METHODS IN MOLECULAR BIOLOGY

Series Editor
John M. Walker
School of Life Sciences
University of Hertfordshire
Hatfield, Hertfordshire, AL10 9AB, UK

For further volumes:
http://www.springer.com/series/7651

Glycosaminoglycans

Chemistry and Biology

Edited by

Kuberan Balagurunathan

Department of Medicinal Chemistry, University of Utah, Salt Lake City, UT, USA

Hiroshi Nakato

Department of Genetics, Cell Biology and Development, The University of Minnesota, Minneapolis, MN, USA

Umesh R. Desai

Department of Medicinal Chemistry, Virginia Commonwealth University, Richmond , VA, USA

 Humana Press

Editors
Kuberan Balagurunathan
Department of Medicinal Chemistry
University of Utah
Salt Lake City, UT, USA

Umesh R. Desai
Department of Medicinal Chemistry
Virginia Commonwealth University
Richmond, VA, USA

Hiroshi Nakato
Department of Genetics, Cell Biology
 and Development
The University of Minnesota
Minneapolis, MN, USA

ISSN 1064-3745   ISSN 1940-6029 (electronic)
ISBN 978-1-4939-1713-6   ISBN 978-1-4939-1714-3 (eBook)
DOI 10.1007/978-1-4939-1714-3
Springer New York Heidelberg Dordrecht London

Library of Congress Control Number: 2014949860

Humana Press is a brand of Springer
Springer is part of Springer Science+Business Media (www.springer.com)

Preface

Glycosaminoglycans are structurally the most complex biopolymers among the many naturally occurring polysaccharides. Complexity arises from the variation in the sugar residue-type, interglycosidic linkage, chain length, sulfation level, and sulfation position. There are six types of glycosaminoglycans including heparan sulfate, heparin, chondroitin sulfate, dermatan sulfate, keratan sulfate, and hyaluronic acid. With the exception of hyaluronic acid, all glycosaminoglycans are covalently attached to proteins. Glycosaminoglycans have been shown to play significant roles in many model organisms starting from early growth period, through development and beyond. These complex carbohydrate polymeric chains regulate numerous biological and pathological processes such as angiogenesis, morphogenesis, organogenesis, neurogenesis, stem cell differentiation, cell proliferation, cell migration, left-right axis induction, synaptic plasticity, synapse formation, neuronal guidance and growth, hemostasis, infection, and many others. Naturally, there is a considerable interest in understanding the structural basis for their numerous biological actions and in identifying protein ligands that orchestrate these functions through direct interaction with glycosaminoglycan chains. There has been a steady growth in the appreciation for the roles of glycosaminoglycan in many scientific disciplines ranging from developmental biology, chemical biology, organic synthesis, structural biology, biochemistry, cell signaling, drug discovery, stem cell biology, tissue engineering, bioinformatics, and computational glycobiology as seen by the large number of glycosaminoglycan-related papers published in the last 10 years.

In this volume of *Methods in Molecular Biology*, we provide robust methods for studying chemistry and biology of glycosaminoglycans. The volume emphasizes several areas of glycosaminoglycan research including structural analysis of GAGs using a variety of approaches, chemical and enzymatic synthesis of GAGs for therapeutic purposes; biophysical and biochemical methods for studying GAG–protein interactions; molecular approaches for modulating and defining GAG biosynthesis; informatics approaches for deciphering GAG code; computational approaches for establishing specific and/or nonspecific interactions; and genetic and biochemical tools for manipulating/visualizing glycosaminoglycan expression and studying their functions in a variety of model organisms. This volume has a primary goal of providing practical guidance for the chemist to carry out biological experiments, for the biologist to perform chemical/biochemical studies, and for the applied scientist to harness therapeutic application possibilities using chemical, biochemical, genetic, and computational tools. Overall, it is our expectation that this volume will serve as a valuable manual for cutting-edge methodologies and practical (hard to find from the primary literature) tips to overcome any obstacles with experimentation pertaining to chemistry and biology of glycosaminoglycans.

Salt Lake City, UT, USA   *Kuberan Balagurunathan*
Minneapolis, MN, USA   *Hiroshi Nakato*
Richmond, VA, USA   *Umesh R. Desai*

Contents

Contributors

NIKOLAOS A. AFRATIS • *Laboratory of Biochemistry, Department of Chemistry, University of Patras, Patras, Greece*

RAMI A. AL-HORANI • *Department of Medicinal Chemistry, Institute for Structural Biology and Drug Discovery, Virginia Commonwealth University, Richmond, VA, USA*

MD. FERDOUS ANOWER-E-KHUDA • *Advanced Medical Research Center, Aichi Medical University, Nagakute, Aichi, Japan; Cellular and Molecular Medicine, University of California San Diego, La Jolla, CA, USA*

SUNEEL S. APTE • *Department of Biomedical Engineering, Lerner Research Institute, Cleveland Clinic, Cleveland, OH, USA*

SAILAJA ARUNGUNDRAM • *Academic Programs, South Seattle College, Seattle, WA, USA*

ATHANASIA P. ASIMAKOPOULOU • *Laboratory of Biochemistry, Department of Chemistry, University of Patras, Patras, Greece*

MATTHEW ATTREED • *Department of Genetics, Albert Einstein College of Medicine, Bronx, NY, USA; Department of Neuroscience, Albert Einstein College of Medicine, Bronx, NY, USA*

PONNUSAMY BABU • *Glycomics and Glycoproteomics, Centre for Cellular and Molecular Platforms, NCBS-TIFR, Bangalore, Karnataka, India*

SOMESH BARANWAL • *Division of Hematology and Oncology, Department of Medicine, Virginia Commonwealth University, Richmond, VA, USA*

CONSUELO N. BEECHER • *Department of Chemistry, University of California – Riverside, Riverside, CA, USA*

MARIKA BOGDANI • *Matrix Biology Program, Benaroya Research Institute at Virginia Mason, Seattle, WA, USA*

RIO S. BOOTHELLO • *Department of Medicinal Chemistry, Institute for Structural Biology and Drug Discovery, Virginia Commonwealth University, Richmond, VA, USA*

TIMOTHY BOWEN • *Department of Nephrology, Cardiff University School of Medicine, Cardiff University, Cardiff, UK*

SPENCER BROWN • *Department of Medicinal Chemistry, University of Utah, Salt Lake City, UT, USA; Department of Bioengineering, University of Utah, Salt Lake City, UT, USA*

HANNES E. BÜLOW • *Department of Genetics, Albert Einstein College of Medicine, Bronx, NY, USA; Department of Neuroscience, Albert Einstein College of Medicine, Bronx, NY, USA*

ISHAN CAPILA • *Momenta Pharmaceuticals, Cambridge, MA, USA*

CHRISTINA K. CHAN • *Matrix Biology Program, Benaroya Research Institute at Virginia Mason, Seattle, WA, USA*

QIUFANG CHENG • *Department of Pathology, Microbiology and Immunology, Vanderbilt University, Nashville, TN, USA*

LESLIE K. CORTES • *Department of Pediatrics, University of Chicago, Chicago, IL, USA*

MAURICIO CORTES • *Department of Pediatrics, University of Chicago, Chicago, IL, USA*

ANDERS DAGÄLV • *Department of Medical Biochemistry and Microbiology, Science for Life Laboratory, Uppsala University, Uppsala, Sweden*

ANTHONY J. DAY • *Wellcome Trust Centre for Cell-Matrix Research, Faculty of Life Sciences, University of Manchester, Manchester, UK*

UMESH R. DESAI • *Department of Medicinal Chemistry, Institute for Structural Biology and Drug Discovery, Virginia Commonwealth University, Richmond, VA, USA*

TABEA DIERKER • *Department of Medical Biochemistry and Microbiology, Science for Life Laboratory, Uppsala University, Uppsala, Sweden*

AMIT K. DUTTA • *Department of Biochemistry and Molecular Biology, The University of Texas Medical Branch, Galveston, TX, USA*

COLIN D. EICHINGER • *Department of Bioengineering, University of Utah, Salt Lake City, UT, USA*

JANE R. ENGLER • *Department of Neurological Surgery, Brain Tumor Research Center, University of California, San Francisco, CA, USA*

INGER ERIKSSON • *Department of Medical Biochemistry and Microbiology, Science for Life Laboratory, Uppsala University, Uppsala, Sweden*

CHERYL M. ETHEN • *Department of Enzyme, R&D Systems, Minneapolis, MN, USA*

BEATA FILIPEK-GÓRNIOK • *Department of Medical Biochemistry and Microbiology, Science for Life Laboratory, Uppsala University, Uppsala, Sweden*

SIMON J. FOULCER • *Department of Biomedical Engineering, Lerner Research Institute, Cleveland Clinic, Cleveland, OH, USA*

LEWIS J. FREY • *Biomedical Informatics Center, Medical University of South Carolina, Charleston, SC, USA*

DAVID GAILANI • *Department of Pathology, Microbiology and Immunology, Vanderbilt University, Nashville, TN, USA; Hematology/Oncology, Vanderbilt University, Nashville, TN, USA*

DINESH R. GARUD • *Department of Chemistry, Sir Parashurambhau College, Pune, India*

ALAN GRODZINSKY • *Department of Biological Engineering, MIT, Cambridge, MA, USA; Department of Electrical Engineering and Computer Science, MIT, Cambridge, MA, USA; Department of Mechanical Engineering, MIT, Cambridge, MA, USA*

HEUNG SIK HAHM • *Department of Biomolecular Systems, Max Planck Institute of Colloids and Interfaces, Potsdam, Germany*

VINCENT C. HASCALL • *Lerner Research Institute, Cleveland Clinic, Cleveland, OH, USA*

VLADIMIR HLADY • *Department of Bioengineering, University of Utah, Salt Lake City, UT, USA*

JOHN HOGWOOD • *National Institute for Biological Standards and Control, Hertfordshire, UK*

REBECCA J. HOLLEY • *Wellcome Trust Centre for Cell Matrix Research, Faculty of Life Sciences, University of Manchester, Manchester, UK*

TONY W. HSIAO • *Department of Bioengineering, University of Utah, Salt Lake City, UT, USA*

IVAN S. IVANOV • *Department of Pathology, Microbiology and Immunology, Vanderbilt University, Nashville, TN, USA*

PAMELA Y. JOHNSON • *Matrix Biology Program, Benaroya Research Institute at Virginia Mason, Seattle, WA, USA*

APRIL JOICE • *Department of Medicinal Chemistry, University of Utah, Salt Lake City, UT, USA; Department of Bioengineering, University of Utah, Salt Lake City, UT, USA*

PREM RAJ B. JOSEPH • *Department of Biochemistry and Molecular Biology, Sealy Center for Structural Biology and Molecular Biophysics, The University of Texas Medical Branch, Galveston, TX, USA*

MAUSAM KALITA • *Department of Medicinal Chemistry, University of Utah, Salt Lake City, UT, USA*

NIKOS K. KARAMANOS • *Laboratory of Biochemistry, Department of Chemistry, University of Patras, Patras, Greece*

EVGENIA KAROUSOU • *Department of Surgical and Morphological Sciences, University of Insubria, Varese, Italy*

RAJESH KARUTURI • *Department of Medicinal Chemistry, Institute for Structural Biology and Drug Discovery, Virginia Commonwealth University, Richmond, VA, USA*

GANESH V. KAUNDINYA • *Momenta Pharmaceuticals, Cambridge, MA, USA*

KOJI KIMATA • *Advanced Medical Research Center, Aichi Medical University, Nagakute, Aichi, Japan*

LENA KJELLÉN • *Department of Medical Biochemistry and Microbiology, Science for Life Laboratory, Uppsala University, Uppsala, Sweden*

MAMORU KOKETSU • *Department of Chemistry and Biomolecular Science, Faculty of Engineering, Gifu University, Gifu, Japan*

BALAGURUNATHAN KUBERAN • *Department of Medicinal Chemistry, University of Utah, Salt Lake City, UT, USA; Department of Bioengineering, University of Utah, Salt Lake City, UT, USA; Interdepartmental Program in Neuroscience, University of Utah, Salt Lake City, UT, USA*

TOIN H. VAN KUPPEVELT • *Department of Biochemistry, Radboud Institute for Molecular Life Sciences, Radboud University Medical Centre, Nijmegen, The Netherlands*

FOTINI N. LAMARI • *Department of Pharmacy, University of Patras, Patras, Greece*

CYNTHIA K. LARIVE • *Department of Chemistry, University of California – Riverside, Riverside, CA, USA*

MARK E. LAUER • *Lerner Research Institute, Cleveland Clinic, Cleveland, OH, USA*

MIROSLAW LECH • *Momenta Pharmaceuticals, Cambridge, MA, USA*

AIYE LIANG • *Department of Physical Sciences, Charleston Southern University, North Charleston, SC, USA*

CHIEN-FU LIANG • *Department of Biomolecular Systems, Max Planck Institute of Colloids and Interfaces, Potsdam, Germany*

JACQUELINE LOFTIS • *Lerner Research Institute, Cleveland Clinic, Cleveland, OH, USA*

MEGAN S. LORD • *Graduate School of Biomedical Engineering, The University of New South Wales, Sydney, NSW, Australia*

ANDERS LUNDEQUIST • *Department of Medical Biochemistry and Microbiology, Science for Life Laboratory, Uppsala University, Uppsala, Sweden*

MIRANDA MACHACEK • *Department of Enzyme, R&D Systems, Minneapolis, MN, USA*

NOBUAKI MAEDA • *Tokyo Metropolitan Institute of Medical Science, Tokyo, Japan*

CHRISTINA MALAVAKI • *Laboratory of Biochemistry, Department of Chemistry, University of Patras, Patras, Greece*

CAITLIN MENCIO • *Interdepartmental Program in Neuroscience, University of Utah, Salt Lake City, UT, USA*

C.L.R. MERRY • *Stem Cell Glycobiology Group, School of Materials, University of Manchester, Manchester, UK*

ADAM C. MIDGLEY • *Department of Nephrology, Cardiff University School of Medicine, Cardiff University, Cardiff, UK*

LUCA MONTI • *Unit of Biochemistry, Department of Molecular Medicine, University of Pavia, Pavia, Italy*

Philip D. Mosier • *Department of Medicinal Chemistry, Institute for Structural Biology and Drug Discovery, Virginia Commonwealth University, Richmond, VA, USA*

Carol de la Motte • *Lerner Research Institute, Cleveland Clinic, Cleveland, OH, USA*

Barbara Mulloy • *Institute of Pharmaceutical Science, King's College London, London, UK*

Nadine Nagy • *Matrix Biology Program, Benaroya Research Institute at Virginia Mason, Seattle, WA, USA*

Hiroshi Nakato • *Department of Genetics, Cell Biology and Development, University of Minnesota, Minneapolis, MN, USA*

Thao Kim Nu Nguyen • *Institute of Microbiology and Biotechnology, Vietnam National University Hanoi, Hanoi, Vietnam*

Hadi Tavakoli Nia • *Department of Mechanical Engineering, MIT, Cambridge, MA, USA*

Akiko Jinno • *Division of Respiratory Diseases, Children's Hospital, Harvard Medical School, Boston, MA, USA*

Christine Ortiz • *Department of Materials Science and Engineering, MIT, Cambridge, MA, USA*

Pyong Woo Park • *Children's Hospital, Harvard Medical School, Boston, MA, USA*

Alberto Passi • *Department of Surgical and Morphological Sciences, University of Insubria, Varese, Italy*

Bhaumik B. Patel • *Division of Hematology and Oncology, Department of Medicine, Virginia Commonwealth University, Richmond, VA, USA*

Nirmita Patel • *Hunter Holmes McGuire VA Medical Center, Richmond, VA, USA*

Mauro S.G. Pavão • *Instituto de Bioquímica Médica Leopoldo de Meis, Hospital Universitário Clementino Fraga Filho, Universidade Federal do Rio de Janeiro, Cidade Universitária, Rio de Janeiro, Brazil*

Joanna J. Phillips • *Department of Neurological Surgery, Brain Tumor Research Center, University of California, San Francisco, CA, USA; Department of Pathology, University of California, San Francisco, CA, USA*

Claire E. Pickford • *Stem Cell Glycobiology Group, School of Materials, University of Manchester, Manchester, UK*

Krishna Mohan Poluri • *Department of Biochemistry and Molecular Biology, Sealy Center for Structural Biology and Molecular Biophysics, The University of Texas Medical Branch, Galveston, TX, USA*

Fabienne E. Poulain • *Department of Biological Sciences, University of South Carolina, Columbia, SC, USA*

Brittany Prather • *Department of Enzyme, R&D Systems, Minneapolis, MN, USA*

Hong Qiu • *Complex Carbohydrate Research Center, University of Georgia, Athens, GA, USA*

Maritza V. Quintero • *Department of Medicinal Chemistry, University of Utah, Salt Lake City, UT, USA*

Krishna Rajarathnam • *Department of Biochemistry and Molecular Biology, Sealy Center for Structural Biology and Molecular Biophysics, The University of Texas Medical Branch, Galveston, TX, USA*

Karthik Raman • *Department of Bioengineering, University of Utah, Salt Lake City, UT, USA*

Jörg Rösgen • *Department of Biochemistry and Molecular Biology, Penn State College of Medicine, Hershey, PA, USA*

ANTONIO ROSSI • *Unit of Biochemistry, Department of Molecular Medicine, University of Pavia, Pavia, Italy*

NEHRU VIJI SANKARANARAYANAN • *Department of Medicinal Chemistry, Institute for Structural Biology and Drug Discovery, Virginia Commonwealth University, Richmond, VA, USA*

AURIJIT SARKAR • *Department of Medicinal Chemistry, Institute for Structural Biology and Drug Discovery, Virginia Commonwealth University, Richmond, VA, USA*

RAM SASISEKHARAN • *Department of Biological Engineering, Koch Institute of Integrative Cancer Research, Massachusetts Institute of Technology, Cambridge, MA, USA*

NANCY B. SCHWARTZ • *Department of Pediatrics, University of Chicago, Chicago, IL, USA*

PETER H. SEEBERGER • *Department of Biomolecular Systems, Max Planck Institute of Colloids and Interfaces, Potsdam, Germany*

KRISHNA MOHAN SEPURU • *Department of Biochemistry and Molecular Biology, Sealy Center for Structural Biology and Molecular Biophysics, The University of Texas Medical Branch, Galveston, TX, USA*

ZACHARY SHRIVER • *Department of Biological Engineering, Koch Institute of Integrative Cancer Research, Massachusetts Institute of Technology, Cambridge, MA, USA*

CHARMAINE SIMEONOVIC • *Diabetes/Transplantation Immunobiology Laboratory, Department of Immunology, The John Curtin School of Medical Research, The Australian National University, Canberra, ACT, Australia*

RAYMOND A. SMITH • *Stem Cell Glycobiology Group, School of Materials, University of Manchester, Manchester, UK*

MARCUS SOLIAI • *Department of Medicinal Chemistry, University of Utah, Salt Lake City, UT, USA*

VIMAL P. SWARUP • *Department of Medicinal Chemistry, University of Utah, Salt Lake City, UT, USA; Department of Bioengineering, University of Utah, Salt Lake City, UT, USA*

MASAHIKO TAKEMURA • *Department of Genetics, Cell Biology and Development, University of Minnesota, Minneapolis, MN, USA*

TIMOTHY TATGE • *Department of Enzyme, R&D Systems, Minneapolis, MN, USA*

ACHILLEAS D. THEOCHARIS • *Laboratory of Biochemistry, Department of Chemistry, University of Patras, Patras, Greece*

VY M. TRAN • *Department of Neurological Surgery, Brain Tumor Research Center, University of California, San Francisco, CA, USA*

KENJI UCHIMURA • *Department of Biochemistry, Nagoya University Graduate School of Medicine, Nagoya, Aichi, Japan*

JAMES A. VASSIE • *Graduate School of Biomedical Engineering, The University of New South Wales, Sydney, NSW, Australia*

STEPHEN VERESPY III • *Chemical Biology Program, Department of Chemistry, Virginia Commonwealth University, Richmond, VA, USA*

XYLOPHONE V. VICTOR • *Department of Medicinal Chemistry, University of Utah, Salt Lake City, UT, USA; Department of Bioengineering, University of Utah, Salt Lake City, UT, USA*

DAVIDE VIGETTI • *Department of Surgical and Morphological Sciences, University of Insubria, Varese, Italy*

MANUELA VIOLA • *Department of Surgical and Morphological Sciences, University of Insubria, Varese, Italy*

ANNA WADE • *Department of Neurological Surgery, Brain Tumor Research Center, University of California, San Francisco, CA, USA*

LIANCHUN WANG • *Complex Carbohydrate Research Center, University of Georgia, Athens, GA, USA*

ELS M.A. VAN DE WESTERLO • *Department of Biochemistry, Radboud Institute for Molecular Life Sciences, Radboud University Medical Centre, Nijmegen, The Netherlands*

JOHN M. WHITELOCK • *Graduate School of Biomedical Engineering, The University of New South Wales, Sydney, NSW, Australia*

THOMAS N. WIGHT • *Matrix Biology Program, Benaroya Research Institute at Virginia Mason, Seattle, WA, USA*

ZHENGLIANG L. WU • *Department of Enzyme, R&D Systems, Minneapolis, MN, USA*

WENYUAN XIAO • *Complex Carbohydrate Research Center, University of Georgia, Athens, GA, USA*

TOMIO YABE • *Department of Applied Life Science, Faculty of Applied Biological Sciences, Gifu University, Gifu, Japan*

HAIXIAO YU • *Department of Enzyme, R&D Systems, Minneapolis, MN, USA*

JINGWEN YUE • *Department of Biochemistry and Molecular Biology, Complex Carbohydrate Research Center, University of Georgia, Athens, GA, USA*

VASSILIKI ZAFEIROPOULOU • *Laboratory of Biochemistry, Department of Chemistry, University of Patras, Patras, Greece*

Part I

Structure

Chapter 1

Automated Synthesis of Chondroitin Sulfate Oligosaccharides

Chien-Fu Liang, Heung Sik Hahm, and Peter H. Seeberger

Abstract

Glycosaminoglycans (GAGs) are important sulfated carbohydrates prevalently found in the extracellular matrix that serve many biological functions. The synthesis of structurally diverse but defined GAGs is extremely challenging as one has to account for the various sulfation patterns. Described is the automated synthesis of two chondroitin sulfate hexasaccharides. The oligosaccharides are prepared on a solid support that is equipped with a photolabile linker. The linker cleavage from the resin is performed in a continuous-flow photoreactor under chemically mild conditions. The described approach will serve as a general scheme to systematically access oligosaccharides of all GAG families.

Key words Glycosaminoglycans, Automated solid-phase oligosaccharide synthesis, Photolabile linker, Chondroitin sulfate, N-acetyl-β-D-galactosamine, β-D-Glucuronic acid

1 Introduction

The most ubiquitous biopolymers playing critical roles in various cellular processes are the carbohydrates that are classified based on their conjugation partners such as proteins and lipids. Typical conjugates are glycoproteins, glycolipids, glycosylphosphatidylinositol (GPI) anchors, and proteoglycans that contain glycosaminoglycans (GAGs) [1]. GAGs are linear, however, structurally highly complex polysaccharides with complicated sulfation patterns. Chondroitin sulfates (CSs), one subclass of GAGs, are of particular interest due to their critical role in cell recognition [2], antithrombin III-meditated anticoagulant activity [3], and cell division processes [4]. In order to advance biological research, it is important to procure sufficient amounts of structurally well-defined chondroitin sulfate oligosaccharides with specific sulfation sequences. Although chemical synthesis is preferred to obtaining large amounts of defined oligosaccharides, conventional solution-phase oligosaccharide synthesis [5, 6] requires time-consuming and labor-intensive

Kuberan Balagurunathan et al. (eds.), *Glycosaminoglycans: Chemistry and Biology*, Methods in Molecular Biology, vol. 1229, DOI 10.1007/978-1-4939-1714-3_1, © Springer Science+Business Media New York 2015

repetitive processes in order to achieve glycosylation, deprotection, and purification procedures.

The focus in the development of automated solid-phase oligosaccharide protocols has been on iterative glycosylations. The first successful proof-of-principle process was disclosed in 2001 [7]. Automated synthesis streamlines the repetitive process that is traditionally utilized and is beginning to shift carbohydrate chemistry from solution phase to automated solid-phase reactions. Ever longer and more complex glycans such as a 30-mer mannose [8] and a 12-mer beta-glucan [9] are now coming within reach. Even cis-glycosidic linkages that cannot benefit from directing neighboring group participation have been installed stereoselectively as exemplified by mannuronate [10]. Fragments of hyaluronan, a non-sulfated member of the GAG family, were recently synthesized [11].

Here, we describe a synthetic approach to prepare CS-A and CS-C employing automated solid-phase synthesis. For that purpose, differentially protected building blocks were prepared, and a photocleavable linker that tolerates acidic as well as basic conditions was developed [12]. A new automated solid-phase oligosaccharide synthesizer [13] that performs fully automated, computer-controlled glycan-coupling cycles was further improved to carry out automated sulfation reactions on the solid support [14].

2   Materials

All chemicals used were of reagent grade, except where noted. All reactions were performed in oven-dried glassware under an argon atmosphere. All solvents used were purified in a cycle-trainer solvent delivery system. Analytical thin-layer chromatography (TLC) was performed on Merck silica gel 60 F254 plates (0.25 mm). Compounds were visualized by UV irradiation, or by dipping the plate either in a cerium sulfate ammonium molybdate solution or a 1:1 mixture of H_2SO_4 (2N) and resorcine monomethylether (0.2 %) in ethanol. Column chromatography was carried out using forced flow of the indicated solvent on Fluka Kieselgel 60 (230–240 mesh). All automated glycosylations were performed on a prototype automated oligosaccharide synthesizer using anhydrous solvents of the cycle-trainer solvent delivery system. LCMS chromatograms were recorded on an Agilent 1100 Series spectrometer. Preparative HPLC purifications were performed on an Agilent 1200 Series. The linker loading of functionalized resins was determined using a Shimadzu UV-MINI-1240 UV spectrometer.

Fig. 1 Automated solid-phase oligosaccharide synthesizer. (1) Computer; (2) chiller; (3) argon manifold; (4) solvents; (5) solenoid valves; (6) reagents (aqueous unit); (7) reaction vessel; (8) syringe pumps; (9) fraction collector

2.1 Automated Solid-Phase Oligosaccharide Synthesizer

See Fig. 1.

2.2 Continuous-Flow Reactor

See Fig. 2.

2.3 Preparation of Linker-Bound Resin (Fig. 3)

Merrifield resin (1 eq.) is swollen for 1 h in *N*,*N*-dimethylformamide (DMF) (4 mL/g of resin) in a flask while being gently shaken on a rotary evaporator. Compound **1** (4 eq.) is added to the flask followed by Cs_2CO_3 (4 eq.) and *tetra*-butylammonium iodide (TBAI) (0.5 eq.). The solution is stirred overnight on the rotary evaporator at 60 °C. The resin is washed successively with DMF/water (1/1), DMF, methanol (MeOH), dichloromethane (DCM), MeOH, and DCM (six times each) and then swollen in DCM for 1 h. The swollen resin is placed in a flask with DMF (4 mL/g of resin) and cesium acetate (CsOAc) (4 eq.). The suspension is stirred overnight on a rotary evaporator at 60 °C. The resin is then washed successively with DMF/water (1/1), DMF, MeOH, DCM, MeOH, and DCM (six times each), the solvent is drained, and the resin is dried under vacuum. Loading on resin is quantified with Fmoc quantification method [15].

Fig. 2 Continuous-flow photoreactor. (1) 450 W Power supply; (2) FEP tubing inlet connected to disposable syringe on a syringe pump (4); (3) FEP tubing outlet; (5) UV box fitted with two fans (7); (6) a quartz glass cooling system connected to a chiller (9); (8) UV filter (Pyrex, 50 % transmittance at 305 nm)

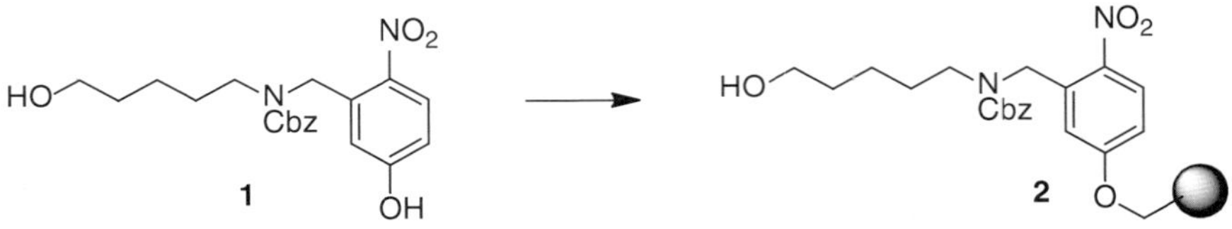

Fig. 3 Preparation of photolabile linker-bound resin

2.4 Preparation of Building Block Solutions (Fig. 4)

Purified building blocks (BBs) [14] are co-evaporated three times with toluene, left under high vacuum overnight, transferred to the building block vials (*see* **Note 1**) under Ar atmosphere, and connected to the corresponding ports (3,2 for compound **4**, and **5**), and (3,3 for compound **3**) in the automated solid-phase oligosaccharide synthesizer.

2.5 Preparation of Reagent Solutions

2.5.1 Activator Solution

In a dry bottle under Ar atmosphere, trimethylsilyl trifluoromethanesulfonate (900 μL) is dissolved in anhydrous DCM (40 mL) and the bottle is connected to the corresponding port (6,2) on the automated solid-phase oligosaccharide synthesizer.

2.5.2 Fmoc Removal Solution

In a dry bottle under Ar atmosphere, piperidine (20 mL) is dissolved in anhydrous DMF (80 mL). This reagent bottle is connected on the corresponding port (4,2) on the automated solid-phase oligosaccharide synthesizer.

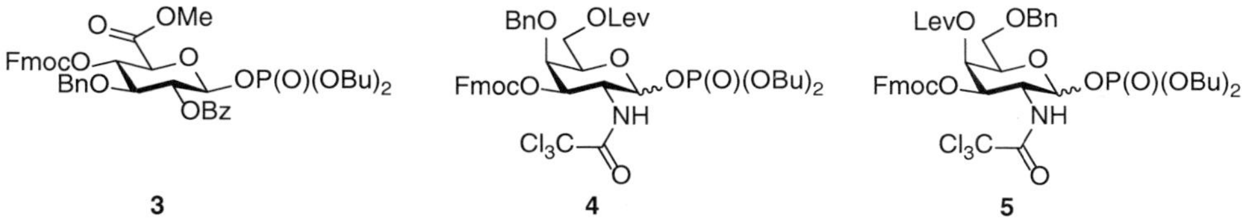

Fig. 4 Building blocks for chondroitin sulfate oligosaccharide synthesis

2.5.3 Acetic Anhydride Solution	In a dry bottle under Ar atmosphere, this reagent bottle with neat acetic anhydride (20 mL) is connected to the corresponding port (5,3) on the automated solid-phase oligosaccharide synthesizer.
2.5.4 Levaloyl (Lev) Removal Solution	In a dry bottle under Ar atmosphere, hydrazine hydrate (1.4 mL) is dissolved in anhydrous pyridine (30 mL), and acetic acid (20 mL). This reagent bottle is connected to the corresponding port (5,2) on the automated solid-phase oligosaccharide synthesizer.
2.5.5 Sulfation Solution	In a dry bottle under Ar atmosphere, dissolve sulfur trioxide pyridine complex (3.2 g) in pyridine (20 mL), and DMF (20 mL). This reagent bottle is connected on the corresponding port (4,4) in automated solid-phase oligosaccharide synthesizer.

3   Methods

3.1 Operation Modules *3.1.1 Module 1 (Washing and Glycosylation)*	The resin is washed three times with DMF, tetrahydrofuran (THF), DCM each (with 2 mL for 25 s at room temperature), and once with 0.375 mL of trimethylsilyl trifluoromethanesulfonate (TMSOTf) in DCM (at –20 °C). The resin is swollen in 2 mL DCM and the temperature of the reaction vessel is adjusted to 25 °C. For glycosylations, the DCM is drained and a solution of the respective building block (3 eq. in 1.0 mL DCM) is delivered to the reaction vessel. After the desired temperature is reached, the reaction is started by the addition of 1 mL of TMSOTf (3 eq. in 1.0 mL DCM) solution. The glycosylation is performed for 45 min at –15 °C and for 15 min at 0 °C. After the reaction the solution is drained and the resin is washed with DCM (six times with 2 mL for 15 s).
3.1.2 Module 2: Fmoc Removal	The resin is washed with DMF (six times with 2 mL for 25 s) and swollen in 2 mL DMF and the temperature of the reaction vessel is adjusted to 25 °C. For fluorenylmethyloxycarbonyl (Fmoc) deprotection the DMF is drained and 2 mL of a solution of 20 % piperidine in DMF is delivered to the reaction vessel. After 5 min the reaction solution is collected in the fraction collector of the oligosaccharide synthesizer. This procedure is repeated three times.

3.1.3 Module 3: Acetylation of Free Hydroxyl Groups

After the last Fmoc deprotection, the resin is washed with DMF (six times with 3 mL for 25 s), DCM, and pyridine (six times each with 2 mL for 25 s) and swollen in 2 mL pyridine and the temperature of the reaction vessel is adjusted to 25 °C. The reaction is started by addition of 1 mL of acetic anhydride to the reaction vessel. After 60 min the reaction solution is drained and the resin is washed with pyridine (six times with 2 mL for 25 s). This acetylation procedure is performed three times.

3.1.4 Module 4: Lev Deprotection

The resin is washed with DCM (three times with 2 mL for 25 s) and swollen in 1.3 mL DCM and the temperature of the reaction vessel is adjusted to 25 °C. For levulinic ester (Lev) deprotection, 0.8 mL of a 0.56 m solution of hydrazine hydrate in pyridine/acetic acid (1.5:1) is added. After 30 min the reaction solution is drained and the resin is washed with 0.2 m acetic acid in DCM and DCM (six times each with 2 mL for 25 s). The entire procedure is performed three times.

3.1.5 Module 5: Sulfation

The resin is washed with DMF and pyridine (six times each with 2 mL for 25 s) and swollen in 2 mL pyridine and the temperature of the reaction vessel is adjusted to 50 °C. For sulfation, 2 mL of a 0.5 M solution of sulfur trioxide pyridine complex in DMF/pyridine (1:1) is added. After 3 h, the reaction solution is drained and the resin is washed with DMF and pyridine (six times each with 2 mL for 25 s). The entire procedure is performed three times.

3.2 Fmoc Quantification (Equation)

A 100 µL aliquot from the Fmoc removal reaction solution is taken for Fmoc quantification. This aliquot, in 10 mL of a volumetric flask, is diluted with DMF, and the UV absorption at $\lambda = 301$ nm is determined. The coupling efficiency is calculated according to the following formula:

$$X\left[\text{Amount of deprotected Fmoc}\right] = \frac{\text{Absorbtion} \times \text{dilution} \times \text{volume}\,(\text{mL})}{\text{Extinction coefficient}}$$

$$= \frac{\text{Absortion} \times 100 \times 6}{7{,}800}\left[\text{mmol}\right]$$

$$Y\left[\text{Amount of reactive sites on resin}\right] = \text{Loading on resin}\left[\frac{\text{mmol}}{\text{g}}\right] \times \text{Resin amount}\,(\text{g})$$

$$\text{Coupling efficiency}\,[\%] = \frac{X}{Y} \times 100 = \frac{\text{Absorbtion} \times 100 \times 6}{7{,}800 \times \text{Loading on resin} \times \text{resin amount}\,(\text{g})} \times 100$$

3.3 Linker-Cleavage in a Continuous-Flow Reactor

To prepare the photoreactor, the FEP tubing is washed with 20 mL DCM using a flow rate of 4 mL/min. The oligosaccharide-bound resin is transferred with 20 mL of DCM (*see* **Note 2**) into a disposable syringe (20 mL). For the cleavage, the resin is slowly injected from the disposable syringe into the reactor and pushed through the tubing (flow rate: 0.6 mL/min). To shrink and wash out remaining resin, the tubing is washed with 20 mL DCM/MeOH, 1:1 (flow rate: 0.6 mL/min for 10 mL and 2 mL/min for 10 mL), and finally with 20 mL MeOH (flow rate: 4 mL/min). The reaction mixture flows through a filter where the resin is filtered out and washed with DCM/MeOH, 1:1, MeOH, and DCM. The tubing is re-equilibrated with 15 mL DCM using a flow rate of 4 mL/min. The entire procedure is performed twice. The resulting solution is evaporated in vacuo (*see* **Note 3**).

3.4 Analytical HPLC and Preparative HPLC

The crude product is analyzed by analytical HPLC (column: C18-Nucleodur (5×250 mm; 5 μm); flow rate: 1 mL/min; eluents: 0.01 M NH_4HCO_3 in water/MeCN), and purified by preparative HPLC (column: C18-Nucleodur (21×250 mm; 5 μm); flow rate: 10 mL/min; eluents: 0.01 M NH_4HCO_3 in water/MeCN). Gradient: 45 % (5 min) → 60 % (in 40 min) → 100 % (in 5 min); detection: 210 and 280 nm.

4 Notes

1. Whenever building blocks are co-evaporated, one should use a pear-shaped flask rather than a round-bottom flask to minimize loss of building block. Each building block vial should be dried in a drying oven for an hour, and cooled down to room temperature under an argon atmosphere. After the building block solution is transferred into the building block vial it is connected to the right port immediately.

2. A syringe pump is adjusted at an almost vertical angle while injecting the oligosaccharide-containing resin, because the resin, when swollen in DCM, is floating on top in the syringe. When other solvents are used to inject the resin, the resin remains on the bottom of the syringe.

3. The crude product was dissolved in MeOH and treated with Amberlite IR-120 (Na) ion-exchange resin to convert it to the more stable sodium form.

Acknowledgment

We gratefully thank the Max Planck Society and the European Research Council (ERC Advanced Grant AUTOHEPARIN to P.H.S.) for financial support.

References

1. Varki A, Cummings R, Esko J, Freeze H, Stanley P, Bertozzi C, Hart G, Etzler M (eds) (2009) Essentials of glycobiology, 2nd edn. Cold Spring Harbor Laboratory Press, Cold Spring Harbor

2. Jalkanen S, Jalkanen M (1992) Lymphocyte CD44 binds the COOH-terminal heparin-binding domain of fibronectin. J Cell Biol 116:817–825

3. Aikawa J, Isemura M, Munakata H (1986) Isolation and characterization of chondroitin sulfate proteoglycans from porcine thoracic aorta. Biochim Biophys Acta 883:83–90

4. Mizuguchi S, Uyama T, Kitagawa H et al (2003) Chondroitin proteoglycans are involved in cell division of Caenorhabditis elegans. Nature 423:443–448

5. Gama C, Tully S, Sotogaku N et al (2006) Sulfation patterns of glycosaminoglycans encode molecular recognition and activity. Nat Chem Biol 2:467–473

6. Jacquinet J, Lopin-Bon C, Vibert A (2009) From polymer to size-defined oligomers: a highly divergent and stereocontrolled construction of chondroitin sulfate A, C, D, E, K, L, and M oligomers from a single precursor: part 2. Chemistry 15:9579–9595

7. Plante O, Palmacci E, Seeberger P (2001) Automated solid-phase synthesis of oligosaccharides. Science 291:1523–1527

8. Calin O, Eller S, Seeberger P (2013) Automated polysaccharide synthesis: assembly of a 30mer mannoside. Angew Chem Int Ed 52:5862–5865

9. Weishaupt M, Matthies S, Seeberger P (2013) Automated solid-phase synthesis of a β-(1,3)-glucan dodecasaccharide. Chemistry 19:12497–12503

10. Walvoort M, Elst H, Plante O et al (2012) Automated solid-phase synthesis of β-mannuronic acid alginates. Angew Chem Int Ed 51:4393–4396

11. Walvoort M, Volbeda A, Reintjens N (2012) Automated solid-phase synthesis of hyaluronan oligosaccharides. Org Lett 14:3776–3779

12. Nicolaou KC, Winssinger N, Pastor J et al (1997) A general and highly efficient solid phase synthesis of oligosaccharides. Total synthesis of a heptasaccharide phytoalexin elicitor (HPE). J Am Chem Soc 119:449–450

13. Krock L, Esposito D, Castagner B et al (2012) Streamlined access to conjugation-ready glycans by automated synthesis. Chem Sci 3:1617–1622

14. Eller S, Collot M, Yin J, Hahm HS, Seeberger P (2013) Automated solid-phase synthesis of chondroitin sulfate glycosaminoglycans. Angew Chem Int Ed 52:5858–5861

15. Gude M, Ryf J, White P (2002) An accurate method for the quantitation of Fmoc-derivatized solid phase supports. Lett Pept Sci 9:203–206

Chapter 2

Enzymatic Synthesis of Heparan Sulfate and Heparin

April Joice, Karthik Raman, Caitlin Mencio, Maritza V. Quintero, Spencer Brown, Thao Kim Nu Nguyen, and Balagurunathan Kuberan

Abstract

Heparan sulfate (HS) polysaccharide chains have been shown to orchestrate distinct biological functions in several systems. Study of HS structure-function relations is, however, hampered due to the lack of availability of HS in sufficient quantities as well as the molecular heterogeneity of naturally occurring HS. Enzymatic synthesis of HS is an attractive alternative to the use of naturally occurring HS, as it reduces molecular heterogeneity, or a long and daunting chemical synthesis of HS. Heparosan, produced by *E. coli* K5 bacteria, has a structure similar to the unmodified HS backbone structure and can be used as a precursor in the enzymatic synthesis of HS-like polysaccharides. Here, we describe an enzymatic approach to synthesize several specifically sulfated HS polysaccharides for biological studies using the heparosan backbone and a combination of recombinant biosynthetic enzymes such as C5-epimerase and sulfotransferases.

Key words Glycosaminoglycan, Proteoglycan, Heparin, Heparan sulfate, Enzymatic synthesis, Epimerase, Sulfotransferase

1 Introduction

Heparan sulfate (HS) is a negatively charged, linear polysaccharide that regulates many biological systems. Some of these include blood coagulation, pathogenesis of infectious disease, cell growth, and developmental processes (as reviewed in refs. 1–3). The primary obstacle with isolating HS from natural sources for use in research is the molecular heterogeneity that occurs within the polysaccharide chain, resulting in various regions with differing sulfation patterns. In order to overcome the challenge associated with the naturally occurring molecular heterogeneity, HS has been produced previously using chemical methods. This chemical method has led to the production of an ATIII-binding pentasaccharide [4, 5]; however, the production of HS using chemical approaches requires as many as 60 steps and the final structure is obtained in very low yields. Enzymatic synthesis of HS-like polysaccharides significantly reduces the molecular heterogeneity of the

Kuberan Balagurunathan et al. (eds.), *Glycosaminoglycans: Chemistry and Biology*, Methods in Molecular Biology, vol. 1229, DOI 10.1007/978-1-4939-1714-3_2, © Springer Science+Business Media New York 2015

naturally occurring chains, is more efficient than chemical synthesis, and also results in higher yields.

In 1973, heparin was conclusively shown to be specifically accelerate the rate of antithrombin III (ATIII) mediated Factor Xa/Thrombin inhibition [6]. Since then the structure-function relationship of HS and ATIII has been well characterized, making this anticoagulant an attractive target for enzymatic synthesis [7–10]. In fact, the first enzymatically synthesized HS pentasaccharide was reported in 2003 and studied for its ATIII-binding properties [7, 8]. However, the possibilities for enzymatically synthesizing HS extend far beyond the production of anticoagulants, as the importance of HS has been identified in numerous biological processes.

Heparan sulfate is composed of a repeating disaccharide unit, consisting of a uronic acid that is linked to an N-acetyl-D-glucosamine. This repeating disaccharide unit is also termed heparosan and can be produced in an *Escherichia coli* K5 strain [11]. After production of the starting material, various enzymatic modifications can be performed (Fig. 1). C5-epimerase converts the D-glucuronic acid

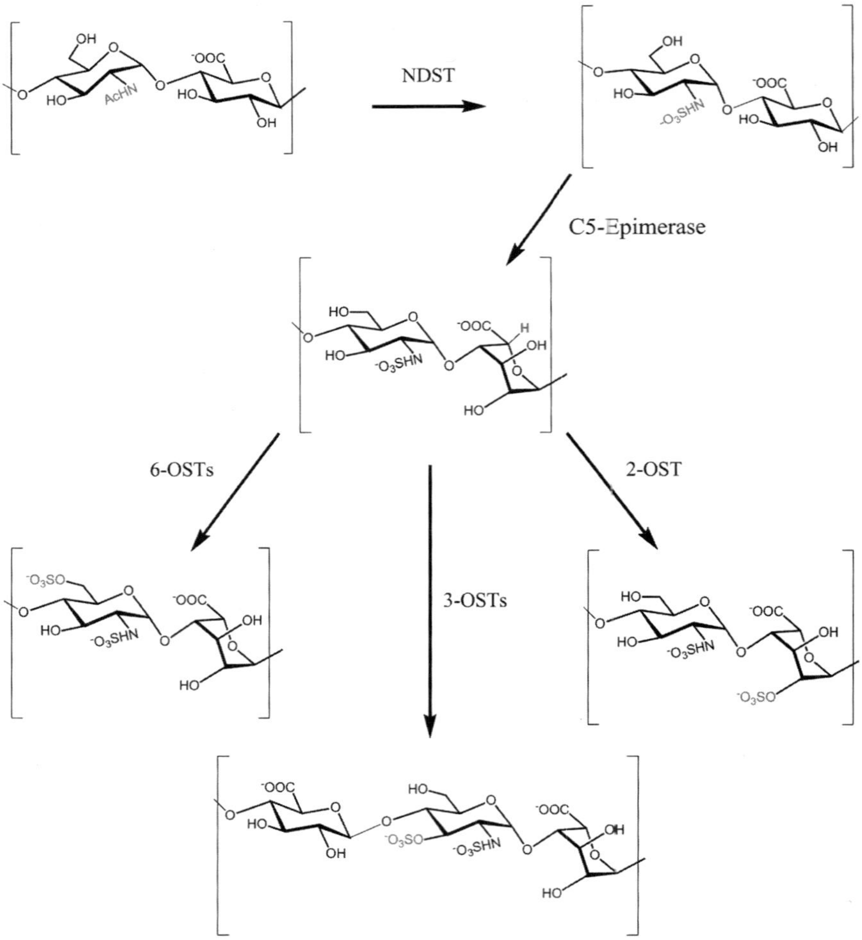

Fig. 1 Schematic representation of modifications involved in the enzymatic synthesis

residues of the heparosan backbone to L-iduronic acid residues, and N-deacetylation/N-sulfation is carried out by N-deacetylase-N-sulfotransferase. 2-O-sulfotransferase modifies the uronic acid residues at C2 position, while 3-O-sulfotransferases and 6-O-sulfotransferases modify the glucosamine residues at C3 and C6 positions, respectively. Many of the HS biosynthetic enzymes exist in multiple isoforms with the exception of C5-epimerase and 2-O-sulfotransferase.

Once the enzymatic synthesis reactions are complete, the polymer product must be analyzed for disaccharide composition to determine the sulfation group(s) present (2-O-sulfation, N-sulfation, 6-O-sulfation, and 3-O-sulfation). Analysis of disaccharides can be performed by strong anion exchange, mass spectrometry, and capillary electrophoresis. A strong anion exchange column coupled with a DAD or UV detector can determine the exact composition of the disaccharides, for example N-sulfated and 6-O-sulfated versus N-sulfated and 3-O-sulfated, when compared to a known disaccharide standard. Mass spectrometry, on the other hand, gives you the exact MW of the components, but cannot be used to obtain the quantitative information on disaccharides [12]. Ideally, it is best to confirm the disaccharide sulfation using both strong anion exchange and mass spectrometry. Using the heparosan backbone and a combination of various modification enzymes mentioned above, many different specifically sulfated HS polysaccharides can be obtained to study and establish the structure-function relationship.

2 Materials

2.1 Heparosan Production (See Note 1)

1. LB medium: 10 g/L Bactotryptone, 5 g/L yeast extract, 10 g/L NaCl.

2. Growth media for high-yield *E. coli* K5 cultures: 20 g/L Casamino acids, 10 g/L yeast extract, 4.3 g/L NaH_2PO_4, 5.3 g/L K_2PO_4 dibasic, 4.2 g/L KH_2PO_4 monobasic, 500 mg/L $MgSO_4$, 18 mg/L $FeSO_4$ heptahydrate, 1 g/L NaCl, 0.25 g/L Na_2SO_4, add water to a total volume of 1 L.

3. Trace element mixture for high-yield *E. coli* K5 cultures: 50 mg H_3PO_4, 2.5 mg $CuSO_4$, 0.2 g $ZnCl_2$, 1.2 g $CoCl_2$ hexahydrate, 0.25 g $AlCl_3$, 0.2 g Na_2MoO_4, Add 1 mL of HCl to salt mixture while stirring to dissolve, then q.s. with water up to 100 mL.

4. Nutrient buffer for high-yield *E. coli* K5 cultures: 20 % glycerol w/v, 100 g glucose, 50 g $(NH_4)_2SO_4$, 100 mL trace element mix, 680 mL H_2O.

5. Vitamin buffer: 210 mg Riboflavin, 2.7 g pantothenic acid, 3.0 g niacin, 0.7 g pyridoxin, 30 mg biotin, 20 mg folic acid, q.s. with water to 500 mL, store protected from light.

6. Econo-column.

7. DEAE-Sepharose Resin.

8. DEAE wash buffer: 0.1 M NaCl, 0.02 M NaOAc, 0.01 % Triton X-100, pH 6.0.

9. DEAE elution buffer: 1 M NaCl, 0.02 M NaOAc, 0.01 % Triton X-100, pH 6.0.

10. 1,000 MWCO centrifugal filters.

11. 1,000 MWCO dialysis membrane.

2.2 Expression and Purification of O-Sulfotransferases, N-Sulfotransferase, and Other Heparan Sulfate Biosynthetic Enzymes (See Note 2)

1. SF9 cells.

2. SF900 II SFM media.

3. Toyopearl AF-heparin column resin.

4. Nickel Sepharose Fast Flow column.

5. PCG buffer A: 10 mM PIPES, 0.1 % CHAPS, 1 % glycerol, 50 mM NaCl, pH 7, ice-cold.

6. PCG buffer B: 10 mM PIPES, 0.1 % CHAPS, 1 % glycerol, 1 M NaCl, pH 7, ice-cold.

7. Ni buffer A: 100 mM Tris-HCl 0.5 M NaCl, 5 % glycerol, 5 mM beta mercaptoethanol, 20 mM imidazole, pH 7, ice cold.

8. Ni buffer B: 100 mM Tris-HCl 0.5 M NaCl, 5 % glycerol, 5 mM beta mercaptoethanol, 500 mM imidazole, pH 7, ice cold.

9. Enzyme storage buffer: 10 mM PIPES, 5 % glycerol, pH 7, ice cold.

10. 10,000 MWCO centrifugal filters.

2.3 Enzymatic Synthesis Reactions

1. 5× MES buffer: 125 mM MES (2-(N-morpholino)ethanesulfonic acid hemisodium salt), 2.5 % (v/v) Triton X-100, 12.5 mM $MgCl_2$, 6.25 mM $CaCl_2$, 0.5 mg/mL BSA, pH 8.0.

2. Heparosan polymer.

3. HS biosynthetic enzyme.

4. 3′-Phosphoadenosine-5′-phosphosulfate (PAPS).

2.4 Purification of Enzymatically Synthesized Reaction Product

1. DEAE wash buffer: 0.1 M NaCl, 20 mM sodium acetate, 0.01 % Triton X-100, pH 6.8.

2. DEAE elution buffer: 1 M NaCl, 20 mM sodium acetate, pH 6.8.

3. DEAE-Sepharose Resin.

4. Disposable columns.

5. 3,000 MWCO centrifugal filter.

2.5 Digestion of Polymer

1. 10× Heparitinase buffer: 33 mM Calcium acetate, 400 mM ammonium acetate, 1 mg/mL BSA, pH 7.0.

2. Heparin lyase I/II/III mixture.

2.6 Disaccharide Analysis by Strong Anion Exchange HPLC	1. HPLC buffer A: dH_2O, pH 3.5 Filtered and degassed.

2. HPLC buffer B: 1 M NaCl, pH 3.5 Filtered and degassed.

3. Analytical Carbopac column (*see* **Note 3**).

4. Disaccharide standards.

3 Methods

3.1 Heparosan Production

1. Growth of *E. coli* K5 with D-glucose (*see* **Note 4**).

   (a) *E. coli* K5 bacteria are established overnight in 100 mL of LB medium containing 0.5 % w/v D-glucose.

   (b) In a flask combine 1 L autoclaved growth medium, 10 mL nutrient buffer, 100 mL K5 overnight culture, and 3 mL vitamin solution.

   (c) Incubate for 28 h at 37 °C, shaking at 200 rpm.

   (d) Then add an extra 5 mL of nutrient buffer and incubate for 22 h at 37 °C, shaking at 200 rpm.

   (e) Bacterial cultures are autoclaved for 30 min.

   (f) The autoclaved solution is centrifuged at $6238 \times g$ for 1 h to pellet lysed bacteria.

   (g) The supernatant is then purified over DEAE-Sepharose.

2. Purification of *E. coli* K5 Heparosan through DEAE-Sepharose.

   (a) First a purification column is set up with a bed volume of 2 mL filled with DEAE-Sepharose (*see* **Note 5**).

   (b) The column is washed carefully with 5 column volumes (10 mL) of nano pure H_2O.

   (c) After the beads settle, the column is equilibrated with 5 column volumes of DEAE wash buffer.

   (d) The column is then loaded with the supernatant while the valve is closed. As the beads are slowly allowed to settle, the valve is opened.

   (e) The flow through is loaded one more time to maximize binding of heparosan to the beads.

   (f) The column is then washed with 30 column volumes of DEAE wash buffer.

   (g) The bound heparosan is then eluted with 6 column volumes (12 mL) of DEAE elution buffer.

   (h) The elution is then concentrated through a 1,000 MWCO Amicon centrifugal filter column and desalted by concentration and dilution at least seven times (*see* **Note 5**).

3.2 Expression and Purification of O-Sulfotransferases, N-Sulfotransferase, and Other Heparan Sulfate Biosynthetic Enzymes

1. Expression of epimerase

    (a) SF9 insect cells were grown in SF900 II SFM media at 25 °C, shaking at 150 rpm.

    (b) 1 L of SF900 II SFM media was inoculated with SF9 cells to a final cell count of 1×10^6 cells/mL in a sterile 3 L plastic flask (*see* **Note 6**).

    (c) 30 mL of Epimerase virus with 30–45 pfu/cell titer was then added to the plastic flask (*see* **Note 7**).

    (d) Cells were swirled once and left stationary in the hood for 10 min.

    (e) Subsequently, cells were shaken at 150 rpm overnight at 25 °C.

    (f) After 48 h, the cells were centrifuged at $250 \times g$, 4 °C, for 30 min.

2. Purification of epimerase over heparin and nickel columns

    (a) The supernatant, after centrifugation, was chilled in an ice-cold container and 1 M PIPES was added till a final concentration reaches 10 mM PIPES. The pH was then adjusted to pH 6.0 (*see* **Note 8**).

    (b) A solution of PMSF in 10 mL of isopropanol was added to the supernatant till a final concentration of 0.1 mg/mL and the supernatant was kept stationary for 1 h at 4 °C.

    (c) Subsequently, the supernatant was centrifuged at $3993 \times g$ at 4 °C for 1 h.

    (d) 600 µL of ice-cold Toyopearl Heparin resin was loaded into a 5 mL syringe containing a filter.

    (e) The heparin resin was then equilibrated with 5 mL of ice-cold PCG A buffer.

    (f) The supernatant containing the protein was then loaded onto the heparin resin carefully (in a 4 °C fridge) and flow through was reloaded twice.

    (g) The column was then washed with 30 column volumes of PCG A.

    (h) The purified protein was then eluted with 6 column volumes of PCG B (collected in 1 mL fractions).

    (i) The eluant was then diluted tenfold with PCG A.

    (j) Similar to the heparin column, 600 µL of nickel resin was loaded into a 5 mL syringe column.

    (k) The resin was equilibrated with 5 mL of Ni buffer A.

    (l) The eluant from the heparin column was then loaded onto the nickel column.

    (m) The nickel column was washed with 30 column volumes of Ni buffer A.

(n) The protein was then eluted with 6 column volumes of Ni buffer B. Add 1 mL at a time to minimize disturbance of the bed and maximize elution.

(o) The purified enzyme was then concentrated and desalted through a 10,000 MWCO filter at 4 °C.

(p) The resulting desalted enzyme was then stored in liquid nitrogen in enzyme storage buffer until needed.

3.3 Enzymatic Synthesis Reaction

1. Mix the following (*see* **Note 9**):

   (a) 200 µg starting material (heparosan or other polymer, such as *N*-sulfated polymer).

   (b) 20 µL 5× MES buffer.

   (c) 200 µg of PAPS.

   (d) Appropriate amount of enzyme for modification. (This will need to be optimized based on the desired sulfation density.)

   (e) Bring total volume up to 100 µL with water.

2. Incubate the reaction mixture for 24–72 h with rotation at 37 °C.

3.4 Purification of Enzymatic Product

1. A 200 µL column volume of DEAE-sepharose resin is loaded onto the disposable column (*see* **Note 10**).

2. The column is equilibrated with 5 column volumes of DEAE wash buffer.

3. The enzymatic synthesis reaction is loaded three times onto the column.

4. The column is washed with 30 column volumes of DEAE wash buffer.

5. The product is eluted with 6 column volumes of DEAE elution buffer.

6. The sample is then concentrated and filter-desalted by addition of water five times in a centrifugal filter unit (3,000 MWCO).

3.5 Digestion of Enzymatic Product

1. It is important to filter-desalt all samples before proceeding with this procedure.

2. Mix the following (*see* **Note 11**):

   (a) 100 µL of sample.

   (b) 20 µL of 10× heparin lyase buffer.

   (c) Adjust the volume to 195 µL with dH_2O.

   (d) Add 5 µL of heparin lyase I/II/III mixture.

3. Incubate overnight at 37 °C.

4. Boil the sample for 15–45 s.

5. Centrifuge at $13400 \times g$ for 20–30 min and save supernatant.

3.6 Disaccharide Analysis by Strong Anion Exchange HPLC

1. The disaccharide standards may be co-injected to determine the elution profile of specific disaccharides (*see* **Note 12**). The disaccharides can be detected at 232 nm.

2. The disaccharides are eluted at 1 mL/min using the following buffer profile:
   (a) 100 % HPLC buffer A for 5 min.
   (b) Linear gradient of 0–100 % buffer B for 45 min.
   (c) 100 % HPLC buffer B for 10 min.
   (d) 100 % HPLC buffer A for 15 min.

---

4   Notes

1. All biochemicals, disposable plasticwares, and chromatographic resins/columns can be easily obtained from many commercial vendors.

2. All heparan sulfate biosynthetic enzymes are produced using the same protocol. C5-epimerase is shown here as an example.

3. Other strong anion exchange columns may also be used.

4. *E. coli* K5 is a pathogenic bacterial strain. Do not ingest or breathe in. Wear 0.2 μm pore size masks when growing this strain as an additional safety precaution.

5. This procedure is an outline for purifying small amounts of heparosan. Always check the loading capacity of DEAE-Sepharose resin to maximize recovery yield based on the batch size that is to be purified. Dialysis and centrifugal concentration are interchangeable techniques to desalt proteins and carbohydrates.

6. SF9 cells have a typical doubling time of 20 h. If the cells seem to grow significantly faster or slower by 3–4 h, then they may be too old and enzyme preparation is not recommended using these aberrant cells.

7. It is advisable to do a viral plaque assay with the C5-epimerase virus prior to doing large-scale infections for reproducible results.

8. Use plastic to minimize binding of the protein to the sides of the flask or any tube that is used.

9. This reaction is scalable.

10. Increase the column volume as necessary for the reaction scale.

11. This reaction is also scalable. The amount of disaccharide you need for analysis will vary depending on your method of analysis.

12. The amount of disaccharide used will depend on the detection abilities of the instrument. Typically, 20 μg of disaccharide is sufficient for detection at 232 nm by DAD or UV detector.

Acknowledgement

This work was supported in part by NIH grants (P01HL107152 and R01GM075168) to B.K. and by the NIH fellowship F31CA168198 to K.R.

References

1. Bernfield M, Götte M, Park P et al (1999) Functions of cell surface heparan sulfate proteoglycans. Annu Rev Biochem 68:729–777

2. Nakato H, Kimata K. (2002) Heparan sulfate fine structure and specificity of proteoglycan functions. Biochim Biophys Acta 1573:312–318

3. Forsberg E, Kjellén L. (2001) Heparan sulfate: lessons from knockout mice. J Clin Invest 108:175–180

4. Sinaÿ P, Jacquinet J, Petitou M et al (1984) Total synthesis of a heparin pentasaccharide fragment having high affinity for antithrombin III. Carbohydr Res 132:C5–C9

5. Petitou M, Hérault J, Bernat A et al (1999) Synthesis of thrombin-inhibiting heparin mimetics without side effects. Nature 398:417–422

6. Rosenberg R, Damus P (1973) The purification and mechanism of action of human antithrombin-heparin cofactor. J Biol Chem 248:6490–6505

7. Kuberan B, Miroslaw L, Beeler D et al (2003) Enzymatic synthesis of antithrombin III-binding heparan sulfate pentasaccharide. Nat Biotechnol 21:1343–1346

8. Kuberan B, Beeler D, Lawrence R et al (2003) Rapid two-step synthesis of mitrin from heparosan: a replacement for heparin. J Am Chem Soc 125:12424–12425

9. Lindahl U, Li JP, Kusche-Gullberg M et al (2005) Generation of "neoheparin" from E. coli K5 capsular polysaccharide. J Med Chem 48:349–352

10. Xu Y, Sayaka M, Takieddin M et al (2011) Chemoenzymatic synthesis of homogeneous ultralow molecular weight heparins. Science 334:498–501

11. Vann W, Schmidt M, Jann B et al (2005) The structure of the capsular polysaccharide (K5 antigen) of urinary-tract-infective Escherichia coli 010:K5:H4: a polymer similar to desulfoheparin. Eur J Biochem 116:359–364

12. Kuberan B, Lech M, Zhang L et al (2002) Analysis of heparan sulfate oligosaccharides with ion pair-reverse phase capillary high performance liquid chromatography-microelectrospray ionization time-of-flight mass spectrometry. J Am Chem Soc 124:8707–8718

Production of Size-Defined Heparosan, Heparan Sulfate, and Heparin Oligosaccharides by Enzymatic Depolymerization

Spencer Brown and Balagurunathan Kuberan

Abstract

Glycosaminoglycans (GAG) are most commonly isolated as large polymers from various animal origins, the functional units of which are oligosaccharides, which bind their target proteins to induce conformational changes, compete with other ligands, or facilitate the formation of signaling complexes. One example, the extensively studied heparin pentasaccharide sequence—which binds antithrombin-III, inducing a conformational change that increases its serpin protease activity by 1,000-fold—is unique in that no other specific GAG-protein structure-function relations have been described to the same degree. Thus, production of heparan sulfate (HS) oligosaccharides is critical for obtaining specific structural information regarding the binding interactions of GAG and their ligands (typically proteins). Purely synthetic methods of oligosaccharide synthesis are possible, but the cost, time requirement, and difficulty of their preparation prohibit library synthesis in significant amounts. Herein, the use of bacterial heparin lyases for the production of HS oligosaccharides via enzymatic depolymerization of HS polymers is discussed. The separation and purification of these oligosaccharides by liquid chromatography are also described.

Key words Glycosaminoglycans, Heparan sulfate, Heparosan, Heparin, Oligosaccharides, Enzymatic depolymerization

1 Introduction

Heparan sulfate (HS) represents a major class of glycosaminoglycans (GAG), which is composed of repeating disaccharide units of glucosamine and uronic acid. These linear polysaccharides are heavily modified in the Golgi apparatus by a series of sulfotransferases, deacetylases, and epimerase to yield highly heterogeneous, acidic, biologically-active, complex molecules [1]. Heparin, the most densely sulfated HS-GAG and its specific interaction with antithrombin III have been extensively characterized; yet many of the putative structure-activity relationships between HS and GAG-binding proteins remain elusive [2, 3]. This paucity of specific data is due both to the complexities of multimeric interactions dominated by electrostatic forces as well as difficulties in producing adequate quantities of specific HS oligosaccharides for study.

In common laboratory practice, HS-GAGs may be produced either by chemical degradation using nitrous acid or by enzymatic cleavage using bacterial heparin lyases (also called heparitinases) by unique mechanism of random endolytic followed by exolytic depolymerization. Thus, heparitinase first binds along the HS or heparin polysaccharide and then cleaves the chain to produce smaller oligosaccharide chains. Heparitinase may then remain bound and proceed down the HS/heparin chain, processively catalyzing depolymerization into HS disaccharide units, each composed of glucosamine and uronic acid [4].

Application of nitrous acid depolymerization of HS has been described in depth for a variety of pH conditions with or without hydrazinolysis [5]. However, its application for oligosaccharide production, rather than complete degradation, requires significant optimization and sample work-up, and yields a product devoid of UV absorbance. Enzymatic depolymerization, on the other hand, produces oligosaccharides with UV absorbance at 232 nm, which can easily be monitored and separated after a single-step enzymatic digestion with no work-up with a detection limit of 50 pmol [6].

Three major commercially available heparin lyase enzymes are commonly employed in the study of heparan sulfate, each of which varies slightly in its substrate specificity. Therefore, in the production of glycosaminoglycan oligosaccharides, the choice of a specific heparin lyase depends on the sulfation pattern and density as well as degree of epimerization of the chosen substrate [1, 7]. Heparan sulfate is more structurally diverse than heparin, with alternating domains of low (N-acetyl or NA domains) and high (N-sulfated or NS domains) degrees of sulfation. Heparin, in contrast, is highly N-sulfated with a dominating tri-sulfated disaccharide motif (GlcNS6S α(1–4) IdoA2S). Another GAG, N-acetylheparosan, has been used extensively as an HS precursor and NA domain analog in chemical and enzymatic synthesis of HS-GAGs for biological studies. This low-sulfated substrate is preferred by heparin lyase III [8]. Table 1 shows the distinct specificities of the three heparin lyases and the corresponding choice of substrate for enzymatic oligosaccharide preparation [1].

Originally, methods for obtaining oligosaccharides relied on calculations for heparin lyase activity and determination of enzyme kinetics that give the optimal yield of an oligosaccharide of a given length by halting the reaction at a certain point by heat inactivation of the enzyme [7]. Herein is described a simpler method of depolymerization combined with filtration which proves more robust since oligosaccharides are removed from the reaction as soon as they are cleaved from the polysaccharide chain. This effectively avoids the disproportionate overproduction of HS disaccharide and enhances the yield of larger oligosaccharide sizes of biological significance.

Table 1
Substrate specificity of heparitinases

Enzyme isoform	Preferred cleavage site	Preferred GAG substrate(s)
Heparin lyase I (heparinase, heparitinase III)	GlcNS6S-α(1–4)-IdoA2S	Heparin > heparan sulfate
Heparin lyase II (heparitinase II)	GlcNAc/NS+/−6S-α(1–4)-IdoA+/−2S	Heparan sulfate > heparin
Heparin lyase III (heparitinase I)	GlcNAc/NS-α(1–4)-GluA/IdoA	Heparosan, N-sulfated heparosan, NA HS domains

Adapted from ref. 1

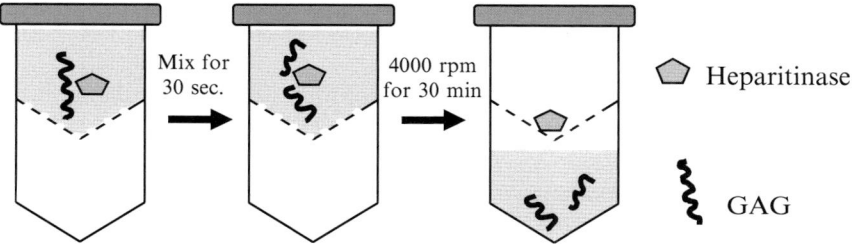

Fig. 1 Schematic of GAG depolymerization by centrifugal filtration

In this method, the reaction mixture is transferred into a centrifugal filter and spun in a desktop centrifuge until the heparitinase enzyme and the remaining undigested polysaccharides are concentrated in the retentate and an oligosaccharide mixture is collected in the flow-through (as shown in Fig. 1). The exact pore size required for optimal oligosaccharide yield is difficult to calculate since GAGs are linear polymers with a large volume of hydration, and not globular with a defined radius like the proteins for which centrifugal filter tubes are designed. However, anywhere from 5,000 to 20,000 MWCO membranes may be effective and have been used in centrifugal depolymerization as well as ultrafiltration methods. Larger pore sizes will allow too much enzyme and undigested polysaccharide through the filter, while smaller pore sizes will retain the larger oligosaccharides.

Unlike heparanase, which catalyzes random depolymerization of heparan sulfate by hydrolysis of glycosidic bonds, heparitinases catalyzes an *elimination* reaction at glycosidic bonds, which produces a Δ^{4-5} unsaturated bond when the protonated C4-hydroxyl of the uronic acid is expelled as a leaving group. This double bond present on this nonreducing terminal uronic acid residue has a UV absorption maximum at 232 nm, which is utilized during separation and purification of these oligosaccharides [9].

Although other methods may be used such as low-pressure chromatography through a Biogel P-10 column, complete separation of oligosaccharides is best achieved by preparative

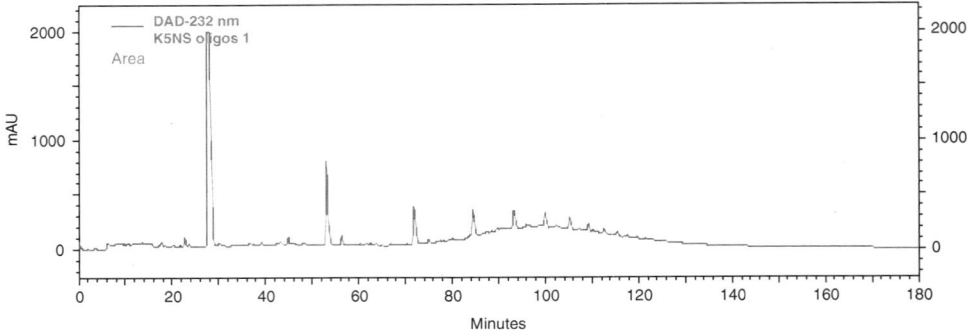

Fig. 2 HPLC chromatogram of UV absorbance at 232 nm shows effective separation of *N*-sulfoheparosan oligosaccharides, dp2–dp24, on a semi-preparative (9 × 250 mm) CarboPac PA1 HPLC column with a linear 0–1.0 M NaCl gradient

high-performance liquid chromatography (HPLC) through a strong anion exchange column. One particular anion exchange column that gives excellent resolution and reproducibility of oligosaccharide peaks, including *N*-acetylheparosan, is the CarboPac PA1 column, which is stable over a pH range from 0 to 14 and at all concentrations of buffer salts. Excellent oligosaccharide separation can be achieved with a simple gradient from 0 to 1 M NaCl (*see* **Note 1**) with larger and more negatively charged oligosaccharides eluting at a higher salt concentration as shown in Fig. 2 [6]. Although the high ionic strength of the eluent precludes the use of this method in tandem with mass-spectrometry analysis, anion exchange chromatography is superior for separation of larger oligosaccharides with higher yields.

Each of the peaks observed in the chromatogram in Fig. 2 correlates to a specific N-sulfated heparosan oligosaccharide. These oligosaccharides are commonly described by their degree of polymerization (dp), which refers to the total number of monomer units in that molecule. Since heparosan is composed of repeating disaccharide units of glucuronic acid and glucosamine, a dp6 oligosaccharide would consist of three disaccharide units, each of which has a glucuronic acid and a glucosamine, for a total of six carbohydrate monomers. Thus, in Fig. 2, the absorbance peaks at 28, 53, 72, and 85 min may be called K5NS dp2, K5NS dp4, K5NS dp6, K5NS dp8, and so on. Initially, it might appear as though there is a much lower yield of high-molecular-weight oligosaccharides than of disaccharide, but a direct comparison of peak areas under the curve (AUC) is not accurate for weight equivalents. Note that dp10 has five times the mass of dp2, yet each molecule has only one double bond, or absorptive functional group. Therefore, if the AUC of a dp2 is five times greater than the dp10 peak, the actual yield of those two structures will be nearly equal by weight.

Each of the elution fractions corresponding to these peaks may be collected and individually desalted by gel-filtration chromatography through a hand-packed bead media such as Sephadex G-10 or Biogel P-2 with a running buffer of 1–200 mM ammonium bicarbonate. Some researchers tend to avoid Sephadex gel for carbohydrate applications since its dextran-based scaffold may potentially contaminate eluate, complicating structural and functional analysis. Fractions collected from the desalting column are then analyzed in a microplate reader or UV spectrophotometer at 232 nm in order to identify the oligosaccharide-containing fractions. These fractions may then be pooled together and lyophilized to yield a pure oligosaccharide. An optimized digestion of 10 mg of heparosan may yield up to 200–500 µg of each oligosaccharide size, ranging from dp2 up to dp24 (*see* **Note 2**).

Isolation of distinct heparan sulfate and heparin oligosaccharides from enzymatic digestion (*see* **Note 3**) is more challenging because of their heterogenous sulfation patterns. With homogenous GAG such as heparosan or *N*-sulfoheparosan, differently sized oligosaccharides will always have different retention times by anion exchange chromatography because the size and charge are directly related. This is not the case with heparan sulfate and heparin. Thus, resolution of HS/heparin oligosaccharide mixtures may require a size-exclusion chromatography (SEC) step (often done using a low-pressure Biogel P-10 column), in which oligosaccharides are separated by size without any regard to charge density (*see* **Note 4**). This is followed by charge density-based separation of size-specific fractions by strong anion exchange (SAX) HPLC of distinct structures [10]. Classical methodologies also include an antithrombin affinity separation step, but this may be unnecessary when investigating non-anticoagulant activities of HS. One drawback of enzymatic oligosaccharide preparation is that it only allows collection of naturally occurring HS sequences. Thus, although enzymatic depolymerization and separation can yield significant amounts of many oligosaccharides, production of unnatural oligosaccharides may require alternative synthetic methods.

2   Materials

1. Heparosan
   (a) Heparosan may also be produced in-house by purifying capsular polysaccharide from cultured K5 *Escherichia coli*.

2. Heparan sulfate.

3. Heparin.

4. Heparin lyase enzyme I/II/III (aliquot into 20 µL aliquots of 1–5 units each and store at –20 °C).

5. 10× Heparin enzyme buffer (33 mM calcium acetate, 400 mM ammonium acetate, 1 mg/mL bovine serum albumin, filtered, deionized water, adjust to pH 7.0 with NaOH/HCl).

6. Centrifugal filter tubes 10,000 Da MWCO.

7. Low-pressure chromatography column (1×50 cm).

8. BioGel P-2 and P-10 size-exclusion gel.

9. Strong anion exchange chromatography column.

10. HPLC equipped with binary pump, UV detector with 232 nm range, and an attached fraction collector.

3 Methods

3.1 Enzymatic Depolymerization of HS-GAG by Centrifugal Filtration

1. Wash centrifugal filter according to the manufacturer's instructions, typically by adding deionized water to the tube above the filter and spinning at high speed for 5–10 min.

2. Combine the following reaction mixture in the washed centrifugal filter in the following order:

   (a) 500 μL 10× heparin lyase buffer.

   (b) 4,280 μL milli-Q water.

   (c) 1–5 units of heparin lyase (*see* **Notes 3, 5, 6**).

   (d) 200 μL HS-GAG 50 mg/mL (10 mg).

3. Tightly cap the reaction mixture and mix gently by inverting the tube several times for 30 s (*see* **Note 7**).

4. Place the centrifugal filter tube in a desktop centrifuge, and after balancing the rotor immediately spin for 60 min at 4,000 rpm or until <5 % of the reaction mixture remains in the retentate (*see* **Note 8**). This flow-through sample contains a mixture of oligosaccharides, each of which is capped by a terminal 4,5-unsaturated uronic acid residue at the nonreducing end as a result of the heparitinase cleavage.

5. Collect the flow-through (below the filter), lyophilize, and reconstitute in an appropriate volume for separation by HPLC.

3.2 Separation of HS-GAG Oligosaccharides by SAX-HPLC

1. Prime HPLC system with filtered, degassed, purified water, pH 3.5, as buffer A and 1.0 M NaCl (2.0 M NaCl for heparin oligosaccharides, *see* **Notes 1, 9**) pH 3.5 as buffer B. HPLC method should include a linear gradient of 45–90 min (depending on sample size) for adequate separation.

2. Inject the sample to HPLC fitted with a strong anion exchange column, UV absorbance detector, and fraction collector.

3. Collect and combine HPLC fractions corresponding to oligosaccharide peaks and lyophilize if necessary to reduce sample volume.

4. Remove salt and buffer by dialysis through <1,000 MWCO membrane (dialysis may have excessive sample loss for smaller oligosaccharides) over 3 days or alternatively by low-pressure, size-exclusion chromatography on a Biogel P-2 column.

5. Eluted fractions from Biogel P-2 column are then read by UV spectrometry at 232 nm and peaks—corresponding to desalted oligosaccharide—may then be pooled and lyophilized to obtain dry, purified oligosaccharide.

4 Notes

1. Sodium chloride gradient separation should be tailored to the oligosaccharides being separated. For example, heparin oligosaccharides are the most negatively charged and are best separated on a gradient from 1.0 to 2.0 M NaCl, whereas heparosan oligosaccharides are much less negatively charged and can therefore be separated more easily on a gradient of 0–0.5 M NaCl. Furthermore, nonlinear gradients can also be used to achieve greater separation between similar species in complex mixtures.

2. The amount of heparin lyase used in each depolymerization reaction may need to be optimized by performing several digestions with varying amounts of heparin lyase simultaneously in separate reaction mixtures and comparing overall oligosaccharide yield as well as relative disaccharide composition.

3. Choice of heparin lyase depends on the HS-GAG being depolymerized. As discussed in the introduction, lyase activity varies based on the degree of substrate sulfation. Thus, although a mixture of all three lyases may be used for any depolymerization, heparin lyase III (heparitinase I) is best for heparosan and N-sulfoheparosan, and heparin lyase I (heparitinase III) is best for heparin. Heterogeneously O-sulfated HS-GAG (heparan sulfate) should be depolymerized by heparin lyase II (heparitinase II) or by a combination of all three lyase enzymes.

4. In contrast to the centrifugal filtration method described, enzymatic depolymerization of heterogenous HS can be accomplished alternatively via uncontrolled digestion using a single heparin lyase to digest undesired sequences and procure the remaining oligosaccharides. For example, overnight digestion of heparin with heparin lyase III removes the low-sulfated, NA domains as disaccharides, leaving highly sulfated oligosaccharides for purification.

5. Heparin lyase activity varies significantly between manufacturers based on the source and mode of production (e.g., *Bacteroides* or *Flavobacterium*). Enzyme substrate specificity may also vary slightly (e.g., activity on iduronic acid or glucuronic acid residues). One unit of heparin lyase is defined as the amount of enzyme that will liberate 1.0 μmol unsaturated oligosaccharides from porcine mucosal heparin per minute at 30 °C and pH 7.0 in a total reaction volume of 100 μL, but definitions vary between manufacturers.

6. Given that manufacturer and inter-user variability may be significant, attempts of HS oligosaccharide production should begin with smaller reactions (~0.1–1.0 mg polysaccharide substrate) and several different enzyme concentrations. Comparison of oligosaccharide peak AUCs serves as an aid in adjusting enzyme/substrate ratios. A dp2/dp4 AUC ratio >10–20 denotes excessive enzyme. Low overall yields and disproportionate yields of oligosaccharides >dp14 indicate insufficient enzyme.

7. To avoid excessive production of disaccharides, the GAG polysaccharide should be added last, and then the tube is capped tightly, and briefly mixed for 15–30 s by rotating the tube to ensure a homogenous mixture. Next, the tube should be immediately centrifuged at 4,000 rpm (assuming that this will not compromise the filter membrane, check the manufacturer's instructions).

8. Required centrifugation time will depend on the purity of the sample and capacity of the filter membrane. Typically 60 min at 4,000 rpm is sufficient to collect the majority of the solution. Once >90 % of the reaction has collected in the flow-through, the flow-through is lyophilized until a total volume has been reduced to ~1.0 mL (or whatever volume may be completely injected into the HPLC in a single run).

9. Extensive use of sodium chloride gradients can lead to corrosion of HPLC machinery. Special care should be taken to rinse the instrument flow path thoroughly following each run to remove excess salt and avoid blockage.

Acknowledgement

This work was supported in part by NIH grants (P01HL107152 and R01GM075168) to B.K. and by the Eccles fellowship to S.B.

References

1. Esko JD, Kimata K, Lindahl U (2009) Proteoglycans and sulfated glycosaminoglycans. In: Varki A, Cummings RD, Esko JD, Freeze HH, Stanley P, Bertozzi CR, Hart GW, Etzler ME (eds) Essentials of glycobiology, 2nd edn. Cold Spring Harbor Laboratory Press, Cold Spring Harbor, NY, pp 229–248

2. Casu B, Lindahl U (2001) Structure and biological interactions of heparin and heparan sulfate. Adv Carbohydr Chem Biochem 57: 159–206

3. Capila I, Linhardt RJ (2002) Heparin-protein interactions. Angew Chem Int Ed Engl 41(3): 391–412

4. Conrad HE (2001) Nitrous acid degradation of glycosaminoglycans. Curr Protoc Mol Biol. Chapter 17:Unit17.22A

5. Hovingh P, Linker A (1970) The enzymatic degradation of heparin and heparitin sulfate. 3. Purification of a heparitinase and a heparinase from flavobacteria. J Biol Chem 245(22): 6170–6175

6. Pervin A, Gallo C, Jandik KA, Han XJ, Linhardt RJ (1995) Preparation and structural characterization of large heparin-derived oligosaccharides. Glycobiology 5(1):83–95

7. Babu P, Kuberan B (2010) Fluorescent-tagged heparan sulfate precursor oligosaccharides to probe the enzymatic action of heparitinase I. Anal Biochem 396(1):124–132

8. Rusnati M, Oreste P, Zoppetti G, Presta M (2005) Biotechnological engineering of heparin/heparan sulphate: a novel area of multi-target drug discovery. Curr Pharm Des 11(19):2489–2499

9. Linhardt RJ, Galliher PM, Cooney CL (1986) Polysaccharide lyases. Appl Biochem Biotechnol 12(2):135–176

10. Jones CJ, Beni S, Limtiaco JF, Langeslay DJ, Larive CK (2011) Heparin characterization: challenges and solutions. Annu Rev Anal Chem 4:439–465

Chapter 4

Chemical Modification of Heparin and Heparosan

Karthik Raman, Balagurunathan Kuberan, and Sailaja Arungundram

Abstract

Heparin is a potent clinically used anticoagulant. It is a heterogeneous mixture of polymers that contain a variety of sulfation patterns. However, only 3-O sulfonated heparin pentasaccharide units have been proven to bind to antithrombin and elicit an anticoagulant response. Heparins with other sulfation patterns are able to bind to a variety of other proteins such as FGF, VEGF, and CXCL-3. By modulating heparin's sulfation pattern, it is possible to generate polymers that can regulate biological processes beyond hemostasis. Here we describe a variety of simple chemical modification methods, N-acetylation, N-deacetylation, N-sulfation, O-sulfation, 2-O desulfation, and complete desulfation, to prepare heparin-like polymers with distinct sulfation patterns.

Key words Heparin, Desulfation, O-sulfation, Deacetylation, N-sulfation

1 Introduction

Glycosaminoglycans (GAGs) are composed of repeating disaccharide units of hexuronic acid and hexosamine sugar. Heparin is primarily composed of a trisulfated disaccharide units (Ido2S-GlcNS6S) whereas heparosan, a precursor derived from *E. coli* K5, is primarily composed of unsulfated disaccharide units (GlcA-GlcNHAc). Utilizing these two polymers, it is possible to design GAGs with any sulfation pattern from a top-down (desulfation of heparin) or bottom-up (O- and N-sulfation of heparosan) approach. By varying the content of epimerized residues, sulfation pattern, sulfation density, and chain length, it is possible to significantly affect the biological functions of GAGs. It is known that heparin-FGF interactions can be modulated by N-desulfation of heparin, 2-O desulfation of heparin's iduronic acid residues, and the removal of 6-O sulfate groups from heparin's glucosamine residues [1–3]. Additionally, N-sulfate and 3-O sulfate groups are critical to heparin's anticoagulant activity [4, 5].

In the past decades several research groups have modified heparin's sulfation pattern using a variety of simple techniques.

Kuberan Balagurunathan et al. (eds.), *Glycosaminoglycans: Chemistry and Biology*, Methods in Molecular Biology, vol. 1229, DOI 10.1007/978-1-4939-1714-3_4, © Springer Science+Business Media New York 2015

It is possible to synthesize *N*-acetylated amino sugars from heparin or heparosan [6]. Heparin may be subjected to *N*-deacetylation and *N*-sulfation to increase its binding affinity for FGFs and other growth factors [7]. It is also possible to utilize dimethyl sulfoxide and high temperature to solvolytically desulfate heparin to make it devoid of any anticoagulant activity [8]. A key component of this mixture is the small content of water and methanol which facilitates hydrolysis of the sulfate esters. More recently, some research groups have developed a technique to persulfonate heparin to increase its ability to bind growth factors and also to modulate coagulation [9]. The pyridine:sulfur trioxide complex selectively reacts with heparin-like polymers to generate *O*-sulfo groups in the following order of preference: 6-*O* sulfo groups on glucosamine > 2-*O* sulfo groups on iduronic acid > 2-*O* sulfo groups on glucuronic acid. It is also possible to selectively remove the 2-*O* sulfate groups of heparin using basic conditions under lyophilization and minimize chain scission [10]. The 6-*O* sulfate groups on the glucosamine residues of heparin can be selectively removed using *N*-methyltrimethylsilyl-trifluoroacetamide [11]. An enormous variety of chemically modified heparin polymers can be prepared using techniques from these seminal studies.

2   Materials

2.1   Components for Chemical Modifications

Amberlite H⁺ ion exchange resin.

Exchange solution (10 % v/v tributylammonium in 200 proof ethanol).

Deacetylation buffer (2.5 M NaOH).

Neutralization buffer (2.5 M acetic acid in water).

Precipitation buffer (saturated NaOAc in 200 proof ethanol).

Sulfur trioxide:triethylamine salt.

Centrifugal filters, 1,000 MWCO.

Dialysis membrane, 1,000 MWCO.

Sulfur trioxide:pyridine salt.

N-methyltrimethylsilyl-trifluoroacetamide (MTS-TFA).

3   Methods

3.1   Production of the Pyridinium Salt of Heparin/Heparosan Polymer

1. Dissolve 100 mg of polymer in 10 ml of water.

2. Take a 5 ml column.

3. Fill the column with Amberlite H⁺ ion exchange resin.

4. Wash the column with 20 ml of water or until the orange coloring disappears from the eluate.

5. Load the polymer solution onto the column and collect the eluate in a beaker on ice.

6. Elute with additional 10 ml of water to ensure that all the polymer passes through the column.

7. Add pyridine until the pH of the eluate reaches pH 9 and stir the solution for 30 min on ice.

8. Subsequently evaporate the pyridine and water by using a rotary evaporator to yield the pyridinium salt of polymer.

3.2 Production of the Tributyl-ammonium Salt of Heparin/Heparosan Polymer

1. Dissolve 100 mg of polymer in 10 ml of water.

2. Take a 5 ml column.

3. Fill the column with Amberlite H^+ ion exchange resin.

4. Wash the column with 20 ml of water or until the orange coloring disappears from the eluate.

5. Load the polymer solution onto the column and collect the eluate in a beaker on ice.

6. Elute with additional 10 ml of water to ensure that all the polymer passes through the column.

7. Add exchange solution until the eluant reaches pH 9 and stir the solution for 30 min on ice.

8. Subsequently extract the souliton with ether: 1 volume of diethyl ether is added to the solution and the aqueous phase is separated from the ether phase using a separatory funnel. This is repeated three times.

9. Finally, the aqueous phase containing the polymer is lyophilized to dryness.

3.3 N-Deacetylation of Heparosan

1. Heparosan is deacetylated by dissolving 1 mg of heparosan in 1 ml of deacetylation buffer and heating at 55 °C overnight (*see* **Note 1**).

2. After 16 h, the solution is neutralized to pH 7.0 using neutralization buffer.

3. Subsequently, the product is precipitated by adding 6 volumes of precipitation buffer and letting the container stand at −20 °C overnight.

4. The precipitate is centrifuged at $4,000 \times g$ at 4 °C and the pellet is redissolved in water.

5. This solution, containing deacetylated heparosan, can be purified through DEAE-sepharose, desalted, and finally analyzed using an HPLC after heparitinase treatment (*see* **Note 2**).

3.4 N-Reacetylation of Heparosan

1. 60 mg of heparosan carrying free amine is dissolved in ice-cold 10 % methanol containing 0.05 M Na_2CO_3.

2. To this mixture 1 ml of ice-cold acetic anhydride is added for 1 h at 10-min intervals and the reaction pH is maintained

between 7 and 7.5 with 10 % methanol saturated with Na_2CO_3 on ice.

3. Subsequently, the polymer is precipitated using precipitation buffer overnight at −20 °C.

4. The precipitate is then centrifuged (as above), redissolved in water, purified through DEAE-sepharose, dialyzed, and lyophilized.

3.5   N-Sulfation of Heparosan

1. To a solution of 1 mg of deacetylated heparosan, 2.5 mg of Na_2CO_3 and 2.5 mg of sulfur trioxide:triethylamine complex are added and stirred overnight while heating at 48 °C. The pH is maintained at 9.5 by using neutralization buffer and sodium hydroxide.

2. The next day, an additional 2.5 mg of Na_2CO_3 and 2.5 mg of sulfur trioxide:triethylamine are added and stirred overnight. The pH is checked once again to maintain the reaction at pH 9.5.

3. Subsequently, the solution is neutralized using neutralization buffer.

4. The product is purified through DEAE-sepharose and is dialyzed through a 1,000 MWCO membrane for 3 days. The water is changed five times in the 3-day period at 1 h, 3 h, 24 h, 2 days, and 3 days (*see* **Note 3**).

5. The retentate (*N*-sulfoheparosan) is then lyophilized to dryness.

3.6   O-Sulfation of Heparosan

1. 10 mg of the tributylammonium salt of heparosan is dissolved in 5 ml of DMF.

2. Three equivalents of the sulfur trioxide:pyridine complex is added and the solution is stirred at 4 °C for 4 h.

3. Additional three equivalents of sulfur trioxide:pyridine complex is added and the solution is stirred at 4 °C for another 4 h.

4. Another three equivalents of sulfur trioxide:pyridine complex is added and the solution is stirred overnight at room temp.

5. Subsequently, the solution is neutralized with 1 volume of ice-cold neutralization buffer.

6. Finally, the sulfated heparosan is precipitated by addition of 6 volumes of ice-cold ethanol saturated with sodium acetate (precipitation buffer) at 20 °C overnight.

7. The resulting precipitate is centrifuged, resuspended in water, digested with heparitinase, and analyzed by HPLC (*see* **Note 4**).

3.7   Complete Desulfation of Heparin

1. Dissolve 100 mg of the pyridinium salt of heparin in 10 ml of a 9:1 DMSO:methanol mixture.

2. Stir overnight at 100 °C under reflux conditions.

3. Check the extent of desulfation at various time points by taking out a 100 μl aliquot, diluting with water, desalting, digesting with heparitinase, and checking on via HPLC analysis.

4. Once the reaction is complete, dilute it with 10 volumes of water, dialyze through a 1,000 MWCO membrane, and lyophilize to dryness (*see* **Note 5**).

3.8 2-O Desulfation of Heparin

1. 10 mg of heparin is mixed with 1 mg of NaBH$_4$ in 10 ml of 0.4 N NaOH.

2. This mixture is then frozen in a –80 °C refrigerator and lyophilized.

3. The resulting crusty yellow solid is subsequently redissolved in 5 ml of water and neutralized to pH 7 with ice-cold neutralization buffer.

4. This solution is then dialyzed for 3 days and lyophilized.

3.9 6-O Desulfation of Heparin

1. 100 mg of the pyridium salt of heparin is dissolved in 10 ml of pyridine.

2. Next, 2 ml of N-methyltrimethylsilyl-trifluoroacetamide (MTS-TFA) is added to the reaction and the reaction is stirred at 110 °C for 60 min.

3. This mixture is concentrated on a rotary evaporator at 40 °C for 30 min.

4. Subsequently the mixture is quenched with 5 ml of water and precipitated by addition of 6 volumes of precipitation buffer overnight at –20 °C.

5. Next, the precipitate is dissolved in water, dialyzed for 3 days, and then lyophilized.

4 Notes

1. NaOH reduces the overall size of the polymers because it slowly degrades the reducing end. Do not leave the polymer in these harsh conditions for too long.

2. It is advisable to constantly monitor how much of the reaction has completed using HPLC analysis at several time points. It is possible that some of these chemical reactions may need different reaction times to reach completion depending on the scale of the reaction and the ability to heat/stir effectively. DEAE-sepharose purification and HPLC analysis are described in the chapter about enzymatic synthesis.

3. To rapidly desalt within 1 h instead of waiting for several days using dialysis membranes, it is possible to use centrifugal filters with the defined MWCO.

4. If the polymer is highly sulfated, it will not be cleaved by heparitinase. NMR is a good alternative to analyze these polymers.

5. The DMSO in this reaction will damage dialysis membranes. Make sure to dilute it to ≤ 5 % v/v in water. Alternatively, one can precipitate this polymer prior to dialysis using the precipitation buffer.

Acknowledgement

This work was supported in part by NIH grants (P01HL107152 and R01GM075168) to B.K. and by the NIH fellowship F31CA168198 to K.R.

References

1. Guimond S, Maccarana M, Olwin BB, Lindahl U, Rapraeger AC (1993) Activating and inhibitory heparin sequences for FGF-2 (basic FGF). Distinct requirements for FGF-1, FGF-2, and FGF-4. J Biol Chem 268(32):23906–23914

2. Lundin L, Larsson H, Kreuger J, Kanda S, Lindahl U, Salmivirta M, Claesson-Welsh L (2000) Selectively desulfated heparin inhibits fibroblast growth factor-induced mitogenicity and angiogenesis. J Biol Chem 275(32): 24653–24660

3. Sugaya N, Habuchi H, Nagai N, Ashikari-Hada S, Kimata K (2008) 6-O-sulfation of heparan sulfate differentially regulates various fibroblast growth factor-dependent signalings in culture. J Biol Chem 283(16):10366–10376

4. Atha DH, Lormeau JC, Petitou M, Rosenberg RD, Choay J (1987) Contribution of 3-O- and 6-O-sulfated glucosamine residues in the heparin-induced conformational change in antithrombin III. Biochemistry 26(20):6454–6461

5. Danishefsky I, Ahrens M, Klein S (1977) Effect of heparin modification on its activity in enhancing the inhibition of thrombin by antithrombin III. Biochim Biophys Acta 498(1):215–222

6. Levvy GA, McAllan A (1959) The N-acetylation and estimation of hexosamines. Biochem J 73:127–132

7. Lloyd AG, Embery G, Fowler LJ (1971) Studies on heparin degradation. I. Preparation of (35 S) sulphamate derivatives for studies on heparin degrading enzymes of mammalian origin. Biochem Pharmacol 20(3):637–648

8. Nagasawa K, Inoue Y, Kamata T (1977) Solvolytic desulfation of glycosaminoglycuronan sulfates with dimethyl sulfoxide containing water or methanol. Carbohydr Res 58(1):47–55

9. Ogamo A, Metori A, Uchiyama H, Nagasawa K (1989) Reactivity toward chemical sulfation of hydroxyl-groups of heparin. Carbohydr Res 193:165–172

10. Ishihara M, Kariya Y, Kikuchi H, Minamisawa T, Yoshida K (1997) Importance of 2-O-sulfate groups of uronate residues in heparin for activation of FGF-1 and FGF-2. J Biochem 121(2):345–349

11. Ishihara M, Takano R, Kanda T, Hayashi K, Hara S, Kikuchi H, Yoshida K (1995) Importance of 6-O-sulfate groups of glucosamine residues in heparin for activation of FGF-1 and FGF-2. J Biochem 118:1255–1260

Synthesis of Sulfur Isotope-Labeled Sulfate Donor, 3′-Phosphoadenosine-5′-Phosphosulfate, for Studying Glycosaminoglycan Functions

Caitlin Mencio, Vimal P. Swarup, Marcus Soliai, and Balagurunathan Kuberan

Abstract

The biological activity of glycosaminoglycans (GAGs) depends greatly on the sulfation pattern present within the GAG chain. Chemical biology of GAGs can be further advanced by preparation of sulfur-isotope-enriched sulfated GAGs. 3′-Phosphoadenosine-5′-phosphosulfate (PAPS) serves as a universal sulfate donor in the sulfation of GAGs by sulfotransferases. Therefore, synthesis of PAPS carrying sulfur isotopes is critical in the preparation of labeled GAGs for biochemical studies. Here we describe a robust in vitro enzymatic synthesis of sulfur isotope-enriched PAPS which allows for heavy- or radio-isotope labeling of GAG chains.

Key words PAPS, APS kinase, Sulfurylase, PAPS synthase, Radioactive sulfate, Glycosaminoglycans

1 Introduction

The sulfation of glycosaminoglycans is essential for their involvement in a variety of biological activities such as cell signaling [1], coagulation [2], and development [3]. To achieve this sulfation in vivo, cells produce 3′-phosphoadenosine-5′-phosphosulfate (PAPS) that serves as the sulfuryl donor for sulfation by sulfotransferases [4]. PAPS synthesis in vivo (Fig. 1) has been well characterized and consists of a series of enzyme-catalyzed reactions. First inorganic sulfate is adenylated to form adenosine 5′-phosphosulfate (APS) through a reversible reaction catalyzed by an ATP sulfurylase. Next through the activity of APS kinase, an adenylylsulfate kinase, APS is converted to PAPS. Both reactions utilize magnesium ATP as substrate [5–7]. In vitro, the production of PAPS is modulated by the hydrolysis of inorganic pyrophosphate by a pyrophosphatase and the rate of conversion by APS kinase [6]. ATP sulfurylase and APS kinase can exist as separate proteins, as is the case with plants/

Kuberan Balagurunathan et al. (eds.), *Glycosaminoglycans: Chemistry and Biology*, Methods in Molecular Biology, vol. 1229, DOI 10.1007/978-1-4939-1714-3_5, © Springer Science+Business Media New York 2015

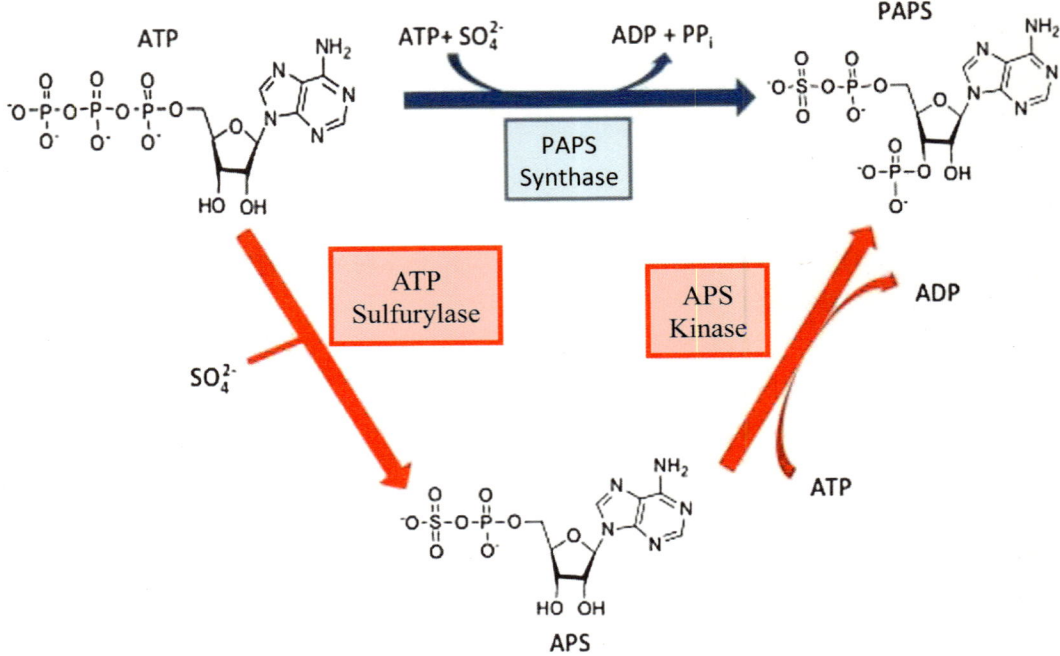

Fig. 1 The schematic for the production of PAPS in biological systems. The red arrows correspond to the pathway used in plants, bacteria and some animals. This pathways requires two separate enzymes: APS kinase and ATP sulfurylase. The blue pathway represents the bifunctional enzyme, PAPS synthase, present in humans that allows direct conversion of ATP into PAPS

fungi, but in animals and some bacteria these two enzymes have been found in a single polypeptide chain resulting in a bifunctional enzyme, PAPS synthase [8–10]. The discovery of PAPS synthase has allowed for a less complicated in vitro reaction for the production of PAPS. The production of radiolabeled PAPS (PAPS[35]) as well as heavy isotope-labeled PAPS (PAPS[33], PAPS[34]) can enable one to assess sulfotransferase activity and specificity, and also protein-GAG binding affinity. This chapter provides a detailed protocol for the production and characterization of radioactive S[35]-labeled PAPS. Additionally this protocol can also be used to prepare PAPS labeled with nonradioactive sulfur isotopes.

2 Materials

Prepare all solutions using ultrapure water, either autoclaved D.I. water or D.I. water sterilized through a 0.22 μm filter, and analytical grade reagents. Prepare and store all solutions at room temperature unless otherwise indicated. Dispose of all waste in accordance with the rules and requirements of your research facility.

2.1 Materials for the Synthesis of PAPS

1. MOPS buffer: 50 mM MOPS, pH 7.0. Adjust pH using NaOH.
2. MgATP solution: 100 mM MgATP.
3. Magnesium chloride solution: 100 mM $MgCl_2$ in 50 mM Tris.
4. 10% Triton X-100 solution.
5. BSA solution: 100 mg/mL BSA.
6. PEP solution: 150 mM PEP.
7. $Na_2S^xO_4$: 10 mCi/mL or 3 mg/mL x = isotope value of sulfur. Method described has been verified for production using S^{35} and S^{34}. See **Note 1**.
8. IPP solution: Inorganic pyruvate phosphate: 2000 U/mL.
9. PAPS synthase: 0.5 mg/mL.
10. Pyruvate kinase: 10,000 U/mL.

2.2 Materials for the Purification of PAPS

1. DEAE-Sephacel beads.
2. Disposable 5 mL polypropylene column.
3. Wash buffer: 200 mM TEA, adjust to pH 8 using CO_2.
4. Elution buffer: 400 mM TEA, adjust to pH 8 using CO_2.

2.3 Materials for TLC Analysis of PAPS[35]

1. Ethanol solution: 100 % EtOH.
2. Lithium chloride solution: 1 M LiCl.
3. PEI-cellulose TLC plates.

3 Methods

3.1 In Vitro Synthesis of PAPS

1. Combine in a 2 mL microcentrifuge tube:
   (a) 500 µL MOPS buffer.
   (b) 20 µL MgATP solution.
   (c) 25 µL Magnesium chloride.
   (d) 5 µL 10 % Triton X-100.
   (e) 1 µL BSA solution.
   (f) 100 µL PEP solution.
   (g) 200 µL $Na_2S^xO_4$ solution.
   (h) 5 µL IPP solution.
   (i) 20 µL PAPS synthase.
   (j) 10 µL Pyruvate kinase.
   (k) Adjust to 1 mL with dH_2O.
2. Place the reaction vial in 37 °C water bath for 12 h or overnight.
3. Remove the reaction vial from water bath. See **Note 1**.

**3.2 Purification
of the PAPS Reaction**

1. Set up a purification column using the disposable polypropylene columns with a bed volume of 0.5 mL DEAE-sephacel beads. *See* **Note 2**.

2. Wash the column with 1 mL of dH_2O.

3. Equilibrate the gel with 1 mL of wash buffer. *See* **Note 2**.

4. Dilute the reaction mixture with an equal volume of dH_2O.

5. Load the diluted mixture onto the column three times by collecting the flow-through and re-loading the filtrate. *See* **Note 3**.

6. Wash with 30 column volumes of wash buffer.

7. Elute with 6 column volumes of elution buffer into 6 fractions of 500 μL each.

8. To remove the TEA and exchange it with water, place all fractions in the speed-vac until all liquid is removed. Rehydrate each fraction with 1 mL of water and speed-vac again. Repeat this step five times. *See* **Note 1**.

9. Remove from speed-vac and rehydrate each fraction in 200 μL water.

**3.3 Analysis of PAPS
Production by TLC or
Mass Spectrometry**

1. Set up TLC to check fractions for PAPS[35] or run PAPS[34] on mass spectrometry to identify the fractions containing PAPS[34].

2. TLC analysis.

   (a) Draw a line in pencil about ¼ of the way from one end of the PEI-cellulose TLC plate.

   (b) Place ethanol solution into a beaker large enough to accommodate TLC plate standing on end. Fill the beaker with about a quarter inch of ethanol.

   (c) Stand the TLC plate in the beaker with the pencil line toward the bottom of the beaker.

   (d) Allow the ethanol to run up the plate until about ¾ of the plate is wet. Remove from the beaker and let air-dry completely. Pour the ethanol out of the beaker. *See* **Note 4**.

   (e) After the paper is completely dry, draw a few small dots on the pencil line, one for each fraction and all controls.

   (f) Place 2 μL of a fraction on a dot. Repeat for all fractions and controls. Allow the spots to dry completely.

   (g) Put about a quarter inch of LiCl solution into the beaker.

   (h) Place the spotted chromatography paper in the beaker with the dots toward the bottom.

   (i) Allow the solution to run up the paper until about ¾ of the paper is wet. Remove from the beaker and allow to completely air-dry. Pour out the LiCl solution. *See* **Note 4**.

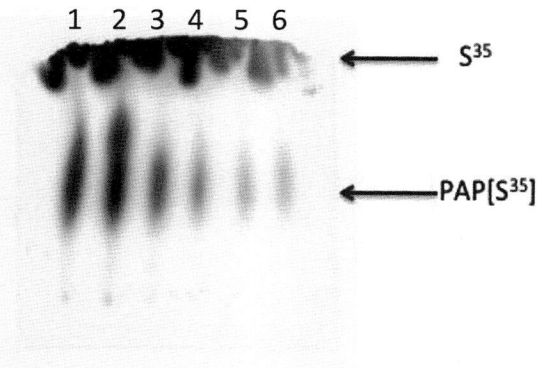

Fig. 2 Confirmation of radiolabeled PAP[S³⁵] production by TLC analysis. Each lane represents a sample of radiolabeled PAP[S³⁵] run on the PEI cellulose in a TLC. The first 2 lanes are neat samples from fraction 1-3 and 4-6 of the elutions. The subsequent lanes are dilutions of the original fractions

(j) After the paper is dry, position the paper on the phospho-imaging cassette. Exposure time can vary, but usually 60–120 min is sufficient. This is dependent on the amount of radioactive PAPS synthesized.

(k) Expose the cassette using the phosphoimager (Fig. 2).

3. After the fractions containing PAPS are identified, they are combined and the PAPS is aliquoted and stored at −80 °C.

4. If the PAPS was made using sulfate enriched with S^{34} or S^{33} isotopes, then analyze the fractions using mass spectrometry to identify fractions containing PAPS.

4 Notes

1. If you are using radioactive sodium sulfate, it is important to use the method of disposal required by your institution. It is likely that the tubes will be radioactive when they are removed from the speed-vac and therefore, it is important to check your hands often when handling radiochemicals.

2. If your columns come with a small disk/filter pad, this can be used as a first step in creating the purification column. If there is no disk, glass wool can be substituted but be careful as skin irritation can occur if exposed to bare skin. If you have the disk/filter pad, it is important to prevent air bubbles from getting trapped between the filter and the outlet of the column. To do this, plug the bottom of the column and fill it with water. Place the filter on the top and then push through the

water until it reaches the bottom of the column. This will displace water out the top of the column, so it is best done over a sink. After the filter is settled at the bottom, remove the plug and let the remaining water drained down. This pre-loading of water prevents air bubble formation.

3. The number of times the sample is loaded may need to be optimized based on individual experience.

4. Both the ethanol and LiCl solutions can be reused. Label and store for future TLCs.

Acknowledgement

This work was supported in part by NIH grants (P01HL107152 and R01GM075168) to B.K.

References

1. Pan J, Qian Y, Zhou X et al (2010) Chemically oversulfated glycosaminoglycans are potent modulators of contact system activation and different cell signaling pathways. J Biol Chem 285(30):22966–22975

2. Lindahl U, Backstrom G, Thunberg L, Leder I (1980) Evidence for a 3-O-sulfated D-glucosamine residue in the antithrombin-binding sequence of heparin. Proc Natl Acad Sci U S A 77(11):6551–6555

3. Burkart T, Wiesmann U (1987) Sulfated glycosaminoglycans (GAG) in the developing mouse brain. Quantitative aspects on the metabolism of total and individual sulfated GAG in vivo. Dev Biol 120(2):447–456

4. Gay S, Fribourgh J, Donohoue P et al (2009) Kinetic properties of ATP sulfurylase and APS kinase from *Thiobacillus denitrificans*. Arch Biochem Biophys 489:110–117

5. Yu Z, Lemongello D, Segel I, Fisher A (2008) Crystal structure of saccharomyces cerevisiae 3'-phosphoadenosine-5'-phosphosulfate reductase complexed with adenosine 3',5'-bisphosphate. Biochemistry 47(48):12777–127786

6. Segel I, Renosto F, Seubert P (1987) Sulfate-activating enzymes. In: Jakoby W, Griffith O (eds) Methods in enzymology, vol 143. Academic, New York

7. Sugahara K, Schwartz N (1979) Defect in 3'-phosphoadenosine 5'-phosphosulfate formation in brachymorphic mice. Proc Natl Acad Sci U S A 76(12):6615–6618

8. Deyrup A, Krishnan S, Singh B, Schwartz N (1999) Activity and stability of recombinant bifunctional rearranged and monofunctional domains of ATP-sulfurylase and adenosine 5'-phosphosulfate kinase. J Biol Chem 274(16):10751–10757

9. Venkatachslam K, Akita H, Strott C (1998) Molecular cloning, expression and characterization of human bifunctional 3'-phosphoadenosine 5'-phosphosulfate synthase and its functional domains. J Biol Chem 273(30):19311–19320

10. Satishchandran C, Markham G (1989) Adenosine-5'-phosphosulfate kinase from Escherichia coli K12. Purification, characterization and identification of a phosphorylated enzyme intermediate. J Biol Chem 264(25):15012–15021

Chapter 6

Preparation of Isotope-Enriched Heparan Sulfate Precursors for Structural Biology Studies

Xylophone V. Victor, Vy M. Tran, Balagurunathan Kuberan, and Thao Kim Nu Nguyen

Abstract

Heparan sulfate (HS) plays numerous important roles in biological systems through their interactions with a wide array of proteins. Structural biology studies of heparan sulfate are often challenging due to the heterogeneity and complexity of the HS molecules. Radioisotope metabolic labeling of HS in cellular systems has enabled the elucidation of HS structures as well as the interactions between HS and proteins. However, radiolabeled structures are not amenable for advanced structural glycobiology studies using sophisticated instruments such as nuclear magnetic resonance (NMR) spectroscopy and mass spectrometry (MS). The utilization of stable isotope-enriched HS precursors is an appealing approach to overcome these challenges. The application of stable isotope-enriched HS precursors has facilitated the HS structural analysis by NMR spectroscopy and mass spectrometry. Herein we describe a simple method to prepare isotopically enriched HS precursors.

Key words Heparan sulfate, Heparin, Stable isotope, Radioisotope, Heparosan, Nuclear magnetic resonance, Mass spectrometry

1 Introduction

Heparan sulfate (HS) is a linear, sulfated polysaccharide, involved in many cellular, physiological, and pathological processes through its interactions with various proteins [1]. Therefore, it is essential to elucidate the structure-function relationships of HS chains. However, the elucidation of HS structures is very challenging due to the structural complexity of HS molecules, arising from their highly variable length, sulfation pattern, epimer content, domain organization, and chain valency [2, 3]. Such structural complexity is the factor that contributes to the diverse functions of HS in biological systems.

Various techniques such as nuclear magnetic resonance (NMR) spectroscopy, mass spectrometry (MS), liquid chromatography (LC), and capillary electrophoresis (CE), have been developed for

Kuberan Balagurunathan et al. (eds.), *Glycosaminoglycans: Chemistry and Biology*, Methods in Molecular Biology, vol. 1229, DOI 10.1007/978-1-4939-1714-3_6, © Springer Science+Business Media New York 2015

structural biology studies of HS. Together with the development of these techniques, stable isotopes and radioisotopes are also employed to advance the structural and functional elucidation of the complex HS molecules. In this chapter, we describe two methods for the preparation of stable isotope-enriched HS precursors and the radioisotope metabolic labeling of HS.

The first method to be discussed is for the preparation of stable isotope-enriched HS precursors, which is a very efficient facilitation for NMR spectroscopy. NMR spectroscopy has been widely used for the structural elucidation of heparin and heparan sulfate but limited to small oligosaccharides or disaccharides due to the heterogeneity of the polymer. However, even 1D ^{1}H NMR spectroscopy of such small fragments is still complicated due to the extensive signal overlap and the requirement of large quantity for analysis. Heteronuclear NMR spectroscopy with isotopic enrichment has been developed to decrease the signal overlap and provide higher resolution and sensitivity [2, 4, 5]. Even though uniformly ^{13}C-labeled HS precursors can facilitate the assignments of ^{1}H and ^{13}C chemical shifts using multidimensional NMR spectroscopy, the analysis is still complicated due to the one-bond ^{13}C-^{13}C couplings between adjacent carbon atoms. Meanwhile, atom-specific ^{13}C labels at the anomeric carbons can minimize this problem [2].

Together with stable isotope labeling, radioisotope metabolic labeling of HS in cell culture is a robust approach to label the intact HS chains for structural and functional biology studies. The commonly used radioisotope precursors are [^{3}H] glucose, [^{3}H] glucosamine, [^{3}H] galactose, [^{14}C] glucosamine, and [^{35}S] sodium sulfate [6–8]. Radioisotope labeling allows miniscule detection limits and can be employed in liquid chromatography as well as gel electrophoresis. Therefore, radioisotope labeling has been utilized widely in the elucidation of HS structures, characterization of HS-modifying enzymes, and investigation of HS-protein interactions. Here, we describe the radioisotope metabolic labeling of HS in cellular system by [^{35}S]-Na$_2$SO$_4$ and D-[6-^{3}H]-glucosamine to label the sulfate groups and the sugar backbone of the HS polymer, respectively.

2   Materials

2.1 Production of ^{15}N- and ^{13}C-Enriched Heparosan (See Note 1)

1. *Escherichia coli* K5 (*see* **Note 2**).

2. LB medium: 10 g/L Bactotryptone, 5 g/L yeast extract, 10 g/L NaCl.

3. Minimal medium: Add 40 ml M9 minimal salt solution to 960 ml water, and then autoclave. Before use, add 2 ml "O" solution and 1.2 ml vitamin solution.

4. M9 minimal salt solution: 16.5 g KH_2PO_4, 87.5 g KH_2PO_4, 18.25 g NaCl, 5 g NH_4Cl, add water up to 500 ml, then autoclave.

5. Vitamin solution: 210 mg Riboflavin, 2.7 g pantothenic acid, 3.0 g niacin, 0.7 g pyridoxin, 30 mg biotin, 20 mg folic acid, add water up to 500 ml, filter, and store in the dark.

6. "O" stock solution: 5 g $FeSO_4 \cdot 7H_2O$, 184 mg $CaCl_2 \cdot 2H_2O$, 64 mg H_3BO_3, 40 mg $MnCl_2 \cdot 4H_2O$, 18 mg $CoCl_2 \cdot 6H_2O$, 4 mg $CuCl_2 \cdot 2H_2O$, 340 mg $ZnCl_2$, 605 mg $Na_2MoO_4 \cdot 2H_2O$. Add 8 ml of concentrated HCl, add water up to 100 ml, dissolve overnight, and autoclave.

7. $[1,2,3,4,5,6-{}^{13}C_6]$-D-glucose or $[1-{}^{13}C]$-D-glucose, $({}^{15}NH_4)_2SO_4$ or ${}^{15}NH_4Cl$.

8. *Streptomyces griseus* protease type IV.

9. Econo-column.

10. DEAE Sepharose.

11. Wash buffer: 0.1 M NaCl, 0.02 M NaOAc, 0.01 % Triton X-100, pH 6.0.

12. Elution buffer: 1 M NaCl, 0.02 M NaOAc, 0.01 % Triton X-100, pH 6.0.

2.2 Radioisotope Metabolic Labeling of Heparan Sulfate in Cell Culture (See Note 1)

1. $[{}^{35}S]$-Na_2SO_4 or D-$[6-{}^3H]$glucosamine.

2. Cell lines and appropriate medium.

3. 6× pronase solution: *Streptomyces griseus* protease type IV (1 mg/ml).

4. DEAE Sepharose.

5. 6-Well plate, disposable 3 ml syringe.

6. Wash buffer: 0.1 M NaCl, 0.02 M NaOAc, 0.01 % Triton X-100, pH 6.0.

7. Elution buffer: 1 M NaCl, 0.02 M NaOAc, 0.01 % Triton X-100, pH 6.0.

8. Amicon centrifugal filter 3,000 MWCO.

9. Chondroitinase ABC.

3 Methods

3.1 Production of ${}^{15}N$- and ${}^{13}C$-Enriched Heparosan

1. *E. coli* K5 bacteria are grown in 100 ml of LB medium for 24 h at 37 °C with shaking (250 rpm).

2. The 100 ml primary culture is used to inoculate 1 L of minimal medium in a 3 L Erlenmeyer flask containing 4 g/L $[1,2,3,4,5,6-{}^{13}C_6]$-D-glucose or 1 g/L $[1-{}^{13}C]$-D-glucose as the principal carbon source and 2 g/L $({}^{15}NH_4)_2SO_4$ or ${}^{15}NH_4Cl$ as the principal nitrogen source (*see* **Note 3**).

3. Incubate at 37 °C for 48 h with shaking (250 rpm).

4. The bacterial culture is autoclaved, adjusted to pH 7.0, and treated with protease (20 mg/L) at 37 °C for 12 h with shaking (50 rpm).

5. The solution is centrifuged at $6,000 \times g$ for 30 min, and the supernatant is then purified by DEAE-sepharose.

3.2 Purification of ^{15}N- and ^{13}C-Enriched Heparosan

1. A 50-ml DEAE-Sepharose column is prepared, washed with 5 column volumes of ddH$_2$O, and equilibrated with 5 column volumes of wash buffer (*see* **Note 4**).

2. The sample is loaded onto the column.

3. The column is washed with 30 column volumes of wash buffer.

4. The heparosan polysaccharide is eluted with 6 column volumes of elution buffer.

5. The eluate is adjusted to 1 M NaCl, 4 volumes of 99 % ethanol is added, and the sample is kept at 4 °C for 24 h.

6. The sample is centrifuged at $6,000 \times g$ for 30 min at 4 °C, and the heparosan pellet is allowed to air-dry overnight (*see* **Note 5**). The structures of the ^{13}C- and ^{15}N-enriched HS precursor chains are shown in Fig. 1.

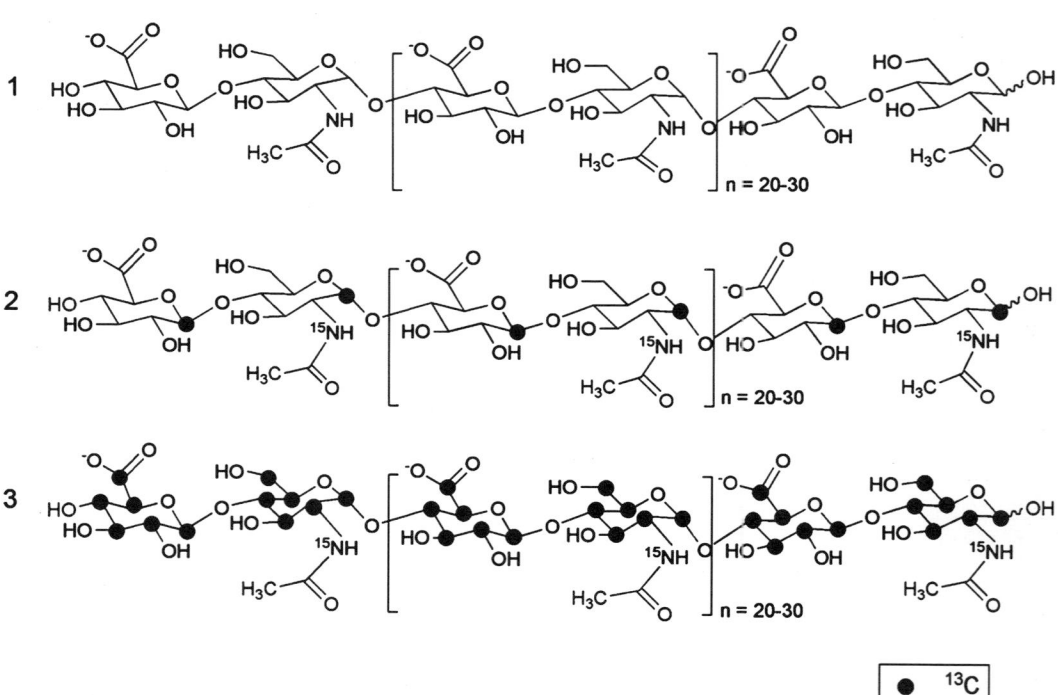

Fig. 1 Structures of the HS precursor chains prepared without isotope enrichment (**1**) or with ^{13}C and ^{15}N enrichment (**2**, **3**). Polysaccharide **2** is atom-specific ^{13}C labeled, polysaccharide **cp** is uniformly ^{13}C labeled. The positions of ^{13}C isotopes are indicated by the *filled dark circles*

3.3 ^{35}**S- or**
^{3}**H-Metabolic Labeling**
of Heparan Sulfate
in Cell Culture

1. Approximately 4×10^5 cells are plated per well in a 6-well plate containing the appropriate medium (*see* **Note 6**).

2. The plate is incubated at 37 °C in a 5 % CO_2, humidified incubator for 24 h to reach a confluency of about 50 %.

3. The cells are washed with sterile PBS and 2 ml of appropriate medium containing 10 % dialyzed FBS is added (*see* **Note 7**).

4. 200 μCi of [^{35}S]Na_2SO_4 and/or D-[6-^{3}H]glucosamine is added (*see* **Note 8**).

5. The plate is incubated at 37 °C in a 5 % CO_2, humidified incubator for 24–72 h.

3.4   Purification
of Radioisotope-
Labeled Heparan
Sulfate

1. 400 μl of 6× pronase solution is added, and the plate is then incubated at 37 °C overnight.

2. The cell solution is transferred to an Eppendorf tube and centrifuged at $16,000 \times g$ for 5 min.

3. The supernatant is collected, and half a volume of 0.016 % Triton X-100 is added.

4. 200 μl DEAE-Sepharose column is prepared, washed with 5 column volumes of ddH_2O, and equilibrated with 5 column volumes of wash buffer.

5. The supernatant is loaded onto the column.

6. The column is washed with 30 column volumes of wash buffer.

7. The glycosaminoglycan is eluted with 6 column volumes of elution buffer.

8. The eluate is then concentrated and filter-desalted five times with water in a 3,000 MWCO filter.

9. Chondroitin sulfate in the total glycosaminoglycan is digested by chondroitinase ABC at 37 °C overnight.

10. The digested chondroitin sulfate disaccharides are removed by filtering five times with water in a 3,000 MWCO filter.

4   Notes

1. All products were purchased from commercial sources unless otherwise stated.

2. *E. coli* K5 is a pathogenic bacterial strain. Use biosafety level 2 procedures when handling this strain.

3. [1,2,3,4,5,6-^{13}C$_6$]-D-glucose is used to generate uniformly ^{13}C-labeled polysaccharide and [1-^{13}C]-D-glucose is used to generate anomeric carbon-specific ^{13}C-labeled polysaccharide.

4. The column volume is scalable depending on the amount of heparosan loaded onto the column.

5. The ^{15}N- and ^{13}C-enriched heparosan is further modified by various approaches in order to be utilized in structural biology studies. The isotope-enriched heparosan can be depolymerized into size-defined oligosaccharides [5], modified by enzymes to generate heparan sulfate structures [4], or degraded into disaccharide to generate internal disaccharide standards for mass spectrometry [9].

6. The number of cells can be varied depending on the types of cells.

7. The concentration of glucose in the medium should be 1 mM in the case of labeling with D-[6-^{3}H] glucosamine. The medium should be sulfate deficient in the case of labeling with [^{35}S] sulfate.

8. This experiment is scalable.

Acknowledgement

This work was supported in part by NIH grants (P01HL107152 and R01GM075168) to B.K. T.K.N. acknowledges a graduate fellowship support from Vietnam Education Fund.

References

1. Bishop JR, Schuksz M, Esko JD (2007) Heparan sulphate proteoglycans fine-tune mammalian physiology. Nature 446:1030–1037

2. Nguyen TK, Tran VM, Victor XV, Skalicky JJ, Kuberan B (2010) Characterization of uniformly and atom-specifically (13)C-labeled heparin and heparan sulfate polysaccharide precursors using (13)C NMR spectroscopy and ESI mass spectrometry. Carbohydr Res 345: 2228–2232

3. Tran VM, Nguyen TK, Raman K, Kuberan B (2011) Applications of isotopes in advancing structural and functional heparanomics. Anal Bioanal Chem 399:559–570

4. Zhang Z, McCallum SA, Xie J, Nieto L, Corzana F, Jimenez-Barbero J, Chen M, Liu J, Linhardt RJ (2008) Solution structures of chemoenzymatically synthesized heparin and its precursors. J Am Chem Soc 130:12998–13007

5. Sigulinsky C, Babu P, Victor XV, Kuberan B (2010) Preparation and characterization of (15)N-enriched, size-defined heparan sulfate precursor oligosaccharides. Carbohydr Res 345:250–256

6. Zhang L, Lawrence R, Schwartz JJ, Bai X, Wei G, Esko JD, Rosenberg RD (2001) The effect of precursor structures on the action of glucosaminyl 3-O-sulfotransferase-1 and the biosynthesis of anticoagulant heparan sulfate. J Biol Chem 276:28806–28813

7. Victor XV, Nguyen TK, Ethirajan M, Tran VM, Nguyen KV, Kuberan B (2009) Investigating the elusive mechanism of glycosaminoglycan biosynthesis. J Biol Chem 284:25842–25853

8. Hagner-McWhirter A, Li JP, Oscarson S, Lindahl U (2004) Irreversible glucuronyl C5-epimerization in the biosynthesis of heparan sulfate. J Biol Chem 279:14631–14638

9. Zhang Z, Xie J, Liu H, Liu J, Linhardt RJ (2009) Quantification of heparan sulfate disaccharides using ion-pairing reversed-phase microflow high-performance liquid chromatography with electrospray ionization trap mass spectrometry. Anal Chem 81:4349–4355

Chapter 7

Synthesis of Glycosaminoglycan Mimetics Through Sulfation of Polyphenols

Rami A. Al-Horani, Rajesh Karuturi, Stephen Verespy III, and Umesh R. Desai

Abstract

In nearly all cases of biological activity of sulfated GAGs, the sulfate group(s) are critical for interacting with target proteins. A growing paradigm is that appropriate small, sulfated, nonsaccharide GAG mimetics can be designed to either mimic or interfere with the biological functions of natural GAG sequences resulting in the discovery of either antagonist or agonist agents. A number of times these sulfated NSGMs can be computationally designed based on the parent GAG–protein interaction. The small sulfated NSGMs may possess considerable aromatic character so as to engineer hydrophobic, hydrogen-bonding, Coulombic or cation–pi forces in their interactions with target protein(s) resulting in higher specificity of action relative to parent GAGs. The sulfated NSGMs can be easily synthesized in one step from appropriate natural polyphenols through chemical sulfation under microwave-based conditions. We describe step-by-step procedures to perform microwave-based sulfation of several small polyphenol scaffolds so as to prepare homogenous NSGMs containing one to more than 10 sulfate groups per molecule in high yields.

Key words Allosteric inhibitors, Glycosaminoglycans, Library, Microwave, Mimetics, Sulfation, Synthesis

1 Introduction

Glycosaminoglycans (GAGs) are ubiquitous molecules that are present on cell surfaces (as a part of proteoglycans (PGs)), in extracellular matrix (ECM) and in secreted form as components of plasma and other biological fluids. Common GAGs include heparin/heparan sulfate (H/HS), chondroitin sulfate (CS), dermatan sulfate (DS), and keratan sulfate (KS) [1]. The GAG backbone is composed of repeating hexosamine–hexuronic acid (or hexose) disaccharide units that extend some 50–500 residues generating a fairly long polymeric chain. The hexosamine unit could be either D-glucosamine or D-galactosamine, while the uronic acid unit could be either L-glucuronic acid or L-iduronic acid (or hexose could be D-galactose) (*see* Chaps. 2 and 3 of this volume).

Kuberan Balagurunathan et al. (eds.), *Glycosaminoglycans: Chemistry and Biology*, Methods in Molecular Biology, vol. 1229, DOI 10.1007/978-1-4939-1714-3_7, © Springer Science+Business Media New York 2015

Despite the apparent simplicity in the types of building blocks, GAG sequences are extraordinarily diverse. The primary reason for the structural diversity of GAGs is their template-free enzymatic biosynthesis [2, 3]. A number of sulfotransferases and epimerases are involved in the biosynthetic process, many of which are spatio-temporally expressed [2, 3], resulting in a highly variable structure. A secondary reason for structural diversity stems from the conformational variability of iduronic acid residues that may constitute the chain. Whereas glucosamine, galactosamine, and glucuronic acid residues exhibit only one conformational preference, iduronic acid exhibits multiple forms, especially the ^2S_O and $^1C_4$ forms [4, 5]). Finally, another reason for GAG diversity is the structural fine-tuning introduced by the action of sulfatases, which may remove select group of sulfate groups on the GAG chain, in response to environment cues [6]. These configurational and conformational variations generate a large number of distinct structures that may exhibit unique protein-binding characteristics. A simple calculation indicates the level of structural diversity possible in GAG chains. For an HS oligomer consisting of six repeating disaccharide units, approximately 0.8 billion distinct sequences are theoretically possible, if one assumes that every possible modification at the 2-, 3-, and 6-positions of glucosamine and uronic acid. In comparison, a hexameric peptide may have one of the 64 million, while a hexameric nucleotide can exhibit one of the 4,096 sequences. This implies that the chemical space of GAGs is way greater than that of peptides or nucleic acids, although it is important to note that not all GAG sequences may occur naturally [7].

The phenomenal structural diversity of GAGs translates into myriad biological responses including modulation of coagulation, inflammation, angiogenesis, cell growth and signaling, morphogenesis, and host cell defense [1, 8]. In nearly all cases studied so far, the sulfate groups ($-OSO_3^-$) of GAGs facilitate interactions with target proteins [7, 9]. For the case of heparin–antithrombin interaction, molecular modeling indicates that the scaffold appears to have minimal enthalpic contribution, other than optimal three-dimensional positioning of key sulfate groups [10]. While extensive studies are not available, this conclusion can also be drawn from most GAG–protein crystal structures that show sulfate–Arg/Lys interactions as the primary contributors to recognition. A sulfate group on a GAG sequence can interact with Arg/Lys residues through ionic (electrostatic) forces and/or nonionic (hydrogen bonding) forces [9]. The hydrogen-bonding capability of sulfate groups implies that highly directional intermolecular bonds are possible, which can afford high selectivity of interaction.

2   Rationale for Nonsaccharide Mimetics of GAGs

To date, GAG sequences identified to display high selectivity in recognition of target proteins have been not many. These include the pentasaccharide and octasaccharide sequences, which appear to be naturally engineered in polymeric H/HS to specifically interact with antithrombin and viral glycoprotein D, respectively [11–14]. The pentasaccharide–antithrombin interaction results in an anticoagulant response, whereas the octasaccharide–glycoprotein D interaction results in an antiviral response. However, specificity features are in general poorly understood for GAG–protein interactions. Several investigators have pursued molecular modeling-based approaches to elucidate specificity characteristics of these interactions with marginal success [15–19] (also *see* Chap. 24 in this volume). A possible reason for this state of art with regard to modeling is that GAGs are intrinsically flexible, which arises from the gyrational motions of sulfate groups [7]. This translates into huge conformational search space that is not easy to cover exhaustively.

GAG specificity elements can also be theoretically elucidated through the use of a library of homogenous GAG sequences. Several attempts have been made to synthesize GAG sequences. Majority of these syntheses require a large number of steps [20–27], which is difficult to apply toward the synthesis of a library of reasonable diversity. For example, heparin octasaccharides that antagonize HSV entry into cells were synthesized in about 50 steps [26] but the approach cannot be exploited for all possible combinations of individual monosaccharide units. Alternative approaches have relied on microarray technology [28, 29]. However, identification of GAG sequences that show activity has proved to be quite challenging.

Another attempt to discover GAG-like modulators has relied on the idea that a limited number of anionic groups (e.g., sulfate, carboxylate, phosphate) on a smaller saccharide scaffold may overcome the difficulties of working with GAG polymers. Examples of these include oligosaccharides containing sulfate [27, 30–33] and phosphate [34] groups, aptamers [35], sulfated linked cyclitols [36], dendritic polyglycerol sulfates [37], and steroid-modified sulfated oligosaccharides [38]. There is substantial promise in this approach as indicated by the initiation of several clinical trials studying PI-88, a pentaphosphate oligosaccharide.

We have embarked on a new paradigm that sulfated, nonsaccharide GAG mimetics (NSGMs) can be designed to modulate the biological functions of their parent GAGs. This hypothesis gradually developed following our effort to design an antithrombin activator that mimics three residues of the heparin pentasaccharide [39, 40]. Starting from this first generation activator epicatechin sulfate, which activated the serpin ~8-fold, to the third generation molecule, a tetrasulfated tetrahydroisoquinoline, which induced ~80-fold activation, we developed structure-based and pharmacophore-based approaches to design NSGMs (Fig. 1)

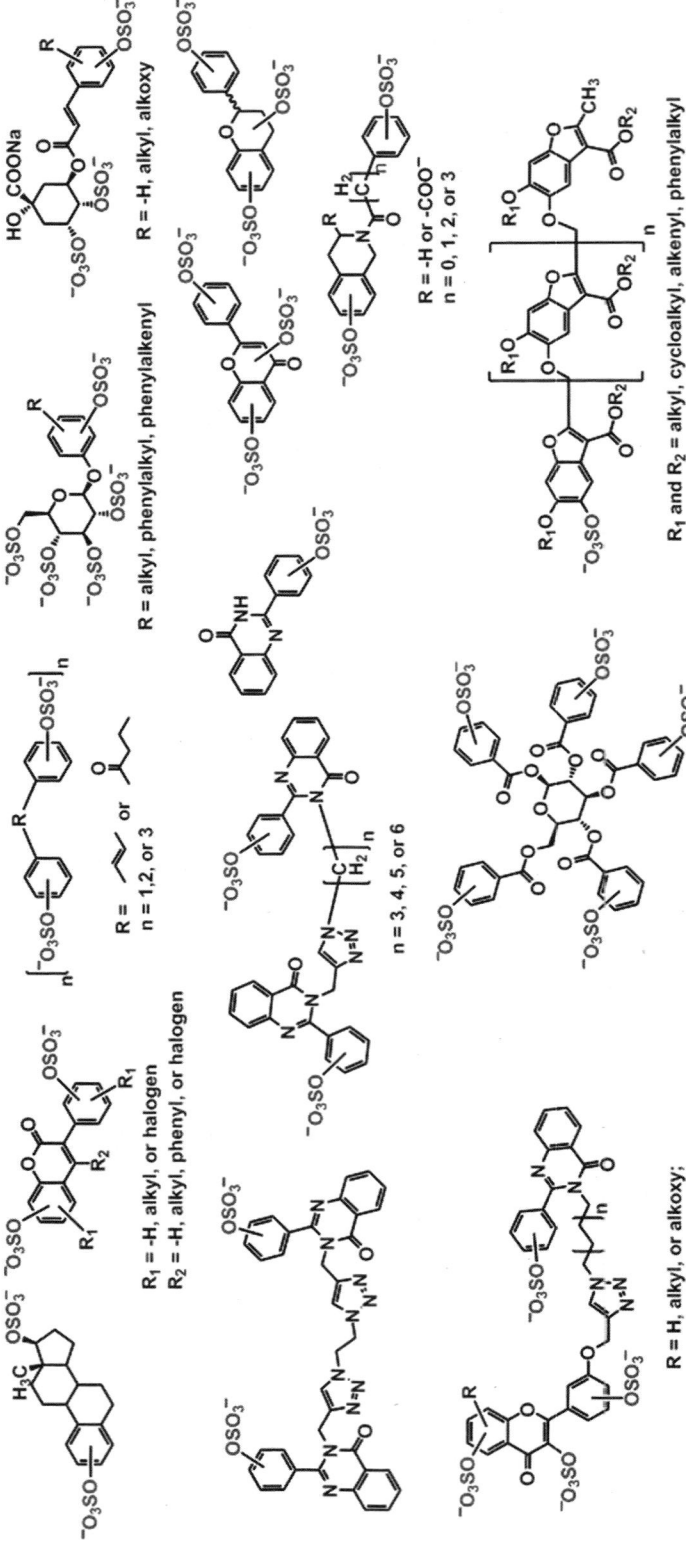

Fig. 1 The diverse library of glycosaminoglycan (GAG) mimetics

[39–42]. Likewise, sulfated benzofuran-based NSGMs were designed to interfere with the binding of heparin to anion-binding exosite II of thrombin [43, 44]. These sulfated NSGMs were designed to mimic or interfere with the function of GAGs although their structural similarity with the natural biopolymers was minimal. But a major advance was realized recently when highly sulfated pentagalloyl glucopyranoside (SPGG) was synthesized as a structural mimetic of polydisperse heparin [45]. A particular species of this molecule containing nearly ten sulfate groups was found to selectively inhibit human factor XIa by binding to the heparin-binding site [45].

The sulfated NSGMs can be computationally designed. These possess considerable aromatic character so as to possibly introduce hydrophobic, hydrogen-bonding, Coulombic and cation–pi forces in their interactions, which could engineer enhanced specificity of binding to protein targets relative to GAGs. The sulfated NSGMs would preferentially bind to GAG-binding sites on proteins, which implies that typically these molecules would utilize an allosteric modulation mechanism of action. Finally, the availability of polyphenols in nature allows the easy synthesis of the sulfated GAG mimetics in a homogenous manner, which is a major advance from the difficulties encountered in the synthesis of GAG sequences.

3 Sulfation in the Synthesis of GAG Mimetics

The discovery of multiple functional roles of sulfated NSGMs suggests a strong possibility of developing novel sulfated pharmaceutical agents. These are typically synthesized by sulfation of an appropriate polyphenolic scaffold. Hence, several chemical sulfation protocols have been introduced as to quantitatively synthesize sulfated NSGMs (*see* **Note 1**). Literature reveals that sulfation used to be carried out with H_2SO_4 in the early part of the twentieth century [9]. Another reagent is sulfamic acid, a modified form of sulfuric acid, which has been used for synthesis of saturated monohydric alcohol sulfates and carbohydrate sulfates [9]. To improve the yields, DCC along with H_2SO_4 has been suggested. Yet, strong acid (H_2SO_4) is typically not appropriate because aromatic polyphenols may be sensitive to high acidity. Another approach is to use sulfur trioxide; however its direct use has been fraught with problems of polymerization. This issue is typically avoided by using SO_3 as an adduct with amine, amide, ether, or phosphate containing molecules. These complexes have been used for sulfating a variety of scaffolds containing alcoholic, phenolic, amine, thiol, and other functional groups [46–50]. Complexes of SO_3 with organic bases including pyridine, trimethylamine $((CH_3)_3N)$, and triethylamine $(CH_3CH_2)_3N)$, or amides such as DMF have typically found wide

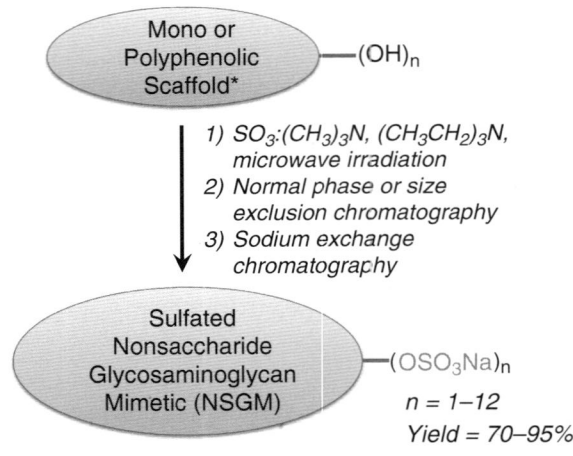

Fig. 2 Generic microwave-assisted sulfation of diverse nonsaccharide mono- or polyphenolic scaffolds including benzofurans, flavonoids, coumarins, quinazolinones, chalcones, galloyl glycosides, steroids, tetrahydroisoquinolines

use. Furthermore, a low temperature, triflic acid-catalyzed sulfation reaction was also reported to quantitatively conclude the synthesis of several polysulfated scaffolds [51].

Considering that GAG mimetics present a major source of discovery of GAG function modulating molecules, we have developed a library of sulfated NSGMs by per-sulfation of an appropriate synthetic or natural polyphenol. Per-sulfation of a polyphenolic scaffold is typically difficult because of anionic overcrowding as the reaction proceeds. Hence, a microwave-based protocol has been developed (Fig. 2) [52]. In this protocol, acetonitrile (CH_3CN) was used as the solvent for its microwave-friendliness and ease of evaporation. The presence of free base in the reaction mixture appears to promote the per-sulfation reaction. Typically, optimal yields were achieved by using SO_3-amine complex (−NR_3 or −Py, where R=CH_3 or CH_2CH_3) at approximately 6–10 times molar proportion per phenolic group in the presence of microwaves at 90–100 °C. Using these conditions, per-sulfated products could be isolated in 70–95 % yield.

The microwave-assisted sulfation protocol seems to tolerate a range of functional groups such as amides, esters, aldehydes, as well as alkenes. Besides, the high yield of the products makes the method convenient for construction of sulfated compounds library. The molecules synthesized so far have hypothetically been similar to partial sequences of GAGs including fully sulfated, partially sulfated, and un-sulfated domains. The method can be applied uniformly toward the synthesis of mono- to dodeca-sulfated compounds, which is important considering that repulsive intramolecular forces are thought to limit polysulfation resulting in a mixture of partially sulfated products (*see* **Note 2**). Both alcoholic and phenolic hydroxyl groups can be sulfated equally well and the method appears to provide the per-sulfated product in high purity using a one step normal phase flash chromatography or an aqueous G-10 size exclusion chromatography. The method is convenient for quantitative isolation of 10–200 mg of the per-sulfated products and may be possible for a scale up.

The microwave protocol has been applied successfully in synthesis of several sulfated GAG mimetics belonging to different chemical classes modulating different GAG–protein interactions (Fig. 1). In addition to the examples discussed above, sulfated low molecular weight lignins (LMWLs) have been synthesized. Sulfated LMWLs are polymeric mimics of heparin with inhibitory action on several serine proteases in coagulation cascade [53–55] and inflammation [56]. LMWLs exhibited potent antiviral activity against HSV and HIV [57]. Also, sulfated chalcones and steroids have demonstrated a reasonable level of angiogenic modulation in an in vitro tube formation assay [58].

4 Materials

4.1 Chemicals

1. Anhydrous CH_3CN.
2. D_2O.
3. DMSO-d_6.
4. High purity water for chromatography.
5. HPLC grade solvents (acetonitrile and formic acid).
6. Sulfating agent SO_3–$(CH_3)_3N$.
7. Base $(CH_3CH_2)_3N$.
8. Coumarin.
9. Trihydroxy flavonoid.
10. Pentagalloylglucoside and quinazolinone dimer (synthesized according to previous reports [59, 60]).
11. n-Hexylamine for ion-pairing UPLC.

4.2 Instruments

Reaction system—Sulfation is to be performed using a chemical synthesis microwave (we used CEM discover® SP system from CEM, Matthews, NC) in sealed tubes (e.g., 10 mL). Heating rate is usually 2–6 °C per second and electromagnetic stirring with adjustable speed is to be employed. All reaction parameters including temperature, pressure, and microwave power are typically controlled by the instrument software (in this case the Synergy™ software).

Purification systems—Two different purification schemes are to be used to obtain high purity (>95 %) sulfated NSGMs. Molecules with sufficient hydrophobicity should be purified using normal phase flash chromatography followed by sodium exchange, whereas those having less hydrophobicity should be purified by size exclusion followed by sodium exchange chromatographies.

Normal phase flash chromatography is performed using a combiflash system (we used the Teledyne ISCO (Lincoln, NE) Combiflash RF system) and disposable normal silica cartridges of 30–50 μm particle size, 230–400 mesh size, and 60 Å pore size. The flow rate of the mobile phase is in the range 18–35 mL/min and mobile phase gradients of CH_2Cl_2/CH_3OH are used to elute sulfated compounds.

Size exclusion (Sephadex G10) and *sodium exchange (SP Sephadex C-25) chromatographies* are to be performed using regular glass chromatography columns with dimensions of 40×2.5 cm or Flex columns of dimensions 170×1.5 cm for size exclusion and 75×1.5 cm for sodium exchange. For regeneration of the cation exchange column, 1 L of 2 M NaCl solution is used. Water is used as eluent in both chromatographies.

Drying systems—A rotary evaporator equipped with water bath and oil-free high vacuum pump (65 L/min) is used for removal of acetonitrile and initial drying. An effective lyophilizer equipped with high vacuum pump is used to remove water from samples.

Structural Characterization of Sulfated Molecules—Sulfated NSGMs are characterized using ^1H and ^{13}C NMR spectroscopy, UPLC-MS, and/or capillary electrophoresis.

NMR spectroscopy is performed on a standard 400 or higher MHz spectrometer (e.g., Bruker) in either D_2O or DMSO-d_6. In this work, signals, in part per million (ppm), are reported relative either to the internal standard (tetramethyl silane, TMS) or to the residual peak of the solvent. The NMR data are reported as chemical shift (ppm), multiplicity of signal (s = singlet, d = doublet, t = triplet, q = quartet, dd = doublet of doublet, m = multiplet), coupling constants (Hz), and integration. Samples for NMR are prepared by dissolving 5–10 mg of sulfated NSGM in an appropriate NMR solvent (0.5 mL) and the spectra recorded at ambient temperature.

UPLC-MS experiments are performed using a standard system (we used a Waters Acquity H-class UPLC system (Milford, MA) equipped with a photodiode array detector and triple quadrupole mass spectrometer). In this work, a reversed-phase Waters BEH C18 column of particle size 1.7 μm and 2.1×50 mm dimensions at 30 ± 2 °C was used for separation of sulfated species components, if any. Solvent A consisted of 25 mM *n*-hexylamine in H_2O containing 0.1 % (v/v) formic acid, while solvent B consisted of 25 mM *n*-hexylamine in $CH_3CN–H_2O$ mixture (3:1 v/v) containing 0.1 % (v/v) formic acid. A flow rate of 500 μL/min and a linear gradient of 3 % solvent B per min over 20 min (initial solvent B proportion was 20 % v/v) was used. The sample was first monitored for absorbance in the range of 190–400 nm and then directly introduced into the mass spectrometer. ESI-MS detection was performed in positive ion mode for which the capillary voltage was 4 kV, cone voltage was 20 V, desolvation temperature was 350 °C, and nitrogen gas flow was maintained at 650 L/h. Mass scans were collected in the range of 1,000–2,048 amu within 0.25 s and several of these added to enhance signal-to-noise ratio.

For higher solution MS (HR-MS) measurements, a TOF-MS spectrometer is used. We used a Perkin–Elmer AxION 2 TOF MS in negative ion mode. Ionization conditions were optimized for each compound to maximize the ionization of the parent ion. Generally, the extractor voltage was set to 3 V, the Rf lens voltage was 0.1 V, the source block temperature was set to 150 °C, and the desolvation temperature was about 250 °C.

Capillary electrophoresis (CE) experiments are performed using a standard CE system. We used a Beckman P/ACE MDQ system (Fullerton, CA). Electrophoresis was performed at 25 °C and a constant voltage of 8 kV or a constant current of 75 µA using an uncoated fused silica capillary (ID 75 µm) with the total and effective lengths of 31.2 and 21 cm, respectively. A sequential wash of 1 M HCl (10 min), water (3 min), 1 M NaOH (10 min), and water (3 min) at 20 psi was used to activate the capillary. Before each run, the capillary was rinsed with the run buffer; 50 mM sodium phosphate buffer of pH = 3, for 3 min at 20 psi. Sulfated compounds injected at the cathode (0.5 psi for 4 s) and detected at the anode (214 nm).

5 Methods

5.1 The Sulfation Protocol

In our endeavor to establish a diverse library of sulfated molecules, several polyphenolic chemical scaffolds were considered for sulfation. The diverse library was planned to span enough chemical space in order to identify functionally active GAGs mimetics, which may yield drug-like candidates for modulating a GAG–protein interaction. Figure 1 shows representative structures that belong to different chemical classes, while Schemes 1, 2, 3, and 4 display specific examples to perform mono-sulfation (**1** to **1S**), di-sulfation (**2** to **2S**), tri-sulfation (**3** to **3S**), and poly-sulfation (**4** to **4S**). For simplicity, a step-by-step protocol for the synthesis of mono-sulfated 3-(3′-chloro-phenyl)-7-hydroxy-4-phenylcoumarin (**1S**) [sodium 3-(3-chlorophenyl)-2-oxo-4-phenyl-*2H*-chromen-7-yl sulfate] is presented (*see* **Notes 3–5**).

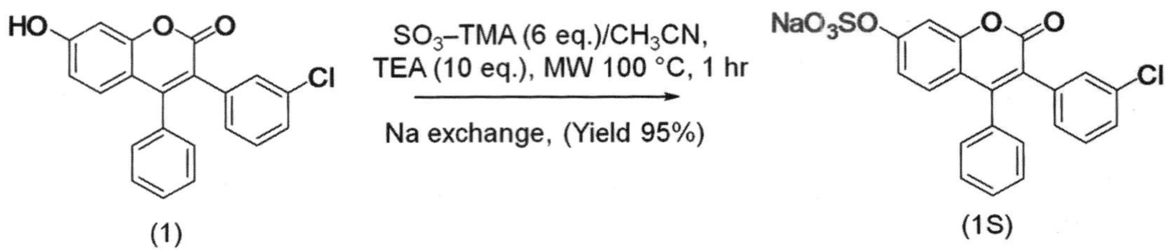

Scheme 1 Synthesis of mono-sulfated NSGM derivatives. A representative example of the coumarin class of NSGMs is shown

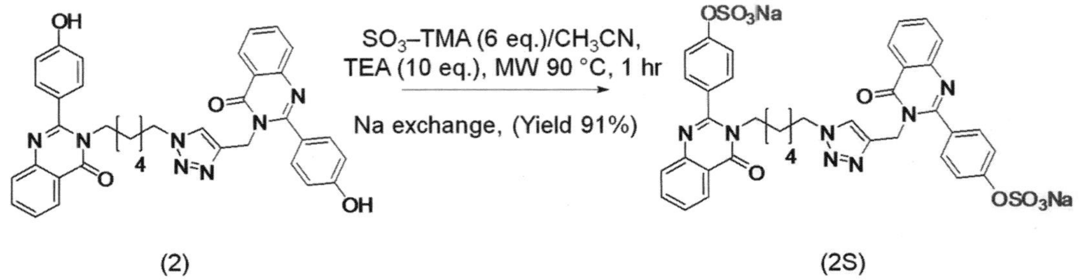

Scheme 2 Synthesis of di-sulfated NSGM derivatives. A representative example of the quinazolinone class of NSGMs is shown

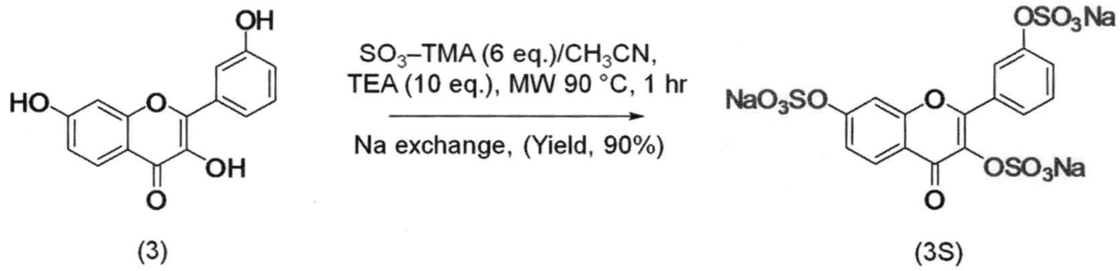

Scheme 3 Synthesis of tri-sulfated NSGM derivatives. A representative example of the flavonoid class of NSGMs is shown

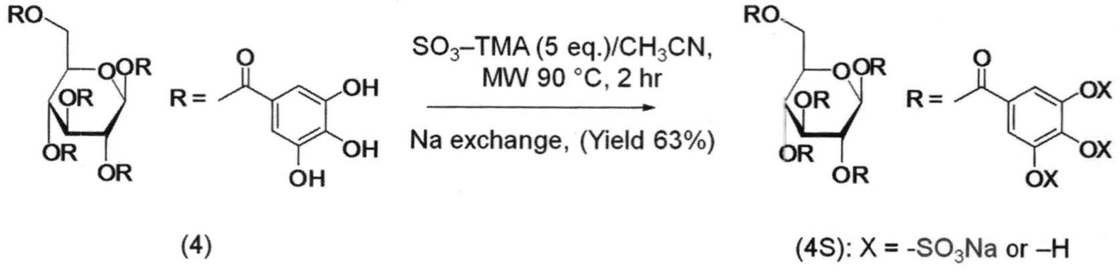

Scheme 4 Synthesis of poly-sulfated NSGM derivatives. A representative example of the galloloid class of NSGMs is shown

5.1.1 Synthesis of Mono-sulfated NSGM Derivative (1S), a Representative Hydrophobic Scaffold

Perform sulfation reaction as follows:

1. Add 3-(3′-chlorophenyl)-7-hydroxy-4-phenylcoumarin (**1**) (1 equivalent, 100 mg, 0.346 mmol) to a 10 mL microwave reaction vessel containing a magnetic stir bar.

2. Add SO_3–$(CH_3)_3N$ complex (6 equivalents/–OH, 288.9 mg, 2.076 mmol) to the same reaction vessel.

3. Add 3 mL of anhydrous CH_3CN to the reaction vessel to obtain a suspension and stir for 2 min.

4. Add anhydrous triethylamine (10 equivalents/–OH, 0.5 mL, 3.46 mmol) to the reaction vessel and continue stirring for another 2 min. Then, seal the vessel and place it in the microwave reactor and run the reaction at temperature of 100 °C for 1 h (*see* **Note 6**).

5. After 1 h, let the reaction vessel cool down, remove magnetic stir bar, and wash contents into a round bottom flask using CH_2Cl_2.

6. Ensure that the coumarin starting material has been consumed and the reaction has reached completion using TLC (e.g., use UNIPLATE silica gel GHLF, 250 μm precoated plates from ANALTECH (Newark, DE)) with CH_2Cl_2:CH_3OH (90:10) as the mobile phase.

7. Transfer the reaction mixture into 100 mL round bottom flask, concentrate it in vacuo (to remove CH_3CN), and resuspend the resulting reaction concentrate with minimal volume of CH_2Cl_2.

8. To obtain the desired product in pure form, add dry silica powder (~20 g) into the reaction crude to make slurry. Concentrate the resulting slurry under vacuum to afford sulfated product bound silica (dry loading). Then, pour compound–silica mixture into RediSep® loading cartridge and plug it with a silica wafer.

9. Purify the loaded sulfated compound by Combiflash® flash chromatography system using 4 g RediSep® normal-phase silica flash column and a mobile phase gradient of CH_2Cl_2:CH_3OH starting from (100:0) to (70:30) over 18 min. Monitor the elution of desired products through absorbance at 254 and 280 nm.

10. Collect appropriate fractions having the sulfated coumarin derivative (**1S**) into 18 mL tubes, pool the desired fractions into 250 mL round bottom flask, and reduce the volume in vacuo to 3–5 mL.

11. Dissolve the sulfated product in as little high purity water as possible, load the resulting solution onto a sodium exchange column (SP Sephadex C-25, prepared as instructed by manufacturer and regenerated as described above), and collect fractions by monitoring on TLC (*see* **Note 6**).

12. Combine fractions having the sulfated coumarin derivative (**1S**) and freeze it in –80 °C refrigerator for approximately 1 h.

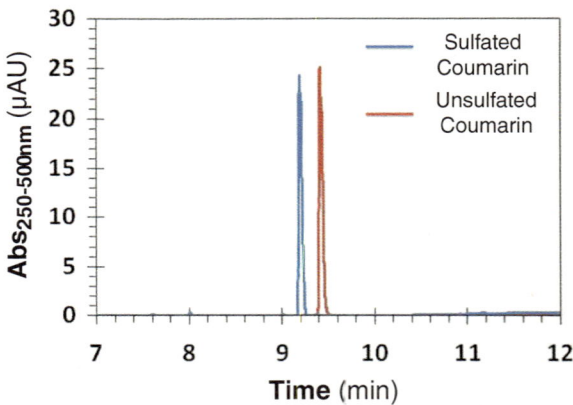

Fig. 3 Characteristic difference in UPLC profile of unsulfated (**1**) and sulfated coumarin (**1S**)

Then place the frozen product on lyophilizer for 24–48 h to afford the desired sulfated product (**1S**) as a fluffy solid powder in yield of 90–95 % (*see* **Note 7**).

13. Confirm the chemical structure of the mono-sulfated coumarin (**1S**) using ^{1}H NMR, ^{13}C NMR, and UPLC-MS. For reference the **1S** spectral data are as follows: ^{1}H NMR (DMSO-d_6, 400 MHz): 7.32–7.25 (m, 4H), 7.18–7.12 (m, 5H), 7.03–7.01 (m, 2H), 6.91–6.89 (d, J=8 Hz, 1H). ^{13}C NMR (DMSO-d_6, 100 MHz) 177.67, 170.10, 160.22, 156.83, 153.39, 151.66, 151.23, 136.60, 134.13, 132.00, 130.39, 129.37, 129.22, 128.96, 128.23, 127.89, 127.19, 122.85, 116.62, 115.08, 106.80. MS (ESI) calculated for $C_{21}H_{12}ClNaO_6S$, [M–Na]$^-$, m/z 427.83, found for [M–Na]$^-$, m/z 427.004. The UPLC profile is shown in Fig. 3.

5.1.2 Synthesis of Di-sulfated NSGM Derivative (2S), Another Representative Hydrophobic Scaffold

The synthesis of this sulfated NSGM essentially follows that described in **steps 1** through **13** above except for the following modifications.

Mix the diphenolic dimer (**2**, 1 equivalent, 40 mg, 0.0625 mmol), triethylamine (10 equivalents/–OH, 0.17 mL, 1.25 mmol), and SO_3–$(CH_3)_3N$ complex (6 equivalents/–OH, 104.3 mg, 0.75 mmol) in anhydrous CH_3CN (2 mL) at room temperature and microwave the resulting mixture for 1 h at 90 °C. Purify the product using normal phase flash chromatography in $CH_2Cl_2/$$CH_3OH$ (85:15) solvent system followed by sodium exchange chromatography to obtain the desired di-sulfated molecule (typical yield 91 % or 48 mg). Confirm the structure of the di-sulfated quinazolinone dimer (**2S**) using ^{1}H NMR, ^{13}C NMR, and UPLC-MS, as reported previously [60].

5.1.3 Synthesis of Tri-sulfated NSGM Derivative (3S), a Representative Scaffold with Intermediate Polarity

The synthesis of this sulfated NSGM follows that described above (*see* **steps 1** through **13**) except for the following modifications.

Mix the trihydroxy flavonoid (**3**, 1 equivalent, 40 mg, 0.0625 mmol), triethylamine (10 equivalents/–OH, 0.619 mL, 4.44 mmol), and SO_3–$(CH_3)_3N$ complex (6 equivalents/–OH, 317 mg, 2.66 mmol) in anhydrous CH_3CN (2 mL) at room temperature and microwave the resulting mixture for 1 h at 90 °C. Purify using normal phase flash chromatography in CH_2Cl_2/CH_3OH (85:15) solvent system followed by sodium exchange chromatography to obtain the desired tri-sulfated molecule in good yields (90 % or 77 mg). Confirm the structure of the tri-sulfated flavonoid (**3S**) using ^1H NMR, ^{13}C NMR, and UPLC-MS, as reported previously [58].

5.1.4 Synthesis of Poly-sulfated NSGM Derivative (4S), a Representative Scaffold with High Polarity

The synthesis of this sulfated NSGM follows that described in **steps 1** through **13** above except for the following modifications.

Mix PGG (**4**, 1 equivalent, 25 mg, 0.027 mmol) and SO_3–$(CH_3)_3N$ complex (5 equivalents/–OH, 281 mg, 2.03 mmol) in anhydrous CH_3CN (2 mL) at room temperature and microwave the resulting mixture for 2 h at 90 °C. No base is to be added. Perform G-10 size exclusion chromatography using high purity water as the mobile phase followed by sodium exchange to obtain the desired poly-sulfated SPGG (Yield 63 % or 42 mg). Confirm the structure of SPGG (**4S**) using ^1H and ^{13}C NMR, as reported previously [45].

SPGG is a mixture of differentially sulfated species and hence will require additional compositional analysis. This is typically performed using high resolution SEC analysis. In this work, we measured the average molecular weight (M_R) and composition of SPGG using reversed-phase ion-pairing (RP-IP) UPLC–MS (Fig. 4), which is especially useful to identify and calculate the proportion of different sulfated components of complex sulfated molecules. RP-IP UPLC-MS of SPGG utilized *n*-hexylamine as the ion-pairing agent, which is introduced in the mobile phase so as to replace sodium cations present on each sulfate group and impart considerable hydrophobicity to the molecule. Resolution arises from the varying hydrophobicity of the ion-paired constituents. The UPLC profile of SPGG shows the presence of six major nearly baseline resolved peaks, labeled *p1* through *p6* (Fig. 4a), each of which was found to further containing multiple peaks. The ESI–MS profile of each peak, observed between 1,000 and 2,048 *m/z* range, was found to contain a doubly charged molecular ion. For example, peaks *p3*, *p4*, and *p5* displayed molecular ions at 1,388.43, 1,478.99. and 1,569.60 *m/z*, respectively, corresponding to doubly charged SPGG species containing 9, 10, and 11 sulfate groups with 11, 12, and 13 *n*-hexylamines, respectively, as ion-pairs. A similar behavior was observed for peaks *p1*, *p2*, and *p6*, which corresponded to SPGG species with 7, 8, and 12 sulfate groups,

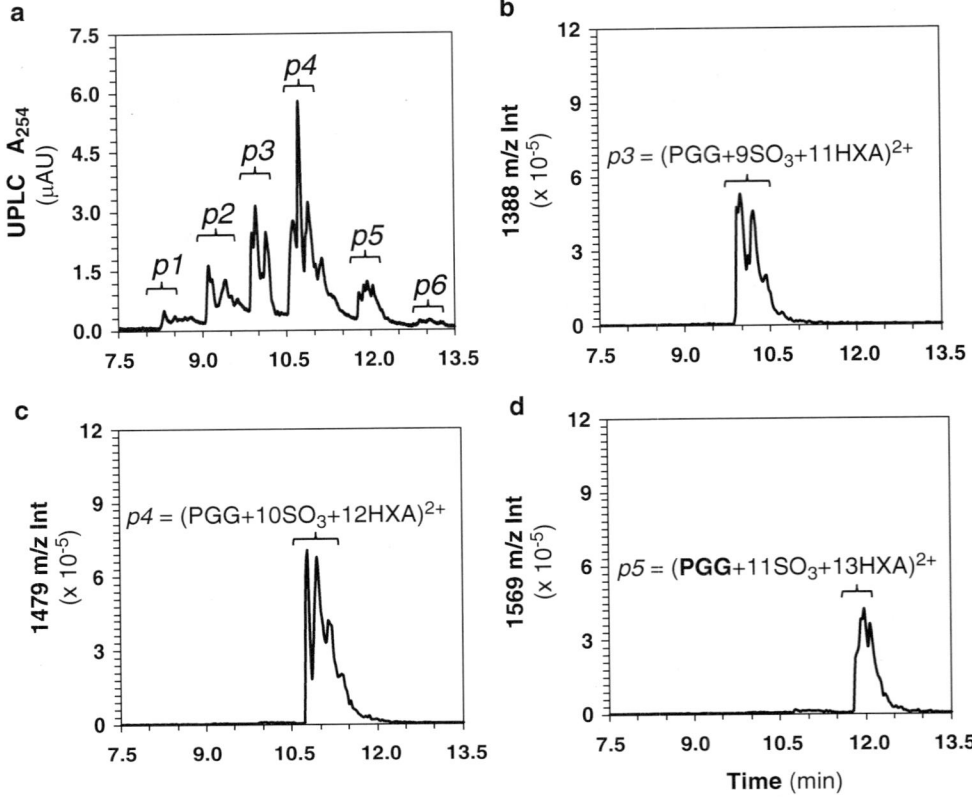

Fig. 4 (**a**) UPLC resolution of SPGG (**4S**) into six peaks (*p1–p6*), which arise from variable sulfation of the PGG scaffold. (**b**)–(**d**) show the ESI-MS analysis of individual components of SPGG using selective ion recording (SIR) monitoring at 1,388, 1,479, and 1,569 *m/z*, respectively. Peaks *p3–p6* correspond to 9, 10, and 11 sulfated PGG species (see inset text for ions formed). Similar SIR profiles were measured for 7, 8, and 12 sulfated species. *PGG* pentagalloyl glucopyranoside, *HXA* n-hexylamine. (Reprinted from ref. 45 with permission of the author)

respectively. In addition to the molecular ions, the MS also displayed several other ions corresponding to the loss of one or more hexylamine-paired sulfonate groups further confirming the identity of the parent sulfated species.

Further, RP-IP UPLC-MS technique can also help identify the structures of components present in the sample. For this purpose, selective ion recording (SIR)–MS was exploited to better characterize peaks *p1* through *p6*. In this approach, the spectrometer was tuned to monitor a specific ion, e.g., 1,478.99 *m/z* corresponding to decasulfated SPGG ion, resulting in the identification of all peaks that contain this molecular ion. Figure 4b–d shows three SIR profiles of SPGG. Monitoring at 1,388.43 *m/z* gave an SIR profile that essentially mimicked *p3* of the UV chromatogram suggesting that each component present in the *p3* peak contained nine sulfate groups. More importantly, the ion corresponding to 1,388.43 was not present in any peak other than *p3*. Likewise, monitoring at 1,478.99 or 1,569.90 *m/z* resulted in a profile equivalent to chromatographic peaks *p4* or *p5*, respectively. This was also found to be the case for peaks *p1*, *p2*, and *p6*.

Thus, RP-IP UPLC–MS coupled with SIR analysis indicated that SPGG was a mixture of septa- (*p1*), octa- (*p2*), nona- (*p3*), deca- (*p4*), undeca- (*p5*), and dodeca- (*p6*) sulfated species, which further contain several subspecies, each with an identical number of sulfate groups. Analysis of the UPLC profile gave a composition of 6, 17, 21, 45, 11, and 3 % for peaks *p1* through *p6*, respectively. Using these peaks and their associated molecular weights, the M_R of SPGG was calculated to be 2,178 (Na$^+$ form).

6 Notes

1. Chemical synthesis of sulfated small molecules is quite challenging, although sulfation appears simple because it is a one-step chemical reaction. The introduction of a sulfate group drastically alters the physicochemical properties of a small molecule; hence an experimenter should pay close attention to potential multidimensional complications described below.

    (a) All sulfated molecules are inherently water soluble, which complicates their isolation in highly pure form. A good lyophilizer is critical for removal of significant quantities of water.

    (b) Because the purification stage of sulfated molecules is performed in aqueous medium, the presence of inorganic salts is a considerable problem and may lead to significant inconsistencies in weight measurements. It is important to perform good size exclusion G-10 chromatography or nonaqueous combiflash to ensure the remove of inorganic salts.

    (c) The stability of sulfate groups on aromatic scaffolds to acidic conditions and high temperatures is also a concern, as described by Liang et al. [61].

    (d) Very few functional group transformations can be successfully performed in the presence of a sulfate group due to the relatively limited solubility of sulfated NSGMs in organic solvents. This essentially forces the synthetic scheme to include sulfation as the final step.

2. Although the synthetic approach for a mono-sulfated scaffold has been extended to a poly-sulfated scaffold, in general this is practically difficult because gradual generation of negative charges on a small scaffold slows down further sulfation considerably. The major challenge here is to drive the reaction to completion to sulfate all available reactive functional groups. Generally, several partially sulfated side products can be anticipated. Moreover, lack of regioselectivity can be also expected to become a dominant issue with poly-functional substrates undergoing sulfation.

3. Based on acid-base balance, SO_3–$(CH_3)_3N$ and SO_3–$(CH_3CH_2)_3N$ complexes appear to be well suited for sulfation of alcoholic groups present in carbohydrates, steroids, and aliphatic or alicyclic scaffolds. Scaffolds based on phenolic structures and containing more acidic–OH groups can be easily sulfated with SO_3 complexes with weaker bases such as pyridine and DMF. Yet, additional considerations also play a role including the stability of the resulting quaternary salt, the ease of the product purification, and the simplicity of the sulfating complex preparation. In fact, SO_3–Py complex has been most often used for sulfation of carbohydrate scaffolds.

4. Although the reaction temperature spans a limited range (90–100 °C), the reaction time spans a wider range of 0.5–8 h. This seems to be highly dependent on the scaffold, the relative position of phenolic groups to be sulfated, and the number of sulfate groups to be introduced. It appears that mono- to tri-sulfated products can be quantitatively obtained within 1–2 h, tetra- to octa-sulfated products require 2–5 h, and higher sulfated species need extended microwave sulfation reaction time. Although the use of triethylamine $(CH_3CH_2)_3N$ is highly dependent on the stability of polyphenolic precursors, its use is recommended because it decreases the time needed to obtain per-sulfated molecules.

5. Precautions should be strictly undertaken with respect to the temperature and pH over the course of reaction, purification, and storage. Our chemical stability studies of sulfated small molecules revealed that they are essentially stable under neutral and basic conditions in a manner similar to the heparins, but are considerably unstable under acidic conditions in contrast to heparins [61]. In addition, temperatures during handling sulfated small molecule GAGs mimetics are preferentially required to be less than 40 °C to assure maximum chemical stability.

6. When setting up the reaction in the microwave tube, it can be seen that in some cases not all solid material dissolves even after stirring for few minutes. This should not be an issue as the reactants will eventually go into solution under heat from the microwave reactor. In addition, during sodium exchange stage, some mono-sulfated compounds will not be readily soluble in high purity water. Therefore, sonication and/or addition of small amount of methanol along with stirring are highly recommended as that will likely enhance solubility.

7. It is highly recommended to purify the resulting reaction mixture right after the end of reaction. However, if there likely to be a significant time gap between completion of reaction and subsequent chromatographic purification, store the reaction vessel at 4 °C. Low temperature may cause precipitation, but warming up the reaction to ambient temperature along with continuous stirring will again dissolve the reaction mixture components.

References

1. Gandhi NS, Mancera RL (2008) The structure of glycosaminoglycans and their interactions with proteins. Chem Biol Drug Des 72: 455–482

2. Silbert JE, Sugumaran G (2002) Biosynthesis of chondroitin/dermatan sulfate. IUBMB Life 54:177–186

3. Kreuger J, Kjellén L (2012) Heparan sulfate biosynthesis. Regulation and variability. J Histochem Cytochem 60:898–907

4. Khan S, Fung KW, Rodriguez E, Patel R, Gor J, Mulloy B, Perkins SJ (2013) The solution structure of heparan sulfate differs from that of heparin, Implications for function. J Biol Chem 288:27737–27751

5. Khan S, Gor J, Mulloy B, Perkins SJ (2010) Semi-rigid solution structures of heparin by constrained X-ray scattering modeling: new insight into heparin-protein complexes. J Mol Biol 395:504–521, Erratum in (2013) J Mol Biol 425:1847

6. Gallagher JT (2012) Heparan sulphate: a heparin in miniature. Handb Exp Pharmacol 207:347–360

7. Desai UR (2013) The promise of sulfated synthetic small molecules as modulators of glycosaminoglycan function. Future Med Chem 5:1363–1366

8. Whitelock JM, Iozzo RV (2005) Heparan sulfate: a complex polymer charged with biological activity. Chem Rev 105:2745–2764

9. Al-Horani RA, Desai UR (2010) Chemical sulfation of small molecules – advances and challenges. Tetrahedron 66:2907–2918

10. Desai UR (2004) New antithrombin-based anticoagulants. Med Res Rev 24:151–181

11. Olson ST, Björk I, Bock SC (2002) Identification of critical molecular interactions mediating heparin activation of antithrombin: implications for the design of improved heparin anticoagulants. Trends Cardiovasc Med 12:198–205

12. Petitou M, van Boeckel CA (2004) A synthetic antithrombin III binding pentasaccharide is now a drug! What comes next? Angew Chem Int Ed 43:3118–3133

13. Shukla D, Liu J, Blaiklock P, Shworak NW, Bai X, Esko JD, Cohen GH, Eisenberg RJ, Rosenberg RD, Spear PG (1999) A novel role for 3-O-sulfated heparan sulfate in herpes simplex virus 1 entry. Cell 99:13–22

14. Copeland R, Balasubramaniam A, Tiwari V, Zhang F, Bridges A, Linhardt RJ, Shukla D, Liu J (2008) Using a 3-O-sulfated heparin octasaccharide to inhibit the entry of herpes simplex virus type 1. Biochemistry 47:774–5783

15. Raghuraman A, Mosier PD, Desai UR (2010) Understanding dermatan sulfate-heparin cofactor II interaction through virtual library screening. ACS Med Chem Lett 1:281–285

16. Raghuraman A, Mosier PD, Desai UR (2006) Finding a needle in a haystack: development of a combinatorial virtual screening approach for identifying high specificity heparin/heparan sulfate sequence(s). J Med Chem 49:3553–3562

17. Pichert A, Samsonov SA, Theisgen S, Thomas L, Baumann L, Schiller J, Beck-Sickinger AG, Huster D, Pisabarro MT (2012) Characterization of the interaction of interleukin-8 with hyaluronan, chondroitin sulfate, dermatan sulfate and their sulfated derivatives by spectroscopy and molecular modeling. Glycobiology 22:134–145

18. Bitomsky W, Wade RC (1999) Docking of glycosaminoglycans to heparin-binding proteins: validation for aFGF, bFGF, and antithrombin and application to IL-8. J Am Chem Soc 121:3004–3013

19. Grootenhuis PDJ, van Boeckel CAA (1991) Constructing a molecular model of the interaction between antithrombin III and a potent heparin analog. J Am Chem Soc 113: 2743–2747

20. Sears P, Wong CH (1998) Enzyme action in glycoprotein synthesis. Cell Mol Life Sci 54: 223–252

21. Sears P, Wong CH (2001) Toward automated synthesis of oligosaccharides and glycoproteins. Science 291:2344–2350

22. Seeberger PH, Haase WC (2000) Solid-phase oligosaccharide synthesis and combinatorial carbohydrate libraries. Chem Rev 100: 4349–4393

23. Heidlas JE, Williams KW, Whitesides GM (1992) Nucleoside phosphate sugars: syntheses on practical scales for use as reagents in enzymatic preparation of oligosaccharides and glycoconjugates. Acc Chem Res 25:307–314

24. Karst NA, Linhardt RJ (2003) Recent chemical and enzymatic approaches to the synthesis of glycosaminoglycan oligosaccharides. Curr Med Chem 10:1993–2031

25. Borgia JA, Malkar NB, Abbasi HU, Fields GB (2001) Difficulties encountered during glycopeptide syntheses. J Biomol Tech 12:44–68

26. Hu YP, Lin SY, Huang CY, Zulueta MM, Liu JY, Chang W, Hung SC (2011) Synthesis of 3-O-sulfonated heparan sulfate octasaccharides that inhibit the herpes simplex virus type 1 host-cell interaction. Nat Chem 3:557–563

27. Sheng GJ, Oh YI, Chang S-K, Hsieh-Wilson LC (2013) Tunable heparan sulfate mimetics

for modulating chemokine activity. J Am Chem Soc 135:10898–10901

28. de Paz JL, Seeberger PH (2008) Deciphering the glycosaminoglycan code with the help of microarrays. Mol Biosyst 4:707–711

29. Yin J, Seeberger PH (2010) Applications of heparin and heparan sulfate microarrays. Methods Enzymol 478:197–218

30. Petitou M, Nancy-Portebois V, Dubreucq G, Motte V, Meuleman D, de Kort M, van Boeckel CA, Vogel GM, Wisse JA (2009) From heparin to EP217609: the long way to a new pentasaccharide-based neutralisable anticoagulant with an unprecedented pharmacological profile. Thromb Haemostas 102:804–810

31. Saxena K, Schieborr U, Anderka O, Duchardt-Ferner E, Elshorst B, Gande SL, Janzon J, Kudlinzki D, Sreeramulu S, Dreyer MK, Wendt KU, Herbert C, Duchaussoy P, Bianciotto M, Driguez PA, Lassalle G, Savi P, Mohammadi M, Bono F, Schwalbe H (2010) Influence of heparin mimetics on assembly of the FGF · FGFR4 signaling complex. J Biol Chem 285:26628–26640

32. Sarrazin S, Bonnaffé D, Lubineau A, Lortat-Jacob H (2005) Heparan sulfate mimicry: a synthetic glycoconjugate that recognizes the heparin binding domain of interferon-gamma inhibits the cytokine activity. J Biol Chem 280:37558–37564

33. Lubineau A, Lortat-Jacob H, Gavard O, Sarrazin S, Bonnaffé D (2004) Synthesis of tailor-made glycoconjugate mimetics of heparan sulfate that bind IFN-γ in the nanomolar range. Chem Eur J 10:4265–4282

34. Ferro V, Dredge K, Liu L, Hammond E, Bytheway I, Li C, Johnstone K, Karoli T, Davis K, Copeman E, Gautam A (2007) PI-88 and novel heparan sulfate mimetics inhibit angiogenesis. Semin Thromb Hemostas 33:557–568

35. Nimjee SM, Oney S, Volovyk Z, Bompiani KM, Long SB, Hoffman M, Sullenger BA (2009) Synergistic effect of aptamers that inhibit exosites 1 and 2 on thrombin. RNA 15:2105–2111

36. Freeman C, Liu L, Banwell MG, Brown KJ, Bezos A, Ferro V, Parish CR (2005) Use of sulfated linked cyclitols as heparin sulfate mimetics to probe the heparin/heparan sulfate binding specificity of proteins. J Biol Chem 280:8842–8849

37. Türk H, Haag R, Alban S (2004) Dendritic polyglycerol sulfates as new heparin analogues and potent inhibitors of the complement system. Bioconjug Chem 15:162–167

38. Ferro V, Liu L, Johnstone KD, Wimmer N, Karoli T, Handley P, Rowley J, Dredge K, Li

CP, Hammond E, Davis K, Sarimaa L, Harenberg J, Bytheway I (2012) Discovery of PG545: a highly potent and simultaneous inhibitor of angiogenesis, tumor growth, and metastasis. J Med Chem 55:3804–3813

39. Gunnarsson GT, Desai UR (2002) Designing small, nonsugar activators of antithrombin using hydropathic interaction analyses. J Med Chem 45:1233–1243

40. Gunnarsson GT, Desai UR (2002) Interaction of designed sulfated flavanoids with antithrombin: lessons on the design of organic activators. J Med Chem 45:4460–4470

41. Raghuraman A, Liang A, Krishnasamy C, Lauck T, Gunnarsson GT, Desai UR (2009) On designing non-saccharide, allosteric activators of antithrombin. Eur J Med Chem 44:2626–2631

42. Al-Horani RA, Liang A, Desai UR (2011) Designing nonsaccharide, allosteric activators of antithrombin for accelerated inhibition of factor Xa. J Med Chem 54:6125–6138

43. Abdel Aziz MH, Sidhu PS, Liang A, Kim JY, Mosier PD, Zhou Q, Farrell DH, Desai UR (2012) Designing allosteric regulators of thrombin. Monosulfated benzofuran dimers selectively interact with Arg173 of exosite 2 to induce inhibition. J Med Chem 55: 6888–6897

44. Sidhu PS, Abdel Aziz MH, Sarkar A, Mehta AY, Zhou Q, Desai UR (2013) Designing allosteric regulators of thrombin. Exosite 2 features multiple sub-sites that can be targeted by sulfated small molecules for inducing inhibition. J Med Chem 56:5059–5070

45. Al-Horani RA, Ponnusamy P, Mehta AY, Gailani D, Desai UR (2013) Sulfated Pentagalloylglucoside is a potent, allosteric, and selective inhibitor of factor XIa. J Med Chem 56:867–878

46. Dusza JP, Joseph JP, Bernstein S (1985) The preparation of estradiol-17 beta sulfates with triethylamine-sulfur trioxide. Steroids 45:303–315

47. Kitagawa K, Aida C, Fujiwara H, Yagami T, Futaki S, Kogire M, Ida J, Inoue K (2001) Facile solid-phase synthesis of sulfated tyrosine-containing peptides: total synthesis of human big gastrin-II and cholecystokinin (CCK)-39. J Org Chem 66:1–10

48. Lee JC, Lu XA, Kulkarni SS, Wen YS, Hung SC (2004) Synthesis of heparin oligosaccharides. J Am Chem Soc 126:476–477

49. Tully SE, Mabon R, Gama CI, Tsai SM, Liu X, Hsieh-Wilson LC (2004) A chondroitin sulfate small molecule that stimulates neuronal growth. J Am Chem Soc 126:7736–7737

50. Young T, Kiessling LL (2002) A strategy for the synthesis of sulfated peptides. Angew Chem Int Ed Engl 41:3449–3451

51. Krylov VB, Ustyuzhanina NE, Grachev AA, Nifantiev NE (2008) Efficient acid-promoted per-O-sulfation of organic polyols. Tetrahedron Lett 49:5877–5879

52. Raghuraman A, Riaz M, Hindle M, Desai UR (2007) Rapid and efficient microwave-assisted synthesis of highly sulfated organic scaffolds. Tetrahedron Lett 48:6754–6758

53. Monien BH, Henry BL, Raghuraman A, Hindle M, Desai UR (2006) Novel chemo-enzymatic oligomers of cinnamic acids as direct and indirect inhibitors of coagulation proteinases. Bioorg Med Chem 14:7988–7998

54. Henry BL, Monien BH, Bock PE, Desai UR (2007) A novel allosteric pathway of thrombin inhibition: exosite II mediated potent inhibition of thrombin by chemo-enzymatic, sulfated dehydropolymers of 4-hydroxycinnamic acids. J Biol Chem 282:31891–31899

55. Henry BL, Abdel Aziz M, Zhou Q, Desai UR (2010) Sulfated, low-molecular-weight lignins are potent inhibitors of plasmin, in addition to thrombin and factor Xa: novel opportunity for controlling complex pathologies. Thromb Haemost 103:507–515

56. Saluja B, Thakkar JN, Li H, Desai UR, Sakagami M (2013) Novel low molecular weight lignins as potential anti-emphysema agents: in vitro triple inhibitory activity against elastase, oxidation and inflammation. Pulm Pharmacol Ther 26:296–304

57. Raghuraman A, Tiwari V, Zhao Q, Shukla D, Debnath AK, Desai UR (2007) Viral inhibition studies on sulfated lignin, a chemically modified biopolymer and a potential mimic of heparan sulfate. Biomacromolecules 8:1759–1763

58. Raman K, Karuturi R, Swarup VP, Desai UR, Kuberan B (2012) Discovery of novel sulfonated small molecules that inhibit vascular tube formation. Bioorg Med Chem Lett 22:4467–4470

59. Khanbabaee K, Lötzerich K (1997) Efficient total synthesis of the natural products 2,3,4,6-tetra-O-galloyl-D-glucopyranose, 1,2,3,4,6-penta-O-galloyl-β-D-glucopyranose and the unnatural 1,2,3,4,6-penta-O-galloyl-α-D-glucopyranose. Tetrahedron 53:10725–10732

60. Karuturi R, Al-Horani RA, Mehta SC, Gailani D, Desai UR (2013) Discovery of allosteric modulators of factor XIa by targeting hydrophobic domains adjacent to its heparin-binding site. J Med Chem 56:2415–2428

61. Liang A, Thakkar JN, Desai UR (2010) Study of physico-chemical properties of novel highly sulfated, aromatic, mimetics of heparin and heparan sulfate. J Pharm Sci 99:1207–1216

Synthesis of Selective Inhibitors of Heparan Sulfate and Chondroitin Sulfate Proteoglycan Biosynthesis

Caitlin Mencio, Dinesh R. Garud, Balagurunathan Kuberan, and Mamoru Koketsu

Abstract

Glycosaminoglycan (GAG) side chains of proteoglycans are involved in a wide variety of developmental and pathophysiological functions. Similar to a gene knockout, the ability to inhibit GAG biosynthesis would allow us to examine the function of endogenous GAG chains. However, ubiquitously and irreversibly knocking out all GAG biosynthesis would cause multiple effects making it difficult to attribute a specific biological role to a specific GAG structure in spatiotemporal manner. Reversible and selective inhibition of GAG biosynthesis would allow us to examine the importance of endogenous GAGs to specific cellular, tissue, or organ systems. In this chapter, we describe the chemical synthesis and biological evaluation of 4-deoxy-4-fluoro-xylosides as selective inhibitors of heparan sulfate and chondroitin/dermatan sulfate proteoglycan biosynthesis.

Key words Glycosaminoglycan biosynthesis, 4-Deoxy-4-fluoro-xylosides, Proteoglycan inhibitors, Heparan sulfate, Chondroitin sulfate, Dermatan sulfate

1 Introduction

Proteoglycan (PG) is composed of a core protein and one or more glycosaminoglycan (GAG) side chains. These side chains can be divided into several different classes with the two largest being chondroitin sulfate and heparan sulfate GAGs. The first step of PG biosynthesis is the xylosylation of certain serine residues on the core protein [1]. After the addition of a xylose residue, a tetrasaccharide linker is assembled before the repeating disaccharide units are attached to extend the GAG chain [2, 3]. Many of the biological activities of PGs are dependent on the interaction of proteins with the sulfated GAG side chains [4–6]. The ability to manipulate, enhance, or inhibit GAG biosynthesis would allow for a better understanding of the functions of endogenous GAGs.

Several different biochemical approaches have been utilized to manipulate GAG biosynthesis. The two most common include the

Kuberan Balagurunathan et al. (eds.), *Glycosaminoglycans: Chemistry and Biology*, Methods in Molecular Biology, vol. 1229, DOI 10.1007/978-1-4939-1714-3_8, © Springer Science+Business Media New York 2015

use of sodium chlorate which is a bleaching agent and Brefeldin A (BFA), a fungal isoprenoid metabolite [7–10]. These substances have different mechanisms of action but result in a loss of sulfated GAG chains. Sodium chlorate prevents formation of 3′-phosphoadenosine-5′-phosphosulfate (PAPS), the sulfate donor required by the sulfotransferases for the sulfation of GAG chains [9]. BFA disrupts the secretory transport of Golgi vesicles. GAG biosynthesis occurs in the Golgi, so preventing the transport of Golgi vesicles disrupts proteoglycan formation [8, 10]. While effective, these approaches are not ideal. Sodium chlorate treatment requires high concentrations, up to 30 mM, to be effective. These high concentrations are not suitable for in vivo studies. BFA disruption is not proteoglycan specific. Disruption of Golgi vesicle secretory transport also disrupts the biosynthesis of other glycoconjugates resulting in outcomes that may not be attributed solely to GAG function [11]. A better method to disrupt specifically GAG biosynthesis is needed.

GAGs have a common tetrasaccharide linkage region prior to extension by repeating disaccharide units specific to GAG type. The first step to GAG biosynthesis is xylosylation. Xylose residues are added to specific serine residues on the core protein, followed by the rest of the linkage region [1, 2]. Exploiting this specific method for biosynthesis has been attempted. The use of 4-deoxy-glucosamine analogs has been examined as a method for the inhibition of GAG biosynthesis; however, GAG chains were still produced although they were smaller in size than endogenous GAGs. These analogs also affected other glycoconjugates as well [12].

Using the formation of this linkage region as a starting place and previous knowledge about xylosides and their ability to serve as decoys for GAG biosynthesis, we proposed and synthesized a series of 4-deoxy-4-fluoro-xylosides using click chemistry on 2,3-di-O-benzoyl-4-deoxy-4-fluoro-D-xylopyranosyl azide [13] followed by deprotection of hydroxyl groups as shown in Scheme 8.1. These molecules, unlike traditional xylosides, contain a fluoro group at the fourth position where extension of GAG chains would normally

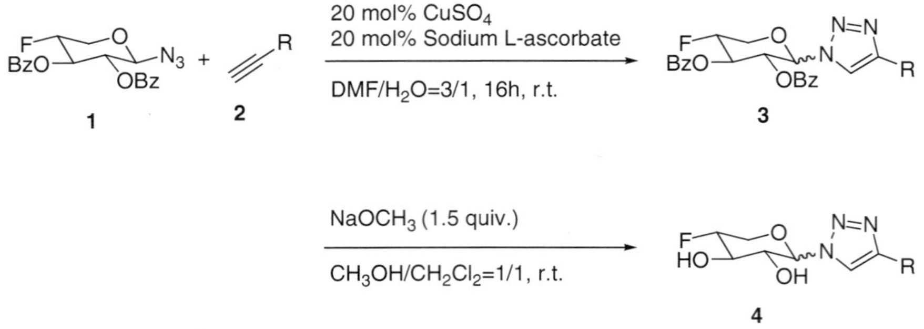

Scheme 1 Formation of 4-deoxy-4-fluoro-xyloside

occur [14, 15]. A number of these molecules were shown to inhibit GAG biosynthesis efficiently and selectively. Here we describe the method of synthesis of these molecules and their use in vitro.

2 Materials

2.1 Materials for the Synthesis of 4-Deoxy-4-fluoro-xylosides

1. Chemicals were obtained from many commercial vendors.
2. Reactions were carried out under a nitrogen atmosphere in oven-dried glassware using standard syringe and septum techniques.

2.2 Materials for the Confirmation and Characterization of 4-Deoxy-4-fluoro-xylosides

1. IR spectra were recorded on a JASCO FTIR-460 Plus.
2. The ^{1}H NMR, ^{13}C NMR, and ^{19}F NMR spectra were measured on JEOL:JNM ECX-400P, JEOL:JNM ECA-500, JEOL:JNM ECA-600, and Bruker 400-MHz spectrometers. Chemical shifts of protons are reported in δ values referred to TMS as an internal standard.
3. MS were recorded using a JOEL JMS-700/GI spectrometer. High-resolution mass spectrometry was performed using a Finnigan LCQ mass spectrometer in either positive or negative ion mode.
4. Column chromatography was performed on silica gel N-60 (40–50 μm) and, TLC on pre-coated plates of 0.25 mm thick silica gel 60F254 aluminum sheets. Chromatograms were observed under short- and long-wavelength UV light and were visualized by heating plates that were dipped in a solution of ammonium (VI) molybdate tetrahydrate (12.5 g) and cerium (IV) sulfate tetrahydrate (5.0 g) in 10 % aqueous sulfuric acid (500 mL).
5. The α- and β-isomers of 4-deoxy-4-fluoro-xylosides were purified by HPLC using a C18 column, Wakosil-II 5C18 HG prep (20.0 mm × 250 mm) (*see* **Note 1**).

2.3 Materials for In Vitro Analysis of 4-Deoxy-4-fluoro-xylosides Using CHO Cells

1. Wild-type K1 Chinese hamster ovary (CHO) cells.
2. Radiolabeled sodium sulfate, ^{35}S-Na$_2$SO$_4$; 10 mCi/mL solution.
3. Growth medium: Add 50 mL of cell culture-grade, sterile fetal bovine serum and 5 mL of cell culture-grade penicillin/streptomycin to 500 mL of Hams-F12 media.
4. Screening media: Dialyze 100 mL of cell culture-grade fetal bovine serum in dialysis tubing (1,500 MWCO). Add 50 mL of dialyzed and filter-sterilized serum and 5 mL of penicillin/streptomycin, cell culture grade, to 500 mL of Hams-F12 media.
5. Sterile phosphate-buffered saline.

6. Stock solution of xylosides. Weigh out the appropriate amount of xyloside based on molecular weight. Add calculated amount of DMSO or water, depending on the solubility of the compound to make a 10 mM solution.

7. Sterile cell culture-grade 24-well plate.

2.4 Materials for GAG Analysis of CHO Cells After 4-Deoxy-4-fluoroxyloside Treatment

1. 6× pronase solution: Prepared from *Streptomyces greisus* protease type XIV at a concentration of 1 mg/mL.

2. Microcentrifuge tubes.

3. DEAE-sepharose beads.

4. Disposable polypropylene columns.

5. Wash buffer solution: 20 m MNaOAc, pH 6, with 0.1 M NaCl and 0.01 % Triton X-100.

6. Elution buffer solution: 20 m MNaOAC, pH 6 with 1 M NaCl and 0.01 % Triton X-100.

7. Flow scintillation cocktail, Ultima-FloAP.

8. DEAE Buffer A for HPLC: 0.01 M KH_2PO_4, 0.001 % w/v CHAPS, pH 6.4 (*see* **Note 2**).

9. DEAE Buffer B for HPLC: 0.01 M KH_2PO_4, 1 M NaCl, 0.001 % w/v CHAPS, pH 6.4 (*see* **Note 2**).

10. Tskgel DEAE-3SW column.

3 Methods

3.1 Synthesis of 4-Deoxy-4-fluoroxylosides

1. Add copper (II) sulfate pentahydrate (20 mol%) and sodium L-ascorbate (20 mol%) to a stirred solution of 2,3-di-*O*-benzoyl-4-deoxy-4-fluoro-D-xylopyranosyl azide (1.0 equivalent) and azide-reactive triple bond-containing hydrophobic agents (1.5 equivalents) in DMF/water = 3/1.

2. Continue the reaction for additional 16 h with stirring at room temperature.

3. After the completion of reaction, concentrate the reaction mixture and purify over a silica gel column (ethyl acetate/*n*-hexane = 1/2) to yield inseparatable α:β-isomers of 2,3-di-*O*-benzoyl-4-deoxy-4-fluoro-D-xylopyranosyl azide derivatives (*see* Table 1).

4. Add sodium methoxide (1.5 equivalents) to a solution of 2,3-di-*O*-benzoyl-4-deoxy-4-fluoro-D-xylopyranosyl azide (1.0 equivalent) in 10 mL of 1:1 dichloromethane/methanol at room temperature.

5. Continue the reaction for an additional 3 h with stirring at room temperature.

Table 1
Structures of 2,3-di-*O*-benzoyl-4-deoxy-4-fluoro-D-xylopyranosyl azide derivatives, 3

Entry	R	Entry	R
3a		3l	
3b		3m	
3c		3n	
3d		3o	
3e		3p	
3f		3q	
3g		3r	
3h		3s	
3i		3t	
3j		3u	
3k			

6. Neutralize the reaction mixture by addition of Dowex (H⁺) resins, filter, and concentrate to give a crude mixture of the α:β-isomers of 4-deoxy-4-fluoro-xylosides **4**.

7. Separate a mixture of α:β-isomers of 4-deoxy-4-fluoro-xyloside **4** derivatives by reverse-phase chromatography of HPLC system using a C18 column, Wakosil-II 5C18 HG prep (20.0 mm×250 mm). HPLC conditions: eluting solvent: aqueous CH_3CN, flow rate: 5 mL/min, column temperature: 40 °C, UV detection: 256–340 nm, respectively (for example *see* Fig. 1).

3.2 Characterization and Confirmation of 4-Deoxy-4-fluoro-xylosides

1. The structures of α- and β-isomers of 4-deoxy-4-fluoro-xylosides **4** were confirmed by IR, ¹H NMR, ¹³C NMR, ¹⁹F NMR, and optical rotation and MS spectra. For example:

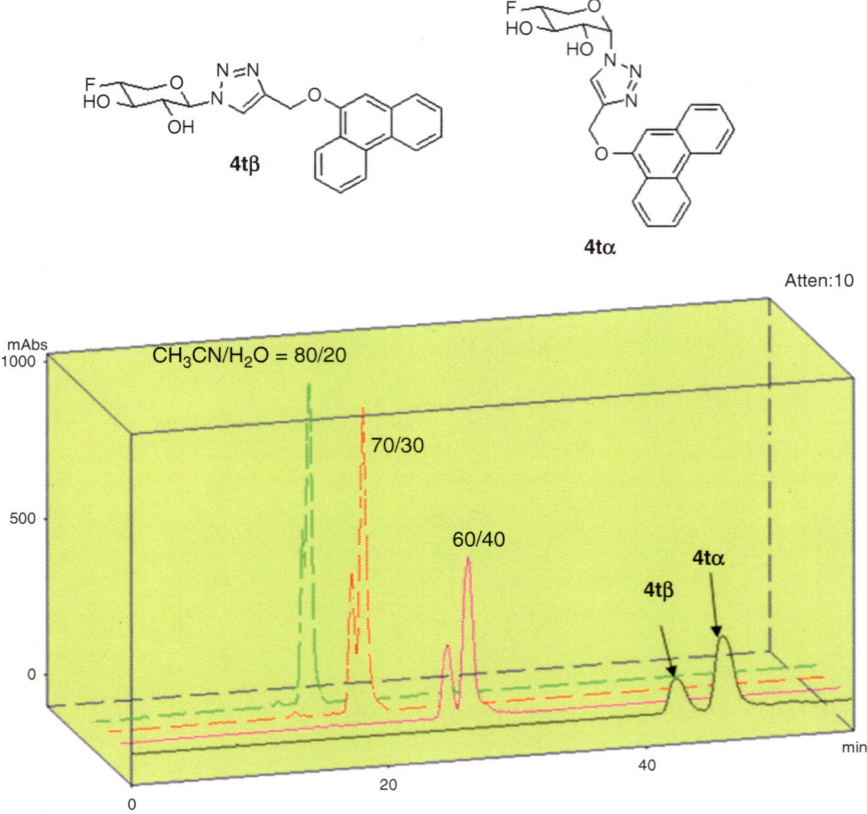

Fig. 1 Separation of α- and β-form of 1-(4-deoxy-4-fluoro-D-xylopyranosyl)-4-(9-phenanthryloxymethyl)-1,2,3-triazole 4t by HPLC ODS column. HPLC conditions: Column: Wakosil-II 5C18 HG prep (φ 20.0 mm × 250 mm); flow rate: 5 mL/min; column temperature: 40 °C; detection: UV 256 nm

(a) α-Isomer of 1-(4-deoxy-4-fluoro-D-xylopyranosyl)-4-(9-phenanthryloxymethyl)-1,2,3-triazole **4tα**. $[\alpha]_D^{25} + 14.1°$ (c 1.0, $(CH_3)_2CO$); ^{1}H NMR (600 MHz, $(CD_3)_2CO$): δ 3.96–3.99 (m, 1H, H-5b), 4.08–4.19 (m, 2H, H-3, H-2), 4.40–4.43 (m, 2H, H-5a, OH), 4.51–4.61 (m, $J_{4,F}$=46.8 Hz, 1H, H-4), 5.16 (d, J=3.4 Hz, 1H, OH), 5.49 (s, 2H), 6.20 (d, J=3.4 Hz, 1H, H-1), 7.42 (s, 1H), 7.49 (t, J=6.9 Hz, 1H), 7.53 (t, J=6.8 Hz, 1H), 7.59 (t, J=8.3 Hz, 1H), 7.67 (t, J=6.9 Hz, 1H), 7.86 (d, J=8.9 Hz, 1H) 8.29 (d, J=8.3 Hz, 1H), 8.44 (s, 1H), 8.66 (d, J=8.2 Hz, 1H), 8.75 (d, J=8.2 Hz, 1H); ^{13}C NMR (150 MHz, $(CD_3)_2CO$): δ 62.7, 66.3 ($^2J_{FC}$=23.1 Hz, C5), 70.2 ($^3J_{FC}$=21.7 Hz, C2), 71.0 (C3), 85.7 (C1), 89.1 ($^1J_{FC}$=177.8 Hz, C4), 104.3, 123.3, 123.4, 123.5, 125.4, 125.6, 127.3, 127.4, 127.8, 128.1, 128.4, 132.2, 133.8, 143.7, 153.0; ^{19}F NMR ($(CD_3)_2CO$): δ –193.4. MS (FAB): m/z=410 $[M+H]^+$.

(b) β-Isomer of 1-(4-deoxy-4-fluoro-D-xylopyranosyl)-4-(9-phenanthryloxymethyl)-1,2,3-triazole **4tβ**. $[\alpha]_D$ ^{25}-51.6° (c 1.0, $(CH_3)_2CO$); ^{1}H NMR (600 MHz, $(CD_3)_2CO$): δ 3.74 (dt, J=4.2, 11.0 Hz, 1H, H-5b), 3.85–3.91 (m, 1H, H-3), 4.12–4.18 (m, 2H, H-2, H-5a), 4.49–4.62 (m, $J_{4,F}$=49.4 Hz, 1H, H-4), 4.99 (d, J=4.8 Hz, 1H, OH), 5.06 (d, J=4.1 Hz, 1H, OH), 5.48 (s, 2H), 5.71 (d, J=9.6 Hz, 1H, H-1), 7.41 (s, 1H), 7.49 (t, J=8.2 Hz, 1H), 7.54 (t, J=8.3 Hz, 1H), 7.59 (t, J=6.9 Hz, 1H), 7.68 (t, J=8.3 Hz, 1H), 7.86 (d, J=6.8 Hz, 1H) 8.30 (d, J=8.2 Hz, 1H), 8.43 (s, 1H), 8.67 (d, J=8.3 Hz, 1H), 8.75 (d, J=8.2 Hz, 1H); ^{13}C NMR (150 MHz, $(CD_3)_2CO$): δ 62.8, 66.1 ($^2J_{FC}$=30.4 Hz, C5), 73.1 ($^3J_{FC}$=8.7 Hz, C2), 76.3 ($^2J_{FC}$=18.8 Hz, C3), 89.1 (C1), 89.9 ($^1J_{FC}$=180.6 Hz, C4), 104.3, 123.3, 123.4, 123.5, 123.9, 125.4, 127.2, 127.3, 127.4, 127.9, 128.2, 128.4, 132.2, 133.8, 144.4, 153.0; ^{19}F NMR ($(CD_3)_2CO$): δ –201.1. MS (FAB): m/z=410 [M+H]$^+$.

3.3 In Vitro Analysis of 4-Deoxy-4-fluoro-xylosides Using CHO Cells

1. Plate approximately 1×10^5 cells per well in a 24-well cell culture plate containing 0.5 mL of growth media per well. Incubate the plate for 24 h at 37 °C with 5 % CO_2 to reach a confluence of about 50 %.

2. Remove the media and wash the cells with sterile PBS and replace it with 495 μL of screening media. Add 5 μL of xyloside stock solution for a final concentration of 100 μM (or PBS for control) and 50 μCi of ^{35}S-Na_2SO_4. The radioactive sulfate will label the GAG chains that are produced by the cells.

3. Incubate the plate for 24 h at 37 °C with 5 % CO_2.

3.4 Analysis of GAGs of CHO Cells After 4-Deoxy-4-fluoro-xyloside Treatment

4. Add 83.3 μL of 6× pronase to each well and incubate overnight at 37 °C with 5 % CO_2.

5. Transfer the entire contents of the wells to a microcentrifuge tube and spin at $16,000 \times g$ for 5 min.

6. Transfer the supernatant to a fresh tube and add 1 mL of 0.016 % Triton X-100.

7. Assemble a purification column with a 200 μL bed volume of DEAE-sepharose beads (*see* **Note 3**).

8. Wash the column with 1 mL dH$_2$O.

9. Equilibrate the gel with 10 column volumes of wash buffer.

10. Load the diluted supernatant onto the DEAE-sepharose column.

11. Wash the column with 30 column volumes of wash buffer.

12. Elute the bound GAG chains using 6 column volumes into a clean collection tube.

13. Place 5 mL of scintillation fluid into a scintillation vial for each well purified.

14. Add 50 μL of purified GAG to the vial and shake to mix. Place the vial in the scintillation counter and compare the scintillation counts between the control and the treated samples.

15. To remove salt from the elution, filter through a 3,000 MWCO spin filtration unit at $<14{,}000 \times g$ for 10 min each spin. Discard flow through in appropriate containers for radioactivity. Add sample until total elutions have been spun through the column. After the sample, add water and spin. Repeat the addition of water step at least six times.

16. In a fresh column, flip the filter unit over and spin again to remove all liquid from the filter unit.

17. Dilute the sample to 300–1,000 μL and place in vial for HPLC analysis. Vial will vary based on machine and size of sample.

18. Sample is then run on the HPLC using linear gradient transitioning from DEAE Buffer A to DEAE Buffer B. Samples that produce GAG will result in two peaks in a DEAE profile. The first is the less sulfated GAGs, often the heparan sulfate, and the second is the more highly sulfated GAGs, more commonly the chondroitin sulfate. 4-Deoxy-4-fluoro-xylosides block production of these molecules and thus you will see reduction or complete loss of these peaks in a DEAE profile (*see* Fig. 2 for an example).

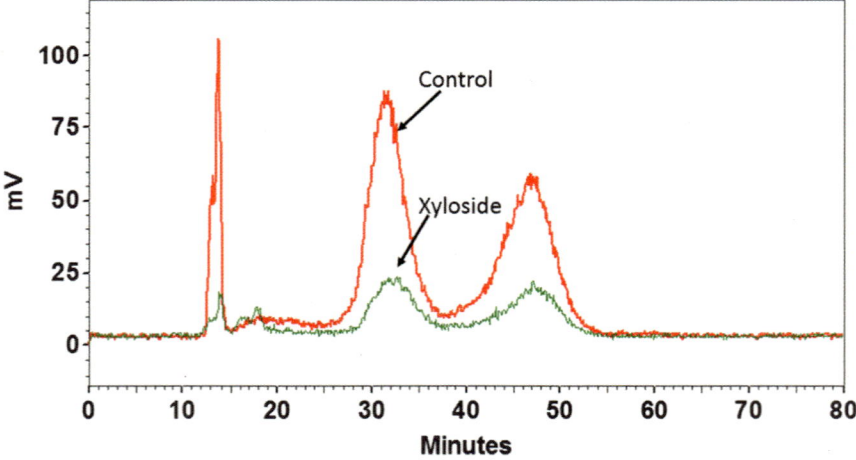

Fig. 2 HPLC DEAE profile showing reduction of sulfated glycosaminoglycans after treatment with 4-deoxy-4-fluoro-xyloside

4   Notes

1. Please note that for isolation of pure α- and β-isomers the % of eluting solvent and UV detection for the HPLC changes with the change in the functional group at triazolyl ring. Increase in the polarity gives good separation of α:β-isomers.

2. Please note that CHAPS can damage your pH probe. It is, therefore, advisable to adjust the pH of the buffer prior to the addition of CHAPS. Furthermore, addition of CHAPS may generate bubbles that require degassing each buffer well before use for HPLC analysis.

3. If your columns come with a small disk/filter pad, this can be used as the first step to create the purification column. If there is no disk, glass wool can be substituted but be careful as skin irritation can occur if exposed to bare skin. If you have the disk/filter pad it is important to prevent air bubbles from getting trapped between the filter and the outlet of the column. To do this plug the bottom of the column and fill it with water. Place the filter on the top and then push through the water until it reaches the bottom of the column. This will displace water out the top of the column, so it is best done over a sink. After the filter is settled at the bottom, remove the plug and the water left should run through. This pre-loading of water prevents air bubble formation.

Acknowledgement

This work was supported in part by NIH grants (GM075168 and HL107152 to B.K.), Human Frontier Science Program grant (RGP0044/2006 to B.K.), and KAKENHI grant (No. 23590005 to M.K.).

References

1. Gotting C, Kuhn J, Zahn R et al (2000) Molecular cloning and expression of human UDP-D-xylose:proteoglycan core protein β-D-xylosyltransferase and its first isoform XT-II. J Mol Biol 304:517–528

2. Lindahl U, Roden L (1966) The chondroitin 4-sulfate protein linkage. J Biol Chem 241: 2113–2119

3. Muir H (1958) The nature of the link between protein and carbohydrate of a chondroitin sulphate complex from hyaline cartilage. Biochem J 69:195–204

4. Mikami T, Kitagawa H (2013) Biosynthesis and function of chondroitin sulfate. Biochim Biophys Acta 1830:4719–4733

5. Lin X (2004) Functions of heparin sulfate proteoglycans in cell signaling during development. Development 131:6009–6021

6. Bernfield M, Gotte M, Park P et al (1999) Functions of cell surface heparin sulfate proteoglycans. Annu Rev Biochem 68: 729–777

7. Greve H, Cully Z, Blumberg P, Kresse H (1988) Influence of chlorate on proteoglycan

biosynthesis by cultured human fibroblasts. J Biol Chem 263:12886–12892

8. Calabro A, Hascall V (1994) Effects of brefeldin A on aggrecan core protein synthesis and maturation in rat chondrosarcoma cells. J Biol Chem 269(36): 22771–22778

9. Safaiyan F, Kolset S, Prydz K et al (1999) Selective effects of sodium chlorate treatment on the sulfation of heparin sulfate. J Biol Chem 274:36267–36273

10. Uhlin-Hansen L, Yanagishita M (1993) Differential effect of brefeldin A on the biosynthesis of heparin sulfate and chondroitin/dermatan sulfate proteoglycans in rat ovarian granulosa cells in culture. J Biol Chem 268: 17370–17376

11. Sherwood A, Holmes E (1992) Brefeldin A induced inhibition of de novo globo- and neolacto-series glycolipid core chain biosynthesis in human cells. Evidence for an effect on beta 1→4 galactosyltransferase activity. J Biol Chem 267:25328–25336

12. Berkin A, Szarek W, Kisilevsky R (2005) Biological evaluation of a series of 2-acetoamido-2-deoxy-D-glucose analogs towards cellular glycosaminoglycan and protein synthesis in vitro. Glycoconj J 22:443–451

13. Wicki J, Schloegl J, Tarling CA, Withers SG (2007) Recruitment of both uniform and differential binding energy in enzymatic catalysis: xylanases from families 10 and 11. Biochemistry 46:6996–7005

14. Garud DR, Tran VM, Victor XV, Koketsu M, Kuberan B (2008) Inhibition of heparan sulfate and chondroitin sulfate proteoglycan biosynthesis. J Biol Chem 283:28881–28887

15. Tsuzuki Y, Nguyen TK, Garud DR, Kuberan B, Koketsu M (2010) 4-Deoxy-4-fluoro-xyloside derivatives as inhibitors of glycosaminoglycan biosynthesis. Bioorg Med Chem Lett 20: 7269–7273

Ascidian (Chordata-Tunicata) Glycosaminoglycans: Extraction, Purification, Biochemical, and Spectroscopic Analysis

Mauro S.G. Pavão

Abstract

Sulfated polysaccharides with unique structures of the chondroitin/dermatan and heparin/heparan families of sulfated glycosaminoglycans have been described in several species of ascidians (Chordata-Tunicata). These unique sulfated glycans have been isolated from–ascidians and characterized by biochemical and spectroscopic methods. The ascidian glycans can be extracted by different tissues or cells by proteolytic digestion followed by cetylpyridinium chloride/ethanol precipitation. The total glycans are then fractionated by ion-exchange chromatography on DEAE-cellulose and/or Mono Q (HR 5/5) columns. Alternatively, precipitation with different ethanol concentrations can be employed. An initial analysis of the purified ascidian glycans is carried out by agarose gel electrophoresis on diaminopropane/acetate buffer, before or after digestion with specific glycosaminoglycan lyases or deaminative cleavage with nitrous acid. The disaccharides formed by exhaustive degradation of the glycans is purified by gel-filtration chromatography on a Superdex-peptide column and analyzed by HPLC on a strong ion exchange Sax-Spherisorb column. 1H or 13C nuclear magnetic resonance spectroscopy in one or two dimensions is used to confirm the structure of the intact glycans.

Key words Glycosaminoglycans, Dermatan sulfate, Heparin, Agarose gel electrophoresis, Polyacrylamide gel electrophoresis, Ion-exchange chromatography, Gel filtration chromatography, 1H-NMR, 13C-NMR

1 Introduction

Marine invertebrates are a rich source of glycosaminoglycans containing unique sulfation patterns. The extraction, purification, and initial characterization of these molecules are an important step for a more detailed analysis of the fine structure by spectroscopic methods such as nuclear magnetic resonance and the study of the biological activity of these polymers.

Dermatan sulfate glycosaminoglycans with high content of disulfated disaccharide units and unique sulfation patterns have been described in ascidians from the orders Phlebobranchia and Stolidobranchia. An oversulfated dermatan sulfate consisting of

Kuberan Balagurunathan et al. (eds.), *Glycosaminoglycans: Chemistry and Biology*, Methods in Molecular Biology, vol. 1229, DOI 10.1007/978-1-4939-1714-3_9, © Springer Science+Business Media New York 2015

[IdoA2S-GalNAc6S] (major) and [IdoA-GalNAc6S] disaccharide units occurs in the organs of the phlebobranch *Phallusia nigra* [1–3]. Highly sulfated dermatan sulfates with the same core structure, [IdoA2-GalNAc], but sulfated at carbon 4 of the GalNAc residues have been detected in the extracellular matrix of the stolidobranchs *Styela plicata* and *Halocynthia pyriformis* [3, 4]. Interestingly, low content of GlcAc residues were detected in the dermatan sulfates from all ascidians. Using these unique oversulfated dermatan sulfates, it was possible to show the relationship between the position of sulfate and heparin cofactor II activity. Thus, the dermatan sulfate from *S. plicata* and *H. pyriformis*, composed mainly by the disulfated disaccharide [IdoA2S-GalNAc4S], has high heparin cofactor II activity [1]. On the other hand, the dermatan sulfate obtained from *P. nigra*, possessing the same sulfation degree but composed mainly by [IdoA2S-GalNAc6S] disaccharide units [1], has very low heparin cofactor II activity [3, 5], as a result of a discernible binding to the inhibitor. Overall, these results indicate that binding of oversulfated dermatan sulfate polymers to heparin cofactor II requires a specific sulfation pattern on the glycans, composed by [IdoA2S-GalNAc4S]-enriched sequences.

Heparin-like glycosaminoglycans have been described only in the stolidobranchia ascidian *S. plicata*. Heparins composed mainly by [IdoA2S-GlcNS,6S] and minor amount of [HexA-GlcNS,6S] disaccharide units have been detected in intracellular granules of mast cell-like cells, the test cells of the oocytes [6, 7]. In the same ascidian, a heparin composed mainly by [IdoA2S-GlcNS] and [IdoA2S-GlcNS,6S] disaccharide units and lower amounts of GlcNAc3S-containing units [IdoA2S-GlcNS,3S] and [IdoA2S-GlcNS,3S,6S] has been detected in intracellular granules of a basophil-like cell, which circulates in the hemolymph [8].

The oocyte test cell heparin has anticoagulant and antithrombin-mediated thrombin inhibition activities 10 times and 20 times lower than mammalian heparin, respectively [6]. On the other hand the heparin isolated from *S. plicata* basophil-like cells has an antithrombin-mediated thrombin inhibition activity similar to that of porcine intestinal mucosa heparin [8].

In this work, we describe the methods for extraction, purification, and biochemical and structural analysis of glycosaminoglycan-like polymers isolated from ascidians.

2   Materials

All reagents should be of analytical grade and properly stored either at room temperature or at 4 °C, when specified. Animals should be collected and kept in an aerated aquarium containing filtered (0.22 μm) seawater until use. Alternatively, the animals can be placed immediately in 70 % ethanol at room temperature. The sulfated glycans can be obtained from different starting materials (*see* **Note 1**).

2.1 Obtaining the Starting Material

1. Oocyte and hemocyte washing buffer 1: 0.22 μm filtered seawater (pH 6.0).

2. Oocyte and hemocyte washing buffer 2: 0.22 μm filtered seawater (pH 8.0).

3. Dehydration agent: Acetone.

4. Marine anticoagulant (MAC): 0.45 M NaCl, 100 mM glucose, 1.5 mM trisodium citrate, 13 mM citric acid, 10 mM EDTA in 1 l distilled water, pH 7.0).

5. MAC, containing 0.45 M sodium chloride, 0.1 M glucose, 0.01 M disodium citrate, 0.01 M citric acid, and 0.001 EDTA (pH 7.0).

2.2 Extraction of the Sulfated Glycosaminoglycans

1. Digestion buffer: 0.1 M sodium acetate buffer (pH 5.5), 5 mM EDTA, and 5 mM cysteine, containing 100 mg papain (6,000 USP-U/mg, Merck) (*see* **Note 2**).

2. First precipitation buffer: Cetylpyridinium chloride (CPC) (0.5 % final concentration).

3. Washing buffer: CPC solution (0.05 % final concentration).

4. Sulfated glycosaminoglycan-CPC complex dissociation buffer: 2 M NaCl in 95 % ethanol (100:15, v/v).

5. Final precipitation solution: 2 volumes of 95 % ethanol (*see* **Note 3**).

2.3 Fractionation of the Sulfated Glycosaminoglycans

Columns used to fractionate the sulfated glycans:

1. DEAE-cellulose.

2. Mono-Q (HR 5/5) (GE Healthcare Biosciences AB—Uppsala).

3. Q-Sepharose (GE Healthcare Biosciences AB—Uppsala).

4. Elution buffers: 0.5 M sodium acetate (pH 5.0) and 20 mM Tris/HCl (pH 8.0), under increasing NaCl concentration (0–2.0 M) (*see* **Note 4**).

2.4 Biochemical Analysis by Electrophoresis

2.4.1 Agarose Gel

1. Agarose gel: 0.5 % agarose gel (agarose, molecular biology grade, Bio Agency—São Paulo) in 0.05 M 1,3-diaminopropane/acetate (pH 9.0).

2. Running buffer: 0.05 M 1,3-diaminopropane/acetate (pH 9.0).

3. Fixation buffer: 0.1 % cetylmethylammonium bromide solution; staining buffer: 0.1 % toluidine blue in acetic acid/ethanol/water (0.1:5:5, v/v).

4. The electrophoresis system is described by Jaques, 1967 [9], and modified by Dietrich, 1977 [10].

2.4.2 Polyacrylamide Gel
*(See **Note** 5)*

1. Polyacrylamide gel: 1-mm-thick 6 % polyacrylamide (acrylamide/bis-acrylamide 30 % solution, Sigma-Aldrich, Saint Louis) slab gel.

2. Running buffer: 0.06 M sodium barbital (pH 8.6).

3. Staining buffer: 0.1 % toluidine blue in 1 % acetic acid.

4. Washing buffer: 1 % acetic acid.

5. Polyacrylamide gel electrophoresis system: Mini-PROTEAN® Tetra System from Bio Rad.

2.5 Enzymatic Analysis

*2.5.1 Chondroitinase AC or Chondroitin AC Lyase (EC# 4.2.2.5) (See **Note** 6)*

1. Incubation buffer: 0.1 ml 50 mM Tris–HCl (pH 8.0), containing 5 mM EDTA and 15 mM sodium acetate.

*2.5.2 Chondroitinase ABC or Chondroitin ABC Lyase (EC# 4.2.2.4) (See **Note** 7)*

1. Incubation buffer: 0.1 ml 50 mM Tris–HCl (pH 8.0), containing 5 mM EDTA and 15 mM sodium acetate.

*2.5.3 Heparinase I or Heparin Lyase I (EC# 4.2.2.7) (See **Note** 8)*

1. Incubation buffer: 20 mM Tris–HCl, pH 7.5, containing 0.1 mg/ml BSA and 4 mM calcium acetate or 100 mM sodium acetate buffer (pH 7.0), containing 10 mM calcium acetate.

*2.5.4 Heparinase III or Heparin Lyase III or Heparitinase I (EC# 4.2.2.8) (See **Note** 9)*

1. Incubation buffer: 20 mM Tris–HCl, pH 7.5, containing 0.1 mg/ml BSA and 4 mM calcium acetate or 100 mM sodium acetate buffer (pH 7.0), containing 10 mM calcium acetate.

2.6 Deaminative Cleavage with Nitrous Acid (See Note 10)

2.6.1 Generation of Fresh Nitrous Acid

Dissolve 56 mg of $BaNO_2$ in 500 µl of ice-cold distilled water. Then add to this solution 500 µl of ice-cold H_2SO_4 and mix thoroughly. Remove the turbidity formed by centrifugation and recover the resulting nitrous acid, pH 1.5 in the supernatant [11].

2.7 Preparation and Isolation of the Disaccharides Formed by the Action of Glycosaminoglycan Lyases

*2.7.1 Fractionation of the Digestion Products (See **Note** 11)*

1. Fractionation column: Superdex Peptide column (GE Healthcare Biosciences AB—Uppsala) linked to an HPLC.

2. Elution buffer: Distilled water:acetonitrile:trifluoroacetic acid (80:20:0.1, v/v).

2.7.2 Disaccharide
*Analysis (See **Note 12**)*

1. Fractionation column: SAX-HPLC analytic column (250 × 4.6 mm, Sigma-Aldrich), under a linear NaCl gradient in the running buffer.

2. Running buffer: Distilled water adjusted to pH 3.5 with HCl, under a linear NaCl gradient.

2.8 NMR
Spectroscopy
(See Note 13)

NMR equipment: Bruker DRX 600 with a triple-resonance probe. The ascidian resolution buffer: glycans are dissolved in 99.9 % D2O (CIL).

3 Methods

The procedures for the extraction, purification, and biochemical/ structural analysis of the ascidian glycans are summarized in the scheme below as well as in Fig. 1, and detailed in the next paragraphs.

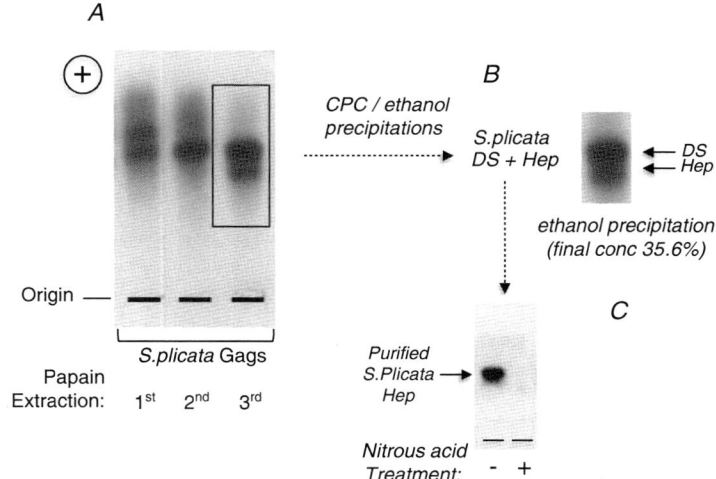

Fig. 1 Alternative method to purify the sulfated glycans from ascidians using ethanol precipitation. The solution of the glycans obtained after the third proteolytic extraction of the glycans from the ascidian *Styela plicata* contains a mixture of dermatan sulfate and heparin, which can be identified by agarose gel electrophoresis (**a**). To isolate the glycans mix the third extraction solution with CPC (0.5 % final concentration) overnight, at room temperature, wash the pellet formed with CPC (0.05 % final concentration), and dissolve the pellet with 2 M NaCl in 95 % ethanol overnight at 4 °C. The precipitate formed is collected by centrifugation and dissolved in 50 ml of distilled water. The aqueous solution of ascidian dermatan sulfate + heparin mixture (**b**) is mixed with absolute ethanol (final concentration of 35.6 %) overnight at 4 °C. The precipitate formed contains the purified *S. plicata* heparin (**c**). The ascidian dermatan sulfate can be obtained precipitating the pellet with ethanol at a final concentration of 95 %

Procedures for extraction, fractionation, biochemical and spectroscopic analysis of ascidian glycans

1- Extraction of the ascidians glycans

      1.1- Papain extraction
      1.2- CPC/ethanol precipitation

      OBS: A preliminary analysis of the total glycans can be performed at this time by agarose gel electrophoresis

2- Purification of the ascidian glycans

      2.1- DEAE-cellulose
      2.2- Mono Q (HR 5/5)
      2.3- Alternatively Ethanol precipitation

3- Gel electrophoresis of the purified glycans

      3.1- Agarose
      3.2- Polyacrylamide

4- Biochemical analysis

      4.1- Enzymatic (Chase AC/ABC; Heparinase/heparitinase)
      4.1- Chemical (Deaminative cleavage by nitrous acid

      OBS: intact or degraded glycans can be analyzed by agarose and/or polyacrylamide gel.

5- Disaccharide analysis

      5.1- Exhaustive digestion with glycan lyases
      5.2- Disaccharide isolation by gel filtration chromatography (Superdex peptide column)
      5.3- Disaccharide fractionation by Ion exchange (SAX-Spherisorb column)

6- NMR spectroscopy

      6.1- 1H-Proton
      6.2- 13C-Carbon

3.1 Extraction of the Ascidian Glycans

3.1.1 Organs

After isolating the organs (*see* **Note 1**), remove the excess of acetone and dry the pieces in an oven at 60 °C. Then, suspend the dried material (~25 g) in the digestion buffer: 20 ml of 0.1 M sodium acetate buffer (pH 5.5), containing 100 mg papain, 5 mM EDTA, and 5 mM cysteine, and incubate at 60 °C overnight. Centrifuge the incubation mixture (2,000×g for 10 min at room temperature), separate the supernatant, and place the pellet in the digestion buffer. Incubate the pellet for an additional 12-h period at 60 °C. This procedure is repeated two more times, as described above. The clear supernatants from the extractions are combined, and mixed with a solution of cetylpyridinium chloride (CPC) (0.5 % final concentration) overnight at room temperature to precipitate the glycans. Wash the precipitate formed with CPC (0.05 % final concentration) and suspend it with 2 M NaCl in 95 % ethanol (100:15, v/v) to dissolve the complex CPC-glycans. Mix the solution with 2 volumes of 95 % ethanol and kept overnight at 4 °C. The precipitate formed (containing the total glycans) is collected by centrifugation (2,000×g for 10 min at room temperature), dried, and dissolved in distilled water.

3.1.2 Test Cells

Immerse the ascidian eggs in acetone and keep for 24 h at 4 °C. Then, remove the excess of acetone and dry the pieces in an oven at 60 °C. Suspend the dried eggs (1 g) in 20 ml of 0.1 M sodium acetate buffer (pH 5.5), containing 100 mg of papain, 5 mM EDTA, and 5 mM cysteine and incubate at 60 °C, overnight. Then, centrifuge the incubation mixture (2,000×g for 10 min at room temperature) and add another 100 mg of papain in 20 ml of the same buffer, containing 5 mM EDTA and 5 mM cysteine to the precipitate. Incubate the precipitate for an additional overnight period. The clear supernatant from the two extractions is combined and the glycans precipitated with 2 volumes of 95 % ethanol and maintained at 4 °C for 24 h. Collect the precipitates formed by centrifugation (2,000×g for 10 min at room temperature), combine them, freeze-dry, and dissolve in 2 ml of distilled water. This material is called "pool 1" and contains a non-glycosaminoglycan polysaccharide. In order to remove the heparin from the test cells, it is necessary to subject the pellet to the extraction procedure with papain two more times, as described earlier. The material obtained is named "pool 2" and contains the test cell heparin [8].

3.1.3 Hemocytes

Immerse the hemocytes in acetone and keep them for 24 h at 4 °C. Then, remove the excess of acetone and dry the material in an oven at 60 °C. Suspend the dried hemocytes (1 g) in 20 ml of 0.1 M sodium acetate buffer (pH 5.5), containing 100 mg of papain, 5 mM EDTA, and 5 mM cysteine and incubate the digestion solution at 60 °C for 24 h. The digestion mixture is then centrifuged (2,000×g for 10 min at room temperature), the supernatant is separated, and the precipitate is incubated with papain two more times, as described above. The clear supernatants from the extractions are combined, and the polysaccharides are precipitated with 2 volumes of 95 % ethanol and maintained at 4 °C for 24 h. The precipitate formed is collected by centrifugation (2,000 × g for 10 min at room temperature) and freeze-dried.

3.2 Purification of the Sulfated Glycosaminoglycans from Different Tissues of the Ascidians

3.2.1 Organs

To purify the sulfated polysaccharides extracted from the body of the ascidians, apply 200 mg of the crude glycans to a DEAE-cellulose column (10 × 1.5 cm), equilibrated with 0.5 M sodium acetate (pH 5.0). Elute the column stepwise with 50 ml each of 0.5 M and 1.0 M NaCl in the same buffer. Adjust the flow rate of the column to 8.0 ml/h. To identify the sulfated glycans assay each fraction by metachromasia using 1,9-dimethylmethylene blue [12] and by the carbazole reaction for hexuronic acid [13]. The fractions eluted with 1.0 M NaCl are pooled, dialyzed against distilled water, and lyophilized.

The DEAE-cellulose purified glycan (60 mg) is then applied to a Mono Q (HR 5/5)-FPLC column, equilibrated with 20 mM Tris/HCl (pH 8.0), containing 0.75 M NaCl. Develop the column by a linear gradient of 0.75–2.0 M NaCl in the same buffer. Adjust the flow rate of the column to 0.45 ml/min, and collect fractions of 0.5 ml. Each fraction is assayed by metachromasia using 1,9-dimethylmethylene blue [12] and by the carbazole reaction for hexuronic acid [13]. The fractions under the major peak eluted between 1.5 and 2.0 M NaCl contain the dermatan sulfate (as identified by the positive metachromatic and hexuronic acid assays). Pool the fractions, dialyze against distilled water, and lyophilize the material. This should yield ~10 mg (dry weight) of ascidian dermatan sulfate. This sample is then reapplied to a Mono Q (HR 5/5)-FPLC column and developed as described above, yielding ~4.0 mg (dry weight) [3].

Fractionation with Ethanol

Alternatively, the ascidian glycans from the organs can be separated by differential ethanol precipitation. The solution of the glycans obtained after the third extraction with papain contains a mixture of ascidian dermatan sulfate and heparin. To isolate the glycans, mix the third extraction solution with a cetylpyridinium chloride solution (0.5 % final concentration) overnight, at room temperature. Then wash the pellet formed with cetylpyridinium chloride (0.05 % final concentration) and dissolve the pellet with 2 M NaCl in 95 % ethanol (100:15, v/v). The solution is mixed with 2 volumes of 95 % ethanol and kept overnight at 4 °C. The precipitate formed is collected by centrifugation (2,000 × g for 10 min at room temperature), dried, and dissolved in 50 ml of distilled water. Mix the aqueous solution of ascidian dermatan sulfate + heparin mixture with absolute ethanol to achieve a final concentration of 35.6 % and maintain it at 4 °C overnight. The precipitate formed, containing the purified *S. plicata* heparin, is collected by centrifugation and freeze-dried. To obtain the ascidian dermatan sulfate, the supernatant should be precipitated with ethanol to a final concentration of 95 % (Fig. 1).

3.2.2　Test Cells

The test cells contain only one glycosaminoglycan, heparin. To purify the test cell heparin apply ~20 mg of the glycans, obtained from the third and fourth extractions with papain (pool 2, see above) to a Mono Q (HR 5/5)-FPLC column, equilibrated with 20 mM Tris/HCl buffer (pH 8.0). Elute the column with a linear gradient of 0–2.0 M NaCl (10 ml), using a flow rate of 0.45 ml/min. Collect 0.5 ml fractions and check them by their metachromatic properties using 1,9-dimethyl methylene blue [12]. The fractions containing heparin are pooled, dialyzed against distilled water, and lyophilized.

3.2.3　Hemocytes

The hemocytes of the ascidian *S. plicata* contain five types of cells [8]. Among these five cells, only the granulocyte or basophil-like cell contains a glycosaminoglycan of the heparin type. To isolate the heparin, apply ~2 mg of the glycans extracted from the hemocytes with papain to a Mono Q (HR 5/5)-FPLC column. Equilibrate the column with 20 mM Tris/HCl buffer (pH 8.0) and elute the glycans with a linear gradient of 0–2.0 M NaCl (45 ml), at a flow rate of 0.5 ml/min. Collect 0.5 ml fractions and check them for their metachromatic properties, using 1,9-dimethylmethylene blue. Pool the fractions under the peak eluted between 1.4 and 1.5 M NaCl that contain the granulocyte heparin, dialyze the glycan solution against distilled water, and lyophilize.

3.3　Analysis of the Ascidian Glycans by Electrophoresis

3.3.1　Agarose Gel

The crude or purified glycosaminoglycans extracted from the ascidians can be analyzed by agarose gel electrophoresis, where the negative charges of the glycans interact with the positive charges of the amines in the running buffer. As a result, each glycan has a specific migration profile in the gel.

Prepare the agarose gel (0.5 %, w/v) in 0.05 M 1,3-diaminopropane/acetate (pH 9.0), on 7.5 × 5.0 cm glass plates, according to the instructions described previously [9, 10]. Then, apply ~15 μg of the glycans in a 0.5 cm cut made with the edge of a cutter in the agarose gel. Run the electrophoresis in 0.05 M 1,3 diaminopropane/acetate (pH 9.0) for 1 h at

120 V. After the running fix the glycosaminoglycans in the gel with 0.1 % *N*-acetyl-*N*,*N*,*N*-trimethylammonium bromide in water overnight at room temperature. The glycans in the gel are then stained with 0.1 % toluidine blue in acetic acid/ethanol/water (0.1:5:5, v/v). After staining, wash the gel for about 15 min in acetic acid/ethanol/water (0.1:5:5, v/v).

3.3.2 Polyacrylamide Gel

The molecular mass of the glycosaminoglycans can be estimated by polyacrylamide gel electrophoresis, comparing the migration in the gel with those from molecular weight standards. Apply the glycan samples (10 µg) to a 1-mm-thick 6 % polyacrylamide slab gel in 0.06 M sodium barbital (pH 8.6), and run in the Mini-PROTEAN® Tetra System from Bio Rad, at 100 V for 1 h in the same buffer. After running, the gel is stained with 0.1 % toluidine blue in 1 % acetic acid for 30 min and destained in 1 % acetic acid.

3.4 Enzymatic and Deaminative Cleavage with Nitrous Acid Followed by Analysis of the Ascidian Glycans by Agarose Gel Electrophoresis

A qualitative analysis of the glycosaminoglycans isolated from the ascidian tissues or cells can be obtained by agarose gel electrophoresis before and after digestion of the glycans with specific lyases or deaminative cleavage with nitrous acid.

3.4.1 Enzymatic Treatments with Chondroitin AC or ABC Lyases

Incubate ~200 µg of the isolated glycans with 0.1 unit of chondroitin AC lyase or chondroitin ABC lyase in 300 µl of 50 mM Tris/HCl (pH 8.0), containing 5 mM EDTA and 15 mM sodium acetate for 12 h, at 37 °C. Aliquots containing the enzyme-resistant glycosaminoglycans in the reaction mixtures are analyzed by agarose gel electrophoresis, as described above.

3.4.2 Enzymatic Treatments with Heparan Sulfate and Heparin Lyases

Incubate ~50 µg (as dry weight) of the glycans isolated from the ascidian tissues or cells with −0.05 units of either heparan sulfate lyase or heparin lyase in 100 µl of 100 mM sodium acetate buffer (pH 7.0), containing 10 mM calcium acetate for 17 h at 37 °C. At the end of the incubation period, aliquots containing the enzyme-resistant glycosaminoglycans in the reaction mixtures are analyzed by agarose gel electrophoresis, as described above.

3.4.3 Deaminative Cleavage with Nitrous Acid

Deamination by nitrous acid is performed at pH 1.5. The purified glycan isolated from ascidian tissues or cells (~20 µg) is incubated with 5 µL of freshly generated HNO_2 at room temperature for 1.5 h. The reaction mixtures are then neutralized with 1.0 M Na_2CO_3 [11]. Intact and nitrous acid-degraded glycans are analyzed by agarose gel electrophoresis.

3.5 Determination of the Disaccharide Composition of the Ascidian Glycans

In order to determine the disaccharide composition, the ascidian glycans are exhaustively digested simultaneously with both heparin- and heparan sulfate-lyases or with chondroitin ABC lyase, as described earlier. The products of the enzymes are applied to a Superdex peptide column, linked to an HPLC system from Shimadzu (Tokyo, Japan). The column is eluted with distilled water:acetonitrile:trifluoroacetic acid (80:20:0.1, v/v), at a flow rate of 0.5 ml/min. Fractions of 0.25 ml are collected and monitored for UV absorbance at 232 nm, which detects the double bond introduced between carbons 4 and 5 of the hexuronic acid in the disaccharides by the action of the lyases. Fractions corresponding to disaccharides (~90 % of the degraded material) are pooled, freeze-dried, and stored at 20 °C. This disaccharide preparation and standard compounds are subjected to a SAX-Spherisorb HPLC analytical column (250×4.6 mm, Sigma-Aldrich), as follows. After equilibration in the mobile phase (distilled water adjusted to pH 3.5 with HCl) at 0.5 ml/min, samples are injected and disaccharides eluted with a linear gradient of NaCl from 0 to 1.0 M over 45 min in the same mobile phase. The eluant is collected in 0.5-ml fractions and monitored for UV absorbance at 232 nm for comparison with lyase-derived disaccharide standards.

3.6 NMR Spectroscopy

1H and 13C spectra are recorded using a Bruker DRX 600 with a triple-resonance probe. About 2 mg purified ascidian glycans is dissolved in 0.5 ml of 99.9 % D2O (CIL). All spectra are recorded at 60 °C with HOD suppression by presaturation. COSY, TOCSY, and 1H/13C heteronuclear correlation (HMQC) spectra are recorded using states-time proportion-phase incrementation for quadrature detection in the indirect dimension. TOCSY spectra are run with 4,096×400 points with a spin-lock field of about 10 kHz and a mixing time of 80 ms. HMQC is run with 1,024×256 points and globally optimized alternating phase rectangular pulses for decoupling. Nuclear Overhauser effect spectroscopy spectra are run with a mixing time of 100 ms. All chemical shifts are relative to external trimethylsilylpropionic acid and [^{13}C]-methanol (Fig. 2).

4 Notes

1. *Obtaining the starting material.*
   All ascidians are covered by a thick external tunic, which protects the animal. The tunic contains high- and low-molecular-weight sulfated polysaccharides. The high-molecular-weight polymers are very homogeneous, composed mainly by L-galactose and sulfate [14], whereas the low-molecular-weight polysaccharides are heterogeneous polymers, composed by different proportions of galactose, mannose, and fucose, with different

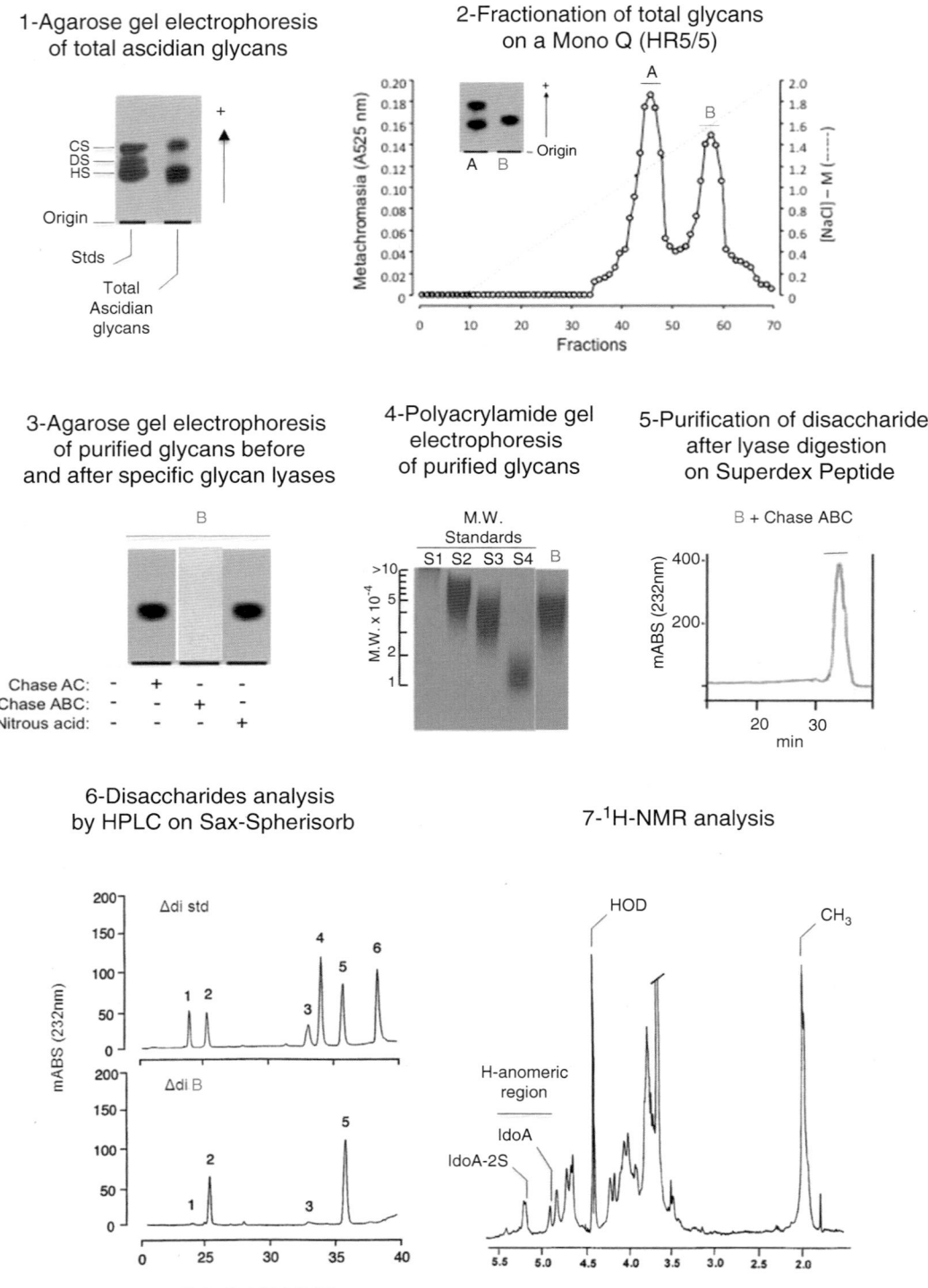

Fig. 2 Schematic representation of the steps involved in the fractionation and biochemical and structural analysis of ascidian glycosaminoglycans. (*1*) After proteolytic extraction, the total glycans are analyzed by agarose gel electrophoresis, comparing their migration with those of standard chondroitin sulfate (CS),

degrees of sulfation [14]. No glycosaminoglycan-like compounds have been detected in the tunic of all ascidian studies so far. Therefore, the first step to obtain the sulfated glycosaminoglycans from ascidians is to physically separate the tunic from the internal organs to avoid contamination with the tunic sulfated polysaccharides.

The sulfated glycosaminoglycans can be obtained from organs, test cells of the oocyte, and cells present in the hemolymph, the hemocytes. These can be isolated as follows:

(a) *Organs*: To isolate the internal organs carefully cut the external tunic longitudinally from the siphons to the base of the animal, with the help of a scissors or a sharp blade. Then, remove the internal organs using tweezers. Cut the isolated organs in approximately 3-mm pieces and place them in acetone at 4 °C overnight or until use.

(b) *Test cells*: Ascidian test cells are accessory cells that surround the oocytes and are the only cells in the whole structure of the oocyte that contain sulfated glycosaminoglycans. Therefore, the oocytes are the source of the sulfated glycans of test cells.

To obtain the oocytes, carefully separate the gonads from other tissues, under magnifying lenses. Then, stimulate the release of the oocytes by steering and cutting the gonads in a beaker containing filtered seawater (pH 6.0). To isolate the oocytes filter the solution of the gonads through a 0.5-mm-diameter net. The oocytes are recovered in the filtrate, free of contaminating tissues. Then, wash the oocytes with 500 ml of filtered seawater (pH 8.0) and centrifuge at $2,000 \times g$. Place the isolated oocytes in acetone and keep at 4 °C until use.

(c) *Hemocytes*: The hemolymph of ascidians contains different types of circulating blood cells, called hemocytes. Only one type of these cells, the granulocyte or basophil-like cell, contains glycosaminoglycan-like compounds [8].

Fig. 2 (continued) dermatan sulfate (DS), and heparan sulfate (HS). (*2*) The total glycans are fractionated by ion-exchange chromatography on a Mono Q (HR 5/5) column, and the fractions under the peaks eluted with different salt concentrations are analyzed by agarose gel electrophoresis to access purity. (*3*) An initial biochemical analysis of the purified glycans is carried out by agarose gel electrophoresis before and after specific enzyme digestion or deaminative cleavage with nitrous acid, pH 1.5. (*4*) The molecular weight of the purified glycans can be estimated by polyacrylamide gel electrophoresis comparing the migration of the sample with those of standard molecular weight markers. (*5*) The disaccharides formed by an exhaustive enzymatic digestion of the glycans can be isolated by gel filtration chromatography on a Superdex Peptide column. (*6*) The disaccharide units are then analyzed by HPLC on a strong anion-exchange column, such as SAX-Spherisorb. (*7*) A final analysis of the intact purified glycan is obtained by 1H and 13C NMR on one and two dimensions

To obtain the hemolymph, remove a fragment of the tunic to expose the mantle and the internal organs. Then, insert the needle of a syringe containing marine anticoagulant (MAC: 0.45 M NaCl, 100 mM glucose, 1.5 mM trisodium citrate, 13 mM citric acid, 10 mM EDTA in 1 l distilled water, pH 7.0) directly into the pericardial cavity of the ascidian. Collect the liquid with the syringe and pour it in plastic tubes containing an equal volume of marine anticoagulant, containing 0.45 M sodium chloride, 0.1 M glucose, 0.01 M disodium citrate, 0.01 M citric acid, and 0.001 EDTA (pH 7.0). After harvesting, the hemocytes are separated from plasma by centrifugation ($130 \times g$ for 10 min at room temperature). Place the isolated hemocytes in acetone and keep at 4 °C until use.

2. *Solubilization of the sulfated glycosaminoglycans*: This is achieved by proteolytic digestion of the material (organs or cells).

3. After solubilization in the digestion buffer, the sulfated glycosaminoglycans are precipitated with cetylpyridinium chloride (CPC) and ethanol.

4. The purification of the sulfated glycosaminoglycans from ascidian organs and cells can be achieved by ion-exchange chromatography on a variety of columns. Additionally, the sulfated glycosaminoglycans, mainly dermatan sulfate and heparin, can be separated by precipitation using different ethanol concentrations.

5. An estimation of the average molecular weight of the sulfated glycosaminoglycans can be obtained by polyacrylamide gel electrophoresis in sodium barbital buffer, comparing the migration with those of molecular weight standards.

6. *Chondroitinase AC or chondroitin AC lyase (EC# 4.2.2.5)*: This enzyme acts as an eliminase, degrading chondroitin sulfates A and C, but not chondroitin sulfate B. The enzyme cleaves 1,4-glycosidic bounds between hexosamine and glucuronic acid residues in sulfated and non-sulfated glycans. The reaction yields mainly disaccharides, containing unsaturated uronic acids that can be detected by UV spectroscopy at 232 nm.

7. *Chondroitinase ABC or chondroitin ABC lyase (EC# 4.2.2.4)*: This enzyme acts as an eliminase. The enzyme cleaves 1,4-β-D-hexosaminyl and 1,3-β-D-glucuronosyl or 1,3-α-L-iduronosyl linkages on chondroitin 4-sulfate, chondroitin 6-sulfate, and dermatan sulfate.

8. *Heparinase I or heparin lyase I (EC# 4.2.2.7)*: Heparinase I cleaves glycosidic linkages between hexosamines and O-sulfated iduronic acids on heparin and heparan sulfate (relative activity about 3:1), yielding mainly disaccharides.

9. *Heparinase III or heparin lyase III or heparitinase I (EC# 4.2.2.8)*: This enzyme cleaves the 1–4 linkages between hexosamine and

glucuronic acid residues in heparan sulfate, yielding mainly disaccharides. The enzyme is not active towards heparin or low-molecular-weight heparins.

10. Deamination by nitrous acid at pH 1.5 will cleave *N*-acetyl glucosamine residues where *N*-sulfate group replaces the *N*-acetyl group. The *N*-sulfated glucosamine residues are found only in heparin and in the *N*-sulfated domains in heparan sulfate and are unique in their susceptibility to cleavage by nitrous acid at pH 1.5.

11. The products obtained by the exhaustive enzymatic digestion are fractionated by gel-filtration chromatography.

12. *Disaccharide analysis*. The purified disaccharides obtained from the Superdex Peptide column are analyzed by a strong ion-exchange column.

13. The structural analysis of the ascidian glycans is completed by 1H and 13C nuclear magnetic resonance analysis.

Acknowledgements

This work was supported by grants from CNPq, FAPERJ, Fundação do Câncer. M.S.G.P. is a research fellow from CNPq, FAPERJ.

References

1. Pavao MS, Mourao PA, Mulloy B, Tollefsen DM (1995) A unique dermatan sulfate-like glycosaminoglycan from ascidian. Its structure and the effect of its unusual sulfation pattern on anticoagulant activity. J Biol Chem 270: 31027–31036

2. Mourao PA, Pavao MS, Mulloy B, Wait R (1997) Chondroitin ABC lyase digestion of an ascidian dermatan sulfate. Occurrence of unusual 6-O-sulfo-2-acetamido-2-deoxy-3-O-(2-O-sulfo-alpha-L-idopyranosyluronic acid)-beta-D-galactose units. Carbohydr Res 300:315–321

3. Pavao MS, Aiello KR, Werneck CC, Silva LC, Valente AP, Mulloy B, Colwell NS, Tollefsen DM, Mourao PA (1998) Highly sulfated dermatan sulfates from Ascidians. Structure versus anticoagulant activity of these glycosaminoglycans. J Biol Chem 273:27848–27857

4. Gandra M, Kozlowski EO, Andrade LR, de Barros CM, Pascarelli BM, Takiya CM, Pavao MS (2006) Collagen colocalizes with a protein containing a decorin-specific peptide in the tissues of the ascidian Styela plicata. Comp Biochem Physiol B Biochem Mol Biol 144: 215–222

5. Vicente CP, He L, Pavao MS, Tollefsen DM (2004) Antithrombotic activity of dermatan sulfate in heparin cofactor II-deficient mice. Blood 104:3965–3970

6. Cavalcante MC, Allodi S, Valente AP, Straus AH, Takahashi HK, Mourao PA, Pavao MS (2000) Occurrence of heparin in the invertebrate styela plicata (Tunicata) is restricted to cell layers facing the outside environment. An ancient role in defense? J Biol Chem 275: 36189-6

7. Cavalcante MC, de Andrade LR, Du Bocage Santos-Pinto C, Straus AH, Takahashi HK, Allodi S, Pavao MS (2002) Colocalization of heparin and histamine in the intracellular granules of test cells from the invertebrate Styela plicata (Chordata-Tunicata). J Struct Biol 137: 313–321

8. de Barros CM, Andrade LR, Allodi S, Viskov C, Mourier PA, Cavalcante MC, Straus AH, Takahashi HK, Pomin VH, Carvalho VF, Martins MA, Pavao MS (2007) The Hemolymph of the ascidian Styela plicata (Chordata-Tunicata) contains heparin inside basophil-like cells and a unique sulfated galactoglucan in the plasma. J Biol Chem 282:1615–1626

9. Jaques LB, Ballieux RE, Dietrich CP, Kavanagh LW (1968) A microelectrophoresis method for heparin. Can J Physiol Pharmacol 46:351–360

10. Dietrich CP, McDuffie NM, Sampaio LO (1977) Identification of acidic mucopolysaccharides by agarose gel electrophoresis. J Chromatogr 130:299–304

11. Conrad HE (1993) Dissection of heparin – past and future. Pure Appl Chem 65:787–791

12. Farndale RW, Buttle DJ, Barrett AJ (1986) Improved quantitation and discrimination of sulphated glycosaminoglycans by use of dimethylmethylene blue. Biochim Biophys Acta 883:173–177

13. Bitter T, Muir HM (1962) A modified uronic acid carbazole reaction. Anal Biochem 4: 330–334

14. Pavao MS, Albano RM, Lawson AM, Mourao PA (1989) Structural heterogeneity among unique sulfated L-galactans from different species of ascidians (tunicates). J Biol Chem 264:9972–9979

Chapter 10

Human Blood Glycosaminoglycans: Isolation and Analysis

Md. Ferdous Anower-E-Khuda and Koji Kimata

Abstract

Glycosaminoglycans (GAGs) are linear polysaccharides having disaccharide building blocks consisting of an amino sugar (*N*-acetylglucosamine, or *N*-acetylgalactosamine) and a uronic acid (glucuronic acid or iduronic acid) or galactose. Glycosaminoglycans have sulfated residues at various positions except for hyaluronan, and those sulfated residues regulate the biological functions of a wide variety of proteins, primarily through high-affinity interactions mediated by specific patterns/densities of sulfation and sugar sequences. Alteration of GAG structure is associated with a number of disease conditions and therefore the analyses of GAG structures and their sulfation patterns are important for the development of disease biomarkers and for treatment options. Extensive structural and quantitative analyses of GAGs from human blood are largely unexplored which may be due to the exhaustive isolation process because of the presence of too much interfering proteins and lipids such as serum albumin. Therefore we established a new GAG isolation method using the least amount (~200 µl) of human blood, consisting of a combination of proteolytic digestion and selective ethanol precipitation of GAGs, digestion of GAGs recovered on the filter cup by direct addition of GAG lyase reaction solution, and subsequent high-pressure liquid chromatography of unsaturated disaccharide products that enable to analyze GAG structures and contents. This isolation method offers an 80 % recovery of GAGs and can be applied to analyze a minute GAG content (≥1 nmol) from the least amount of biological fluids. Hence the method could be useful for the development of disease biomarkers.

Key words Blood, Chondroitin sulfate, Chondroitinase, Disaccharides, Glycosaminoglycans, Heparan sulfate, Heparitinase, High-pressure liquid chromatography (HPLC), Plasma, Proteoglycans

1 Introduction

Glycosaminoglycans (GAGs) are a family of polysaccharides which consist of repeating disaccharide units of β-D-glucuronate (GlcA) linked to either α-D-*N*-acetylglucosamine (GlcNAc) or β-D-*N*-acetylgalactosamine (GalNAc) with sulfated residues at various positions in the units except for hyaluronan [1]. Human tissues including blood contain a number of glycosaminoglycans which differ in molecular structure and the degree of sulfation. Glycosaminoglycans are implicated in many clinical conditions and are involved in the regulation of many growth factor interactions

Kuberan Balagurunathan et al. (eds.), *Glycosaminoglycans: Chemistry and Biology*, Methods in Molecular Biology, vol. 1229, DOI 10.1007/978-1-4939-1714-3_10, © Springer Science+Business Media New York 2015

and many protein interactions and in host cell invasion by a variety of microorganisms [1–3].

The major sulfated GAG components in human body are heparan sulfate/heparin and chondroitin sulfate/dermatan sulfate having a variety of sulfation patterns and different disaccharide content. In heparin/heparan sulfate, sulfation largely occurs at position 2 of iduronic acid (IdoA) and the *N-* and position 6 of hexosamine (and rarely at position 3). In chondroitin/dermatan sulfate it can occur at position 4 or position 6 of GalNAc residue as well as position 2 of IdoA residue [2, 4]. These variations give rise to a complex GAG structure that differs in composition and sequence among tissues and cells [5].

The biological functions of GAGs are closely related to those sulfation patterns, and structural diversities which result in the different abilities to interact with a number of cell surface and extracellular molecules including growth factors, cytokines, chemokines, proteases, protease inhibitors, coagulant and anticoagulant proteins, complement proteins, lipoproteins, lipolytic enzymes, and other extracellular matrix molecules [1, 5]. However, those interactions make their extraction and purification of GAGs quite difficult, especially GAGs in blood.

Many efforts have been made for the precise isolation of glycosaminoglycans from human blood such as trichloroacetic acid or heating precipitation, gel electrophoretic separation, proteolytic enzyme digestion, lipid extraction, and dialysis [6, 7]. But still to date none of the above techniques has yielded efficient extraction of heparan sulfate (HS) and chondroitin sulfate (CS) from blood. For this reason we established a precise GAG isolation technique involving several ethanol precipitation, desalting GAGs in the elution solution from the ion exchange column and GAG lyase digestion of the desalted GAGs over filtering membrane by direct addition of either heparitinase or chondroitinase reaction solution, and the detection of digested disaccharides by fluorometric post-column reverse-phase high-pressure liquid chromatography (HPLC) technique. The method would be available for only a small amount of blood samples (the least 200 μl). The same isolation method is equally applicable for the GAG isolation and analysis from culture cells, tissues, neo-cartilages, and other biological fluids (*see* **Note 10**).

2 Materials

2.1 General Materials

95 % (v/v) Ethanol containing 1.3 % (w/v) potassium acetate (abbreviated as EtOH/KAc).

20 mM Tris–HCl pH 7.5 containing 0.1 % Triton X-100 (v/v), Buffer A.

20 mM Tris–HCl pH 7.5 containing 0.2 M NaCl (v/v) and 0.1 % Triton X-100, Buffer B.

20 mM Tris–HCl pH 7.5 containing 0.2 M NaCl, Buffer C; 20 mM Tris–HCl pH 7.5 containing 2 M NaCl, Buffer D.

50 mM Tris–HCl pH 7.2 containing 1 mM $CaCl_2$ (v/v) and 5 μg bovine serum albumin (w/v), Buffer E.

50 mM Tris–HCl pH 7.2 containing 1 mM $CaCl_2$.

DEAE Sephacel.

Proteinase-K (*Tritirachium album*).

Potassium acetate.

Acetic acid.

Ethylenediaminetetraacetic acid (EDTA).

Deoxyribonuclease I from bovine pancreas.

Centrifuged filtering unit (5 kDa molecular weight cutoff filtering unit).

Micro bio-spin chromatography columns.

Heparitinase I, II, and III (*Flavobacterium heparinum*).

Chondroitinase ABC (*P. vulgaris*).

Heparin/heparan sulfate disaccharide standards: ΔDi-0S, -HexA-GlcNAc-; ΔDi-*NS*, -HexA-GlcNS-; ΔDi-6S, -HexAGlcNAc(6S)-; ΔDi-(*N*,6)diS, -HexA-GlcNS(6S)-; ΔDi-(*N*,2)diS, -HexA(2S)-GlcNS-; and ΔDi-(*N*,6,2)triS, -HexA(2S)-GlcNS(6S)-.

Chondroitin sulfate disaccharide standards: ΔDi-0S, -HexA-GalNAc-; ΔDi-4S, -HexA-GalNAc(4S)-; ΔDi-6S, -HexAGalNAc(6S)-; ΔDi-diS$_E$, -HexA-GalNAc(4,6)diS-; and ΔDi-diS$_D$, -HexA(2S)-GalNAc(6S)-.

2.2 HPLC Apparatus

The chromatographic equipment included two high-pressure pumps, a controller, an autosampler, a heated (125 °C) chemical reaction box, and a fluorescence detector (RF-10AxL, Shimadzu, Kyoto, Japan) with a flow rate of 1.1 ml/min on a Docosil HPLC column (column size and temperature: 4.6 mm × 150 mm and 50 °C). For post-column fluorometric detection unsaturated disaccharides in the effluent were labeled with the reaction with 0.25 M sodium hydroxide and 0.5 % 2-cyanoacetoamide. The fluorescent-labeled disaccharides were monitored with the excitation of 345 nm and emission at 410 nm. For this experiment maximum injection volume was 90 μl (HPLC apparatus maximum injection 100 μl).

3 Methods

3.1 Glycosamino-glycan Isolation

Collected blood was separated to plasma and cellular fractions, subjected to the systematic isolation of GAGs, and then their structure was analyzed by a high-pressure liquid chromatography technology.

A schematic diagram of the method is presented in Fig. 1 and below is the detailed description. This isolation method gives high GAG yield as we observed by adding known amount of heparan sulfate and chondroitin sulfate to human plasma and cellular fractions and obtained ~80 % recovery [2]. Our isolation procedure is highly sensitive and can even detect ≥1 nmol unsaturated disaccharide unit from human plasma and/or cellular fractions because of the adaptation of post-column fluorometric detection technique with 2-cyanoacetoamide [2].

1. Freshly drawn human blood was collected in heparin-free and EDTA-coated collection tubes.

2. Immediately after collection of human blood, it was transferred into 15 ml centrifuge conical tube and was centrifuged at 1,500×g for 15 min at 4 °C. After centrifugation the upper plasma and the lower cellular fraction were harvested separately.

3. About 100 µl plasma and 100 µl cellular fraction from 200 µl blood in 2 ml centrifuge tubes, both were processed at the same time.

4. Both the plasma and cellular fractions were subjected to ethanol precipitation by adding three volumes of cold EtOH/KAc to each fraction and vortexed several times. Ethanol-mixed plasma and cellular fractions were kept on ice (0 °C) for 30 min and were centrifuged at 12,000×g for 20 min at 4 °C. Supernatant was discarded and 400 µl deionized water was added to the pellet and was well mixed. In the mixed solution three volumes of EtOH/KAc was added, iced, and centrifuged as the same process as described above and were repeated three times. The final precipitate was suspended in alkali (*see* **Notes 1** and **2**).

5. The pellet was treated with alkali for inducing the β-elimination reaction. 1 ml of 0.4 M potassium hydroxide was added to the precipitate of each plasma and cellular fractions and was kept for 12 h (usually overnight) at room temperature.

6. Alkali was neutralized by a mild acid (acetic acid).

7. Genomic DNA was removed by the treatment of deoxyribonuclease I (25 µg DNA/ml of human blood [8]) for 20 h at 37 °C. 300 µg and 50 µg deoxyribonuclease I was added for the cellular fraction and plasma fraction, respectively (*see* **Notes 3** and **4**).

8. Protein is one of the most interfering molecules during GAG isolation because more than 95 % of plasma GAGs form complexes with plasma proteins. So both the plasma and cellular fractions were treated with 2.6 mg and 3.5 mg Proteinase-K (final concentration of Proteinase-K in the solution was 1 mg/ml), respectively, for 20 h at 45 °C. Further 1.3 mg and 1.8 mg of Proteinase-K were added to the plasma and cellular fractions, respectively, and were incubated for another 20 h at 50 °C (*see* **Note 5**).

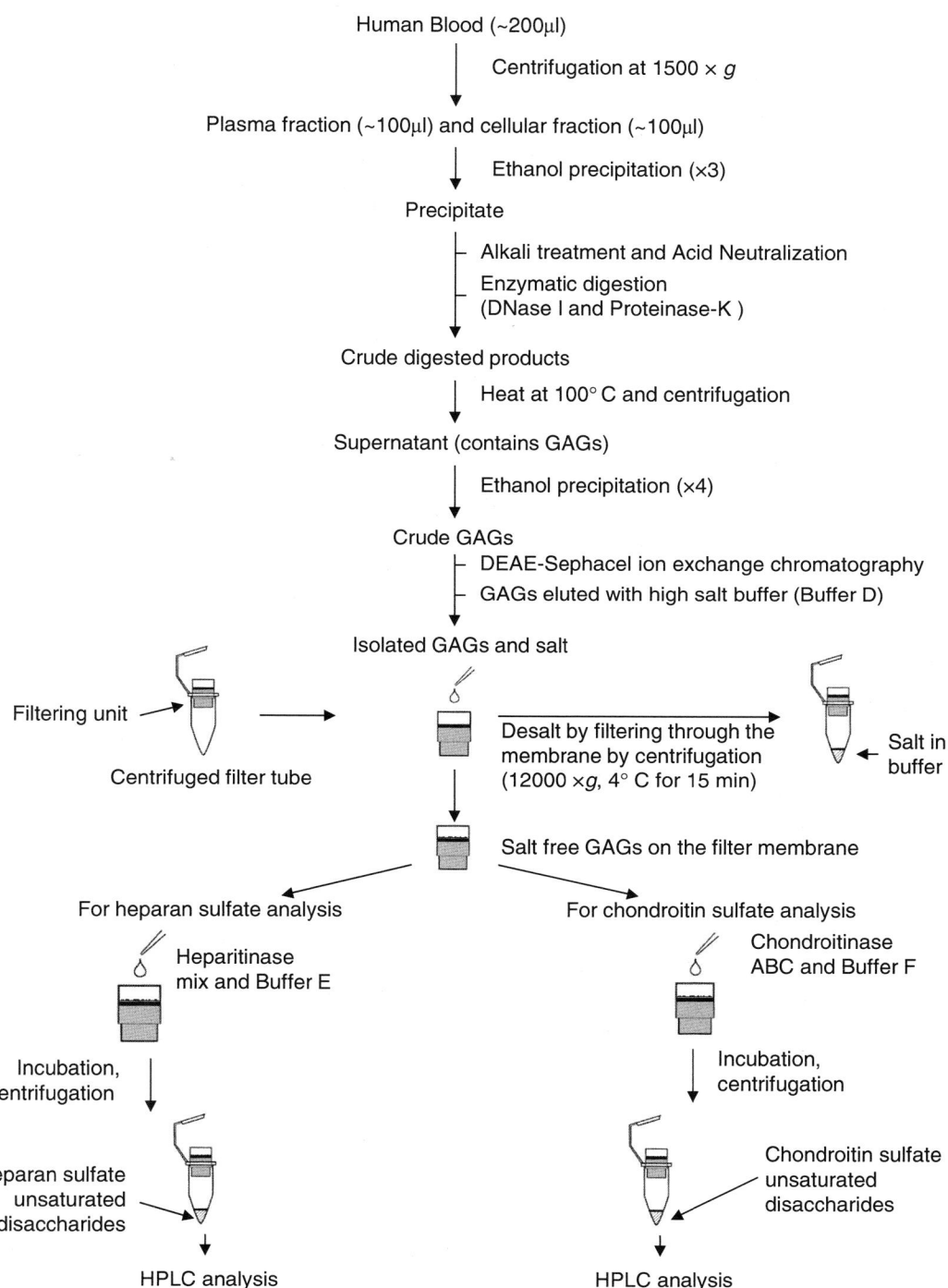

Fig. 1 A schematic presentation of GAG isolation method from human blood

9. Proteinase-K activity was destroyed by heating the entire tubes at 100 °C for 5–6 min (*see* **Note 6**).

10. After centrifugation at $12,000 \times g$ for 20 min at 4 °C the supernatant was collected.

11. The supernatant was then subjected to ethanol precipitation (*see* **Note 9**) by adding three volumes of EtOH/KAc, mixed, and incubated for 30 min at 0 °C following centrifugation at $12,000 \times g$ for 20 min at 4 °C. Ethanol precipitation was repeated four times and the final precipitate was dissolved in 100 µl Buffer A.

12. The final precipitate was purified by DEAE-Sephacel ion exchange chromatography. Column volume of DEAE-Sephacel was 300 µl and equilibrated with Buffer A.

13. Samples were applied over DEAE-Sephacel column and were washed several times with Buffer B. The same column was further washed with Buffer C.

14. Finally crude GAG samples were eluted from DEAE-Sephacel column by adding four column volumes of Buffer D.

15. A portion (40 %) of the eluate was applied to filtering unit (5 kDa molecular weight cutoff filtering unit) to desalt the crude GAGs and centrifuged at $12,000 \times g$ for 15 min at 4 °C. Desalted crude GAGs on the membranes were directly digested by GAG lyase as described below (*see* **Notes 7** and **8**).

16. To obtain heparan sulfate from the crude GAGs, 50 µl Buffer E was added over filtering membrane and 5 µl heparitinase mix (0.2 mU of heparitinase I, 0.1 mU of heparitinase II, and 0.2 mU of heparitinase III) was added with the Buffer E. The solution over filtering unit was briefly vortexed and incubated at 37 °C for 2 h.

17. To obtain chondroitin sulfate from the crude GAGs, 50 µl Buffer F was added over filtering membrane and 5 mU of chondroitinase ABC enzyme was mixed with the Buffer F. The solution over filtering unit was briefly vortexed and incubated at 37 °C for 2 h.

18. Purified unsaturated disaccharide products were obtained in the bottom tubes of the filtering units by the centrifugation of the filtering units at $13,800 \times g$ for 10 min at 4 °C.

19. The unsaturated disaccharide products were analyzed by a standard HPLC technology.

3.2 Structure Analysis of Glycosaminoglycans from HPLC

1. Glycosaminoglycan structure was analyzed through a parallel run of standard disaccharides. Figures 2 and 3 show representative chromatogram of heparan sulfate and chondroitin sulfate structure, respectively, from healthy human blood.

2. Quantity of the respective GAGs can also be calculated from the sample peak height with the respective GAG standard peak

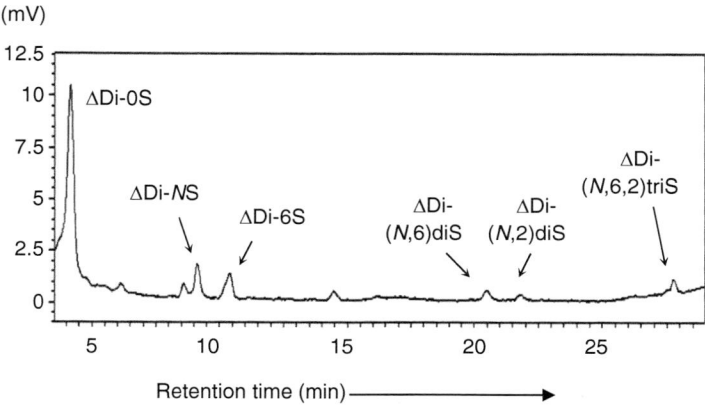

Fig. 2 A representative chromatogram illustrates heparan sulfate structure from the cellular fractions of healthy human blood

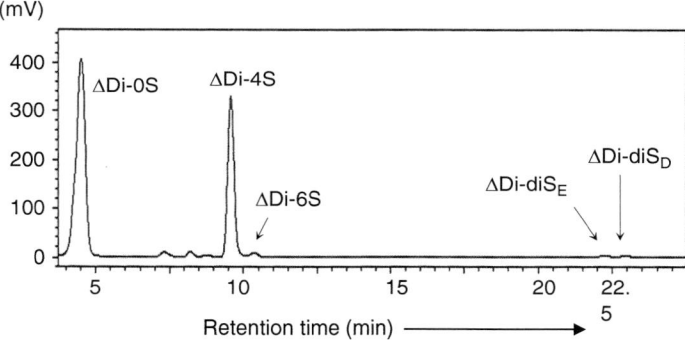

Fig. 3 A representative chromatogram illustrates chondroitin sulfate structure from the plasma fractions of healthy human blood

height (applied known amounts (1 nmol) of the respective GAG) which are obtained through the chromatogram.

4 Notes

1. To obtain a good yield of GAGs, several rounds of ethanol precipitation is required. The mixing of three volumes of EtOH/KAc with the sample containing glycosaminoglycans should be complete. Repeated ethanol precipitation is essential to reduce the amount of contaminants such as lipids, lipid derivatives, and low-molecular-weight materials. This is also an effective way to condense GAG molecules from the sample solution.

2. When the cellular fraction is mixed with EtOH/KAc, it tends to coagulate. Immediate and several vortexing can reduce the

formation of large coagulation particles. Coagulation should keep less by several vortexing to obtain higher GAG molecules.

3. After 8–10 h of alkali treatment, cellular fraction seems to be almost coagulated. This is mostly due to genomic DNA in cells. Treatment of deoxyribonuclease I at 37 °C makes such coagulation less.

4. Deoxyribonuclease I has an absolute requirement for divalent metal cations and we used magnesium at a final concentration of 10 mM to activate deoxyribonuclease I.

5. Calcium is required for prolonged activation of Proteinase-K and we added calcium at a final concentration of 10 mM to the solution.

6. Inactivation of Proteinase-K is one of the key points to be done completely. Otherwise further enzymatic reactions (heparitin-ase and/or chondroitinase digestion) will not work and unsat-urated disaccharides will not be obtained. Heating the entire solution at 100 °C for 5–6 min will work fine.

7. GAG lyase enzymatic digestion of GAGs over the filtering membranes is another key point of this method which can reduce further GAG loss. During the incubation, several brief vortexing may help increase substrate contact with the enzymes and hence the GAG yield.

8. Filtering unit should be pretreated once with 70 % ethanol and twice with HPLC-graded water to remove any fluorescence materials that might interfere with the results.

9. We investigated whether dialysis instead of ethanol precipita-tion would improve GAG yield from plasma and cellular frac-tions after Proteinase-K digestion. We observed that solvent transfer was limited and almost blocked during dialysis, which might be due to the high content of lipid and protein particles in plasma and cellular fractions.

10. We isolated GAGs from culture cells and tissue using the same method. For 1×10^5 cells, 200 μl 0.2 M KOH can induce β-elimination reaction. All the reagents should be scaled down accordingly.

References

1. Esko JD, Kimata K, Lindahl U (2009) Proteoglycans and sulfated glycosaminoglycans. In: Varki A, Cummings R (eds) Essentials of gly-cobiology, 2nd edn. Cold Spring Harbor Laboratory Press, New York

2. Anower-E-Khuda MF, Matsumoto K, Habuchi H et al (2013) Glycosaminoglycans in the blood of hereditary multiple exostoses patients: half reduction of heparan sulfate to chondroitin sul-fate ratio and the possible diagnostic application. Glycobiology 23:865–876

3. Rostrand KS, Esko JD (1997) Microbial adher-ence to and invasion through proteoglycans. Infect Immun 65:1–8

4. Anower-E-Khuda MF, Habuchi H, Nagai N et al (2013) Heparan sulfate 6-O-sulfotransferase isoform-dependent regulatory effects of heparin on the activities of various proteases in mast cells and the biosynthesis of 6-O-sulfated heparin. J Biol Chem 288: 3705–3717

5. Bulow HE, Hobert O (2006) The molecular diversity of glycosaminoglycans shapes animal development. Annu Rev Cell Dev Biol 22: 375–407

6. Higgins MK (2008) The structure of a chondroitin sulfate binding domain important in placental malaria. J Biol Chem 283:21842–21846

7. Mulloy B, Hart GW, Stanley P (2009) Structural analysis of glycans. In: Varki A, Cummings R (eds) Essentials of glycobiology, 2nd edn. Cold Spring Harbor Laboratory Press, New York

8. Adell K, Ogbonna G (1990) Rapid purification of human DNA from whole blood for potential application in clinical chemistry laboratories. Clin Chem 36:261–264

Chapter 11

Chromatographic Molecular Weight Measurements for Heparin, Its Fragments and Fractions, and Other Glycosaminoglycans

Barbara Mulloy and John Hogwood

Abstract

Glycosaminoglycan samples are usually polydisperse, consisting of molecules with differing length and differing sequence. Methods for measuring the molecular weight of heparin have been developed to assure the quality and consistency of heparin products for medicinal use, and these methods can be applied in other laboratory contexts. In the method described here, high-performance gel permeation chromatography is calibrated using appropriate heparin molecular weight markers or a single broad standard calibrant, and used to characterize the molecular weight distribution of polydisperse samples or the peak molecular weight of monodisperse, or approximately monodisperse, heparin fractions. The same technology can be adapted for use with other glycosaminoglycans.

Key words Gel permeation chromatography, Heparin, Heparan sulfate, Molecular weight

1 Introduction

Glycosaminoglycans (GAGs) and their fragments have biological properties that are often molecular weight dependent. An example is the minimum length of 18 monosaccharides for a heparin molecule needed to bind to antithrombin and thrombin at the same time, important for the anticoagulant activity of heparin [1]. It is therefore useful to be able to characterize the molecular weight distribution of a polydisperse GAG sample or to determine the molecular weight of monodisperse fragments. Unfortunately, GAG oligosaccharides are not as readily amenable to analysis by mass spectrometry as are peptides [2], so determination of the size of GAG molecules relies on other electrophoretic and chromatographic techniques [3–9].

Gel permeation chromatography can be used in both preparative and analytical modes to separate GAGs by molecular weight. The method described here uses analytical high-performance gel

Kuberan Balagurunathan et al. (eds.), *Glycosaminoglycans: Chemistry and Biology*, Methods in Molecular Biology, vol. 1229, DOI 10.1007/978-1-4939-1714-3_11, © Springer Science+Business Media New York 2015

permeation chromatography for molecular weight analysis, and relies on calibration of the chromatographic system with heparin molecular weight markers of low polydispersity or a single broad standard heparin molecular weight calibrant. Depending on the choice of columns and calibrant(s) it is possible to characterize unfractionated heparin samples as well as low-molecular-weight heparins and heparin oligosaccharides.

The use of refractive index detection limits the sensitivity of the technique, but once the system is calibrated, samples of interest may be analyzed using UV or fluorimetric detection. While the use of light scattering detection has been described for the successful analysis of both unfractionated and low-molecular-weight heparin [4, 5, 10], its technical challenges for these relatively small molecules, and dependence on adjustable parameters and specialist expertise, make it impractical for many laboratories.

The experimental part of this method is very simple, and barely requires detailed description to a chromatographer. Calibration by means of a broad standard, on the other hand, may not be so familiar; it is most conveniently performed using specialist software aimed at polymer analysis but can be managed using a spreadsheet, as described below.

2   Materials

All solutions should be prepared with ultrapure water, such as 18 MΩ deionized water. Reagents should be of analytical grade. Stock and working solutions may be stored at room temperature except where otherwise indicated.

2.1   Stock Solutions and Mobile Phase

1. Stock solution, 1 M ammonium acetate: Weigh 77.1 g ammonium acetate into a large weighing boat. Wash into a 1 L measuring cylinder, dissolve with stirring if necessary, and make up to 1 L with water.

2. Stock solution sodium azide (if used—see **Note 1**): Weigh 1 g sodium azide into a large weighing boat. Wash into a 100 mL measuring cylinder, dissolve with stirring if necessary, and make up to 100 mL with water.

3. Mobile-phase solution: Measure out 100 mL stock ammonium acetate solution into a 1 L measuring cylinder, add 20 mL sodium azide stock solution, and make up to 1 L with water.

4. Filter the mobile phase through a 0.22 μm membrane before use.

2.2   Sample Preparation

1. Any fraction or fragment of heparin consisting entirely of material with molecular weight less than 18,000 can be analyzed by this method (see **Note 2**). Polydisperse low-molecular-weight heparins generally contain only a tiny fraction of material above this value, and so can also be analyzed in this way [2].

2. Sample solution: Weigh out 50 mg sucrose (*see* **Note 3**) and make up to 50 mL with mobile-phase solution.

3. Samples may be presented as solids or solutions. For polydisperse samples, dissolve or dilute the sample to 5 mg/mL with sample solution; for monodisperse oligosaccharides, dissolve or dilute the solution to 1 mg/mL with sample solution (*see* **Note 4**). Filter all sample solutions through a 0.22 μm membrane before use (*see* **Note 5**).

4. Broad standard calibrant solution: Dissolve the contents of one ampoule of the 2nd International Standard Low Molecular Weight Heparin for Molecular Weight Calibration (*see* **Note 6**) (or 25 mg of your own Broad Standard) in 5 mL sample solution. Filter calibrant solutions through a 0.22 μm membrane before use.

2.3 Chromatography Equipment and Columns

1. HPLC equipment including the following components: isocratic HPLC pump; autosampler (optional but desirable); valve injector, with suitable sized loop; refractive index (RI) detector (*see* **Note 7**); analogue-to-digital data converter; computer with HPLC software.

2. Chromatography columns—(*see* **Note 8**): Guard column, such as Tosoh TSK SWxl guard column; analytical columns in series such as Tosoh TSK G3000 SWxl 7.8 mm × 30 cm and TSK G2000 SWxl 7.8 mm × 30 cm. Other column types must be used for glycosaminoglycans of high molecular weight (*see* **Note 9**).

3. Filters: Disposable filter units, 1 L, 500 mL, 250 mL; syringe filters about 10 mm diameter, all with 0.22 μm membranes. A pre-column filter, e.g., Upchurch A315, can be installed between HPLC pump and column if necessary.

4. Consumables: Syringes: single-use 1 mL or 2 mL syringes for sample filtration; HPLC tubing and fittings, steel or PEEK, for connections; autoinjector vials.

3 Method

3.1 System Setup

1. Switch on the chromatography equipment and the autosampler; turn on the PC and start up the control software allowing connection to the instrument. Ensure that the columns are fitted in the correct sequence (inlet side—Guard—G3000—G2000—outlet side) (*see* **Note 10**).

2. Set up the instrument control software according to the manufacturer's instructions. Prime the pump, and set running at 0.2 mL/min to fit in-line filter and columns. Once the columns are fitted, set the pump to 0.5 mL/min and allow the columns and detector to equilibrate for at least 2 h (*see* **Note 11**).

3. Software setup: Set the chromatography software up to handle the recording of data for each sample in turn, and to transfer the data

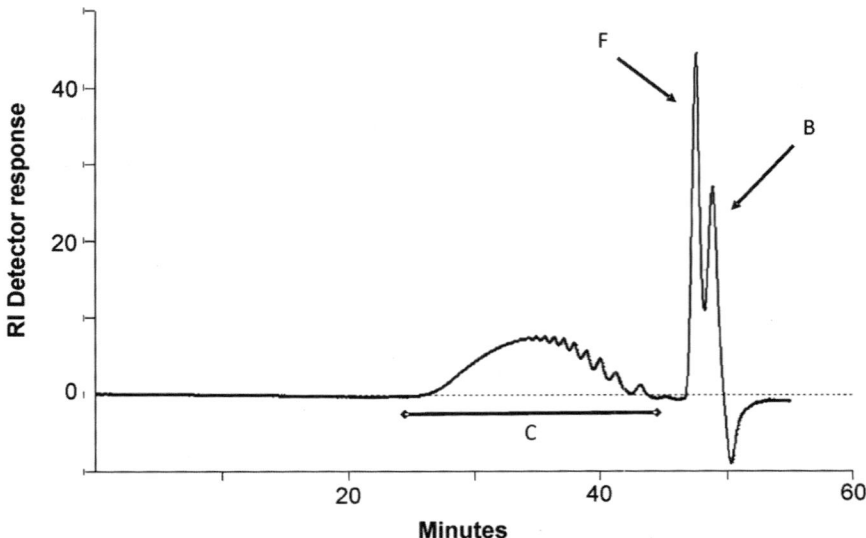

Fig. 1 A chromatogram of the 2nd International Standard Low Molecular Weight Heparin Molecular Weight Calibrant. The broad peak *underlined* as C is the calibrant peak. Partly resolved sub-peaks corresponding to shorter oligosaccharides are ignored for the purposes of calibration. Peak F is the sucrose flow rate marker peak, used optionally to align chromatograms. Peak B arises from buffer salts, and is followed by a negative fluctuation at the total column volume characteristic of RI detection

to the computer, according to the manufacturer's instructions. The chromatography time should be just over 50 min (Fig. 1).

3.2 Chromatography

1. Sample injection may be performed by hand or, more conveniently, an autosampler can be set up. In that case, set the autosampler to inject 20 μl of sample and wait for at least 50 min before the next injection. Make sure that one vial of calibrant is included with every set of samples, and run duplicates of each sample (*see* **Note 12**).

2. Initiate the injection and data collection using the software or by hand.

3.3 Data Storage and Analysis Using Specialized Software

1. Data from the RI detector is automatically digitized and transferred to the computer, and stored there in a format determined by the software used.

2. Data analysis is also dependent on the software, but the process will consist of a calibration on the basis of the calibrant chromatogram followed by analysis of individual samples (*see* **Note 13**).

3. Values calculated for polydisperse samples are most commonly the peak molecular weight (M_p), the number average molecular weight ($\bar{M}_n$), the weight average molecular weight ($\bar{M}_w$), and the polydispersity (PD) (*see* **Note 14**).

3.4 Data Analysis Using a Spreadsheet

1. A spreadsheet may be used to derive a calibration curve for the chromatographic system, linking retention time (RT) to molec-

ular weight (M). Calculation of average molecular weights, for polydisperse samples, and of peak molecular weights for monodisperse samples, can also be performed using a spreadsheet.

2. A typical session consists of one calibrant (or a set of molecular weight markers) and several samples, for which chromatograms have been saved in a digital format.

3. First, process the calibrant chromatogram(s) (Fig. 1): If a set of molecular weight markers is used, tabulate the retention times of the narrow peaks against the corresponding values for M and $\log(M)$, and go directly to **step 7** in this list. For a single polydisperse broad standard proceed as follows: use chromatography software with manual adjustment if necessary to assign a baseline and integrate the broad peak from the calibrant. If the calibrant peak is divided into multiple sub-peaks, as in Fig. 1, ignore them. Export the resulting file to a format that can be read into a spreadsheet.

4. Open the spreadsheet and import the file. Regardless of other information now in the spreadsheet, you will use only the retention time (column A in Fig. 2a) and the response (column B in Fig. 2a).

a

A	B	C	D	E	F	G	H	
RT (mins)	Response (mV)	Cumulative response			% cum. resp.			
			D(n-1) + B(n-1)		Dn*100/D1046		M from table	
25.3167	0.140906		0		0			
25.3333	0.14457		0.140906		0.002195			
25.35	0.149759		0.285476		0.004446			
25.3667	0.152812		0.435235		0.006779			
25.3833	0.159374		0.588047		0.009159			
25.4	0.164411		0.747421		0.011642			
25.4167	0.171279		0.911832		0.014202			
27.2833	2.85239		134.0523		2.087963	%>M		
27.3	2.88977		136.9047		2.132391	from table:		
27.3167	2.92548		139.7944		2.177401			
27.3333	**2.96256**		**142.7199**		**2.222967**	**2.23**	**18000**	
27.35	3.00132		145.6825		2.269111			
27.3667	3.0352		148.6838		2.315859			
27.3833	3.07105		151.719		2.363135			
41.7	2.04445		6389.554		99.52198	%>M		
41.7167	1.89675		6391.598		99.55382	from table:		
41.7333	1.75332		6393.495		99.58337			
41.75	**1.61446**		**6395.248**		**99.61068**	**99.6**	**600**	
41.7667	1.48507		6396.863		99.63582			
41.7833	1.36819		6398.348		99.65895			
41.8	1.26367		6399.716		99.68027			

Fig. 2 (a) Three blocks taken from a spreadsheet for calibration of the chromatographic system using the broad standard method, with the 2nd International Standard Low Molecular Weight Heparin for Molecular

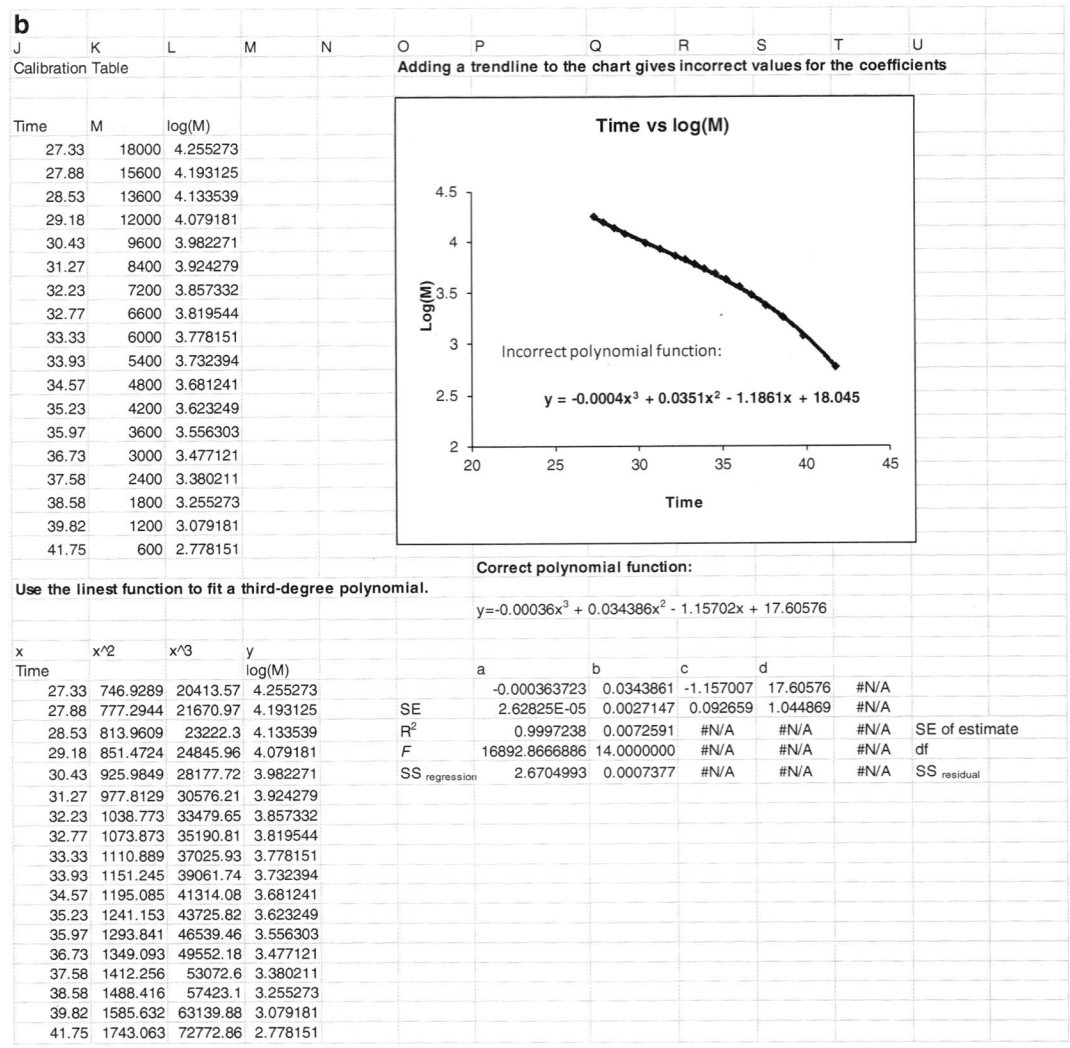

Fig. 2 (continued) Weight Calibration as calibrant and the Broad Standard Table as listed in Table 1. Retention time and response are taken from the chromatogram; cumulative response expressed as a percent of the total (column F) is calculated from the response (column B). Values in column F closest to those in Table 1 (column G) are used to assign M from Table 1 to the corresponding retention time in column A. (**b**) The calibration table assembled as shown in (**a**) is listed in columns J–L in the *top left*, and the Excel function "linest" has been used to calculate the polynomial coefficients a, b, c, and d using the input table in columns J–M, *lower left*. The "linest" output appears as shown in the *lower right* quadrant. Ignore everything except the polynomial coefficients. Under no circumstances use the "trendline" facility in an Excel chart (*upper right quadrant*) to derive these coefficients, as they will be incorrect, as shown

5. In another column (column D in Fig. 2a), calculate the cumulative response, which is the rolling total of column B. The last entry in column D is therefore the total integral of the chromatogram. In a fourth column (column F in Fig. 2a) calculate the cumulative response as a percentage of the total (the last entry in column D; in our example this is cell D1046).

Table 1
Broad Standard Table for the 2nd IS Low Molecular Weight Heparin for Molecular Weight Calibration

Point	M	Log(M)	Cum. % < M	Cum. % > M
1	600	2.78	0.40	99.60
2	1,200	3.08	3.87	96.13
3	1,800	3.26	8.94	91.06
4	2,400	3.38	14.49	85.51
5	3,000	3.48	20.68	79.32
6	3,600	3.56	27.20	72.80
7	4,200	3.62	33.89	66.11
8	4,800	3.68	40.49	59.51
9	5,400	3.73	46.83	53.17
10	6,000	3.78	52.92	47.08
11	6,600	3.82	58.59	41.41
12	7,200	3.86	63.89	36.11
13	8,400	3.92	72.96	27.04
14	9,600	3.98	80.09	19.91
15	12,000	4.08	89.21	10.79
16	13,600	4.13	92.96	7.04
17	15,600	4.19	95.95	4.05
18	18,000	4.26	97.77	2.23

6. The Broad Standard Table (Table 1) links cumulative percent integrals (% > M) to M. Our spreadsheet links cumulative percent integrals (in column F) to RT (in column A). We can therefore now tabulate molecular weight M against RT, by looking down column F (Fig. 2a) to find the values of % > M (column G in Fig. 2a) listed in Table 1, and listing the corresponding RT and M (columns A and H, respectively) in a calibration table (columns J, K, and L in Fig. 2b).

7. The next step is to fit a calibration function to the points in the calibration table. A third-degree polynomial $y = ax^3 + bx^2 + cx + d$, where y is log(M) and x is retention time, is recommended. In Microsoft Excel, the function "linest" should be used to determine the coefficients a, b, c, and d for the polynomial (*see* **Note 15** and Fig. 2b).

8. A separate spreadsheet is used to calculate average and peak molecular weights of the samples (*see* **Note 14** and Fig. 3).

		Calibration function:		y = Ax3 + Bx2 + Cx + D							where y = log(M)		
											x = RT		
		a	b	c	d						Mn	Mw	
	Coefficients	-0.00036	0.0343861	-1.15701	17.6057578						5112	6152	
Chromatogram of unknown													
A	B	C	D	E	F	G	H	I	J	K	L	M	N
RT (mins)	RI response				log(M)	M	cum. %	M*RI	RI/M		$Mn = \Sigma(RI)/\Sigma(RI/M)$		B851/J851
								(G*B)	(B/G)		$Mw = \Sigma(RI*M)/\Sigma(RI)$		I851/B851
26.2333	0												
26.25	0												
26.2667	0												
26.2833	0.000732				4.346059522	22185		16.23909	3.29946E-08				
26.3	0.004791				4.344337012	22097.19		105.8606	2.168E-07				
26.3167	0.007934				4.342617674	22009.88		174.6229	3.60467E-07				
26.3333	0.012145				4.340911767	21923.59		266.2643	5.53974E-07				
26.35	0.011931				4.339198726	21837.29		260.5494	5.46377E-07				
26.3667	0.012175				4.337488827	21751.48		264.8308	5.59746E-07				
26.3833	0.015929				4.335792272	21666.68		345.1263	7.3518E-07				

Fig. 3 A spreadsheet for calculation of average molecular weights from a chromatogram of an unknown polydisperse sample of low-molecular-weight heparin. Retention time and response (columns A and B) are imported from the chromatogram. The polynomial function from the calibration spreadsheet (Fig. 2) is used to calculate log(M) (column F) and M in column G. The number average molecular weight $\overline{M}_n$ and the weight average molecular weight $\overline{M}_w$ are both then calculated as described in **Note 14** (and *see* columns L and M)

9. As was done for the calibrant, use chromatography software to assign baseline, integrate, and export a sample chromatogram file in a format that can be imported into a spreadsheet (Fig. 3 shows an extract from such a spreadsheet for a low-molecular-weight heparin sample). In the spreadsheet, import the file and discard everything but the columns of retention time and response (columns A and B, respectively, in Fig. 3).

10. Calculate the total integrated response (in our example, in cell B851).

11. Using the retention times in column A and the calibration polynomial function, calculate $\log(M)$ at each point in the chromatogram (column F in Fig. 3) and from that M (in column G). Columns G and B are then used to calculate $M \times RI$ ($G \times B$) and RI/M (B/G) (columns I and J).

12. The final, total values in columns B, I, and J are used to calculate $\bar{M}_n$ and $\bar{M}_w$ as shown in Fig. 3 (and **Note 14**), $\bar{M}_n = B851/J851$ and $\bar{M}_w = I851/B851$ in our example.

13. Peak molecular weights (M_p) are simple; the calibration function derived in **step 7** can be used to calculate the peak molecular weight from the retention time for the maximum response (*see* **Note 16**).

4 Notes

1. Sodium azide is very toxic. It does not contribute to the chromatographic properties of the mobile phase and is only used to prevent the growth of microorganisms in the dilute ammonium acetate solution. Its use can therefore be avoided, as long as the mobile phase is changed frequently and containers such as mobile-phase reservoirs thoroughly cleaned between uses. If sodium azide is used, all local safety advice must be implemented with respect to use, spillages, and disposal.

2. Monodisperse, or almost monodisperse, fragments of heparin, can be prepared by depolymerization of heparin using either chemical methods such as deamination with nitrous acid [11] or by treatment with enzymes such as heparinase [12]. Heparan sulfate can also be enzymatically degraded [13, 14]. Careful preparative gel filtration chromatography can be used to extract almost monodisperse fragments from the digest [12, 15]. It is in principle also possible to examine chondroitin sulfate and dermatan sulfate fragments using the same technique; calibration with heparin will give molecular weight results that are not necessarily accurate, but are likely to be closer to the mark than calibration with neutral polysaccharide markers. The

chondroitin sulfates can also be depolymerized using chemical [16, 17] or enzymatic [18] methods.

3. Sucrose gives rise to a sharp monodisperse peak just before the salt peak (Fig. 1). This can be used in some chromatography software programs as a flow rate marker, to align sample chromatograms with the calibrant chromatogram. This function is particularly useful when the autosampler is used for a long chromatography session, for example over a weekend, so that drift in the properties of the chromatography system can be compensated for. Its use is not absolutely necessary, especially when chromatography sessions are kept short (less than 24 h).

4. The volume of sample solution prepared will depend on the amount of sample available, and on the volume and depth of autosampler vials, if used. The sample concentration can be much lower for monodisperse samples; 1 mg/mL is usually adequate.

5. Filtration of the samples through a fine membrane may sometimes be difficult, in which case a short spin in a bench centrifuge can be used to pull down particulate matter. In that case care must be taken not to disturb the sediment when transferring sample from the centrifuge tube.

6. The 2nd International Standard Low Molecular Weight Heparin for Molecular Weight Calibration is available from NIBSC (www.nibsc.org). It is intended as a primary standard, and can easily be used to calibrate a laboratory standard, or a set of molecular weight markers, for everyday use.

7. The RI detector is the only piece of specialist equipment in this method. It is not the most sensitive mode of detection, but it may be used for molecules that have no useful UV absorbance, and the RI response is proportional to the mass of sample rather than the number of chromophores in the sample, for the calculations described later. The refractive index of the mobile phase is much more sensitive to changes in composition and temperature than, for example, UV absorbance, so topping up the mobile-phase reservoir part way through a run is out of the question, and a steady temperature is essential. RI detectors are as prone as any others to bubbles.

8. Other manufacturers also produce suitable columns. In choosing a column, it is important to note that molecular weight ranges defined for globular proteins do not apply to glycosaminoglycans [19].

9. These columns cannot be used for unfractionated heparin, or any other unfractionated glycosaminoglycan. For unfractionated heparin a combination of TSK G4000 SWxl and TSK

G3000 SWxl, preceded by a guard column, works well, with a mobile-phase flow rate of 0.6 mL/min. For unfractionated heparin, at the time of writing the only available broad standard is the USP Heparin Sodium Molecular Weight Calibrant, available directly from USP (http://www.usp.org/store/products-services/reference-standards). This reference standard is for use with the molecular weight method described in the USP monograph for heparin sodium [6], and is supplied with its own Broad Standard Table. If a flow rate marker is needed, α-cyclodextrin can be used.

10. When fitting columns that have been stored for some time, and may contain a little air, initially fit each column the wrong way round; when liquid emerges, refit in the correct orientation. This avoids pushing bubbles through the columns.

11. Equilibration of the system will be faster in a steady temperature. Two hours is achievable for a system already turned on and pre-equilibrated with a temperature-controlled room. The RI detector is more sensitive to temperature than the columns. It can take more than 1 day to reach equilibrium, and a flat baseline, from a completely cold (or hot) start.

12. Frequent calibration makes for repeatable measurements. Some drift in retention times is expected, as silica columns very slowly deteriorate over time. In the experience of the authors, it is usually necessary to recalibrate at least once every 24 h in order to ensure comparability of the sample and calibrant chromatograms.

13. Automation of this process is possible to a certain extent, but baseline allocation and determination of integration start and stop points may have to be performed, or at least adjusted, manually.

14. Definitions of number average molecular weight and weight average molecular weight:

    The number average molecular weight, $\bar{M}_n$, is defined as follows:

$$\bar{M}_n = \frac{\sum_i N_i M_i}{\sum_i N_i}$$

where N_i is the number of molecules at molecular weight M_i.

The weight average molecular weight, $\bar{M}_w$, is defined as

$$\bar{M}_w = \frac{\sum_i g_i M_i}{\sum_i g_i}$$

where g_i is the weight of the sample at molecular weight M_i.

As $g_i = N_i M_i$, this may also be written as

$$\bar{M}_w = \frac{\sum_i N_i M_i^2}{\sum_i N_i M_i}$$

The number average and weight average of a polydisperse sample are different, and the ratio between them, the polydispersity PD, is a measure of the breadth of the molecular weight distribution:

$$PD = \frac{\bar{M}_w}{\bar{M}_n}$$

PD has a value of 1 for a monodisperse sample, for which $\bar{M}_n = \bar{M}_w$.

For polydisperse polymers PD will be greater than 1.

Translating these equations into terms which can be derived from the RI chromatogram:

The RI signal is proportional to mass, i.e., $N_i M_i$. We can derive N_i by dividing RI by M_i. The equation for $\bar{M}_n$ then becomes $\Sigma(RI)/\Sigma(RI/M_i)$.

We can also derive $N_i M_i^2$ by multiplying RI by M_i. The equation for $\bar{M}_w$ becomes $\Sigma(RI \times M_i)/\Sigma(RI)$.

Once the system is calibrated, in principle it is also possible to use a UV or fluorimetric trace for a compound in which every molecule bears a single chromophore or fluorophore. As the resulting detector response will then be proportional to N_i, for a UV trace the equation for $\bar{M}_n$ becomes $\Sigma(UV \times M_i)/\Sigma(UV)$, and that for $\bar{M}_w$ becomes $\Sigma(UV \times M_i^2)/\Sigma(UV \times M_i)$. No example is given for this, as we have never implemented it ourselves.

15. Fitting a trendline gives coefficients for the polynomial function that are not correct. This results in substantial errors in the calculated average molecular weights. Use of the "linest" function in this context is explained in the Excel help facility, though not at the beginning; scroll down to find the information you need.

16. A rapid estimate of M_p for small, monodisperse oligosaccharides can be found by aligning the unknown peak with partially resolved peaks in the calibrant chromatogram, as shown in Fig. 4. This technique also works for partially enzyme-depolymerized chondroitin and dermatan oligosaccharides, using the parent digest mixture as a comparator. The sub-peak with the longest retention time is the disaccharide.

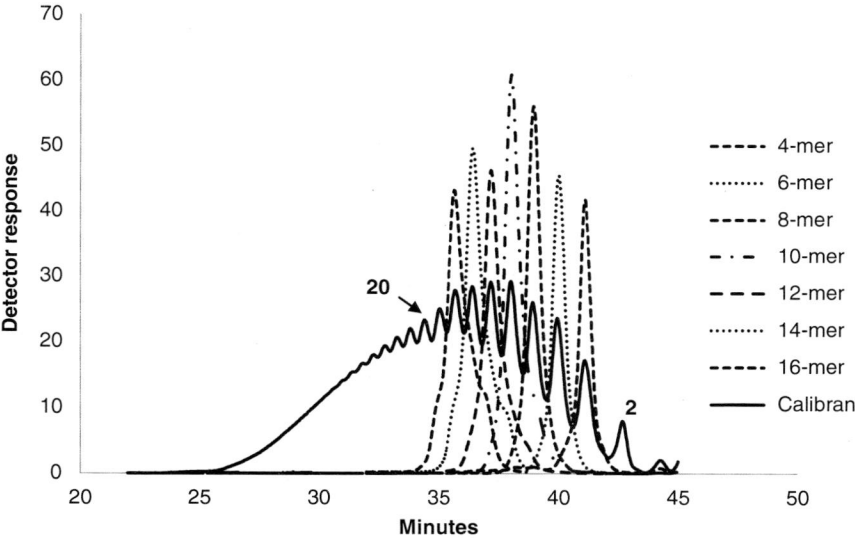

Fig. 4 Overlaid chromatograms of seven almost monodisperse oligosaccharides derived from heparin by digestion with heparinase I, and a low-molecular-weight heparin calibrant. The sub-peaks in the calibrant corresponding to the disaccharide (labeled **2**) and the eicosasaccharide (labeled **20**) are indicated. The oligosaccharide peaks match up well with the sub-peaks in the calibrant (which was also derived by heparinase treatment of an unfractionated heparin sample). Assigning a molecular weight of 600 per disaccharide [2], the approximate peak molecular weights of the oligosaccharides can be obtained without complex calculations

References

1. Al Dieri R, Wagenvoord R, van Dedem GW, Beguin S, Hemker HC (2003) The inhibition of blood coagulation by heparins of different molecular weight is caused by a common functional motif—the C-domain. J Thromb Haemost 1:907–914

2. Kailemia MJ, Li L, Xu Y, Liu J, Linhardt RJ, Amster IJ (2013) Structurally informative tandem mass spectrometry of highly sulfated natural and chemoenzymatically synthesized heparin and heparan sulfate glycosaminoglycans. Mol Cell Proteomics 12:979–990

3. Volpi N, Maccari F, Suwan J, Linhardt RJ (2012) Electrophoresis for the analysis of heparin purity and quality. Electrophoresis 33:1531–1537

4. Sommers CD, Ye H, Kolinski RE, Nasr M, Buhse LF, Al-Hakim A, Keire DA (2011) Characterization of currently marketed heparin products: analysis of molecular weight and heparinase-I digest patterns. Anal Bioanal Chem 401:2445–2454

5. Mulloy B, Gee C, Wheeler SF, Wait R, Gray E, Barrowcliffe TW (1997) Molecular weight measurements of low molecular weight heparins by gel permeation chromatography. Thromb Haemost 77:668–674

6. Mulloy B, Heath A, Shriver Z, Jameison F, Al-Hakim A, Morris TS, Szajek A (2014) Development of a compendial method for the chromatographic determination of molecular weight distributions for unfractionated heparin. Anal Bioanal Chem 460:4815–4823

7. Beirne J, Truchan H, Rao L (2011) Development and qualification of a size exclusion chromatography coupled with multiangle light scattering method for molecular weight determination of unfractionated heparin. Anal Bioanal Chem 399:717–725

8. Bertini S, Bisio A, Torri G, Bensi D, Terbojevich M (2005) Molecular weight determination of heparin and dermatan sulfate by size exclusion chromatography with a triple detector array. Biomacromolecules 6:168–173

9. Mulloy B (2002) Gel permeation chromatography of heparin. In: Volpi N (ed) Analytical techniques to evaluate the structure and function of natural polysaccharides, glycosaminoglycans. Research Signpost, Trivandrum

10. Knobloch JE, Shaklee PN (1997) Absolute molecular weight distribution of low-molecular-weight heparins by size-exclusion chromatography with multiangle laser light scattering detection. Anal Biochem 245:231–241

11. Huckerby TN, Sanderson PN, Nieduszynski IA (1986) N.M.R. studies of oligosaccharides obtained by degradation of bovine lung heparin with nitrous acid. Carbohydr Res 154:15–27

12. Khan S, Gor J, Mulloy B, Perkins SJ (2010) Semi-rigid solution structures of heparin by constrained X-ray scattering modelling: new insight into heparin-protein complexes. J Mol Biol 395:504–521

13. Khan S, Fung KW, Rodriguez E, Patel R, Gor J, Mulloy B, Perkins SJ (2013) The solution structure of heparan sulfate differs from that of heparin: implications for function. J Biol Chem 288:27737–27751

14. Murphy KJ, Merry CL, Lyon M, Thompson JE, Roberts IS, Gallagher JT (2004) A new model for the domain structure of heparan sulfate based on the novel specificity of K5 lyase. J Biol Chem 279:27239–27245

15. Hasan J, Shnyder SD, Clamp AR, McGown AT, Bicknell R, Presta M, Bibby M, Double J, Craig S, Leeming D, Stevenson K, Gallagher JT, Jayson GC (2005) Heparin octasaccharides inhibit angiogenesis in vivo. Clin Cancer Res 11:8172–8179

16. Lauder RM, Huckerby TN, Nieduszynski IA, Sadler IH (2011) Characterisation of oligosaccharides from the chondroitin/dermatan sulphates: (1)H and (13)C NMR studies of oligosaccharides generated by nitrous acid depolymerisation. Carbohydr Res 346:2222–2227

17. Toida T, Sato K, Sakamoto N, Sakai S, Hosoyama S, Linhardt RJ (2009) Solvolytic depolymerization of chondroitin and dermatan sulfates. Carbohydr Res 344:888–893

18. Pomin VH, Park Y, Huang R, Heiss C, Sharp JS, Azadi P, Prestegard JH (2012) Exploiting enzyme specificities in digestions of chondroitin sulfates A and C: production of well-defined hexasaccharides. Glycobiology 22:826–838

19. Volpi N, Bolognani L (1993) Glycosaminoglycans and proteins: different behaviours in high-performance size-exclusion chromatography. J Chromatogr 630:390–396

Chapter 12

Mass Spectrometric Methods for the Analysis of Heparin and Heparan Sulfate

Miroslaw Lech, Ishan Capila, and Ganesh V. Kaundinya

Abstract

Glycosaminoglycans like heparin and heparan sulfate exhibit a high degree of structural heterogeneity. This structural heterogeneity results from the biosynthetic process that produces these linear polysaccharides in cells and tissues. Heparin and heparan sulfate play critical roles in normal physiology and pathophysiology; hence it is important to understand how their structural features may influence overall activity. Therefore, high-resolution techniques like mass spectrometry represent a key part of the suite of methodologies available to probe the fine structural details of heparin and heparan sulfate. This chapter outlines the application of techniques like LC-MS and LC-MS/MS to study the composition of these polysaccharides, and techniques like GPC-MS that allow for an analysis of oligosaccharide fragments in these mixtures.

Key words Glycosaminoglycans, Heparin, Heparan sulfate, Ion-pairing reversed-phase liquid chromatography (IPRP-HPLC), Low molecular weight heparins (LMWHs), Gel permeation chromatography (GPC), Mass spectrometry (MS)

1 Introduction

Heparin and heparan sulfate are complex, linear polysaccharides, which are predominantly found in the extracellular matrix or on the surface of cells, attached to a protein core. They belong to the family of glycosaminoglycans (GAGs) and at a basic level are made up of repeating units of N-acetylglucosamine α 1–4 linked to a glucuronic acid residue. The biosynthetic process that produces these polysaccharides consists of a series of enzymatic modifications that are introduced in a sequential manner on this basic disaccharide repeating unit along the polysaccharide chain. Each of these enzymatic reactions is dependent, to some extent, on the previous step, as products of one reaction can act as substrate for the subsequent steps. Furthermore, the fact that not all these biosynthetic reactions go to completion leads to the introduction of significant structural heterogeneity in these polysaccharides.

Kuberan Balagurunathan et al. (eds.), *Glycosaminoglycans: Chemistry and Biology*, Methods in Molecular Biology, vol. 1229, DOI 10.1007/978-1-4939-1714-3_12, © Springer Science+Business Media New York 2015

At a disaccharide level the N-position of the glucosamine can either be acetylated or sulfated; in rare cases the free amine has also been reported. Additionally, the enzymatic steps can lead to the introduction of sulfate groups at either the 6-O-position of the glucosamine or the 2-O-position of the uronic acid, and rarely also the 3-O-position of the glucosamine. Lastly the glucuronic acid can also be epimerized to the iduronic acid. This leads to a very large number of possible polysaccharide structures, which can be dynamically remodeled in the extracellular matrix or on the cell surface, thereby driving important biological processes.

Understanding the structural diversity of these polysaccharides therefore is critical to understanding how changes in structure may lead to modulation of biological processes. One of the key analytical technologies to facilitate this investigation into the fine structure of heparin and heparan sulfate is mass spectrometry. Glycosaminoglycans present a couple of unique challenges to mass spectrometrists; specifically they are tough to detect as molecular ions, as they are predominantly complexed to positively charged metal ions like sodium, and also they tend to lose their sulfate fairly readily, thereby leading to loss of structural information, and also leading to extensive fragmentation. Nevertheless, over the years, a number of approaches have been developed to analyze glycosaminoglycans by MS. These include both electrospray ionization (ESI) and matrix-assisted laser desorption ionization (MALDI) techniques that allow for the structural elucidation of disaccharide and oligosaccharide fragments of these GAGs. In conjunction, there were also different approaches applied to the fragmentation of these disaccharides and oligosaccharides. The coupling of chromatographic separation methods in-line to ESI-MS represented a key breakthrough in the field and led to a greater level of structural inquiry into these polysaccharides. In the subsequent sections we discuss two of these methodologies that have allowed investigation into the shorter saccharides that constitute the compositional building blocks of these polysaccharides, as well as longer oligosaccharide chains that could represent intact chains in the mixture, or could result from partial enzymatic digestion of the mixture.

1.1 Ion-Pairing Reversed-Phase High-Performance Liquid Chromatography-Mass Spectrometry (IPRP-HPLC-MS)

Ion-pairing reversed-phase high-performance liquid chromatography (IPRP-HPLC) coupled to electrospray ionization mass spectrometry (ESI-MS) is an indispensable tool for analysis of sulfated glycosaminoglycans including heparin/heparan sulfate. The IPRP-HPLC-MS method is based on the methodology that was originally developed at MIT [1]. This technique represents a very sensitive methodology that can detect building block structures present at very low levels. In addition, building block analysis by LC-MS represents an independent confirmation of the presence and structural assignment of peaks as determined by building block analysis using other analytical techniques including capillary electrophoresis and IPRP-HPLC. This method is suitable for separation of short-chain oligosaccharides, from di- to hexasaccharides (dp2–dp6). Therefore, this method is mainly used for building block analysis of unfractionated heparin (UFH)/ low molecular weight heparins (LMWHs)/heparan sulfate (HS)/ dermatan sulfate (DS)/chondroitin sulfates (CS) prepared by complete enzymatic digestion using a cocktail of heparinases. IPRP-HPLC separates on the basis of hydrophobicity of the stationary phase, concentration of the ion-pairing reagent, size, charge, and isomerization.

1.2 Gel Permeation Chromatography-Mass Spectrometry (GPC-MS)

Gel permeation chromatography (GPC) coupled to electrospray ionization mass spectrometry (ESI-MS) is a powerful technique that is suitable for separation of oligosaccharides from di- to octadecasaccharides (dp2–dp18) and higher without any chemical or enzymatic pretreatment of the sample prior to analysis. Since ion-pairing reversed-phase high-performance liquid chromatography (IPRP-HPLC) is mainly used for analysis of completely depolymerized glycosaminoglycans (GAGs), as it demonstrates high resolving power for short-chain-size oligosaccharides, GPC can be used for analysis of longer oligosaccharides (Fig. 1a). In addition, the in-line MS provides us with a mass (and thereby composition) for species present within the major size-fractionated components of the analyzed mixture (Fig. 1b, c). This methodology is based upon a study published in 2004 that outlined an in-line size-exclusion chromatography/mass spectrometry approach for analyzing low-molecular-mass heparin [2]. GPC separates mostly on the basis of size. The described GPC-MS method allows the determination of total number of monosaccharide units, and number of sulfate and acetyl groups of the components present in the mixtures; however, it does not allow the distinction of isomers with respect to the position of the sulfate and acetyl groups.

2 Materials

2.1 IPRP-HPLC-MS

1. Prepare all solutions using ultrapure water (prepared by purifying deionized water to attain a sensitivity of 18.2 MΩ cm at 25 °C) and LCMS-grade reagents. Prepare and store all reagents at room temperature (unless indicated otherwise).

2. LC Buffer "A": Prepare 100 % aqueous solution containing 8 mM dibutylammonium acetate as an ion-pairing reagent by diluting 1:62.5 0.5 M stock solution of dibutylammonium acetate in water (see **Note 1**).

3. LC Buffer "B": Prepare 70 % aqueous methanol solution containing 8 mM dibutylammonium acetate as an ion-pairing reagent by diluting 1:62.5 0.5 M stock solution of dibutylammonium acetate in water.

4. A capillary C18 column (0.3 mm × 250 mm) (see **Note 2**).

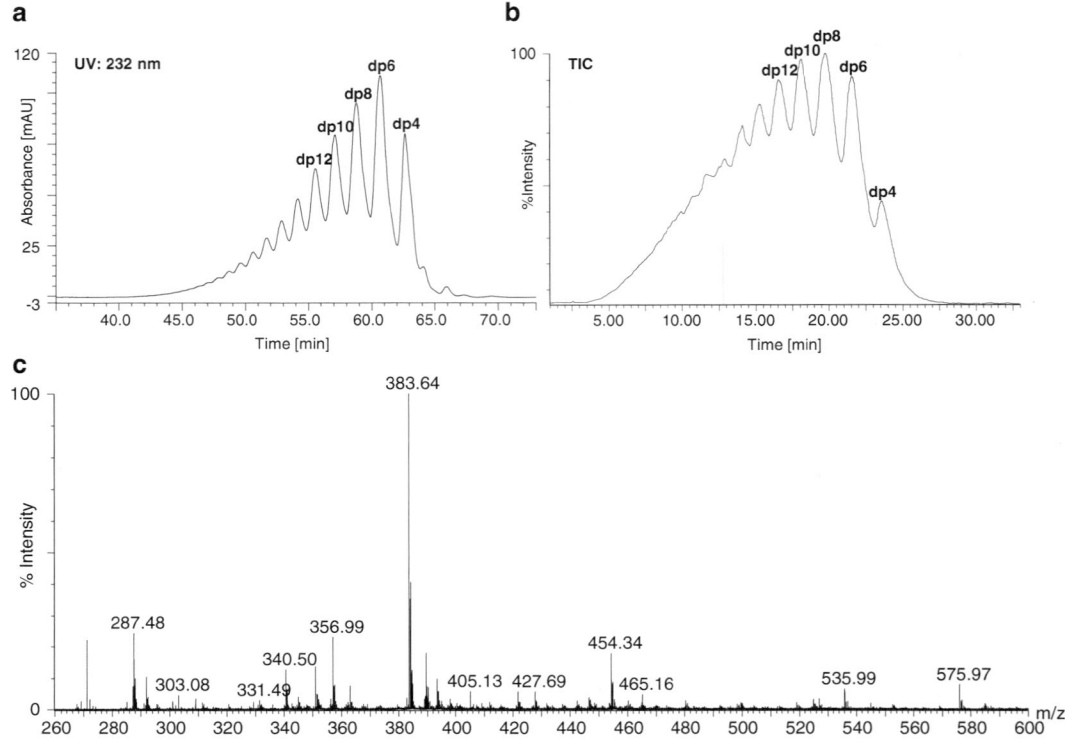

Fig. 1 (**a**) GPC separation of the entire low molecular weight heparin (LMWH) mixture. (**b**) The total ion chromatogram (TIC) based on GPC separation of the entire LMWH mixture. (**c**) Negative mass spectra observed within the tetrasaccharides. Mass spectra were acquired on an electrospray ionization (ESI) quadrupole time-of-flight (QTOF) mass spectrometer (Waters). An Ultimate 3000 HPLC workstation (Dionex) was used for GPC separation

2.2   GPC-MS

1. Prepare all solutions using ultrapure water (prepared by purifying deionized water having specific resistance of 18.2 MΩ, or greater, at 25 °C) and LCMS-grade reagents. Prepare and store all reagents at room temperature.

2. An electrospray ionization high-resolution mass spectrometer.

3. A HPLC pump system capable to deliver the flow rate at 100 μl/min.

4. A UV detector, if desired (*see* **Note 3**).

5. Flow splitter: Only if desired (*see* **Note 4**).

6. Anion Electrolytically Regenerated Suppressor 500 desalter (e.g., Dionex AERS 500 2 mm) (*see* **Note 5**).

7. A water delivery system such as the ERS 500 Pressurized Bottle Kit from Dionex P/N 038018.

8. Power supply: The Reagent-Free Controller (Dionex RFC-10 P/N 060335).

9. LC buffer: Prepare 100 % aqueous solution containing 100 mM ammonium acetate by diluting 1:10, 1.0 M stock solution of ammonium acetate in water.

10. HPLC columns: Two 4.6 mm × 300 mm, 4 µm gel permeation TOSOH TSKgel SuperSW2000 columns.

3   Methods

3.1   IPRP-HPLC-MS

3.1.1   Capillary HPLC Microseparation of Lyase-Digested Sulfated Glycosaminoglycans

Perform HPLC separations on a 0.3 mm × 250 mm C18 column thermostatically controlled at 17 °C, with the flow rate set at 5 µl/min using a step gradient elution and a binary solvent system composed of water (mobile phase A) and 70 % aqueous methanol (mobile phase B), both containing 8 mM dibutylammonium acetate. Set up the following elution profile: 0 % B for 4 min, 9 % B for 20 min, 20 % B for 21 min, 32 % B for 19 min, 48 % B for 17 min, and 63 % B for 14 min. After each separation run, wash the column for 12 min with 90 % methanol/water, and equilibrate with 100 % A for 20 min (*see* **Note 6**).

3.1.2   Mass Spectrometry of Lyase-Digested Sulfated Glycosaminoglycans

Perform mass spectrometry in the negative-ion mode using an electrospray ionization high-resolution instrument such as time-of-flight mass spectrometer. Conditions for ESI-MS can vary depending on the manufacturer of the instrument. The main concern with mass spectrometric analysis of highly sulfated saccharides is the potential loss of sulfates, thereby resulting in a loss of structural information. In general, use low capillary and cone voltage potential as well as very low collision energy. The other settings including source temperature, desolvation temperature, cone gas flow rate, and desolvation gas flow rate are set up as per the manufacturer's recommendations. The recommended settings for the SYNAPT G2 HDMS MS system (Waters Corporation) are provided below. Acquire mass spectra in negative "V" ion mode with a typical resolving power of 20,000 FWHM (full width half maximum) and mass range of 50–1,200 Da. Externally calibrate the TOF (time of flight) analyzer using a NaCsI mixture from m/z 50 to 1,200. A reference compound (Leu-enkephalin, Waters) can be used as a LockSpray. Deliver the reference compound at 5 µl/min flow rate and sample it for three scans every 60 s. Set up other instrument parameters as follows: capillary voltage 2.00 kV, sampling cone voltage 8 V, extraction cone voltage 2 V, source temperature 90 °C, desolvation temperature 250 °C, cone gas flow 50 L/h, desolvation gas flow 800–1,000 L/h, trap collision energy 2.0 V, transfer collision energy 0.4 V, trap gas flow 1.00 mL/min, trap DC Bias 0.0, IMS Bias 0.0, trap wave height 0.0 V, transfer wave height 0.0 V.

3.1.3   Tandem Mass Spectrometry

Precursor ions with negative charges equal to the number of sulfate groups are preferred for the tandem MS analysis, as they produce larger number of fragments (glycosidic-bond and cross-ring cleavages) and the sulfate losses are minimized [2]. Product ions are assigned on the basis of cleavage patterns previously observed

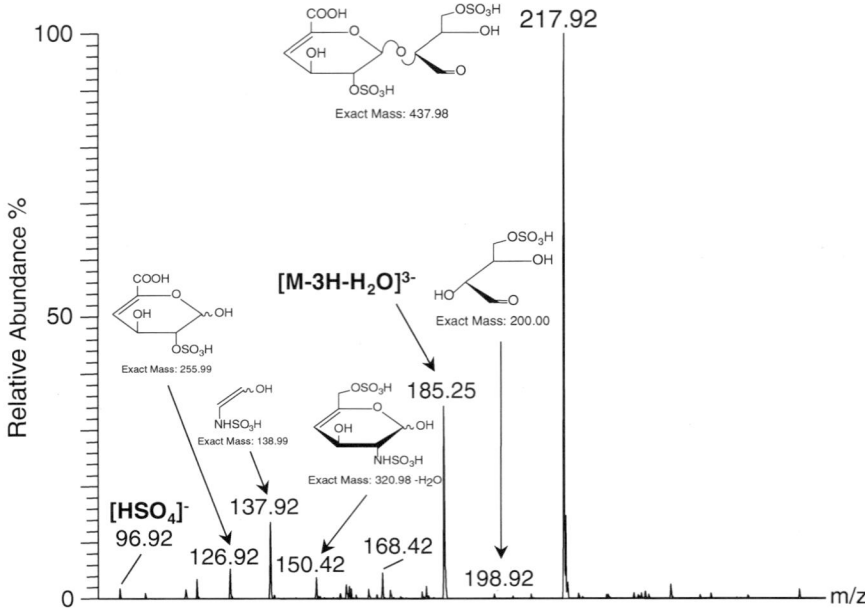

Fig. 2 Negative tandem mass spectra of trisulfated disaccharide of the formula ΔU_{2S}–H_{NS6S} (577 Da) using the precursor ion at m/z 191.33 [M-3H]$^{3-}$

in HS disaccharides [3, 4]. In this document the negative-ion mode mass spectra are acquired using an electrospray ionization source (ESI) on the quadrupole ion trap mass spectrometer LTQ XL from Thermo Scientific. ESI-MS and collision-induced dissociation (CID) MS2 experiments are performed on the precursor ion at m/z 191.33 [M-3H]$^{3-}$ of trisulfated disaccharide with the formula ΔU_{2S}–H_{NS6S} (577 Da) (Fig. 2) using data-dependent triple-play acquisition. The MS to MS2 switch criteria are based on the peak intensity. The first scan event is the MS scan of full scan type generated by scanning the range of m/z 120–1,400, followed by the data-dependent ZoomScan and MS2 scan. For MS2 experiments, an isolation window is set at 2 Da and the normalized collision energy at 25.

3.2 GPC-MS

3.2.1 GPC Separation of the Intact Mixture (Partially Depolymerized Unfractionated Heparin (UFH)/Heparan Sulfate (HS)/Dermatan Sulfate (DS)/Chondroitin Sulfates (CS) Including Low Molecular Weight Heparins (LMWHs))

Perform GPC separations using two 4.6 mm × 300 mm, 4 μm gel permeation TOSOH TSKgel SuperSW2000 columns placed in series and thermostatically controlled at 25 °C, with the flow rate set at 100 μl/min and using an isocratic elution composed of 100 mM ammonium acetate in water. To obtain good MS signal and suppress formation of adducts it is necessary to remove the excess of ammonia; therefore, the columns are followed by the Anion Electrolytically Regenerated Suppressor 500 (AERS 500) operating in the AutoSuppression External Water Mode. This mode incorporates an external source of deionized water pneumatically supplied as the regenerant. Electrolysis of water in the AERS 500 is driven by the Dionex RFC-10 controller using 50 mA current. Eluent from the AERS enters the mass spectrometer.

3.2.2 Mass Spectrometry Parameters

Perform mass spectrometry in the negative-ion mode using an electrospray ionization high-resolution instrument such as time-of-flight mass spectrometer. Conditions for ESI-MS can vary depending on the manufacturer of the instrument. The main concern with mass spectrometric analysis of highly sulfated saccharides is the potential loss of sulfates, thereby resulting in a loss of structural information. In general, use low capillary and cone voltage potential as well as very low collision energy. The other settings including source temperature, desolvation temperature, cone gas flow rate, and desolvation gas flow rate are set up as per the manufacturer's recommendations. The recommended settings for the SYNAPT G2 HDMS MS system (Waters Corporation) are provided below. Acquire mass spectra in negative "V" ion mode with a typical resolving power of 20,000 FWHM (full width half maximum) and mass range of 50–1,200 Da. Externally calibrate the TOF (time of flight) analyzer using a NaCsI mixture from m/z 50 to 1,200. A reference compound (Leu-enkephalin, Waters) can be used as a LockSpray. Deliver the reference compound at 5 μl/min flow rate and sample it for three scans every 60 s. Set up other instrument parameters as follows: capillary voltage 2.00 kV, sampling cone voltage 8 V, extraction cone voltage 2 V, source temperature 90 °C, desolvation temperature 250 °C, cone gas flow 50 L/h, desolvation gas flow 800–1,000 L/h, trap collision energy 2.0 V, transfer collision energy 0.4 V, trap gas flow 1.00 mL/min, trap DC Bias 0.0, IMS Bias 0.0, trap wave height 0.0 V, and transfer wave height 0.0 V.

4 Notes

1. The acidic substances are ionized under the neutral conditions; they form an ion-pair with ion-pair reagents in the mobile phase to become electrically neutral. The increase in hydrophobic character of the ion pair results in a greater affinity for the reverse stationary phase and leads to sample resolution. The UV absorption of sodium alkane sulfonates and quaternary ammonium salts (tetrabutylammonium compounds being one example) is minimal so that these reagents are commonly used as ion-pairing agents for IP-RP-HPLC [5]. However, direct coupling of ion-pair chromatography to mass spectrometry is made difficult by the contamination of the interface by the nonvolatile tetraalkylammonium salts as well as their tendency to form cluster ions with anionic analytes and lower the ionization efficiency when being used as mobile-phase modifiers. The dibutylammonium acetate is a highly volatile ion-pair reagent for acidic samples supplied as 0.5 M aqueous solution adjusted to pH 7.5 allowing for continuous LC-MS analysis without contaminating the interfaces. In addition, the same reagent can be used for the UV absorption. The heparin

lyase-derived saccharide fragments contain the unsaturated double bond on the uronic acid at the nonreducing end and in conjugation with the carboxyl group at C-5 they can be easily monitored by UV absorption at 232 nm. The dibutylammonium acetate solution can be used as a neutral mobile phase after dilution with the LC solvents (acetonitrile/water or methanol/water) to 5–8 mM. The dibutylammonium acetate is mainly used for analysis of completely depolymerized heparin/HS/DS/CS samples, as it demonstrates high resolving power for short oligosaccharides. For analysis of longer oligosaccharides including the analysis of a low molecular weight heparin (LMWH) 8 mM pentylammonium acetate can be used as the ion-pairing reagent. A combination of the nonlinear %-gradient (step gradient curves) and the ramp gradient provides high-resolution separation for a broad size (dp) range of sulfated oligosaccharides.

2. The analyses using ion-pairing reversed-phase liquid chromatography is performed on C18 capillary polymeric silica column made by Vydac. There are other commercially available columns that can replace Vydac column as needed. For example the PROTO 300 C18 made by Higgins Analytical, Inc. can be used as an alternative. After several adjustments to the elution gradient PROTO 300 can give very similar separation of the sulfated saccharide building blocks. Moreover, the PROTO C18 column showed stronger retention, a future that could help analysis of the samples that have large amounts of the non-sulfated species.

3. The heparin lyase-derived oligosaccharides as well as some of the chemically derived oligosaccharides including some LMWHs contain the unsaturated double bond on the uronic acid at the nonreducing end and, therefore in addition to MS, they can also be monitored by UV absorption at 232 nm (Fig. 3a).

4. Most of the modern electrospray ionization (ESI) interfaces in use today are pneumatically assisted, usually using nitrogen as a desolvation gas, and can handle flow rates up to 1 mL/min. This means that our gel filtration LC columns (4.6 mm i.d.) do not require post-column splitting when used in combination with the AERS 500 2 mm, which can handle the maximum eluent flow rate of 1 mL/min. Nonetheless a post-column adjustable flow splitter can provide the desired split ratio range and back pressure.

5. The Anion Electrolytically Regenerated Suppressor 500 is designed to be run on a stand-alone controller such as the Dionex RFC-10.

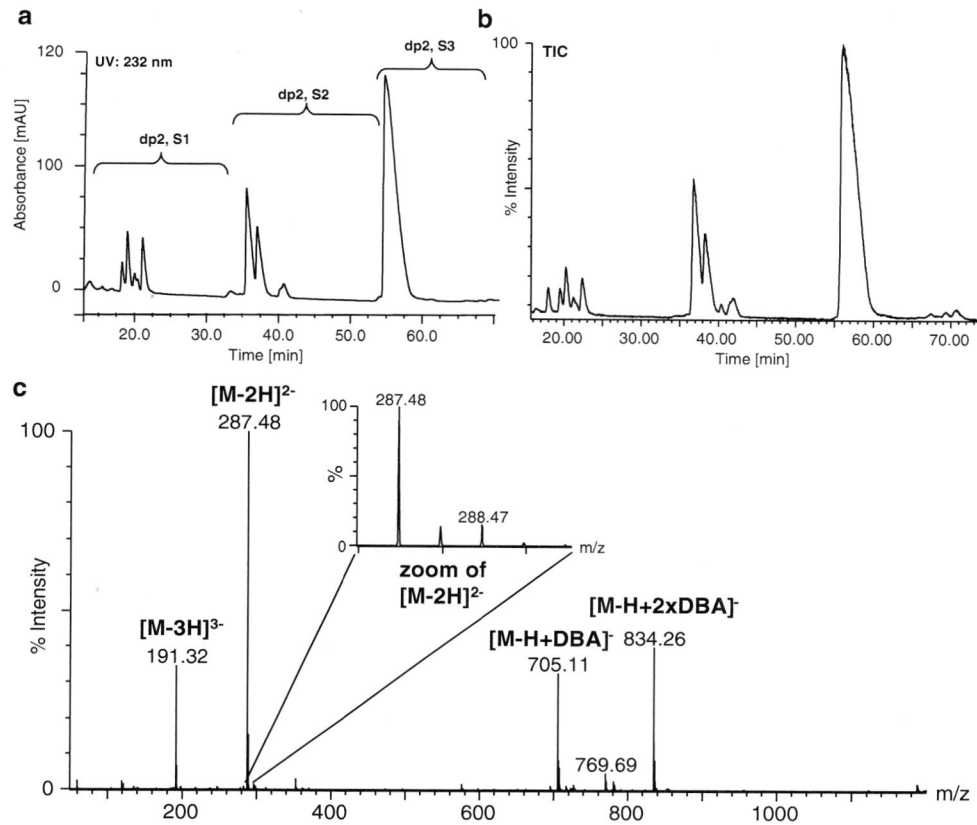

Fig. 3 (**a**) Capillary IPRP-HPLC of disaccharides from depolymerized heparin by heparin lyases I, III, and IV from *Bacteroides thetaiotaomicron*. (**b**) The total ion chromatogram (TIC) of disaccharides from depolymerized heparin by heparin lyase digestion. (**c**) Negative mass spectra of trisulfated disaccharide of the formula $\Delta U_{2S}-H_{NS6S}$ (577 Da). Mass spectra were acquired on the SYNAPT G2 HDMS instrument (Waters). An Ultimate 3000 capillary HPLC workstation (Dionex) was used for IPRP microseparation

6. A step gradient elution is used to maximize the separation of the oligosaccharides based on the degree of sulfation and sulfation pattern. The IPRP-LC not only resolves the oligosaccharides that vary by the number of sulfates and/or acetyl groups, but also the numerous isomeric structures of the same molecular weight. The chromatographic conditions are optimized for disaccharides to hexasaccharides (dp2–dp6). Briefly, non-sulfated disaccharides elute with 100 % A, single-sulfated disaccharides elute with 9 % B (first step), double-sulfated disaccharides elute with 20 % B (second step), triple-sulfated disaccharides elute with 32 % B (third step), tetrasaccharides elute with 48 % B (fourth step), and finally hexasaccharides elute with 63 % B (fifth step).

References

1. Kuberan B, Lech M, Zhang L, Wu ZL, Beeler DL, Rosenberg RD (2002) Analysis of heparan sulfate oligosaccharides with ion pair-reverse phase capillary high performance liquid chromatography-microelectrospray ionization time-of-flight mass spectrometry. J Am Chem Soc 124(29):8707–8718

2. Zaia J, Costello CE (2003) Tandem mass spectrometry of sulfated heparin-like glycosaminoglycan oligosaccharides. Anal Chem 75(10): 2445–2455

3. Saad OM, Leary JA (2004) Delineating mechanisms of dissociation for isomeric heparin disaccharides using isotope labeling and ion trap tandem mass spectrometry. J Am Soc Mass Spectrom 15(9):1274–1286

4. Domon B, Costello CE (1988) Structure elucidation of glycosphingolipids and gangliosides using high-performance tandem mass spectrometry. Biochemistry 27(5): 1534–1543

5. Henriksen J, Ringborg LH, Roepstorrf P (2004) On-line size-exclusion chromatography/mass spectrometry of low molecular mass heparin. J Mass Spectrom 39: 1305–1312

Chapter 13

Validated Capillary Electrophoretic Assays for Disaccharide Composition Analysis of Galactosaminoglycans in Biologic Samples and Drugs/Nutraceuticals

Athanasia P. Asimakopoulou, Christina Malavaki, Nikolaos A. Afratis, Achilleas D. Theocharis, Fotini N. Lamari, and Nikos K. Karamanos

Abstract

Capillary electrophoresis is a separation technique with high resolving power and sensitivity with applications in glycosaminoglycan analysis. In this chapter, we present validated protocols for determining the variously sulfated chondroitin or dermatan sulfate-derived disaccharides. These approaches involve degradation of the polysaccharides with specific chondro/dermato-lyases and electrophoretic analysis with capillary zone electrophoresis in a low pH operating buffer and reversed polarity. This methodology has been applied to drug/nutraceutical formulations or to biologic samples (blood serum, lens capsule) and has been validated. Analysis of biologic tissue samples is often more demanding in terms of detection sensitivity, and thus concentration pretreatment steps and/or a derivatization step with 2-aminoacridone are often advisable.

Key words Disaccharides, Capillary electrophoresis, 2-Aminoacridone, Enzymic treatment, Glycosaminoglycans, Validation

1 Introduction

1.1 Glycosaminoglycans (GAGs)

Glycosaminoglycans (GAGs) are ubiquitous acidic linear hetero-polysaccharides containing repeating disaccharide units composed of uronic acids and hexosamines. GAGs, except hyaluronan, are covalently bound into protein core of proteoglycans and the heterogenicity, such as relative molecular mass, charge density, and physicochemical properties, determines targeting of PGs to proper cellular or extracellular environment. GAGs are distinguished in four main categories: hyaluronan, chondroitin sulfate (CS) and dermatan sulfate (DS), heparan sulfate and heparin, and keratan sulfate [1]. CS and DS contain as hexosamine galactosamine (GalN) and thus are known as galactosaminoglycans.

Kuberan Balagurunathan et al. (eds.), *Glycosaminoglycans: Chemistry and Biology*, Methods in Molecular Biology, vol. 1229, DOI 10.1007/978-1-4939-1714-3_13, © Springer Science+Business Media New York 2015

**1.2 Chondroitin
and Dermatan Sulfate
Structure**

Chondroitin/dermatan sulfate (CS/DS), as part of proteoglycans, are an essential component of the extracellular and cellular surface matrix, and determine their biological properties and pharmacological targeting in disease [2, 3]. Galactosaminoglycans are composed of repeating disaccharide units of D-glucuronic acid (GlcA) and N-acetyl-galactosamine (GalNAc). This basic unit ($\rightarrow$4GlcAβ1$\rightarrow$3GalNAcβ1$\rightarrow$) can be substituted with O-sulfo groups at the C-4 or C-6 positions of GalNAc and also at the C-2 of IdoA in the case of dermatan sulfate (DS). The IdoA, a C-5 epimer of GlcA, imparts conformational flexibility to the DS chain [4]. The variability in substitution creates a number of differently sulfated disaccharide units; more than 20 different disaccharides have been reported but this variability is significantly reduced in mammals (Fig. 1).

CS is the main GAG component of blood and contains disaccharides mainly sulfated at C-4 of GalNAc (Di-4S) (40–60 %), whereas the rest of the polysaccharide is mainly non-sulfated [5–8]. In blood, CS interacts with protease inhibitors, coagulation factors, lipoproteins, and complement proteins, thus participating in several biological events. CS in blood serum originates from many different tissues and cells (endothelial, platelets, malignant ones, etc.) and its final quantity and structure reflect modifications in proteoglycan biosynthesis in the tissue of origin and in the plasma by GAG-modifying enzymes. Thus, analysis of blood CS is considered as a tool that may contribute to the discovery of useful biomarkers.

GAGs, and especially CS, are implicated in several pathological conditions like osteoarthritis, atherosclerosis, cancer, and ophthalmologic disease. The pseudoexfoliation (PSX) syndrome, an age-related systemic disease of the extracellular matrix, is characterized by the presence of amyloid-like fibrillar deposits on the anterior lens capsule. The pathological deposits (PSX material) can obstruct aqueous outflow leading to increased intraocular pressure that in turn can result in glaucoma [9]. The determination of CS structure in lens tissues offers an insight into the pathophysiological changes of extracellular matrix.

In recent years, CS from various animal sources and, thus, with different structural composition are used in pharmaceutical formulations and dietary supplements. Supplements of CS are mainly used for improving symptoms and arresting or reversing the degenerative process occurring in the development of osteoarthritis [10, 11]. Besides, CS is sometimes used for ophthalmologic diseases and there is preliminary evidence that CS may help in treatment of psoriasis [12]. Therefore, a proper quality control of CS formulations in terms of amount, purity, and structural characteristics (i.e., disaccharide composition) is necessary since they affect effectiveness [13].

Fig. 1 Structural characteristics of Δ-disaccharides derived from CS/DS by chondroitinase digestions. The IdoA- or GlcA-derived disaccharides acquire the same structure after digestion with chondro/dermato-lyases. The common disaccharides in mammals are those in *bold*. Rear glycoforms of CS/DS chains have been determined only in marine invertebrates. Adapted from Karamanos et al. [14] with permission of Elsevier

No	Name	Other names	Hexuronic acid				Galactosamine		
			R^2	R^3	R^6_G	R^6_I	R^2	R^4	R^6
1	**$\Delta di\text{-}nonS_{GlcA}$**	**$\Delta di\text{-}0S_{CS}$**	H	H	COOH	H	Ac	H	H
2	**$\Delta di\text{-}nonS_{IdoA}$**		H	H	H	COOH	Ac	H	H
3	**$\Delta di\text{-}mono4S_{GlcA}$**	**$\Delta di\text{-}4S$/ A unit**	H	H	COOH	H	Ac	SO_3H	H
4	**$\Delta di\text{-}mono4S_{IdoA}$**		H	H	H	COOH	Ac	SO_3H	H
5	**$\Delta di\text{-}mono6S_{GlcA}$**	**$\Delta di\text{-}6S$/ C unit**	H	H	COOH	H	Ac	H	SO_3H
6	**$\Delta di\text{-}mono6S_{IdoA}$**		H	H	H	COOH	Ac	H	SO_3H
7	**$\Delta di\text{-}mono2S_{GlcA}$**		SO_3H	H	COOH	H	Ac	H	H
8	**$\Delta di\text{-}mono2S_{IdoA}$**		SO_3H	H	H	COOH	Ac	H	H
9	$\Delta di\text{-}mono3S_{GlcA}$		H	SO_3H	COOH	H	Ac	H	H
10	$\Delta di\text{-}mono3S_{IdoA}$		H	SO_3H	H	COOH	Ac	H	H
11	$\Delta di\text{-}monoNS_{IdoA}$		H	H	H	COOH	SO_3H	H	H
12	**$\Delta di\text{-}di(2,6)S_{GlcA}$**	**$\Delta di\text{-}diS_D$/ D unit**	SO_3H	H	COOH	H	Ac	H	SO_3H
13	**$\Delta di\text{-}di(2,6)S_{IdoA}$**		SO_3H	H	H	COOH	Ac	H	SO_3H
14	**$\Delta di\text{-}di(2,4)S_{GlcA}$**		SO_3H	H	COOH	H	Ac	SO_3H	H
15	**$\Delta di\text{-}di(2,4)S_{IdoA}$**	**$\Delta di\text{-}diS_B$ / B unit**	SO_3H	H	H	COOH	Ac	SO_3H	H
16	**$\Delta di\text{-}di(4,6)S_{GlcA}$**	**$\Delta di\text{-}diS_E$ / E unit**	H	H	COOH	H	Ac	SO_3H	SO_3H
17	**$\Delta di\text{-}di(4,6)S_{IdoA}$**	**$\Delta di\text{-}diS_H$ / iE unit**	H	H	H	COOH	Ac	SO_3H	SO_3H
18	$\Delta di\text{-}di(3,6)S_{GlcA}$	$\Delta di\text{-}diS_L$ / L unit	H	SO_3H	COOH	H	Ac	H	SO_3H
19	$\Delta di\text{-}di(2,3)S_{IdoA}$		SO_3H	SO_3H	H	COOH	Ac	H	H
20	$\Delta di\text{-}di(3,4)S_{GlcA}$	$\Delta di\text{-}diS_K$ / K unit	H	SO_3H	COOH	H	Ac	SO_3H	H
21	$\Delta di\text{-}di(2,N)S_{IdoA}$		SO_3H	H	H	COOH	SO_3H	H	H
22	**$\Delta di\text{-}tri(2,4,6)S_{GlcA}$**	**$\Delta di\text{-}triS$**	SO_3H	H	COOH	H	Ac	SO_3H	SO_3H
23	**$\Delta di\text{-}tri(2,4,6)S_{IdoA}$**	**$\Delta di\text{-}triS_{iT}$ / iT unit**	SO_3H	H	H	COOH	Ac	SO_3H	SO_3H
24	$\Delta di\text{-}tri(3,4,6)S_{GlcA}$	$\Delta di\text{-}triS_M$ / M unit	H	SO_3H	COOH	H	Ac	SO_3H	SO_3H
25	$\Delta di\text{-}tri(2,6,N)S_{IdoA}$		SO_3H	H	H	COOH	SO_3H	H	SO_3H

1.3 Analytical Approach

High-performance capillary electrophoresis (HPCE or CE) is a miniaturization technique, which has been used for the analysis and structural characterization of various carbohydrate species, including galactosaminoglycans. Analysis of intact polysaccharides may give information on the identity, charge polydispersity, and maybe the molecular size of GAG chains, but information on structural

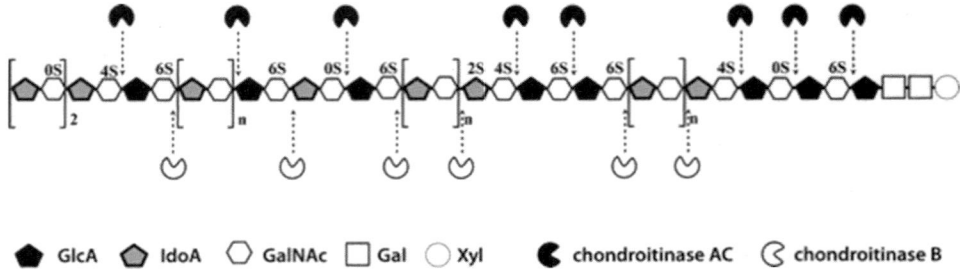

Fig. 2 Schematic representation of specific cleavage sites of CS/DS chains after chondroitinase treatment. The number of disaccharide units *n* is equal to or larger than 4. Adapted from Karamanos et al. [15] with permission of Elsevier

organization is established after utilization of specific depolymerization enzymes. Enzymic degradation via lyases produces an unsaturated hexuronic moiety at the nonreducing terminal of the resultant saccharides. The unsaturated uronic acid endows them with high UV molar absorptivity ($5,500 \ M^{-1} \ cm^{-1}$) and enables their sensitive ultraviolet detection at 232 nm [16]. Galactosaminoglycan degradation enzymes are chondroitinase ABC, which cleaves the *1β→4* glycosidic bonds between the galactosamine and the uronic acids (either GlcA or IdoA), chondroitinase ACII which cleaves only those between galactosamine and GlcA, and chondroitinase B between GalNAc and IdoA [17, 18]. It is well documented that the complete degradation of intact CS to CS Δ-disaccharides by the use of commercial enzymes is achieved by the combined action of both chondroitinase ABC and ACII [19]. This is because of the necessity of the presence in the reaction of both the lyases, chondroitinase ABC which cleaves intact CS to tetra- and disaccharides and also the chondroitinase AC II, which cleaves to tetra- and hexa-saccharide. The combination with both enzymes results in yields more than 92 % in $\Delta^{4,5}$ disaccharides (Figs. 1 and 2).

The detection of Δ-disaccharides with higher specificity and sensitivity than ultraviolet spectroscopy can be held by fluorescence detection. For this purpose, we propose the derivatization with 2-aminoacridone (AMAC), a fluorescent neutral molecule with $\lambda_{ex} = 425$ nm and $\lambda_{em} = 520$ nm. The AMAC derivatives are analyzed in CE and detected either with UV at 245 nm or with special laser-induced fluorescent (LIF) detectors [20].

2   Materials

2.1   Biologic Samples

Determination of CS in biologic samples (blood serum and lens capsule) with validated methods

1. Blood serum is obtained according to the rules imposed by the local and national bioethics committees, centrifuged, and directly stored at –20 °C.

2. Lens capsules from healthy volunteers and patients with exfoliation syndrome (XFS) are obtained from the University Hospital according to the procedures provisioned by the local and national bioethics committees. The central anterior lens capsule is excised during the capsulorhexis step in routine cataract surgery in the form of a thin circular disc approximately 5.5 mm in diameter. During the cataract surgery a pharmaceutical formulation of CS and HA is used to prevent injury. Samples were stored at –80 °C until treatment.

2.2 CS Standards

1. Avian CS (average molecular size 18 kDa) was kindly provided by Pierre Fabre Laboratories (Castres, France), CS-B (dermatan sulfate) from porcine intestinal mucosa, CS-C from shark cartilage, CS-D and CS-E from squid cranial cartilage are obtained from Sigma-Aldrich [21].

2. Unsaturated Chondro-Disaccharide Kit and unsaturated Dermato/Hyaluro-disaccharide Kit (e.g., C-Kit, Cat No. 400571-1 and D-Kit, Cat No. 400572-1 from Seikagaku Corporation (Tokyo, Japan)): These kits contain Δdi-HA, Δdi-0S, Δdi-4S, Δdi-6S, Δdi-diS$_E$, Δdi-diS$_D$, Δdi-diS$_B$, and Δdi-triS. Prepare stock solutions of standard CS Δ-disaccharides by dissolving 250 ng of each disaccharide in 250 µL water for injection and store at -20 ± 5 °C.

2.3 Enzymes

1. Protease from *Streptomyces griseus* (EC Number 232-909-5) is dissolved in 50 mM Tris–HCl buffer, pH 8.00, and stored at -20 ± 5 °C (*see* **Note 1**).

2. Chondroitinases ABC (EC 4.2.2.4) from *Proteus vulgaris* and AC II (EC 4.2.2.6) from *Arthrobacter aurescens* (e.g., from Seikagaku Corporation (Tokyo, Japan)).

3. A chondroitinase mixture solution is prepared. Chondroitinase ABC (10u) is dissolved in 250 µL of 50 mM Tris–HCl buffer pH 7.5 and chondroitinase AC II (5u) is dissolved in 125 µL of 50 mM Tris–HCl buffer pH 7.5. Then stock solutions containing the two chondroitinases are prepared by mixing 1:1 of each chondroitinase solution and aliquots are stored at -20 ± 5 °C. Aliquots are diluted in the same buffer at 1:10 whenever needed.

2.4 Capillary Electrophoresis

Uncoated fused-silica capillary tube (50 µm i.d., 64.5 cm total length, and 56 cm effective length) is used. Many companies offer uncoated fused-silica capillaries with extended light path for higher sensitivity, and those offer analysis with higher sensitivity albeit they are a bit more expensive. Those extended light-path capillaries should be preferred for the analysis of biologic samples, especially if the samples are not derivatized. The capillary electrophoresis experiments can be performed on any CE instrument (we used HP³DCE from Agilent Technologies (Waldbronn, Germany) instrument with a built-in diode array detector).

3 Methods

3.1 Blood Serum Pretreatment

1. Add 20 μL of 50 mM Tris–HCl buffer (pH 8.0) and 5 μL of protease solution (0.04 units/5 μL) to 150 μL serum sample in an Eppendorf tube and incubate it capped overnight at 37 °C (*see* **Notes 2** and **3**).

2. Add 20 μL of a 5 M NaCl solution in water to the sample solution. Heat the mixture in a boiling water bath for 2 min and cool on ice (*see* **Note 4**). Centrifuge at $10,000 \times g$ for 5 min.

3. Transfer supernatant to an Eppendorf tube and add 4.5 volumes of ethanol (95 %) saturated in sodium acetate (*see* **Note 5**).

4. Store all samples at $+4 \pm 2$ °C for 2 h (*see* **Note 5**).

5. Centrifuge at $10,000 \times g$ for 5 min and remove the supernatant (*see* **Note 6**).

6. Dissolve the precipitate in 45 μL of 50 mM Tris–HCl buffer, pH 7.5 (*see* **Note 7**).

7. Add 5 μL of a mixture solution containing chondroitinases ABC and ACII (0.01 units each) for 3 h at 37 °C (*see* **Note 8**).

8. Heat samples in a boiling water bath for 1 min to terminate the reaction and centrifuge at $10,000 \times g$ for 5 min.

9. Transfer supernatant in an Eppendorf tube with cap and store at -20 ± 5 °C until analysis (*see* **Note 9**) (Fig. 3).

3.2 Lens Capsule Pretreatment

1. Place gently lens capsule of approximately 5.5 mm in diameter, in an Eppendorf tube with cap. Add 290 μL of a 50 mM Tris–HCl buffer, pH 8.0, and 10 μL of protease solution (0.08 units). Mix gently and incubate overnight at 37 °C (*see* **Note 10**).

2. Add 40 μL of a 5 M NaCl solution in water (final concentration in sample 0.5 M) to the sample and heat in a boiling water bath for 2–3 min. Cool sample in ice and centrifuge at $10,000 \times g$ for 5 min.

3. Transfer supernatant in an Eppendorf tube and add 4.5 volumes of ethanol solution saturated in sodium acetate. Store sample at 4 °C for 2 h.

4. Centrifuge at $10,000 \times g$ for 5 min.

5. Remove supernatant immediately and place the Eppendorf tube slightly tilted on filter paper at room temperature for 10 min, so as the remaining ethanol is evaporated.

6. Dissolve the precipitate in 95 μL 50 mM Tris–HCl buffer (pH 7.5) and add 5 μL of a chondroitinase solution containing 0.01 units of chondroitinase ABC and 0.01 units of chondroitinase AC II.

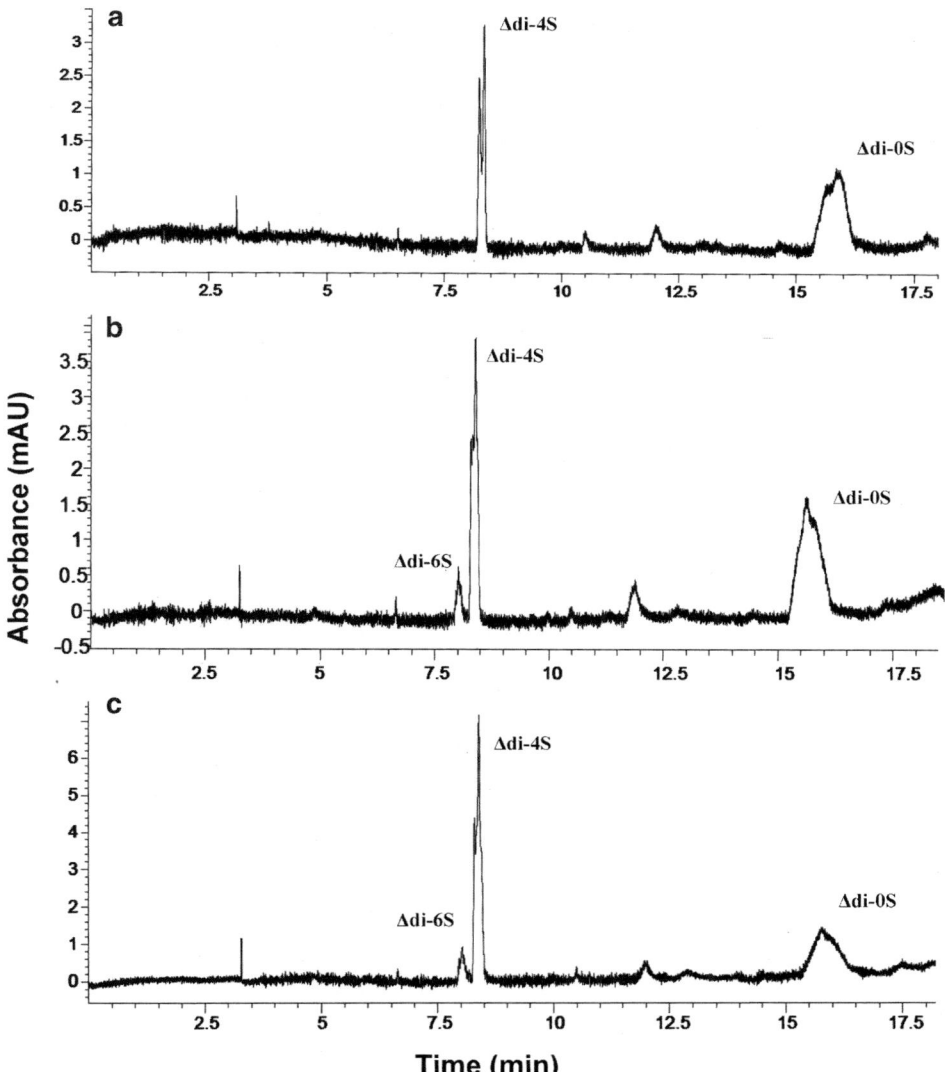

Fig. 3 Representative electropherograms showing the Δdi-4S, Δdi-6S, and Δdi-0S disaccharides in blood serum (**a**), serum spiked with Δ-disaccharide at concentration levels of 5 μg/mL (**b**), and serum spiked with 20 μg/mL avian CS (**c**)

7. Place sample in a boiling water bath for 1 min and after cooling the sample on ice, centrifuge at $10,000 \times g$ for 5 min.

8. Remove carefully the supernatant and transfer it to an Eppendorf tube with cap.

9. Store sample at –20 °C until CE analysis (Fig. 4).

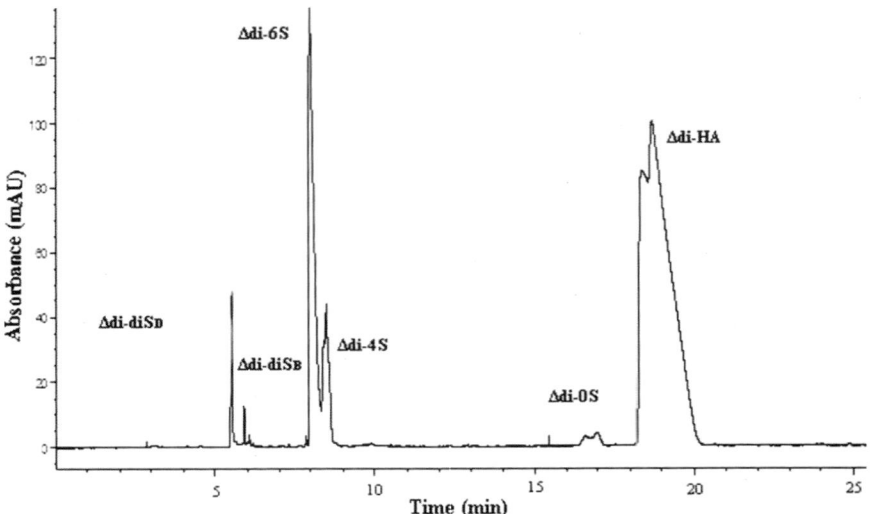

Fig. 4 Typical electropherogram showing the separation of Δdi-diS_D, Δdi-diS_B, Δdi-6S, Δdi-4S, Δdi-0S, and Δdi-HA disaccharides present in tissue of lens capsule from patients with exfoliation syndrome

3.3 Formulations and Preparations

1. Weigh intact CS, raw materials, tablets, and the content of hard capsules and dissolve in water for injection (*see* **Note 11**). Dilute liquid formulations using water for injection.

2. Centrifuge all aqueous CS preparations at $10,000 \times g$ for 5 min.

3. Digest preparations with chondroitinases ABC and AC II at 37 °C overnight in 50 mM Tris–HCl (pH 7.5), using 0.01 units/10 μg of uronic acid (*see* **Note 12**).

4. Boil for 2–3 min in a boiling water bath to terminate the digestion.

5. Centrifuge mixtures at $10,000 \times g$ for 5 min (Fig. 5).

3.4 Derivatization with 2-Aminoacridone

1. Concentrate to dryness a volume of sample or standard solution containing approximately 1–10 nmol or more than 10 nmol of Δ-disaccharides (in case of LIF or UV detection, respectively) in a speed vac microcentrifuge.

2. Add 5 μL of a 0.1 M AMAC solution in glacial acetic acid/ DMSO 3:17 % (v/v) and 5 μL of 1 M NaCNBH_3 in each sample (*see* **Note 13**).

3. Incubate the mixture at 45 °C for 2 h.

4. After evaporation to dryness in a speed vac instrument, reconstitute in 10 μL of 50 % DMSO (Fig. 6).

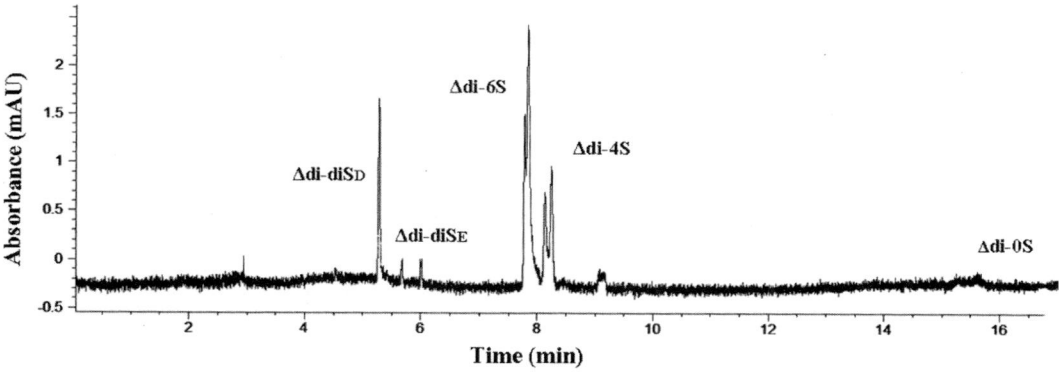

Fig. 5 Representative electropherogram of CSC 90 µg/mL analysis after treatment with ABC and AC II lyases

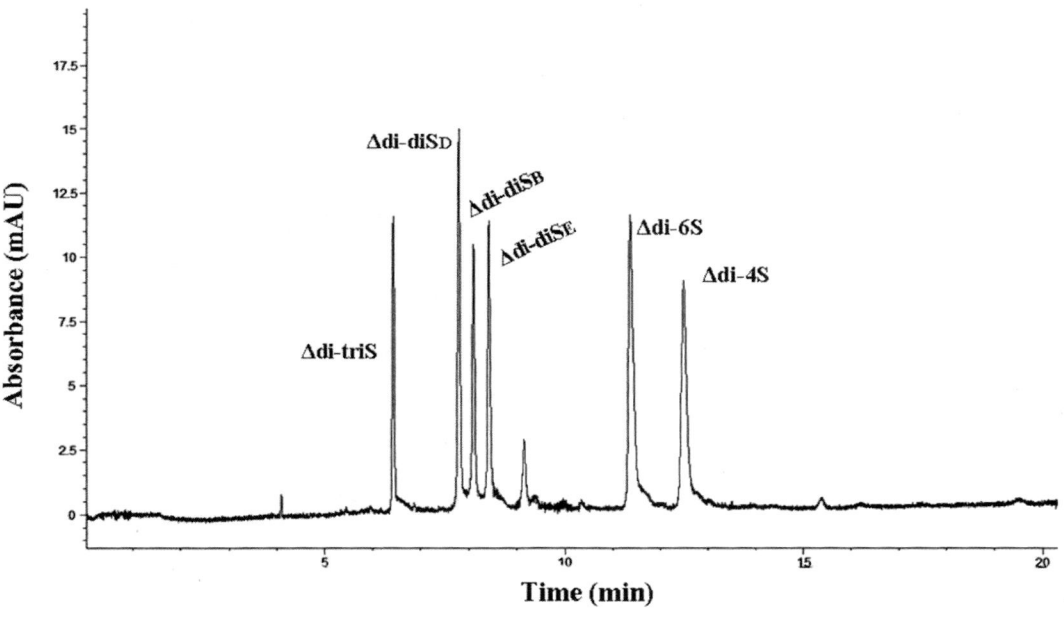

Fig. 6 Representative electropherogram of CS Δ-disaccharides derivatized with AMAC

3.5 Capillary Electrophoresis Analysis

1. Prepare the operating buffer of 50 mM sodium phosphate buffer, pH 3.00, produced by dissolving sodium dihydrogen phosphate in ¾ of the required water, adjusting the pH with a 3 M H_3PO_4 solution and bringing to the final estimated with water so that the concentration is 50 mM (*see* **Note 14**).

2. Filter operating buffer through 0.2 µm membrane filter and degas with agitation in an ultrasonic bath.

3. Before the first use, wash the fused silica capillary with 1.0 M NaOH for 5 min, with H_2O for 5 min, and with the operating buffer for 10 min (*see* **Note 15**).

4. Analysis of CS disaccharides was performed at 30 kV, 25 °C, using reversed polarity in 50 mM phosphate buffer, pH 3.0.

5. Sample introduction is performed hydrodynamically (500 mbar × s) using the pressure injection mode.

6. UV detection is performed at 232 nm for non-derivatized Δ-disaccharides or at 245 nm for AMAC derivatives.

7. After each analysis, the capillary was post-conditioned with H_2O for 2 min.

8. Peak areas are recorded and evaluated using the HP Kayak XA software system HP³ᴰCE ChemStation.

4   Notes

1. Degradation of the core protein of the proteoglycans is necessary in order to liberate the free GAG chains. Many proteases of general specificity can be used. In the past we have used papain, but in the last years we prefer this *Streptomyces* protease because of its effectiveness at a low incubation temperature (37 °C) and of the high purity of the final samples.

2. In order to construct calibration curves which enable the quantitation, calibration samples are necessary. These are prepared by spiking blood serum samples with standard Δ-disaccharides at the concentration levels of 5–100 μg/mL Δ-disaccharides (<10 μL) before **step 8**, and the steps thereafter are omitted. In order to estimate CS recovery, serum samples are spiked with intact CS at concentration in the range of 10–100 μg/mL. Small volumes (<10 μL) of standard CS (0.3 or 3.0 mg/mL) are added in blood serum before **step 1**.

3. This buffer can be stored in aliquots of 1,500 μL in Eppendorf tubes with caps at –20 ± 5 °C for 5 months. Incubation takes place in a water bath.

4. Samples were boiled in the presence of high salt concentration in order to fully dissociate any interacting molecules and precipitate the bulk of proteins in serum.

5. Stir the mixture gently after ethanol addition. When ethanol solution is added, formation of a white precipitate is observed; it certainly depends on the GAG concentration—at low concentrations the solution might just turn cloudy. Alcohol addition in the presence of sodium acetate leads to precipitation of the liberated GAGs. Samples are stored at low temperature in order to facilitate faster precipitation.

6. Place Eppendorf tubes containing the precipitates slightly tilted on a filter paper at room temperature so that ethanol completely evaporates from samples. Avoid leaving overnight, because it is difficult to dissolve the precipitate.

7. The precipitate is hard to dissolve. Leave sample at room temperature for a few minutes and stir regularly until it is dissolved. Then follow **step 8** of the pretreatment procedure.

8. With this procedure the final volume of the injected sample is the 1/3 of the original. This serves as a preconcentration step, but should then be taken into account in all calculations.

9. Transfer supernatant carefully and immediately after centrifugation, because parts of the precipitate might be transferred as well.

10. Lens capsule before treatment is a thin transparent disc. After overnight incubation at 37 °C it is totally dissolved in 50 mM Tris–HCl buffer solution, containing protease.

11. The content of hard capsules is removed from Eppendorf tubes whereas tablets are pulverized. Stock solutions of 10 mg/mL are prepared by dissolving the weighed amounts in water for injection and stored at -20 ± 5 °C.

12. In other tissues, sample uronic acid content is determined by Bitter and Muir procedure with lactone as a standard [22].

13. 0.1 M AMAC solution and 1 M NaCNBH$_3$ must be freshly prepared.

14. This buffer can be stored at $+4 \pm 2$ °C for 2 weeks and is used at CE analysis with a replenishment mode. During the CE sequence analysis, replenishment bottle is filled with buffer (~200 mL). After CE analysis the buffer can be transferred to a measuring flask and stored at $+4 \pm 2$ °C. This step is of utmost importance since it determines the migration time of the analytes and especially of the non-sulfated ones. The 15 mM phosphate buffer also works as well.

15. Before each run, the capillary tube was washed with 0.1 M NaOH for 1 min, with H$_2$O for injection for 2 min, and with the operating buffer for 5 min.

5 Concluding Remark

Assays described above could easily be adapted to different tissues in order to record the galactosaminoglycan disaccharide composition in a low sample volume/amount. When the knowledge on the disaccharide composition is not enough, and it is important to know whether there are more than one GAG types and the molecular size of the GAG chain, then purification using anion chromatography and gel filtration is performed. More information on the structural organization of an isolated galactosaminoglycan is derived by separate treatment of each polysaccharide with chondroitinase AC or B.

References

1. Afratis N, Gialeli C, Nikitovic D, Tsegenidis T, Karousou E, Theocharis AD, Pavao MS, Tzanakakis GN, Karamanos NK (2012) Glycosaminoglycans: key players in cancer cell biology and treatment. FEBS J 279(7):1177–1197.doi:10.1111/j.1742-4658.2012.08529.x

2. Lamari FN, Theocharis AD, Asimakopoulou AP, Malavaki CJ, Karamanos NK (2006) Metabolism and biochemical/physiological roles of chondroitin sulfates: analysis of endogenous and supplemental chondroitin sulfates in blood circulation. Biomed Chromatogr 20(6–7):539–550. doi:10.1002/bmc.669

3. Karamanos NK, Tzanakakis GN (2012) Glycosaminoglycans: from "cellular glue" to novel therapeutical agents. Curr Opin Pharmacol 12(2):220–222. doi:10.1016/j.coph.2011.12.003, S1471-4892(11)00238-4 [pii]

4. Lamari FN, Karamanos NK (2006) Structure of chondroitin sulfate. Adv Pharmacol 53:33–48. doi:10.1016/S1054-3589(05)53003-5, S1054-3589(05)53003-5 [pii]

5. Imanari T, Toyoda H, Yamanashi S, Shinomiya K, Koshiishi I, Oguma T (1992) Study of the measurement of chondroitin sulphates in rabbit plasma and serum. J Chromatogr 574(1):142–145

6. Imanari T, Toida T, Koshiishi I, Toyoda H (1996) High-performance liquid chromatographic analysis of glycosaminoglycan-derived oligosaccharides. J Chromatogr A 720(1–2):275–293

7. Volpi N, Maccari F (2005) Microdetermination of chondroitin sulfate in normal human plasma by fluorophore-assisted carbohydrate electrophoresis (FACE). Clin Chim Acta 356(1–2):125–133. doi:10.1016/j.cccn.2005.01.016, S0009-8981(05)00067-7 [pii]

8. Sakai S, Onose J, Nakamura H, Toyoda H, Toida T, Imanari T, Linhardt RJ (2002) Pretreatment procedure for the microdetermination of chondroitin sulfate in plasma and urine. Anal Biochem 302(2):169–174. doi:10.1006/abio.2001.5545, S0003269701955459 [pii]

9. Schlotzer-Schrehardt U, Dorfler S, Naumann GO (1992) Immunohistochemical localization of basement membrane components in pseudoexfoliation material of the lens capsule. Curr Eye Res 11(4):343–355

10. Deal CL, Moskowitz RW (1999) Nutraceuticals as therapeutic agents in osteoarthritis. The role of glucosamine, chondroitin sulfate, and collagen hydrolysate. Rheum Dis Clin North Am 25(2):379–395

11. Volpi N (2004) The pathobiology of osteoarthritis and the rationale for using the chondroitin sulfate for its treatment. Curr Drug Targets Immune Endocr Metabol Disord 4(2):119–127

12. Verges J, Montell E, Herrero M, Perna C, Cuevas J, Perez M, Moller I (2005) Clinical and histopathological improvement of psoriasis with oral chondroitin sulfate: a serendipitous finding. Dermatol Online J 11(1):31

13. Malavaki CJ, Asimakopoulou AP, Lamari FN, Theocharis AD, Tzanakakis GN, Karamanos NK (2008) Capillary electrophoresis for the quality control of chondroitin sulfates in raw materials and formulations. Anal Biochem 374(1):213–220. doi:10.1016/j.ab.2007.11.006, S0003-2697(07)00726-9 [pii]

14. Karamanos NK, Syrokou A, Vanky P, Nurminen M, Hjerpe A (1994) Determination of 24 variously sulfated galactosaminoglycan- and hyaluronan-derived disaccharides by high-performance liquid chromatography. Anal Biochem 221(1):189–199

15. Karamanos NK, Vanky P, Syrokou A, Hjerpe A (1995) Identity of dermatan and chondroitin sequences in dermatan sulfate chains determined by using fragmentation with chondroitinases and ion-pair high-performance liquid chromatography. Anal Biochem 225(2):220–230

16. Karamanos NK, Hjerpe A (1998) A survey of methodological challenges for glycosaminoglycan/proteoglycan analysis and structural characterization by capillary electrophoresis. Electrophoresis 19(15):2561–2571. doi:10.1002/elps.1150191504

17. Yamagata T, Saito H, Habuchi O, Suzuki S (1968) Purification and properties of bacterial chondroitinases and chondrosulfatases. J Biol Chem 243(7):1523–1535

18. Michelacci YM, Dietrich CP (1975) A comparative study between a chondroitinase B and a chondroitinase AC from Flavobacterium heparinum: isolation of a chondroitinase AC-susceptible dodecasaccharide from chondroitin sulphate B. Biochem J 151(1):121–129

19. Hamai A, Hashimoto N, Mochizuki H, Kato F, Makiguchi Y, Horie K, Suzuki S (1997) Two distinct chondroitin sulfate ABC lyases. An endoeliminase yielding tetrasaccharides and an exoeliminase preferentially acting on

oligosaccharides. J Biol Chem 272(14): 9123–9130

20. Lamari F, Theocharis A, Hjerpe A, Karamanos NK (1999) Ultrasensitive capillary electrophoresis of sulfated disaccharides in chondroitin/dermatan sulfates by laser-induced fluorescence after derivatization with 2-aminoacridone. J Chromatogr B Biomed Sci Appl 730(1):129–133

21. Hjerpe A, Engfeldt B, Tsegenidis T, Antonopoulos CA, Vynios DH, Tsiganos CP (1983) Analysis of the acid polysaccharides from squid cranial cartilage and examination of a novel polysaccharide. Biochim Biophys Acta 757(1):85–91

22. Bitter T, Muir HM (1962) A modified uronic acid carbazole reaction. Anal Biochem 4(4): 330–334

Fast Screening of Glycosaminoglycan Disaccharides by Fluorophore-Assisted Carbohydrate Electrophoresis (FACE): Applications to Biologic Samples and Pharmaceutical Formulations

Evgenia Karousou, Athanasia P. Asimakopoulou, Vassiliki Zafeiropoulou, Manuela Viola, Luca Monti, Antonio Rossi, Alberto Passi, and Nikos Karamanos

Abstract

Hyaluronan (HA), chondroitin sulfate (CS), and heparan sulfate (HS) are glycosaminoglycans (GAGs) with a great importance in biological processes as they participate in functional cell properties, such as migration, adhesion, and proliferation. A perturbation of the quantity and/or the sulfation of GAGs is often associated with pathological conditions. In this chapter, we present valuable and validated protocols for the analysis of HA-, CS-, and HS-derived disaccharides after derivatization with 2-aminoacridone and by using the fluorophore-assisted carbohydrate electrophoresis (FACE). FACE is a well-known technique and a reliable tool for a fast screening of GAGs, as it is possible to analyze 16 samples at the same time with one electrophoretic apparatus. The protocols for the gel preparation are based on the variations of the acrylamide/bisacrylamide and buffer concentrations. Different approaches for the extraction and purification of the disaccharides of various biologic samples and pharmaceutical preparations are also stressed.

Key words Glycosaminoglycans, Disaccharides, Polyacrylamide gel electrophoresis, 2-Aminoacridone, Enzymic treatment

1 Introduction

Glycosaminoglycans (GAGs) are long polyanionic chains that, with the exception of hyaluronan (HA), are carried on the protein core of proteoglycans (PGs) positioned on cell plasma membrane or secreted in the extracellular matrix [1]. GAGs play key roles in many physiological and pathological conditions, such as the stabilization of the fibrillar extracellular matrix, the control of hydration, and the tissue and organism development, as well as in progression of cancer [2–4]. Because GAGs are involved in the induction of cell signaling through their ligation to specific plasma membrane receptors, they are considered an important target for the discovery and improvement of drugs. The establishment and improvement of accurate and sensitive techniques for the examination of the quality of pharmaceutical products/formulations containing GAGs and for the diagnosis by analyzing biologic liquids and tissues are of high importance. In this chapter, we describe improved methods and validated protocols for the analysis of GAGs by using the fast screening technique of fluorophore-assisted carbohydrate electrophoresis/polyacrylamide gel electrophoresis of fluorescent saccharides FACE/PAGEFS [5, 6]. In fact, the anionic feature of the polymeric GAG chains is suitable for their separation in an electric field, with the appropriate techniques. For validation of the results obtained using FACE analysis and for achieving higher sensitivities when very low concentration of GAGs are present in biologic samples, such as in the case of urine, we present also the respective HPLC methodology where the same derivatization protocol with 2-aminocridone is used [7–9].

The constitution of GAGs is characterized by the repetition of a disaccharide unit containing one hexosamine and one uronic acid (UA) or the neutral hexose galactose (Gal) which are linked together with proper glycosidic bonds. There are four subfamilies of GAGs which mainly depend on the monosaccharide composition and the linkage structure: chondroitin/dermatan sulfate (CS/DS), heparin/heparan sulfate (Hep/HS), hyaluronan (HA), and keratan sulfate (KS). The CS/DS disaccharides are composed of a UA linked with a mono- or disulfated N-acetyl-galactosamine (GalNAc). In both DS and HS/Hep chains the UA can be an iduronic acid (IdoA) or glucuronic acid (GlcA). The Hep/HS disaccharides are made up by UA and mono- or disulfated glucosamine (GlcN) substituted with N-acetyl or N-sulfonyl groups (GlcNAc or GlcNS). It is well established that the GAG chains carried on PGs are frequently present as copolymer of HS and Hep or CS and DS disaccharides [10, 11]. The HA is the heteropolymer of the disaccharide composed of GlcA and GlcNAc. HA is neither esterified with sulfate groups nor covalently bound to a core protein. The main variable aspect of HA is not due to its chemical feature but mainly to its length and as a consequence to its physical and mechanical properties [12, 13]. The molecular dimension of the HA polymer has also important implications in tissue behavior, both in physiological and pathological condition.

The first step of the GAG analysis is the enzymatic treatment to obtain a mixture of the disaccharides constituting the chains. Degradation of GAGs into disaccharides is a critical step for the analysis of GAGs since it permits the detection of the total amount of GAGs avoiding any possible interference of the intact GAG with other matrix proteins. The availability of specific glycosidases allows the identification of each GAG type and depending on their substrate target they can also disrupt only partially the polymer, releasing the disaccharide units that contain only monosaccharides with particular substitutions. In this chapter, for the complete digestion of CS/DS chains we treat the CS/DS-containing samples with the chondroitin lyase ABC (chondroitinase ABC) as well as the chondroitin lyases AC (chondroitinase AC) that release only GlcA-containing disaccharides from specific cluster of the chains [14, 15]. The digestion of Hep and HS is performed by using a mix of the three heparinases I, II, and III [8, 16, 17]. Although the HA can be digested by hyaluronidases acting as hydrolases or exoglucosidases, in the methods presented here we use the hyaluronidase SD (from *Streptococcus dysgalactiae*) as it is the most efficient hyaluronidase [18].

The subsequent labeling with a fluorescent dye useful for increasing the sensitivity of the methods is actually a further step for the disaccharide mixture resolution. Derivatizing agents can favor or impede the resolution of the disaccharides depending on their sulfation degree. Here, we use the neutral and hydrophobic 2-aminoacridone (AMAC) with $\lambda_{ex} = 425$ nm and $\lambda_{em} = 520$ nm. The amino group of AMAC reacts with the free reducing aldehyde of the saccharide to form a Schiff base that is stabilized by reduction with cyanoborohydride [6, 19]. The derivatization procedure is relatively simple and fast.

The pretreatment procedure for isolation of GAGs in biologic samples, such as blood plasma and urine, is a key requirement in the development of a rapid, accurate, and sensitive method for the determination of GAG sulfation. These procedures are also presented in detail.

2 Materials

Prepare all solutions using ultrapure water (MilliQ or prepared by purifying deionized water to attain a sensitivity of 18 MΩ cm at 25 °C) and analytical grade reagents.

2.1 Standard HA and CS Disaccharides and Intact Hep/HS

1. 2-Acetamido-2-deoxy-3-*O*-(4-deoxy-a-L-*threo*-hex-4-enopyranosyluronic acid)-D-glucose [Δdi-nonSHA], dissolved at 2 nmol/10 μL in water.

2. 2-Acetamido-2-deoxy-3-*O*-(4-deoxy-2-*O*-sulfo-a-L-*threo*-hex-4-enopyranosyluronic acid)-4-*O*-sulfo-D-galactose [Δdi-di(2,4)S], dissolved at 2 nmol/10 μL in water.

3. 2-Acetamido-2-deoxy-3-*O*-(4-deoxy-a-L-*threo*-hex-4-enopyranosyluronic acid)-4-*O*-sulfo-D-galactose [Δdi-mono4S], dissolved at 2 nmol/10 μL in water.

4. 2-Acetamido-2-deoxy-3-*O*-(4-deoxy-a-L-*threo*-hex-4-enopyranosyluronic acid)-6-*O*-sulfo-D-galactose [Δdi-mono6S], dissolved at 2 nmol/10 μL in water.

5. 2-Acetamido-2-deoxy-3-O-(4-deoxy-2-O-sulfo-a-L-*threo*-hex-4-enopyranosyluronic acid)-D-galactose [Δdi-mono2S], dissolved at 2 nmol/10 μL in water.

6. 2-Acetamido-2-deoxy-3-O-(4-deoxy-a-L-*threo*-hex-4-enopyranosyluronic acid)-D-galactose [Δdi-nonSCS], dissolved at 2 nmol/10 μL in water.

7. Commercial heparin and heparan sulfate at 10 mg/mL in 100 mM ammonium acetate, pH 7.0.

2.2 Enzymes for Extraction and Degradation

1. Protease K (EC 3.4.21.64) for the degradation of proteins and protein cores of proteoglycans.

2. Hyaluronidase SD (EC 3.2.1.35) for digestion of hyaluronan to disaccharides.

3. Chondroitinase ABC (EC 4.2.2.4) and chondroitinase ACII for the digestion of chondroitinase to disaccharides and oligosaccharides.

4. A mix of heparinase I (EC 4.2.2.7), heparinase II (no EC number), and heparinase III (EC 4.2.2.8) for the digestion of heparin and heparin sulfate.

2.3 CS Disaccharide Extraction

2.3.1 Urine Samples

1. Urine specimens from human volunteers or from mouse, frozen at –20 °C.

2. Microcentrifuge to achieve centrifugation at least at $13,000 \times g$.

3. Eppendorf microcentrifuge tubes, 2.0 mL.

4. 10 % w/v cetylpyridinium chloride (CPC) in water.

5. Potassium acetate 10 %.

6. 96 % ethanol.

7. 0.1 M ammonium acetate, pH 7.35.

8. Chondroitinase ABC and chondroitinase ACII, stock solution of 10 units each/mL.

9. Incubator or oven at 37 °C.

10. Lyophilizer or speed-vac.

2.3.2 Serum, Capsule Lens, Aqueous Humor, and Cell Supernatants

1. Serum sample, capsule lens obtained during a surgical procedure in 290 μL Viscoat®, aqueous humor samples of 15–20 μL and 13–14 mL of cell supernatants.

2. Vivaspin filters (3 kDa cutoff).

3. Tris–HCl buffer (50 mM, pH 8.0).

4. Protease solution (0.08 U/10 μL) in 50 mM Tris–HCl buffer (pH 8.0).

5. Eppendorf tube, 1.5 mL.

6. Hood or incubator at 37 °C.

7. Sodium chloride (NaCl) solution (final concentration in sample 0.6 M).

8. Incubator at 99 °C or a boiling water bath.

9. Ice.

10. Ethanol (95 %).

11. Sodium acetate.

12. Chondroitinase ABC and chondroitinase ACII lyases, stock solution of 10 units each/mL.

2.3.3 Pharmaceutical Formulations

1. Stock solutions of 1 mg/mL in 75 μL buffer 50 mM Tris–HCl pH 7.5.

2. Chondroitinase ABC and chondroitinase ACII, stock solution of 2 units each/mL.

3. Incubator at 37 °C.

4. Incubator or water bath at 99 °C.

5. Ice.

6. Eppendorf tubes, 1.5 mL.

7. Lyophilizer.

2.3.4 Hep and HS Extraction

1. Fresh rat kidney and commercial Hep/HS.

2. Incubator at 60 °C.

3. Ammonium acetate buffer (100 mM, pH 7.0).

4. Proteinase K (final concentration 20 U/mL).

5. Incubator or water bath at 99 °C.

6. 96 % ethanol.

7. Centrifuge.

8. Heparinases I, II, and III, stock solution of 1 unit each/mL.

9. Lyophilizer.

2.4 Derivatization Materials

1. Acetic acid/dimethyl-sulfoxide (DMSO) solution in a proportion of 3/17 v/v.

2. 2-Aminoadridone (AMAC) dissolved at 0.1 M or 12.5 mM in glacial acetic acid/DMSO (3:17 v/v).

3. Sodium cyanoborohydrate (NaBH$_3$CN) dissolved at 1 M or 1.25 M in water.

4. Incubator at 37 °C or 45 °C.

2.5 Gel Materials

1. 1.5 M Tris–borate, pH 8.8.

2. 1.5 M Tris–HCl, pH 8.8.

3. 0.5 M Tris–HCl, pH 6.8.

4. A solution containing 25 mM Tris–HCl and 192 mM glycine, pH 8.3.

5. Acrylamide and N,N'-methylenebisacrylamide solutions of T 50 %/C 5 %, T 50 %/C 7.5 %, and T 50 %/C 15 % (*see* **Note 1**).

6. Butanol.

7. Tetramethyl-ethylene-diamide (TEMED) from BDH Chemicals (Poole, England).

8. Ammonium persulfate (10 % w/v in water, freshly prepared or store in –20 °C in aliquots).

9. Glycerol for the samples.

10. Bromophenol blue as a marker.

2.6 Gel Imaging and Product Quantitation

1. UV-light box using a CCD camera and an imaging system.

2. Analysis by QuantityOne system (BioRad) or another similar system of quantification of bands.

2.7 Reversed-Phase HPLC

1. Chromatograph system with fluorophore detector set at $\lambda_{ex} = 442$ nm and $\lambda_{em} = 520$ nm for the identification of AMAC and a gradient elution system using a binary solvent system.

2. Reversed-phase column C-18, 4.6×150 mm or 4.6×200 mm at room temperature.

3. Eluent A: 0.1 M ammonium acetate, pH 7.0, filtered through a 0.45 mm membrane filter.

4. Eluent B: Acetonitrile.

3   Methods

3.1 Extraction and Digestion of GAGs

3.1.1 Analysis of CS Disaccharide in Urine Samples (Fig. 1a)

1. Urine specimens obtained from normal human volunteers should be frozen immediately and stored for short term (1–3 months) at –20 °C. For long-term storage (3–12 months), the –80 °C should be preferred.

2. Thaw samples and centrifuge in Eppendorf tubes at 12,000 rpm ($13,000 \times g$) for 5 min to remove any insoluble material. In 1.7 mL of the urine supernatant, GAGs are precipitated with 34 µL of 10 % w/v CPC (final concentration 0.2 %) at 4 °C overnight.

3. Centrifuge the samples for 15 min at 12,000 rpm at 4 °C, suspend the pellet in 500 µL of 10 % w/v potassium acetate in 96 % ethanol, and immediately centrifuge for 10 min at $13,000 \times g$ at room temperature to remove CPC. Repeat this step two additional times with 10 % potassium acetate and three times with 500 µL of 96 % ethanol to remove potassium acetate.

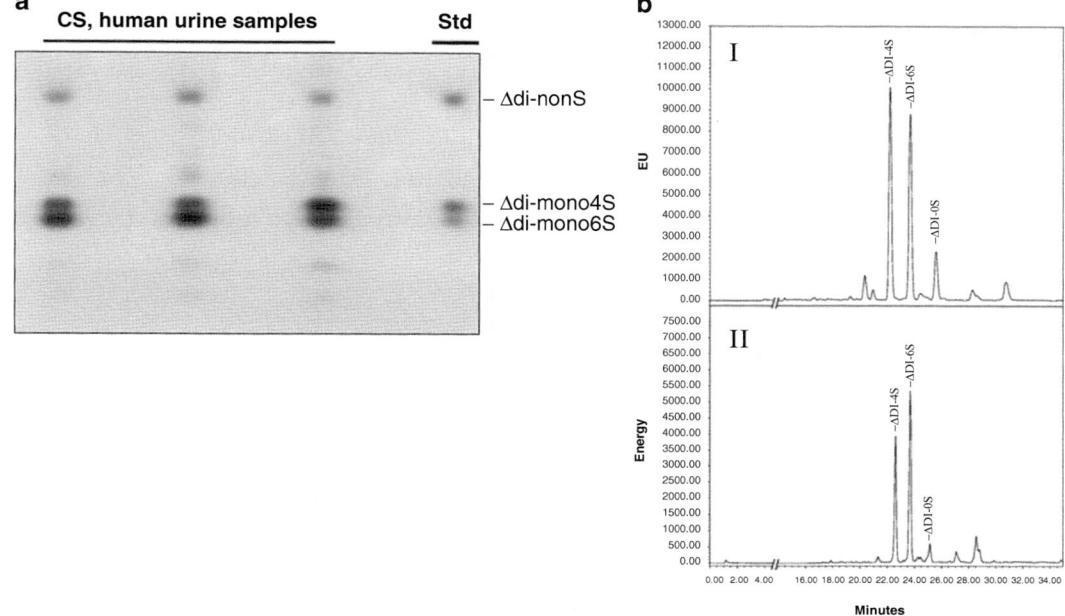

Fig. 1 (**a**) Analysis of AMAC-derived CS disaccharides by FACE/PAGEFS, using the optimized conditions of acrylamide and gel and running buffers specific for CS/DS analysis, as described by Viola et al. [9]. (**b**) RP-HPLC chromatogram of Δ-disaccharides from human (I) and mouse (II) urine. Analyses were performed using a C18 column 4.6 × 200 mm and AMAC derivatives were detected with a fluorescence detector ($\lambda_{ex} = 425$ and $\lambda_{em} = 525$ nm). Mixtures containing at least 10 pmol each were well separated and detected with a PMT gain in the range of 1–10

4. Dry the pellet at RT, and dissolve in 247 μL of 0.1 M ammonium acetate, pH 7.35, adding 3 μL of the stock solution of chondroitinase ABC and chondroitinase ACII (0.03 units each).

5. Incubate at 37 °C overnight.

6. Spin down the samples to remove any insoluble material, freeze at –80 °C, and lyophilize.

7. Derivatize the lyophilized samples, as described in Subheading 3.2.

8. Analyze by FACE, as described in Subheading 3.3.

3.1.2 Analysis of CS in Mouse Urine (Fig. 1)

The following procedure is a developed protocol to purify and digest chondroitin sulfate GAGs starting from the reduced volume of 100 μL of urine samples from young or diseased mice.

1. Centrifuge mouse urine for 5 min at 12,000 rpm ($13,000 \times g$) to clarify the sample; precipitate the supernatant that contains GAGs 2 μL of 10 % CPC (final concentration 0.2 %) at 4 °C overnight.

2. Centrifuge the samples at $13,000 \times g$ for 15 min at 4 °C and wash the pellet three times with 10 % w/v potassium acetate in

96 % ethanol and with 96 % ethanol, respectively, as described for human urine.

3. Dissolve the dried pellet in 197 μL of 0.1 M ammonium acetate, pH 7.35, adding 3 μL of the stock solution of chondroitinase ABC and chondroitinase ACII (0.03 units each).

4. Incubate at 37 °C overnight.

5. Spin down the samples to remove any insoluble material, freeze at −80 °C, and lyophilize.

6. Derivatize the lyophilized samples, as described in Subheading 3.2.

7. Analyze by FACE and HPLC, as described in Subheadings 3.3 and 3.4.

3.1.3 Serum Pretreatment Procedure (Fig. 2)

1. Add 40 μL of 50 mM Tris–HCl buffer (pH 8.0) and 10 μL of protease solution (0.08 U) to 300 μL of serum samples in an Eppendorf tube and incubate at 37 °C overnight.

2. Add 40 μL of NaCl solution to the sample solution (final concentration in each sample 0.5 M NaCl), heat the mixture in a boiling water bath for 1 min, and cool on ice.

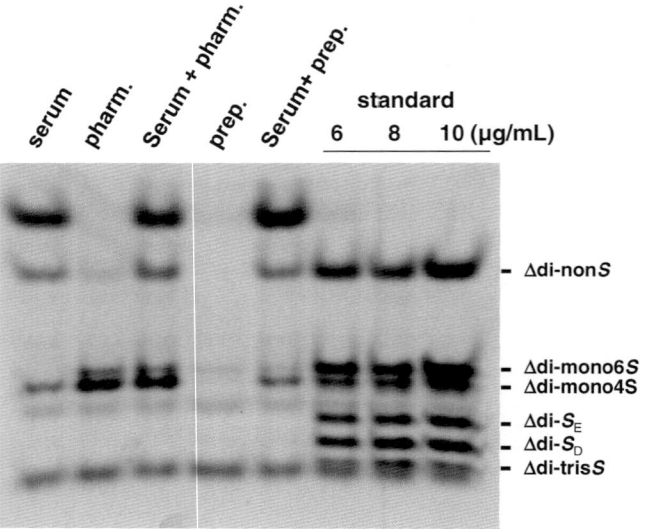

Fig. 2 Analysis of CS disaccharides obtained from serum, pharmaceutical preparations (pharm.), commercial CS preparations (prep.), and mixes of serum with pharmaceutical or commercial CS preparations. The initial volumes of each sample were 50 μL, and the mixes contained 50 μL of serum and 50 μL of each preparation. The obtained disaccharides were derivatized with AMAC and analysis was performed on a 25 % T/7.5 % C polyacrylamide gel with a stacking gel of 5 % T/15 % C, at 400 V, at 4 4 oC for 45 min. The electrophoresis buffer was 0.15 M Tris–borate, pH 8.8; the resolving gel buffer was a mix of 187.5 mM Tris–borate and 187.5 mM Tris–HCl, pH 8.8; and the stacking gel buffer was 0.36 M Tris–HCl, pH 8.8, as described by Karousou et al. [7]

3. After centrifugation at $13,000 \times g$ for 5 min, transfer the supernatant to an Eppendorf tube and add 4.5 volumes of the supernatant of a 96 % ethanol solution saturated in sodium acetate.

4. Keep the samples at 4 °C for 2 h and then centrifuge at $13,000 \times g$ for 5 min.

5. Throw the supernatant and dissolve the precipitate in 95 μL 50 mM Tris–HCl buffer, pH 7.5, adding 5 μL of the stock mixture of chondroitinase ABC and ACII (0.1 units each).

6. Incubate for 3 h at 37 °C.

7. Freeze samples at –80 °C and lyophilize.

8. Derivatize the lyophilized samples, as described in Subheading 3.2.

3.1.4 Capsule Lens Pretreatment Procedure

The capsule lenses are obtained during a surgical procedure in the presence of Viscoat®.

1. Add 290 μL of 50 mM Tris–HCl buffer (pH 8.0) and 10 μL of protease solution (0.08 units/10 μL) in 50 mM Tris–HCl buffer (pH 8.0) to sample (tissue sample—capsule lens) in an Eppendorf tube, and incubate at 37 °C overnight.

2. Follow the same steps as with serum samples and as a final one proceed to derivatization of the lyophilized material.

3.1.5 Aqueous Humor Pretreatment Procedure and Cell Supernatant Pretreatment Procedure

Aqueous humor and cell supernatant samples were treated according to the same protocol used for serum. The initial volume for the aqueous humor is 70–80 μL, whereas for the cell supernatants samples initial volumes of 13–14 mL are concentrated using vivaspin filters (3 kDa cutoff) to volumes up to 300 μL.

3.1.6 Pharmaceutical Formulations of CS (Fig. 3)

1. Stock solutions of 1 mg/mL are dissolved in water.

2. Add 75 μL buffer 50 mM Tris–HCl pH 7.5 and 5 μL of the stock solution of chondroitinase ABC/ACII (0.01 units each) to 20 μL of stock sample.

3. Incubate samples at 37 °C overnight.

4. Heat the mixture in a boiling water bath for 1 min and cool in ice.

5. After centrifugation at $13,000 \times g$ for 5 min, the supernatant is transferred to an Eppendorf tube, frozen, and lyophilized.

3.1.7 Commercial Heparin and Heparan Sulfate and Extraction from Rat Kidney

1. A 300 μL solution of commercial Hep or HS dissolved at 10 mg/mL in 100 mM ammonium acetate, pH 7.0 or fresh rat kidney in 300 μL of 100 mM ammonium acetate buffer, pH 7.0 is digested by 20 U/mL of Proteinase K at 60 °C for 2 h.

2. The enzymatic treatment is terminated by boiling for 5 min.

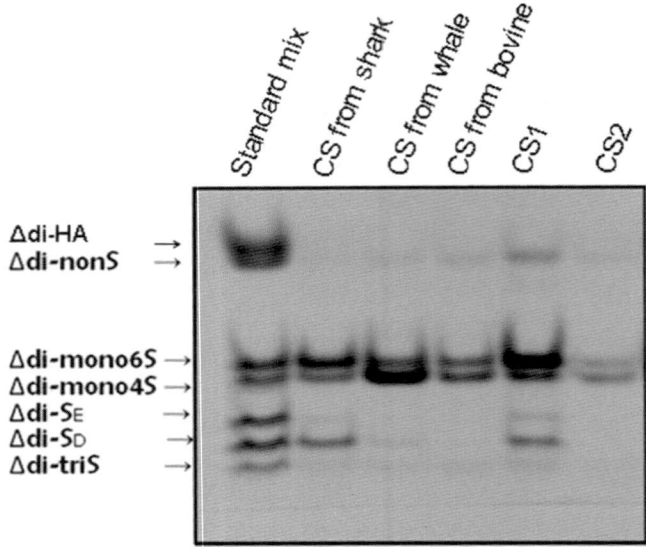

Fig. 3 Analysis of AMAC-derived CS unsaturated disaccharides by FACE/PAGEFS, as described by Karousou et al. [7]. Samples were commercial CS from shark, whale, and bovine; the CS1 and CS2 are two preparations among the various commercially available CS-containing food supplements which show different sulfation pattern

3. Add 1,200 μL of 96 % ethanol (four volumes *per* sample volume) and precipitate the GAGs in the mixture at –20 °C overnight.

4. Centrifuge the ethanol-precipitated GAGs at 11,000 ×*g* at 4 °C for 30 min.

5. Dry the obtained pellets, dissolve in 100 mL of 100 mM ammonium acetate, pH 7.0, containing a mix of heparinases I, II, and III (0.005 units each), and digest at 37 °C for 16–18 h.

6. Add 400 μL of 96 % ethanol (four volumes *per* sample volume) and precipitate the GAGs in the mixture at –20 °C overnight.

7. The unsaturated disaccharides (Δ-disaccharides) from HS are recovered in the supernatant and lyophilized.

3.2 Derivatization of the Disaccharides

For the derivatization of disaccharides with AMAC, use either an aqueous solution containing 5–10 nmol of standard disaccharides or the quantity of the obtained disaccharides from the various biological materials or pharmaceutical preparations (*see* **Note 2**). The volume of these reagents should be equal to the total number of samples plus one more for the pipeting error.

1. Evaporate in a speed-vac 10 nmol of each standard D-disaccharide in a microcentrifuge tube at 11,000 ×*g* at room temperature.

2. Prepare a solution of glacial acetic acid/DMSO (3:17 v/v) (*see* **Note 3**).

3. Dissolve 123.36 mg of AMAC in 40 μL of glacial acetic acid/ DMSO (prepared in **step 2**) to achieve the concentration 12.5 mM, and incubate samples for 10–15 min at room temperature (*see* **Note 4**).

4. Dissolve 2.51 mg of NaBH$_3$CN in 40 μL of water for a final concentration of 1.25 M solution and add to each sample (*see* **Note 5**).

5. Incubate overnight at 37 °C (*see* **Note 6**). Centrifuge the samples at 4,400 × g for 3 min and use immediately or store at –20 °C.

Alternatively the derivatization may be performed as follows:

1. Add 5 μL of 0.1 M AMAC and 5 μL of 1 M NaBH$_3$CN to the lyophilized samples and incubate the mixtures at 45 °C for 4 h.

2. Centrifuge at 4,400 × g for 3 min and dilute the derivatized products with 30 mL of 50 % DMSO v/v.

3. Centrifuge the samples at 4,400 × g for 3 min and use immediately or store at –20 °C.

3.3 FACE or PAGEFS Analysis

The following procedure describes the analysis of HA and CS disaccharides using acrylamide gel electrophoresis. Analysis of various sulfated CS disaccharides or of Hep/HS disaccharides follows the same procedure, but it utilizes different concentrations of acrylamide and gel or running buffer. Therefore, for particular details, *see* also Table 1.

Table 1
Optimized final concentrations of acrylamide, gel buffer, and running buffer solutions for the analysis of AMAC-derived HA, CS, or Hep/HS disaccharides using the FACE/PAGEFS technique

	Acrylamide % T/% C		Gel buffer		
	Resolving gel	Stacking gel	Resolving gel	Stacking gel	Running buffer
HA/CS References [7, 18]	25 % T/7.5 % C	5 % T/15 % C	187.5 mM Tris–borate and 187.5 mM Tris–HCl, pH 8.8	360 mM Tris–HCl, pH 8.8	0.15 M Tris–borate, pH 8.8
CS/DS Reference [9]	33 % T/10 % C	20 % T/15 % C	375 mM Tris–HCl pH 8.8	375 mM Tris–HCl, pH 8.8	25 mM Tris–HCl and 192 mM glycine, pH 8.3.
Hep/HS Reference [8]	30 % T/5 % C	5 % T/5 % C	375 mM Tris–HCl pH 8.8	120 mM Tris–HCl, pH 6.8	25 mM Tris–HCl and 192 mM glycine, pH 8.3.

*3.3.1 Preparation
of the Stock Buffers
and Solutions*

1. For the preparation of stock solution of 50 % T/15 % C use 42.5 g acrylamide and 7.5 g bisacrylamide and dilute them on 80 mL of water. Then mix them in magnetic stirrer until to be dissolved and diluted up to 100 mL. Store at 4 °C. Low temperature creates precipitate, so before the use mix again for 20 min with magentic stirrer at room temperature.

2. For the preparation of 100 mL stock solution of 50 % T/7.5 % C and 50 % T/5 % C use 42.5 g acrylamide/3.75 g and 42.5 g acrylamide/2.5 g bisacrylamide, respectively.

3.3.2 Gel Preparation

1. Wash well and dry with pure ethanol two pairs of plates of 7.2 cm hight, with 0.75 mm spacer and wells of 0.5 cm (*see* **Note 7**).

2. Prepare the resolving gel of 25 % T/7.5 % C in 187.5 mM Tris–borate and 187.5 mM Tris–HCl buffer solution (final concentrations). For two gels, prepare a 10 mL final volume of acrylamide solution. Add in 2.5 mL of water, 5 mL of 50 % T/7.5 % C acrylamide solution, 1.25 mL of 1.5 M Tris–borate, pH 8.8, and 1.25 mL of 1.5 M Tris–HCl, pH 8.8; mix well and degas in a bath of ultrasound for 5 min.

3. Add 5 μL volume of TEMED for the creation of the radicals of the acrylamide and 50 μL volume of 10 % w/v ammonium persulfate for the catalysis of the acrylamide.

4. Mix rapidly the solution and rapidly place it between the glass plates, avoiding air bubbles. Overlay the non-polymerized gel with butanol or ethanol. Polymerization is exothermic and occurs within 20 min. Throw the butanol or ethanol from the upper surface of the resolving gel and rinse with the stacking gel buffer 0.15 M Tris–HCl, pH 8.8, prepared by a 1/10 dilution of the stock solution.

5. Prepare 5 mL of 5 % T/15 % C acrylamide stacking gel in 0.36 M Tris–HCl buffer solution. In 3.3 mL of pure water, add 0.5 mL of 50 % T/15 % C acrylamide stock solution and 1.2 mL of 1.5 M Tris–HCl, pH 8.8. Mix well by pipeting few times.

6. Add 10 μL of TEMED and then 50 μL of 10 % (w/v) ammonium persulfate.

7. Pour immediately the solution on the top of the resolving gel and insert the well-forming comb. The polymerization requires approximately 45 min and the gel has a white color. The gel can be used immediately or stored at 4 °C for 1–2 weeks (*see* **Note 8**).

*3.3.3 Electrophoresis
of AMAC-Disaccharides*

1. Prepare the electrophoresis buffer by diluting 100 mL of 1.5 M Tris–borate, pH 8.8, stock buffer, in 900 mL water, in order to obtain a 0.15 M Tris–borate.

2. Immediately before electrophoresis, rinse thee wells with the electrophoresis buffer.

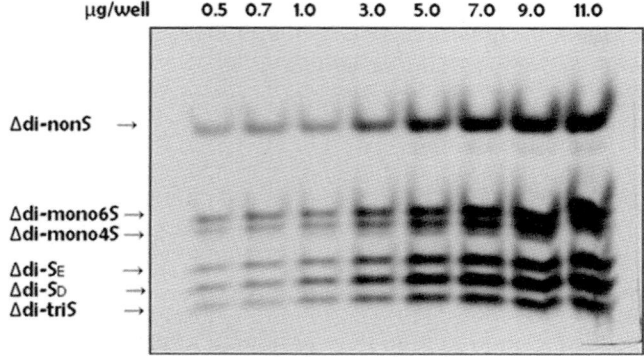

Fig. 4 Analysis of different quantities of the mix of AMAC-derivatized unsaturated CS disaccharides after separation by FACE/PAGEFS. Mixtures containing from 0.5 to 11 μg/well of each AMAC derivative were separated by FACE/PAGEFS, as described in Subheading 3.2. The gel was then imaged using a CCD camera, and the images analyzed using a Gel Doc program

3. Perform a pre-run of the gel at 400 V and 4 °C for 10 min in order to stabilize the current. Load 5 mL of each sample, supplemented with glycerol in a final concentration of 20 % v/v, in each well. Prefer the wells in the middle of the gel. For a ten-well comb, you may use a maximum of eight wells for the samples. Use the wells at the edges of the gel for the mix of standard disaccharides and a marker sample containing 0.02 % (w/v) bromophenol blue, as an indicator for the term of the run. In the remaining empty wells of the gel, load 5 μL of a sample prepared by electrophoresis buffer supplemented with 20 % of glycerol.

4. Perform the electrophoresis at 400 V and 4 °C or in cold room; electrophoresis is terminated when the marker dye is 1.2 mm from the bottom of the gel (running time approximately 35 min).

3.3.4 Gel Imaging and Product Quantification (Fig. 4)

1. Scan gels in a UV-light box using a CCD camera.

2. Identification and quantification of sample bands are performed by comparing their migration and the pixel density with those of standard Δ-disaccharides, running in the same gel.

3.4 HPLC Analysis of AMAC-Derivatives

The following outlines the details and conditions of analysis of AMAC derivatives of Δ-disaccharides using a chromatograph system with a fluorophore detector ($\lambda_{ex} = 442$ nm and $\lambda_{em} = 520$ nm). For details, *see* also Table 2. The mobile phase is composed of two eluents: (a) 0.1 M ammonium acetate, pH 7.0, filtered through a 0.45 mm membrane filter and (b) acetonitrile (*see* **Note 9**). The flow rate is at 1 mL/min and the sample injection volume of all the biological samples is set to 10 μL, which should correspond to 100 pmol of each AMAC-derived standard disaccharide (*see* **Note 10**).

Table 2
HPLC conditions and gradient details for the optimum separation of HA/CS or CS from urine or Hep/HS AMAC-derived unsaturated disaccharides

Column	Reversed-phase C-18, 4.6×150 mm or 4.6×200 mm		
Mobile phase	Eluent A: 0.1 M ammonium acetate, pH 7.0, filtered through a 0.45 mm membrane filter; eluent B: acetonitrile		
Flow rate	1 mL/min		
Sample injection volume	10 µL		
Gradient	HA/CS (references [7, 18]) 100 % eluent A for 5 min 0–30 % B for 30 min 30–50 % B for 5 min	CS for urine samples (reference [9]) 100 % A for 3 min 0–15 % B for 3 min 15–25 % B for 24 min Wash step with 60 % B for 10 min Equilibration in 100 % A for 8 min	Hep/HS (reference [8]) 100 % eluent A for 5 min 0–12.5 % B for 30 min 12.5–50 % B for 10 min Equilibration in 100 % A for 10 min

3.4.1 HPLC for HA and CS Analysis

1. Column details and separation:
   - Column: reversed-phase C-18, 4.6×150 mm.
   - Start with 100 % A for 30 min.
   - Gradient—Isocratic elution with 100 % eluent A for 5 min:
     - Gradient elution from 0 to 30 % eluent B for 30 min.
     - Gradient elution from 30 to 50 % eluent B for 5 min.
     - Wash step for 20 min.

3.4.2 HPLC for CS Extracted from Human or Mouse Urine

1. Column details and separation:
   - Column: Prontosil 120-3-C18-ace EPS 3 µm, 4.6×200 mm.
   - Start with 100 % A for 10 min and then 50 % A/50 % B for 20 min.
   - Gradient—Isocratic elution with 100 % eluent A for 3 min:
     - Gradient elution from 0 to 15 % eluent B for 3 min.
     - Gradient elution from 15 to 25 % eluent B for 24 min.
     - Wash step with 60 % B for 10 min.
     - Equilibration in 100 % eluent A for 8 min.
   - The photomultiplier (PMT) gain of the detector was set to 1 for analysis of human urine and to 5–10 for murine urine.

3.4.3 HPLC for Hep and HS

1. Column details and separation:
   - Column: C-18, 4.6×150 mm, Bischoff 3.0 mm.
   - Start with 100 % A for 10 min and then 50 % A/50 % B for 20 min.

- Gradient—Isocratic elution with 100 % eluent A for 5 min:
  - Gradient elution from 0 to 12.5 % eluent B for 30 min.
  - Gradient elution from 12.5 to 50 % eluent B for 10 min.
  - Equilibration in 100 % eluent A for 10 min.

3.4.4 Qualitative and Quantitative Analysis of HPLC Results

1. Identify the peaks corresponding to AMAC-derived disaccharide comparing the retention time of peaks with the chromatogram of the standard mix and measure the peak area (A) of the standard sample ($A_{standard}$ with concentration $C_{standard}$: 100 pmols/10 μL) and the unknown sample ($A_{unknown}$) by using the dedicated software by integration for quantitative analysis (*see* **Note 11**).

2. Calculate the amount X ($C_{unknown}$: Xpmol/10 μL) of AMAC-derived disaccharides of the samples by using the formula $C_{unknown} = (A_{unknown} \times C_{standard})/A_{standard}$.

4 Notes

1. For the acrylamide stock solutions, % T refers to the total concentration (w/v) of acrylamide monomer (i.e., acrylamide plus methylenebisacrylamide) dissolved in water; % C refers to the concentration (w/w) of cross-linker relative to the total monomer. Stock solutions are stored at 4 °C; before use, agitate the solution at room temperature for 30 min using a magnetic agitator.

2. The AMAC derivatization should take place in well-dried samples; be sure that lyophilization is complete.

3. For the derivatization of X number of samples, you need to prepare $V = (X+1) \times 40$ μL of acetic acid/DMSO (3/17). For example, if $X = 9$, then $V = 400$ μL, which means that you need to mix 60 μL of acetic acid with 340 μL of DMSO.

4. Prepare the AMAC solution fresh each time, protected from light. To dissolve AMAC, mix with vortex vigorously and spin down taking care to avoid use of the possible pellet of insoluble reagent.

5. Prepare the NaBH$_3$CN solution fresh each time. This is a very toxic compound; weight, suspend, and use under fume hood.

6. Wrap the tubes with an aluminum foil to protect against light.

7. Mark with a pen in a distance of 0.5 cm below the bottom of the well-forming comb to indicate the level of the resolving gel solution.

8. Store the prepared gels without taking off the combs in a humid environment; wrap the gels with wet papers and keep them in plastic-sealed bags.

9. Avoiding gas bubbles in the HPLC is critical to remove gasses from HPLC buffers. Typically degassing can be made by vacuum or using the inline degassers of certain chromatography systems.

10. All the samples should be centrifuged at $11,000 \times g$ for 5 min at room temperature and transferred to a new tube, to avoid any insoluble material.

11. Every five runs, we typically re-inject a derivatized standard mix to check the stability of AMAC-derived disaccharide peak retention times. In case of difficult peak identification co-inject sample and standard and compare the peak retention time with a run containing the sample only.

References

1. Iozzo RV (2005) Basement membrane proteoglycans: from cellar to ceiling. Nat Rev Mol Cell Biol 6:646–656

2. Afratis N, Gialeli C, Nikitovic D, Tsegenidis T, Karousou E, Theocharis AD, Pavao MS, Tzanakakis GN, Karamanos NK (2012) Glycosaminoglycans: key players in cancer cell biology and treatment. FEBS J 279:1177–1197

3. Mizumoto S, Sugahara K (2013) Glycosaminoglycans are functional ligands for receptor for advanced glycation end-products in tumors. FEBS J 280:2462–2470

4. Karousou E, Stachtea X, Moretto P, Viola M, Vigetti D, D'Angelo ML, Raio L, Ghezzi F, Pallotti F, De Luca G, Karamanos NK, Passi A (2013) New insights into the pathobiology of Down syndrome—hyaluronan synthase-2 overexpression is regulated by collagen VI alpha2 chain. FEBS J 280:2418–2430

5. Karousou EG, Porta G, De Luca G, Passi A (2004) Analysis of fluorophore-labelled hyaluronan and chondroitin sulfate disaccharides in biological samples. J Pharm Biomed Anal 34:791–795

6. Karousou EG, Viola M, Vigetti D, Genasetti A, Rizzi M, Clerici M, Bartolini B, De Luca G, Passi A (2008) Analysis of glycosaminoglycans by electrophoretic approach. Curr Pharmaceut Anal 4:78–89

7. Karousou EG, Militsopoulou M, Porta G, De Luca G, Hascall VC, Passi A (2004) Polyacrylamide gel electrophoresis of fluorophore-labeled hyaluronan and chondroitin sulfate disaccharides: application to the analysis in cells and tissues. Electrophoresis 25:2919–2925

8. Viola M, Vigetti D, Karousou E, Bartolini B, Genasetti A, Rizzi M, Clerici M, Pallotti F, De Luca G, Passi A (2008) New electrophoretic and chromatographic techniques for analysis of heparin and heparan sulfate. Electrophoresis 29:3168–3174

9. Viola M, Karousou EG, Vigetti D, Genasetti A, Pallotti F, Guidetti GF, Tira E, De Luca G, Passi A (2006) Decorin from different bovine tissues: study of glycosaminoglycan chain by PAGEFS. J Pharm Biomed Anal 41:36–42

10. Sugahara K, Kitagawa H (2002) Heparin and heparan sulfate biosynthesis. IUBMB Life 54:163–175

11. Sugahara K, Mikami T, Uyama T, Mizuguchi S, Nomura K, Kitagawa H (2003) Recent advances in the structural biology of chondroitin sulfate and dermatan sulfate. Curr Opin Struct Biol 13:612–620

12. Heldin P (2003) Importance of hyaluronan biosynthesis and degradation in cell differentiation and tumor formation. Braz J Med Biol Res 36:967–973

13. Vigetti D, Viola M, Karousou E, Rizzi M, Moretto P, Genasetti A, Clerici M, Hascall VC, De Luca G, Passi A (2008) Hyaluronan-CD44-ERK1/2 regulate human aortic smooth muscle cell motility during aging. J Biol Chem 283:4448–4458

14. Karamanos NK, Syrokou A, Vanky P, Nurminen M, Hjerpe A (1994) Determination of 24 variously sulfated galactosaminoglycan- and hyaluronan-derived disaccharides by high-performance liquid chromatography. Anal Biochem 221:189–199

15. Karamanos NK, Vanky P, Syrokou A, Hjerpe A (1995) Identity of dermatan and chondroitin sequences in dermatan sulfate chains determined by using fragmentation with chondroitinases

and ion-pair high-performance liquid chromatography. Anal Biochem 225:220–230

16. Karamanos NK, Vanky P, Tzanakakis GN, Hjerpe A (1996) High performance capillary electrophoresis method to characterize heparin and heparan sulfate disaccharides. Electrophoresis 17:391–395

17. Karamanos NK, Vanky P, Tzanakakis GN, Tsegenidis T, Hjerpe A (1997) Ion-pair high-performance liquid chromatography for determining disaccharide composition in heparin and heparan sulphate. J Chromatogr A 765: 169–179

18. Karousou EG, Viola M, Genasetti A, Vigetti D, Luca GD, Karamanos NK, Passi A (2005) Application of polyacrylamide gel electrophoresis of fluorophore-labeled saccharides for analysis of hyaluronan and chondroitin sulfate in human and animal tissues and cell cultures. Biomed Chromatogr 19:761–765

19. Lamari FN, Kuhn R, Karamanos NK (2003) Derivatization of carbohydrates for chromatographic, electrophoretic and mass spectrometric structure analysis. J Chromatogr B Analyt Technol Biomed Life Sci 793: 15–36

Capillary Electrophoretic Analysis of Isolated Sulfated Polysaccharides to Characterize Pharmaceutical Products

Zachary Shriver and Ram Sasisekharan

Abstract

Capillary electrophoresis is a powerful methodology for quantification and structural characterization of highly anionic polysaccharides. Separation of saccharides under conditions of electrophoretic flow, typically achieved under low pH (Ampofo et al., Anal Biochem 199:249–255, 1991; Rhomberg et al., Proc Natl Acad Sci U S A 95:4176–4181, 1998), is charge-based. Resolution of components is often superior to flow-based techniques, such as liquid chromatography. During the heparin contamination crisis, capillary electrophoresis was one of the key methodologies used to identify whether or not heparin lots were contaminated (Guerrini et al., Nat Biotechnol 26:669–675, 2008). Here we describe a method for isolation of sulfated heparin/heparan sulfate saccharides from urine, their digestion by deployment of heparinase enzymes (Ernst et al., Crit Rev Biochem Mol Biol 30:387–444, 1995), resolution of species through use of orthogonal digestions, and analysis of the resulting disaccharides by capillary electrophoresis.

Key words Heparin, Heparan sulfate, Pentosan polysulfate, Capillary electrophoresis, Heparinase, Alcian Blue

1 Introduction

Sulfated polysaccharides have been shown to be potentially important therapies for a number of diseases, including thrombosis [5], cancer [6], and interstitial cystitis [7]. However, while pharmacodynamic markers are available for some indications (such as anti-Xa activity), a systematic understanding of the pharmacology of complex polysaccharides, particularly highly sulfated sugars, such as heparin, pentosan polysulfate, and heparin derivatives are necessary for understanding tissue distribution, pharmacokinetics, and elimination pathways. In addition, a convergence of–omics-based analysis of biological samples (a bottom-up approach) with a structure-based analysis of protein-glycan interactions (a top-down approach) [8] is enabling the more specific description of these drugs and enabling the creation of optimized therapeutic modalities for a range of diseases, including antivirals [9], anti-inflammatories [10], and antithrombotics [11].

Kuberan Balagurunathan et al. (eds.), *Glycosaminoglycans: Chemistry and Biology*, Methods in Molecular Biology, vol. 1229, DOI 10.1007/978-1-4939-1714-3_15, © Springer Science+Business Media New York 2015

Most recently, techniques such as capillary electrophoresis [12], nuclear magnetic resonance [13], and mass spectrometry [2, 14] have been applied to the structural analysis of anionic complex polysaccharides, including to define structure-activity relationships, for comparison of products, and for the detection of contaminants, such as oversulfated chondroitin sulfate [15] (Fig. 1). Implementation of these methods is critical for understanding the biological roles of complex polysaccharides and elucidating new uses for them in a range of diseases. Additionally, these methods enable development of true assessments of pharmacokinetics since the composition of complex polysaccharides, such as heparin, is known to be altered after in vivo administration because of differential filtration in the kidney [16].

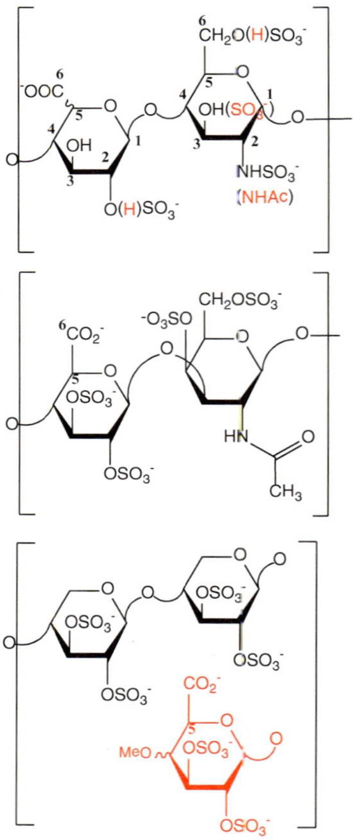

Fig. 1 Disaccharide units of representative complex polysaccharides. (*Top*) Disaccharide unit of heparin/heparan sulfate. Differential *O*-sulfonation and the presence of either *N*-acetylation or sulfation yield significant structural diversity. (*Middle*) Oversulfated chondroitin sulfate (OSCS). OSCS, a contaminant in heparin contains primarily a tetrasulfated disaccharide unit [3]. (*Bottom*) Pentosan polysulfate is a 1 → 4-linked β-D-xylopyranose with laterally substituted 4-methylglucopyranosyluronic acid units glycosidically linked to the 2 position of the main chain at every 10th xylopyranose unit on average

The current method employs isolation of heparin/heparin sulfate from urine, quantification of the amount of material by Alcian Blue, digestion of the material into its components by the use of heparinases, and analysis by capillary zone electrophoresis in an uncoated fused silica capillary. Under conditions of low pH, separation is dictated by analyte electrophoretic mobility almost exclusively [1, 2]. Due to the fact that all saccharide components have a net negative charge, arising from the presence of carboxylate and sulfate moieties, separation is conducted under reverse polarity. In addition, supplementation of the low pH (pH 2.5) buffer with dextran sulfate and tris prevents nonspecific adsorption of analytes, enabling symmetrical peak shapes and accurate quantification.

The net result is that separation under these conditions in capillary electrophoresis is orthogonal to more traditional separation techniques, such as HPLC. In capillary electrophoresis, the most highly sulfated species migrate through the capillary the fastest and are detected first. While many reports have indicated that this technology enables the resolution of the basic saccharide structures in heparin and related materials, there are a number of considerations:

1. Due to the potential dynamic range of the components in the sample (i.e., 50 % of the sample for the principal component compared to 0.1 % for some of the minor components), one needs to take care with regard to the amount of sample that is placed on the capillary to ensure quantitative results.

2. There is little/no resolution of α and β anomers in CE.

3. Accurately separating and quantifying constituent saccharides require the implementation of distinct digests, increasing the potential for the introduction of error. The introduction of an internal standard is essential.

Due to the presence of multiple isomeric disaccharide units within acidic polysaccharides, the resolving power of CE can be amplified through the use of polysaccharide lyases, such as the heparinases. Digestion of heparin-based materials [4], followed by CE analysis, enables exact quantification, compositional analysis, and determination of the presence of contaminants [17], even if highly related to heparin (Fig. 1). Taken together, capillary electrophoresis is an important approach to the identification and analysis of complex, anionic polysaccharides.

2 Materials

2.1 Isolation of HS from Urine

1. 10× Benzonase buffer (500 mM Tris–Cl [8.0], 200 mM NaCl, 20 mM MgCl$_2$) (*see* **Note 1**).

2. Benzonase (25 u/µl).

3. Bovine kidney heparan sulfate (*see* **Note 2**).

4. Proteinase K.

5. 0.1 M CaCl$_2$ (10× dilution in ultrapure water of 1 M solution).

6. Vivapure Q mini H columns (*see* **Note 3**).

7. 50 mM Ammonium acetate: Dissolve 385.4 mg dissolved in 90 ml of ultrapure water. Q.S. to 100 ml.

8. 2 M NaCl: 11.66 g dissolved in 75 ml of ultrapure water. Adjust volume to 100 ml.

9. Vivaspin 5000 (MWCO 3,000).

10. Speedvac.

2.2 Alcian Blue Dot Blot to Quantify Yield

1. Nitrocellulose membrane (0.45 μM pore size; 8 × 11.5 cm).

2. Easy Titer ELIFA System.

3. CTAB solution (1 % (w/v) hexadecyltrimethylammonium bromide): 30 % (v/v) 1-propanol. Add 30 ml 1-propanol to 70 ml of ultrapure water and mix well. To 50 ml of 30 % 1-propanol add 500 mg of CTAB, and mix well with a stir bar.

4. Wash buffer (50 mM Tris–acetate/150 mM NaCl, pH 7.4)— 10 ml of 0.5 M Tris-acetate stock (6.06 g Tris base dissolved in 80 ml of water, pH adjusted to 7.4 with acetic acid and volume adjusted to 100 ml with ultrapure water) + 7.5 ml of 2 M NaCl + 82.5 ml of ultrapure water for a total of 100 ml.

5. 8 M Guanidine chloride: To make an 8 M solution in water, heat the solution to 35 °C for approximately 30 min. Q.S. to 100 ml with ultrapure water.

6. 18 mM Sulfuric acid/0.25 % Triton X-100—Dissolve 2.5 ml of 10 % Triton X-100 in 100 ml of ultrapure water. Add 0.1 ml of concentrated sulfuric acid.

7. Staining solution: 5 ml Alcian Blue stock + 5 ml 8 M guanidine chloride + 90 ml 18 mM sulfuric acid/0.25 % Triton X-100. This solution should be prepared fresh for every use.

2.3 Digestion of Heparin/Heparan Sulfate by Heparinase Cocktail

1. Heparinase I (4.2.2.7), heparinase II (4.2.2.7), and heparinase III (4.2.2.8) from *Pedobacter heparinus* are purified as recombinant enzymes from *E. coli* [18–20] (*see* **Note 4**).

2. 2-*O* sulfatase and D^{4,5} glycuronidase from *Pedobacter heparinus* are purified as recombinant enzymes from *E. coli* [21, 22] (*see* **Note 5**).

3. Enzyme cocktail buffer 10× stock (250 mM NaOAc, 10 mM CaOAc$_2$): 2 g of anhydrous sodium acetate and 158.16 mg calcium acetate (anhydrous) are dissolved in 80 ml water. The pH is adjusted to 7.0 with 1 N acetic acid. Bring up volume to 100 ml with water.

2.4 Analysis by Capillary Electrophoresis

1. Agilent 3D Capillary Electrophoresis or equivalent.
2. Extended light path capillary, 75 μm I.D., 72 cm.
3. Water, CE grade.
4. 1 N Sodium hydroxide, CE grade.
5. Phosphoric acid, CE grade.
6. Tris base.
7. Dextran sulfate.
8. Heparin disaccharide standards (*see* **Note 6**).

3 Methods

3.1 HS Isolation from Urine

An optional pre-step is to heat inactivate the sample through incubation at 52 °C for 1.5 h. This inactivates many viruses, if present. HS isolation should occur in a BSL-2 hood, regardless.

1. Nuclease treatment: To 360 μl of urine sample add 40 μl of 10× benzonase buffer (*see* Subheading 2).
2. Incubate overnight at 37 °C. In parallel set up a control reaction with 50 ng of bovine kidney HS in PBS in place of sample as positive control.
3. Protease treatment: To the above samples add 20 μl proteinase K (20 mg/ml) and 20 μl $CaCl_2$ (0.1 M).
4. Incubate at 55 °C for 2 h (*see* **Note 7**).
5. After digestion, enrichment of HS is completed using ion exchange spin columns. Equilibrate Vivapure Q mini H columns with 200 μl of 50 mM ammonium acetate. Spin at $2,000 \times g$ for 2 min. Load samples (approximately 450 μl) onto the column and spin at $2,000 \times g$ for 2 min. Wash columns with 200 μl 50 mM ammonium acetate, three times, each time spinning at $2,000 \times g$ for 2 min and discarding the flow-through. Elute bound HS with 200 μl of 2 M NaCl (*see* **Note 8**).
6. Desalting: Vivaspin 5000 (MWCO 3,000) filters are rinsed with 500 μl water 3× (each time accompanied by a spin at $15,000 \times g$ 20 min, discarding the flow-through). The crude HS prep is loaded and spun at $13,500 \times g$. Add 500 μl of 50 mM ammonium acetate and spin at $15,000 \times g$ for approximately 35 min; repeat the step five times.
7. Collect the *retentate* after each spin (× 5) in a separate Eppendorf tube; do not pool the ammonium acetate elutions.
8. Speedvac dry each collection (*see* **Note 9**).
9. Resuspend pellet, if present, in 100 μl of 50 mM ammonium acetate.

10. Sample and standard formulation: Prepare purified HS standards (45, 22.5, 11.25, 5.62, 2.81, 1.4, 0.7, 0 µg/ml) in 50 mM ammonium acetate. After isolation of HS material from urine, prepare duplicates of three dilutions for each sample: neat, 1:3, and 1:10. Dilution should be made using ultrapure water. Total volume for each should be 250–300 µl (enough to comfortably transfer 100 µl for sample/dot analysis). Load samples in duplicates. Prepare standards of heparin via serial dilution of stock (4.5 mg/ml): 45, 22.5, 11.25, 5.62, 2.81, 1.4, 0.7, and 0 µg/ml. Dilutions should be made using ultrapure water. Total volume for each should be 250–300 µl (enough to comfortably transfer 100 µl for sample/dot analysis). Load standards in duplicates.

3.2 Dot Blot Assay to Quantify HS

1. Activate nitrocellulose membrane (0.4 µM) filter via addition of CTAB buffer for 5 min. Add enough solution to cover the membrane.

2. Wash 3× with wash buffer (50 mM Tris–acetate/150 mM NaCl, pH 7.4). For each wash, allow the membrane to soak for 15 min.

3. Mount activated membrane in the Easy Titer ELIFA system. Transfer 100 µl sample or std to blot. Apply vacuum until membrane sample completely passes through. Remove membrane from ELIFA system and stain with Alcian Blue solution (see Subheading 2). Allow the membrane to stand for 15–60 min on a shaker. Destain membrane in wash buffer (3× 5–10 min per wash/shaker). Dry membrane, scan, and determine intensity of staining (Fig. 2) (see **Note 10**).

3.3 Heparinase Digestion

1. Enzyme cocktail preparation: Mix Hep I (approximately 800 mU), Hep II (approximately 400 mU), and Hep III (approximately 600 mU) with 10× cocktail buffer (250 mM NaOAc, 10 mM CaOAc$_2$) to a total volume of 50 µl (see **Note 11** and **12**).

2. Digestion: The digestion reaction is set up in a total volume of 60 µl containing the following—3 µl of enzyme cocktail (Table 1), 6 µl of 10× cocktail buffer, 6 µl of 10 mg/ml heparin/heparan sulfate sample, and 45 µl of ultrapure water mixed in a 1.5 ml centrifuge tube. The centrifuge tube should be closed tightly and incubated at 30 °C for 16 h.

3. An aliquot of the digested material is set aside, while the rest of the digest undergoes subsequent enzymatic treatment with either 2-O-S or $\Delta^{4,5}$ glycuronidase (3 µl, Table 1). The setup for this enzymatic modification is as follows—17 µl of the digest from **step 2** is added to each of the three reactions.

4. Incubate each reaction for 6 h at 30 °C.

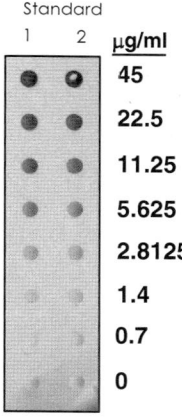

Standard

1   2   µg/ml

45

22.5

11.25

5.625

2.8125

1.4

0.7

0

Fig. 2 Representative dot blot for the quantification of acidic polysaccharides isolated from urine. Standards run simultaneously enable the creation of a standard curve (via densitometry), allowing for a (semi-)quantitative estimate of the amount of material isolated

Table 1
Orthogonal digestions

Digest	2-O Sulfatase	Δ4,5 Glycuronidase	Water
1	–	–	3
2	3	–	–
3	–	3	–

5. Nickel spin column cleanup: Digested material is prepared for CE analysis by passing through a nickel spin column (Qiagen) equilibrated with water. In brief, add 600 µl of water to nickel spin column. Spin at 2,000 rpm (~400 × g) for 2 min. Replace flow-through tube (make sure that no residual water is present). Add sample to the column. Spin at 2,000 rpm for 2 min to elute. His-tagged enzyme should bind to the column and be removed.

3.4 Capillary Electrophoresis

1. All analyses are performed on an Agilent 3D CE system or equivalent equipped with a photodiode array detector.

2. Electrophoretic separations are completed using an uncoated fused-silica capillary with a length of 72 cm (75 µm diameter), at 25 °C.

3. Prior to daily analysis, wash capillary with ultrapure water for 15 min.

4. Ensure removal of residual material on the capillary (to ensure smooth electrophoretic flow) by washing capillary with 0.1 N NaOH. Remove NaOH via washing with water.

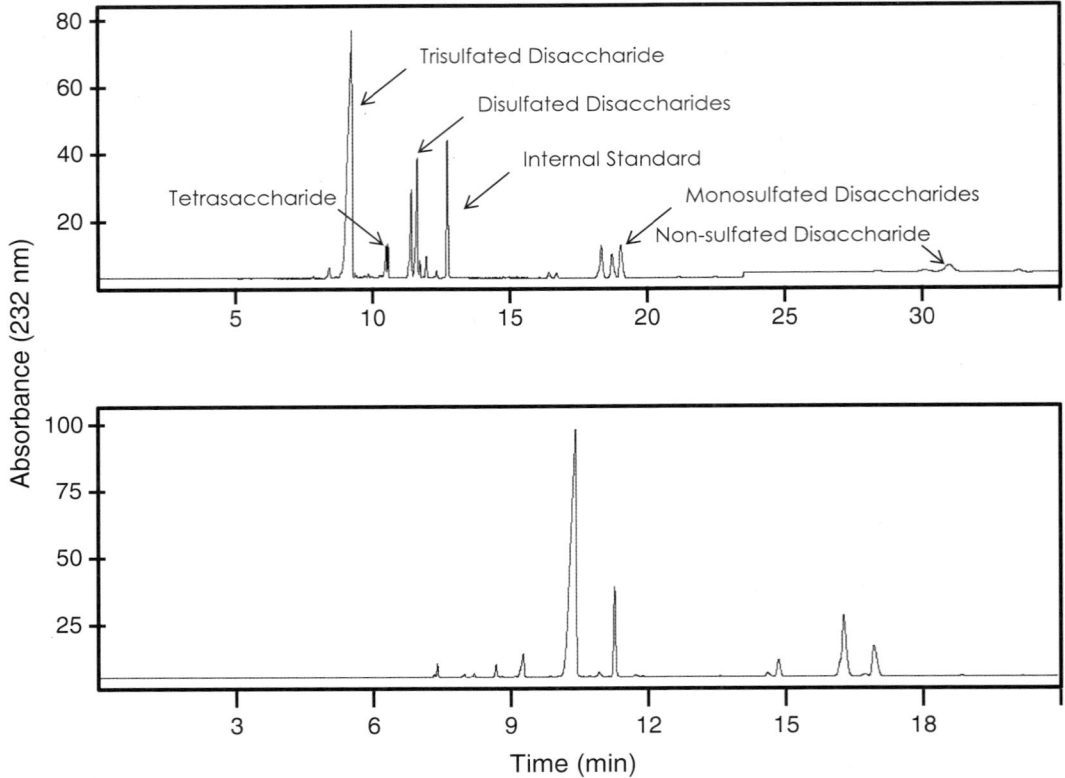

Fig. 3 Representative electropherograms of disaccharide analysis of heparin by capillary electrophoresis. (*Top*) Digestion with the three heparinases yields primarily trisulfated disaccharide with decreasing amounts of other disaccharide products. A tetrasaccharide, containing a 3-O sulfate, is observed between the major disaccharide products. (*Bottom*) Digestion with the 2-O sulfatase moves the major disaccharide peak to a longer migration time, enabling the resolution of minor products

5. Wash capillary with running buffer (50 mM Tris, 10 µM dextran sulfate (10,000 MW) at pH 2.5) for 15 min (*see* **Note 13**).

6. The sample for analysis is injected at a pressure of 20 mbar for 30 s (*see* **Note 14**).

6. Separations are carried out at 30 kV in negative polarity by applying the sample at the cathode.

7. Detection is conducted by monitoring absorbance at 232 nm. Digestion with the heparinases results in the formation of a UV-detectable, $\Delta^{4,5}$ double bond (digest 1, Fig. 3). Treatment with the 2-O sulfatase results in a shift to longer migration times of those species that contain a 2-O sulfate [21] (digest 2, Fig. 3). Treatment with the $\Delta^{4,5}$ glycuronidase results in the disappearance of species that do not contain a 2-O sulfate [22] (digest 3).

8. Detection and quantification of the individual species are completed by injection of standards and through integration of the area under the curve. The general migration time for disaccharide

species is trisulfated < disulfated < monosulfated < non-sulfated. In addition, sulfate positioning affects migration with 2-O-sulfate < N-sulfate < 6-O-sulfate. Finally tetrasaccharide species tend to migrate between the trisulfated disaccharide and disulfated disaccharides (*see* **Note 15**).

4 Notes

1. Prepare all solutions using ultrapure water and analytical grade reagents. Prepare and store all reagents at room temperature, unless otherwise indicated.

2. Standard selected should be as close as possible in composition to that of samples.

3. Column is from Santarus; equivalent materials are available and should be strong anion exchange.

4. The three heparinases have overlapping substrate specificities. Employing all three enzymes results in cleavage of 90- to >99 % of the glycosidic bonds of heparin/heparan sulfate. The enzymes are lyases, meaning cleavage results in the formation of a $D^{4,5}$ double bond that is UV absorbing [$\lambda_{max} = 232$ nm].

5. $\Delta^{4,5}$ glycuronidase cleaves only $\Delta^{4,5}$ double bonds that do not contain a 2-O sulfate.

6. Heparin disaccharide standards are commercially available or can be isolated from heparin or chemically synthesized.

7. Precipitation may be observed in the sample. If this is the case, spinning of the sample to remove precipitate is warranted.

8. An optional step is to complete a second 2 M elution to ensure that all HS material is collected. In our experience, this is not required. Also, addition of urea after **step 4** can help ensure that HS bound to proteins is released.

9. Visible material should be white to off-white.

10. Quantification is by number of sulfate groups. Thus, ensuring as accurate of quantification as possible requires the use of a standard with a similar sulfate-to-carboxylate ratio.

11. Heparinase enzymes used here are His-tagged for easy removal. Enzymes can be left in the reaction mixture but may affect baseline on CE analysis.

12. An example calculation is as follows—

    4.4 μl Hep I (1.2 mg/ml, specific activity 154.6 μmoles/min × mg)

    9.2 μl Hep II (5.6 mg/ml, specific activity 7.8 μmoles/min × mg)

    1.9 μl Hep III (3.5 mg/ml, specific activity 88 μmoles/min × mg)

25 µl of EC buffer

9.5 µl of ultrapure water ($V_t = 50$ µl)].

13. If a large (>10) number of samples are being run, **steps 3–5** should be repeated to ensure removal of residual material adsorbed to the capillary surface. Failure to do so will interfere with the electrophoretic separation.

14. Pressure injection ensures that all digested material, regardless of charge, enters the capillary.

15. Tetrasaccharide species typically migrate between the trisulfated disaccharide and the disulfated disaccharides.

Acknowledgements

This work was funded in part by National Institutes of Health (R37 GM057073-13) to R.S.

References

1. Ampofo SA, Wang HM, Linhardt RJ (1991) Disaccharide compositional analysis of heparin and heparan sulfate using capillary zone electrophoresis. Anal Biochem 199:249–255

2. Rhomberg AJ, Ernst S, Sasisekharan R, Biemann K (1998) Mass spectrometric and capillary electrophoretic investigation of the enzymatic degradation of heparin-like glycosaminoglycans. Proc Natl Acad Sci U S A 95: 4176–4181

3. Guerrini M et al (2008) Oversulfated chondroitin sulfate is a contaminant in heparin associated with adverse clinical events. Nat Biotechnol 26:669–675

4. Ernst S, Langer R, Cooney CL, Sasisekharan R (1995) Enzymatic degradation of glycosaminoglycans. Crit Rev Biochem Mol Biol 30:387–444

5. Petitou M et al (1999) Synthesis of thrombin-inhibiting heparin mimetics without side effects. Nature 398:417–422

6. Zacharski LR, Ornstein DL (1998) Heparin and cancer. Thromb Haemost 80:10–23

7. Anderson VR, Perry CM (2006) Pentosan polysulfate: a review of its use in the relief of bladder pain or discomfort in interstitial cystitis. Drugs 66:821–835

8. Robinson LN et al (2012) Harnessing glycomics technologies: integrating structure with function for glycan characterization. Electrophoresis 33:797–814

9. Shukla D et al (1999) A novel role for 3-O-sulfated heparan sulfate in herpes simplex virus 1 entry. Cell 99:13–22

10. Nelson RM et al (1993) Heparin oligosaccharides bind L- and P-selectin and inhibit acute inflammation. Blood 82:3253–3258

11. Sundaram M et al (2003) Rational design of low-molecular weight heparins with improved in vivo activity. Proc Natl Acad Sci U S A 100:651–656

12. Schirm B, Benend H, Watzig H (2001) Improvements in pentosan polysulfate sodium quality assurance using fingerprint electropherograms. Electrophoresis 22:1150–1162

13. Yates EA et al (1996) 1H and 13C NMR spectral assignments of the major sequences of twelve systematically modified heparin derivatives. Carbohydr Res 294:15–27

14. Kuberan B et al (2002) Analysis of heparan sulfate oligosaccharides with ion pair-reverse phase capillary high performance liquid chromatography-microelectrospray ionization time-of-flight mass spectrometry. J Am Chem Soc 124:8707–8718

15. Nemes P, Hoover WJ, Keire DA (2013) High-throughput differentiation of heparin from other glycosaminoglycans by pyrolysis mass spectrometry. Anal Chem 85:7405–7412

16. Kandrotas RJ (1992) Heparin pharmacokinetics and pharmacodynamics. Clin Pharmacokinet 22:359–374

17. Aich U, Shriver Z, Tharakaraman K, Raman R, Sasisekharan R (2011) Competitive inhibition of heparinase by persulfonated glycosamino-glycans: a tool to detect heparin contamination. Anal Chem 83:7815–7822

18. Godavarti R et al (1996) Heparinase III from Flavobacterium heparinum: cloning and recombinant expression in Escherichia coli. Biochem Biophys Res Commun 225:751–758

19. Ernst S et al (1996) Expression in Escherichia coli, purification and characterization of heparinase I from Flavobacterium heparinum. Biochem J 315(Pt 2):589–597

20. Sasisekharan R, Bulmer M, Moremen KW, Cooney CL, Langer R (1993) Cloning and expression of heparinase I gene from Flavobacterium heparinum. Proc Natl Acad Sci U S A 90:3660–3664

21. Myette JR et al (2003) The heparin/heparan sulfate 2-O-sulfatase from Flavobacterium heparinum. Molecular cloning, recombinant expression, and biochemical characterization. J Biol Chem 278:12157–12166

22. Myette JR et al (2002) Molecular cloning of the heparin/heparan sulfate delta 4,5 unsaturated glycuronidase from Flavobacterium heparinum, its recombinant expression in Escherichia coli, and biochemical determination of its unique substrate specificity. Biochemistry 41:7424–7434

Chapter 16

Methods for Measuring Exchangeable Protons in Glycosaminoglycans

Consuelo N. Beecher and Cynthia K. Larive

Abstract

Recent NMR studies of the exchangeable protons of GAGs in aqueous solution, including those of the amide, sulfamate, and hydroxyl moieties, have demonstrated potential for the detection of intramolecular hydrogen bonds, providing insights into secondary structure preferences. GAG amide protons are observable by NMR over wide pH and temperature ranges; however, specific solution conditions are required to reduce the exchange rate of the sulfamate and hydroxyl protons and allow their detection by NMR. Building on the vast body of knowledge on detection of hydrogen bonds in peptides and proteins, a variety of methods can be used to identify hydrogen bonds in GAGs including temperature coefficient measurements, evaluation of chemical shift differences between oligo- and monosaccharides, and relative exchange rates measured through line shape analysis and EXSY spectra. Emerging strategies to allow direct detection of hydrogen bonds through heteronuclear couplings offer promise for the future. Molecular dynamic simulations are important in this effort both to predict and confirm hydrogen bond donors and acceptors.

Key words Glycosaminoglycan, NMR, Chemical exchange, Hydrogen bond, Temperature coefficient, Chemical shift difference, Activation energy, EXSY

1 Introduction

Glycosaminoglycans (GAGs) are a family of long-chain polysaccharides including chondroitin sulfate (CS), hyaluronan (HA), dermatan sulfate (DS), keratan sulfate (KS), heparin (Hp), and heparan sulfate (HS). The inherent microheterogeneity of GAGs is introduced during biosynthesis as various patterns of O-sulfation, C5 epimerization, and N-acetylation or sulfation. While this microheterogeneity enables the GAGs to mediate a variety of biological processes through their ability to bind to many different proteins it also makes the molecular level characterization of intact GAGs challenging, though advances in mass spectrometry hold promise for GAG sequencing [1, 2]. While disaccharide analysis can be useful for characterization of the overall composition, most information about GAG microstructure is obtained by depolymerization

Kuberan Balagurunathan et al. (eds.), *Glycosaminoglycans: Chemistry and Biology*, Methods in Molecular Biology, vol. 1229, DOI 10.1007/978-1-4939-1714-3_16, © Springer Science+Business Media New York 2015

through enzymatic or chemical means and isolation of individual oligosaccharides using chromatographic methods [3]. Advances in the synthesis of GAG oligosaccharides is emerging as an efficient method to obtain well-defined sequences and promises to eliminate the bottleneck of oligosaccharide isolation from GAG digests [4]. Whether the oligosaccharide is isolated from a GAG digest or produced synthetically, its primary structure must be determined and mass spectrometry and NMR methods are well established for this purpose [5]. In contrast, methods for understanding the solution structure and dynamics of GAG oligosaccharides are still evolving.

Experiments to evaluate solution-state GAG secondary structure have primarily relied on analysis of coupling constants and detection of NOEs [6]. Though GAG oligosaccharides exhibit a great deal of flexibility in solution, elements of primary structure such as iduronic acid residues (IdoA) are well known to limit conformational flexibility. In this chapter we describe strategies to identify new elements of GAG secondary structure by characterization of the exchangeable protons, which can uniquely report on the presence of intramolecular hydrogen bonds. These exchangeable protons are located on the GAG amine (NH_3^+), amide ($NHCOCH_3$), sulfamate ($NHSO_3^-$), and hydroxyl (OH) groups. These NH and OH protons are called "exchangeable" because they exchange chemically with the aqueous solvent. The process of chemical exchange can be described by the following reversible reaction, using an N-sulfoglucosamine ($GlcNHSO_3^-$) sulfamate proton as an example:

$$HOH + GlcNHSO_3^- \rightleftharpoons HOH + GlcNHSO_3^-$$

In an NMR measurement, the exchanging protons, indicated in the chemical reaction above in red, are labeled with different frequencies (chemical shifts) that reflect their local chemical environment before exchange. There are three possible exchange regimes that can be considered in an NMR measurement: slow, fast, and intermediate, defined by the exchange rate (reactions per second) and the difference in frequency (in Hz) between the two chemical environments (or sites) experienced by the exchanging protons [7]. Because of the dependence of the spectrum on the chemical shift difference of the exchanging species, the appearance of the spectrum can be dramatically affected if it is measured at different magnetic field strengths. In slow exchange, the rate of exchange is much less than the frequency difference of the two sites and the protons will be detected in the NMR spectrum as two distinct resonances. In the fast exchange regime, the exchange rate is much greater than the frequency difference and a single exchange-averaged resonance is detected at a chemical shift weighted by the populations of the two species. In a typical solution of a GAG polymer or oligosaccharide, the water protons are present at such a high concentration (55 M) that the chemical shift is essentially that of the water protons. Lying between these two extremes in the intermediate exchange regime, the exchange rate is on the same order of magnitude as the chemical shift difference between the exchanging species and an increase in the resonance line widths (or line broadening) is observed in the NMR spectrum.

If instead of exchange rates we think about the lifetime of the exchanging proton on the time scale of the NMR measurement, these exchange phenomena become easier to understand. In the slow exchange limit, the lifetime of the water and $GlcNHSO_3^-$ protons is long compared to the time required to measure the spectrum. Almost all of the protons initially labeled with the water or $GlcNHSO_3^-$ frequency at the start of acquisition remain in that environment for the duration of the acquisition (typically 1–5 s). If the exchange rate is increased, for example by heating the solution, the lifetimes of the protons in the two sites are reduced and the reaction can move into the intermediate exchange regime. In intermediate exchange, a proton that is in a $GlcNHSO_3^-$ molecule at the start of acquisition may exchange with water while the spectrum is being recorded. When the whole population of protons in the sample is considered, the exchange reaction introduces an uncertainty in the chemical shift of the exchanging protons which is reflected by an increase in the resonance line width. In the fast exchange limit, the protons exchange rapidly between $GlcNHSO_3^-$ and water such that many exchanges occur as the spectrum is being measured, producing a single resonance at a chemical shift that represents a weighted average of the two environments.

Evidence for all three exchange regimes is observed in Fig. 1, which shows the pH dependence of the $GlcNHSO_3^-$ sulfamate protons [8]. At the low and high pH extremes, no resonance is detected for the sulfamate protons. Exchange with water is fast on the NMR time scale and the average chemical shift of the exchanging protons is that of the water resonance because of the high concentration of water in the solution relative to that of the GlcNS monosaccharide. As the pH is raised from 6.67 to 8.54 the resonances of the NH protons appear in Fig. 1. Chemical exchange between the α- and β-anomers is slow on the NMR time scale and a separate resonance is detected for the NH resonance of each anomeric form. The resonance of the α-anomer NH appears at a chemical shift of 5.34 ppm while that of the β-anomer is at 5.79 ppm. It is interesting to note that there is apparently a difference in the exchange rate of the protons in the α- and β-anomers; the resonance of the α-anomer NH can be detected over a wider pH regime than the β-anomer. Between pH 7.49 and 7.97 the exchange rate of the α-anomer NH becomes sufficiently slow that the coupling to the adjacent C2 proton becomes observable. Outside of this pH optimum the resonances broaden as the exchange rate increases.

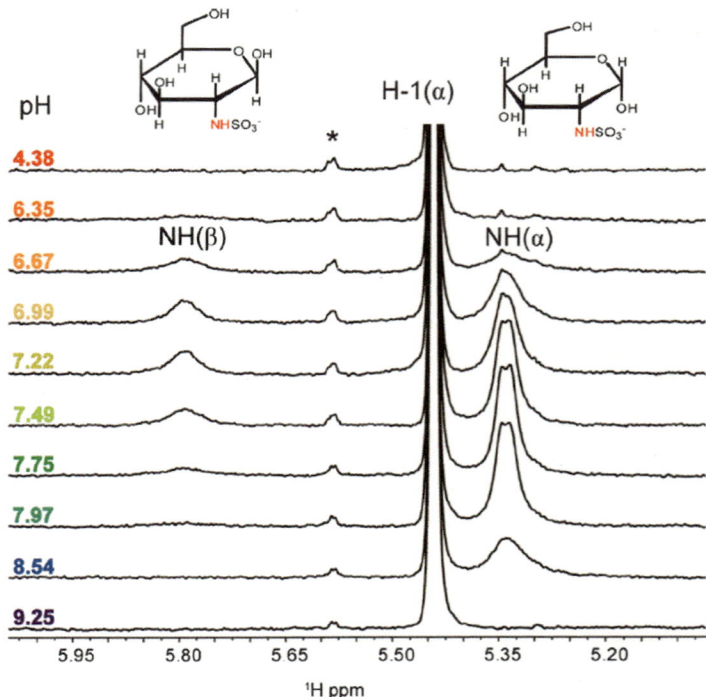

Fig. 1 The line widths of the sulfamate protons of GlcNS vary with solution pH. The exchange rate of the protons is reduced between pH 7.2 and 8.0 allowing them to be observed by NMR [8]. Copyright 2011 American Chemical Society

Various NMR experiments can be used to measure exchange rates, several of which are discussed in this chapter. Comparison of the relative exchange rates of protons in different chemical environments can highlight unusually slow exchange, which is typically attributed to participation by the proton in a hydrogen bond. Experiments to detect hydrogen bonding in GAGs are based on the extensive body of work to identify hydrogen bonds involving the amide protons of proteins [9–11]. While the hydrogen bond donors can be unambiguously identified through NMR measurements, identifying the hydrogen bond acceptor is more difficult; therefore molecular dynamics (MD) simulations are useful both to predict the existence of hydrogen bonds and to identify the hydrogen-bonding pairs. For example, MD simulations by Langeslay et al. for the synthetic Hp pentasaccharide Arixtra (Fig. 2) provided additional evidence for the hydrogen bond involving the central $GlcNHSO_3^-$ NH and identified the hydrogen bond acceptor as the adjacent 3-O-sulfo group [12]. This simulation identified the hydrogen bond in structures in which the 2-O-sulfo-iduronic acid (IdoA2S) residue is in either the ^2S_O (Fig. 2a) or $^1C_4$ (Fig. 2b) conformation.

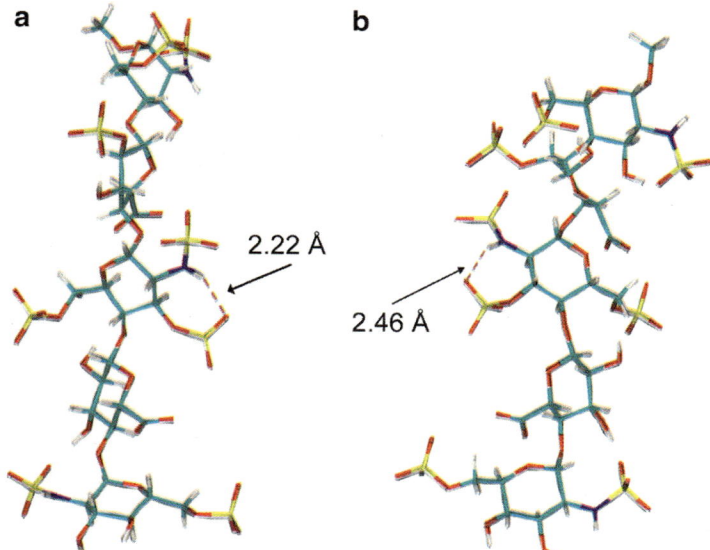

Fig. 2 MD predictions of a sulfamate proton hydrogen bond in Arixtra with the IdoA2S(IV) in the $^2S_0$ (**a**) and $^1C_4$ (**b**) conformations [12]. In each IdoA conformation, the hydrogen bond occurs between the $NHSO_3^-$ and the adjacent 3-O-sulfate group. Published with permission of Oxford University Press

1.1   Observing
Exchangeable Protons
by NMR

To be able to characterize the exchangeable protons of GAGs in solution, they must be detectable by NMR. The different types of exchangeable GAG protons undergo chemical exchange with the aqueous solvent at different rates and therefore different solution conditions may be required for their detection. For example, there are currently no conditions known that allow detection of GAG primary amines in aqueous solution because their exchange is too rapid. As is the case for peptides and protons, the GAG amide protons exchange fairly slowly with water and can be detected over a wide range of pH values, with 90 % H_2O/10 % D_2O and pH 6.0 typical of conditions used for measurement of NMR spectra. Although lower temperatures are often used to slow the exchange rate and sharpen the resonances of exchangeable protons, amide protons, like those of HA, can be detected well above room temperature [13, 14]. Recently, Langeslay and co-workers reported solution conditions that slow the rate of solvent exchange of the Hp and HS sulfamate protons and allow their detection by NMR [8]. The optimum solution pH for detection of the sulfamate protons is between 7.5 and 8.5, with more negatively charged oligosaccharides preferring a more basic pH [8, 12, 15]. The resonances of the sulfamate protons can be detected at room temperature, but become sharper at lower temperatures because the solvent exchange rate is reduced.

The GAG hydroxyl protons exchange more rapidly than either the amide or sulfamate protons. Avoiding buffer anions and removing dissolved CO_2, both of which catalyze the solvent exchange of hydroxyl protons, are essential. Hydroxyl proton resonances are observable by NMR at lower temperatures and are sharpest below the freezing point of water [16–19]. This requires the addition of deuteroacetone or another solvent to reduce the freezing point of water, ideally without affecting its solution structure [16]. With sufficient oligosaccharide or salt content, however, solutions could be supercooled without the addition of aprotic solvent providing solution conditions that are more biologically relevant [19].

An important consideration in the measurement of NMR spectra for exchangeable protons is the selection of a method for solvent suppression. Spectra should be acquired with the minimum amount of D_2O required to maintain a stable lock signal, typically 5 or 10 %. Suppression of the intense H_2O resonance should be accomplished using a method such as WATERGATE [20] or excitation sculpting [21] that avoids transfer of saturation from the water resonance to the signals of the exchangeable protons. Resonance assignments for the exchangeable protons can be accomplished using standard homonuclear two-dimensional NMR experiments such as COSY, TOCSY, NOESY, and ROESY; however care should again be taken in the choice of the solvent suppression method employed to avoid saturation of the resonances of interest.

1.2 Temperature Coefficient Measurements

A well-established and simple method to identify protons involved in hydrogen bonds is through the measurement of the temperature dependence of their chemical shift ($\Delta\delta/\Delta T$), a parameter known as the temperature coefficient [12–14, 17, 18]. Exchangeable protons with smaller temperature coefficients are considered to be protected from solvent exchange, most likely through participation in a hydrogen bond. Langeslay et al. used temperature coefficients to provide evidence for a hydrogen bond involving the internal sulfamate proton of Arixtra (Fig. 3a) [12]. As shown in Fig. 3b, the Arixtra sulfamate NH resonances broaden and shift upfield towards the water resonance as the solution temperature is raised. The sulfamate protons at the reducing (V) and nonreducing (I) ends have similar temperature coefficients, while the NH resonance labeled III is hardly affected by the increase in temperature, suggesting its involvement in a persistent hydrogen bond. Blundell et al. used temperature coefficients of the amide ^1H and ^{15}N resonances of HA to demonstrate that the amide groups participate in transient rather than persistent hydrogen bonds as previously proposed [13, 14]. In addition, Nestor et al. used temperature coefficients as evidence for structural stabilization of HA through the participation of hydroxyl protons in hydrogen bonds [18]. Though a reduced temperature coefficient is consistent with the presence of a hydrogen bond, this parameter alone does not provide definitive evidence for its existence and most studies rely on multiple lines of evidence.

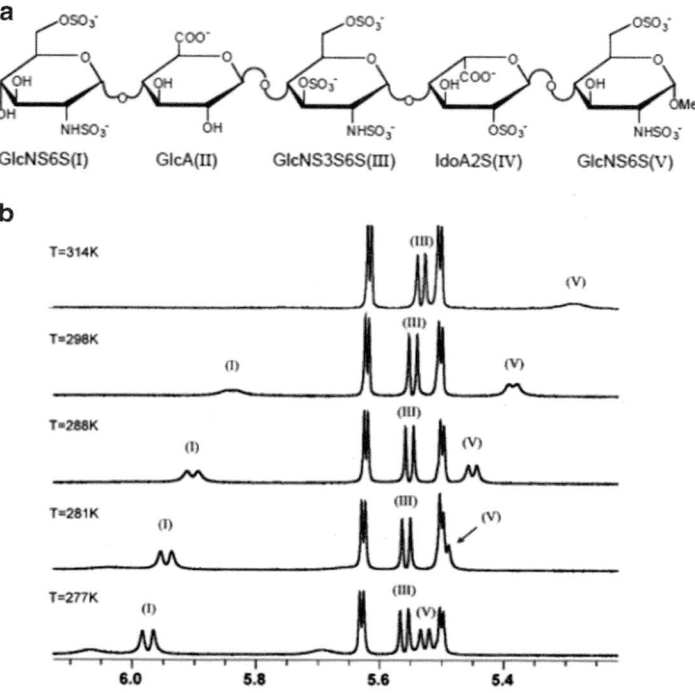

Fig. 3 (a) Arixtra residues are labeled with a Roman numeral, starting with the nonreducing end. **(b)** The sulfamate protons of Arixtra broaden and experience an upfield shift with increasing temperature [12]. At the lowest temperature shown here, the hydroxyl protons are visible as very broad resonances. Published with permission of Oxford University Press

1.3 Chemical Shift Difference Measurements

Comparison of the chemical shifts of the exchangeable protons of a residue within an oligosaccharide to their value in the corresponding monosaccharide, expressed as $\Delta\delta = \delta_{\text{oligo}} - \delta_{\text{mono}}$, can also be an indication of hydrogen bonding. Exchangeable protons experience a downfield shift when they are in close proximity to an electronegative atom. The larger and more negative the value of $\Delta\delta$, the closer the proton is to the electronegative atom and the more likely it is participating in a hydrogen bond [17, 18]. This parameter has been used as evidence for weak hydrogen bonds involving the HA hydroxyl protons [18]. Although $\Delta\delta$ values can provide evidence for hydrogen bonds, they do not provide conclusive evidence of a hydrogen bond because the differences in chemical shift reflect a balance of hydration and proximity to an electronegative atom [17, 22]. Chemical shift differences should be considered together with other experimental results, such as temperature coefficients and exchange rates.

1.4   Line Shape Analysis for the Evaluation of Exchange Rates and Determination of Activation Energies

As shown in Fig. 3b, as the solution temperature is raised, the NH resonances of Arixtra broaden due to an increase in the rate of the solvent exchange reaction. Line shape analysis allows this resonance broadening to be quantified. Figure 4 illustrates this process using least squares fitting in Mathematica to simulate the measured spectrum with a Lorentzian and determine the width at half height of the Arixtra sulfamate protons. The resonance width at half height is proportional to the exchange rate constant, k, and can be used to calculate the energy barrier, $\Delta G^{\ddagger}$, for chemical exchange using the Eyring-Polanyi equation shown below [7, 12, 23, 24]:

$$k = \frac{k_b T}{h} e^{-\frac{\Delta G^{\ddagger}}{RT}}$$

A smaller value of $\Delta G^{\ddagger}$ suggests a low barrier to solvent exchange. A larger value of $\Delta G^{\ddagger}$ reflects the energetic penalty necessary to break a hydrogen bond to allow solvent exchange to occur. The energy difference typically expected for an NH or OH proton involved in a hydrogen bond is roughly 2–3 kcal/mol. Comparison of the $\Delta G^{\ddagger}$ values determined for the three sulfamate protons of Arixtra by Langeslay et al. provided conclusive evidence for the involvement of the internal glucosamine sulfamate proton in a hydrogen bond, the first $NHSO_3^-$ hydrogen bond observed in aqueous solution [12].

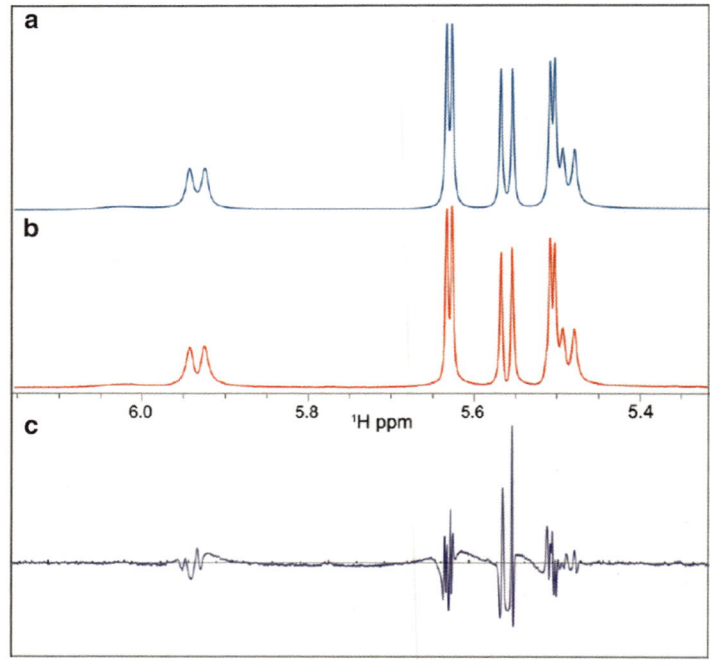

Fig. 4 Example of Arixtra sulfamate proton resonance peak fitting using Mathematica; (**a**) simulated spectrum, (**b**) measured spectrum, and (**c**) residuals shown at a 20 times the intensity of the spectra [12]. Published with permission of Oxford University Press

1.5 EXSY Measurements for Evaluation of Relative Exchange Rates

The faster exchange rates of the hydroxyl protons allow for evaluation of their relative exchange rates through measurement of EXSY spectra [25, 26]. The EXSY spectra measured for the α- and β-glucuronic acid (GlcA) hydroxyl protons as a function of mixing time are shown in Fig. 5. The cross peaks at the chemical shift of the water resonance are observed to increase as the mixing time is incremented. A slower rate of cross peak intensity buildup indicates that the proton is protected from solvent exchange through participation in a hydrogen bond. Sandström and co-workers used this method to identify hydrogen bonds in 3,4-disubstituted methyl α-D-galactopyranosides [17].

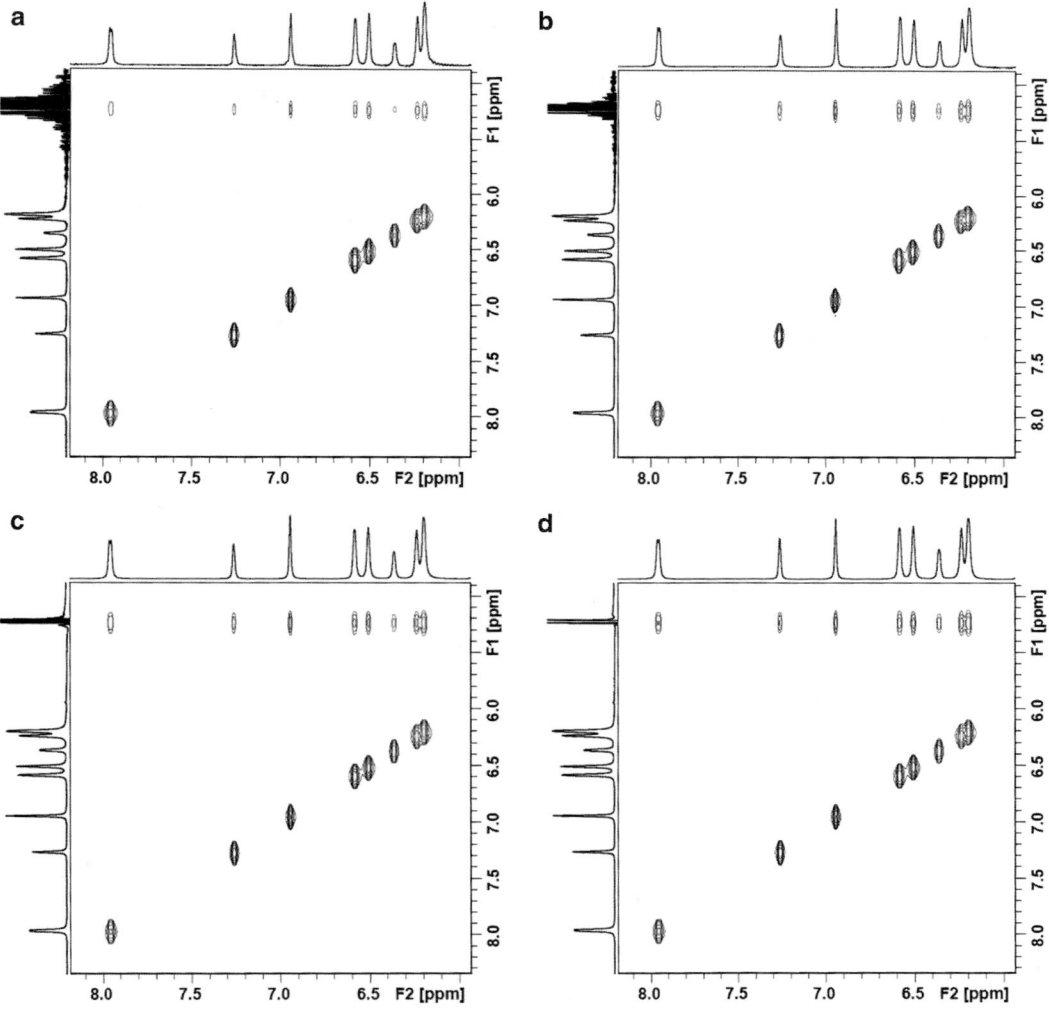

Fig. 5 EXSY plots of the α- and β-GlcA hydroxyl protons at mixing times of (**a**) 6 ms, (**b**) 12 ms, (**c**) 18 ms, and (**d**) 24 ms. As the mixing time increases, the intensity of the exchange cross peaks between the hydroxyl protons and the water resonance increases

1.6 Direct Detection of Hydrogen Bonds Through Heteronuclear Coupling

Battistel et al. identified hydroxyl proton hydrogen bonds in aqueous solutions of sucrose, a topic that has been previously debated in the literature [16, 27], through direct observation of heteronuclear coupling [19]. A fully aqueous solution containing no organic modifiers was used to measure [^{1}H, ^{13}C] HSQC-TOCSY cross peaks (Fig. 6) for 300 mM sucrose at natural abundance

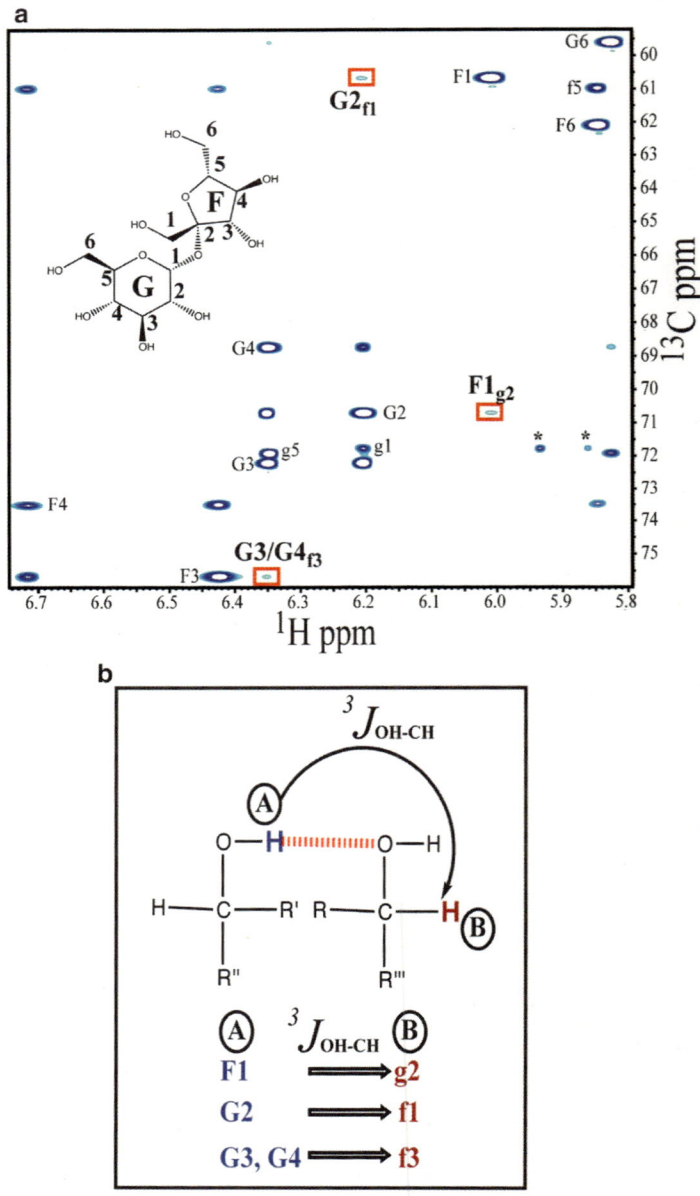

Fig. 6 (**a**) An HSQC-TOCSY spectrum of 300 mM sucrose. The *red squares* indicate those cross peaks corresponding to the hydrogen bonds shown in (**b**) [19]. Copyright 2011 American Chemical Society

levels of ^{13}C, unambiguously identifying the hydrogen bond donors (^{1}H) and acceptors (^{13}C). MD simulations confirmed the presence of intramolecular interring hydrogen bonds between the fructose OH1 and glucose OH2, which act as both donors and acceptors in "flip-flop" hydrogen bonds. MD simulations also confirmed the presence of intermolecular hydrogen bonds between both glucose OH3 (donor) and fructose OH3 (acceptor), which was also detected by cross peaks arising from the ^{2}J$_{OH-OH}$ coupling using selective COSY experiments, demonstrating the potential of this experiment to study interactions between GAGs. Though this work used concentrations (50 mM and above) that are not generally feasible for GAG oligosaccharides, this limitation could be overcome through the use of isotopically labeled compounds [28].

2   Materials

1. HA amide buffer: 5 % (v/v) D_2O, 0.02 % (w/v) NaN_3, pH 6.0 [14].

2. Sulfamate proton buffer: 20 mM sodium phosphate buffer, 2 mM EDTA-*d16*, 1 mM DSS in 90 % H_2O/10 % D_2O at pH 8.2 [8].

3. Hp hydroxyl proton solution: 85 % H_2O/15 % acetone-*d6*, pH 8.1.

4. Methanol standard for temperature measurements.

5. Program for calculating line widths and exchange rates (i.e., Mathematica).

3   Methods

3.1   Preparing Solutions

3.1.1   NH Protons

1. Dissolve oligosaccharides in 600 μL of 90 % H_2O/10 % D_2O in an Eppendorf tube.

2. Measure pH directly in the Eppendorf tube using a microelectrode.

3. Make pH adjustments with small amounts of 0.1 M HCl and 0.1 M NaCl in 90 % H_2O/10 % D_2O to pH 8.2 (*see* **Note 1**).

3.1.2   OH Protons

1. Boil HPLC-grade water to remove any dissolved CO_2. Allow water to cool to room temperature in a sealed container (*see* **Note 2**).

2. All solution preparation must be performed in a glovebox to prevent absorption of CO_2.

3. Adjust pH of the solution to 8.1 using degassed 0.1 M HCl and 0.1 M NaOH (*see* **Note 3**).

4. Add acetone-*d6* to prevent solution from freezing (*see* **Note 4**).

5. Transfer solution to an NMR tube in the glovebox and cap tightly (*see* **Notes 5** and **6**).

3.2 Resonance Assignments

1. Record TOCSY, COSY, and ROESY spectra to assign the carbon-bound proton resonances [29]. This measurement can be performed using a D_2O solution to allow detection of protons that are close in chemical shift to the water resonance.

2. Measure TOCSY and COSY spectra in aqueous solution using a solvent suppression method such as WATERGATE [20] or excitation sculpting [21] that avoids saturation of the exchangeable protons.

3. The TOCSY spectrum will show cross peaks between the exchangeable proton and the other carbon-bound protons within the same ring system.

4. Cross peaks arising from 3-bond couplings between the exchangeable proton and neighboring carbon-bound protons will be observed in the COSY spectrum, confirming the resonance assignment.

3.3 Temperature Coefficient Measurements

1. Cool the NMR sample temperature to 5 °C for amide and sulfamate protons and −15 °C for hydroxyl protons.

2. Measure the initial temperature using the methanol standard. Allow the NMR tube to sit in the probe for at least 10 min for the solution to come to a uniform temperature.

3. Measure the spectrum and calculate the temperature based on the Van Geet equation (*see* **Note 7**) [30].

4. Place the sample in the probe and allow it to come to a uniform temperature for at least 10 min.

5. Measure the spectrum and determine the chemical shift of the exchangeable protons. For doublets, the chemical shift is calculated by measuring the chemical shifts of the two resonances that make up the doublet and calculating the average.

6. Spectra should be recorded for at least seven different temperatures to accurately determine the temperature coefficient value.

3.4 Chemical Shift Difference Measurements

1. Prepare separate solutions of the oligosaccharide and of each corresponding monosaccharide contained within the larger oligosaccharide sequence.

2. Measure the chemical shifts of the exchangeable protons at the temperature of interest.

3. Calculate the chemical shift differences by subtracting the chemical shift of the proton within the monosaccharide from the chemical shift of the corresponding proton within the oligosaccharide ($\Delta\delta = \delta_{oligo} - \delta_{mono}$).

3.5 Line Shape Analysis for the Evaluation of Exchange Rates and Determination of Activation Energies

1. Repeat procedure for temperature coefficient measurements. These calculations require 15–20 different temperature points for accurate fits. Spectra should be recorded with high digital resolution. All spectra must be acquired and processed using exactly the same parameters (*see* **Note 8**).

2. For TopSpin users, type the command "tojdx" in the command line to extract as a JCAMP file for Mathematica. Use the parameters "JCAMP-FIX," "RSPEC," and version "5.0" for each ^1H NMR spectrum.

3. Determine the resonance line widths of the exchangeable protons by nonlinear least-squares fitting of a Lorentzian peak shape to the experimental data using a program such as Mathematica (*see* **Note 9**) [12, 31, 32].

4. Plot these resonance line widths as a function of temperature to the Eyring-Polanyi equation to calculate the activation energy of proton exchange, $\Delta G^{\ddagger}$ using a program such as Mathematica [7, 23, 24].

3.6 EXSY Measurements for Evaluation of Relative Exchange Rates

1. Cool NMR sample temperature to –15 °C to –10 °C to obtain sharper hydroxyl proton resonances.

2. Acquire EXSY spectra using the NOESY pulse sequence with excitation sculpting. Use the largest anticipated mixing time (e.g., 24 ms) to set the receiver gain (RG).

3. Using the same RG for all spectra, acquire EXSY spectra varying the mixing time from 0 to 21 ms.

4. Measure the cross peak volume at the water chemical shift for each of the exchangeable protons at each mixing time.

5. Normalize the cross peak volumes of each resonance to the corresponding diagonal peak volume at zero mixing time.

6. Plot the normalized cross peak volume as a function of mixing time.

7. Calculate the initial buildup exchange rate constant using the modified Bloch equations [33].

4 Notes

1. The pH that minimizes the exchange rate is compound dependent. More highly charged oligosaccharides give sharper peaks in a somewhat more basic pH than neutral compounds.

2. For measurements of the OH resonances, remove the CO_2 from the solvent used to prepare solutions, including the NaOH and HCl solutions used to adjust the solution pH, by boiling and/or bubbling with N_2.

3. Prepare solutions and perform pH adjustments inside a glovebox filled with N_2 to avoid absorption of CO_2.

4. If sufficient compound is available it is possible to supercool the solution without adding aprotic solvents.

5. Parafilm around the NMR tube will help keep CO_2 from absorbing into the solution during the measurements.

6. Between experiments store NMR tubes containing solutions for OH measurements in a jar full of N_2 in the refrigerator to reduce CO_2 absorption and to help avoid evaporation of deuteroacetone.

7. When calibrating the temperature using the methanol standard, only one scan is needed with the lock and sweep turned off. Measure the chemical shift between the two methanol resonances and convert to temperature using a calibration plot [30].

8. Acquiring spectra at a large number temperature (e.g., 15–20) will provide more accurate fits to the Eyring-Polanyi equation.

9. Including ^1J coupling constants for the NH protons measured at low temperature in the spectral simulation produces more accurate fits of the experimental spectra when the resonances become very broad.

Acknowledgements

This work was supported by the National Science Foundation grant CHE-1213845 to C.K.L. C.B. acknowledges support through a UCR GRMP fellowship and the US Department of Education, GAANN Award #P200A120170.

References

1. Ly M, Leach FE, Laremore TN et al (2011) The proteoglycan bikunin has a defined sequence. Nat Chem Biol 7:827–833

2. Jones CJ, Larive CK (2011) Carbohydrates: cracking the glycan sequence code. Nat Chem Biol 7:758–759

3. Jones CJ, Beni S, Limtiaco JFK et al (2011) Heparin characterization: challenges and solutions. Ann Rev Anal Chem 4:439–465

4. Xu YM, Pempe EH, Liu J (2012) Chemoenzymatic synthesis of heparin oligosaccharides with both anti-factor Xa and anti-factor IIa activities. J Biol Chem 287:29054–29061

5. Limtiaco JFK, Beni S, Jones CJ et al (2011) The efficient structure elucidation of minor components in heparin digests using microcoil NMR. Carbohydr Res 346:2244–2254

6. Guerrini M, Guglieri S, Beccati D et al (2006) Conformational transitions induced in heparin octasaccharides by binding with antithrombin III. Biochem J 399:191–198

7. Bain AD (2003) Chemical exchange in NMR. Prog Nucl Magn Reson Spectrosc 43:63–103

8. Langeslay DJ, Beni S, Larive CK (2011) Detection of the H-1 and N-15 NMR resonances of sulfamate groups in aqueous solution: a new tool for heparin and heparan sulfate characterization. Anal Chem 83:8006–8010

9. Ohnishi M, Urry DW (1969) Temperature dependence of amide proton chemical shifts—secondary structures of gramicidin S and valinomycin. Biochem Biophys Res Commun 36:194–202

10. Englander SW, Kallenbach NR (1983) Hydrogen-exchange and structural dynamics of proteins and nucleic-acids. Q Rev Biophys 16:521–655

11. Andersen NH, Neidigh JW, Harris SM et al (1997) Extracting information from the temperature gradients of polypeptide NH chemical shifts. 1. The importance of conformational averaging. J Am Chem Soc 119:8547–8561

12. Langeslay DJ, Young RP, Beni S et al (2012) Sulfamate proton solvent exchange in heparin oligosaccharides: evidence for a persistent hydrogen bond in the antithrombin-binding pentasaccharide Arixtra. Glycobiology 22:1173–1182

13. Blundell CD, Deangelis PL, Almond A (2006) Hyaluronan: the absence of amide-carboxylate hydrogen bonds and the chain conformation in aqueous solution are incompatible with stable secondary and tertiary structure models. Biochem J 396:487–498

14. Blundell CD, Almond A (2007) Temperature dependencies of amide 1H- and 15N-chemical shifts in hyaluronan oligosaccharides. Magn Reson Chem 45:430–433

15. Langeslay DJ, Beni S, Larive CK (2012) A closer look at the nitrogen next door: H-1-N-15 NMR methods for glycosaminoglycan structural characterization. J Magn Reson 216:169–174

16. Adams B, Lerner L (1992) Observation of hydroxyl protons of sucrose in aqueous-solution—no evidence for persistent intramolecular hydrogen-bonds. J Am Chem Soc 114:4827–4829

17. Sandstrom C, Baumann H, Kenne L (1998) NMR spectroscopy of hydroxy protons of 3,4-disubstituted methyl α-D-galactopyranosides in aqueous solution. J Chem Soc Perkin Trans 2:809–815

18. Nestor G, Kenne L, Sandstrom C (2010) Experimental evidence of chemical exchange over the β-(1→3) glycosidic linkage and hydrogen bonding involving hydroxy protons in hyaluronan oligosaccharides by NMR spectroscopy. Org Biomol Chem 8:2795–2802

19. Battistel MD, Pendrill R, Widmalm G et al (2013) Direct evidence for hydrogen bonding in glycans: a combined NMR and molecular dynamics study. J Phys Chem B 117:4860–4869

20. Piotto M, Saudek V, Sklenar V (1992) Gradient-tailored excitation for single-quantum NMR-spectroscopy of aqueous-solutions. J Biomol NMR 2:661–665

21. Hwang TL, Shaka AJ (1995) Water suppression that works—excitation sculpting using arbitrary wave-forms and pulsed-field gradients. J Magn Reson 112:275–279

22. Bekiroglu S, Kenne L, Sandström C (2004) NMR study on the hydroxy protons of the Lewis X and Lewis Y oligosaccharides. Carbohydr Res 339:2465–2468

23. Eyring H (1935) The activated complex in chemical reactions. J Chem Phys 3:107–115

24. Pechukas P (1981) Transition state theory. Annu Rev Phys Chem 32:159–177

25. Jeener J, Meier BH, Bachmann P et al (1979) Investigation of exchange processes by 2-dimensional NMR-spectroscopy. J Chem Phys 71:4546–4553

26. Dobson CM, Lian LY, Redfield C et al (1986) Measurement of hydrogen-exchange rates using 2D NMR-spectroscopy. J Magn Reson 69:201–209

27. Poppe L, Vanhalbeek H (1992) The rigidity of sucrose—just an illusion. J Am Chem Soc 114:1092–1094

28. Battistel MD, Shangold M, Trinh L et al (2012) Evidence for helical structure in a tetramer of α2-8 sialic acid: unveiling a structural antigen. J Am Chem Soc 134:10717–10720

29. Langeslay DJ, Beecher CN, Dinges MM et al (2013) Glycosaminoglycan structural characterization. eMagRes 2:205–214

30. Van Geet AL (1968) Calibration of the methanol and glycol nuclear magnetic resonance thermometers with a static thermistor probe. Anal Chem 40:2227–2229

31. Wolfram S (1991) Mathematica: a system for doing mathematics by computer, 2nd edn. Addison-Wesley Publishing Co., Reading, MA

32. Olsen RA, Liu L, Ghaderi N et al (2003) The amide rotational barriers in picolinamide and nicotinamide: NMR and ab initio studies. J Am Chem Soc 125:10125–10132

33. Chen JH, Mao XA (1998) Measurement of chemical exchange rate constants with solvent protons using radiation damping. J Magn Reson 131:358–361

Chapter 17

Heparan Sulfate Structure: Methods to Study *N*-Sulfation and NDST Action

Anders Dagälv, Anders Lundequist, Beata Filipek-Górniok,
Tabea Dierker, Inger Eriksson, and Lena Kjellén

Abstract

Heparan sulfate proteoglycans are important modulators of cellular processes where the negatively charged polysaccharide chains interact with target proteins. The sulfation pattern of the heparan sulfate chains will determine the proteins that will bind and the affinity of the interactions. The *N*-deacetylase/ *N*-sulfotransferase (NDST) enzymes are of key importance during heparan sulfate biosynthesis when the sulfation pattern is determined. In this chapter, metabolic labeling of heparan sulfate with [^{35}S]sulfate or [^{3}H]glucosamine in cell cultures is described, in addition to characterization of polysaccharide chain length and degree of *N*-sulfation. Methods to measure NDST enzyme activity are also presented.

Key words Heparan sulfate biosynthesis, *N*-deacetylase/*N*-sulfotransferase, *N*-sulfation, ELISA, JM-403

1 Introduction

1.1 Importance of HS N-Sulfation

Heparan sulfate (HS) proteoglycans at the surface of cells and in the extracellular matrix interact with growth factors and morphogens, thereby influencing key processes in embryonic development and homeostasis [1]. The HS glycosaminoglycan chains are synthesized in the Golgi compartment, where *N*-acetylglucosamine (GlcNAc) residues and glucuronic acid residues are added in alternating sequence to a linkage tetrasaccharide attached to specific serine residues in the proteoglycan core protein [2]. At the same time as the chain is growing, sulfotransferases and an epimerase modify the polysaccharide backbone (Fig. 1). The final structure of the heparan sulfate chain depends on the cell responsible for its synthesis and possibly also the core protein to which the heparan sulfate chain is attached [3]. Therefore, HS structure is different in different tissues and varies during embryogenesis and cell differentiation [4]. While polymerization

Kuberan Balagurunathan et al. (eds.), *Glycosaminoglycans: Chemistry and Biology*, Methods in Molecular Biology, vol. 1229, DOI 10.1007/978-1-4939-1714-3_17, © Springer Science+Business Media New York 2015

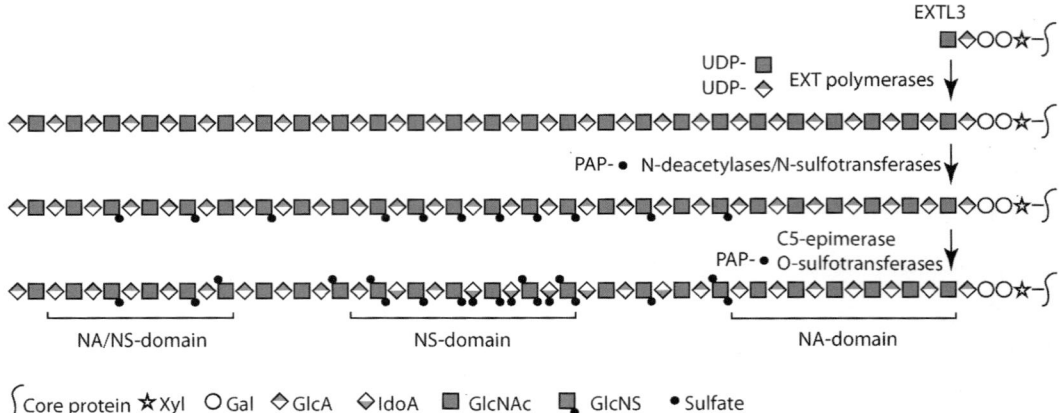

Fig. 1 Heparan sulfate biosynthesis. After synthesis of the tetrasaccharide linkage region and addition of the first GlcNAc residue by EXTL3, NDST enzymes catalyze *N*-deacetylation of GlcNAc residues followed by *N*-sulfation of the generated free amino groups. Subsequent modification by the C5-epimerase and *O*-sulfotransferases occurs mostly in previously *N*-sulfated regions of the polysaccharide. The final product contains NS-domains with a high degree of sulfation, non-sulfated NA-domains, and NA/NS mixed domains

of the HS chains mainly is carried out by the HS-copolymerase EXT1/EXT2, several enzymes are responsible for their modification. The *N*-deacetylase/*N*-sulfotransferases (NDSTs) which remove *N*-acetyl groups and replace them with sulfate groups are responsible for the overall design of the HS chain, since subsequent *O*-sulfation and epimerization of glucuronic to iduronic acid mostly occurs in close proximity to *N*-sulfoglucosamine residues. Four NDST genes are expressed in mammals, where NDST1 and NDST2 seem to be the major ones, present in most tissues and cells. As shown for these two NDSTs [5], they can work in a processive manner creating stretches of fully *N*-sulfated domains (NS-domains) interspersed with *N*-acetylated regions (NA-domains). At the border of NA- and NS-domains, mixed (NA/NS) domains are found. The sulfation pattern will determine the potential of the HS chain to interact with target proteins, the domain structure being of particular importance [6].

In the following sections, methods to study chain length, domain structure, and *N*-sulfation of HS are outlined. These will depend on metabolic labeling with ^{35}S-sulfate or ^{3}H-glucosamine. In addition, assays to determine *N*-deacetylase and *N*-sulfotransferase activities are described.

1.2 Metabolic Labeling of Cells

Any cells kept in culture can be used for the experiments, but their capacity to synthesize glycosaminoglycans will vary greatly; for example fibroblasts will produce large amounts of both HS and CS, primary hepatocytes make nearly only HS, and wild-type HEK 293 cells, often used for transfection studies, synthesize only low levels of either HS or CS. The advantage with [^{35}S]sulfate is that nearly all macromolecules labeled are glycosaminoglycans. This is not the case for [³H]glucosamine which to a large extent also is incorporated into glycoproteins and glycolipids. After purification it is important to check that all recovered radioactive macromolecules are glycosaminoglycans. Sulfate-free medium is sometimes used to increase recovery of labeled products, but should be avoided since the sulfation degree may be altered at low sulfate concentrations [7].

1.3 Isolation and Characterization of Heparan Sulfate

For solubilization of cells, Triton X-100 is often used since it is an uncharged detergent, not interfering with subsequent ion exchange chromatography. If the salt concentration is kept low in the extraction buffer, most of the nucleic acids will reman in the nuclei, remaining associated with the extracellular matrix left at the bottom of the cell culture dish. The glycosaminoglycan present in the Triton X-100 extract will be a mixture of newly synthesized glycosaminoglycans and glycosaminoglycans undergoing degradation. The longer the labeling times, the more of the labeled glycosaminoglycans will be heparanase fragments, present in endosomes and lysosomes. To determine HS chain lengths, it is important to remove the degradation products before the HS chains are released from their core proteins and analyzed. Preparative gel chromatography on, e.g., Superose 6 can be used for this purpose. After recovery of the high-molecular-weight proteoglycan peak, the glycosaminoglycan chains are released from the core proteins by alkali treatment, and their size distribution determined by gel chromatography on Superose 6 after chondroitinase ABC treatment to degrade contaminating CS [8].

The best way to get a good yield as well as pure products from a cell extract is to load the samples on the ion exchange column at pH 7–8, wash the column in the loading buffer, and then lower the pH to 4 for a second wash before elution at a high-salt concentration. A vast majority of proteins initially bound to the ion exchange column will lose their negative charges at pH 4, leaving glycosaminoglycans and nucleic acids bound to the column. To load the column at a low pH is not a good idea since many proteins will then be positively charged and interact with the glycosaminoglycans, preventing them from binding to the resin. For isolation of [³⁵S]glycosaminoglycans, gradient elution of the column is not necessary. However, if the HS produced by the investigated cells is low sulfated, elution with a salt gradient may be needed to separate ³H-labeled hyaluronan from the sulfated [³H]glycosaminoglycans.

To calculate the degree of *N*-sulfation, [³H]HS is treated with nitrous acid at pH 1.5, resulting in cleavage of the polysaccharide chain at *N*-sulfated residues. Since all oligosaccharides generated previously contained one *N*-sulfate group each, the degree of depolymerization is proportional to the *N*-sulfate content.

Gel chromatography on Bio-Gel P-10 takes long time but is the best method to get a good separation of the generated oligosaccharides (*see* ref. 9).

1.4 NDST Enzyme Activity

Since NDSTs are bifunctional, both *N*-deacetylase and *N*-sulfo transferase activities can be measured as well as the combined action of the two. The *N*-deacetylase ELISA assay described relies on the monoclonal JM-403 antibody which recognizes the deacetylated substrate (*see* also ref. 10). We use the *E. coli* K5 polysaccharide as a substrate, which has the same backbone structure as unmodified HS (heparosan). For the *N*-sulfotransferase assay, the substrate is deacetylated K5 polysaccharide which is incubated with the enzyme together with ^{35}S-labeled 3′-phosphoadenosine 5′-phosphosulfate, [^{35}S]PAPS. To measure the combined action of the two enzyme activities, the K5 substrate without prior *N*-deacetylation is used together with [^{35}S]PAPS and the sample.

2 Materials

2.1 Metabolic Labeling and Isolation of ^{3}H- or ^{35}S-Labeled Heparan Sulfate

1. Na$_2$^{35}SO$_4$, 25 mCi/ml (NEX 041H PerkinElmer).

2. D-[6^{3}H] Glucosamine hydrochloride, 1 mCi/ml (NET 190A005MC PerkinElmer).

3. Poly-prep chromatography columns (Biorad).

4. DEAE-Sephacel (GE Healthcare Life Sciences).

5. Superose 6 10/300 GL column (GE Healthcare Life Sciences).

6. PD10 desalting columns (GE Healthcare Life Sciences).

7. Sephadex G50 fine (GE Healthcare Life Sciences).

8. Cell lysis buffer: 50 mM Tris–HCl, pH 7.5, 0.15 M NaCl, 1 % Triton X-100. Add complete protease inhibitor cocktail (Roche) from 100× stock just before use.

9. DEAE wash buffer 1: 50 mM Tris–HCl, pH 7.5, 0.2 M NaCl, 0.1 % Triton X-100.

10. DEAE wash buffer 2: 50 mM acetate buffer, pH 4.0, 0.2 M NaCl, 0.1 % Triton X-100.

11. DEAE elution buffer: 50 mM acetate buffer, pH 4.0, 2 M NaCl, 0.1 % Triton X-100.

12. Superose running buffer: 50 mM Tris–HCl, pH 7.4, 1 M NaCl, 0.1 % Triton X-100.

13. 5× Chondroitinase ABC buffer: 250 mM Tris–HCl, pH 8.0, 150 mM sodium acetate.

14. Chondroitinase ABC: Dissolve 10 units chondroitinase ABC (Amsbio, Abingdon, UK) to a final concentration of 0.01 units/μl in 1 ml of 1× chondroitinase ABC buffer containing 0.1 % bovine serum albumin. Store in small aliquots at ≤−20 °C.

2.2 Characterization of ^{3}H- or ^{35}S-Labeled Heparan Sulfate	1. 0.5 M Ba(NO$_2$)$_2$.
	2. 0.5 M H$_2$SO$_4$.
	3. 2 M NaCO$_3$.
	4. Bio-Gel P-10 (fine grade; Bio-Rad).
	5. 0.2 M NH$_4$HCO$_3$.

2.3   *N*-Deacetylase ELISA Assay

Prepare all the solutions with deionized water and store at room temperature unless otherwise indicated.

1. 96-Well plate: Nunc MaxiSorp (Affymetrix eBioscience).

2. K5 polysaccharide (Amsbio or other supplier): 50 µg/ml in PBS. Store at –20 °C until use.

3. Solubilization buffer: 50 mM Tris–HCl, pH 7.5, 1 % Triton X-100, complete protease inhibitor cocktail, EDTA free (Roche). Add protease inhibitor from the aliquoted stock solution kept at –20 °C prior to use.

4. *N*-Deacetylase reaction mix: 50 mM MES, pH 6.3, 10 mM MnCl$_2$, 1 % Triton X-100, and 25 µg/ml of polybrene. Store a master mix containing 100 mM MES, pH 6.3, 2 % Triton X-100, and 50 µg/ml polybrene at –20 °C. A 0.2 M MnCl$_2$ solution is prepared fresh every time. Mix sample, master mix, MnCl$_2$, and H$_2$O. 100 µl will contain 50 µl master mix, 5 µl 0.2 M MnCl$_2$, and 45 µl sample + H$_2$O.

5. TBS-T: 0.02 M Tris–HCl, pH 7.6, 0.15 M NaCl, and 0.05 % Tween 20. Store at 4 °C.

6. Gelatin solution: Prepare a 1 % solution of gelatin in TBS (0.02 M Tris–HCl, pH 7.6, 0.15 M NaCl) by heating the solution to approximately 90 °C until the gelatin is completely dissolved. Store at –20 °C.

7. JM 403 antibody (5 mg/ml; Amsbio): Dilute the antibody 1:10,000 in TBS-T. Store at 4 °C (short time) or at –20 °C (for storage).

8. IgM-peroxidase (0.5 mg/ml): Dilute the antibody 1:500 in TBS-T. Store at 4 °C.

9. Peroxidase substrate mix: 10 µg tetramethylbenzidine (TMB) in DMSO (store in the dark), 110 mM NaAc, and 0.3 % H$_2$O$_2$. All can be stored at 4 °C.

10. 2 M H$_2$SO$_4$.

11. A spectrophotometer with a plate reader, for example Titertec.

2.4   *N*-Sulfotransferase Assay

1. 5× Reaction buffer: 250 mM HEPES pH 7.4, 50 mM MgCl$_2$, 25 mM CaCl$_2$, 17.5 µM NaF.

2. Freshly prepared 0.2 M MnCl$_2$.

3. 10 % Triton X-100.

4. 0.2 mg/ml deacetylated K5 polysaccharide, obtained after incubation of K5 polysaccharide (Amsbio) in 2 M NaOH (10 mg/ml) at 60 °C for 12–15 h. After neutralization with 6 M HCl, the deacetylated polysaccharide is desalted on a PD-10 column (GE Healthcare Life Sciences) eluted in H_2O.

5. ^{35}S-labeled 3′-phosphoadenosine 5′-phosphosulfate [^{35}S] PAPS.

6. 1.3 % NaAc in 95 % ethanol.

7. 10 mg/ml heparin.

8. Sephadex G-25 Superfine (GE Healthcare) in 0.2 M NH_4HCO_3 (*see* **Note 5**).

9. Glass fiber filters (Whatman Grade GF/F).

2.5 Combined N-Deacetylase/ N-Sulfotransferase Assay

Same materials as for the sulfotransferase assay except:

1. 5× Combined reaction buffer: 250 mM MES pH 6.3, 50 mM $MgCl_2$, 25 mM $CaCl_2$, 17.5 µM NaF. (Replaces 5× reaction buffer in the sulfotransferase assay.)

2. 0.2 mg/ml K5 polysaccharide. (Replaces deacetylated K5 polysaccharide in the sulfotransferase assay.)

3   Methods

3.1 Metabolic Labeling and Isolation of ^{3}H- or ^{35}S-Labeled Heparan Sulfate

1. Grow cells in T75 flasks in 15 ml of their favorite medium (*see* **Note 1**).

2. Change to 5 ml fresh medium containing 50–100 µCi $Na_2^{35}SO_4$ or [6-^{3}H] glucosamine–HCl and incubate for 2–24 h (*see* **Note 2**).

3. Carefully remove the radioactive medium by manual pipetting and wash twice with PBS (*see* **Note 3**).

4. Solubilize the cells by adding 4 ml of ice-cold lysis buffer followed by incubation at 4 °C for 30–60 min on a rocking table. Collect the lysate in a centrifuge tube and spin for 5 min at $13,000 \times g$ at 4 °C. Collect the supernatant (*see* **Note 4**).

5. Prepare a small column with 0.3 ml DEAE-Sephacel. Wash the resin with at least 2 ml (6× column volume) of DEAE wash buffer 1.

6. Load the supernatant on the column. Wash with 3 ml (10× column volume) DEAE wash buffer 1 and then with 3 ml (10× column volume) DEAE wash buffer 2.

7. Elute the column with 6 × 0.3 ml DEAE elution buffer. Analyze a small aliquot of each fraction (5–10 µl) in a β-counter. Pool fractions containing radioactivity.

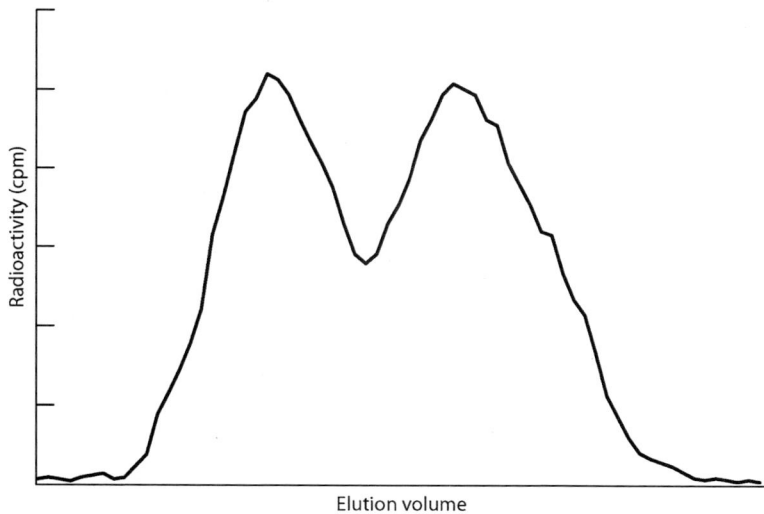

Fig. 2 Superose 6 chromatography. In this step, intact proteoglycans eluting in the first peak are separated from heparanase degradation products present in the second peak

8. Separate labeled proteoglycans from lysosomal/endosomal degradation products on a Superose 6 column. Run the column at a flow rate of 15 ml/h. Collect 50 fractions of 0.5 ml. Analyze a small aliquot of each fraction (10–50 µl) in a β-counter. Pool fractions eluting in the first peak containing the proteoglycans (Fig. 2).

9. Release the HS and CS chains from their core proteins by alkali treatment: Calculate the amount of 10 M NaOH needed to achieve a final concentration of 0.5 M:
$$\frac{\text{Volume of the sample in ml} \times 0.5}{9.5} = \text{Volume NaOH in ml}$$ and add that to the sample. Incubate at 4 °C overnight. Neutralize with the same volume 10 M HCl.

10. Desalt the sample on a PD10 column eluted in H_2O (*see* **Note 5**).

11. Chondroitinase ABC digestion followed by gel chromatography is used to remove chondroitin sulfate from the sample. Mix 20 µl of 5× chondroitinase ABC buffer, 2 µl of chondroitinase ABC, and desalted sample and add H_2O to achieve a final concentration of 100 µl. Incubate for 18 h at 37 °C. Apply the digested sample to a Sephadex G-50 fine column (0.5×90 cm), eluted in 0.2 M NaCl at a flow rate of 0.05 ml/min. Collect fractions of 0.5 ml and analyze a small aliquot of each fraction (10–50 µl) in a scintillation counter. Pool the first peak eluting in the void volume which contains the purified HS chains (Fig. 3).

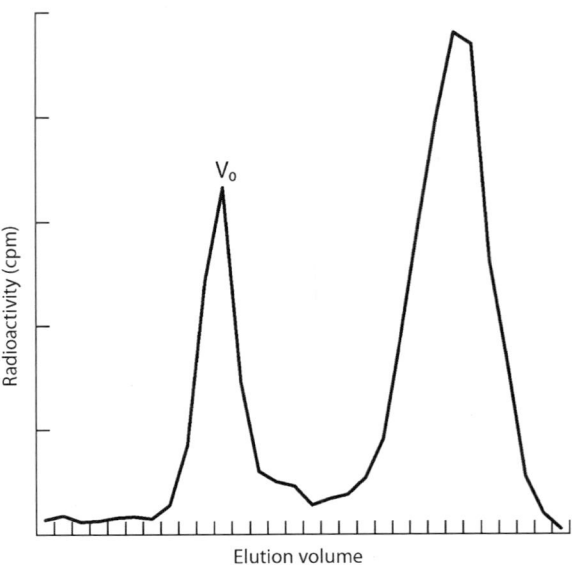

Fig. 3 Sephadex G50 chromatography. Here chondroitin sulfate degradation products (*second peak*) are separated from intact HS chains eluting in the void volume of the column

3.2   Characterization of ³H- or ³⁵S-Labeled Heparan Sulfate

1. To determine the average chain length, analyze 5,000 cpm of the purified ^{3}H- or ^{35}S-labeled HS chains by gel chromatography on Superose 6 as described above. Analyze the entire fractions for radioactivity and calculate Kav using the equation $K_{av} = \dfrac{V_e - V_0}{V_t - V_0}$. V_e=elution volume of the sample, V_0=void volume, and V_t=total volume of column. By comparing the K_{av} values of the purified ^{3}H- or ^{35}S-labeled HS chains with those of known standards (*see* ref. 8 where heparin and hyaluronan fragments of known molecular weight were analyzed), the apparent molecular weight can be calculated.

2. To determine the degree of *N*-sulfation, ^{3}H-labeled HS (20,000 cpm or more) is deaminated with nitrous acid at pH 1.5 followed by gel chromatography on Bio-Gel P-10. The nitrous acid reagent is prepared by mixing in an Eppendorf tube 0.5 ml freshly prepared 0.5 M Ba(NO$_2$)$_2$ with 0.5 ml 0.5 M H$_2$SO$_4$ on ice in a fume hood [11]. A white precipitate of BaSO$_4$ is formed. After centrifugation for 5 min at 10,000×*g*, the supernatant, the nitrous acid reagent, is collected. Within 10 min, add 200 µl reagent to 50 µl sample. Leave at room temp for 10 min. Neutralize by adding 2 M Na$_2$CO$_3$ (20–25 µl). Check that pH is 7–8 with pH paper.

3. Apply the sample to a Bio-Gel P-10 column (1×140 cm) eluted in 0.2 M NH$_4$HCO$_3$. Collect fractions of 0.8 ml at a flow rate of 1.6 ml/h.

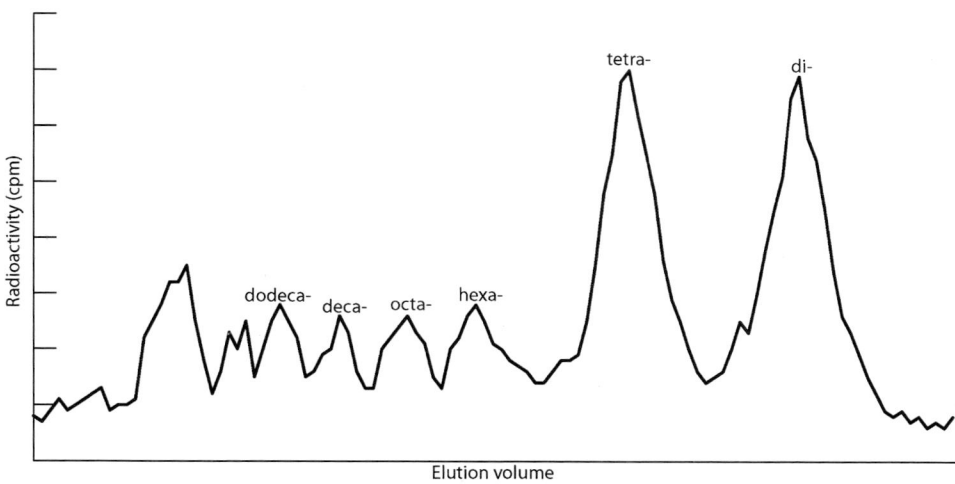

Fig. 4 Bio-Gel P-10 chromatography. The oligosaccharides generated after nitrous acid treatment are separated according to size

4. Analyze the entire fractions by scintillation counting and plot the data into a graph (Fig. 4). The percentage of GlcN residues carrying *N*-sulfate groups is determined using weighted integration of the oligosaccharide peak areas according to the equation

$$100 \times \left(\frac{\begin{array}{c} \text{cpm in dis.} + \dfrac{\text{cpm in tetras.}}{2} + \dfrac{\text{cpm in hexas.}}{3} + \dfrac{\text{cpm in octas.}}{4} \\[2ex] + \dfrac{\text{cpm in decas.}}{5} + \dfrac{\text{cpm in dodecas.}}{6} \end{array}}{\text{total cpm}} \right)$$

(*see* **Note 6**).

3.3 N-Deacetylase ELISA Assay

Here, a sensitive method to analyze the *N*-deacetylase activity of NDST enzymes is described. It is a simplified variant of a previously published method (ref. 10).

1. Coat a 96 flat-bottom well plate overnight at room temperature with 100 μl/well of K5 polysaccharide (*see* **Note 7**).

2. Solubilize the cells as described above (Subheading 3.1, **step 4**). However, remember to use the protease inhibitor mix without EDTA (*see* **Note 8**).

3. Wash the wells three times with 200 μl TBS-T.

4. Mix your samples with the *N*-deacetylase reaction mix and add to the wells in a total volume of 100 μl (*see* **Notes 9** and **10**).

5. Incubate the plate at 37 °C for 30–60 min.

6. Wash the wells three times with 200 μl TBS-T.

7. Block the wells with 1 % gelatin in TBS for 2 h at room temperature (200 μl/well).

8. Incubate the wells with the JM 403 antibody (100 μl/well) for 1 h at room temperature. The antibody will recognize the free amino groups generated when the enzyme has worked on the polysaccharide substrate.

9. Wash the wells three times with 200 μl TBS-T.

10. Incubate the wells with the goat anti-mouse IgM-peroxidase solution, 100 μl/well, for 1 h at room temperature.

11. Wash the wells three times with 200 μl TBS-T.

12. Incubate the wells with the peroxidase substrate mix, 100 μl/well, for 10–15 min, or until a color change can be observed (max. 30 min of incubation).

13. Stop the reaction with 100 μl/well of 2 M H_2SO_4.

14. Measure the absorbance at 450 nm in the spectrophotometer plate reader (*see* **Note 11**).

3.4 N-Sulfotransferase Assay

In this part, an *N*-sulfotransferase assay is described, where incorporation of [^{35}S]sulfate from [^{35}S]PAPS into K5 polysaccharide is measured.

Day 1

1. Mix 20 μl 5× reaction buffer, 5 μl freshly made 0.2 M $MnCl_2$, 10 μl 10 % Triton X-100, and 10 μl 0.2 mg/ml deacetylated K5 polysaccharide with 2 μCi [^{35}S]PAPS and the enzyme sample to be analyzed. Add H_2O to a final volume of 100 μl. Prepare a "blank" without enzyme sample.

2. Vortex briefly and incubate at 37 °C for 30 min.

3. Place tubes on ice and stop the reaction by adding 300 μl 1.3 % NaAc in 95 % ethanol.

4. Add 50 μl heparin (10 mg/ml), mix by vortexing, and place the samples at −20 °C overnight.

Day 2

5. Prepare one spin column per sample with Sephadex G-25 Superfine (*see* **Note 12**):

   (a) Punch/cut out glass fiber filters that will fit inside the 2 ml syringes.

   (b) Remove the plungers and place the cut out glass fiber filters inside the syringes.

   (c) Put the syringes inside 15 ml tubes and add Sephadex G-25 Superfine slurry to the syringes until the settled gel bed is 2 ml.

(d) Spin the 15 ml tubes holding the 2 ml syringes with Sephadex G-25 for 5 min at $800 \times g$.

(e) Discard the flow-through.

6. Take out your samples from –20 °C and spin them for 10 min in a tabletop centrifuge at $16,000 \times g$.

7. Remove the supernatants completely and resuspend the pellet in 200 μl H_2O (*see* **Note 13**).

8. Apply the resuspended samples to individual spin columns and spin the columns for 5 min at $800 \times g$.

9. Transfer the eluates to scintillation vials.

10. Add 2 ml of scintillation cocktail and measure radioactivity in a β-counter.

3.5 Combined N-Deacetylase/ N-Sulfotransferase Assay

In this assay, incorporation of [^{35}S]sulfate from [^{35}S]PAPS into the K5 polysaccharide depends on both *N*-deacetylation and *N*-sulfation. It is performed in the same way as the *N*-sulfotransferase assay but K5 polysaccharide is used as substrate instead of the deacetylated counterpart. The reaction buffer for the combined assay also contains MES instead of HEPES and has a pH of 6.3 instead of 7.5.

4 Notes

1. For metabolic labeling of HEK 293 cells, cultures at 90 % confluency, grown in DMEM medium containing 10 % fetal bovine serum, are used.

2. Shorter labeling times result in less total radioactivity incorporated into total GAGs, but the relative amounts of HS as well as intact proteoglycans are greater.

3. The medium fraction could of course also be analyzed in the same way as the cell-associated HS proteoglycans.

4. This procedure will solubilize the cell surface and intracellular pools of HS, but will only partially release proteoglycans associated with the extracellular matrix.

5. The sample volume can then be reduced using for example a speedvac concentrator.

6. Remember that the contribution of the larger oligosaccharides to the number of *N*-sulfate groups is limited. Therefore, there is no need to panic if oligosaccharides larger than octasaccharides did not separate well in the run. Just skip the last terms of the equation.

7. Alternatively, coating with the K5 polysaccharide can be carried out at 4 °C for 24 h.

8. Using this assay we have successfully assayed *N*-deacetylase activity in whole zebrafish embryos (24 h post-fertilization), minced in 1.5 ml tubes in solubilization buffer, using a pellet pestle (Sigma-Aldrich, Fischer Scientific, or other).

9. Start by performing serial dilutions of your sample. Sometimes even logarithmic dilutions are needed to find optimal concentrations of the sample. The assay is quite sensitive but the linear range of the assay is rather narrow. Run samples in triplicates.

10. The *N*-deacetylase reaction mix contains polybrene which is a cationic polymer crucial for NDST2 but not for NDST1 activity.

11. The assay is only relative since it lacks a standard. Use your own!

12. To prepare Sephadex G-25 Superfine (GE Healthcare) in 0.2 M NH_4HCO_3, the dried substance is mixed with 0.2 M NH_4HCO_3 for at least 3 h prior to use to let the gel swell completely.

13. It is important to completely remove the supernatant as residual radioactivity in the supernatant will influence the end result.

References

1. Bishop JR, Schuksz M, Esko JD (2007) Heparan sulphate proteoglycans fine-tune mammalian physiology. Nature 446:1030–1037

2. Kreuger J, Kjellen L (2012) Heparan sulfate biosynthesis: regulation and variability. J Histochem Cytochem 60:898–907

3. Li F, Shi W, Capurro M, Filmus J (2011) Glypican-5 stimulates rhabdomyosarcoma cell proliferation by activating Hedgehog signaling. J Cell Biol 192:691–704

4. Ledin J, Staatz W, Li JP, Gotte M, Selleck S, Kjellen L et al (2004) Heparan sulfate structure in mice with genetically modified heparan sulfate production. J Biol Chem 279:42732–42741

5. Carlsson P, Presto J, Spillmann D, Lindahl U, Kjellen L (2008) Heparin/heparan sulfate biosynthesis: processive formation of N-sulfated domains. J Biol Chem 283:20008–20014

6. Kreuger J, Spillmann D, Li JP, Lindahl U (2006) Interactions between heparan sulfate and proteins: the concept of specificity. J Cell Biol 174:323–327

7. Humphries DE, Silbert CK, Silbert JE (1986) Glycosaminoglycan production by bovine aortic endothelial cells cultured in sulfate-depleted medium. J Biol Chem 261:9122–9127

8. Pikas DS, Eriksson I, Kjellen L (2000) Overexpression of different isoforms of glucosaminyl N-deacetylase/N-sulfotransferase results in distinct heparan sulfate N-sulfation patterns. Biochemistry 39:4552–4558

9. Ledin J, Ringvall M, Thuveson M, Eriksson I, Wilen M, Kusche-Gullberg M et al (2006) Enzymatically active N-deacetylase/N-sulfotransferase-2 is present in liver but does not contribute to heparan sulfate N-sulfation. J Biol Chem 281:35727–35734

10. van den Born J, Pikas DS, Pisa BJ, Eriksson I, Kjellen L, Berden JH (2003) Antibody-based assay for N-deacetylase activity of heparan sulfate/heparin N-deacetylase/N-sulfotransferase (NDST): novel characteristics of NDST-1 and -2. Glycobiology 13:1–10

11. Shively JE, Conrad HE (1976) Formation of anhydrosugars in the chemical depolymerization of heparin. Biochemistry 15:3932–3942

Chapter 18

Analysis of Hyaluronan Synthase Activity

Davide Vigetti, Evgenia Karousou, Manuela Viola, and Alberto Passi

Abstract

Hyaluronan (HA) is a component of the extracellular matrix that is involved in many physiological and pathological processes. As HA modulates several functions (i.e., cell proliferation and migration, inflammation), its presence in the tissues can have positive or negative effects. HA synthases (HAS) are a family of three isoenzymes located on the plasma membrane that are responsible for the production of such polysaccharide and, therefore, their activity is critical to determine the accumulation of HA in tissues. Here, we describe a nonradioactive method to quantify the HAS enzymatic activity in crude cellular membrane preparation.

Key words Enzymatic activity, UDP sugars, Glycosaminoglycan synthesis, Hyaluronan, Hyaluronan synthase

1 Introduction

HA is a glycosaminoglycan (GAG) that is ubiquitously distributed in the human body. In contrast to the other GAGs that can be heavily chemically modified, HA is a linear repetition of glucuronic acid (GlcUA) and N-acetylglucosamine (GlcNac) disaccharide. The polymer length can greatly vary reaching a molecular mass of 4–6 million Dalton [1]. This heterogeneity causes the peculiar rheological properties of HA, but also determines the different effects on cell and tissues. In fact, low-molecular-weight HA induces angiogenesis, NF-kB, and inflammation whereas high-molecular-weight HA has anti-inflammatory properties and inhibits angiogenesis [2].

HA is synthesized by three different isoenzymes named hyaluronan synthases (HASes) 1, 2, and 3 that are located on the plasma membrane [3]. HASes use cytosolic UDP-GlcNAc and UDP-GlcUA as precursors to polymerize the HA chain out of the cell without the necessity of a primer or an anchor protein or lipid [4]. HAS1 and 2 produce HA of high molecular weight whereas HAS3 synthesizes a shorter polysaccharide [5]. In vivo, short active HA

Kuberan Balagurunathan et al. (eds.), *Glycosaminoglycans: Chemistry and Biology*, Methods in Molecular Biology, vol. 1229, DOI 10.1007/978-1-4939-1714-3_18, © Springer Science+Business Media New York 2015

oligosaccharides can be also produced by the action of hyaluronidases and oxidative stress [6]. Moreover, an alteration of UDP-sugar substrate availability can modify HA metabolism [7–9] by modifying the length of HA synthesized by HASes and also HAS dysregulation could alter the size of HA chains [10].

Although the 3D structure of HASes is still unknown, topological studies revealed the presence of two cytosolic loops that contain critical residues for catalytic activity [4]. Interestingly, these two loops are accessible to intracellular modulators as kinases or other transferases. Human HAS2 seems to be the most regulated at posttranslational level and it has been reported to be phosphorylated at T110 by AMP-activated protein kinase (AMPK) [11], S221 is reported to be modified by O-GlcNAc [12], and K190 is known to be ubiquitinated [13]. Moreover, HAS1, 2, and 3 are the substrates of ERK activity [14]. These modifications can affect many properties of the enzyme as activity, turnover, and localization.

The determination of HAS enzymatic activity is critical to assay the functions of enzyme modifications. In mammals HASes are measurable in membrane preparations only since their purification inactivates the enzymes (with the exception of mouse HAS1 [15]) while streptococcal HASes can be studied in a soluble form [16]. Experimental results indicate that AMPK and O-GlcNacylation inhibit and activate HAS2 activity, respectively, ERK increases HAS functionality, and ubiquitination mediates HAS dimerization [11–14].

The canonical protocol to measure the catalytic activity of HASes in membranes expects the use of ^{14}C- or ^{3}H-labeled UDP precursors that produce a radioactive HA [17]. After the removal of the excess of UDP sugars, the newly synthesized HA is easily quantifiable by liquid scintillation count. Although this method is widely used and sensitive, it requires radioactive manipulations along with expert training and special facilities. Therefore, here we report a nonradioactive method to determine HAS activity in cellular membranes (microsomes) maintaining a comparable sensitivity with the radioactive assay. This protocol (Fig. 1) uses nonradioactive UDP-sugar precursors, and the newly synthesized HA is then digested to obtain unsaturated disaccharides which are quantified by HPLC.

2 Materials

All the solutions are made with ultrapure water. Cell culturing was performed in a laminar flood hood using filtered media and sterilized pipette tips. For HPLC solutions, all the reagents are of HPLC purity grade, and solutions are filtered with 0.45 μm membranes and degassed by helium purging.

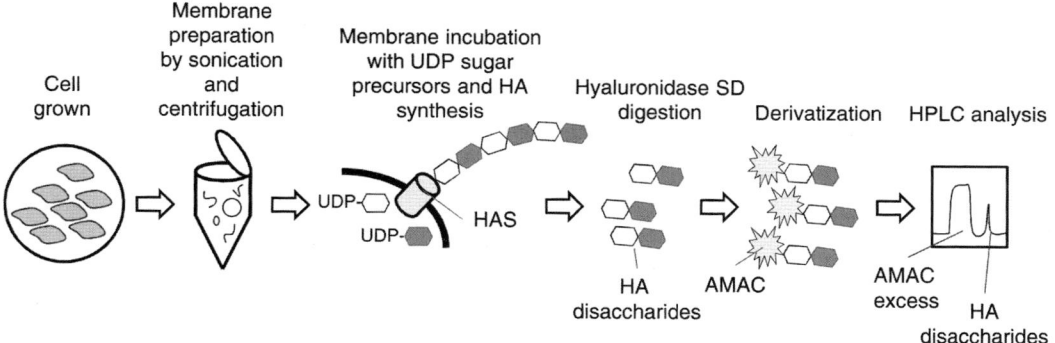

Fig. 1 Schematic representation of the method to obtain and measure HAS enzymatic activity from cell membranes. Cells were grown in Petri dishes in the appropriate medium with or without additional treatments; cells were removed by scraping and the membranes prepared by sonication and centrifugation; HA synthesis is allowed by incubating with UDP-GlcUA and UDP-GlcNAc precursors; newly synthesized HA was digested to disaccharides by hyaluronidase SD; disaccharides were derivatized with a fluorophore (AMAC); the tagged disaccharides were quantified in HPLC

2.1 Microsome Preparations	1. Lysis buffer: 10 mM KCl, 1.5 mM MgCl$_2$, and 10 mM Tris–HCl (pH 7.4). Store at 4 °C (*see* **Note 1**).
	2. Bradford reagents for protein quantification in membrane preparation.
2.2 HAS Enzymatic Assay	1. HA synthase buffer 10×: 50 mM DTT, 150 mM MgCl$_2$, and 250 mM HEPES, pH 7.1.
	2. UDP-GlcNac 100× solution: 100 mM UDP-GlcNAc in water. Prepare small aliquots and store at −20 °C.
	3. UDP-GlcUA 100× solution: 5 mM UDP-GlcUA in water. Prepare small aliquots and store at −20 °C.
2.3 HA Digestion	1. Hyaluronidase SD 100× solution: 0.01 U/μL of hyaluronidase SD (form *Streptococcus dysgalactiae*) in water. Prepare small aliquots and store at −20 °C (*see* **Note 2**).
	2. Stop solution 10×: 10 % SDS in water. Store at room temperature to avoid precipitation (*see* **Note 3**).
2.4 Unsaturated Disaccharide Derivatization	1. AMAC solution: 0.1 M 2-aminoacridone (AMAC) dissolved in 85 % DMSO/15 % acetic acid (*see* **Note 4**).
	2. NaBH$_3$CN solution: 1 M of Sodium cyanoborohydride dissolved in water (*see* **Note 5**).
	3. 50 % DMSO solution: Dilute pure DMSO with water in 1:1 proportion.
2.5 HPLC Quantification of Derivatized Disaccharides	Separation and analysis of AMAC derivatives of disaccharides were performed using an HPLC system coupled with a fluorescence detector (λ_{ex} = 425 nm and λ_{em} = 525 nm) as previously described [18].

Chromatography is carried out using a C-18 reverse-phase column (4.6 × 150 mm) at room temperature.

1. Eluent A: 0.1 M ammonium acetate buffer, pH 7.0 in water, filtered through a 0.45 mm membrane filter.

2. Eluent B: Acetonitrile HPLC grade.

3. HA disaccharide standard: Dissolve 10 nmol of commercially available unsaturated HA disaccharide in 50 μL of water. Keep at –20 °C.

3 Methods

3.1 Purify Cellular Membranes

1. Grow the cells in Petri dishes in appropriate medium and conditions up to 2×10^6 cells (*see* **Note 6**).

2. Add ice-cold PBS to wash the cell layer and carefully aspirate all the PBS. Repeat three times.

3. Add ice-cold lysis buffer (*see* **Note 7**).

4. With a rubber cell scraper carefully remove the cell layer (*see* **Note 8**) and transfer the lysate into a 1.5 ml plastic tube.

5. Incubate the lysate in ice for 15 min.

6. Disrupt the cells by using three strokes of 6 s each at 80 % microns amplitude using a high-intensity ultrasonic liquid processor maintaining the tube in ice.

7. To pellet the nuclei, spin at $5,000 \times g$ for 5 min in a refrigerated centrifuge. Save the supernatant.

8. To pellet microsomes containing vesiculated fragments of the plasma membranes [19], centrifuge the supernatant obtained in the previous step at $100,000 \times g$ for 40 min at 4 °C. Discard the supernatant (*see* **Note 9**).

9. Resuspend the pellet obtained in the previous step in 100 μL of lysis buffer with protease inhibitors and transfer into a 1.5 ml plastic tube. Maintain on ice.

10. Quantify the protein content by using the Bradford method. Proceed to the next point or store at –20 °C.

3.2 HAS Enzymatic Assay and HA Digestion

1. In a new 1.5 ml plastic tube, a volume of microsomes containing 20 μg of proteins, 10 μL of HA synthase buffer 10×, 1 μL of UDP-GlcNac 100× solution, and 1 μL of UDP-GlcUA 100× solution, and brought to a volume of 100 μL with lysis buffer with protease inhibitors (*see* **Note 10**).

2. Incubate for 60 min at 37 °C.

3. To obtain the unsaturated HA disaccharides, digest the newly synthesized HA by adding 1 μL of hyaluronidase SD 100× solution and at 37 °C for 3 h.

4. Stop the reaction by adding 10 µL of stop solution and heating at 100 °C for 5 min.

5. Spin the tube to collect drops from the lid and freeze at –80 °C for 30 min. Proceed to the next point or store at –80 °C.

3.3 Unsaturated Disaccharide Derivatization

The derivatization of unsaturated HA disaccharides is carried out as previously described [20, 21].

1. Remove the tubes containing the unsaturated HA disaccharide derived from the previous step and the standard HA disaccharide (2 nmol) from the freezer at –80 °C and quickly lyophilize under vacuum without thawing the samples (see **Note 11**).

2. Add 10 µL of freshly made AMAC solution to the lyophilized disaccharides, suspend by vortexing, and then spin down (see **Note 12**).

3. Centrifuge 3 min at room temperature at $16,000 \times g$.

4. Add 10 µL of $NaBH_3CN$ solution, vortex mix, spin down, and then incubate in the dark for 4 h at 45 °C.

5. Add 60 µL of 50 % DMSO to stop the reaction and store at –20 °C.

3.4 HPLC Quantification of Derivatized Disaccharides

Set the gradient as follows: equilibrate column as follows: 100 % eluent A for 20 min, isocratic elution with 100 % eluent A for 5 min, gradient elution to 30 % eluent B for 30 min and from 30 to 50 % for 5 min. Flow rate 1 ml/min. Before HPLS runs, extensively equilibrate the column with filtered and degassed eluent A until the base line is stable (see **Note 13**).

1. Inject 10 µL of derivatized standard HA (containing 100 pmol of HA standard) or derivatized samples from microsomes (see **Note 14**).

2. Identify the peak corresponding to HA-derivatized disaccharides comparing retention time and measure the peak area (A) by using the dedicated software for integration (see **Note 15**).

3. Calculate the amount (nmol) of newly synthesized HA by using the formula

nmol of newly synthesized $HA = (A_{sample} \times 0.1)/A_{standard}$.

4. Calculate the specific activity of HAS in the sample as follows (see **Note 16**).

HAS-specific units = (nmol of newly synthesized HA)/60/µg of proteins in the sample.

4 Notes

1. Before using completely dissolve protease or phosphatase inhibitors (commercially available as lyophilized tablet cocktails) and keep on ice for 30 min. Please note that in the lysis buffer detergents are not present since these inhibit HAS activity.

2. Hyaluronidase SD is absolutely stringent as it is specific for HA. It cuts HA chains forming unsaturated disaccharides; without this enzyme formation of polysaccharide fragments cannot occur.

3. If SDS solution forms precipitates, incubate at 37 °C with stirring.

4. Prepare the AMAC solution fresh each time, protected from light. To dissolve AMAC, mix with vortex vigorously and spin down taking care to avoid using the possible pellet of undissolved reagent.

5. Prepare the NaBH₃CN solution fresh each time. This is a very toxic compound; weigh, suspend, and use under fume hood.

6. HAS activity can be measured in almost every cell. We measured HAS activity in ECV104 and OVCAR-3 tumor cell line; however, as HASes are very-low-expressed enzymes, we usually overexpress HASes by transfections using HAS coding vector in COS-7 or NIH3T3 cells. In this way we can treat transfected cells with various modulators and study the effect on HAS activity. Alternatively we have generated a stable cell line overexpressing c-myc-HAS2 which can be used for several treatments without considering the efficiency of transfections.

7. Before using lysis buffer add the appropriate inhibitor amount and incubate on ice for 30 min. The amount of lysis buffer to be used depends on the number of cells and needs to be empirically determined.

8. Cell scraping must be done on ice or in a cold room in order to prevent protein degradation.

9. More complex methodologies can be used to separate membrane from different cellular compartments by using cell fractionation as we previously reported (*see* ref. 22).

10. As negative controls we usually incubate the microsomes without UDP sugars.

11. The samples should be frozen during the lyophilization process in order to avoid degradation. An efficient vacuum system is critical for this step.

12. Wrap the tubes with an aluminum foil to protect against light.

13. Avoiding gas bubbles in the HPLC is critical to degas HPLC buffers. Typically degassing can be done by vacuum, helium sparging, or using the inline degassers of certain chromatography systems.

14. Every five runs, we typically re-inject a derivatized standard HA to check the stability of HA peak retention time.

15. In case of difficult peak identification co-inject sample and standard and compare the peak retention time with a run containing the sample only.

16. One unit of HAS is defined as the amount of the enzyme that catalyzes the synthesis of 1 nmol of HA disaccharide in 1 min at 37 °C.

Acknowledgements

The authors acknowledge the "Centro Grandi Attrezzature per la Ricerca Biomedica", Università degli Studi dell'Insubria, for instruments availability and the PhD School in Biological and Medical Sciences.

References

1. Jiang D, Liang J, Noble PW (2011) Hyaluronan as an immune regulator in human diseases. Physiol Rev 91:221–264

2. Jiang D, Liang J, Noble PW (2007) Hyaluronan in tissue injury and repair. Annu Rev Cell Dev Biol 23:435–461

3. Tammi RH, Passi AG, Rilla K, Karousou E, Vigetti D, Makkonen K, Tammi MI (2011) Transcriptional and post-translational regulation of hyaluronan synthesis. FEBS J 278:1419–1428

4. Weigel PH, DeAngelis PL (2007) Hyaluronan synthases: a decade-plus of novel glycosyltransferases. J Biol Chem 282:36777–36781

5. Itano N, Sawai T, Yoshida M, Lenas P, Yamada Y, Imagawa M, Shinomura T, Hamaguchi M, Yoshida Y, Ohnuki Y, Miyauchi S, Spicer AP, McDonald JA, Kimata K (1999) Three isoforms of mammalian hyaluronan synthases have distinct enzymatic properties. J Biol Chem 274:25085–25092

6. Stern R, Asari AA, Sugahara KN (2006) Hyaluronan fragments: an information-rich system. Eur J Cell Biol 85:699–715

7. Jokela TA, Jauhiainen M, Auriola S, Kauhanen M, Tiihonen R, Tammi MI, Tammi RH (2008) Mannose inhibits hyaluronan synthesis by down-regulation of the cellular pool of UDP-N-acetylhexosamines. J Biol Chem 283:7666–7673

8. Jokela TA, Makkonen KM, Oikari S, Karna R, Koli E, Hart GW, Tammi RH, Carlberg C, Tammi MI (2011) Cellular content of UDP-N-acetylhexosamines controls hyaluronan synthase 2 expression and correlates with O-GlcNAc modification of transcription factors YY1 and SP1. J Biol Chem 286:33632–33640

9. Vigetti D, Ori M, Viola M, Genasetti A, Karousou E, Rizzi M, Pallotti F, Nardi I, Hascall VC, De Luca G, Passi A (2006) Molecular cloning and characterization of UDP-glucose dehydrogenase from the amphibian Xenopus laevis and its involvement in hyaluronan synthesis. J Biol Chem 281:8254–8263

10. Jing W, DeAngelis PL (2004) Synchronized chemoenzymatic synthesis of monodisperse hyaluronan polymers. J Biol Chem 279:42345–42349

11. Vigetti D, Clerici M, Deleonibus S, Karousou E, Viola M, Moretto P, Heldin P, Hascall VC, De Luca G, Passi A (2011) Hyaluronan synthesis is inhibited by adenosine monophosphate-activated protein kinase through the regulation of HAS2 activity in human aortic smooth muscle cells. J Biol Chem 286:7917–7924

12. Vigetti D, Deleonibus S, Moretto P, Karousou E, Viola M, Bartolini B, Hascall VC, Tammi M, De Luca G, Passi A (2012) Role of UDP-N-acetylglucosamine (GlcNAc) and O-GlcNAcylation of hyaluronan synthase 2 in the control of chondroitin sulfate and hyaluronan synthesis. J Biol Chem 287:35544–35555

13. Karousou E, Kamiryo M, Skandalis SS, Ruusala A, Asteriou T, Passi A, Yamashita H, Hellman U, Heldin CH, Heldin P (2010) The activity of hyaluronan synthase 2 is regulated by dimerization and ubiquitination. J Biol Chem 285:23647–23654

14. Bourguignon LY, Gilad E, Peyrollier K (2007) Heregulin-mediated ErbB2-ERK signaling activates hyaluronan synthases leading to CD44-dependent ovarian tumor cell growth and migration. J Biol Chem 282:19426–19441

15. Yoshida M, Itano N, Yamada Y, Kimata K (2000) In vitro synthesis of hyaluronan by a single protein derived from mouse HAS1 gene and characterization of amino acid residues essential for the activity. J Biol Chem 275:497–506

16. Tlapak-Simmons VL, Baron CA, Gotschall R, Haque D, Canfield WM, Weigel PH (2005) Hyaluronan biosynthesis by class I streptococcal hyaluronan synthases occurs at the reducing end. J Biol Chem 280:13012–13018

17. Spicer AP (2001) In vitro assays for hyaluronan synthase. Methods Mol Biol 171:373–382

18. Raio L, Cromi A, Ghezzi F, Passi A, Karousou E, Viola M, Vigetti D, De Luca G, Bolis P (2005) Hyaluronan content of Wharton's jelly

in healthy and Down syndrome fetuses. Matrix Biol 24:166–174

19. Caride AJ, Filoteo AG, Enyedi A, Verma AK, Penniston JT (1996) Detection of isoform 4 of the plasma membrane calcium pump in human tissues by using isoform-specific monoclonal antibodies. Biochem J 316(Pt 1):353–359

20. Karousou EG, Viola M, Genasetti A, Vigetti D, Luca GD, Karamanos NK, Passi A (2005) Application of polyacrylamide gel electrophoresis of fluorophore-labeled saccharides for analysis of hyaluronan and chondroitin sulfate in human and animal tissues and cell cultures. Biomed Chromatogr 19:761–765

21. Camenisch TD, Spicer AP, Brehm-Gibson T, Biesterfeldt J, Augustine ML, Calabro A Jr, Kubalak S, Klewer SE, McDonald JA (2000) Disruption of hyaluronan synthase-2 abrogates normal cardiac morphogenesis and hyaluronan-mediated transformation of epithelium to mesenchyme. J Clin Invest 106:349–360

22. Vigetti D, Genasetti A, Karousou E, Viola M, Clerici M, Bartolini B, Moretto P, De Luca G, Hascall VC, Passi A (2009) Modulation of hyaluronan synthase activity in cellular membrane fractions. J Biol Chem 284:30684–30694

Chapter 19

A Rapid, Nonradioactive Assay for Measuring Heparan Sulfate C-5 Epimerase Activity Using Hydrogen/Deuterium Exchange-Mass Spectrometry

Ponnusamy Babu, Xylophone V. Victor, Karthik Raman, and Balagurunathan Kuberan

Abstract

Heparin and heparan sulfate (HS) glycosaminoglycans have important roles in anticoagulation, human development, and human diseases. HS C5-epimerase, which catalyzes the epimerization of GlcA to IdoA, is a crucial enzyme involved in the biosynthesis of heparin-related biomolecules. Here, we describe a detailed method for measuring the total activity of HS C5-epimerase that involves the following steps: H/D exchange upon epimerization of the substrate with HS C5-epimerase, low-pH nitrous acid treatment of the substrate, the separation of low-pH nitrous acid-cleaved disaccharides using HPLC, and mass spectrometry analysis. This nonradioactive method is rapid and sensitive and, importantly, allows us to study the reversible nature of HS C5-epimerase.

Key words Glycosaminoglycans, Heparan sulfate, C5-epimerase, H/D exchange, Mass spectrometry

1 Introduction

Mass spectrometry, with recent advancements in ionization, resolution, accuracy, sensitivity, and speed, has become an invaluable tool not only for characterization of chemicals and biomolecules but also in the field of biology and biomedical applications [1, 2]. Special interest has been in using mass spectrometric techniques for enzyme kinetics and assay development, which conventionally requires radioactive or chromogenic labeling of substrates [3, 4]. The use of radioactive methods requires preparation of radioactive substrate, creates hazardous waste, and often uses time-consuming protocols. More importantly, part of the information may be lost due to mere measurement of the radioactive isotope. Use of spectroscopic methods becomes redundant because of lack of chromophores present in natural substrates and because modification of substrate may alter the enzyme kinetics. Here, we describe

Kuberan Balagurunathan et al. (eds.), *Glycosaminoglycans: Chemistry and Biology*, Methods in Molecular Biology, vol. 1229, DOI 10.1007/978-1-4939-1714-3_19, © Springer Science+Business Media New York 2015

a detailed method for measuring in vitro activity of heparan sulfate (HS) C5-epimerase using D/H exchange upon epimerization protocol (DEEP) and LC-MS [5].

HS proteoglycans play a major role in development and disease conditions by interacting with various receptors and growth factors [6]. The HS binding specificities, though poorly understood, are dictated by domain organization and subtle changes in the sulfation pattern. HS C5-epimerase, which converts glucuronic acid to iduronic acid, is a key enzyme responsible for generation of differential sulfation patterns. Knockout of C5-epimerase gene in mouse led to early developmental defects and neonatal lethality due to incomplete biosynthesis of HS proteoglycans [7]. Moreover, recently there has been a great deal of interest in the chemo-enzymatic synthesis of heparin-like molecules due to contamination of heparin produced from porcine [8]. For the production of biotechnological heparin-like molecules, HS C5-epimerase is used as a key enzyme because epimerization of glucuronic acid to iduronic acid by chemical means is not currently possible [9, 10]. At present, radioactive ^{3}H-labeled heparosan is used as a substrate for measuring the activity of HS C5-epimerase [11, 12]. However, this method does not truly reflect the total reactivity of C5-epimerase as the enzymatic activity in vitro and in cell-free systems is reversible; that is, GlcA can be converted to IdoA and back to GlcA. However, this irreversible nature of C5-epimerase seems to be depended on the substrate [13]. The LC-MS-based method developed here is not only rapid and sensitive but also able to provide the total reactivity of the C5-epimerase. In this method, first we developed an LC-MS-based protocol to separate and quantify GlcA-$_A$Man$_R$ and IdoA-$_A$Man$_R$ disaccharides obtained from low-pH nitrous acid hydrolysis followed by reduction (Scheme 1, Fig. 1). Secondly, we used HS C5-epimerase to react upon glucuronic acid units present in N-sulfoheparosan in D$_2$O; this allowed us to differentiate the products formed upon epimerization by DEEP-LC-MS. This newly developed method has enabled us to quantify the relative abundance of GlcA-$_A$Man$_R$ and deuterium-labeled ^{D}GlcA-$_A$Man$_R$ and ^{D}IdoA-$_A$Man$_R$ (Fig. 2). By using simple mathematic approach we can calculate the total activity of C5-epimerase.

2   Materials

Prepare all solutions using milliQ water (18 Ω cm at 25 °C).

2.1 Preparation of N-Sulfoheparosan

1. Heparosan: Prepared using a reported procedure or obtained commercially (Iduron, Manchester, UK).

2. NaOH solution (2.5 M): Dissolve 1 g of NaOH in 10 mL of deionized water.

3. Dialysis membrane Spectra/Por 1,000 MWCO.

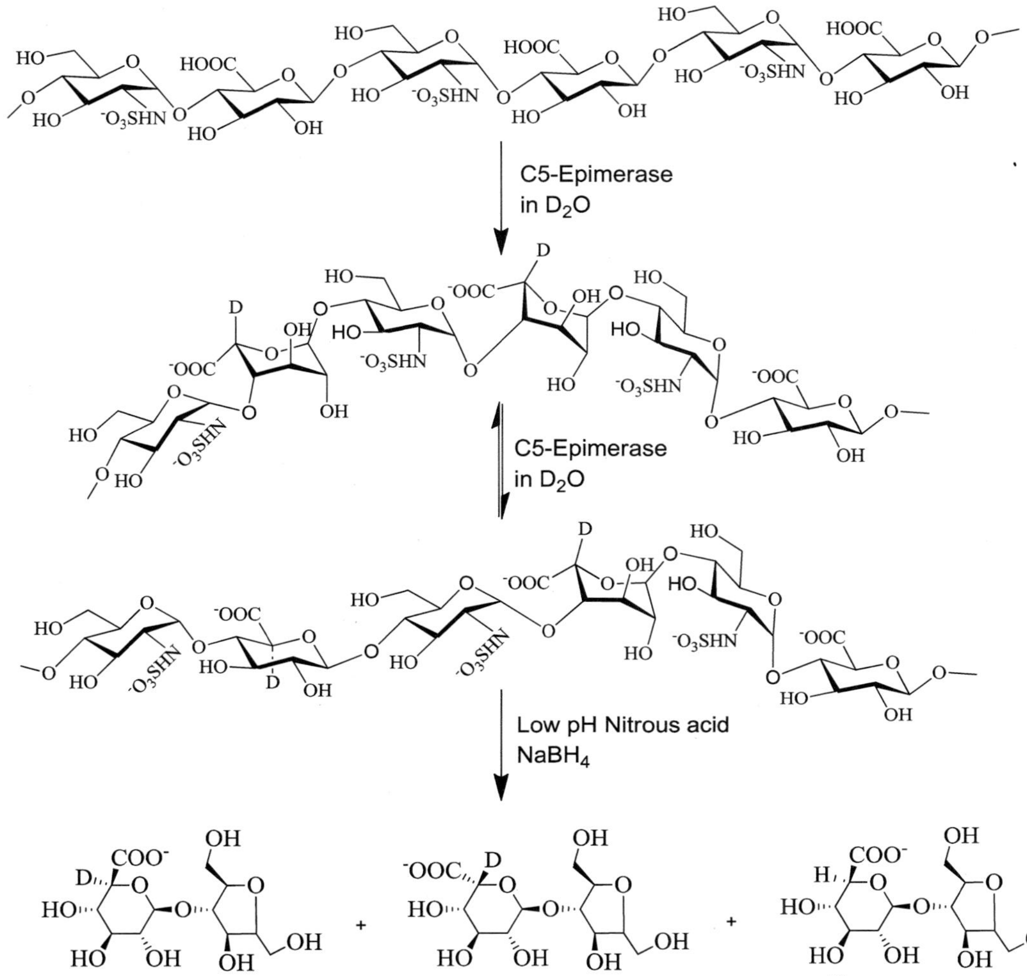

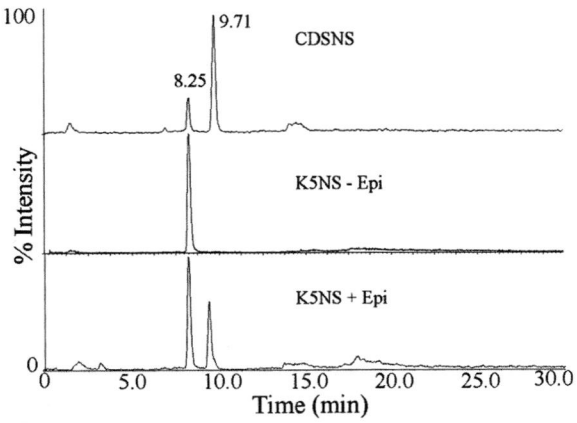

Scheme 1 Heparan sulfate C5-epimerase-catalyzed conversion of GlcA to IdoA in polymeric K5NS and its deaminative-cleaved products

$^DGlcA-_AMan_R$
m/z 340.1

$^DIdoA-_AMan_R$
m/z 340.1

$^HGlcA-_AMan_R$
m/z 339.1

Fig. 1 Extracted ion chromatogram of m/z 339.1 for nitrous acid-cleaved deaminative products of CDSNS-heparin (*top*), K5NS (*middle*), and C5-epimerase-reacted K5NS (*bottom*)

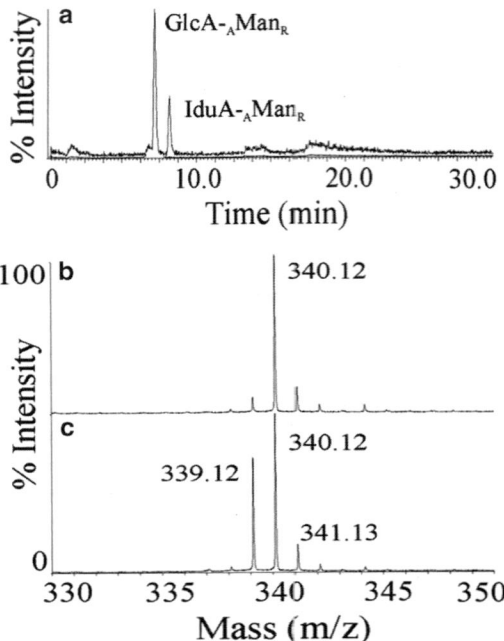

Fig. 2 Total ion chromatogram of epimerase-treated K5NS in D$_2$O. After the reaction with epimerase the product is subjected to low-pH nitrous acid treatment followed by sodium borohydride reduction. The *first* and *second peaks* correspond to GlcA-$_A$Man$_R$ and IdoA-$_A$Man$_R$, respectively (**a**); mass spectrum of the *first peak* corresponding to the mixture of ^{H}GlcA-$_A$Man$_R$ and ^{D}GlcA-$_A$Man$_R$ (**b**) and the *second peak* corresponding to ^{D}IdoA-$_A$Man$_R$ (**c**)

2.2 Expression and Purification of C5-Epimerase

1. Sf9 cells (ATCC, Manassas, VA).

2. C5-epimerase gene was cloned according to literature and recombinant enzyme was expressed using baculovirus system [14].

3. Stock 0.1 M PMSF (100×): Dissolve 0.174 g of PMSF in 10 mL of isopropanol to obtain 0.1 M PMSF stock solution. Store the solution at −20 °C in 1.0 mL aliquots.

4. Nalgene Rapidflow sterile disposable filter units (Cat. no. 168-0045).

5. ToyoPearl AF heparin 650 M (Tosoh Biosciences, USA).

6. PCG-50: 10 mM PIPES, pH 7.0, 2 % glycerol, 0.6 % CHAPS, 50 mM NaCl.

7. PCG-1000: 10 mM PIPES, pH 7.0, 2 % glycerol, 0.6 % CHAPS, 1.0 M NaCl.

8. Amicon YM-10 filters (Cat. No. UFC900308; Millipore, Billerica, MA, USA).

9. AKTAPrime Plus FPLC (GE Life Sciences, USA).

2.3 Measuring Activity of C5-Epimerase

1. Epimerase reaction buffer: 25 mM MES (pH 7.0), 0.02 % Triton X-100, 2.5 mM $MgCl_2$, 2.5 mM $MnCl_2$, 1.25 mM $CaCl_2$, and 0.75 mg/mL BSA.

2. DEAE-Sepharose fast flow gel (Amersham Biosciences, USA).

3. Plastic 2 mL syringe.

4. Wash buffer: 20 mM NaOAc, 0.1 M NaCl, and 0.01 % Triton X-100, pH 6.0.

5. Elution buffer: 20 mM NaOAc, 1 M NaCl, pH 6.0.

6. Scintillation mixture (Cat no. SX18-4, Perkin Elmer, USA).

7. Scintillation counter (Beckman LS 6500, Beckman Coulter Inc., USA).

2.4 Epimer Content Determination of IPRP-LC/MS

1. Sodium nitrite stock solution (5.5 M): Dissolve 0.38 g of sodium nitrite in 1 mL of water.

2. Sulfuric acid (0.5 M): Carefully add 0.5 mL of concentrated sulfuric acid through a glass pipette to 36 mL of deionized water. This solution can be stored in glass bottle for weeks.

3. Chemically desulfated and N-sulfated heparin (CDSNS heparin) (Seikagaku, Cape Cod, MA, USA).

4. N-sulfoheparosan (K5NS): Prepared from K5 *E. coli*; *see* Subheading 3.1.

5. Na_2CO_3 solution (2 M): Dissolve 2.12 g of sodium carbonate in 10 mL of deionized water.

6. NaOH solution (10 mM): Dissolve 4 mg of NaOH in 10 mL of water.

7. Sodium borohydride solution (2 M): Dissolve 75.6 mg of Sodium borohydride in 1 mL of 10 mM NaOH (*see* **Note 1**).

8. Sephadex G-10 gel (Amersham Biosciences, USA).

9. C18 column (4.6×150 mm, Vydac, USA).

10. HPLC system: Aliance 6250 HPLC (Waters, Milford, MA, USA) or DGP3600M HPLC (Dionex, USA).

11. MS system: Micromass LCT Premier MS (Waters, Milford, MA, USA) or micrOTOF-Q II MS (Bruker, Billerica, MA, USA).

12. Eluate A: Mix 0.225 mL of AcOH and 0.420 mL of dibutyl-amine in a reagent bottle. After 5 min add 500 mL of water to make up the solution (*see* **Note 2**).

13. Eluate B: Mix 0.225 mL of AcOH and 0.420 mL of dibutyl-amine in a reagent bottle. After 5 min add 500 mL of 70 % acetonitrile/30 % water to make up the solution.

14. Deuterium oxide (D_2O).

3   Methods

3.1 Preparation of N-Sulfoheparosan

1. Heparosan deacetylation: Dissolve 1 mg of K5 heparosan in 1 mL of 2.5 M NaOH and shake at 500 rpm at 55 °C overnight.

2. After 16 h, neutralize the solution to pH 7.0 using glacial acetic acid.

3. To this solution, add 2.5 mg of Na_2CO_3 and 2.5 mg of sulfur trioxide:triethylamine complex and stir overnight. Maintain the pH at 9.5.

4. Add an additional 2.5 mg of Na_2CO_3 and 2.5 mg of sulfur trioxide:triethylamine and stir overnight. Maintain the pH at 9.5.

5. Subsequently, neutralize the solution using glacial acetic acid.

6. Purify the product by dialysis through a 1,000 MWCO membrane for 3 days. The water is changed five times in the 3-day period at 1 h, 3 h, 24 h, 2 days, and 3 days.

7. Lyophilize the retentate (N-sulfoheparosan) to dryness.

3.2 Expression and Purification of C5-Epimerase

1. Add 15 mL of C5-epimerase viral stock to 2×10^9 Sf9 cells in a 250 mL sterile Erlenmeyer flask and shake at 90 rpm at 28 °C for 4 days (*see* **Note 3**).

2. Centrifuge the infected cell suspension at $1,000 \times g$ for 30 min to pellet cells. Dilute the supernatant containing C5-epimerase to 1:2 with 10 mM PIPES, titrate with 0.2 M NaOH to obtain pH 7.0, and chill the solution on ice for 30 min (*see* **Note 4**).

3. Centrifuge crude C5-epimerase solution at $4,000 \times g$ for 30 min, and then add appropriate volume of 100× phenyl-methylsulfonyl fluoride (0.1 M) stock solutions in 10 mL isopropanol to get a final concentration of 1 mM PMSF (*see* **Note 5**).

4. Filter the solution using sterile one-time use 0.45 μm filters and load onto a 10 mL ToyoPearl AF heparin 650 M column fitted in AKTAPrime Plus FPLC.

5. Wash the column with 250 mL of PCG-50 buffer at a flow rate of 1.0 mL per minute and elute with a 50 mL linear gradient of 50–1,000 mM NaCl in PCG (*see* **Note 6**).

6. Aliquots of selected fractions are analyzed for epimerase activity as described in the following sections (Fig. 3); the positive fractions are then pooled and concentrated using an Amicon YM-10 membrane filter (*see* **Note 7**).

3.3 Determination of Epimer Content by IPRP-LC-MS

1. Low-pH nitrous acid-mediated deaminative cleavage of N-sulfoglycosamine [15]: Add freshly prepared, ice-cold 20 μL of pH 1.5 nitrous acid to 50 μg of N-sulfoheparosan (K5NS) or CDSNS dissolved in 10 μL of water in a 150 μL microcentrifuge

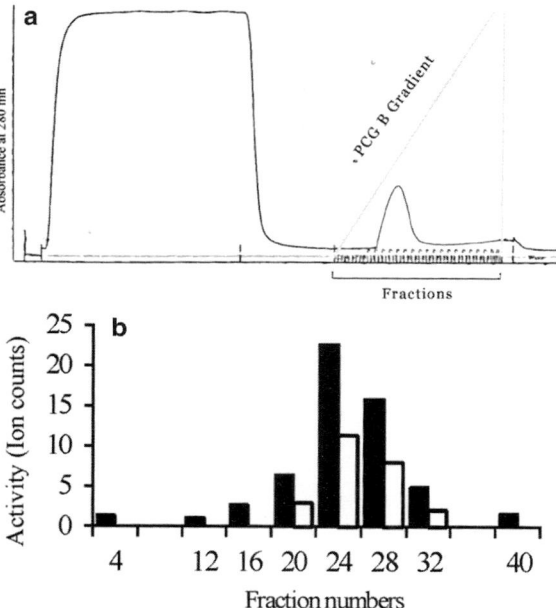

Fig. 3 (**a**) Purification of C5-epimerase from insect cell supernatant medium by heparin-ToyoPearl affinity chromatography. The elution of protein from the affinity column is monitored by online UV detector at 280 nm (*dark trace*). The salt gradient is indicated by the *gray line*. (**b**) DEEP-MS-based assay to measure C5-epimerase activity during enzyme purification: The relative total activity of C5-epimerase is calculated based on the incorporation of deuterium into the substrate by measuring ion counts for ^{D}UA-$_A$Man$_R$ and plotted as a function of eluted fractions

tube cooled in ice bath. Allow the reaction to continue at 4 °C for 20 min.

2. Quench the excess nitrous acid by carefully adding 5 μL aliquots of 2 M Na_2CO_3 until the pH reaches 7.0 (*see* **Note 8**).

3. Centrifuge and add 30 μL of 2 M sodium borohydride in 10 mM NaOH to the reaction mixture and shake (500 rpm) at room temperature for 2 h. Quench excess sodium borohydride using 10 μL of glacial acetic acid. Dry down the reaction mixture in a Savant speed vac.

4. Redissolve the sample in 20 μL of deionized water and load on to a Sephadex G-10 (1×10 cm) mini column pre-equilibrated with deionized water (*see* **Note 9**).

5. Elute the sample with deionized water (3 mL) at a flow rate of 0.2 mL per minute and by collecting 0.5 mL fractions. Concentrate the fractions in speed vac and redissolve in 50 μL of water before LC-MS analysis.

6. Separation of nitrous acid-deaminated epimeric disaccharides (Fig. 1): Epimeric disaccharides are separated on a C18 column

(4.6×150 mm, Vydac, USA) using an Aliance 6250 HPLC system or a Dionex HPLC system. A gradient elution conditions are as follows: 0 min, 95 % A; 25 min, 100 % B; and 30 min, 95 % A at a flow rate of 1.0 mL/min for 30 min, using a binary solvent system composed of water (Eluent A) and 70 % aqueous acetonitrile (Eluent B), both containing 8 mM acetic acid and 5 mM dibutylamine (DBA) as an ion-pairing agent.

7. An HPLC coupled to an electrospray ionization time-of-flight mass spectrometer: In the negative ion mode, the instrument was calibrated according to the manufacturer's instructions. Nitrogen is used as a desolvation gas as well as a nebulizer. Conditions for ESI-MS are as follows: cone gas flow, 50 L/h; nozzle temperature, 130 °C; drying gas (N_2) flow, 450 L/h; spray tip potential, 2.3 kV; and nozzle potential, 35 V. Negative ion spectra were analyzed using either MassLynx V4.1 software (Waters Corporation, Milford, MA) or Data Analysis V4.0 software (Bruker Daltonics, Billerica, MA).

**3.4   Deuterium/
Hydrogen Exchange
upon Epimerization
Protocol
(DEEP)-LC-MS**

Now that LC-MS-based separation and identification of epimeric disaccharides, GlcA-$_A$Man$_R$ and IdoA-$_A$Man$_R$, are developed we used this method for measuring activity of HS C5-epimerase by employing isotope exchange. The basic principle is upon reaction of C5-epimerase with K5NS proton at C5 carbon gets removed, inversion of configuration at C5 followed by addition of proton from reaction solvent (Scheme 1). Moreover, this epimerization is reversible in solution; that is, HS C5-epimerase has been shown to react with both GlcA and IdoA, albeit with higher preference for IdoA.

1. Dissolve 500 μg of N-sulfoheparosan in 5 mL of 1× ST buffer and concentrate to dryness in a speed vac. Redissolve the mixture in 5 mL of D_2O, vortex, and remove the solvent in speed vac. Repeat the same process twice more and finally reconstitute the mixture in 5 mL of D_2O (*see* **Note 10**).

2. Aliquot 0.5 mL of this reaction mixture into ten 1.5 mL microcentrifuge tubes.

3. Add 20 μL of C5-epimerase enzyme fractions obtained from heparin column purification to each of the reaction tubes and incubate at 37 °C.

4. After 24 h, filter the reaction mixtures through a 3 kDa cutoff Microcon (Millipore, Inc.) and wash five times with 1 mL water to remove salts. Freeze-dry the samples and perform low-pH nitrous acid-mediated deaminative cleavage followed by desalting using mini Sephadex-G10 column (1×10 cm), as described in Subheading 3.4 (*see* **Note 11**).

5. HS C5-epimerase fraction numbers 4 (control), 12, 16, 20, 24, 28, and 32 are taken for calculating percentage of deuterium

incorporation in both GlcA and IdoA. Simultaneously, conduct a conventional radioactivity-based assay for C5-epimerase for abovementioned fractions. This way the assay results can be compared between two different methods.

6. Perform LC-MS analysis on deaminative products of C5-epimerase-reacted K5NS as described in Subheading 3.4.

3.5 Measuring Total Activity of C5-Epimerase Using DEEP-LC-MS

1. Calculation of the epimer content: A typical extracted ion chromatogram (XIC) for m/z 339.1 showed only a single peak whereas XIC of m/z 340.1 showed two peaks corresponding to GlcA-$_{AM}$an$_R$ (first peak) and IdoA-$_{AM}$an$_R$ (second peak) disaccharides (*see* **Note 12**).

2. Calculate the relative ratio of the epimeric disaccharides from the total ion counts of the first peak (from m/z 339.1) and the second peak (m/z 340.1). However, since some of the deuterium-incorporated iduronic acids (^{D}IdoA-$_A$Man$_R$) are converted back to glucuronic acids (^{D}GlcA-$_A$Man$_R$), the above calculation does not truly reflect the total activity of C5-epimerase (Fig. 2).

3. Calculate the back-converted product from the increase in the intensity of the second isotopic peak of the spectra for the $^{H/D}$GlcA-$_A$Man$_R$ peak using the following equation: total $[DUA$-$_A Man_R] = [^DIdoA$-$_A Man_R] + \{[^{H/D}GlcA$-$_A Man_R] \times [\text{percent increase in } 340.1]\}$; the percent increase in $340.1 = (\{I_{340.1}/I_{(339.1+340.1)}\} - 15)$ is calculated using the formula.

4. Calculate the total activity of C5-epimerase which is equal to the intensity of deuterium-incorporated disaccharides.

---

4 Notes

1. Prepare this solution freshly before reaction.

2. Do the reaction of AcOH and DBA in a chemical hood as it will produce pungent white smoke. Avoid inhaling this fume.

3. SF9 cells have a typical doubling time of 20 h. If the cells seem to grow significantly faster or slower by 3–4 h, then they are too old and enzyme preparation is not recommended in these aberrant cells.

4. Use a calibrated, glass electrode containing pH meter to measure the pH.

5. PMSF is a protease inhibitor and toxic to humans. When handling PMSF use nitrile gloves. Repeated freeze-thaw of stock solution should be avoided.

6. A simple program was written in AKTAPrime for automated loading, washing, and elution of C5-epimerase.

Equilibrate column with PCG-50 for 5 min and then load the crude enzyme fraction. Wash the column with 250 mL of PCG-50 (25 column volumes) and elute with 5 column volumes (50 mL) of PCG-1000 as a linear gradient.

7. Condition the Amicon filters by filtering 5 mL of PCG-50 solution before concentrating enzyme fractions.

8. Use pH paper to check the pH of the reaction mixture.

9. Void volume of the column was obtained using dextran blue before loading the sample.

10. D_2O is hygroscopic and hence prepare the reaction mixture in D_2O freshly.

11. In the initial experiments, all of the fractions with the exception of void volume fractions, determined by bromophenol blue migration, were subjected to LC-MS analysis. For subsequent experiments, fractions containing sample disaccharides were pooled together and lyophilized. The desalted samples were finally analyzed by directly infusing into mass spectrometry or by IPRP-LCMS.

12. This is primarily due to the fact that the peak for the m/z 340.1 XIC is contributed to by both the first isotopic mass from $^DUA\text{-}_AMan_R$ and the second isotopic mass ($M+1$) of $^HUA\text{-}_AMan_R$, which is approximately 15 % (observed; theoretical value is 13.1 %) of the first isotopic peak (m/z 339.1).

Acknowledgements

This work was supported in part by NIH grants (P01HL107152 and R01GM075168) to B.K. and by the NIH fellowship F31CA168198 to K.R. P.B. thanks Centre for Cellular and Molecular Platforms for support.

References

1. Chen R, Mias GI, Li-Pook-Than J, Jiang L, Lam HY, Miriami E, Karczewski KJ, Hariharan M, Dewey FE, Cheng Y, Clark MJ, Im H, Habegger L, Balasubramanian S, O'Huallachain M, Dudley JT, Hillenmeyer S, Haraksingh R, Sharon D, Euskirchen G, Lacroute P, Bettinger K, Boyle AP, Kasowski M, Grubert F, Seki S, Garcia M, Whirl-Carrillo M, Gallardo M, Blasco MA, Greenberg PL, Snyder P, Klein TE, Altman RB, Butte AJ, Ashley EA, Gerstein M, Nadeau KC, Tang H, Snyder M (2012) Personal omics profiling reveals dynamic molecular and medical phenotypes. Cell 148:1293–1307

2. Walsh GM, Rogalski JC, Klockenbusch C, Kast J (2010) Mass spectrometry-based proteomics in biomedical research: emerging technologies and future strategies. Expert Rev Mol Med 12:e30

3. Liesener A, Karst U (2005) Monitoring enzymatic conversions by mass spectrometry: a critical review. Anal Bioanal Chem 382: 1451–1464

4. Wu J, Takayama S, Wong CH, Siuzdak G (1997) Quantitative electrospray mass spectrometry for the rapid assay of enzyme inhibitors. Chem Biol 4:653–657

5. Babu P, Victor XV, Nelsen E, Nguyen TK, Raman K, Kuberan B (2011) Hydrogen/deuterium exchange-LC-MS approach to characterize the action of heparan sulfate C5-epimerase. Anal Bioanal Chem 401:237–244

6. Esko JD, Selleck SB (2002) Order out of chaos: assembly of ligand binding sites in heparan sulfate. Annu Rev Biochem 71:435–471

7. Li JP, Gong F, Hagner-McWhirter A, Forsberg E, Abrink M, Kisilevsky R, Zhang X, Lindahl U (2003) Targeted disruption of a murine glucuronyl C5-epimerase gene results in heparan sulfate lacking L-iduronic acid and in neonatal lethality. J Biol Chem 278:28363–28366

8. Guerrini M, Beccati D, Shriver Z, Naggi A, Viswanathan K, Bisio A, Capila I, Lansing JC, Guglieri S, Fraser B, Al-Hakim A, Gunay NS, Zhang Z, Robinson L, Buhse L, Nasr M, Woodcock J, Langer R, Venkataraman G, Linhardt RJ, Casu B, Torri G, Sasisekharan R (2008) Oversulfated chondroitin sulfate is a contaminant in heparin associated with adverse clinical events. Nat Biotechnol 26:669–675

9. Naggi A, Torri G, Casu B, Oreste P, Zoppetti G, Li JP, Lindahl U (2001) Toward a biotechnological heparin through combined chemical and enzymatic modification of the Escherichia coli K5 polysaccharide. Semin Thromb Hemost 27:437–443

10. Xu Y, Masuko S, Takieddin M, Xu H, Liu R, Jing J, Mousa SA, Linhardt RJ, Liu J (2011) Chemoenzymatic synthesis of homogeneous ultralow molecular weight heparins. Science 334:498–501

11. Hagner-McWhirter A, Hannesson HH, Campbell P, Westley J, Roden L, Lindahl U, Li JP (2000) Biosynthesis of heparin/heparan sulfate: kinetic studies of the glucuronyl C5-epimerase with N-sulfated derivatives of the Escherichia coli K5 capsular polysaccharide as substrates. Glycobiology 10:159–171

12. Raedts J, Lundgren M, Kengen SW, Li JP, van der Oost J (2013) A novel bacterial enzyme with D-glucuronyl C5-epimerase activity. J Biol Chem 288:24332–24339

13. Sheng J, Xu Y, Dulaney SB, Huang X, Liu J (2012) Uncovering biphasic catalytic mode of C5-epimerase in heparan sulfate biosynthesis. J Biol Chem 287:20996–21002

14. Li J, Hagner-McWhirter A, Kjellen L, Palgi J, Jalkanen M, Lindahl U (1997) Biosynthesis of heparin/heparan sulfate. cDNA cloning and expression of D-glucuronyl C5-epimerase from bovine lung. J Biol Chem 272:28158–28163

15. Shively JE, Conrad HE (1970) Stoichiometry of the nitrous acid deaminative cleavage of model amino sugar glycosides and glycosaminoglycuronans. Biochemistry 9:33–43

Chapter 20

Aggrecan: Approaches to Study Biophysical and Biomechanical Properties

Hadi Tavakoli Nia, Christine Ortiz, and Alan Grodzinsky

Abstract

Aggrecan, the most abundant extracellular proteoglycan in cartilage (~35 % by dry weight), plays a key role in the biophysical and biomechanical properties of cartilage. Here, we review several approaches based on atomic force microscopy (AFM) to probe the physical, mechanical, and structural properties of aggrecan at the molecular level. These approaches probe the response of aggrecan over a wide time (frequency) scale, ranging from equilibrium to impact dynamic loading. Experimental and theoretical methods are described for the investigation of electrostatic and fluid-solid interactions that are key mechanisms underlying the biomechanical and physicochemical functions of aggrecan. Using AFM-based imaging and nanoindentation, ultrastructural features of aggrecan are related to its mechanical properties, based on aggrecans harvested from human vs. bovine, immature vs. mature, and healthy vs. osteoarthritic cartilage.

Key words Aggrecan, Atomic force microscopy, Nanomechanics, Biophysics, Ultrastructure, Cartilage, Extracellular matrix, Dynamic modulus, Elasticity, Poroelasticity, Viscoelasticity

1 Introduction

Proteoglycans make up 5–10 % of the wet weight of human and animal cartilages (~35 % by dry weight) [1]. Aggrecan, the most abundant extracellular proteoglycan in cartilage, is composed of an ~300–400 nm long core protein (225–250 kDa) to which glycosaminoglycan (GAG) side chains are covalently bound [2]. The major GAGs of aggrecan are chondroitin sulfate (CS) and keratansulfate (KS), with sulfation of CS typically at the –4 or –6 position, depending on age. The KS chains (~5 kDa) consist of ~10 repeating disaccharides, while CS chains occupy most of the GAG-rich region of aggrecan and comprise ~95 % of the molecular weight of the entire aggrecan molecule. At physiological pH and ionic strength, the carboxylic acid and the sulfate groups are negatively charged. Macroscopic measurements of cartilage biomechanical properties [3] have shown that the high negative charge density ascribed to aggrecan GAG chains is the major determinant of

Kuberan Balagurunathan et al. (eds.), *Glycosaminoglycans: Chemistry and Biology*, Methods in Molecular Biology, vol. 1229, DOI 10.1007/978-1-4939-1714-3_20, © Springer Science+Business Media New York 2015

cartilage compressive stiffness, responsible for more than 50 % of the equilibrium modulus in compression [4]. In addition, studies have shown that with increasing age, the ratio of CS to KS GAG chains decreases in human and animal cartilages, and the contour length of CS chains, and to a lesser degree the KS chains, becomes shorter with age [5, 6]. The proteolysis and loss of aggrecan from cartilage is one of the earliest events in the initiation of cartilage degradation in osteoarthritis [7].

Here, we review novel approaches to study biophysical and biomechanical properties of aggrecan and its GAG chains at the molecular level using recently developed atomic force microscopy (AFM)-based nanomechanical and imaging methodologies. Such methods enable the quantification of the ultrastructural properties of individual aggrecan molecules and the nanomechanical compressive and shear behavior of assemblages of aggrecan having the same molecular density as that found in native human and animal articular cartilage. These experimental approaches can be coupled to mathematical models to estimate the elastic, electrostatic, and fluid-dependent properties of aggrecan at the nanoscale, and to estimate the contribution of aggrecan to the macroscopic tissue-scale biophysical and biomechanical properties of cartilage in health and disease.

2　Materials

The materials used in the following approaches include mica surfaces, AFM tips, polydimethylsiloxane (PDMS), 3-aminopropyltriethoxysilane, dithiobis sulfosuccinimidyl propionate, dithiothreitol, hydroperoxide, sulfuric acid, 11-mercaptoundecanol, and gold-coated silicon substrate.

Aggrecan from human and animal cartilages can be isolated using the following protocol [8, 9]:

1. Wash cartilage in ice-cold 50 mM sodium acetate, pH 7.0, containing a mixture of protease inhibitors (*see* **Note 1**), and store on ice until further processing.

2. Cut the tissue into small pieces (3×3 mm^2) or microtome into even thinner 20 μm slices and extract with 4 M guanidine hydrochloride (GuHCl) and 100 mM sodium acetate, pH 7.0, with protease inhibitors for 48 h.

3. Separate the unextracted tissue residues by centrifugation, and dialyze the clarified supernatant against two changes of 100 volumes of 0.1 M sodium acetate, pH 7.0, with protease inhibitors [10].

4. Purification of aggrecan is then performed using CsCl density gradient centrifugation. One common approach first utilizes

"associative conditions" with 0.5 M GuHCl to purify aggrecan aggregates, and then "dissociative conditions" with 4 M GuHCl to separate constituents of the aggregates (e.g., link protein and hyaluronan).

5. The simplest purification method isolates aggrecan monomer in one dissociate step (i.e., the D1 fraction): transfer supernatant to a new tube, add CsCl powder until solution density is 1.58 g/ml, and spin at 4 °C for 72 h at $500,000 \times g$ (e.g., Beckman Optima ultracentrifuge with MLA-130 rotor).

6. Alternatively, the supernatant is first dialyzed to associative conditions for 48 h against two changes of 100 volumes of 0.1 M sodium acetate, pH 7.0 containing protease inhibitors. The solution is adjusted with CsCl to a final density of 1.58 g/l and spun for 48 h; aggrecan aggregates are recovered in the bottom 1/8 of the gradient (the A1 fraction), which can be re-centrifuged under the same conditions to obtain the A1A1 fraction. This fraction is readjusted to 4 M guanidine HCl, CsCl added to a density of 1.58 g/l, and spun for 48 h at 4 °C and the resultant A1A1D1 fraction can be centrifuged a second time under dissociative conditions to obtain the A1A1D1D1.

7. Dialyze the purified aggrecan fractions (A1A1D1D1) consecutively against 500 volumes of 1 M NaCl and deionized water to remove excess salts.

8. Determine the aggrecan yield by the dimethyl methylene blue (DMMB) dye-binding assay [11].

9. Make 4 mM hydroxyl-terminated self-assembled monolayer (OH-SAM) solution by dissolving 25 mg 11-mercaptoundecanol ($HS(CH_2)_{11}OH$) powder in 40 ml ethanol (4 mM). Vortex until the powder is fully dissolved.

3 Method

3.1 High-Resolution Imaging and Ultrastructure of Aggrecan Across Age and Species

High-resolution atomic force microscopy (AFM) imaging (Fig. 1a) has been utilized to directly visualize and quantify aggrecan ultrastructural features, including (a) the spatial distribution of GAG chains along the aggrecan core protein and the heterogeneity of GAG chain lengths within individual aggrecan molecules, and (b) differences between aggrecan extracted from cartilages of different animal species and age, including mature and fetal bovine tissue [9], aggrecan extracted from tissue-engineered cartilage formed by bone marrow stromal cells and chondrocytes from immature and mature equines [12, 13], and aggrecan extracted from newborn (Fig. 1c) and adult human cartilage (Fig. 1d) [14]. The following methods are used:

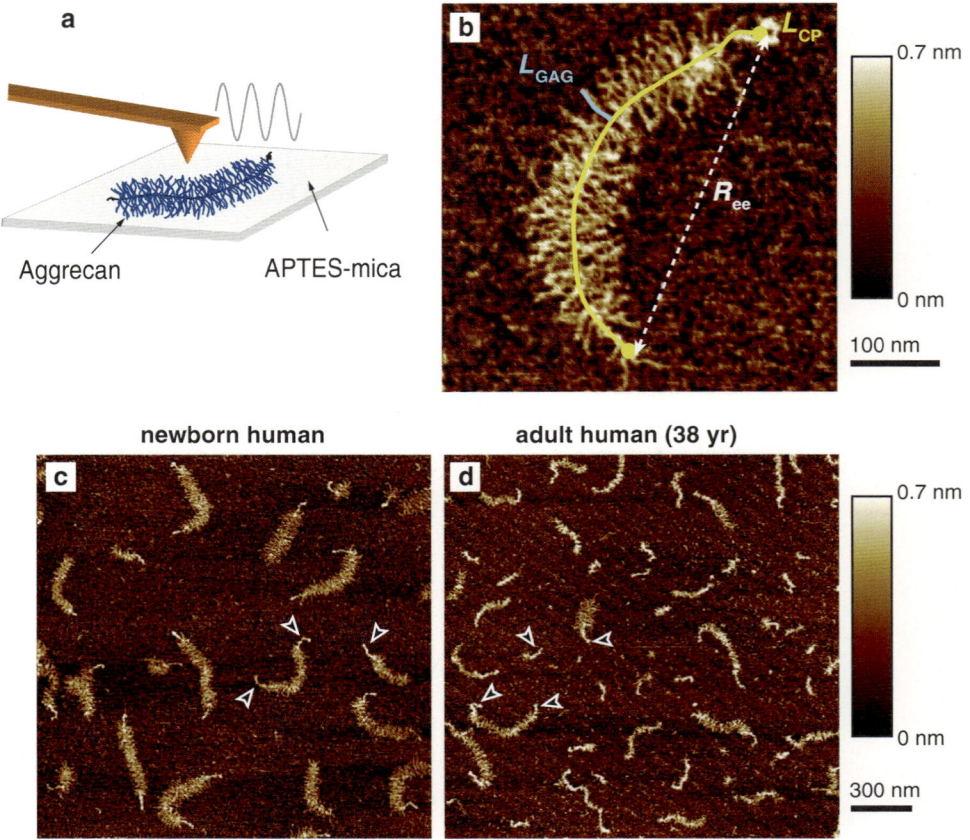

Fig. 1 (**a**) High-resolution imaging of aggrecan is performed using atomic force microscopy (AFM) in tapping mode. (**b**) Important structural features including the trace length of the core protein L_{CP}, the trace length of the GAG chains L_{GAG}, and the end-to-end distance R_{ee} are quantified for aggrecan obtained from different species and ages. As an example, aggrecans from newborn human (**c**) and adult human (38 years) (**d**) are shown in which significant differences in these structural properties are observed. Image **a** is from [13] and images (**b**)–(**d**) are from [14], reproduced by the permission of Elsevier © 2013

1. Dilute the aggrecan solution with DI water to the desired concentration (10–100 µg/ml) and mix well (using DI or MilliQ water avoids imaging salt particles or impurities).

2. Deposit a small aliquot of aggrecan solution (10–50 µl) on 3-aminopropyltriethoxysilane (APTES) freshly treated muscovite mica surfaces and wait for 20–30 min at room temperature.

3. Rinse the mica surface gently with MilliQ water and then blow dry in N_2 gas.

4. Start AFM imaging in tapping mode after the following steps. The AFM vibration isolation needs to be on. We used a Nanoscope IIIA Multimode AFM (Veeco, Santa Barbara, CA);

however most of the following settings are the same for other commercial AFMs:

(a) Mount AFM tip with spring constant $k \sim 2$ N/m, nominal tip radius $R < 10$ nm (AC240TS-2, Olympus), and perform laser alignment and calibration (*see* **Note 2**).

(b) Set the AFM to tapping mode.

(c) Set the following parameters: scan size $\sim \mu$m, aspect ratio 1:1, scan rate: 0.5–1 Hz, sample line: 256 (pre-scan), 512 (high-quality image), integral gain: 0.3–0.5, proportional gain: 0.5, amplitude set point: 0.6–1.0.

5. After digitizing the AFM height images, extract the structural features of the aggrecan including the trace (contour length of core protein L_{CP}, the end-to-end distance, R_{ee}, and the trace lengths of the individual GAG chains L_{GAG} (Fig. 1b)).

3.2 Quasi-Static Compression in the Fully Wet State

AFM has also been utilized to probe the nanomechanical properties of aggrecan and their variation with age, species, and treatment. In this section, the techniques used to probe aggrecan nanomechanics are associated with low-rate (quasi-static) compression of end-grafted aggrecan brush layers (*see* Fig. 2a–d). With quasi-static loading, the loading rate is low enough that the fluid within and surrounding the aggrecan sample does not contribute directly to the measured mechanical properties. These techniques include quasi-static indentation [14–19], shear tests [20], and adhesion tests [21]. To investigate the role of electrostatic interactions between the highly negatively charged GAG chains in the nanomechanical behavior of aggrecan, these tests were performed over a wide range of ionic strengths (IS). The IS dependence of aggrecan interactions is typically associated with salt screening of electrostatic forces.

1. **Steps 1–6** are first performed to obtain thiol-functionalized aggrecan to enable formation of a chemically end-grafted aggrecan layer (as shown schematically in Fig. 2a).

2. Add ~100 μl of dithiobis(sulfosuccinimidyl propionate) (DTSSP, 12 mg/ml) to 2 ml of 1 mg/ml aggrecan aliquot and wait for 1 h.

3. Add ~200 μl of dithiothreitol (DTT, 3 mg/ml) to the aggrecan solution. Wait for 1 h.

4. Filter the aggrecan solution with a 100,000 MW cutoff centrifugal filter device at speed between 3,500 and 4,000 rev/min for ~5–6 h to remove the reactant. Take out the filtrate tube, reverse the device, and centrifuge w/same speed for another 30 min to concentrate the remained aggrecan solution into the sharp-end tube.

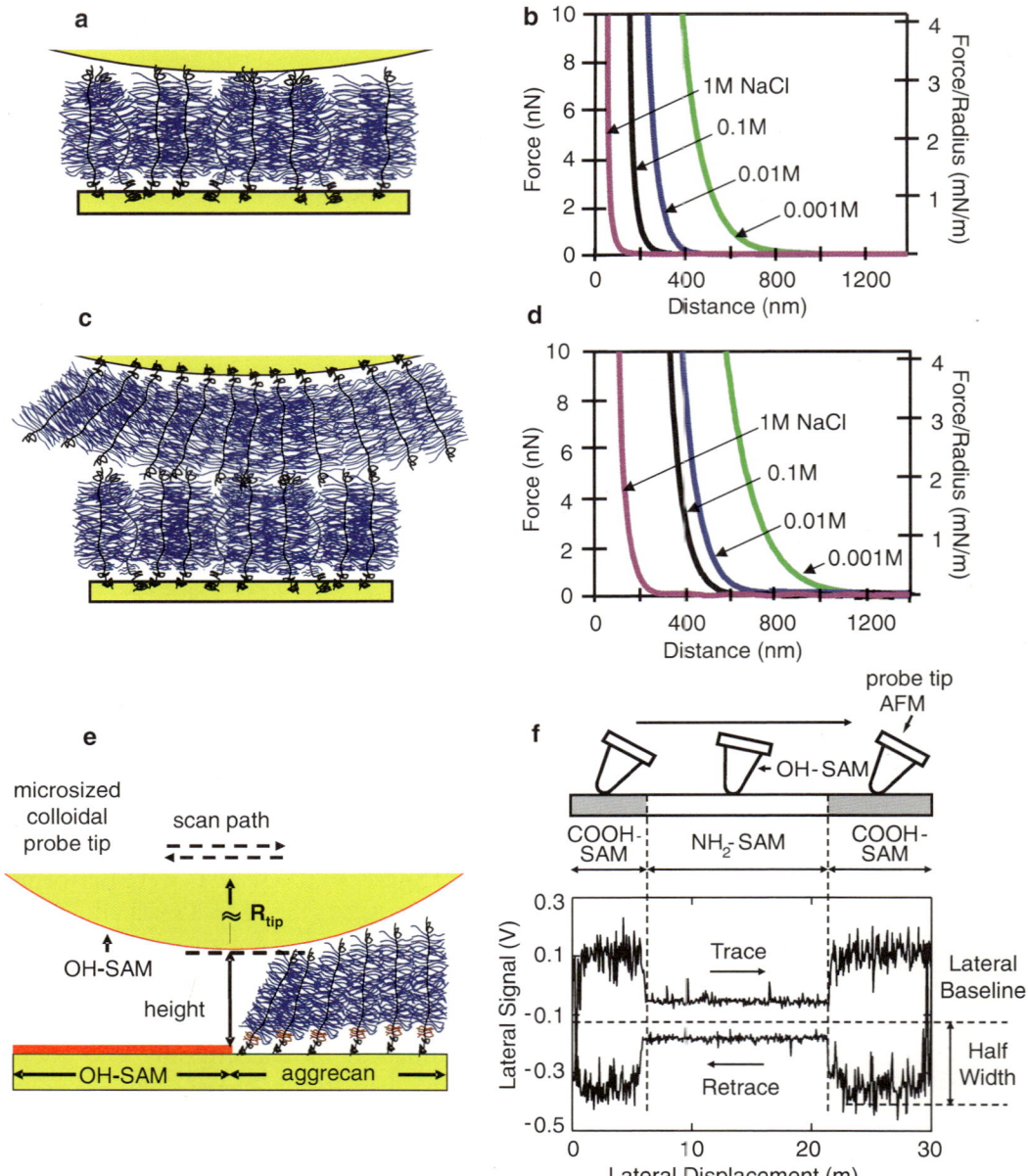

Fig. 2 (**a**) Schematic showing measurement of the compressive properties of aggrecan is explored by AFM-based indentation of a single layer of aggrecan that is end grafted to a gold-coated substrate. (**b**) Nanoindentation curves in the configuration of (**a**) showing the force that the AFM tip experiences as a function of the distance between tip and substrate. The compressive properties of the aggrecan monolayer undergo significant changes with bath ionic strength, implying the key contribution of electrostatic interactions between GAG chains. (**c**) Schematic showing AFM-based indentation performed with aggrecan end grafted to the AFM probe tip in addition to the substrate, in order to study GAG-GAG interactions from two opposing layers of aggrecan. (**d**) The corresponding indentation curves also show that significant ionic strength dependence is intrinsic to aggrecan behavior. (**e**) Configuration and (**f**) examples of data from measurement of aggrecan deformation in shear, studied using AFM-based lateral force microscopy. The AFM tip traces from the region having a COOH-NH2 self-assembled monolayer over to the end-grafted aggrecan. Figures (**a**)–(**d**) from [16]; figures (**e**) and (**f**) from [20], reproduced by the permission of Elsevier © 2013

5. The concentrated aggrecan solution will be very viscous (~200 μl). Dilute the aggrecan solution to desired concentration (~1 mg/ml) with MilliQ water.

6. Clean the gold-coated silicon substrate by dipping it in piranha solution (2 ml of H_2O_2 plus 4 ml H_2SO_4) for 15 min. Rinse the substrates with DI water, and blow dry with N_2 gas. Repeat this step.

7. Place 50 μl of aggrecan solution onto the gold surface. Incubate in a humidity-controlled environment (e.g., airtight box with water) for ~48 h.

8. Mount the AFM probe tip: either a sharp tip (e.g., Veeco, cantilever spring constant, $k \sim 0.06$ N/m, end radius, $R_{tip} \sim 50$ nm measured by SEM), which will interact with only a few aggrecan molecules, or micron-sized colloid (BioForce Nanosciences, $k \sim 0.12$ N/m, $R_{tip} \sim 2.5$ mm), which will interact with a larger assembly of end-grafted aggrecan (~10^4 aggrecan monomers). Do subsequent laser alignment and calibration.

9. Rinse the sample with MilliQ water and place it on the AFM stage. Add a bath solution with desired ionic strength. When changing the ionic strength of the bath in the range from 0.001 M to 1.0 M, start with the lowest ionic strength first. To switch the solutions, blot dry the initial solution with filter paper and add new solution to the stage. Repeat three times.

10. Perform single force-indentation measurements in the desired ionic strength solution using the following adjustments: z-piezo ramp size: 2 μm; ramp velocity: 0.5 μm/s. Capture the deflection and z-piezo signals in order to plot the force-displacement curves (e.g., Fig. 2b, d). The force curve is calculated as deflection times the spring constant of the AFM cantilever. The displacement is calculated as the z-piezo subtracted by the deflection (*see* **Note 3**).

3.3 Microcontact Printing and Aggrecan Height Measurement

A microcontact printing (μCP) technique [15, 22], extensively used in the nanomechanical studies of aggrecan [14–16, 20, 23], consists of the creation of a patterned surface with densely packed, chemically end-grafted aggrecan confined to well-defined micrometer-sized surface area with a hydroxyl-terminated self-assembling monolayer (OH-SAM) filling the rest of the non-grafted surface (Fig. 3a). By using AFM contact mode imaging over boundaries between the aggrecan layer and OH-SAM regions, the height (Fig. 3b–d) and, hence, the compressive deformation of the aggrecan layer can be directly measured as a function of normal compressive load (0–900 nN) in aqueous bath conditions of varying ionic strengths (IS = 0.001–1 M) and as shown in Fig. 3d, e.

1. Fully immerse the micro-printing PDMS stamp in the OH-SAM solution and wait for 30 min.

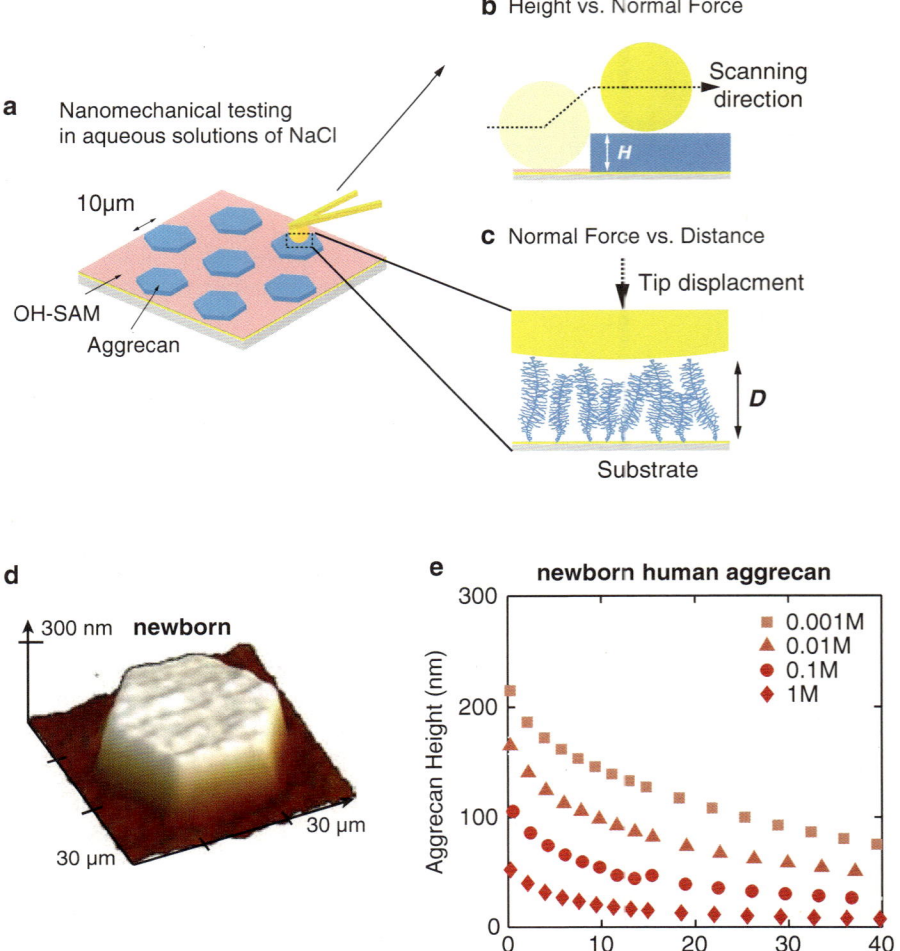

b Height vs. Normal Force

Scanning direction

a Nanomechanical testing in aqueous solutions of NaCl

10μm

OH-SAM

Aggrecan

c Normal Force vs. Distance

Tip displacment

Substrate

d

300 nm **newborn**

30 μm

30 μm

e **newborn human aggrecan**

300

200

Aggrecan Height (nm)

100

0

■ 0.001M
▲ 0.01M
● 0.1M
◆ 1M

0   10   20   30   40

Normal Force F (nN)

Fig. 3 (**a**) Aggrecan molecules are end attached to hexagonal regions using microcontact printing techniques [15]. (**b**, **c**) This methodology enables measurement of the height of the aggrecan monolayer (**b**) versus the distance between the aggrecan layer and substrate (**c**) as the AFM probe tip touches the top surface of the monolayer. (**d**) A typical height map is shown for the case in which the aggrecan monolayer is printed in hexagonal pattern. (**e**) The height of the monolayer can be measured as a function of the applied AFM tip, and in different ionic strength of the solution (IS = 0.001, 0.01, 0.1, and 1.0 M NaCl). Figures (**a**)–(**c**) from [13]; figures (**d**, **e**) from [14], reproduced by the permission of Elsevier © 2013

2. Clean the gold-coated silicon substrate by dipping it in piranha solution (2 ml H_2O_2 plus 4 ml H_2SO_4) for 15 min. Rinse the substrates with DI water, and blow dry with N_2 gas. Repeat this step.

3. Right before piranha clean finishes, take out PDMS micro stamp from OH-SAM. Put it on filter paper with the stamp side facing up. Make sure that there are no water drops on the stamp side.

4. Stamp the gold surface with the micro stamp. Use tweezers to bring the stamp in contact with the gold surface. Once in contact, gently press the stamp to ensure uniform contact. Wait for 2 min.

5. Remove the stamp. Rinse the gold surface with ethanol and DI water. Blow dry with N_2 gas.

6. Place 50 µl aggrecan solution onto the gold surface. Incubate in a humidity-controlled environment (e.g., airtight box with water) for ~48 h.

7. Perform AFM imaging in contact mode with the aspect ratio of the image set to 1:16 (8 lines). Adjust the deflection signal to 0.2 V, and capture the height image for set point voltage: 0.25 V. Increase the set point voltage by 0.1 V, and continue until reaching to 4 V (*see* **Note 4** for converting the voltages to force). Switch the solution with desired ionic strength by blotting dry the old solution with filter paper and dropping new solution without spilling. Repeat the imaging step for the new solution. Save the signals from the height channel.

3.4 Lateral Force Microscopy

Lateral force microscopy [20, 24] has been utilized to explore the role of aggrecan and aggrecan-aggrecan interactions in the shear behavior of cartilage by measuring the lateral resistance to deformation of a monolayer of aggrecan end grafted on a microcontact printed surface in aqueous NaCl solutions (Fig. 2e, f). As the cantilever scans across the surface under a constant applied normal force, it twists in the lateral direction, resulting in a horizontal deflection of the laser spot on a quadrant position-sensitive photodiode that outputs a lateral deflection signal. The microcontact printing technique enables us to distinguish the lateral force of regions with aggrecan (printed in the form of circles or hexagons) from the regions with neutral OH-SAM (Fig. 2e).

1. Repeat **steps 1–6** in Subheading 3.2 for sample preparation.

2. Do the calibration for the lateral microcopy using the wedge approach [25]: first let the probe tip scan on a flat mica sample at a series of normal force, then switch to a tilted mica sample ($\alpha = 20°$), and scan at the top of the tilted region (to avoid the contact of the base of the cantilever to the mica during scanning) at the same normal forces; record the lateral signals in both cases (*see* **Note 5**).

3. Load the calibration sample and AFM tip into the system, adjust the laser position on the cantilever to reach the maximum sum (~7 V), and zero both vertical and horizontal deflections.

4. Repeat **step 2** in Subheading 3.4, and record height, friction in trace, and friction in retrace channels.

3.5 Dynamic Nanomechanics of Aggrecan

Measurements of solid-fluid interactions in intra- and extracellular matrix at the molecular level provide the possibility of increased understanding of the fundamental principles underlying the functional and structural properties of hydrated soft tissues [26], and the effects of pathobiological processes on these functions [23]. However, the conventional biomechanical tests, such as indentation performed as described above, only probe quasi-static (low rate) mechanical responses of porous biomaterials (e.g., the aggrecan monolayer); thus, these tests do not assess the effects of solid-fluid interactions. To involve the frictional losses arising from fluid-solid interactions, a dynamic approach at an appropriate loading rate is required to induce fluid flow through the pores of the solid phase of the sample (i.e., the spaces between aggrecan GAGs in the monolayer). We utilized a novel high-bandwidth AFM-based nanorheology system (Fig. 4a) to probe the full-spectrum dynamic nanomechanical properties of aggrecan monolayers as well as from native intact cartilage over a wide frequency range (1 Hz to 10 kHz) [23, 27–29].

1. Repeat **steps 1–6** in Subheading 3.2 for sample preparation.

2. Mount the AFM colloidal probe tip (using an end radius $R \sim 22.5$ μm), and an additional tip with $R \sim 2.5$ μm (two different radii are used for length-scale studies) attached to cantilevers with nominal spring constant $k \sim 4.0$ N/m, and resonance frequency $f_0 \sim 150$ kHz. Do subsequent laser alignment and calibration.

3. Rinse the sample with MilliQ water and place it on the AFM stage. Add the solution with desired ionic strength. When changing ionic strengths from 0.001 M to 1.0 M, start with the lowest ionic strength. To switch the solutions, blot the old solution with filter paper and drop new solution without spilling. Repeat three times.

4. Design the primary displacement profile to apply an initial indentation with ramp size 2 μm and ramp velocity 0.5 μm/s, followed by a hold with 110 s (this hold time needs to be longer than the dynamic signal designed in **step 10**; e.g., *see* Fig. 4b).

5. Design the secondary dynamic displacement signal: one approach is to use a sinusoidal sweep signal (Fig. 4b) composed of an exponential chirp with the duration of 100 s. The amplitude of the sinusoidal displacement is maintained in the range of 2–15 nm. A second approach using a random binary sequence (RBS) is composed of steps with 2–15 nm amplitude, which are placed randomly in a time series with duration of 20 s (*see* **Note 6** for generation of RBS). The advantage of an RBS is that it has a higher signal-to-noise ratio compared to the sinusoidal sweep.

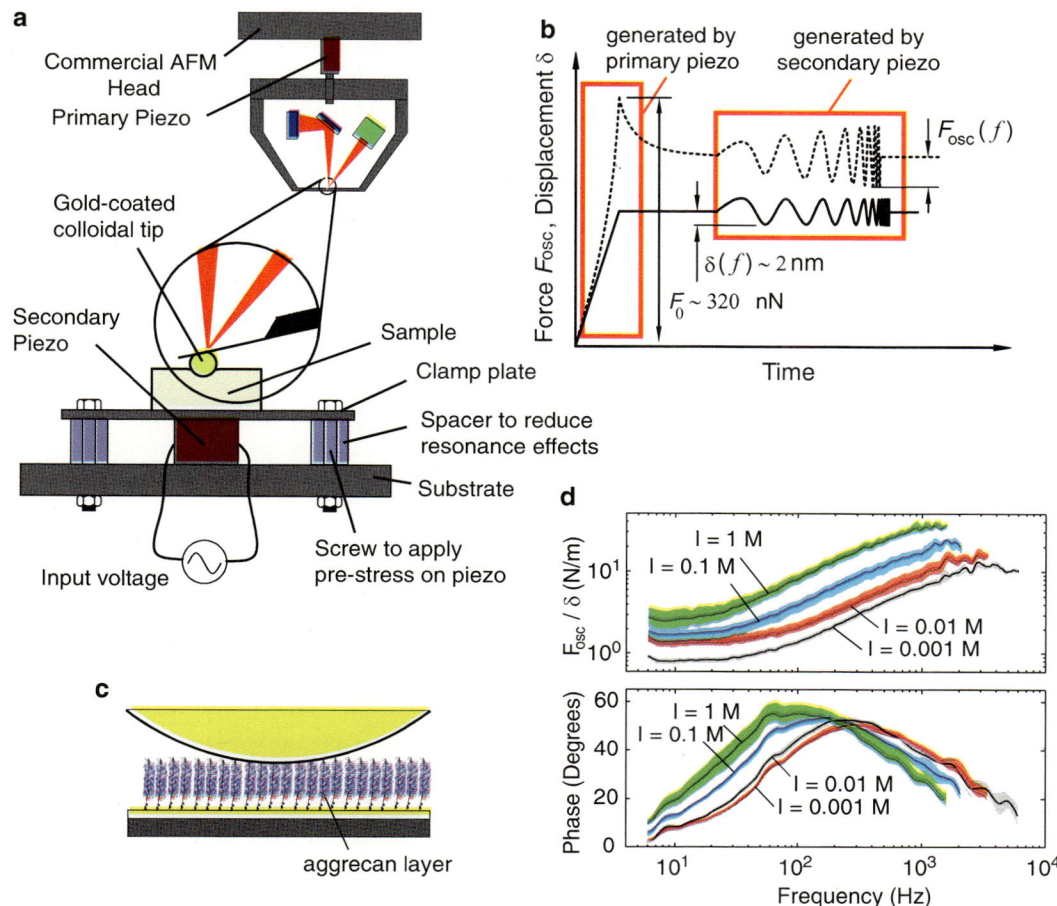

Fig. 4 (**a**) High-frequency nanorheology system coupled to a commercial AFM expands the spectrum of the rheological measurement of single aggrecan layers as well as native tissue samples up to 1 Hz to 10 kHz. This approach involves the formation of a well-defined contact distance between the probe and sample, which can be utilized in extracting key features of solid–fluid interaction at the nanoscale. (**b**) The applied displacement and resulting force profiles can comprise an initial indentation to form the contact distance between the aggrecan monolayer and AFM tip, followed by a series of sinusoidal nanoscale displacement (step size ~2 nm). (**d**) An example of the frequency dependence of the measured force using the protocol of (**b**), and the phase angle of the measured force with respect to the applied displacement, using the configuration of (**c**). Such data can be used to calculate the magnitude and phase of the dynamic modulus of the aggrecan layer over the frequency range of 1 Hz to 10 kHz. In (**d**), the *solid lines* are the mean values of experimental data and the shaded regions represent 95 % confidence intervals. Figures (**a**, **b**) from [23], reproduced by the permission of Elsevier © 2013; Figs. (**c**, **d**) from [33], reproduced by the permission of Orthopaedic Research Society © 2013

6. Apply the primary displacement signal. After the force relaxation that happens in the hold section of the profile, apply the secondary displacement signal. The primary signal is generated by the primary piezo of the commercial AFM, and the secondary signal is generated by the secondary piezo (Fig. 4a, *see* **Note** 7). Capture the deflection and z-piezo signals for

plotting the force-displacement curves. The sampling rate needs to be 100 kHz or higher. The force curve is calculated as deflection times the spring constant of the AFM cantilever. The displacement is calculated as the z-piezo subtracted by the deflection.

7. Calculate the magnitude and phase of the dynamic modulus (shown in Fig. 4d) from the capture force and displacement signals (*see* **Note 8**). From the wide-spectrum dynamic modulus, we quantify the important features of the dynamic nano-mechanics of the sample: the low-frequency modulus E_L, high-frequency modulus E_H, the peak frequency of the phase angle, f_{peak}, and the phase angle at the peak frequency Π_{peak}. These parameters are used as inputs to a finite element model (*see* Subheading 3.6) to extract the intrinsic poroelastic properties of the sample including the Young's modulus, hydraulic permeability, and anisotropy in elasticity of the sample.

3.6   Modeling

3.6.1   Modeling Electrostatic Interaction

In equilibrium, with no fluid flow, a compressive load on aggrecan is balanced by restoring forces that are composed of both electrostatic and nonelectrostatic contributions. The nonelectrostatic contributions include steric interactions between GAG chains, and the electrostatic contribution arises from the Coulomb forces associated with the presence of fixed charges on GAG chains (*see* **Note 9**). To model the electrostatic contribution at both the macroscopic and molecular level, continuum models can be used and have been employed [3]. Macroscopically, the electrostatic contribution has been viewed as a Donnan osmotic swelling pressure. In molecular level, the electrostatic contribution has been associated to repulsive electrostatic forces between GAG chains, which can be estimated using the Poisson-Boltzmann (PB) equations. Based on the PB equations, three models have been used, each of which employs a different charge distribution geometry for the GAG-associated charge groups [30].

In the first model (Fig. 5a), a polyelectrolyte brush layer is represented as a uniform, flat constant surface charge density. The second model (Fig. 5b) approximates the polyelectrolyte brush as a uniform volume charge density. Even though this model takes into account the height of the brush, the molecular shape and charge distribution along the polyelectrolyte GAG chains are not included. The third model (Fig. 5c) represents the time-average space occupied by the individual polyelectrolyte macromolecules in the brush as cylindrical rods of uniform volume charge density and finite height. This approach attempts to account for additional aspects of polymer molecular geometry and nonuniform molecular charge distribution inside the brush. This model is different from the "unit cell" model in which each polyelectrolyte macromolecule is represented as an infinitely long cylinder having a fixed surface charge.

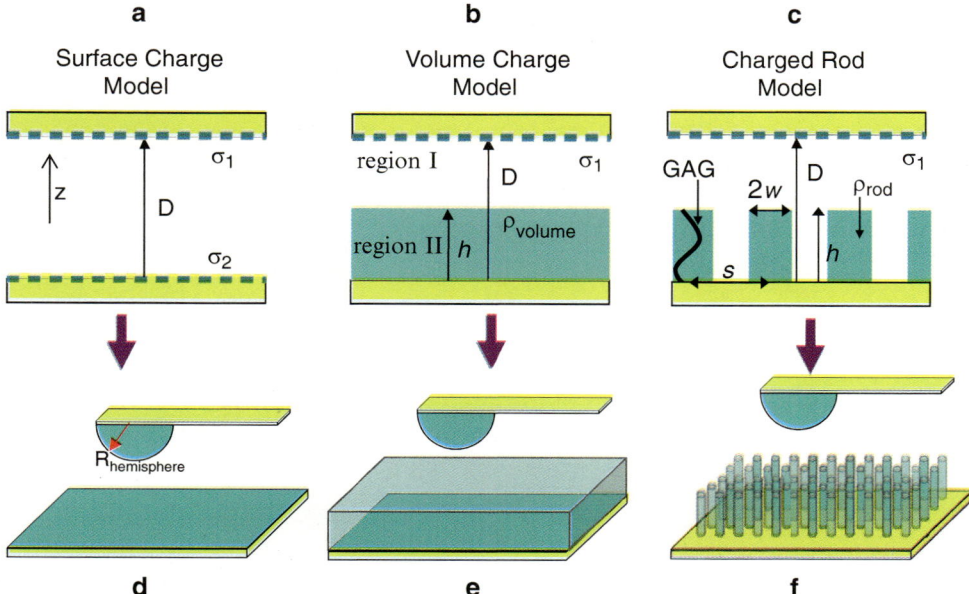

Fig. 5 Schematics of (**a**, **d**) a surface charge model; (**b**, **e**) a volume charge model; and (**c**, **f**) a GAG-rod model that can be used to model the electrostatic contributions of GAG chains in resisting compression. All these models can incorporate the Poisson-Boltzmann equations to estimate electrostatic interactions. Figures (**a**)–(**f**) from [30], reproduced by the permission of American Chemical Society © 2013

To experimentally validate these models, a layer of CS-GAG chains has been end grafted onto a gold-coated substrate and a sulfate-functionalized AFM probe tip has been used to apply compressive force on the GAG layer [30]. Although the total polyelectrolyte charge was the same in all three models, both the rod and volume charge models, which accounted for the height of the brush, predicted much higher forces than the surface charge model at any given separation distance. The comparison between measured and theoretically predicted forces shows that the rod model gives better agreement with the force data over the widest range of distance between indenter and the substrate for reasonable best-fit values of the brush height and rod radius. Changes in the rod radius led to changes in the shape of the predicted force profile. Therefore, in the framework of the PB theory, it appears that molecular level changes in the charge distribution inside polyelectrolyte brush layers, as manifested in the rod model, can significantly change the magnitude and the shape of the resulting force profile. Although the rod model is more general, it is also significantly more computationally intensive than the other two models. In certain experimental regimes, the volume charge model may be sufficient (e.g., when the polyelectrolyte molecules are less than an electrical Debye length apart).

3.6.2 Modeling
Fluid-Solid Interaction

Utilizing the AFM-based high-frequency rheology system, one can induce fluid flow through and between the GAG chains that are structurally part of aggrecan monolayers, and explore the role of frictional solid-fluid interactions at the molecular level. The 2-nm push release of the displacement profile results in a lateral low-Reynolds number fluid flow through the aggrecan layer, traveling back and forth over the contact distance. This flow is resisted by the frictional drag of GAG chains, which results in fluid pressurization between the AFM tip and substrate. By quantifying the magnitude and phase of the force exerted on the AFM tip due to solid stress and fluid pressurization (experimental procedures are explained in Subheading 2.5), we are able to estimate the poroelastic properties of the aggrecan monolayer, and explore the molecular origins of the mechanisms underlying hydraulic permeation, energy dissipation, and self-stiffening of the layer by utilizing a finite element model.

1. The aggrecan is end grafted covalently to the gold-coated substrate, where the substrate confines the aggrecan monolayer from lateral motion at the bottom of the layer (Fig. 4c). This lateral confinement results in higher stiffness in the lateral direction compared to the axial direction.

2. A transversely isotropic model [31] is implemented using the general-purpose commercial finite element software ABAQUS (Version 6.9, SIMULIA, Providence, RI). Because of the symmetry of the problem, the specimen is modeled using axisymmetric, poroelastic elements (CAX4P).

3. The probe tip indenter is modeled as a rigid surface since the spherical tip is much stiffer than the aggrecan monolayer. The probe tip is assigned a displacement history as described in Fig. 4b, and a zero-displacement boundary condition was assumed at the lower aggrecan-substrate interface. The indenter and the substrate surface are assumed to be impermeable to fluid flow, and the indenter-aggrecan contact region is assumed to be frictionless. The pore pressure is set to zero at the top surface of the aggrecan (excluding the indenter contact surface) and the side surfaces of the aggrecan to simulate free draining of the interstitial fluid from the aggrecan at those surfaces.

4. The relevant mechanical properties in this anisotropic model are the Young's modulus in axial direction E_a, the Young's modulus in transverse direction E_t, and the hydraulic permeability, k. Other parameters such as the Poisson's ratios ϑ_a and ϑ_{at} and the shear modulus G_t are assumed to be zero.

4   Notes

1. We use the inhibitor cocktail: Complete, Roche Applied Science, and Indianapolis, IN.

2. The thermal oscillation method was applied to determine the cantilever spring constant for all the AFM experiments [25].

3. Figure 2c depicts compression of an aggrecan-grafted probe tip onto an aggrecan-grafted substrate. In order to end graft aggrecan onto the probe tip, place 30 µl aggrecan solution onto the gold-coated AFM tip (on the side of the chip where the tip is located), and incubate in a humidity-controlled environment (e.g., airtight box with water) for ~48 h.

4. The force that the AFM tip experiences is calculated as (set point voltage – deflection signal) × deflection sensitivity × spring constant of the cantilever.

5. New modifications of the wedge method for calibration of nanosized probe tips directly applicable to such studies are given in [20].

6. To generate such an RBS signal, a sign operator ($\text{sign}(x) = 1$ for $x \geq 0$ and $\text{sign}(x) = -1$ for $x < 0$) is applied to a set of simulated white Gaussian noise data, implemented in LabView (National Instrument Co., Austin, TX). The amplitude of the resulting dataset is then scaled to the maximum allowable excitation given to the secondary piezo actuator. To control the bandwidth of the resulting RBS signal, we applied a low-pass filter to the white Gaussian noise, prior to the application of the sign operator. To control the bandwidth of the resulting RBS signal, we applied a low-pass filter to the white Gaussian noise, prior to the application of the sign operator.

7. The details of construction of the secondary piezo, needed for applications to higher frequencies than those normally possible with commercially available AFM instruments, is described in references [23, 28].

8. A digital Fourier transform can be used to calculate the amplitudes of the fundamental harmonics of both the displacement and force signals. The dynamic modulus is calculated as the ratio of the fast Fourier transform of the dynamic force $F_{osc}(f)$ and displacement signals $\delta(f)$. The magnitude of the dynamic modulus is then calculated as

$$|E^*(f)| = \frac{F_{osc}(f)}{\delta(f)} \frac{1}{2(R\delta_0)^{1/2}}$$

The phase of the dynamic modulus is the phase of the ratio of the force to displacement signal.

9. Such forces are often referred to as electrical double-layer repulsion forces; a derivation is given in [32] (Fields Forces and Flows in Biological Systems, Chapter 4).

Acknowledgements

Supported by Whitaker Foundation Fellowship, National Science Foundation (grant CMMI-0758651), and National Institutes of Health (grant AR060331).

References

1. Maroudas A (1980) Physical chemistry of articular cartilage and the intervertebral disc. The Joints and Synovial Fluid 2:239–291

2. Hardingham T, Fosang A (1992) Proteoglycans: many forms and many functions. FASEB J 6(3):861–870

3. Buschmann MD, Grodzinsky AJ (1995) A molecular-model of proteoglycan-associated electrostatic forces in cartilage mechanics. J Biomech Eng 117(2):179–192

4. Eisenberg SR, Grodzinsky AJ (1985) Swelling of articular cartilage and other connective tissues: electromechanochemical forces. J Orthop Res 3(2):148–159

5. Bayliss MT, Ali SY (1978) Age-related changes in the composition and structure of human articular-cartilage proteoglycans. Biochem J 176:683–693

6. Deutsch AJ, Midura RJ, Plaas AH (1995) Structure of chondroitin sulfate on aggrecan isolated from bovine tibial and costochondral growth plates. J Orthop Res 13(2):230–239

7. Bay-Jensen A-C, Hoegh-Madsen S, Dam E, Henriksen K, Sondergaard BC, Pastoureau P, Qvist P, Karsdal MA (2010) Which elements are involved in reversible and irreversible cartilage degradation in osteoarthritis? Rheumatol Int 30(4):435–442

8. Roughley PJ, White R (1980) Age-related changes in the structure of the proteoglycan subunits from human articular cartilage. J Biol Chem 255(1):217–224

9. Ng L, Grodzinsky AJ, Patwari P, Sandy J, Plaas A, Ortiz C (2003) Individual cartilage aggrecan macromolecules and their constituent glycosaminoglycans visualized via atomic force microscopy. J Struct Biol 143(3):242–257

10. Buckwalter J, Rosenberg L (1982) Electron microscopic studies of cartilage proteoglycans. Direct evidence for the variable length of the chondroitin sulfate-rich region of proteoglycan subunit core protein. J Biol Chem 257(16):9830–9839

11. Farndale RW, Buttle DJ, Barrett AJ (1986) Improved quantitation and discrimination of sulphated glycosaminoglycans by use of dimethylmethylene blue. Biochim Biophys Acta 883(2):173–177

12. Kopesky P, Lee H-Y, Vanderploeg E, Kisiday J, Frisbie D, Plaas A, Ortiz C, Grodzinsky A (2010) Adult equine bone marrow stromal cells produce a cartilage-like ECM mechanically superior to animal-matched adult chondrocytes. Matrix Biol 29(5):427–438

13. Lee H-Y, Kopesky P, Plaas A, Sandy J, Kisiday J, Frisbie D, Grodzinsky A, Ortiz C (2010) Adult bone marrow stromal cell-based tissue-engineered aggrecan exhibits ultrastructure and nanomechanical properties superior to native cartilage. Osteoarthritis Cartilage 18(11):1477–1486

14. Lee H-Y, Han L, Roughley P, Grodzinsky AJ, Ortiz C (2012) Age-related nanostructural and nanomechanical changes of individual human cartilage aggrecan monomers and their glycosaminoglycan side chains. J Struct Biol 181(3):264–273

15. Dean D, Han L, Ortiz C, Grodzinsky AJ (2005) Nanoscale conformation and compressibility of cartilage aggrecan using microcontact printing and atomic force microscopy. Macromolecules 38(10):4047–4049

16. Dean D, Han L, Grodzinsky AJ, Ortiz C (2006) Compressive nanomechanics of

opposing aggrecan macromolecules. J Biomech 39:2555–2565

17. Seog J, Dean D, Plaas A, Wong-Palms S, Grodzinsky A, Ortiz C (2002) Direct measurement of glycosaminoglycan intermolecular interactions via high-resolution force spectroscopy. Macromolecules 35(14):5601–5615

18. Seog J, Dean D, Rolauffs B, Wu T, Genzer J, Plaas AHK, Grodzinsky AJ, Ortiz C (2005) Nanomechanics of opposing glycosaminoglycan macromolecules. J Biomech 38(9): 1789–1797

19. Dean D (2005) Modeling and measurement of intermolecular interaction forces between cartilage ecm macromolecules. Doctoral dissertation, Massachusetts Institute of Technology, retrieved from DSpace. http://dspace.mit.edu/handle/1721.1/30153

20. Han L, Dean D, Ortiz C, Grodzinsky AJ (2007) Lateral nanomechanics of cartilage aggrecan macromolecules. Biophys J 92(4):1384–1398

21. Han L, Dean D, Daher LA, Grodzinsky AJ, Ortiz C (2008) Cartilage aggrecan can undergo self-adhesion. Biophys J 95(10):4862–4870

22. Wilbur JL, Kumar A, Kim E, Whitesides GM (1994) Microfabrication by microcontact printing of self-assembled monolayers. Adv Mater 6(7-8):600–604

23. Nia HT, Bozchalooi IS, Li Y, Han L, Hung H-H, Frank E, Youcef-Toumi K, Ortiz C, Grodzinsky A (2013) High-bandwidth AFM-based rheology reveals that cartilage is most sensitive to high loading rates at early stages of impairment. Biophys J 104(7):1529–1537

24. Han L (2007) Nanomechanics of cartilage extracellular matrix and macromolecules. Doctoral dissertation, Massachusetts Institute of Technology, retrieved from DSpace. http://dspace.mit.edu/handle/1721.1/42134

25. Hutter JL, Bechhoefer J (1993) Calibration of atomic-force microscope tips. Rev Sci Instrum 64:1868

26. Nia HT, Han L, Li Y, Ortiz C, Grodzinsky A (2011) Poroelasticity of cartilage at the nanoscale. Biophys J 101(9):2304–2313

27. Nia HT, Han L, Bozchalooi IS, Roughley P, Youcef-Toumi K, Grodzinsky AJ, Ortiz C, Aggrecan nanoscale solid-fluid interactions are a primary determinant of cartilage dynamic mechanical properties (Submitted)

28. Nia HT, Bozchalooi IS, Youcef-Toumi K, Ortiz C, Grodzinsky AJ, Frank E (2013) High-frequency rheology system. US Patent 8,516,610

29. Nia HT (2013) Nanomechanics of cartilage at the matrix and molecular levels. Doctoral dissertation, Massachusetts Institute of Technology, retrieved from DSpace. http://dspace.mit.edu/handle/1721.1/81709?show=full

30. Dean D, Seog J, Ortiz C, Grodzinsky AJ (2003) Molecular-level theoretical model for electrostatic interactions within polyelectrolyte brushes: applications to charged glycosaminoglycans. Langmuir 19(13):5526–5539

31. Cohen B, Lai WM, Mow VC (1998) A transversely isotropic biphasic model for unconfined compression of growth plate and chondroepiphysis. J Biomech Eng 120:491

32. Grodzinsky AJ (2011) Fields, forces, and flows in biological systems, chapter 4. Garland Science, New York, pp 259–272

33. Nia HT, Han L, Soltani I, Youcef-Toumi K, Grodzinsky A, Ortiz C (2013) Frequency-dependent nanomechanical behavior of aggrecan demonstrates that aggrecan is the dominant constituent responsible for the frequency dependence of cartilage poroelasticity. Orthopedic Research Society, San Antonio, TX, 2013

Use of Flow Cytometry for Characterization and Fractionation of Cell Populations Based on Their Expression of Heparan Sulfate Epitopes

Rebecca J. Holley, Raymond A. Smith, Els M.A. van de Westerlo, Claire E. Pickford, C.L.R. Merry, and Toin H. van Kuppevelt

Abstract

The ability to characterize alterations in heparan sulfate (HS) structure during development or as a result of loss or mutation of one or more components of the HS biosynthetic pathway is essential for broad understanding of the effects these changes may have on cell/tissue function. The use of anti-HS antibodies provides an opportunity to study HS chain composition in situ, with a multitude of different antibodies having been generated that recognize subtle differences in HS patterning, with the number and positioning of sulfate groups influencing antibody binding affinity. Flow cytometry is a valuable technique to enable the rapid characterization of the changes in HS-specific antibody binding in situ, allowing multiple cell types to be directly compared. Additionally fluorescent-activated cell sorting (FACS) allows fractionation of cells based on their HS-epitope expression.

Key words Flow cytometry, FACS, Heparan sulfate, Glycosaminoglycan, Phage-display scFv antibody

1 Introduction

Flow cytometry is a powerful technique, measuring the scattered light or fluorescence associated with cells as they flow in single file past a detector. Commonly, antibodies are used to detect cell surface-displayed epitopes with fluorescent detection via a secondary or tertiary antibody system. The power of the technique comes from the rapid detection of subtle differences in epitope expression between individual cells in a population and the ability to combine reagents to detect co-expression of epitopes. Modern machines can now simultaneously detect up to 20 separate colors using five lasers. Additionally, fluorescent-activated cell sorting (FACS) provides the opportunity to separate cells based on their

Kuberan Balagurunathan et al. (eds.), *Glycosaminoglycans: Chemistry and Biology*, Methods in Molecular Biology, vol. 1229, DOI 10.1007/978-1-4939-1714-3_21, © Springer Science+Business Media New York 2015

antibody-binding, thus enabling subpopulations to be segregated for analysis or further culture.

The glycosaminoglycan (GAG) heparan sulfate (HS) is expressed on the cell surface of virtually all cell types. However the length of the HS chains and the number and distribution of sulfate groups into sulfated and nonsulfated domains are tissue and cell type-specific characteristics [1, 2]. These differences are likely generated by the differential expression and activity of the large number of biosynthetic enzymes that function in concert in the Golgi apparatus to form a functional HS chain. Ultimately it is the number and type of sulfate groups and their distribution into distinct patterns that provides functionality, dictating which of the numerous HS-dependent growth factors, cytokines, enzymes, and extracellular matrix proteins can bind at the cell surface and elicit a cellular response. During development, rapid alterations in HS-structure occur, likely reflecting coordinated changes in the ability of cells to respond to successive waves of morphogens, mitogenic factors, and shifting matrix signals. Thus the ability to identify or isolate a particular cell type based on the type of HS expressed at the cell surface is extremely attractive and provides a mechanism to distinguish between cells with different developmental fates.

A few hybridoma-derived mouse anti-HS antibodies are available, e.g., JM403 and 10E4 [3, 4]; however, the nonimmunogenic character of HS makes this type of antibody difficult to raise. In contrast, the use of phage-display technology has enabled the generation of a large panel of single chain fragment variable (scFv) epitope-specific antibodies recognizing HS and heparin, as well as other GAGs chondroitin sulfate and dermatan sulfate [5–9]. These antibodies were selected from a large antibody phage display library panning against HS and other GAGs isolated from numerous sources, including bovine kidney, human skeletal muscle, and human lung [5], ensuring recognition of an array of different HS-patterns. Characterization of these antibodies has revealed that each antibody is able to recognize distinct features within the HS chain. Rather than recognizing only one specific sequence within the HS chain, they tend to have different binding affinities to the HS chain with preferences for binding dependent on the number and types of sulfation modification [10, 11]. Thus they can potentially bind a single HS chain at multiple sites along the length of the chain (Fig. 1a). When a number of these antibodies are used together as a panel, these provide a unique and highly versatile tool to study the topography, structure, and function of HS (Fig. 1b), by highlighting differences and similarities in binding between cell types (Fig. 2a, b). The antibodies can also be used in combination with other analytical techniques to gain further structural information or to move towards an understanding of how GAG structure influences function. For example, using a modification of the

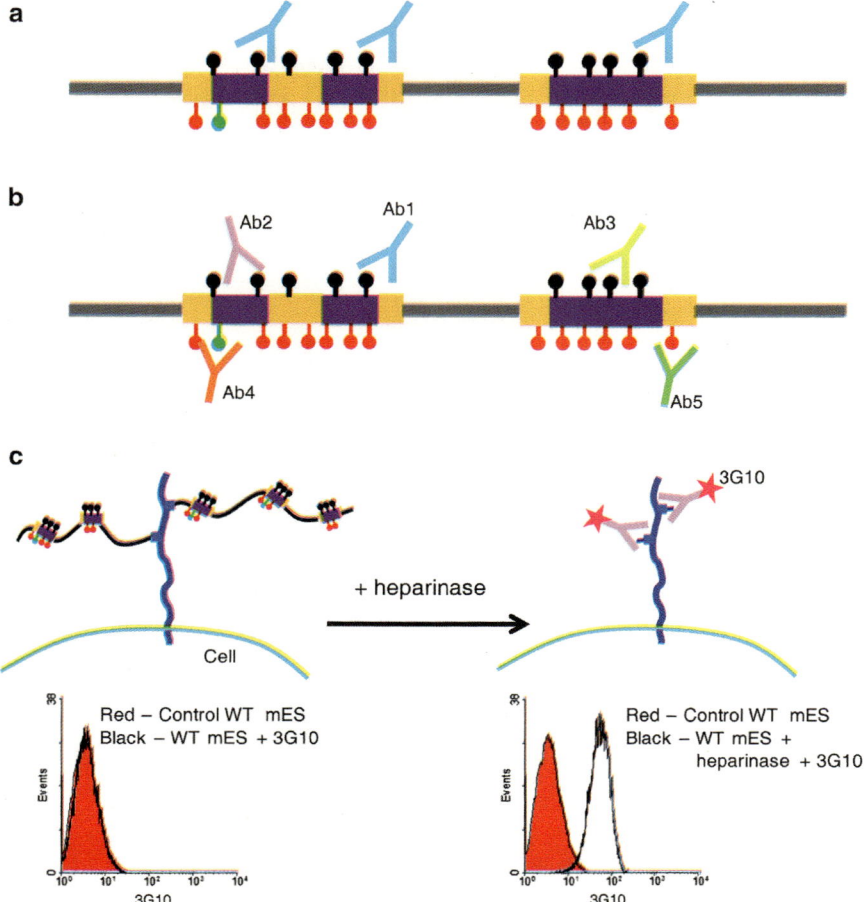

Fig. 1 Schematic showing HS-specific antibody binding to HS chains. (**a**) HS-epitope-specific antibodies have a preference for specific sulfate motifs which can occur at multiple sites along the HS chain. Therefore one antibody may potentially bind a single HS chain multiple times. (**b**) HS epitope-specific antibodies recognize different patterns within the HS chain, enabling them to be used as a panel. (**c**) Antibody 3G10 binds only to the HS stubs produced by heparinase action, allowing comparison of the number of chains per cell. Abbreviations: *Ab* antibody, *mES* mouse embryonic stem cells, *WT* wild type

standard flow cytometric method described here, growth factors can be used as probes to define the proportion of a cell population able to bind to a specific factor [12]. Of course, a further significant benefit is that the non-species-specific expression of GAGs means that reagents created against GAGs from one species are suitable for application to all animals.

An additional tool is the antibody 3G10, which is unique in that it only recognizes HS neo-epitopes created following removal of the HS chain by heparinase enzymes. Addition of bacterial heparinase enzymes to cells prior to antibody addition enables 3G10 to recognize the unsaturated uronic acid residue generated by heparinase enzyme action [3]. Thus 3G10 will bind the remaining HS

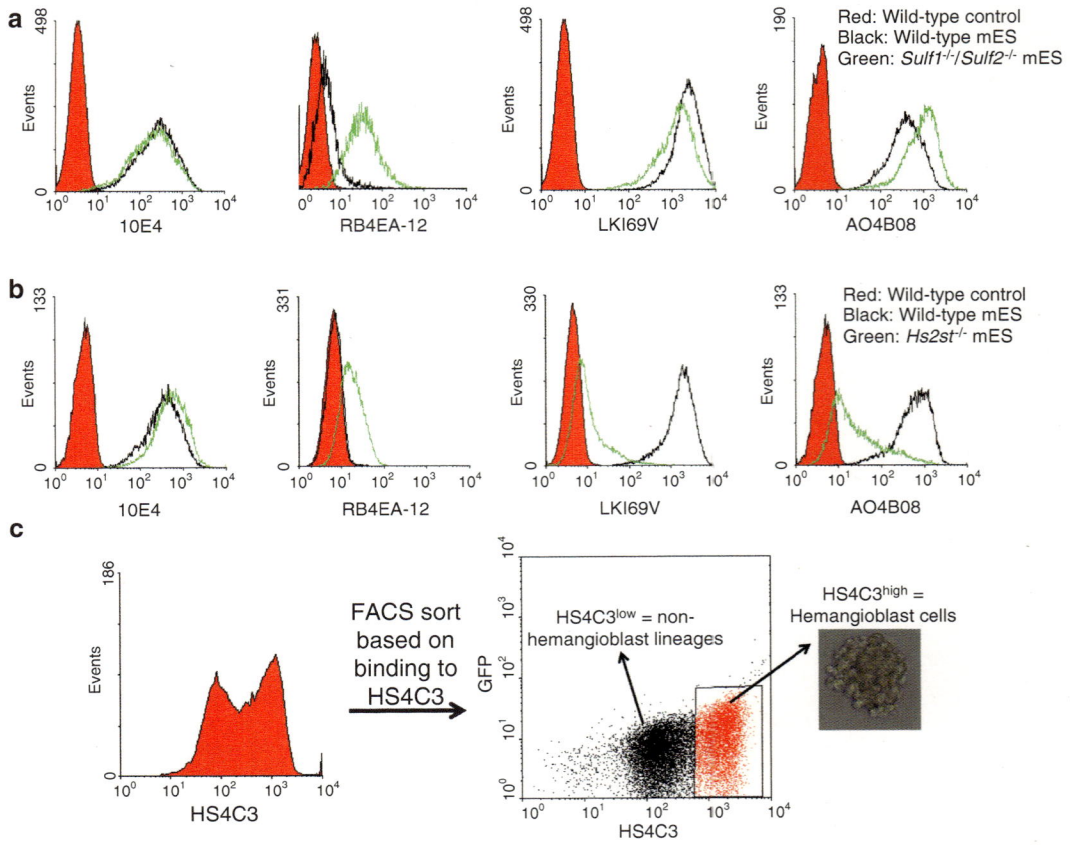

Fig. 2 Use of anti-HS antibodies to distinguish between cell types. (**a/b**) Use of a panel of anti-HS antibodies in wild-type and *Sulf1⁻/⁻/Sulf2⁻/⁻* mouse embryonic stem (mES) cells [17] (**a**) or wild-type and *Hs2st⁻/⁻* mES cells [18] (**b**) to reveal changes in HS patterning. (**c**) Use of anti-HS phage-display antibody HS4C3 to isolate cells with hemangioblast potential from mixed mesodermal lineages

stub attached to a protein core at the cell surface at a ratio of 1:1, detailing the number of chains expressed per cell (Fig. 1c). This is in contrast to pattern-specific HS antibodies that can potentially bind the chain multiple times (Fig. 1a).

The power of using these HS-specific antibodies in flow cytometry studies has been highlighted by numerous groups. The ability to study the shift in HS patterning resulting from the loss or increase in activity of the HS biosynthetic enzymes, either due to mutation or during disease progression provides essential information on the consequences these changes may have to protein binding (e.g., growth factors/cytokines/matrix factors) and therefore cell/tissue function. For example using these antibodies as a panel against different HS-biosynthetic knock-out embryonic stem cells (ES cells) reveals striking differences in the binding of some antibodies, while the binding of others is unaffected (Fig. 2a, b). Also the ability to identify unique shifts in the level or distribution of

HS structural motifs as a consequence of disease offers the potential to use these antibodies as biomarkers [13–15]. HS-epitopes are also rapidly altered during development in a lineage-specific manner, enabling cells to alter their responsiveness to external factors. FACS offers the opportunity to select cells based on their binding to HS-specific antibodies, demonstrated following the in vitro differentiation of ES cells to mesodermal lineages, whereby cells with hematopoietic and endothelial-forming potential (hemangioblasts) could be isolated from other mesodermal lineages purely on the basis of their HS-epitope expression using the antibody HS4C3 (Fig. 2c) [16]. Thus these antibodies provide a powerful tool to distinguish between cells displaying HS chains with subtle variations in sulfation levels and patterning.

2 Materials

Store all reagents at 4 °C under sterile conditions unless stated below. Cells should be grown under appropriate growth conditions as required for chosen cell line.

2.1 Antibodies

1. HS-specific primary antibodies:
   (a) Anti-mouse IgM 10E4 (Amsbio): use at 1:200.
   (b) Anti-mouse IgM JM403 (Amsbio): use at 1:100.
   (c) Anti-mouse IgG_{2b} 3G10 (Amsbio): use at 1:100.
   (d) Anti-HS scFv antibodies, e.g., RB4EA12 and AO4B08. and non-HS binding antibody MPB49 produced in a VSV-G tag version (*see* **Note 1**). Use all at 1:20 (*see* **Note 2**). Working aliquots can be made and stored at −20 °C to avoid repeated freeze-thaw.
   (e) Appropriate isotype controls.

2. Secondary and tertiary antibodies:
   (a) Anti-VSV-G antibody (e.g., Monoclonal Anti-VSV Glycoprotein antibody produced in mouse (IgG1), Sigma, 1:1,000).
   (b) Alexa Fluor or PE-conjugated secondary antibodies, e.g., Goat α-mouse IgM-PE (Santa Cruz Biotechnology, CS-3768; 1/100); Goat α-mouse IgG-PE (Molecular probes, 1/100); Goat α-mouse IgM-488 (Molecular Probes, 1/1,000); Goat α-mouse IgG-488 (Molecular Probes, 1/1,000).

2.2 Equipment

1. Hemocytometer.
2. 15 ml and 50 ml falcon tubes.
3. V-bottomed 96-well plate with lid, non-tissue culture treated (*see* **Note 3**).

4. Centrifuge that can hold 15 ml falcons and 96-well plates (*see* **Note 4**).

5. FACS tube, e.g., 5 ml polystyrene round bottom tube, capped top.

6. FACSCalibur flow cytometer and CellQuest software (both BD) or similar.

7. Cell strainer/sieve.

8. FACS Aria sorter and Diva 6 software (both BD) or similar.

9. WinMDI or similar analysis software.

2.3 Solutions

1. PBS.

2. Cell dissociation buffer, enzyme-free, EDTA, PBS-based (life technologies). Store at room temperature (*see* **Note 5**).

3. Trypan blue, 0.4 % (Sigma).

4. Heparinase I, II, and III (*see* **Note 6**).

5. FACS Buffer: 0.2 % BSA, 0.1 % sodium azide in PBS. Filtered by a 0.22-μm filter.

6. 1 % formaldehyde in PBS.

7. Tissue culture growth media appropriate to cell type with addition of 10 % fetal calf serum.

8. Tissue culture growth media appropriate to cell type with added 25 mM HEPES.

3 Methods

3.1 Cell Preparation

Steps can be carried out either in a Class II Biosafety cabinet or on the bench.

1. Remove media from cells and wash twice with excess PBS.

2. Add enough cell dissociation buffer to completely cover cell layer, e.g., 10 ml for T75 flask (*see* **Notes 7** and **8**).

3. Incubate at 37 °C for 10 min. Rock cell vessel after 5 min to re-distribute dissociation reagent.

4. After 10 min observe cells under the microscope. If cells are beginning to round up, tap flask and mix up and down using a pipette tip to dislodge cells and dissociate into single cells. If necessary replace at 37 °C for 5 min, then repeat pipetting until a single cell suspension is achieved (*see* **Notes 9** and **10**).

5. Pool cell suspension into a 15 ml falcon and top up to 15 ml with cell growth media containing 10 % serum (*see* **Note 11**).

6. Centrifuge at $100 \times g$ for 3 min to pellet cells.

7. Discard supernatant and resuspend cell pellet in appropriate volume of fresh cell growth medium, e.g., 1 ml and mix well.

8. Remove 10 μl cells and mix with 10 μl trypan blue. Using a 10 μl pipette tip, draw up some of the cell suspension containing trypan blue and carefully fill the hemocytometer before determining cell concentration/viability. To obtain an accurate cell concentration, between 50 and 150 cells should be counted per hemocytometer grid. If necessary, dilute original cell stock and repeat this step.

3.2 Antibody Staining of Cells for Flow Cytometric Analysis

1. *Optional step*: heparinase digestions, e.g., for 3G10. Pellet cells at $100 \times g$ for 3 min and remove supernatant. Resuspend at a concentration of 5×10^6 cells/ml in PBS and add 2 mIU each of heparinase I, II, and III per 100 μl of cell suspension, incubating at room temperature for 1 h. Continue to next step.

2. Aliquot between 10^5 and 10^6 cells/well of a V-bottom 96-well plate (*see* **Notes 12** and **13**).

3. Centrifuge 96-well plate at $100 \times g$ for 3 min. Cell pellet should be obvious at the center of the well. Aspirate supernatant without disturbing the cell pellet (*see* **Note 14**).

4. Resuspend pellet in 100 μl FACS buffer containing diluted primary antibody or isotype control. Also include a no antibody control (*see* **Note 15**). Incubate at 4 °C for 1 h.

5. Centrifuge plate to pellet cells and remove supernatant. Wash with 200 μl PBS. Pellet cells and repeat with second wash of PBS.

6. Pellet cells, remove supernatant, and resuspend in 100 μl FACS buffer containing secondary antibody—either fluorescent-conjugated secondary antibody or for the scFv antibodies anti-mouse VSV-G, incubating at 4 °C for 1 h.

7. Pellet cells and wash twice with 200 μl PBS as above.

8. For scFv antibodies a tertiary detection step is necessary. Repeat **step 6/step 7** using appropriate fluorescent conjugated anti-mouse antibody.

9. Resuspend cell pellet in 200 μl 1 % formaldehyde/PBS and transfer to 5 ml polystyrene tube (*see* **Note 16**).

10. Analyze samples using a FACSCalibur (BD) flow cytometer and CellQuest software (BD) or similar. Gate viable cells using forward and side scatter, measuring the fluorescence of this population (*see* **Note 17**).

11. Data analysis can be performed post-run using WinMDI (or similar) software.

3.3 Antibody Staining of Cells for FACS Separation

All steps should be carried out in a Class II Biosafety cabinet and cells should be kept on ice at all times.

1. Harvest single cell suspension as detailed above and determine viable cell count. Remember to remove a small cell sample for controls to set up the FACS machine, using the preparation of samples for flow cytometry protocol as detailed above (*see* **Note 18**).

2. Pool cells for sorting into a 50 ml falcon tube and centrifuge at $100 \times g$ for 3 min to pellet.

3. Prepare primary antibody in the cells normal growth media containing 10 % FCS in a volume to allow resuspension of the cells at 5×10^6 cells/ml. Filter solution through 0.2 μm syringe filter and use to resuspend cells (*see* **Note 19**). Incubate on ice for 1 h.

4. Wash cells by filling remaining volume of 50 ml falcon with growth media/10 % FCS and centrifuge as above to pellet.

5. Repeat wash step.

6. Prepare filtered fluorescent or anti-VSV-G secondary antibody solution as above and use to resuspend cell pellet. Incubate for 1 h on ice.

7. Wash cells twice as above.

8. For phage display antibodies only repeat preparation of fluorescent tertiary antibody and washes as above.

9. Following final wash, resuspend cells at 1×10^7 cells/ml (*see* **Note 20**) in HEPES buffered growth media without serum (*see* **Note 21**).

10. Fix a cell strainer onto a clean FACS tube and prewet by applying 2 ml PBS and allowing it to run through. Discard PBS. Add cell suspension to sieve and collect flow through (*see* **Note 22**).

11. Sort cells using a Becton Dickinson FACS Aria or similar, gating around live cells to separate out debris and dead cells. Collect desired fluorescent region into sterile tubes (*see* **Note 23**).

12. Centrifuge recovered cells to pellet and resuspend in 1 ml fresh growth media/10 % FCS. Perform viable cell count. Use in desired downstream application (*see* **Note 24**).

4 Notes

1. Phage display scFv antibodies are VSV-G tagged. A three-step primary-, secondary-, and tertiary-antibody system gives the best results for flow cytometry. Following primary scFv antibody addition, an anti-VSV-G secondary is used (e.g., mouse

anti-VSV-G), followed by an anti-species-specific fluorescently-labeled tertiary antibody (e.g., Goat α-mouse IgG-PE). A directly labeled anti-VSV-G tag antibody such as Anti-VSV-G-FITC (Abcam ab3863) may be used, however fluorescence is increased if a tertiary detection system is used.

2. Antibodies may need to be titrated for optimal discrimination of epitope-binding dependent on the cell type and detection system (secondary or secondary/tertiary) used. Typically a dilution between 1:10 to 1:200 gives satisfactory results. It is worth noting that many of the HS-epitope antibodies recognize 'preferred' 3-dimensional GAG conformations rather than binding in an 'all or nothing' fashion. Therefore decreasing the abundance of potential binding sites by reducing the cell:antibody ratio can improve separation of cells with different HS patterning by selecting for binding to more 'preferred' sequences.

3. Using V-bottomed plates enables easy capture and visualization of the cell pellet upon centrifugation. U-bottomed plates may also be used. Flat bottomed plates should not be used as there is a high risk of losing the pellet. Use of non-tissue culture-treated plates reduces cell attachment.

4. If a 96-well plate centrifuge is not available the procedure can be completed in eppendorf or 5 ml polystyrene tubes. However, cell loss at each wash/centrifuge step is typically higher in larger vessels. Therefore it is necessary to start with a higher cell number per condition, e.g., 10^6 cells.

5. Enzyme based dissociation methods such as trypsin should not be used because the core proteins to which many GAG chains are attached are trypsin sensitive. Therefore only enzyme-free methods should be used. Dissociation buffer can be made in house if required, using 0.02 % ethylenediamine tetraacetic acid (EDTA) in PBS.

6. If heparinase I, II, and III are not all available, use of heparinase III alone can give satisfactory 3G10 staining results as only one heparinase-mediated cleavage per chain is required for 3G10 reactivity. However heparinase III alone is unlikely to fully remove the cell surface HS chains.

7. For optimal detachment of cells in cell dissociation buffer, cells should be subconfluent prior to use. If cells are in dense monolayer culture additional steps may be required for successful removal (see **Note 9**).

8. Cell dissociation buffer can also be used on cell aggregates such as embryoid bodies. Aggregates need to be pelleted and washed in PBS prior to addition of excess cell dissociation buffer in a 50 ml falcon tube. Incubate aggregates in a 37 °C water bath for 10 min, swirling tube contents every 2–3 min.

After 10 min pipette cells up and down to disperse into a single cell suspension. Repeat incubation (for maximum of 30 min) as required.

9. Leave cells in dissociation buffer for a maximum total time of 30 min to decrease cell damage. If cell detachment is poor (or cell cultures are dense prior to detachment), the protocol can be modified to include gentle use of a cell scraper following the initial 10 min incubation in cell dissociation buffer. Cells should then be incubated for a further approximately 5 min in cell dissociation buffer to enable the cells to round. Pipetting using a 1 ml pipette may be required to ensure cells are detached into a single cell suspension.

10. It is essential that a single cell suspension is achieved to allow successful staining and to prevent blockage of the flow cytometer.

11. Adding media with 10 % FCS at this stage increases the recovery of cells from cell dissociation buffer during the following centrifugation step. However if serum-free culture is required this step can be omitted.

12. Using less than 10^5 cells often results in insufficient cells for meaningful flow analysis. Using higher than 10^6 cells may result in reduced signal intensity due to insufficient antibody to bind all HS-epitopes.

13. It is important if directly comparing two or more cell lines with HS-pattern-specific antibodies that an equal number of cells are used for each cell line per antibody condition and the experiment is performed at the same time. This is essential to allow accurate comparison of the affinity of antibody binding (and therefore fluorescence) between cell types.

14. It is useful to tilt the plate when aspirating and remove solutions down to the start of the 'V' bottom to avoid disrupting the pellet.

15. Controls are absolutely essential for setting up gates, channel voltages, and correct compensation on the flow cytometer and for data interpretation. A sample of unstained cells (for each cell type examined) is absolutely required for each analysis for initial flow cytometer set-up. Nonspecific staining control for each fluorochrome and each cell type is necessary to identify any nonspecific binding of your antibody. Use either appropriate isotype controls or secondary antibody alone (no primary antibody). For the scFv antibodies include secondary/tertiary (no primary) controls as well as using a non-HS binding scFv control (e.g., MPB49). If multiple fluorochromes are used together, single stained cells (with only one antibody-fluorochrome at a time) are necessary to enable correct compensation between channels.

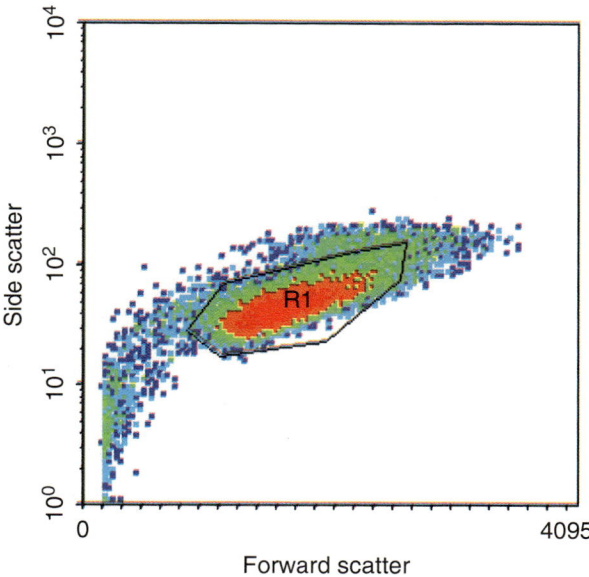

Fig. 3 Plot showing forward scatter/side scatter characteristics allowing the identification of live cells

16. Once fixed in 1 % formaldehyde/PBS at the end of the experiment, cell samples can be stored at 4 °C in the dark for up to 1 month prior to analysis without affecting results, provided samples are not allowed to dry out.

17. Using forward scatter and side scatter characteristics is usually sufficient to allow identification of live cells (*see* Fig. 3). However if necessary a live/dead viability stain such as Propidium Iodide could be used to identify the live cell population.

18. As a minimum an unstained control sample will be required to allow gates and voltages to be set. Ideally the cell sample should be resuspended at the same cell density as for the stained sample (10^7 cells/ml) to aid cytometer set-up.

19. Filtering may not be necessary if the antibody has been prepared under sterile conditions. Filtering should not lead to antibody loss.

20. The cell density may need to be altered depending on the speed/type of FACS machine used.

21. Using a HEPES buffered media may aid in viability by buffering the pH of the sample while sorting.

22. This step is essential to ensure that only single cells are applied to the FACS machine. Cell clumps make accurate sorting difficult and may lead to machine blockage.

23. Adding a small volume of cell growth media/10 % FCS to the bottom of the tube used for sorting into may increase recovered cell viability.

24. It is often necessary following FACS sorting to include penicillin/streptomycin and/or anti-fungicides in the culture media to prevent contamination if cells are to be used for future experiments.

Acknowledgements

This work is supported by a strategic award from the Medical Research Council (UK) and British Heart Foundation (G0902170), the Engineering and Physical Sciences Research Council (EP/H046070/1), the Dutch Cancer Society (2008-4058) and the Netherlands Institute of Regenerative medicine (2.5).

References

1. Ledin J, Staatz W, Li JP et al (2004) Heparan sulfate structure in mice with genetically modified heparan sulfate production. J Biol Chem 279:42732–42741

2. Maccarana M, Sakura Y, Tawada A et al (1996) Domain structure of heparan sulfates from bovine organs. J Biol Chem 271:17804–17810

3. David G, Bai XM, Van der Schueren B et al (1992) Developmental changes in heparan sulfate expression: in situ detection with mAbs. J Cell Biol 119:961–975

4. van den Born J, Salmivirta K, Henttinen T et al (2005) Novel heparan sulfate structures revealed by monoclonal antibodies. J Biol Chem 280:20516–20523

5. Dennissen MA, Jenniskens GJ, Pieffers M et al (2002) Large, tissue-regulated domain diversity of heparan sulfates demonstrated by phage display antibodies. J Biol Chem 277:10982–10986

6. Jenniskens GJ, Oosterhof A, Brandwijk R et al (2000) Heparan sulfate heterogeneity in skeletal muscle basal lamina: demonstration by phage display-derived antibodies. J Neurosci 20:4099–4111

7. Smetsers TFCM, van de Westerlo EMA, ten Dam GB et al (2003) Localization and characterization of melanoma-associated glycosaminoglycans: differential expression of chondroitin and heparan sulfate epitopes in melanoma. Cancer Res 63:2965–2970

8. van de Westerlo EMA, Smetsers TFCM, Dennissen MABA et al (2002) Human single chain antibodies against heparin: selection,

characterization, and effect on coagulation. Blood 99:2427–2433

9. van Kuppevelt TH, Dennissen MA, van Venrooij WJ et al (1998) Generation and application of type-specific anti-heparan sulfate antibodies using phage display technology. Further evidence for heparan sulfate heterogeneity in the kidney. J Biol Chem 273:12960–12966

10. Kurup S, Wijnhoven TJ, Jenniskens GJ et al (2007) Characterization of anti-heparan sulfate phage-display antibodies AO4B08 and HS4E4. J Biol Chem 282(29):21032–21042

11. ten Dam GB, Kurup S, van de Westerlo EM et al (2006) 3-O-sulfated oligosaccharide structures are recognized by anti-heparan sulfate antibody HS4C3. J Biol Chem 281:4654–4662

12. Johnson CE, Crawford BE, Stavridis M et al (2007) Essential alterations of heparan sulfate during the differentiation of embryonic stem cells to Sox1-enhanced green fluorescent protein-expressing neural progenitor cells. Stem Cells 25:1913–1923

13. Hosono-Fukao T, Ohtake-Niimi S, Hoshino H et al (2012) Heparan sulfate subdomains that are degraded by Sulf accumulate in cerebral amyloid ss plaques of Alzheimer's disease: evidence from mouse models and patients. Am J Pathol 180:2056–2067

14. Smits NC, Shworak NW, Dekhuijzen PN et al (2010) Heparan sulfates in the lung: structure, diversity, and role in pulmonary emphysema. Anat Rec (Hoboken) 293:955–967

15. Vallen MJ, Massuger LF, ten Dam GB et al (2012) Highly sulfated chondroitin sulfates, a

novel class of prognostic biomarkers in ovarian cancer tissue. Gynecol Oncol 127:202–209

16. Baldwin RJ, ten Dam GB, van Kuppevelt TH et al (2008) A developmentally regulated heparan sulfate epitope defines a subpopulation with increased blood potential during mesodermal differentiation. Stem Cells 26: 3108–3118

17. Lamanna WC, Baldwin RJ, Padva M et al (2006) Heparan sulfate 6-O-endosulfatases: discrete in vivo activities and functional cooperativity. Biochem J 400:63–73

18. Merry CL, Bullock SL, Swan DC et al (2001) The molecular phenotype of heparan sulfate in the Hs2st−/− mutant mouse. J Biol Chem 276:35429–35434

Chapter 22

A Transgenic Approach to Live Imaging of Heparan Sulfate Modification Patterns

Matthew Attreed and Hannes E. Bülow

Abstract

Heparan sulfate (HS) glycosaminoglycan chains contain highly modified HS domains that are separated by sections of sparse or no modification. HS domains are central to the role of HS in protein binding and mediating protein–protein interactions in the extracellular matrix. Since HS domains are not genetically encoded, they are impossible to visualize and study with conventional methods in vivo. Here we describe a transgenic approach using previously described single chain variable fragment (scFv) antibodies that bind HS in vitro and on tissue sections with different specificities. By engineering a secretion signal and a fluorescent protein to the scFvs and transgenically expressing these fluorescently tagged antibodies in *Caenorhabditis elegans*, we are able to directly visualize specific HS domains in live animals (Attreed et al. Nat Methods 9(5):477–479, 2012). The approach allows concomitant colabeling of multiple epitopes, the study of HS dynamics and, could lend itself to a genetic analysis of HS domain biosynthesis or to visualize other nongenetically encoded or posttranslational modifications.

Key words Heparan sulfate, Single chain variable fragment (scFv) antibody, *Caenorhabditis elegans*, Live imaging, Nongenetically encoded molecules

1 Introduction

Heparan sulfate (HS) is an unbranched polysaccharide of repeating disaccharide units composed of N-acetylglucosamine and glucuronic acid [1]. During synthesis, the disaccharide repeats are modified by a set of type II transmembrane Golgi-resident enzymes which can remove acetyl groups, add or remove sulfate groups, and convert glucuronic acid into the stereoisomeric iduronic acid (Fig. 1a) [4, 5]. The glycan chain is invariably attached via a characteristic tetrasaccharide linker to a serine on core proteins that include the extracellular membrane-attached syndecans and glypicans as well as secreted perlecan and collagen XVIII [1]. Core proteins, together with their glycan chains, are referred to as heparan sulfate proteoglycans (HSPG) [1].

Kuberan Balagurunathan et al. (eds.), *Glycosaminoglycans: Chemistry and Biology*, Methods in Molecular Biology, vol. 1229, DOI 10.1007/978-1-4939-1714-3_22, © Springer Science+Business Media New York 2015

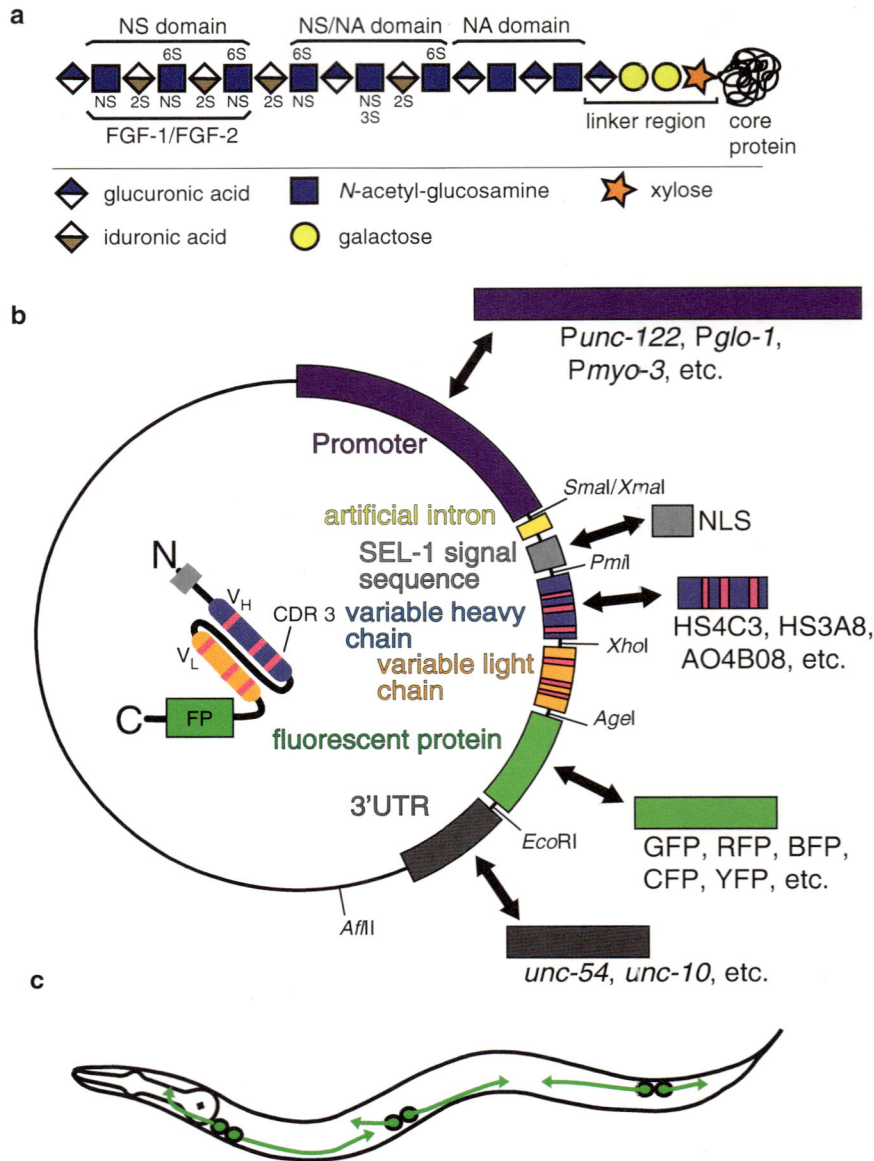

Fig. 1 (**a**) Schematic of the basic HS structure organized into domains with varying degrees of sulfation. Modifications and putative growth factor binding sites are indicated. (**b**) Schematic of the secreted scFv antibody fusion (*inset*) and the basic plasmid construct [2]. The construct was designed in a modular fashion to allow for easy swapping of the heavy chains, targeting sequence, 3′ UTR, fluorophores, and promoters. *Two-headed arrows* indicate alternative modules that could be exchanged with existing modules. Common restriction sites are indicated. *NLS* nuclear localization signal. (**c**) Diagram of a transgenic worm expressing the antibody construct from specific cells (coelomocytes in this example). The antibody is secreted into the pseudocoelom where it diffuses and binds to heparan sulfate domains. Unbound antibody is taken up by the coelomocytes which absorb material from the pseudocoelom [3]

HSPGs have been shown to serve diverse functions in physiology and development [6–8]. Many of these functions are mediated by specific HS modification patterns interacting with a variety of molecules, including growth factors and receptors. At least some of these protein–glycan interactions are highly specific (such as the antithrombin III binding to a characteristic pentasaccharide) whereas other interactions may be less specific [9]. Highly modified regions of the HS chains are interspersed with less modified or unmodified regions where the distance between modified domains is essential for function [10]. The HS domains appear highly specific in their tissue and age distribution [11]. However, without live imaging approaches it has been impossible to study the developmental dynamics of HS domains in vivo, or to pursue genetic approaches to understand HS biosynthesis on a mechanistic level.

Single chain variable fragment (scFv) antibodies are synthetic antibodies based on human heavy and light chain variable regions that have been assembled in vitro [12, 13]. By randomizing the complementarity determining region 3, vast phage display cDNA libraries of scFv antibodies have been created [12, 13]. By panning these libraries with an epitope of interest, scFvs against essentially all types of epitopes have been identified, including glycosaminoglycans such as HS, which are notorious for displaying low antigenicity in traditional approaches [14]. Using a number of different HS/heparin preparations, van Kuppevelt and colleagues identified at least 36 scFv antibodies that bind HS or heparin with varying specificities [14–21]. Some of these scFv antibodies have been characterized as binding to specifically modified heparan sulfate oligosaccharides in vitro (e.g., AO4B08 and HS4E4, [22]) but the majority has not been molecularly characterized in detail.

Caenorhabditis elegans is widely used in genetic and neuroscience research owing to its transparency, the ease of transgenic approaches and the possibility to utilize fluorescent proteins to analyze and characterize cells, organelles, and proteins in high resolution. The *C. elegans* genome encodes one of each class of the heparan sulfate modifying enzymes [7], making this nematode an excellent system to study the genetic underpinnings of heparan sulfate modifications during development and physiology. We have used *C. elegans* and transgenically expressed anti-heparan sulfate-specific scFv antibodies to devise a system for the in vivo study of nongenetically encoded HS modification patterns (Fig. 2) [2].

2   Materials and Equipment

2.1 Nematode Husbandry

1. Nematodes and the OP50 *E. coli* feeding strain can be obtained from the *Caenorhabditis elegans* Genetics Center at the University of Minnesota (www.cbs.umn.edu/cgc).

2. Nematode Growth Medium (NGM) agar plates for growing worms require 60 mm diameter petri plates (e.g., from Tritech

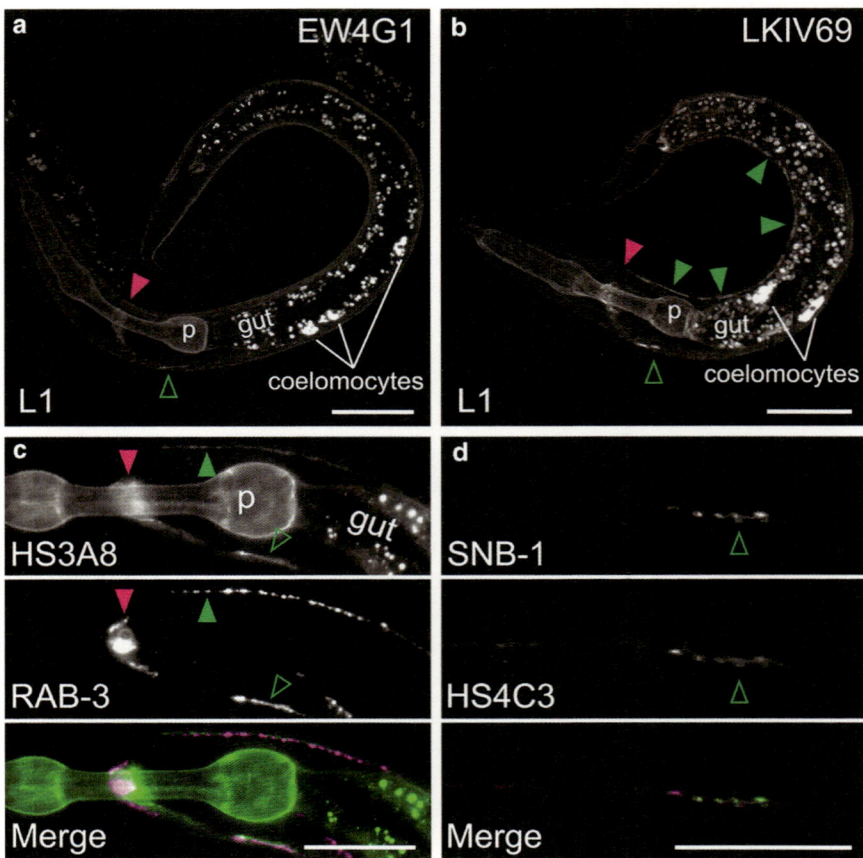

Fig. 2 (**a**, **b**) Epifluorescent micrographs of transgenic animals expressing an EW4G1::GFP (**a**) or LKIV69::GFP fusion (**b**) under control of the coelomocyte-specific *Punc-122* [23] promoter show staining of the nervous system in *C. elegans*. In all panels, *filled* or *open green arrow heads* indicate the dorsal and ventral nerve cords, respectively and, *magenta arrowheads* indicate neuronal staining associated with the *C. elegans* nerve ring. No comparable staining was observed when a control scFv antibody fusion (MPB49::GFP) was expressed under the same conditions [2]. The gut with characteristic vesicular autofluorescence is indicated. *Scale bars* indicate 25 μm in all panels and *p* pharynx. (**c**) Epifluorescent micrographs of transgenic animals expressing an HS3A8::GFP antibody fusion under control of the coelomocyte-specific *Punc-122* (*upper panel*) and a pre-synaptic marker (mCherry::RAB-3 fusion) under control of a promoter specific for GABAergic neurons (*middle panel*) (Attreed and Bülow, unpublished). A merged image (*lower panel*) shows partial colocalization of HS3A8-specific HS epitopes with the presynaptic marker in GABAergic neurons. (**d**) Epifluorescent micrographs of transgenic animals expressing an HS4C3::DsRed2 antibody fusion under control of the coelomocyte-specific *Punc-122* (*upper panel*) and a presynaptic marker (SNB-1::GFP) under control of a promoter specific for GABAergic neurons (*middle panel*) (*juls1*, [24]). A merged image (*lower panel*) shows partial colocalization of HS4C3-specific HS epitopes with the presynaptic marker in GABAergic neurons

Research, www.tritechresearch.com), a peristaltic pump or 50 mL serological pipettes and a pipetting ball.

3. A dissecting microscope with a transmitted light base (e.g., Zeiss Stemi 2000).

4. Platinum wire 99.95 % platinum, 0.05 % iridium 0.01 inches diameter wire (e.g., PT-9901, Tritech Research, www.tritechresearch.com), Glass Pasteur pipets, 1,000 μL pipette tips, hair (from eyelash, pet whisker, or paint brush), cyanoacrylate adhesive "superglue."

2.2 Solutions and Media

1. M9 buffer solution: 3 g KH_2PO_4, 6 g Na_2HPO_4, 5 g NaCl, 1 mL 1 M $MgSO_4$, H_2O to 1 L. Sterilize the solution and store it at room temperature.

2. LB media: 10 g tryptone, 5 g yeast extract, 5 g NaCl, H_2O to 1 L. Autoclave the solution and store it at 4 °C.

3. 40 mM levamisole stock solution: 0.0963 g levamisole hydrochloride, 10 mL ddH_2O. The solution should be aliquoted and stored at –20 °C.

4. 0.6 M 2,3-butanedione monoxime (BDM) stock solution: 0.121 g 2,3-BDM (B0753 Sigma-Aldrich), 2 mL M9 buffer solution. This solution should be prepared fresh each day.

5. 3 % agarose for pads: 3 g agarose, 100 mL ddH_2O. Bring to a boil and keep on a hot plate.

6. LB agar plates:
   (a) Agar solution: 10 g tryptone, 5 g yeast extract, 5 g NaCl, 15 g agar, H_2O to 1 L. Autoclave the solution.
   (b) Cool flask in 55 °C water bath for 15 min.
   (c) Add 1 mL of ampicillin (from stock solution of 100 mg/mL in ddH_2O) before pouring plates if needed.
   (d) Pour approximately 20 mL of agar into 10 cm petri dishes using peristaltic pump or serological pipettes.

7. NGM (*Nematode Growth Medium*) plates:
   (a) Agar solution: 17 g agar, 3 g NaCl, 2.5 g peptone, 975 mL ddH_2O. Combine in a 2 L Erlenmeyer flask, cover with aluminum foil, and autoclave.
   (b) 1 M KPO_4 buffer: 108.3 g KH_2PO_4, 35.6 g K_2HPO_4, ddH_2O to 1 L.
   (c) Cool flask in 55 °C water bath for 15 min.
   (d) Add 1 mL 1 M $CaCl_2$, 1 mL 5 mg/mL cholesterol in ethanol, 1 mL 1 M $MgSO_4$, and 25 mL 1 M KPO_4 buffer. Mix the solution.
   (e) Pour approximately 10 mL into 60 mm petri dishes using peristaltic pump or serological pipettes.

2.3 scFv Constructs

1. Competent bacterial cells for molecular biology.
2. Site-directed mutagenesis kit (e.g., Agilent QuickChange II).
3. LB agar plates with ampicillin.

4. Restriction enzymes (depending on subcloning goals) and materials for standard molecular biology techniques as needed.

2.4 Microinjection and Imaging

1. A standard set-up for microinjection of nematodes is required, including an inverted microscope (e.g., Zeiss Axio Observer), a needle holder (e.g., microINJECTOR Brass Straight-Arm Needle Holder MINJ-4 from Tritech Research, www.tritechresearch.com) combined with a micromanipulator (e.g. Three-Axis Oil Hydraulic Fine Micromanipulator MMO-203, Tritech Research, www.tritechresearch.com), compressed nitrogen gas tank and a pedal-operated valve for injection with pressurized nitrogen (e.g., microINJECTOR System MINJ-1 Tritech Research, www.tritechresearch.com).

2. Microscope slides (75×25 mm) and cover slips (60×24 mm for injection pads and 22×22 mm for imaging) are necessary for microinjection and imaging. Halocarbon Oil 700 (H8898 Sigma-Aldrich) is used to immerse and immobilize the worms on the slides.

3. Needles: Microinjection requires thin glass needles that can be made from thin-walled 1.0 mm outer diameter 0.75 mm inner diameter borosilicate glass capillaries with an internal glass filament (e.g., TW100F-4 World Precision Instruments). These glass capillaries are pulled into two needles using a needle puller such as a Narishige PC-10. The settings of the needle puller vary depending on the make and model as well as the filament age and have to be determined by trial and error. Generally, needles should taper quickly to a fine point. If the needle tapers too slowly or too fast the needle may bend too much to penetrate the cuticle of the worm or could puncture too large of a hole in the cuticle causing the worm to burst.

4. Agarose pads for injection: Prepare a 3 % agarose solution in water by bringing 3 g agarose to a boil in 100 mL ddH$_2$O until completely dissolved. 3 % agarose can be kept on a heat block for a day. Use a glass Pasteur pipette where the tip has been broken to create a wider opening and place around 100 μL of molten agarose onto a glass slide coverslip (50×22 mm). Place another coverslip on top at a 60° angle and lightly compress until the agarose is spread evenly and thinly. The slide can be left to cool while more slides are prepared. Cooled slides are separated and the coverslip with the agarose should be placed in an 80 °C vacuum oven with the agarose facing up for 1 to 4 h. Generally, mounted worms will stick more to a drier pad but they will also desiccate more quickly. Thus, depending on the environmental conditions (humidity etc.) the optimal drying time of agarose pads has to be determined experimentally, but 2 h are a good starting point.

3   Methods

3.1   Designing the scFv Constructs for Transgenic Expression

1. The scFv antibodies contain a variable heavy and a variable light chain (Fig. 1b). For all antibodies in this set, the variable light chain sequences are identical [14]. The heavy chain sequences fall into 13 families. Within a family, all members have identical heavy chain sequences except for the complementarity determining region 3 (CDR3) which is unique for each antibody within the family. Therefore, we initially synthesized all individual heavy chains as well as the common light chain *de novo* with codon usage optimized for *C. elegans*. Unique restriction sites (*see* Fig. 1b, **Note 1**) allowed convenient assembly into a founding member of each class of scFvs that share a common variable heavy chain. Using these constructs as templates, we established all other scFv antibody constructs of this family by site-directed mutagenesis of the CDR3 sequence. For example, HS3A8 was derived from HS4C3 by mutagenesis using primers that replace the CDR3 of HS4C3 with the CDR3 of HS3A8, because both scFvs share the heavy chain of family DP-38 (e.g., HS4C3 CDR3: "…VYYCAR**GRRLKD**WGQGTL…" is altered to HS3A8 CDR 3 "…VYYCAR**GMRPRL**WGQGTL…") [2]. Appropriate controls to demonstrate specificity of any observed staining have to be included (*see* **Note 2**).

2. The scFv heavy and light chains need a secretion signal to target the scFvs to the extracellular space and allow labeling of extracellular matrix components. We have used the secretion signal of *sel-1* [25]. Conceivably, other sequences could be fused to the scFvs instead for targeting to other cellular compartments if so desired, such as for example to the nucleus, the cytosol, or the endoplasmic reticulum. To the C-terminus, we fused in frame a fluorescent protein (Fig. 1b, for a discussion of different fluorescent proteins *see* **Note 3**).

3. To control the level of secretion, an appropriate promoter for transgenic expression needs to be chosen (for a discussion of promoter choice *see* **Note 4**). The level and location of expression both need to be taken into account. Unlike in immunohistochemistry, where the level of antibody can be directly controlled, the level of antibody is mainly determined by the strength of the promoter used. The *unc-122* promoter [23] proved to be an excellent promoter in that it limits expression to the coelomocytes (a set of scavenger cells) (Fig. 1c) and has a moderate level of activity so that background fluorescence is kept low. If specific staining is detected, alternative promoters should be used to confirm results obtained with the coelomocyte-specific promoter, such as the *glo-1* promoter which drives expression in the intestine [26]. Regardless of the promoter used, antibodies will be visible in vesicles of the

coelomocytes which are known to uptake proteins from the pseudocoelom (i.e. the extracellular space) (Fig. 2a, b).

4. All of these components are combined into a single plasmid construct. Sequences for all existing HS-specific scFv constructs are available on request.

3.2 Nematode Husbandry

1. The method in its current form has been established using the nematode *Caenorhabditis elegans*, but has been used in at least one related species (*see* **Note 5**). Worms are maintained using standard procedures and kept at 20 °C [27].

2. Overnight cultures of the uracil auxotroph *Escherichia coli* strain OP50 grown in LB media are used as a food source. Generally, 500 µL of a bacterial overnight culture seeded on a 60 mm NGM plate produces a large bacterial lawn that covers approximately 75 % of the agar surface. The bacterial lawn takes 1 or more days to dry depending on the age of the plates, humidity, etc.. Appropriately dried plates should neither contain remaining liquid culture nor show a cracked agar surface.

3. Worms are moved between plates using a "pick" made from platinum wire fixed to a glass Pasteur pipette. The front of the "pick" should be flattened to make a spade-like shape to aid in the movement of worms between plates. To move worms between plates and a microinjection slide, use a thin but resilient hair such as an animal whisker, human eyelash, or paint brush hair, fixed to a pipette tip with super glue. Cat whiskers can be cut at an angle so that the tip is flatter and can better scoop worms off the plates.

3.3 Transgenesis

1. Here, the transgenic approach using "complex arrays" is discussed (*see* **Note 6**). Prepare the injection mixes for the generation of multicopy arrays as an aqueous solution. Include 5 ng/µL linearized plasmid containing the scFv antibody construct, 5 ng/µL linearized dominant injection marker plasmid such a fluorescent marker for the cells of the gut, and digested genomic DNA from *C. elegans* as a carrier up to a final combined DNA concentration of at least 100 ng/µL. Other concentrations of plasmids are possible but these concentrations are a good start.

2. Pull needles using needle puller and glass capillaries containing an internal filament. If the needle tapers too slowly or too quickly, the needle may bend too much to penetrate the cuticle of the worm or could make a hole in the cuticle, respectively causing the worm to burst.

3. Prepare agarose pads for injection.

4. Arrange all materials for microinjection: injection needles, injection pads, halocarbon oil, M9 buffer, a hair pick, and young, healthy, and well-fed adult *C. elegans*.

5. Load injection needles using a micropipette with 1 μL of injection mix from the back end (let the capillary forces pull in the solution) and mount the loaded needle into a microinjector/needle holder.

6. Prepare the injection pad by placing a drop of halocarbon oil on the dried agarose pad.

7. Prior to mounting worms onto the dried agar pad, position the needle into view and check if the needle is open or if it needs to be broken. Use the edge of the oil drop on the dried agar pad to focus and align the needle in the field of view. Bring the tip of needle into the oil. Test the needle by trying to force out the aqueous injection mix into the oil. If the solution flows freely, there is no need to break the needle. If there is no flow, the needle tip needs to be broken. This can be accomplished by gently dragging the needle tip along the dried agar of the pad while holding the pedal down, i.e. applying gas pressure. Once the tip breaks, the injection mix will flow out driven by the flow of compressed nitrogen. Depending upon the needle, the gas pressure should be between 200 and 350 kPa. Be careful to not break the needle too far from the tip as a wide bore may damage the worms.

8. Worms should be injected as young, healthy, and well-fed adults. These animals can be identified by the number of eggs inside the animal. A young adult will usually contain 6–12 embryos. Pick animals using the hair pick and transfer them into the oil drop on the agarose pad. Try to pick worms without bacteria attached (e.g., by picking worms from outside the bacterial lawn), because attached bacteria may create difficulties during injections (such as variable adhesion to the injection pad). Manipulate worms such that they are lined up in a convenient orientation where the distal arms of the gonad are visible in the field of view (at 200× magnification). Once the worm is in a good orientation, press the animals gently through the oil onto the agarose, where they should begin to stick due to desiccation. At this point, there is a limited amount of time (not more than 10–20 min) before the worms die due to progressive desiccation. Thus, work quickly and deliberately as time may be one of the determining factors in being successful with microinjections.

9. Once the worm(s) are mounted onto the agarose pad, move the cover glass to the microscope for microinjection. A skilled injector may be able to inject as much as 20 animals all lined up on the pad together; a novice should start with 2. Target the needle for microinjection to the distal arm of the gonad, which has a characteristic tubular appearance containing spherical germ cells with large nuclei. This can be clearly seen with differential interference microscopy (DIC). When the needle is

correctly inserted in the distal arm, press the pedal that controls gas flow through the needle; there should be a noticeable flow of injection mix into the distal arm traveling towards the vulval opening. The increased pressure in the gonad sometimes results in the extrusion of an egg. If the needle is incorrectly positioned in the pseudocoelom, the flow will instead push the gut and gonad apart. For an extended discussion of microinjection of nematodes, see ref. [28].

10. Once injected, recover the worms by rescuing them in a drop of M9 buffer which will free the worms from the dried agar pad. Then, use the hair pick or an aspirator to move worms from the pad back to a seeded NGM plate. Once the animals have recovered for an hour, move them to individual NGM plates to lay a brood.

11. After eggs are laid and grow up to the late larval or early adult stages, screen the progeny (F1 generation) for expression of the dominant injection marker that was used. Move these transgenic F1 animals to new individual plates and keep until their progeny (F2 generation) are old enough to screen for germline transmission of the transgene. Often, about 1 in 10 transgenic F1 animals produces transgenic progeny, i.e., result in animals with germ line transmission. The ratio between the total number of F1 animals and those that result in germline transmission depends upon the injector and the injection mix (complex array mixes sometimes have a higher rate of successful transmission). An F1 animal that produces consistent germline transmission is considered a clonal extrachromosomal transgenic line, which has to be maintained by picking transgenic animals.

3.4 Imaging Transgenic Animals

1. Prepare well-fed animals of the desired stage for imaging. Animals can be visualized at any stage but visualization may be easier at younger stages. For example, animals expressing antibodies under control of the *unc-122* promoter are best viewed during the early larval stages as the *unc-122* promoter is most active from late embryonic to mid-larval stages (for a discussion on promoter choice *see* **Note 4**).

2. Prepare anesthetics, 1.5 mL microcentrifuge tubes, imaging slides, and transgenic worms for imaging (for a discussion of anesthesia *see* **Note 7**). Wash worms off the plate into 1.5 mL microcentrifuge tubes using M9 buffer. Gently pellet the worms by centrifugation using a low setting ($500 \times g$) for 2 min. Decrease the volume in the tube to 20 μL and add 20 μL of 16 mM levamisole. Then add 40 μL of 0.6 M 2,3-BDM to the tube and gently mix. Allow worms to settle or pellet using a microcentrifuge at a low setting. Worms will collect at the bottom of the tube and be ready for mounting.

3. Prepare slides one or two at a time as worms anesthetized with this solution will start dying after 30 min. Slides for imaging should be freshly made with 3 % agarose pads prepared in similar fashion as the pads for injection but on microscope slides and without drying. Transfer worms to a slide with an agar pad using a cut pipette tip so as to not damage the worms. Once the worms are on the slide, gently lay a cover glass onto the agar pad over the worms. Mount worms in an appropriate volume of liquid to avoid crushing them (too little liquid) or having them float too freely (too much liquid) (approximately 8 μL for a 25×25 mm cover slip).

4. The fluorescence from the antibodies can be weak. Thus, limit exposure to light, particularly at excitatory wavelength, to avoid bleaching. Increase the camera gain as high as necessary to produce a usable image that limits exposure length and minimizes noise.

4   Notes

1. The modular nature of the constructs allows convenient conversion into different constructs with different specificities. For example, if scFvs against other molecules of interest become available, such as other glycans, small molecules, or against posttranslationally modified molecules (e.g., a phosphorylated or glycanated protein of interest), it should be possible to merely exchange the variable heavy chain as needed. If the heavy chain backbone is available already, this may even be accomplished through simple site-directed mutagenesis.

2. Appropriate controls for transgenic scFv approaches are essential. First, one has to ascertain that the observed staining is specific for the scFv at hand and not the fluorescent protein. For example, in the case of anti-HS scFv antibodies, a control scFv antibody fusion (MPB49::GFP) was expressed under identical conditions which did not result in comparable staining [2]. Second, using genetic methods it has to be tested whether the scFv antibody does indeed recognize the epitope of interest. For HS this has been accomplished in several instances by demonstrating dependence of staining on the presence of certain genes of the HS biosynthetic machinery [2].

3. The fluorescent tag used for visualization is extremely important. The GFP version (which carries a S65T mutation) described here is good starting point. But for experiments that involve detection of multiple epitopes, antibodies fused to different fluorescently labeled proteins are required (Fig. 2c, d). In principle, the antibody constructs can contain any fluorescent marker, but usability is limited by the intensity of the

fluorescent protein. The super folder GFP [29] is a bright alternative to EGFP. DsRed2, tagRFP, and mCherry work, but DsRed2 and mCherry may show some aggregation in the extracellular space. Nonetheless, double labeling experiments with different fluorophores are possible (Fig. 2c, d). Current CFP, YFP, and BFP fluorescent markers appear too weak to be used successfully with this system, but perhaps optimized variants for the extracellular space may prove usable in the future. Beyond fluorophore choice, multimerizing the super folder GFP (a dimer with a short linker sequence) resulted in a brighter signal and allowed better visualization of weaker or less common epitopes. Perhaps producing longer chains of fluorophores may mimic the signal-amplification property of secondary antibodies. Alternatively, nonfluorescently tagged scFv antibodies could be paired with transgenically expressed, fluorescently labeled secondary antibodies to provide signal amplification.

4. It is important to choose an appropriate promoter to drive expression. Keep in mind that the fluorescence of the cells that secrete the antibody may obscure staining on the surface of those cells. Choose cells that are not of interest or unlikely to have staining. Based on these considerations the use of the *unc-122* promoter [23] was an obvious choice and convenient for three reasons. First, it expresses the antibody at a level where it could be visualized without being so high that fluorescently labeled scFv started aggregating in the extracellular space. Second, the coelomocytes are not absolutely fixed to the same position, but are at times found in slightly different positions in the pseudocoelom of *C. elegans* so that no location is always obscured by the source of the antibody. And third, the coelomocytes take up proteins in the pseudocoelom [3] and therefore would eventually contain the fluorescent antibody regardless of whether they produce it. Thus, instead of two sets of cells being "unusable" (because of internal fluorescent signal), only one set is.

Promoters like the *glo-1* promoter [26] provide a slightly weaker level of expression than the coelomocytes, but most structures visible with the *unc-122* promoter are also visible. Other promoters like the *myo-3* or *grd-10* promoters [30, 31] drive expression in muscle or the seam cells so strongly that specific and nonspecific staining cannot be distinguished easily. Initial experiments with *C. elegans* heatshock promoters were not successful, though perhaps with alternative activation lengths or temperatures they might produce usable levels of expression. Chemically inducible systems [32, 33] may provide another alternative approach and may allow expression using stronger promoters (*myo-3* or *grd-10*) by tightly controlling expression before too much antibody is produced. Ideally, a combination that allows both spatial and temporal control of expression may produce the best results such as the recently developed Q-system [33, 34].

5. While the described method uses *C. elegans*, this technique could also be applied to other organisms. To date, the technique has been applied to one other nematode species, *Caenorhabditis briggsae* (Attreed and Bülow, unpublished), but could possibly be applied to any nematode species for which transgenesis methods are available, or even non-nematode model organisms such as flies or fish. A possible concern is that the antibody-fusions are continuously produced and could build up in the extracellular space as an animal ages. One possible path around this is to use heat shock promoters or a chemically induced system (tamoxifen-Cre or quinic acid-Q System) [32, 33]. Another possibility would be to choose a promoter with temporally limited expression. The coelomocytes in *C. elegans* absorb the antibody from the extracellular space and likely play a key role in keeping background levels of staining low by absorbing unbound antibody [35]. Other species have similar cells or organs in charge of this process. Second, *C. elegans* is transparent throughout development. Other organisms, such as Drosophila and zebrafish, have more transparent stages early in development only. For organisms that are more opaque, far-red fluorescent tags have been shown to work in vivo in adult mice [36] and could possibly be employed. Finally, most organs in *C. elegans* are exposed to the pseudocoelom; in other organisms, there is more tissue compartmentalization and a freely diffusing, fluorescently labeled antibody may have a more limited reach. Thus, the use of more region-specific expression of the antibodies may be required to see staining of, for example, specific areas of the nervous system.

6. Several approaches exist to create transgenic animals in *C. elegans* [28], including the generation of multicopy extrachromosomal arrays by microinjection [37], methods for single copy insertion [38] and bombardment [39]. All experiments to date were performed using multicopy arrays, including both "standard arrays" and "complex arrays." For "standard arrays" injection mixes containing circular plasmids where 25 ng/μL should be used for the antibody plasmid and 25–50 ng/μL for the marker plasmid. The rest of the mix, up to a total DNA concentration of 100 ng/μL, should be comprised of generic vector plasmids such as pBluescript or pUC19.

"Complex arrays" mixes only contain linearized DNA. DNA can be linearized by any restriction enzyme that cuts plasmids once in the backbone and cuts genomic DNA approximately every 1–2 kb. Mixes may contain as little as 5 ng/μL of the linearized antibody plasmid along with 5 ng/μL of a dominant injection marker plasmid. The rest of the mix is comprised of digested genomic DNA from *C. elegans* or *E. coli* to a total DNA concentration of 100 ng/μL. The complex array injection mixes have the added benefit of better transmission between generations and better expression. Alternative methods such as single copy insertion [38] and bombardment [39] could also work, although in the case of single copy insertion, the expression may be too weak under a promoter like the *unc-122* promoter. In this case, perhaps a stronger promoter such as the *myo-3* promoter could be used to increase expression. In summary, a variety of transgenic approaches exist and the optimal approach may have to be determined experimentally with "complex arrays" being a convenient starting point.

7. The choice of anesthetic is important. Poorly anesthetized worms may move or twitch during the exposure under the microscope making imaging of weak signals particularly difficult. *C. elegans* tend to twitch when exposed to blue light, which exacerbates this problem. This problem, however, can be managed by using combinations of anesthetics. Sodium azide is a common anesthetic, but tends to weaken fluorescence, particularly fluorescence of extracellular molecules. Levamisole proved to be a better choice because it failed to weaken the fluorescence, though it sometimes lead to morphological changes and "bubbles" near the nerve ring. Since levamisole by itself did not completely inhibit muscle function, and muscles in the pharynx often still twitched, a combination of anesthetics turned out to be the best approach. When levamisole was combined with 2,3-BDM, the worms became completely immobilized, but lived only between 30 and 60 min on the slide before dying.

Acknowledgements

This work was supported by NIH grants F31NS076243 & T32GM07491 (M.A.) and RC1GM090825 & R01GM101313 (H.E.B.).

References

1. Bernfield M, Götte M, Park PW, Reizes O, Fitzgerald ML, Lincecum J, Zako M (1999) Functions of cell surface heparan sulfate proteoglycans. Annu Rev Biochem 68: 729–777

2. Attreed M, Desbois M, van Kuppevelt TH, Bülow HE (2012) Direct visualization of specifically modified extracellular glycans in living animals. Nat Methods 9:477–479

3. Fares H, Greenwald I (2001) Genetic analysis of endocytosis in Caenorhabditis elegans: coelomocyte uptake defective mutants. Genetics 159:133–145

4. Esko JD, Selleck SB (2002) Order out of chaos: assembly of ligand binding sites in heparan sulfate. Annu Rev Biochem 71:435–471

5. Lindahl U, Kusche-Gullberg M, Kjellen L (1998) Regulated diversity of heparan sulfate. J Biol Chem 273:24979–24982

6. Bishop JR, Schuksz M, Esko JD (2007) Heparan sulphate proteoglycans fine-tune mammalian physiology. Nature 446:1030–1037

7. Bülow HE, Hobert O (2006) The molecular diversity of glycosaminoglycans shapes animal development. Annu Rev Cell Dev Biol 22: 375–407

8. Nadanaka S, Kitagawa H (2008) Heparan sulphate biosynthesis and disease. J Biochem 144: 7–14

9. Jakobsson L, Kreuger J, Holmborn K, Lundin L, Eriksson I, Kjellen L, Claesson-Welsh L (2006) Heparan sulfate in trans potentiates VEGFR-mediated angiogenesis. Dev Cell 10: 625–634

10. Sarrazin S, Lamanna WC, Esko JD (2011) Heparan sulfate proteoglycans. Cold Spring Harb Perspect Biol 3

11. Thompson SM, Connell MG, van Kuppevelt TH, Xu R, Turnbull JE, Losty PD, Fernig DG, Jesudason EC (2011) Structure and epitope distribution of heparan sulfate is disrupted in experimental lung hypoplasia: a glycobiological epigenetic cause for malformation? BMC Dev Biol 11:38

12. Hoogenboom HR, Winter G (1992) By-passing immunisation. Human antibodies from synthetic repertoires of germline VH gene segments rearranged in vitro. J Mol Biol 227:381–388

13. Marks JD, Hoogenboom HR, Bonnert TP, McCafferty J, Griffiths AD, Winter G (1991) By-passing immunization. Human antibodies from V-gene libraries displayed on phage. J Mol Biol 222:581–597

14. van Kuppevelt TH, Dennissen MA, van Venrooij WJ, Hoet RM, Veerkamp JH (1998) Generation and application of type-specific anti-heparan sulfate antibodies using phage display technology. Further evidence for heparan sulfate heterogeneity in the kidney. J Biol Chem 273:12960–12966

15. Dennissen MA, Jenniskens GJ, Pieffers M, Versteeg EM, Petitou M, Veerkamp JH, van Kuppevelt TH (2002) Large, tissue-regulated domain diversity of heparan sulfates demonstrated by phage display antibodies. J Biol Chem 277:10982–10986

16. Jenniskens GJ, Oosterhof A, Brandwijk R, Veerkamp JH, van Kuppevelt TH (2000) Heparan sulfate heterogeneity in skeletal muscle basal lamina: demonstration by phage display-derived antibodies. J Neurosci 20: 4099–4111

17. Smits NC, Kurup S, Rops AL, ten Dam GB, Massuger LF, Hafmans T, Turnbull JE, Spillmann D, Li JP, Kennel SJ, Wall JS, Shworak NW, Dekhuijzen PN, van der Vlag J, van Kuppevelt TH (2010) The heparan sulfate motif (GlcNS6S-IdoA2S)3, common in heparin, has a strict topography and is involved in cell behavior and disease. J Biol Chem 285:41143–41151

18. Smits NC, Lensen JF, Wijnhoven TJ, Ten Dam GB, Jenniskens GJ, van Kuppevelt TH (2006) Phage display-derived human antibodies against specific glycosaminoglycan epitopes. Methods Enzymol 416:61–87

19. ten Dam GB, van de Westerlo EM, Smetsers TF, Willemse M, van Muijen GN, Merry CL, Gallagher JT, Kim YS, van Kuppevelt TH (2004) Detection of 2-O-sulfated iduronate and N-acetylglucosamine units in heparan sulfate by an antibody selected against acharan sulfate (IdoA2S-GlcNAc)n. J Biol Chem 279:38346–38352

20. van de Westerlo EM, Smetsers TF, Dennissen MA, Linhardt RJ, Veerkamp JH, van Muijen GN, van Kuppevelt TH (2002) Human single chain antibodies against heparin: selection, characterization, and effect on coagulation. Blood 99:2427–2433

21. Wijnhoven TJ, Lensen JF, Rops AL, van der Vlag J, Kolset SO, Bangstad HJ, Pfeffer P, van den Hoven MJ, Berden JH, van den Heuvel LP, van Kuppevelt TH (2006) Aberrant heparan sulfate profile in the human diabetic kidney offers new clues for therapeutic glycomimetics. Am J Kidney Dis 48:250–261

22. Kurup S, Wijnhoven TJ, Jenniskens GJ, Kimata K, Habuchi H, Li JP, Lindahl U, van Kuppevelt TH, Spillmann D (2007) Characterization of anti-heparan sulfate phage display antibodies AO4B08 and HS4E4. J Biol Chem 282: 21032–21042

23. Loria PM, Hodgkin J, Hobert O (2004) A conserved postsynaptic transmembrane protein affecting neuromuscular signaling in Caenorhabditis elegans. J Neurosci 24: 2191–2201

24. Jorgensen EM, Hartwieg E, Schuske K, Nonet ML, Jin Y, Horvitz HR (1995) Defective recycling of synaptic vesicles in synaptotagmin mutants of Caenorhabditis elegans. Nature 378:196–199

25. Grant B, Greenwald I (1996) The Caenorhabditis elegans sel-1 gene, a negative regulator of lin-12 and glp-1, encodes a predicted extracellular protein. Genetics 143:237–247

26. Hermann GJ, Schroeder LK, Hieb CA, Kershner AM, Rabbitts BM, Fonarev P, Grant BD, Priess JR (2005) Genetic analysis of lysosomal trafficking in Caenorhabditis elegans. Mol Biol Cell 16:3273–3288

27. Brenner S (1974) The genetics of Caenorhabditis elegans. Genetics 77:71–94

28. Evans T (2006) Transformation and microinjection. In: WormBook (ed) The C. elegans

research community, WormBook. doi:10.1895/wormbook.1.108.1, http://www.wormbook.org

29. Pedelacq JD, Cabantous S, Tran T, Terwilliger TC, Waldo GS (2006) Engineering and characterization of a superfolder green fluorescent protein. Nat Biotechnol 24:79–88

30. Hao L, Johnsen R, Lauter G, Baillie D, Burglin TR (2006) Comprehensive analysis of gene expression patterns of hedgehog-related genes. BMC Genomics 7:280

31. Okkema PG, Harrison SW, Plunger V, Aryana A, Fire A (1993) Sequence requirements for myosin gene expression and regulation in Caenorhabditis elegans. Genetics 135:385–404

32. Feil R, Wagner J, Metzger D, Chambon P (1997) Regulation of Cre recombinase activity by mutated estrogen receptor ligand-binding domains. Biochem Biophys Res Commun 237:752–757

33. Potter CJ, Tasic B, Russler EV, Liang L, Luo L (2010) The Q system: a repressible binary system for transgene expression, lineage tracing, and mosaic analysis. Cell 141:536–548

34. Wei X, Potter CJ, Luo L, Shen K (2012) Controlling gene expression with the Q repressible binary expression system in Caenorhabditis elegans. Nat Methods 9:391–395

35. Gottschalk A, Schafer WR (2006) Visualization of integral and peripheral cell surface proteins in live Caenorhabditis elegans. J Neurosci Methods 154:68–79

36. Filonov GS, Piatkevich KD, Ting LM, Zhang J, Kim K, Verkhusha VV (2011) Bright and stable near-infrared fluorescent protein for in vivo imaging. Nat Biotechnol 29:757–761

37. Mello CC, Kramer JM, Stinchcomb D, Ambros V (1991) Efficient gene transfer in C.elegans: extrachromosomal maintenance and integration of transforming sequences. EMBO J 10:3959–3970

38. Frokjaer-Jensen C, Davis MW, Hopkins CE, Newman BJ, Thummel JM, Olesen SP, Grunnet M, Jorgensen EM (2008) Single-copy insertion of transgenes in Caenorhabditis elegans. Nat Genet 40:1375–1383

39. Praitis V, Casey E, Collar D, Austin J (2001) Creation of low-copy integrated transgenic lines in Caenorhabditis elegans. Genetics 157:1217–1226

Part II

Functions

Chapter 23

Informatics Tools to Advance the Biology of Glycosaminoglycans and Proteoglycans

Lewis J. Frey

Abstract

Glycomics researchers have identified the need for integrated database systems for collecting glycomics information in a consistent format. The goal is to create a resource for knowledge discovery and dissemination to wider research communities. This has the potential to extend the research community to include biologists, clinicians, chemists, and computer scientists. This chapter discusses the technology and approach needed to create integrated data resources to empower the broader community to leverage extant glycomics data. The focus is on glycosaminoglycan (GAGs) and proteoglycan research, but the approach can be generalized. The methods described span the development of glycomics standards from CarbBank to Glyco Connection Tables. The existence of integrated data sets provides a foundation for novel methods of analysis such as machine learning for knowledge discovery. The implications of predictive analysis are examined in relation to disease biomarker to expand the target audience of GAG and proteoglycan research.

Key words Glycosaminoglycan, Proteoglycan, Data integration, Machine learning, Data representation, Informatics

1 Introduction

Glycoscience is a challenging multidisciplinary field with a history of advancement via collaborative research [1]. In 2012, the Division on Earth and Life Studies of the National Academies Board on Chemical Sciences and Technology published a roadmap for future directions in glycomics and glycosciences research [2]. The roadmap points to the need to build research systems that support multidisciplinary collaborations that include biology, chemistry, bioinformatics, and computer science. To achieve this, the authors emphasized the collaboration between informatics and glycoscience to create centralized and standardized glycomics databases supported by open-source software ecosystems that advance research in machine learning in glycomics.

Glycans, also referred to as carbohydrates, saccharides, or sugars, can have diverse and complex functions and structures that are

Kuberan Balagurunathan et al. (eds.), *Glycosaminoglycans: Chemistry and Biology*, Methods in Molecular Biology, vol. 1229, DOI 10.1007/978-1-4939-1714-3_23, © Springer Science+Business Media New York 2015

influenced by a range of factors such as cell type, cellular metabolism, and developmental stage. Glycosaminoglycans (GAGs) [3] and proteoglycans are the focus of this paper. GAGs are glycans that contain amino sugars in a repeating sequence with other sugars. GAGs attach to glycosylation sites on proteoglycans, which are amino acid sequences that act as the protein core for GAGs. Proteoglycans are ubiquitous and serve in many roles as secretory, basement-membrane, membrane-bound, and interstitial proteins. Proteoglycans exhibit tremendous diversity in numbers of GAGs and sulfated residues on the GAGs. The variety of interaction creates a complex space that enables regulation of specific cellular process relevant to development, angiogenesis, axonal growth, anticoagulation, microbial pathogenesis, and disease pathogenesis in cancer and immunology [4].

This complexity increases the difficulty of conducting research and of establishing standards. The roadmap points to informatics as serving a critical role in addressing the complexity in glycoscience and widening the community of scientists who are contributing to its advancement. Aoki [5] provides a review of bioinformatics for glycomics research. This chapter focuses on the ecosystem of open source or publically available informatics tools related to data integration of glycomics data sources and machine learning analysis of glycans.

2 Data Integration

Data integration involves combining or linking data from multiple sources to enable data sharing, expanded data sets, secondary analysis/reuse of data, and broadening multidisciplinary collaborations. The development of community curated formats and standards, data repositories, and open-source technologies are central to data integration. In addition to software and open source code, there are web page utilities that interface with databases, provide structure similarity matching and format conversion. The web tools tend to be hosted in conjunction with glycomic databases. The use of standards in databases enables the development of software that can interoperate. Through the standards, application program interfaces (APIs) result in new functionality that enhances the scientific community's access to the data and knowledge derived from it.

2.1 Consortium for Functional Glycomics (CFG) Database and CFG Standard

Frequently, glycans are depicted graphically using the Consortium for Functional Glycomics (CFG) [6, 7] notation (*see* Fig. 1). In this notation, glycans are represented as a two dimensional tree with the reducing monosaccharide being at the root of the tree. The tree is on its side with the root on the right and the leaves or nonreducing end on the left. The children branch out from the

Subcomponent of a Heparin Molecule Without Sulfation

β4   α4

α4 and β4 correspond to the one to four
linkages between monosaccharides

the symbols correspond to:

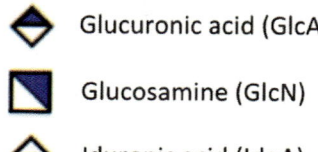

Glucuronic acid (GlcA)

Glucosamine (GlcN)

Iduronic acid (IdoA)

Fig. 1 Graphical nomenclature from CFG for a subcomponent of a Heparin molecule without sulfation. *Diamonds* correspond to acids and *squares* correspond to amines. The β4 and α4 on the links indicate that the bonds go from the first carbon on one monosaccharide to the fourth carbon on the attached monosaccharide with the bonds being β and α

root with nodes representing monosaccharides and the edges representing glycosidic linkages. As an example, a subcomponent of the GAG Heparin without sulfation is presented in Fig. 1. Heparin is used widely in medicine as an anticoagulant. A consistent representation is necessary for recall of unique GAGs, such as Heparin, and methods to support a unified format are described.

Although a useful communication mechanism, graphical standards are difficult for machines to read and thus, limit informatics development. To address this, the glycomics scientific community has developed other standards that support a machine-readable format [8]. This chapter will provide a brief overview of five dominant machine-readable standards: International Union of Pure and Applied Chemistry [9] (IUPAC); Linear Notation for Unique Description of Carbohydrate Sequences (LINUCS) [10]; KEGG Chemical Function (KCF) [11]; GLYcan Data Exchange-II (GLYDE-II) [12, 13]; Glyco Connection Table (GlycoCT) [14]. These standards are described to provide a context for the informatics approaches used within glycoscience to generate consistent representations that uniquely identify glycans.

The range of standards developed after a funding cut in 1997 for CarbBank [15, 16], the first major complex carbohydrate database. Multiple new standards evolved driven by different data systems and propriety approaches. Herget et al. [14] provides an excellent comparison of these as well as other standards and describes GlycoCT that can be used as a convergent format. To illustrate the differences and similarities between the standards,

[non-reducing end] D-GlcpA β—4 D-GlcpN α—4 L-IdopA [reducing end]

Fig. 2 CFG nomenclature for a subcomponent of a heparin molecule without sulfation. D-GlcpA has a 1–4 beta bond with D-GlcpN which has a 1–4 alpha bond with root of the tree, L-IdopA. The *D* and *L* indicate the configuration of the monosaccharide, with either D- or L-glyceraldehyde attached. The *p* indicates the ring size of pyranose

the subcomponent of a Heparin molecule without sulfation depicted in Fig. 1 will be converted to each of the five machine-readable formats.

The database for the CFG web site is a repository for multiple files related to glycans and proteoglycans. The CFG web site provides search tools for researchers to retrieve molecules [17]. The structure of a queried molecule can be entered through a graphical interface. The tool follows IUPAC nomenclature for describing monosaccharides chains and the example Heparin subcomponent is expressed in CFG's IUPAC format in Fig. 2. All glycans have reducing and nonreducing ends, which gives directionality to the molecule. The reducing end is place at the root at the right hand side of the notation according to IUPAC conventions. The CFG format formats does not handle multiple connections and only partially encodes undetermined terminal units, nonstoichiometric modifications, and alternative residues [14]. CFG's database stores data on matrix-assisted laser desorption/ionization mass spectrometry (MALDI-MS) glycan profiling, glycosyltransferase, glycosylation pathways, glycan binding protein interactions, glyco-gene expression, literature citations, and mouse phenotyping data.

The CFG site has cell- and tissue-specific MALDI mass spectrometry glycan profiles. The CFG site provides summary information for the proteoglycans Brevican, Neurocan, Versican, and Agrecan for human and mouse. The material includes links to data from experiments on binding affinities for the proteoglycans. These are listed under glycan binding proteins molecule home page [18] as proteoglycan core proteins. The data can be searched with a linear IUPAC codes for the carbohydrate structures and downloaded from CFG's site.

2.2 CarbBank Database and IUPAC Standard

The CarbBank Project [15, 16] constructed the Complex Carbohydrate Structure Database (CCSD), which is now referred to as CarbBank. The structures are represented in branched two-dimensional IUPAC format. Figure 3 depicts a CarbBank extended IUPAC ASCII format for a subcomponent of a Heparin molecule without sulfation. The reducing end is at the bottom right of the representation. The CarbBank format can encode glyans in a similar to CFG's format except that it is in ASCII and has different notation for the linkages.

The project's funding ended in 1997 and the data in the system serves as the basis for the other databases on carbohydrate structures.

```
b-D-GlcpA-(1-4)+
               |
       a-D-GlcpN-(1-4)+
                      |
                  L-IdopA
```

Fig. 3 CarbBank extended form of IUPAC nomenclature for a subcomponent of a Heparin molecule without sulfation. The *b* or *a* in front of the monosaccharide indicates a beta or alpha bond. The *D* and *L* indicate the configuration of the monosaccharide, with either D- or L-glyceraldehyde attached. The 1–4 in *parenthesis* indicates the atom locants between monosaccharides

Multiple glycan initiatives replaced CarbBank and resulted in the growth of divergent standards for representing glycan structures.

2.3 Glycoscience Database Portal and LINUCS Standard

The Glycoscience database portal [19] gives a structural interface for cross-linking glycan information. The portal provides a retrieval interface based on structure, substructure, motifs, and carbohydrate components. Tools are available from the site for glycosylation, mass spec, and protein analysis. Applications to perform analysis are available and linked through the portal. The site integrates with Protein Data Bank (PDB) tools for displaying and obtaining information from PDB. Tools for analyzing Nuclear Magnetic Resonance (NMR) data are also included on the site.

The Glycoscience.de has a utility for converting from IUPAC to LINUCS and running queries in LINUCS linear code or interactively through a glycan structure building tool. The exact match is done though a two-dimensional IUPAC file format, similar to the output from CarbBank (discussed in the next section). The LINUCS format is used within the glycoscience.de database for matching glycan structures. It has a LINCUS tool [20] for converting IUPAC representation into the LINUCS unique normalized file format of the submitted glycan structure.

2.4 LINUCS (Linear Notation for Unique Description of Carbohydrate Sequences)

Linear Notation for Unique Description of Carbohydrate Sequences (LINUCS) [10] addressed the need to develop a carbohydrate database that provided a linear, canonical description of carbohydrate structures of glycans that was computer readable and could be used to link to other resources. Figure 4 provides the linear string format of LINCUS for a subcomponent of a Heparin molecule without sulfation. The reducing end of LIdoA is on the left hand side with the alpha and beta links being indicated by the letters a and b. Note that link number go from 4 to 1 within the parenthesis due to the reversed position of the reducing end compared with CFG's linear format.

The linear format of LINUCS makes it unable to handle multiple connections, undetermined terminal units, nonstoichiometric

```
[][LIdoA]{[(4+1)][a-GlcN]{[(4+1)][b-GlcA]{}}}
```

Fig. 4 LINCUS nomenclature for a subcomponent of a Heparin molecule without sulfation. GlcA have a 1–4 beta (b) bond with GlcN which has a 1–4 alpha (a) bond with root of the tree, LIdoA.

modifications, and alternative residues [14]. The method does establish uniqueness for the set of molecules that can be represented within the standard, making it a good candidate as a database key.

2.5 KEGG GLYCAN Database and KCF Standard

KEGG GLYCAN [21] is part of the extensive KEGG ecosystem of web services. The database integrates multiple data resources that are connected through the KEGG ecosystem: KEGG PATHWAY, REACTION, and ENZYME. Glycan structures can be accessed by a keyword search on KEGG's GenomeNet Webpage [22]. The applications KEGG-Draw and DrawRings tools enable users to draw graphical pictures of a glycan and then search the KEGG repository for them. The KEGG web site hosts KEGG-Draw and conversion utilities for converting to KCF from other standards such as LINUCS and IUPAC as well as converting to LINUCS from KCF. KEGG-Draw is a graphical retrieval mechanism based on the tree structures of carbohydrate molecules in the KEGG database. The KEGG Carbohydrate Matcher (KCaM) [11] is used to align structures to an index structure and return similar structures from the KEGG GLYCAN database. Aoki et al. discuss KEGG-Draw [11] as an excellent tool that can be used by beginners as an interface to the GlycomeDB data repository.

2.6 KCF (KEGG Chemical Function)

The KEGG Chemical Function (KCF) format was introduced by the developers of KEGG to serve as a standard format to represent chemical objects. It was the first to use the table structure for representing glycan structures. Figure 5 displays the representation of KCF for a subcomponent of a Heparin molecule without sulfation. The representation gives the number of nodes (three) and the number of edges (two) in a table format. The pattern of the nodes starts with the reducing end node LIdoA. The x and y position in a two dimensional space are specified. The alpha and beta bonds are represented as edges. The alpha bond goes from donor node 2 and donor carbon 1 to acceptor node 1 on acceptor carbon 4. The beta bond goes from node 3 to node 2 connecting GlcA and GlcN.

The connection table format of KCF enables it to handle multiple connections unlike the formats discussed up to this point; although, it still does not encode undetermined terminal units, nonstoichiometric modifications, or alternate residues [14].

2.7 GlycomeDB Database with Glyde II and GlycoCT Standards

The developers, Ranzinger et al. [23], of GlycomeDB created a comprehensive glycome database composed of a union of other major databases in glycomics (i.e., Carbbank, CFG, KEGG, GLYCOSCIENCES.de, BCSDB, GlycoBase (Dublin),

```
ENTRY                               Glycan
NODE          3
              1    LIdoA      8.4    -0.2
              2    GlcN       0.4    -0.2
              3    GlcA      -7.6    -0.2
EDGE          2
              1       2:a1    1:4
              2       3:b1    2:4
///
```

Fig. 5 KCF nomenclature for a subcomponent of a Heparin molecule without sulfation. The nodes give *x* and *y* coordinates along with the name of the monosaccharide. The links specify the bond type along with the donor node and acceptor node. The donor and acceptor carbons are also indicated for the links

and GlycoBase (Lille)). In 2008 GlycomeDB had integrated more than 100,000 data sets resulting in 33,000 unique sequence structures. GlycomeDB move the field forward with an integrated representation that brings together glycan structures from multiple divergent standards into one database: 8,498 glycan structures from CFG's database, 23,402 structures from CarbBank, 23,172 structures from the Glycoscience.de database and 10,966 glycan structures from KEGG GLYCAN. The complete GlycomeDB SQL database, updated weekly, can be downloaded from the GlycomeDB site [24]. The complete set of structures can also be obtained as GlycoCT or Glyde-II files. The breakdown of species for the 33,000 unique sequences is 25 % mammalian, 21 % bacteria, 11 % embryophyta and the remaining 43 % being smaller collections of species. Of the mammalian structures 41 % are human [23].

Software and web page utilities in glycomics research have been build, deployed, and in some cases made publically available as open source code. The GlycomeDB has the graphical interfaces of GlycanBuilder and DrawRing and a wide range of text format inputs for search: GlycoCT, GlydeII, KCF, LINUCS, and CarbBank IUPAC. GlycosWorkbench is an example of an open source tool that is currently available that can serve as a collaborative open source code base for development.

Ceroni et al. [25] present the tools GlycosWorkbench and GlycanBuilder for creating glycan structures. The GlycosWorkbench has modifiable code and GlycanBuilder is an application that is precompiled. The later enables the running of a glycan structure manipulation and rendering system without concern for the source code. The former has a fuller set of functionality and can be expanded by integrating new modules. GlycosWorkbench has mass spec analysis functionality while GlycoBuilder does not. Both tools output GlycosCT and other file formats along with two-dimensional CFG symbolic images of carbohydrate molecules. The tools also input multiple formats, which means they can act as a transformation tool between standards. GlycanBuilder is available as supplemental material [26]. The GlycosWorkbench code along with a zip file for the application can be downloaded from Google Code [27]. With the integration of multiple standards combined with tools connected to an integrated database, i.e., GlycomeDB, the ecosystem of tools supports a convergence of representations. With a way to uniquely identify a wider range of glyans within a single repository or even across multiple repositories using the same standard, new analysis techniques can be developed on an expanding consistent set of data. Such large data sets facilitate research in machine learning by having glycans and data that are annotated and available as training instances for the algorithms.

2.8   Glyde II (GLYcan Data Exchange–II)

Sahoo et al. [12] describe the Glyde format as a directed acyclic graph tree structure expressed in Extensible Markup Language (XML) format. Glyde II expanded the Glyde format to include connection tables [13]. The XML standard supports a service-oriented architecture where applications can be web services that run on sites' remote computers. The ecosystem of services collectively supports the data and analysis requirements for glycomics research. The XML is also expressed in a human readable and machine-readable format. Figure 6 provides the Glyde II representation for a subcomponent of a Heparin molecule without sulfation.

The Glyde II format in Fig. 6 was translated from GlycoCT by GlycoWorkbench. The format has a residue and link structure expressed in XML. PartId one is the iduronic acid that is connected to glucosamine. Note that residue 2 has a substitution of an

```
<GlydeII>
  <molecule subtype="glycan" id="From_GlycoCT_Translation">
    <residue subtype="base_type" partid="1"
ref="http://www.monosaccharideDB.org/GLYDE-II.jsp?G=a-lido-HEX-1:5|6:a" />
    <residue subtype="base_type" partid="2"
ref="http://www.monosaccharideDB.org/GLYDE-II.jsp?G=a-dglc-HEX-1:5" />
    <residue subtype="substituent" partid="3"
ref="http://www.monosaccharideDB.org/GLYDE-II.jsp?G=amino" />
    <residue subtype="base_type" partid="4"
ref="http://www.monosaccharideDB.org/GLYDE-II.jsp?G=b-dglc-HEX-1:5|6:a" />
    <residue_link from="2" to="1">
      <atom_link from="C1" to="O4" from_replace="O1" bond_order="1" />
    </residue_link>
    <residue_link from="3" to="2">
      <atom_link from="N1H" to="C2" to_replace="O2" bond_order="1" />
    </residue_link>
    <residue_link from="4" to="2">
      <atom_link from="C1" to="O4" from_replace="O1" bond_order="1" />
    </residue_link>
  </molecule>
</GlydeII>
```

Fig. 6 GlydeII nomenclature for a subcomponent of a Heparin molecule without sulfation

amino which makes it glucosamine. The links are handled in a similar fashion to other formats with the distinction being the XML nature of the format. The standard does not support alternative residues, unlike GlycoCT, which can handle everything Glyde-II does plus alternative residues.

2.9 GlycoCT (Glyco Connection Table)

Herget et al. [14] as part of the EUROCarbDB project developed the GlycoCT standard as a connective-tables approach to representing complex carbohydrates. They developed GlycoCT to extended glycan standards format to include structurally underdetermined sequences and represent known glycan structures in existing database sources. In addition, they added a consistent naming scheme that can be used by database systems as a controlled vocabulary. Strict sorting rules are used to uniquely determine the order of elements in the sequence. Given this uniqueness quality, the authors propose the representation can serve as a primary key in databases of carbohydrate structures. The GlycoCT standard was also developed to incorporate a number of different standards. The GlycoCT representation for a subcomponent of a Heparin molecule without sulfation is listed in Fig. 7. The RES section is the residue list and the LIN section is the connectivity list.

The condensed GlycoCT format gives a compact representation that flexibly handles glycan structures. The lines in RES start with the node number and then a letter indicating basictype (b) or substitution (s), the name of the monosaccharide, the number of consecutive carbon atoms, the closure start and end of the ring and the position and type of a modifier. The lines in LIN consist of the linkage number, parent linkage number, parent atom replacement, parent linkage number, child linkage number, child residue number, and the child atom replacement. The format is designed to uniquely identify the many different structures of glycans and as such could serve as the primary key in a database. The other formats cover some of the cases across glycan databases, but GlycoCT is designed to account for the full range of structures and to support database functionality such as controlled vocabularies.

```
RES
1b:a-lido-HEX-1:5|6:a
2b:a-dglc-HEX-1:5
3s:amino
4b:b-dglc-HEX-1:5|6:a
LIN
1:1o(4+1)2d
2:2d(2+1)3n
3:2o(4+1)4d
```

Fig. 7 Condensed GLycoCT nomenclature for a subcomponent of a Heparin molecule without sulfation

3   Machine Learning for Modeling Structure

Machine learning is a process by which computer algorithms improve performance on tasks through training on data. The tasks can be supervised learning with previously annotated instances or unsupervised learning with unlabeled instances. The glycomics databases discussed above contain both types of instances that are uniquely identified through structured formats. A task in glycomics is the discovery of machine learning models that predict features for new instances. An example of a system that uses machine learning is IBM's Watson [28]. Besides out performing top human competitors in the game Jeopardy, Watson trains on medical oncology data to discover models that predict cancer types for patients [29]. The algorithms incorporated into Watson integrate the oncology training data and then predict the feature of cancer type for a new instance, in this case a patient presenting with an unknown cancer.

The performance of machine learning algorithms improves with rich sets of annotated data that are indexed. The efforts in glycomics to create a repository of such data will expand the applicability of machine learning in the field. With a centralized data repository each new structure cumulatively increases the ability of machines to identify patterns. Glycoscience research already has a history of using machine learning techniques to predict glycan structure, identify biomarker, and characterize glycan profiles. Below are machine learning and informatics approaches for matching similar glycan substructures, predicting glycan classes, identifying structure from mass spec and predicting glycosylation sites.

3.1   Substructure Similarity and Classification Prediction

Alignment methods are used to identify similar glycan structures using dynamic programming that aligns monosaccharides [30]. The algorithm works in a fashion similar to the Smith-Waterman method [31], except that it searches for the maximum common subtree instead of a linear alignment. In computing a global match of two trees, the largest local alignment is computed first on the largest subtree followed by recursively running the algorithm through the remaining subtrees. The algorithm computes similarity by summing the similarity scores for all alignments of subtrees. The task is a supervised learning task because the model is trained on previously labeled instances.

The probabilistic sibling-dependent tree Markov model (PSTMM) [32] algorithm is based on probability characteristics of rooted, ordered, labeled trees. Glycan are represented as labeled ordered trees with monossacharides having parent child and sibling relationships depending upon the bonds between them. PSTMM trains on the trees that have been assigned to classes in a supervised fashion. The algorithm produces a model based on the probabilistic properties of the parent child and sibling relationships of the glycans' monosaccharides in the tagged class. The model can then be used to assign a probability to a new glycan tree instance to predict which tagged class it belongs to.

PSTMM was applied on real data glycan data from the KEGG GLYCAN database [21]. PSTMM generated a model that distinguished among four glycan classes: N-Glycans, O-Glycans, GAGs, and Sphingolipids. The PSTMM model correctly predicted GAGs better than the other three glycan classes. The approach demonstrates the use of machine learning to identify substructures in glycans and for predicting classes of glycans.

The PSTMM system has the risk of overfitting due to sibling dependencies. An extension to the algorithm called ProfilePSTMM [33] defines types of state transitions to manage dependencies among siblings and between parents and children. The new approach reduced overfitting and also improved algorithm efficiency. The algorithm was trained on glycans and outputs a structure profile that can be used to predict glycan classes. The algorithm had an accuracy of 95.4 % in classifying O-Glycans and Sphingolipids.

Kawano et al. [34] use an alternative approach that uses similarity of expression patterns to predict glycan structure. The approach opened up expression data as a data source for predicting structure. They reason that since glycosyltransferases (GTs) determine the biosynthetic code of glycans, then knowing the GT expression patterns can give information about what glycan structures will be generated. This approach of glyco-gene expression analysis looks at predicting glycan structure through gene expression profiles. Kawano et al. construct a reaction pattern library of GTs used for predicting glycan structure. Frequency of glycans cooccurrence patterns is derived from glycan databases to augment the library.

The authors decompose structure files into reaction pairs of monosaccharides and perform analysis on the pairs. The analysis includes a hierarchically clustered tree of the pairs based on a distance matrix between the pairs. The distance measure was created by cooccurrence of reaction pairs in a database of known glycan structures. Using the distance measure, a distance score was computed for each set of reaction pairs associated with a glycan. They use the association of GTs and a library of reaction pairs to relate expression patterns of GTs and potential glycans in the cell. The accuracy of the algorithm was 81 % on predicting known structures using reaction expression patterns generated from KEGG GLYCAN. The method was improved through generating intermediate glycan that are proposed to exist given their similarity to trees that are known to exist [35]. The authors used leukemia expression profiles to show that the improved method predicted glycan structures that are known to occur in cancers.

3.2 Mass Spec

The prior discussion focused on identifying patterns in known structures or sequences. With this approach accuracy measures can easily be calculated. Another active area of research is identifying structure from measurements of GAG chemical properties with unknown structure. The process of annotating structural measurement data is manual, time-consuming, and tedious. A difficult component of the work is feature identification and construction. Informatics approaches using machine learning have been explored to provide pipelines for speeding glycan structure annotation.

An informatics framework called property-encoded nomenclature (PEN) [36] characterized GAGs through constraint satisfaction algorithms. PEN has been used on MALDI-MS data to sequence a Heparin and Heparan Sulfate like GAGs (HLGAG) [37]. They used the PEN approach to identify the HLGAG being investigated and confirmed the structure with NMR. PEN has also been used on NMR spectroscopy to sequence Heparin and Heparan Sulfate GAG oligosaccharides [38]. They showed that the PEN constraint satisfaction method reduced the required number of experimental constraints to identify the structure.

Maxwell et al. [39] tested another software pipeline for characterizing glycan structure from liquid chromatography/mass spectrometry (LC/MS) data. They apply the pipeline to heparan sulfate (HS) GAGs. The addition of liquid chromatography to mass spectrometry enabled them to sample a fuller set of glycans for analysis. A component of the pipeline is GlycReSoft software provides a workflow consisting of LC/MS data input, which runs the data through feature quantification processes. After the features are identified and constructed, the resulting data are submitted to supervised and unsupervised machine learning algorithms. GlycReSoft generates scores for compound lists along with their predicted class.

Their supervised learning method was evaluated by comparing a generated GlycoReSoft predicted compound list to a "gold standard" compound list composed by an expert. Compounds that are predicted to be annotation and are in the gold standard are positive examples otherwise they are negative examples. Logistic regression is used to identify a model that best separates the negative and positive examples. The supervised and unsupervised methods performed equally well at 80 % of the area under the receiver operating characteristic curve. The author concluded that machine learning improved glycan LC/MS profiling and offers future opportunities to improve feature identification and construction scoring.

Other methods applied machine learning approaches to new types of chemical property data. Li et al. [40] use a hydrophilic interaction chromatography (HILIC) and Fourier transform algorithm to identify the structure of GAGs. The authors used

GlycReSoft to get the structure of the GAGs within 5 ppm mass accuracy [40]. Their method provided a way to characterize Heparin. The results of the approach open up the possibility of identifying different qualities of heparin product through a partially automated process.

3.3 Glycosylation Site Prediction

Another area for the application of machine learning is glycosylation analysis to predict glycosylation sites [41, 42]. Statistical analysis has been used to characterize the amino acids surrounding glycosylation sites. Work on the topic has been done at the German Cancer Research Center and is available through the Glycoscience portal where they deployed GlySeq [43], a query tool for searching the amino acid characteristics around glycosylation sites. Wang et al. [44] developed a systematic method for predicting proteoglycan glycosylation sites. They trained a neuronet machine learning algorithm to predict the sites and achieved 98 % accuracy on their data. The approach demonstrates the effectiveness of using learning algorithms to identify glycosylation sites on proteoglycans. They acquired their data from a C. elegans and proposed their approach as a neural network method for glycosylation site prediction.

4 Future Informatics Opportunities

4.1 Data Integration

Integration of data for GAGs and proteoglycans will speed discovery in biology. The ability to create unique identifiers via GlycoCT that handle a wide range of glycans will enable tracking of specific forms of molecules that have medical relevance to diagnosis and treatment. For examples, systems could be developed so that researchers could submit data to have the GAGs and proteoglycans indexed with relevant unique GlycoCT identifiers associated with the molecules.

With sharable and standardized representations of complex carbohydrate molecules, whole areas of research become feasible. Linking to experiments using unique GlycoCT identifiers can accelerate the area of biomarker identification for GAGs and proteoglycans. Specifically, resources could be built cumulatively to characterize tissues, organs, and disease pathologies in the context of glycan profiles.

Another area is the developed of open-source technology that plug and play with clinical information systems to speed understanding of glycoscience within the medical community and to increase utilization of a standardized repository. Work is currently advancing in this area and a few examples will be highlighted as future areas of advancement within informatics and GAG and proteoglycan research.

4.2 Data Analysis

4.2.1 Identifying Biomarkers and Developing Organ and Tissue Profiles

Shao et al. [45] show that in four normal patients, sulfation patterns of Heparan Sulfate in leukocytes were higher compared with levels in organ tissues from a variety of species. Determining normal levels of Heparan Sulfate profiles is a necessary step before determining changed expression in disease tissue. A needed system is one that records and visualizes tissue differences for distinguishing diseased tissue from healthy tissue. Such tools would also expand GAG sulfation informatics research for diagnostic and prognostic biomarker identification. The authors use a Lawrence Code representation, but the material could also use the unique identifiers of GlycoCT and store it in a consistent format so secondary analysis could link this study with other related studies.

Konishi et al. [46] developed a visualization system that displays glycan profiles across different organs and tissues for humans and mice. The system enables visualization of glycan structure and relates it to organs and tissues in human and the mouse. It displays the glycan profiles on a two-dimensional map of a human and a mouse. The graphic lets the user click on organs to see profiles of glycans or on glycans to highlight the set of organs the glycans are expressed in. The tool has functionality for users to upload their own data and visualize it with the GlycomeAtlas.

Research would benefit from the development of visualization-based system that characterizes GAGs for organs and tissue with functionality to compare profiles. Researches could cumulatively contribute GAG data for both healthy and diseased profiles to a public repository. This could then be used to compare to characterization of GAGs for specific health conditions.

Shi and Zaia [47] demonstrate organ-specific sulfation patterns of the GAG Heparan Sulfate in rat tissue. Using liquid chromatography/mass spectrometry they determined average length, disaccharide composition, and expression of structural epitopes. They show that rat organs exhibit different Heparan Sulfate disaccharide profiles. The sulfation patterns also differ with muscles having a significantly lower average number of sulfations per 100 disaccharide profiles compared with other tissues such as liver, brain, spleen, lung, kidney, and heart. The average chain length was longer in some tissues like spleen compared with heart. With unique identifiers based on representations like GlycoCT, the patterns such as these could be stored in a database and represented in GlycomeAtlas-type visualization. The goal would be to establish normal GAG profiles including sulfation that can be used to discriminate tissue and organ types including comparative analysis on diseased tissue.

4.2.2 Disease Profiles

Informatics research often involves comparative analysis of diseased and normal tissue. Results in glycoscience research support such an informatics approach for GAG profiles. Smetsers et al. [48] conducted experiments where antibodies that recognized epitopes of

Heparan and Chondroitin Sulfate were used to show differential expression between melanoma cells and normal cells. The results showed that GAG expression is distinct in melanoma cells and thus, could be used as potential biomarkers for diagnosis and prognosis.

Suarez et al. [49] demonstrated that Hepararn Sulfate mediates trastuzumab in breast cancer cells and has the potential to be a biomarker for resistance to trastuzumab. The profile of GAGs and proteoglycans determines the efficiency of the drug with Heparan Sulfate and the proteoglycan Syndecan-1 binding with the trastuzumab antibody and facilitating the blocking of the human epidermal growth factor receptor 2 (HER2). Gomez et al. [50] give a review of Heparan Sulfate and Heparanase as modulators of breast cancer. The ability to capture the details of disease-specific results in a centralized database with unique identifiers for relevant GAG components will advance the field's ability to discover new treatments and communicate the importance of the science while reducing the complexity of managing it [51].

5   Conclusion

The convergence of standards, databases, and software in the field is of paramount importance to advancing informatics and computer science applications in the study of glycoscience. Without such a centralized and standardized representation the complexities of interoperability and drift among standards will slow discovery. To achieve a robust and widely used information sharing toolkit, there needs to be an active open source developer community that is engaged in the problem space. To get collaborative buy in from computer scientists and informaticists, research questions relevant to their disciplines need to be defined in the space. Given the complexity of glycomics and the advancement of unique identifiers for structures, opportunities in recall and precision analysis become feasible along with a wide range of algorithmic approaches for comparing and predicting glycan structures and identifying predictive biomarkers. Through the articulation of shared research objectives, the goal of advancing glycoscience widely across biology, chemistry, and computer science is in reach.

References

1. Editorial (2005) Sweet collaborations. Nat Methods 2:799
2. National Research Council (US) Committee on Assessing the Importance and Impact of Glycomics and Glycosciences (2012) Transforming glycoscience: a roadmap for the future. National Academies Press, Washington, DC
3. Raman R, Venkataraman M, Ramakrishnan S, Lang W, Raguram S, Sasisekharan R (2006) Advancing glycomics: implementation strategies at the Consortium for Functional Glycomics. Glycobiology 16(5):82R–90R. doi:10.1093/glycob/cwj080
4. Esko JD, Kimata K, Lindahl U (2009) Proteoglycans and sulfated glycosaminoglycans.

In: Varki A, Cummings RD, Esko JD et al (eds) Essentials of glycobiology, 2nd edn. Cold Spring Harbor Laboratory, Cold Spring Harbor, NY

5. Aoki-kinoshita KF (2008) An introduction to bioinformatics for glycomics research. PLoS Comput Biol 4(5):1–7. doi:10.1371/journal.pcbi.1000075

6. Raman R, Raguram S, Venkataraman G, Paulson JC, Sasisekharan R (2005) Glycomics: an integrated systems approach to structure-function relationships of glycans. Nat Methods 2(11):817–824. doi:10.1038/NMETH807

7. Sasisekharan R, Raman R, Prabhakar V (2006) Glycomics approach to structure-function relationships of glycosaminoglycans. Annu Rev Biomed Eng 8:181–231. doi:10.1146/annurev.bioeng.8.061505.095745

8. Perez S, Mulloy B (2005) Prospects for glycoinformatics. Curr Opin Struct Biol 15:517–524

9. International Union of Pure and Applied Chemistry (1997) Compendium of analytical nomenclature, 3rd edn. Blackwell Science, Oxford, UK, http://www.chem.qmul.ac.uk/iupac/2carb/38.html. ISBN 86542-6155

10. Bohne-lang A, Lang E, Fo T (2001) LINUCS: linear notation for unique description of carbohydrate sequences. Carbohydr Res 336:1–11

11. Aoki-kinoshita K, Yamaguchi A, Ueda N, Akutsu T, Mamitsuka H, Goto S, Kanehisa M (2004) KCaM (KEGG carbohydrate matcher): a software tool for analyzing the structures of carbohydrate sugar chains. Nucleic Acids Res 32:W267–W272

12. Sahoo SS, Thomas C, Sheth A, Henson C, York WS (2005) GLYDE-an expressive XML standard for the representation of glycan structure. Carbohydr Res 340:2802–2807. doi:10.1016/j.carres.2005.09.019

13. York WS, Kochut KJ, Miller JA, Sahoo S, Thomas C, Henson C (2007) GLYDE-II–GLYcan structural data exchange using connection tables. University of Georgia Technical Report

14. Herget S, Ranzinger R, Maass K (2008) GlycoCT–a unifying sequence format for carbohydrates. Carbohydr Res 343:2162–2171. doi:10.1016/j.carres.2008.03.011

15. Doubet S, Albersheim P (1992) CarbBank. Glycobiology 2(6):505

16. Doubet S, Bock K, Smith D, Darvill A, Albersheim P (1989) The complex carbohydrate structure database. Trends Biochem Sci 14(12):475–477

17. Consortium for Functional Glycomics (2013) http://www.functionalglycomics.org/glycomics/molecule/jsp/carbohydrate/carbMoleculeHome.jsp. Accessed 23 Dec 2013

18. Consortium for Functional Glycomics Binding Proteins (2013) http://www.functionalglycomics.org/glycomics/molecule/jsp/gbpMoleculehome.jsp. Accessed 23 Dec 2013

19. Lütteke T, Bohne-lang A, Loss A, Goetz T, Frank M, Lieth CW (2006) GLYCOSCIENCES.de: an Internet portal to support glycomics and glycobiology research. Glycobiology 16(5):71–81. doi:10.1093/gly cob/cwj049

20. Glycoscience.de database (2013) http://www.glycosciences.de/tools/linucs/input.php. Accessed 23 Dec 2013

21. Hashimoto K, Goto S, Kawano S, Aoki-kinoshita KF, Ueda N, Hamajima M et al (2006) REVIEW KEGG as a glycome informatics resource. Glycobiology 16(5):63–70. doi:10.1093/glycob/cwj010

22. KEGG GenomeNet (2013) http://www.genome.jp. Accessed 23 Dec 2013

23. Ranzinger R, Herget S, Wetter T, Lieth CW (2008) GlycomeDB: an integration of open-access carbohydrate structure databases. BMC Bioinformatics 13:1–13. doi:10.1186/1471-2105-9-384

24. GlycomeDB (2013) http://www.glycome-db.org/showMenu.action?major=downloads. Accessed 23 Dec 2013

25. Ceroni A, Dell A, Haslam SM (2007) The GlycanBuilder: a fast, intuitive and flexible software tool for building and displaying glycan structures. Source Code Biol Med 13:1–13. doi:10.1186/1751-0473-2-3

26. Ceroni A, Dell A, Haslam SM (2007) GlycanBuilder. http://www.ncbi.nlm.nih.gov/pmc/articles/PMC1994674/bin/1751-0473-2-3-S1.zip. Accessed 23 Dec 2013

27. GlycosWorkbench (2013) http://code.google.com/p/glycoworkbench/. Accessed 23 Dec 2013

28. IBM Watson (2013) http://www.research.ibm.com/labs/watson/index.shtml. Accessed 23 Dec 2013

29. Murdoch TB, Detsky AS (2013) The inevitable application of big data to health care. JAMA 309(13):5–6

30. Aoki-kinoshita KF (2003) Efficient tree-matching methods for accurate carbohydrate database queries. Genome Inform 143:134–143

31. Smith TF, Waterman MS (1981) Identification of common molecular subsequences. J Mol Biol 147:195–197

32. Ueda N, Aoki-kinoshita KF, Yamaguchi A, Akutsu T (2005) A probabilistic model for mining labeled ordered trees: capturing patterns in carbohydrate sugar chains. IEEE Trans Knowl Data Eng 17(8):1051–1064

33. Aoki-kinoshita KF, Ueda N, Mamitsuka H, Kanehisa M (2006) ProfilePSTMM: capturing tree-structure motifs in carbohydrate sugar chains. Bioinformatics 22(14):25–34. doi:10.1093/bioinformatics/btl244

34. Kawano S, Hashimoto K, Miyama T, Goto S, Kanehisa M (2005) Prediction of glycan structures from gene expression data based on glycosyltransferase reactions. Bioinformatics 21: 3976–3982

35. Suga A, Yamanishi Y, Hashimoto K, Goto S, Kanehisa M (2007) An improved scoring scheme for predicting glycan structures from gene expression data. Genome Inform 18:237–246

36. Venkataraman G, Shriver Z, Raman R, Sasisekharan R (1999) Sequencing complex polysaccharides. Science 286:537–542

37. Shriver Z, Raman R, Venkataraman G, Drummond K, Turnbull J et al (2000) Sequencing of 3-O sulfate containing heparin decasaccharides with a partial antithrombin III binding site. Proc Natl Acad Sci U S A 97: 10359–10364

38. Guerrini M, Raman R, Venkataraman G, Torri G, Sasisekharan R, Casu B (2002) A novel computational approach to integrate NMR spectroscopy and capillary electrophoresis for structure assignment of heparin and heparan sulfate oligosaccharides. Glycobiology 12:713–719

39. Maxwell E, Tan Y, Tan Y, Hu H, Benson G, Aizikov K et al (2012) GlycReSoft: a software package for automated recognition of glycans from LC/MS data. PLoS One 7(9):e45474. doi:10.1371/journal.pone.0045474

40. Li L, Zhang F, Zaia J, Linhardt RJ (2012) Top-down approach for the direct characterization of low molecular weight heparins using LC-FT-MS. Anal Chem 84:8822–8829

41. Lieth CW, Bohne-Lang A, Lohmann KK, Frank M (2004) Bioinformatics for glycomics: status, methods, requirements, and perspectives. Brief Bioinform 5:164–178

42. Lieth CW, Lutteke T, Frank M (2006) The role of informatics in glycobiology research with special emphasis on automatic interpretation of MS spectra. Biochim Biophys Acta 1760:568–577

43. Lütteke T, Frank M, von der Lieth CW (2005) Carbohydrate structure suite (CSS): analysis of carbohydrate 3D structures derived from the PDB. Nucleic Acids Res 33:D242–D246

44. Wang H, Julenius K, Hryhorenko J et al (2007) Systematic analysis of proteoglycan modification sites in caenorhabditis elegans by scanning mutagenesis. J Biol Chem. doi:10.1074/jbc.M609193200

45. Shao C, Shi X, White M, Huang Y, Hartshorn K, Zaia J (2013) Comparative glycomics of leukocyte glycosaminoglycans. FEBS J 280: 2447–2461. doi:10.1111/febs.12231

46. Konishi Y, Aoki-kinoshita KF (2012) The GlycomeAtlas tool for visualizing and querying glycome data. Bioinformatics 28(21):2849–2850. doi:10.1093/bioinformatics/bts516

47. Shi X, Zaia J (2009) Organ-specific heparan sulfate structural phenotypes. J Biol Chem 284:11806–11814

48. Smetsers TFCM, Westerlo EMA, Dam GB et al (2003) Localization and characterization of melanoma-associated glycosaminoglycans: differential expression of chondroitin and heparan sulfate epitopes in melanoma localization and characterization of melanoma-associated glycosaminoglycans. Cancer Res 63:2965–2970

49. Suarez ER, Paredes-gamero EJ, Giglio AD, Luis I, Nader HB, Aparecida M et al (2013) Heparan sulfate mediates trastuzumab effect in breast cancer cells. BMC Cancer 13(1):444. doi:10.1186/1471-2407-13-444

50. Gomes AM, Stelling MP, Pavao MSG (2013) Heparan sulfate and heparanase as modulators of breast cancer progression. Biomed Res Int. 11 pgs.http://dx.doi.org/10.1155/2013/852093

51. Packer NH, von der Lieth CW, Aoki-Kinoshita KF, Lebrilla CB, Paulson JC et al (2008) Frontiers in glycomics: bioinformatics and biomarkers in disease. Proteomics 8:8–20

Chapter 24

Designing "High-Affinity, High-Specificity" Glycosaminoglycan Sequences Through Computerized Modeling

Nehru Viji Sankaranarayanan, Aurijit Sarkar, Umesh R. Desai, and Philip D. Mosier

Abstract

The prediction of high-affinity and/or high-specificity protein–glycosaminoglycan (GAG) interactions is an inherently difficult task, due to several factors including the shallow nature of the typical GAG-binding site and the inherent size, flexibility, diversity, and polydisperse nature of the GAG molecules. Here, we present a generally applicable methodology termed Combinatorial Library Virtual Screening (CVLS) that can identify potential high-affinity, high-specificity protein–GAG interactions from very large GAG combinatorial libraries and a suitable GAG-binding protein. We describe the CVLS approach along with the rationale behind it and provide validation for the method using the well-known antithrombin–thrombin–heparin system.

Key words Glycosaminoglycan (GAG), Docking, Virtual library, Virtual screening (VS), Genetic algorithm (GA), Molecular interaction, Specificity

1 Docking of GAGs

Carbohydrates are being increasingly recognized as key players in a number of biological responses. Glycosaminoglycans (GAGs) are a special class of carbohydrates consisting of repeating hexosamine sugar and uronic acid/galactose units that are at the center of a number of these functions [1]. Most GAGs are sulfated, as exemplified by heparin (H), heparan sulfate (HS), chondroitin sulfate (CS), and dermatan sulfate (DS). Patterns of sulfation on GAGs are thought to be important for differential recognition of proteins; however, the extent of sequence-specific function has not been fully established [2].

The specificity of interaction between GAGs and proteins (or lack thereof) is the driving force for several pathways. For example, H/HS binds to antithrombin (AT) with high affinity and high

Kuberan Balagurunathan et al. (eds.), *Glycosaminoglycans: Chemistry and Biology*, Methods in Molecular Biology, vol. 1229, DOI 10.1007/978-1-4939-1714-3_24, © Springer Science+Business Media New York 2015

specificity, while its interaction with thrombin (T) is known to be nonspecific [3, 4]. The AT–H/HS-mediated inhibition is a three-step procedure (Fig. 1). It begins with AT recognition by a specific pentasaccharide sequence on a long H/HS chain, known as the DEFGH sequence (or simply H5), leading to formation of a tight AT–H/HS complex, thus immobilizing AT and causing a conformational change that spatially orients AT. This is followed by a low-affinity (mostly ionic) binding of T at a distal site on the long H/HS chain, and interaction that is thought to be largely nonspecific. This low-affinity and low-specificity interaction between H/HS and T allows the latter to "walk" along the polymeric chain and reach AT.

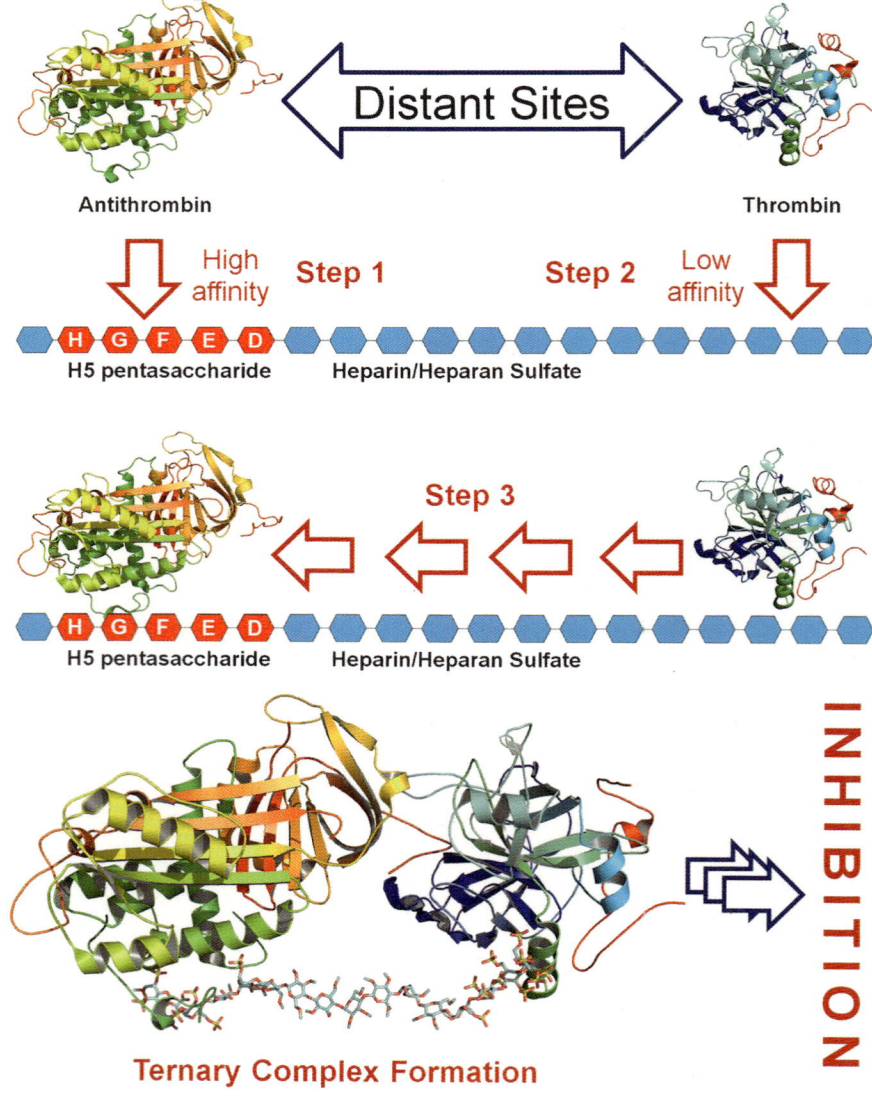

Fig. 1 Bridging mechanism of T inhibition by AT and H/HS (PDB ID:1TB6)

In the final step, the AT–H/HS–T ternary complex results in inactivation of T via a covalent linkage between the AT-reactive center loop (RCL) and S195 of the T catalytic triad. Thus, both the specificity of H/HS–AT interaction and the relative nonspecificity of H/HS–T interaction are necessary for inhibition of T by AT.

With the growing significance of GAGs in biology and the expectation that their specific interactions with proteins are the reason behind physiologic and pathologic responses, it is imperative to develop reliable methodologies for studying these phenomena. While approaches for studying carbohydrate–protein interactions have expanded considerably in the recent past [2], analytical approaches to study GAGs are made difficult by the very sulfate groups that bestow physiological functions to them. The high negative charge density makes it difficult to isolate pure, homogenous GAG sequences, which are mostly isolated as mixtures of chains with varied molecular weights and sulfation patterns. This poses challenges in elucidating their structural biology because techniques such as NMR and X-ray crystallography are highly dependent on the availability of pure homogenous samples. Chemical approaches for GAG synthesis have yielded only a limited number of homogenous GAG sequences. This implies that the plethora of structurally diverse GAG sequences present in nature remain unstudied to date using conventional methodologies.

An atomistic dissection of GAG interactions with proteins is critical to fully elucidate the role of these important biopolymers, which may also lead to GAG sequences as therapeutic agents. Such information is typically provided by theoretical computational studies. Computational molecular modeling methodology typically addresses four major questions: (1) Which GAG sequences are likely to interact with which proteins in a specific manner? (2) What are their likely relative affinities? (3) What is the likely GAG-binding site on proteins? (4) What is the geometry of GAG binding in the binding site? We present the antithrombin–pentasaccharide system as a test case of the methodology developed in the past few years.

The earliest work in prediction of GAG-binding site on proteins was performed by Cardin and Weintraub [5], which was followed by other investigations [6–8]. Analysis of GAG-binding proteins revealed the presence of consensus primary amino acid sequences in some proteins. Other factors were also found to influence protein–GAG binding including the role of secondary and tertiary structure, the spatial separation and type (Arg/Lys) of basic amino acids, and surface exposure of amino acids containing donor nitrogens. In other words, the identification of GAG-binding sites was more related to the 3-D stereoelectronic composition of the protein surface and its complementarity to the GAG topology than its primary amino acid sequence. Automated docking and scoring have been used to predict how GAGs bind to proteins [9–14]. Many different docking routines have been explored including GOLD [15], AutoDock [16], and HADDOCK [17]. However, several challenges have to be addressed to ensure meaningful results. First, the conformational flexibility in GAG chains is much higher than in typical drug-like small molecules. In general, the effectiveness of docking protocols decreases exponentially with the number of rotatable bonds, which implies that such methods do work well with longer GAG sequences. This situation is further complicated by amino acid side chain flexibility and exacerbated by higher order protein structural changes that may occur upon GAG binding. Second, commercially available docking routines are primarily designed to predict how small molecules bind within well-defined deep pockets, whereas GAGs almost exclusively bind to shallow surfaces on proteins. Thus, docking of GAGs almost always begins by reducing the search space. For example, conformational space search is reduced by disallowing rotation around the glycosidic linkages and limiting sampling of ring conformations. Alternatively, an initial rigid body docking may be used as a screen to identify certain oligosaccharide sequences of interest before flexible docking is performed. A distinct advantage of the automated docking paradigm is that it can be made to run reasonably fast, thus allowing simultaneous virtual screening (VS) of many GAG sequences.

Prediction of binding affinities of GAG sequences for proteins may be accomplished either through the use of scoring functions or molecular dynamics (MD) simulations. An in-depth discussion of MD is beyond the scope of this work, but has been addressed elsewhere [18]. The presence of sulfates leads to the almost exclusive interaction of GAGs with positively charged areas of protein surfaces. This is a difficult hurdle in the path towards successful prediction of protein–GAG-binding affinities because GAG-associated sulfates have, as of yet, not been especially well parameterized. Yet, force fields such as GLYCAM/AMBER [19, 20], CHARMM [21–23], and GROMOS96 [24] do exist for carbohydrate MD simulations. Several MD simulations have been conducted on GAGs alone, and also protein–GAG systems [25, 11, 26–30]. Gandhi and Mancera [31] and Pichert et al. [32] have also used MD simulations, combined with MM-PBSA (Molecular Mechanics Poisson Boltzmann Surface Area) and MM-GBSA (Molecular Mechanics General Born Surface Area) methods [33–35] to predict affinities of protein–GAG interactions.

A major advance would be enabled if a methodology that identifies likely "high-affinity, high-specificity" GAG sequences is developed. Here, we present an algorithm for "high-affinity, high-specificity" GAG sequences based on docking and scoring of virtual GAG libraries. We describe the steps involved in using this protocol with supporting rationale. Validation of the method is presented for antithrombin with heparin, the prototypical protein–GAG interaction in which a specific H/HS sequence is recognized.

2 Methodology

2.1 Introduction

The method presented here, known as *CVLS* (*C*ombinatorial *V*irtual *L*ibrary *S*creening), is based loosely on a two-stage virtual screening paradigm (Fig. 2). In the first stage or "filter," the docking is performed relatively quickly, with 10,000 genetic algorithm (GA) iterations per docking run. In this way, many tens or hundreds of thousands of GAG sequences may be screened efficiently (on a large cluster), albeit at the possibility of few false negatives. In the second filter, a small fraction of the highest scoring sequences are then re-docked under more rigorous conditions (100,000 GA iterations) and in triplicate. The top two highest scoring solutions from each of the three docking runs are then compared. If these six solutions form a sufficiently tight cluster (backbone atom RMSD ≤ 2.5 Å; *see* Subheading 2.2.2 for backbone atom definition), the GAG sequence passes the second filter. Using the backbone atoms, which collectively serve as a maximum common subset of atoms common to all GAG sequences, allows sequences with different substituent patterns and conformations to be meaningfully compared.

A key operating principle in CVLS is that if the protein–GAG interaction is specific (meaning that the protein recognizes a specific GAG conformation and pattern of substituents), then the docking routine will return solutions that are consistently docked in a similar fashion (i.e., the RMSD among all solutions is sufficiently low). By this method, then, potentially "high-affinity" and "high-

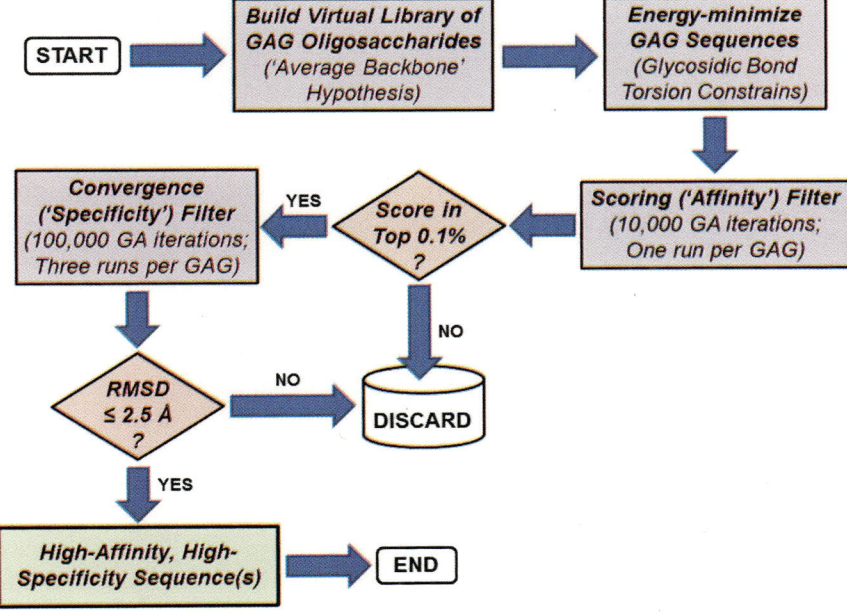

Fig. 2 Overview of the dual-filter CVLS algorithm

specificity" sequences can effectively be identified. It should be noted that if the size of the virtual library is sufficiently small, or the user's computing resources are sufficiently large, the first filter may be effectively removed, thus attenuating the false-negative rate associated with the short-duration docking runs.

To provide an example that validates the method, an H/HS virtual library consisting of 46,656 hexasaccharide sequences is constructed and subjected to CVLS against the activated antithrombin structure from Protein Data Bank (PDB) ID = 1TB6 [36]. It is shown that the method is able to reproduce the heparin pentasaccharide also found in 1TB6. Although a specific example is provided here, it is also important to note that this methodology is potentially applicable to *any* class of GAG and *any* GAG-binding protein with an identified GAG-binding site whose coordinates are determined either experimentally or through homology modeling.

The CVLS algorithm as described here employs two well-established molecular modeling software packages, although other software packages and/or options could be used to achieve similar results. SYBYL-X (Tripos Associates, St. Louis, MO, hereafter referred to as SYBYL), its general-purpose molecular mechanics force field (Tripos Force Field; TFF), its associated scripting language known as SYBYL Programming Language (SPL), and its native file format (mol2) are used in all phases of the CVLS algorithm except for automated docking, which is handled by GOLDSuite (CCDC, Cambridge, UK, hereafter referred to as GOLD) [37] and its native scoring function (GOLDScore). Although the CVLS method is explicitly described in the following subsections, other general considerations are also relevant (*see* **Notes 1–11**).

2.2 Building H/HS Virtual Libraries

2.2.1 Nomenclature of Monosaccharide Units and Disaccharide Building Blocks

H/HS are composed of repeating disaccharide units consisting of β-D-glucuronic acid or α-L-iduronic acid and α-D-glucosamine residues (Fig. 3) [38]. Each monosaccharide unit is variably substituted with *N*-acetyl, *N*-sulfate, and/or *O*-sulfate groups [39], giving rise to sequences that are phenomenally diverse. As H/HS sequences can be long, a simple, intuitive, extensible, and compact naming convention for the sequences is required. Such a system was created based loosely on the symbolic system employed in GLYCAM [19]. The letter "Z" is used for D-glucuronic acid, "u" for L-iduronic acid, and "Y" for D-glucosamine. Similarly, ring conformations are encoded: "a" for $^1C_4$, "b" for $^4C_1$, and "c" for ^2S_O (designations for ring conformation are derived from the Cremer–Pople ring puckering [40–42]). Glucosamine substituents are "H" for an unsubstituted 2-position (NH_3^+), "C" for *N*-acetyl, "2" for *N*-sulfate, "3" for 3-*O*-sulfate, and "6" for 6-*O*-sulfate. Uronic acid substituents: 2 = 2-*O*-sulfate. Anomeric carbon configurations are encoded: "A" for α and "B" for β. These H/HS monosaccharide names used here are shown in Table 1.

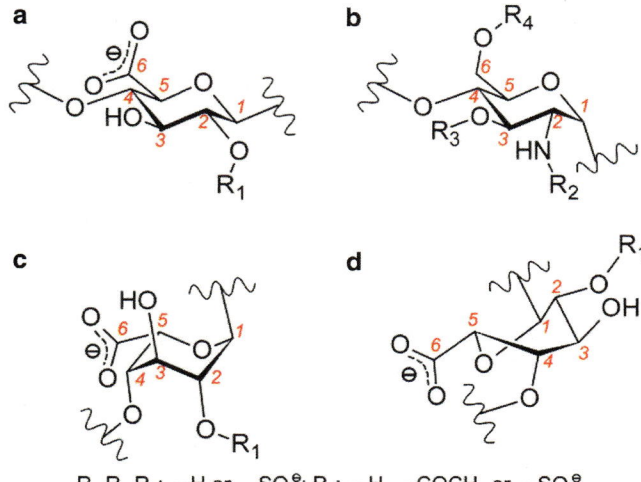

R_1, R_3, R_4: —H or —$SO_3^{\ominus}$; R_2: —H, —$COCH_3$ or —$SO_3^{\ominus}$

Fig. 3 Monosaccharide constituents of H/HS. (**a**) β-D-Glucuronic acid (GlcA), $^4C_1$ conformation. (**b**) α-D-Glucosamine (GlcN), $^4C_1$ conformation. (**c**) α-L-Iduronic acid (IdoA), $^1C_4$ conformation. (**d**) α-L-Iduronic acid (IdoA), $^2S_0$ conformation. Atom numbers are shown in *red*

Table 1
Naming convention for the H/HS monosaccharide building blocks

CVLS library name[a]	Conventional name
ZbB	GlcA ($^4C_1$ conformation, β-anomer)
Zb2B	GlcA(2S) ($^4C_1$ conformation, β-anomer)
uaA	IdoA ($^1C_4$ conformation, α-anomer)
ua2A	IdoA(2S) ($^1C_4$ conformation, α-anomer)
ucA	IdoA ($^2S_0$ conformation, α-anomer)
uc2A	IdoA(2S) ($^2S_0$ conformation, α-anomer)
Yb2A	GlcNS ($^4C_1$ conformation, α-anomer)
Yb23A	GlcNS(3S) ($^4C_1$ conformation, α-anomer)
Yb26A	GlcNS(6S) ($^4C_1$ conformation, α-anomer)
Yb236A	GlcNS(3S,6S) ($^4C_1$ conformation, α-anomer)
YbCA	GlcNAc ($^4C_1$ conformation, α-anomer)
YbC6A	GlcNAc(6S) ($^4C_1$ conformation, α-anomer)
YbHA	GlcN ($^4C_1$ conformation, α-anomer)
YbH3A	GlcN(3S) ($^4C_1$ conformation, α-anomer)
YbH36A	GlcN(3S,6S) ($^4C_1$ conformation, α-anomer)

[a]Symbols used: Z = D-glucuronic acid, u = L-iduronic acid, Y = D-glucosamine. Ring conformations: a = $^1C_4$; b = $^4C_1$; c = $^2S_0$. Glucosamine substituents: H = unsubstituted 2-position (NH_3^+), C = N-acetyl, 2 = N-sulfate, 3 = 3-O-sulfate, 6 = 6-O-sulfate. Uronic acid substituents: 2 = 2-O-sulfate. Anomeric carbon configuration: A = α, B = β

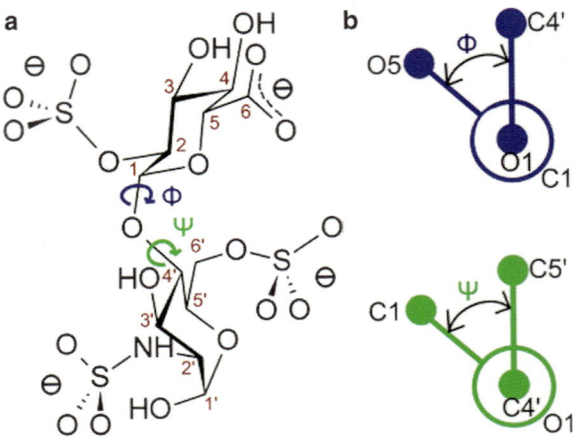

Fig. 4 Inter-glycosidic linkage definitions. (**a**) Location of glycosidic torsion angles phi (φ) and psi (ψ) within a typical H/HS disaccharide, GlcA(2S)–GlcNS(6S). Atom numbers are shown in *red*. (**b**) Newman projection representations of φ and ψ showing the names of the atoms involved in their definition

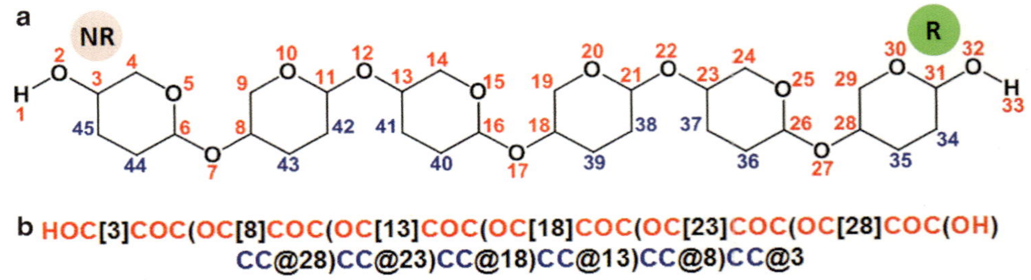

Fig. 5 Representation of the H/HS hexasaccharide backbone atoms. (**a**) Numbering scheme. The reducing (R) and nonreducing (NR) ends are labeled. (**b**) SYBYL Line Notation (SLN) (*see* Subheading 2.2.2 for details)

2.2.2 Atom Numbering Convention

The relative orientation of a pair of adjacent monosaccharide residues is defined by two torsional angles at the glycosidic linkage, denoted by φ and ψ [43]:

$$\varphi = O5 - C1 - O1 - C4' \tag{1}$$

$$\psi = C1 - O1 - C4' - C5' \tag{2}$$

The glycosidic linkage torsion angles are illustrated in Fig. 4. To facilitate the constrained docking of the H/HS oligosaccharides (vide infra), the atoms in the glycosidic linkage are systematically renumbered so that the glycosidic linkage atoms in each disaccharide have consistent numbering; that is, O5 becomes atom number 5, C1 becomes atom number 6, O1 becomes atom number 7, C4′ becomes atom number 8, and C5′ becomes atom number 9 (*see* Fig. 5). In the complete H/HS oligosaccharide sequence, the atoms are analogously renumbered such that the atoms designating the glycosidic linkage span a continuous range from 1 to $n \times 5 + 3$, where n is the number of monosaccharides in the oligosaccharide sequence. The renumbering procedure is facilitated through the use of in-house SPL scripts.

SYBYL Line Notation (SLN), similar to Simplified Molecular-Input Line-Entry System (SMILES) [44], provides a convenient way to specify the unique topology of a molecule via a string of text and is the basis of powerful substructure search functionality within SYBYL. The numbering scheme and SLN used here to represent the H/HS hexasaccharide backbone atoms are shown in Fig. 5. The SLN representation of the backbone for any $1 \rightarrow 4$-linked GAG with n monomers may be created using a loop that runs from $i = 1$ to $(n - 1)$ with an initial string "HOC[3]COC," followed by $n-1$ copies of a second string "(OC[$i \times 5 + 3$]COC," an intermediate string "(OH)," $n-1$ copies of a fourth string "CC@$i \times 5 + 3$," and ending with a terminal string "CC@3," This convention allows facile calculation of the atom numbers in the SLN corresponding to the atoms defining the $n - 1$ glycosidic linkage torsion angles in the GAG (colored red in Fig. 5). Additionally, the atom numbers of other backbone atoms may be generally specified, including the nonreducing-end hydroxyl group oxygen (atom number 2) and hydrogen (atom number 1) and the reducing-end hydroxyl group oxygen (atom number $n \times 5 + 2$) and hydrogen (atom number $n \times 5 + 3$).

2.2.3   Explicit Modeling of Ring Conformations

Ring conformations in the modeled GAGs are kept rigid to greatly simplify the docking of GAGs during the actual docking process. To this end, the pyranose rings of glucuronic acid (GlcA)- and glucosamine (GlcN)-based monosaccharides are assigned the $^4C_1$ conformation, while the rings of iduronic acid (IdoA)-based monosaccharides, which have two primary low-energy conformations, are modeled with either the $^1C_4$ or the ^2S_O conformation.

2.2.4   Atom and Bond Types and Representation of Functional Groups

The proper assignment of atom and bond types for the atoms and bonds of the GAG sequences is very important for the automated docking program to correctly pair the atoms of the protein with the atoms of the GAG. GOLD, the docking program used here, reads files in the SYBYL mol2 format and also uses SYBYL atom and bond types internally; these will be described here. Unless noted otherwise, all bond types are "1" (single bonds).

The atom and bond types of the protein are taken from the input protein mol2 file as assigned automatically by SYBYL and should not normally be changed. For the GAG sequences, the backbone atom types are "C.3" for the carbon atoms and "O.3" for the oxygen atoms. For the free hydroxyl groups, the oxygen atoms are also of type "O.3" and the hydrogen atoms have type "H." Both carboxylate and O-sulfate groups distribute their anionic charge over multiple atoms through delocalization. This in turn affects the electrostatic strength and nature of the hydrogen-bonding ability of these groups. To ensure that the molecular modeling and docking programs correctly represent these interactions, SYBYL atom types are assigned such that the carboxylate carbon atoms have type "C.2," O-sulfate sulfur atoms have type "S.o2," the oxygen atom connecting the O-sulfate group to the GAG backbone has type "O.3," and the terminal oxygen atoms of both carboxylate and O-sulfate groups have type "O.co2." In addition, the bonds connecting either the carboxylate carbon or the O-sulfate sulfur atom to each of the terminal oxygen atoms are of type "ar" (aromatic) to denote the delocalized nature of the electrons in these groups. The sulfur–oxygen bond length should be 1.5 Å, as observed in high-resolution X-ray crystal structures. For primary amine groups, the nitrogen atom will usually be protonated at physiological pH and as such will have type "N.4" with three hydrogen atoms of type "H" attached. The N-acetyl group is an amide: The nitrogen is type "N.am," the carbonyl carbon and oxygen atoms are "C.2" and "O.2," respectively, and the CH_3 group has atoms of type "C.3" and "H." The N-sulfate group nitrogen is "N.3" and the remaining atom and bond types are as described for the O-sulfate group.

2.2.5   The "Average Backbone" Approximation

The H/HS molecules possess both unusual rigidity and flexibility [39]. The unusual rigidity is due to the partial rotational freedom of the glycosidic linkage torsion angles; the majority of the flexibility in H/HS is due to the flexible ring conformation of IdoA, which can easily achieve both $^1C_4$ and ^2S_O conformations at equilibrium. A GAG backbone can be defined as the atoms of the pyranose rings and the glycosidic linkage atoms. On analyzing various available crystal structures [45–47] and also from the reports of various research groups [43], it was found that the glycosidic torsion angles fall within a relatively narrow range and suggest that they remain fairly constant irrespective of a change in substitution pattern around the inter-glycosidic bond. Thus, the "average backbone" approximation, which utilized an average value of torsion for each inter-glycosidic linkage angle φ and ψ (Table 2), was employed. This greatly reduced the number of rotatable bonds in the GAG oligosaccharides, making the otherwise difficult or impossible task of docking of longer H/HS sequences (up to and including hexasaccharides) feasible. Also, in the process of building the H/HS sequences (both disaccharide and longer chains), energy minimization was performed during which the average φ and ψ values were subjected to a restraining force constant of 0.01 kcal/mol/deg^{-1}.

2.2.6   Building Monosaccharide Units

Construction of the monosaccharides (Table 1) may be accomplished in a number of ways, including sketching the structure manually and energy-minimizing or -modifying a previously determined set of atomic coordinates (e.g., from a crystal structure). A typical way of building the monosaccharide units "from scratch" using SYBYL will be briefly explained here. When SYBYL starts its

Table 2
Values used in the "average backbone" approximation

Disaccharide	φ (O5–C1–O1–C4′)	ψ (C1–O1–C4′–C5′)
GlcA(β1→4)GlcN	−81.8	−114.0
IdoA(α1→4)GlcN	−87.7	−128.3
GlcN(α1→4)GlcA	91.1	−151.6
GlcN(α1→4)IdoA	87.4	−132.3

main window is presented. The graphics window contains the menu bar, toolbar icons, and the display area. The console is docked below the display area. The general method of creating a new small-molecule structure is to click "Edit>Molecule>Sketch" to start the sketcher. Draw the structure (heavy atoms only) as a carbon skeleton in Draw mode, and then change the appropriate carbons to the correct heteroatom (nitrogen, oxygen, or sulfur) using the Modify Atom Type mode. Exit the sketcher. Using the Modify Atom Type function from the main menu ("Edit>Atom>Modify Atom Type"), change the SYBYL atom type of the atoms as described in Subheading 2.2.4 where necessary. Do the same for the SYBYL bond types ("Edit>Bond>Modify Type"). Add hydrogen atoms and name the molecule ("Edit>Hydrogens>Add All Hydrogens"). Assign a substructure name to the monosaccharide as shown in the first column of Table 1 (MODIFY SUBSTRUCTURE NAME command). Though not strictly necessary, naming the individual atoms (e.g., S3O for the 3-position *O*-sulfate sulfur atom) uniquely identifies each atom in a meaningful way (MODIFY ATOM NAME command). Energy-minimize the monosaccharide structures using the TFF with Gasteiger–Hückel charges, a fixed dielectric constant of 80 (to simulate an aqueous environment), and a non-bonded cutoff radius of 8 Å. Set the termination criteria for a maximum of 100,000 iterations subject to an energy gradient of 0.05 kcal/(mol-Å). Finally, export the structure as a mol2 file with the name of the monosaccharide contained in the file name.

Building β-D-Glucuronic Acid (GlcA) and Its Derivatives

Although it is possible to build the monosaccharide units "from scratch," it is generally easier and less error prone to start with an experimentally determined structure from the PDB. To build the β-anomer of D-glucuronic acid (GlcA; ZbB) in the $^4C_1$ conformation perform the following steps: Import the 1TB6 crystal structure into SYBYL. Select the substructure named GU1_445 corresponding to the "E" residue of the H5 sequence and copy it to an empty molecule area ("Edit>Copy"). Ensure that the atom and bond types are correct and modify as necessary. Select the atoms of the methyl groups attached to the 2- and 3-position oxygen atoms and delete them ("Edit>Delete>Selected") and then add hydrogens. Select the newly added hydrogen atom corresponding to the C4 oxygen atom and change its atom type to O.3. Add hydrogen atoms again to add the missing hydrogen. Name the substructure "ZbB" and optionally name the atoms as described previously. Export the structure as "ZbB.mol2." For D-glucuronic acid 2-sulfate (GlcA(2S); Zb2B), follow the same procedure except instead of removing both methyl groups at the 2- and 3-positions, remove only the one at the 3-position. Modify the 2-position methyl group carbon atom to type S.o2 and the methyl group hydrogens to type O.co2. Modify the S–O bond types to "ar" if necessary and then add hydrogens.

Building α-L-Iduronic Acid (IdoA) and Its Derivatives

IdoA is capable of adopting equilibrium between the chair ($^1C_4$) and skew-boat (^2S_O) forms, the two most prevalent conformations known in H/HS oligosaccharides. The H/HS library takes into account both conformations explicitly. For the $^1C_4$ conformation of IdoA, import the 1HPN structure from the PDB (model number 2) into SYBYL. Select an IdoA monomer and copy it to an empty molecule area. Continue using process similar to that described above for the GlcA-based structures to generate the uaA and ua2A monosaccharides. To generate the monomer in ^2S_O conformation, import model number 1 from 1HPN, select an IdoA monomer, and copy it to an empty molecule area. Proceed using the same procedure as described for the $^1C_4$ conformation to generate the ucA and uc2A monosaccharides.

Building α-D-Glucosamine (GlcN) and Its Derivatives

To build the α-anomer of D-glucosamine 2,3,6-sulfate (GlcNS(3S,6S); Yb236A) in the $^4C_1$ conformation perform the following steps: From the 1TB6 crystal structure, select the substructure named GU6_446 corresponding to the "F" residue of the H5 sequence and copy it to an empty molecule area ("Edit>Copy"). Modify the atom and bond types as described in Subheading 2.2.4 and then add hydrogens. Select the newly added hydrogen atom corresponding to the C4 oxygen atom and change its atom type to O.3. Add hydrogen atoms again to add the missing hydrogen. Name the substructure "Yb236A" and optionally name the atoms as described previously. Export the structure as "Yb236A.mol2." Continue modifying this structure and exporting for each of the eight remaining GlcN-based monosaccharides listed in Table 1.

When complete, the library should consist of all the monomers listed in Table 1 with appropriate molecule name, and exported as separate SYBYL mol2 files. It is critical when generating the H/HS "building blocks" that the configuration and conformation of the mono- and disaccharide units are correct. Whichever method is employed to generate the monosaccharide structures, ensure that the substitution pattern, configuration, and conformation of the structures are correct either by visual inspection or (preferably)

Table 3
The 36 disaccharide building blocks used to construct virtual H/HS oligosaccharide libraries

GlcA-containing	IdoA-containing		
ZbB-YbCA	uaA-YbCA	ua2A-Yb2A	uaA-Yb236A
ZbB-YbC6A	ucA-YbCA	uc2A-Yb2A	ucA-Yb236A
ZbB-Yb2A	ua2A-YbCA	uaA-Yb26A	ua2A-YbH3A
Zb2B-Yb2A	uc2A-YbCA	ucA-Yb26A	uc2A-YbH3A
ZbB-Yb26A	uaA-YbC6A	ua2A-Yb26A	ua2A-YbH36A
Zb2B-Yb26A	ucA-YbC6A	uc2A-Yb26A	uc2A-YbH36A
ZbB-Yb23A	ua2A-YbC6A	uaA-Yb23A	
Zb2B-Yb23A	uc2A-YbC6A	ucA-Yb23A	
ZbB-Yb236A	uaA-Yb2A	ua2A-Yb23A	
ZbB-YbHA	ucA-Yb2A	uc2A-Yb23A	

using SPL scripts that will facilitate this process with greater simplicity and reliability.

2.2.7 Combining Monosaccharides to Form Disaccharides

A virtual library of the 23 HS disaccharides observed in nature [48] can now be created. The coordinates for the disaccharide sequences can be generated by using the JOIN command in SYBYL where the reducing-end OH group attached to the C1 atom of the first monomer (a uronic acid derivative) is attached to the nonreducing-end OH group attached to the C4 atom of the second monomer (a glucosamine derivative), thereby forming the 1→4 glycosidic linkage and removing H_2O in the process. Use the MODIFY TORSION command in SYBYL to assign the average backbone values to the glycosidic linkage torsions as shown in Table 2. Name the disaccharides so that the names reflect the sequence (e.g., the disaccharide consisting of uc2A and Yb23A would be named uc2A-Yb23A) and save them as separate mol2 files. The completed disaccharide "building block" library should contain all of the naturally occurring disaccharide units as listed in Table 3.

2.2.8 Building Longer Sequences

The disaccharide library has 36 unique sequences [48]. These 36 sequences will be used to generate a combinatorial virtual library of H/HS hexasaccharides. The pseudocode routine shown in Fig. 6 outlines the steps required to create the library and will generate hexasaccharide H/HS sequences for all permutations of disaccharides *r*, *s*, and *t* using steps similar to those described previously through the use of an SPL script. The individual disaccharide units are joined and the glycosidic torsion angles are set to the "average

```
for r=1,N                      # r is first disaccharide
  for s=1,N                    # s is second disaccharide
    for t=1,N                  # t is third disaccharide
      Join r + s = rs          # tetramer formation
      Join rs + t = rst        # hexamer formation
      Renumber                 # renumber glycosidic linkage atoms
      O5-C1-O1-C4' = φ         # φ and ψ modification
      C1-O1-C4'-C5' = ψ
      Minimize energy          # relax structures
      Save mol2 file           # finished hexasaccharide sequence
    endfor
  endfor
endfor
```

Fig. 6 Pseudocode outlining the steps involved in construction of an H/HS hexasaccharide library

backbone" values using a procedure analogous to that described in Subheading 2.2.7. The SLN query in Fig. 5b is then used to identify the atoms of the hexasaccharide backbone and the backbone atoms are renumbered as shown in Fig. 5a, with the atoms of the glycosidic linkages being numbered sequentially. Next, the hexasaccharides are energy-minimized using a two-stage minimization process. In the first stage, the atoms of the backbone are kept rigid while the pendant groups are allowed to move in order to alleviate major repulsive steric interactions among them. Minimization parameters are as described in Subheading 2.2.6. In the second stage, the minimization procedure is repeated but this time all atoms are allowed to move, with only a small torsional constraint of $0.01 \ \mathrm{kcal/mol/deg^{-1}}$ applied to the glycosidic torsion angles. Finally, the structures are named and exported to a database as mol2 files. The virtual library of hexasaccharide sequences combinatorially built from 36 disaccharide building blocks (Table 3) should now include $36 \times 36 \times 36 = 46{,}656$ total sequences.

2.3 Docking of H/HS Sequences

The typical automated molecular docking study consists of the following main steps: (a) preparing the ligands (here the combinatorial virtual library); (b) preparing the protein receptor and setting up the simulation system, which involves identifying the binding site of interest; (c) performing the actual docking simulations; and (d) post-processing and analysis of the results. The steps involved in combinatorial virtual library generation have been discussed in detail in the previous section. The next important step is protein preparation.

2.3.1 Protein Preparation and Identification of GAG-Binding Sites

In this example the crystal structure of the ternary complex of antithrombin–thrombin–heparin (PDB ID:1TB6) is used. To prepare the protein, extract the antithrombin coordinates from this structure using the SYBYL Biopolymer Structure Preparation Tool. Add hydrogen atoms and check for amino acid side chains that need repair.

It is not uncommon for the surface-exposed lysine and arginine side chains that are critical for GAG recognition to be incompletely resolved in the crystal structure, resulting in missing atoms. The SYBYL Mutate Residues feature allows coordinates to be assigned to such missing atoms. For a reasonable docking study it is important to have the residues in the binding site to be in their correct protonation state; this can be done by "Set Protonation State" in the SYBYL structure preparation tool. Then minimize the protein structure with fixed heavy-atom coordinates using the TFF for 1,000 iterations subject to a termination gradient of 0.05 kcal/(mol Å).

In general, the GAG-binding site information can be obtained from the co-crystallized protein–GAG complex deposited in the PDB; if this structural information is not available then it may be inferred by site-directed mutagenesis or through computational prediction of GAG-binding sites. Numerous studies have been reported in identifying the common structural features in H/HS-binding site of proteins, by both experimental studies (X-ray and NMR) and through molecular modeling approaches [49, 50]. The binding site in antithrombin was defined to include all atoms within 14 Å of the K125 Nζ (NZ) atom. This binding site definition covers all basic residues K11, R13, R46, R47, K114, K125, R129, and R132 in the AT pentasaccharide-binding site (PBS).

2.3.2 Docking Algorithm   GOLD is a molecular docking program that uses a GA search strategy that "evolves" docked solutions starting from a set of random ligand conformations, and includes rotational flexibility for selected receptor residues along with full ligand flexibility. The algorithm that GOLD uses to generate docked solutions involves several discrete steps and is outlined in Fig. 7a. The structural state of the ligand (and any flexible receptor side chins) is encoded by a "chromosome," representing its conformation and hydrogen-bonding interactions (Fig. 7b). The "chromosome" is a binary string in which every torsional angle is encoded by a "gene." To generate different ligand conformations, each rotatable torsion angle is allowed to vary between −180° and 180°. A library of preferred conformations is used to assign specific values to the torsion angles. The GA uses two genetic operators: crossover and mutation, as illustrated in Fig. 7b. A GA operator is applied to parent chromosomes that are randomly chosen from the existing population with a bias towards the fittest members (i.e., stronger members are more likely to be parents). Two integer strings encode possible hydrogen bonds between the protein and the ligand. For a given GA iteration, the ligand–receptor hydrogen bonds are subsequently matched using a least-squares fitting protocol to maximize the number of intermolecular hydrogen bonds (i.e., hydrogen bonds between the ligand and the receptor), thus placing the ligand in the binding site and generating a ligand "pose." As a consequence the GA pose generation is biased towards intermo-

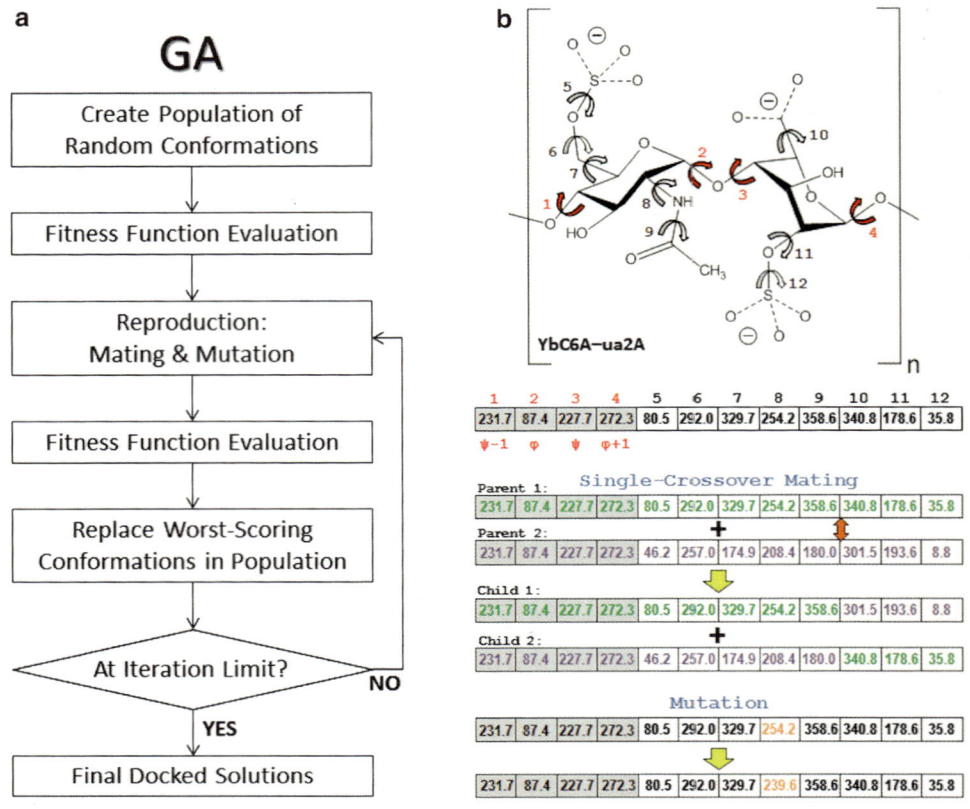

Fig. 7 The genetic algorithm (GA). (**a**) Flow chart outlining the major steps involved. (**b**) Application of the GA to a typical H/HS disaccharide, YbC6A-ua2A. A simplified "chromosome" encoding its 12 rotatable bond torsion angles is shown, with the fixed backbone torsion angles labeled in *red* and the invariant values ("genes") shown in *grey* cells. The evolutionary processes of mating and mutation are also illustrated

lecular hydrogen bonds. The GOLDScore fitness function that evaluates the quality of the docked ligand poses comprises three terms: a hydrogen bond term, a 4–8 intermolecular dispersion potential, and a 6–12 intramolecular dispersion potential energy for the internal energy of the ligand. Although this scoring function in general correlates modestly with the observed free energy of binding, a modified form of the scoring function has been found to be more reliable [12]. This modified GOLD score, which utilizes hydrogen bonding and van der Waals interactions as described above (Eq. 3), can also be used to rank the final docked solutions:

$$GOLDScore = HB_{EXT} + 1.375 \times VDW_{EXT} \qquad (3)$$

Here, HBEXT and VDWEXT are the "external" (i.e., nonbonded interactions taking place between the ligand and receptor) hydrogen bonding and van der Waals terms, respectively. The GOLDScore energy function is designed such that larger (more positive) values are more favorable than negative scores.

2.3.3 Setting Up a GOLD Docking Run

GOLD docking runs can be set up by using the Hermes visualizer, the graphical front end of GOLD. Using Hermes, the input files for docking are prepared in an interactive manner to define the binding site, specify constraints, and set up other parameters for the docking run. In Hermes, the top-level menu has an icon called GOLD; click this icon and there are two options:

1. Setup and Run a Docking.
2. Wizard.

For a new user it is highly recommended to use the *Wizard* icon, as this will walk the user through the entire process of setting up the GOLD docking run. The usual necessary steps in the wizard are the following:

1. *Proteins.* Choose the protein molecule to be used for docking. The protein molecule that was saved after processing through the protein preparation tool in SYBYL is used here.

2. *Define Binding Site.* There are multiple ways to define the binding site, the simplest being to select an atom and an associated radius. Select an atom in the protein; in this case LYS125 NZ from the "I" chain of 1TB6. Specify a radius of 14 Å to include all the atoms within 14 Å of the selected atom.

3. *Select Ligands.* Select the library of mol2 files in the SYBYL database, or concatenate all of the sequences in the library into a single multi-mol2 file. Specify the number of GA runs. For the first CVLS filter (vide infra), set the value to 10; for the second filter set it to 100.

4. *Ligand Flexibility.* To keep the glycosidic linkages fixed during the docking runs, specify the atom numbers of the two atoms joined by each glycosidic bond. Due to the consistent numbering scheme, these will be the same for all of the sequences in the library (*see* Fig. 5a). Check the option to fix ligand rotatable bonds, select "fix specific," and then "Specify Bonds" to enter the atom numbers and "Close" when done.

5. *Fitness and Search Options.* Select GOLDScore.

6. *GA Settings.* Check the "Preset" option. For the first CVLS filter, set the value to 10,000 operations; for the second filter set it to 100,000 operations.

7. *Output Options.* As a very large number of files will be created, select the option to create output sub-directories for each ligand.

8. *Finish.* Save the configuration file (*gold.conf* by default) for further use and to run GOLD outside of the Hermes interface.

2.3.4 Evaluating the Docked Poses and Optimization of Docking Conditions

To ensure that the GAG orientation and the position obtained from the docking studies were likely to represent valid and reasonable binding modes, the GOLD program docking parameters were first used to dock the heparin pentasaccharide (H5) sequence in the crystal structure (native) on AT. The H5 sequence is constrained only at their glycosidic linkage.

Parameters were optimized by performing multiple docking runs and changing various parameters such as the number of GA runs (10–200), the maximum number of iterations, and radius to cover all the atoms in the binding site. The docking procedure was repeated three times, and the results were analyzed and compared with the crystal structure of this complex [36]. The results of the docking study using the optimal parameters indicate that the highest ranked solution is spatially close to the native structure (H5 analog co-crystallized in PDB = 1TB6). Because the crystal structure for this particular GAG–protein interaction is known, a good measure of the reliability and specificity of docking is through an RMSD calculation using the backbone atoms, which for the best examined parameter set is 0.96 Å. In addition, the hydrogen bond and van der Waals interactions were found to be identical, thereby suggesting the docking procedure to be reliable to obtain a specific sequence. These parameters are subsequently used in the second filter of the CVLS algorithm.

2.4 Reproducibility of Observed H5-Binding Mode

To evaluate the ability of the GOLD algorithm to reproduce the observed binding mode of H5 (Fig. 8) with AT using the "average backbone" approximation with fixed glycosidic linkage torsion angles, a small library consisting of only H5-containing sequences

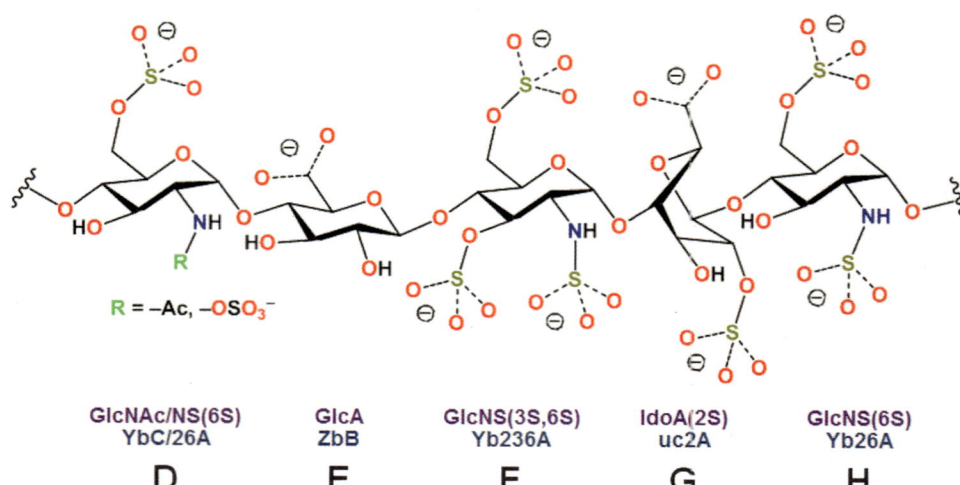

GlcNAc/NS(6S)	GlcA	GlcNS(3S,6S)	IdoA(2S)	GlcNS(6S)
YbC/26A	ZbB	Yb236A	uc2A	Yb26A
D	**E**	**F**	**G**	**H**

R = –Ac, –OSO₃⁻

Fig. 8 Structure and names of the monosaccharide constituents of the AT-specific H5 pentasaccharide sequence

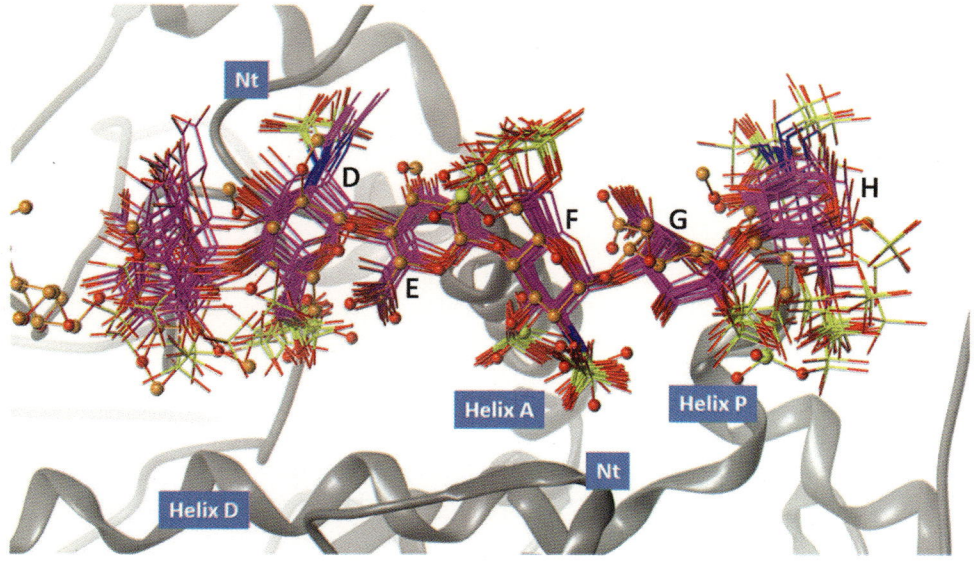

Fig. 9 H5-containing hexasaccharide sequences that reproduced the binding mode found in the PBS of the 1TB6 crystal structure

found in the entire 46,656 sequence hexasaccharide library was constructed and docked to the AT PBS using the 1TB6 crystal structure parameters corresponding to the second CVLS filter. This H5-containing library contained 44 sequences of the form X-Yb[C,2]6A-ZbB-Yb236A-u[a,a2,c,c2]A-Yb26A, where X represents any of the six uronic acid derivatives. GOLD reproduced the observed binding mode for each of the 22 sequences containing iduronic acid in the ^2S_O conformation at monosaccharide position 5. In addition, 21 of these had GOLDScores that placed them in the top 50 % of the database (*see* Fig. 9).

2.5 Application of the CVLS Approach

The previous section outlined how to set up the molecular docking simulation for each of the sequences in the H/HS library for both the first and second CVLS filters. The same GOLD configuration file is used for docking the entire database (46,656 hexasaccharide sequences) against the AT PBS for a given CVLS filter. The GOLD configuration file can be set up in such a way to read every sequence from a single mol2 file or to read the entire set of sequences from a single multi-mol2 file. Taking into account of the "average backbone" approximation, the glycosidic torsions are kept fixed in the GOLD configuration file. The atom numbers of the atoms contributing to the glycosidic linkage in different sequences would normally vary from sequence to sequence, which would complicate the identification and setting up of fixed torsions for all sequences. However, the atom numbering convention discussed above was used to number the atoms in the glycosidic linkage in a consistent manner.

This provides glycosidic linkage atom numbers that are common for all 46,656 sequences in the library. The CVLS docking calculations were performed using an in-house Linux cluster utilizing 120 concurrent processes; the time required to finish the entire hexasaccharide library was 2 days for the first filter and two additional days for the second filter.

2.5.1 Analysis of the Results and Pharmacophore Identification

Two functions (filters) to evaluate the best sequence from the virtual screening experiment were combined: affinity in the form of GOLD score and specificity in the form of consistency of docking, which represents a powerful tool to deduce high-affinity GAG sequences that bind proteins with high specificity. In the first CVLS filter, the entire library of 46,656 sequences was screened for "affinity" using ten GA runs for each sequence with 10,000 GA iterations per GA run. The best ranked solution for each of the 46,656 sequences was selected, and the scores were sorted based on GOLDScore. The sequences corresponding to the top 0.1 % of the database (47 sequences) were selected for further analysis to advance to the second CVLS filter. Of the 47 sequences that passed the first filter, 36 were placed at the same location in the PBS as the H5-like sequence in the 1TB6 crystal structure; the remaining 11 sequences overlapped with the 1TB6 H5-like sequence but were usually shifted by one or two monosaccharide residues towards the extended heparin-binding site (EHBS). Of the 36 sequences that overlapped the 1TB6 H5-like sequence, 33 contained H5 or H5-like sequences. The elements of the classical H5 AT PBS recognition motif (i.e., pharmacophore) were the following: Each sequence contained either ZbB or Zb2B in the third ("E") position and either ucA or uc2A in the fifth ("G") position (*see* Fig. 10a).

In the second CVLS filter, the 47 sequences that passed the first filter were evaluated for "specificity," as measured by docking consistency: Instead of performing a single docking experiment (*see* **Note 11**), the docking experiment was performed in triplicate for each sequence under more rigorous docking conditions (100 GA runs, 100,000 GA iterations per run). This procedure also favors reproducibility and reduces the possibility of false positives. The two highest scoring solutions from each of the three docking experiments (i.e., six docked solutions) were compared and those sequences whose six docked solutions exhibited a backbone atom RMSD of ≤ 2.5 Å were considered for further analysis. Of the 47 sequences that passed the first filter, 20 went on to pass the second filter. Of these 20, 9 reproduced the binding mode exhibited by the 1TB6 H5-like sequence, with the remaining 11 extending into the EHBS. All nine 1TB6-matching sequences shared additional H5 pharmacophoric elements above and beyond those from the first filter: Each contains either Yb26A or YbC6A in the second ("D") position, ZbB or Zb2B in the third ("E") position, either Yb23A or Yb236A in the fourth ("F") position, and either ucA or uc2A in the fifth ("G") position (*see* Fig. 10b).

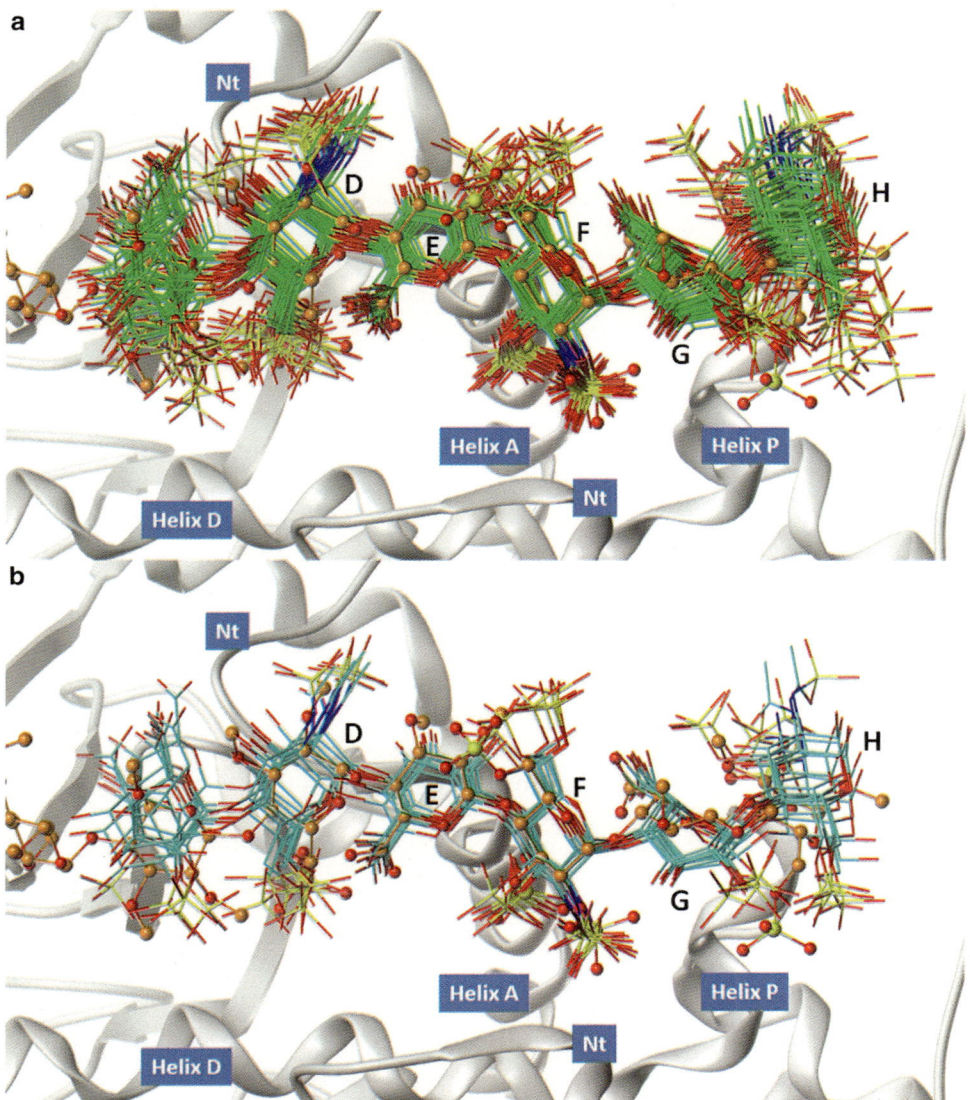

Fig. 10 (**a**) CVLS hits from the first filter. Thirty-three hexasaccharides containing H5 and H5-like hexasaccha-rides (*capped sticks*; green carbon atoms) successfully reproduced the bound conformation observed in the 1TB6 crystal structure (*ball-and-stick*; orange carbon atoms). The N-terminus (Nt) and helices involved in the recognition of the oligosaccharides are labeled, as are GAG monosaccharide units corresponding to residues D–H in the H5 pentasaccharide sequence. (**b**) CVLS hits from the second filter. The nine hexasaccharides (*capped sticks*; cyan carbon atoms) reproducing the 1TB6 H5-like sequence are shown

These results show that the CVLS algorithm may be used to identify pharmacophoric patterns in GAG sequences that specifically recognize GAG-binding sites on proteins. Such information will be invaluable not only in advancing our understanding of biological processes involving GAGs at the molecular level but also in the design of novel therapeutics targeted towards the circumvention of GAG-associated disease states.

3   Conclusions

While GAG-related glycobiology is of immense physiological relevance, it has not caught favor among scientists due to the inherent problems associated with the field. Current experimental methodologies are unable to purify homogenous samples of GAGs, with few exceptions. Synthesis is also difficult due to structural complexity. Under these circumstances, exploratory searches conducted in silico can be of immense use as a preliminary information mining resource to guide experimental efforts. We have presented methodology above that allows such applications, including studying interactions between a specific GAG sequence with a protein, or alternatively identification of specific binders from a combinatorial library of GAG sequences.

There is much room for improvement in the field of computational GAG-related glycoscience. For instance, docking a relatively small conformational library of approximately 100,000 GAG structures takes approximately 2 weeks on a reasonably sized compute cluster. This, too, only samples a very small portion of conformational space due to the excess of rotatable bonds. Luckily, current information seems to suggest that there are some stable conformations adopted by GAG oligosaccharides, which makes the currently presented methodology relevant. However, it is not guaranteed that the same conformations of the same oligosaccharide will interact with every protein. Oligosaccharides exist in multiple interchangeable ensembles of conformations in solution. Thus, first and foremost, the problem of conformational sampling must be solved. If done in the trial and error or brute force way, this will put immense pressure on currently available computational resources and consequently will also increase simulation time by several orders of magnitude. Hence, new technology must be developed to satisfy this need—particularly the development of artificial intelligence-based methodologies must be pursued.

Furthermore, much work is needed to develop force fields and scoring functions that are able to accurately assess the affinity of protein–GAG interactions. While docking and scoring have been useful tools in the case of drug-like molecules, the average success rate is usually on the order of 5–10 %. This is because current scoring functions are unable to distinguish between binders and non-binders unless there is approximately a 1,000-fold difference in affinities. With protein–GAG systems, the accuracy of available force fields falls even further due to the lack of parameterization for sulfated GAGs. Recently, a series of perspectives has suggested that rigorous work is needed for improvement of scoring functions and force fields across the next 25 years, particularly with the aim of identifying drug-like molecules [51]. Perhaps this will serve as impetus in computational glycosciences as well. In the meantime, it is expected that docking and scoring using protocols such as CVLS will contribute towards our understanding of glycobiology.

4 Notes

1. The parameters (and filtering steps) given in Fig. 2 can be adjusted to suit the user's computing resources. For example, if the user's computational resources are sufficiently great, the more rigorous GA settings used in the second filter may be applied in the first filter to reduce the number of false negatives.

2. Alternate SLN numbering schemes other than that described in Fig. 5 could also be used.

3. The energy minimization process can be lengthy for larger libraries; however, it only needs to be performed once.

4. Although the staged energy minimization described above prevents distortion of the pyranose ring and glycosidic linkage conformations, it can occasionally result in the distortion of the sulfate and/or other non-backbone groups due to steric clashes in the conformation of the initially built oligosaccharide. This situation may be remedied by identifying distorted functional groups, readjusting their geometry to an "ideal" state, and re-minimizing the sequence.

5. GAGs are highly complex molecules, and thus analysis tools to assess conformational and configurational integrity are essential in assessing the integrity of the GAG sequences.

6. The mol2 files containing the mono- and disaccharide structures may need to be edited manually in order to, for example, have the substructure label displayed consistently at the same atom position. In our libraries the "ROOT" atom of the substructure (where the substructure label is displayed) corresponds to the anomeric C1 carbon.

7. When building individual GAG sequences, values for the glycosidic linkage torsion angles may also be assigned values corresponding to other defined macroscopic conformations rather than those of the "average backbone." For example, a regular linear H/HS sequence may be generated by assigning values to φ and ψ corresponding to those found in PDB ID = 1HPN [52].

8. Naming of substructures in SYBYL can result in names in which all of the letters are capitalized. Because the naming convention employed here is case sensitive, modification of the saved mol2 files may be necessary to correct this.

9. When docking the H/HS sequences in GOLD, ensure that the "ring corner flipping" option is disabled. Otherwise, the predefined pyranose ring conformations will be modified during the docking process. Additionally, enable the "flip pyrimidal nitrogen" option to allow N-sulfate nitrogen atoms to change conformation.

10. The side chains of the protein may also be made flexible during the docking runs. However, doing so exponentially increases

the size of the conformational search space, resulting in an exponential increase in the number of iterations (and time) required by the docking program to find good docked solutions. Thus, it is not generally recommended for virtual screening applications like CVLS.

11. GOLD starts with a population of 100 arbitrarily ligand conformations, docks and evaluates them using a scoring function (the GA "fitness" function), and improves their average "fitness" by an iterative optimization procedure that is biased towards high scores. As the initial population is selected at random, the "best" docked poses and their associated scores generated during individual GA runs can be different; because of this several such GA runs are required to more reliably predict correct bound conformations. In the second CVLS filter, 100 GA runs are performed with the GOLDScore as the "fitness" function. Collectively, these 100 GA runs are referred to as one docking experiment.

Acknowledgements

This work was supported by the grants HL090586 and HL107152 from the National Institutes of Health and by Award Number S10RR027411 from the National Center for Research Resources. The content is solely the responsibility of the authors and does not necessarily represent the official views of the National Center for Research Resources or the National Institutes of Health.

References

1. Esko JD, Kimata K, Lindahl U (2009) Proteoglycans and sulfated glycosaminoglycans. In: Varki A, Cummings RD, Esko JD et al (eds) Essentials of glycobiology. Cold Spring Harbor Laboratory, Cold Spring Harbor, NY, pp 229–248

2. Blow N (2009) A spoonful of sugar. Nature 457:617–620

3. Dementiev A, Petitou M, Herbert J-M, Gettins PGW (2004) The ternary complex of antithrombin–anhydrothrombin–heparin reveals the basis of inhibitor selectivity. Nat Struct Mol Biol 11:863–867

4. Olson ST, Björk I (1991) Predominant contribution of surface approximation to the mechanism of heparin acceleration of the antithrombin–thrombin reaction. Elucidation from salt concentration effects. J Biol Chem 266:6353–6364

5. Cardin AD, Weintraub HJR (1989) Molecular modeling of protein-glycosaminoglycan interactions. Arterioscler Thromb Vasc Biol 9:21–32

6. Hileman RE, Fromm JR, Weiler JM, Linhardt RJ (1998) Glycosaminoglycan–protein interactions: definition of consensus sites in glycosaminoglycan binding proteins. Bioessays 20: 156–167

7. Margalit H, Fischer N, Ben-Sasson SA (1993) Comparative analysis of structurally defined heparin binding sequences reveals a distinct spatial distribution of basic residues. J Biol Chem 268:19228–19231

8. Sobel M, Soler DF, Kermonde JC, Harris RB (1992) Localization and characterization of a heparin binding domain peptide of human von Willebrand factor. J Biol Chem 267:8857–8862

9. Bitomsky W, Wade RC (1999) Docking of glycosaminoglycans to heparin-binding proteins: validation for aFGF, bFGF, and antithrombin and application to IL-8. J Am Chem Soc 121: 3004–3013

10. Gandhi NS, Coombe DR, Mancera RL (2008) Platelet endothelial cell adhesion molecule 1

(PECAM-1) and its interactions with glycosaminoglycans: 1. Molecular modeling studies. Biochemistry 47:4851–4862

11. Krieger E, Geretti E, Brandner B, Goger B, Wells TN, Kungl AJ (2004) A structural and dynamic model for the interaction of interleukin-8 and glycosaminoglycans: support from isothermal fluorescence titrations. Proteins 54:768–775

12. Raghuraman A, Mosier PD, Desai UR (2006) Finding a needle in a haystack: development of a combinatorial virtual screening approach for identifying high specificity heparin/heparan sulfate sequence(s). J Med Chem 49:3553–3562

13. Raghuraman A, Mosier PD, Desai UR (2010) Understanding dermatan sulfate–heparin cofactor II interaction through virtual library screening. ACS Med Chem Lett 1:281–285

14. Rogers CJ, Clark PM, Tully SE, Abrol R, Garcia C, Goddard WA III, Hsieh-Wilson LC (2011) Elucidating glycosaminoglycan–protein–protein interactions using carbohydrate microarray and computational approaches. Proc Natl Acad Sci U S A 108:9747–9752

15. Cole JC, Nissink JWM, Taylor R (2005) Protein–ligand docking and virtual screening with GOLD. In: Alvarez J, Shoichet B (eds) Virtual screening in drug discovery. Taylor & Francis, Boca Raton, FL, pp 379–416

16. Morris GM, Huey R, Lindstrom W, Sanner MF, Belew RK, Goodsell DS, Olson AJ (2009) AutoDock4 and AutoDockTools4: automated docking with selective receptor flexibility. J Comput Chem 16:2785–2791

17. Dominguez C, Boelens R, Bonvin AMMJ (2003) HADDOCK: a protein–protein docking approach based on biochemical or biophysical information. J Am Chem Soc 125:1731–1737

18. Sapay N, Nurisso A, Imberty A (2013) Simulation of carbohydrates, from molecular docking to dynamics in water. Methods Mol Biol 924:469–483

19. Kirschner KN, Yongye AB, Tschampel SM, González-Outeiriño J, Daniels CR, Foley BL, Woods RJ (2008) GLYCAM06: a generalizable biomolecular force field. Carbohydrates. J Comput Chem 29:622–655

20. Woods RJ, Dwek RA, Edge CE (1995) Molecular mechanical and molecular dynamical simulations of glycoproteins and oligosaccharides. 1. GLYCAM_93 parameter development. J Phys Chem 99:3832–3846

21. Guvench O, Hatcher E, Venable RM, Pastor RW, MacKerell AD Jr (2009) CHARMM additive all-atom force field for glycosidic linkages between hexopyranoses. J Chem Theory Comput 5:2353–2370

22. Huige CJM, Altona C (1995) Force field parameters for sulfates and sulfamates based on ab initio calculations: extensions of AMBER and CHARMm fields. J Comput Chem 16:56–79

23. Mallajosyula SS, Guvench O, Hatcher E, MacKerell AD Jr (2012) CHARMM additive all-atom force field for phosphate and sulfate linked to carbohydrates. J Chem Theory Comput 8:759–776

24. Scott WRP, Hünenberger PH, Tironi IG, Mark AE, Billeter SR, Fennen J, Torda AE, Huber T, Krüger P, van Gunsteren WF (1999) The GROMOS biomolecular simulation program package. J Phys Chem A 103:3596–3607

25. Jin L, Barran PE, Deakin JA, Lyon M, Uhrín D (2005) Conformation of glycosaminoglycans by ion mobility mass spectrometry and molecular modelling. Phys Chem Chem Phys 7:3464–3471

26. Mikhailov D, Linhardt RJ, Mayo KH (1997) NMR solution conformation of heparin-derived hexasaccharide. Biochem J 328:51–61

27. Mikhailov D, Mayo KH, Vlahov IR, Toida T, Pervin A, Linhardt RJ (1996) NMR solution conformation of heparin-derived tetrasaccharide. Biochem J 318:93–102

28. Murphy KJ, McLay N, Pye DA (2008) Structural studies of heparan sulfate hexasaccharides: new insights into iduronate conformational behavior. J Am Chem Soc 130:12435–12444

29. Pol-Fachin L, Becker CF, Guimarães JA, Verli H (2011) Effects of glycosylation on heparin binding and antithrombin activation by heparin. Proteins 79:2735–2745

30. Zhang Z, McCallum SA, Xie J, Nieto L, Corzana F, Jiménez-Barbero J, Chen M, Liu J, Linhardt RJ (2008) Solution structures of chemoenzymatically synthesized heparin and its precursors. J Am Chem Soc 130:12998–13007

31. Gandhi NS, Mancera RL (2009) Free energy calculations of glycosaminoglycan–protein interactions. Glycobiology 19:1103–1115

32. Pichert A, Samsonov SA, Theisgen S, Thomas L, Baumann L, Schiller J, Beck-Sickinger AG, Huster D, Pisabarro MT (2012) Characterization of the interaction of interleukin-8 with hyaluronan, chondroitin sulfate, dermatan sulfate, and their sulfated derivatives by spectroscopy and molecular modelling. Glycobiology 22:134–145

33. Gilson MK, Honig B (1988) Calculation of the total electrostatic energy of a macromolecular system: solvation energies, binding energies, and conformational analysis. Proteins 4:7–18

34. Honig B, Nicholls A (1995) Classical electrostatics in biology and chemistry. Science 268: 1144–1149

35. Tsui V, Case DA (2001) Theory and applications of the generalized Born solvation model in macromolecular simulations. Biopolymers 56:275–291

36. Li W, Johnson DJD, Esmon CT, Huntington JA (2004) Structure of the antithrombin–thrombin–heparin ternary complex reveals the antithrombotic mechanism of heparin. Nat Struct Mol Biol 11:857–862

37. Jones G, Willett P, Glen RC, Leach AR, Taylor R (1997) Development and validation of a genetic algorithm for flexible docking. J Mol Biol 267:727–748

38. Gandhi NS, Mancera RL (2008) The structure of glycosaminoglycans and their interactions with proteins. Chem Biol Drug Des 72:455–482

39. Mulloy B, Forster MJ (2000) Conformation and dynamics of heparin and heparan sulfate. Glycobiology 10:1147–1156

40. Cremer D, Pople JA (1975) A general definition of ring puckering. J Am Chem Soc 97:1354–1358

41. Forster MJ, Mulloy B (1993) Molecular dynamics study of iduronate ring conformations. Biopolymers 33:575–588

42. Rao VSR, Qasba PK, Balaji PV, Chandrasekaran R (1998) Conformation of carbohydrates. Harwood Academic, Amsterdam

43. Pol-Fachin L, Verli H (2008) Depiction of the forces participating in the 2-O-sulfo-α-L-iduronic acid conformational preference in heparin sequences in aqueous solutions. Carbohydr Res 343:1435–1445

44. Weininger D (1988) SMILES, a chemical language and information system. 1. Introduction to methodology and encoding rules. J Chem Inform Comput Sci 28:31–36

45. Jin L, Abrahams JP, Skinner R, Petitou M, Pike RN, Carrell RW (1997) The anticoagulant activation of antithrombin by heparin. Proc Natl Acad Sci U S A 94:14683–14688

46. Johnson DJD, Li W, Adams TE, Huntington JA (2006) Antithrombin–S195A factor Xa–heparin structure reveals the allosteric mechanism of antithrombin activation. EMBO J 25:2029–2037

47. McCoy AJ, Pei XY, Skinner R, Abrahams J-P, Carrell RW (2003) Structure of β-antithrombin and the effect of glycosylation on antithrombin's heparin affinity and activity. J Mol Biol 326:823–833

48. Esko JD, Selleck SB (2002) Order out of chaos: assembly of ligand binding sites in heparan sulfate. Annu Rev Biochem 71:435–471

49. Carter WJ, Cama E, Huntington JA (2005) Crystal structure of thrombin bound to heparin. J Biol Chem 280:2745–2749

50. Faham S, Hileman RE, Fromm JR, Linhardt RJ, Rees DC (1996) Heparin structure and interactions with basic fibroblast growth factor. Science 271:1116–1120

51. Stouch TR (2012) Looking forward into the next 25 years: the 25th anniversary issue of the journal of computer-aided molecular design. J Comp Aided Mol Des 26:1

52. Mulloy B, Forster MJ, Jones C, Davies DB (1993) N.M.R. and molecular-modelling studies of the solution conformation of heparin. Biochem J 293:849–858

Chapter 25

Using Isothermal Titration Calorimetry to Determine Thermodynamic Parameters of Protein–Glycosaminoglycan Interactions

Amit K. Dutta, Jörg Rösgen, and Krishna Rajarathnam

Abstract

It has now become increasingly clear that a complete atomic description of how biomacromolecules recognize each other requires knowledge not only of the structures of the complexes but also of how kinetics and thermodynamics drive the binding process. In particular, such knowledge is lacking for protein–glycosaminoglycan (GAG) complexes. Isothermal titration calorimetry (ITC) is the only technique that can provide various thermodynamic parameters—enthalpy, entropy, free energy (binding constant), and stoichiometry—from a single experiment. Here we describe different factors that must be taken into consideration in carrying out ITC titrations to obtain meaningful thermodynamic data of protein–GAG interactions.

Key words Glycosaminoglycan (GAG), Heparin, Free energy, Thermodynamics, Isothermal titration calorimetry (ITC), Enthalpy, Entropy

1 Introduction

Glycosaminoglycans (GAGs), such as heparan sulfate, are highly sulfated polysaccharides that are ubiquitously expressed by most cell types, and bind a wide range of proteins including growth factors, chemokines, and enzymes. Naturally occurring human mutations and animal knockout mice studies have shown that protein–GAG interactions play crucial roles in processes ranging from development to the host immune response [1, 2]. We refer the reader to other chapters in this book for a description of the GAG structures and properties. A complete understanding of biomacromolecular interactions requires knowledge not only of the structures of the complexes but also of how the kinetics and thermodynamics drive the binding process. At this time, very little is known regarding the thermodynamic basis of protein–GAG interactions.

Kuberan Balagurunathan et al. (eds.), *Glycosaminoglycans: Chemistry and Biology*, Methods in Molecular Biology, vol. 1229, DOI 10.1007/978-1-4939-1714-3_25, © Springer Science+Business Media New York 2015

Isothermal titration calorimetry (ITC) is the only technique that can provide various thermodynamic parameters—enthalpy (ΔH), entropy (ΔS), free energy (ΔG) that is also related to equilibrium binding constant (K_a), and stoichiometry—from a single experiment. The enthalpy of binding provides insight into how favorable hydrophobic, H-bonding, and electrostatic interactions mediate binding, whereas the entropy of binding arises due to changes in restriction/freedom of the backbone and side chain atoms and rearrangement or release of solvent water molecules and ions [3]. Such knowledge is essential to understand the relationship between structure, dynamics, thermodynamics, and function that otherwise can be completely missed from considering structures alone, and also not tractable from conventional mutational studies. ITC studies have been routinely used for studying protein binding to other proteins, peptides, and DNA, but has been challenging for characterizing protein–GAG interactions. We will discuss why the complex interrelationship between factors such as binding-induced precipitation, GAG heterogeneity and size, binding affinities, and sample requirements must be taken into consideration in designing the ITC experiments. We also discuss how to interpret and get the most out of the experimental data to gain thermodynamic insights (*see* **Note 1**).

We have written this chapter on the basis of our ITC experience in characterizing chemokine–heparin interactions. Heparin and heparan sulfate are linear highly sulfated polysaccharides consisting of repeating disaccharide units. Heparin is generally used as a structural and functional surrogate of heparan sulfate, as it can be easily purified in large amounts and readily available. Sequence analysis and binding studies have shown that chemokines bind GAGs such as heparan sulfate expressed on endothelial and epithelial cells and the extracellular matrix, and in vivo studies have shown that GAG binding functions as directional cues and regulates cellular trafficking [4, 5]. Compared to other macromolecular interactions, ITC studies have been few and far between for GAG–protein complexes. This paucity is evident if we consider that a PubMed search for the word "chemokine" results in >70,000 hits, whereas just 1 hit is obtained for the words "chemokine, isothermal titration calorimetry, and heparin" (16th Oct 2013). Even this ITC study was part of a larger study, and merely describes the binding affinities and stoichiometry of one chemokine binding to various commercially available heparins with no discussion of the enthalpy or entropy of binding [6].

2   Materials

Most commercially available heparins are from porcine intestinal mucosa or bovine lungs. Various size heparins ranging from a disaccharide and various oligosaccharides to high molecular weight polysaccharides are available from popular vendors such as Sigma and Calbiochem, and specialized vendors such as Neoparin and Iduron. Moreover, heparin is structurally more homogeneous than heparan sulfate, and not surprisingly, most structural and biophysical studies reported in literature have used heparin. Our discussion on experimental design will be for heparin but is applicable to heparan sulfate and most likely for all other GAGs.

The concentrations of protein and GAG solutions should be determined as accurately as possible. Weigh the protein and GAG accurately using a microbalance (*see* **Notes 2** and **3**). Determine the concentration of the protein solution using UV–vis spectrophotometer and/or bicinchoninic acid (BCA) or similar assay. The extinction coefficients can be calculated from the number of tryptophans, tyrosines, and disulfide bonds [7], and if necessary, precise protein concentrations can be obtained from amino acid analysis from which the experimental extinction coefficient can be obtained. We recommend preparing stock solutions of the protein and GAG and then diluting as required for the ITC experiments.

Make sure that the buffer conditions of the protein and GAG solutions match. Buffer mismatch may not be a problem if the commercially purchased GAGs were dialyzed against water and then lyophilized. We have also observed that titration of commercial heparins into a buffer results in negligible heat release, indicating that heparin dilution does not complicate the titration. If required, dialyze the samples into the required buffer to prevent solvent mismatch. In this case, it is important to measure the concentrations before and after to account for any loss during the dialysis procedure.

3   Methods

3.1   *Instrument Setup and Experimental Parameters*

1. In a typical ITC instrument, a reference cell and a measurement cell are placed in an adiabatic jacket. The reference cell is filled with 1 % azide solution whereas the measurement cell is filled with the protein solution. Both cells are maintained at a constant temperature by using a controller. The equipment has a syringe through which heparin (ligand) is injected into the measurement cell. The volume of the VP-ITC cell is 1.45 mL, whereas the injection volume is around 100 μL (25×4-μL injections). The solution in the measurement cell is constantly stirred for rapid mixing (*see* **Notes 4–6**).

2. The sample temperature should be just below the experimental temperature so that the equilibrium in the ITC instrument is attained quickly. We generally carry out our experiments at 25 °C. The protein and GAG solutions must be degassed to remove air bubbles as they can interfere with the feedback circuit and lead to poor baseline.

3. If necessary, run a blank titration of ligand into the buffer and buffer into protein to correct for any heat change due to contribution from dilution of the protein and/or GAG samples.

4. An important aspect of an ITC experiment is choosing the proper concentration for protein and ligand. The choice of protein concentration depends on the "c" value as given by Eq. 1:

$$c = nP_t / K_d \qquad (1)$$

where, P_t is the protein concentration in the measurement cell; "n" is the number of binding sites per protein molecule. Please note that multiple binding sites are present usually on the GAG and not on the protein. Thus the value of n could be <1. The shape of the binding isotherm depends on the c value. The c value should be between 20 and 100 to obtain a sigmoidal shape of the binding isotherm in order to get a good fit to estimate K_d, ΔH, ΔS, and n. If the c value is very high (~1,000), one can get an estimate of ΔH and n, but not of the other parameters. If the c value is <5, then the shape of the binding isotherm does not permit accurate estimation of the thermodynamic parameters unless one of them is already known (such as the stoichiometry) and is not varied during the fitting process.

A simulation plot of heat released vs. mole ratio explains the importance of c value (Fig. 1). If the K_d and n are not

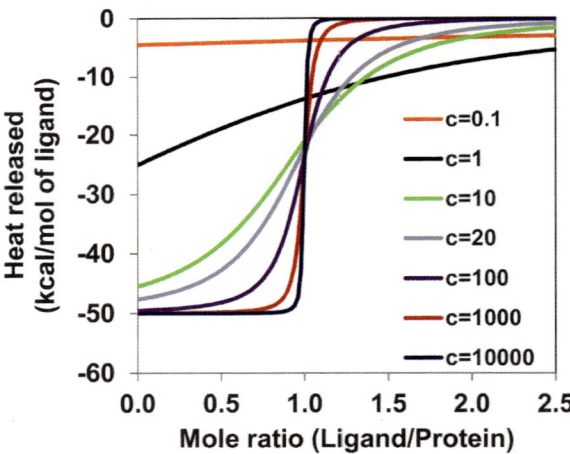

Fig. 1 Simulated binding isotherm for a 1:1 system to show the importance of c value. The data were generated using the model proposed by Wiseman et al. [10] using $\Delta H = -50$ kcal/mol. For low c values (<5), the binding isotherm loses the S-shape and thus determination of the thermodynamic parameters is not possible. At high c values (~1,000), ΔH and stoichiometry (n) can be inferred from the shape of the binding isotherm but the data cannot be fitted to determine the other thermodynamic parameters

known, assume that the K_d is between 1 mM and 1 nM. Start with a protein concentration of 10 µM. The shape of a step jump would indicate a high c value, and in which case reduce the protein concentration by tenfold. A hyperbolic shape would indicate a low c value, and in which case increase the protein concentration by tenfold. Optimize the protein concentration until a good shape for the isotherm is obtained (*see* **Notes 7–9**).

5. If an estimate of the dissociation constant (K_d) is known, then choose a protein concentration in the cell that is 30 times K_d. The ligand concentration should be $\sim n \times 30$ times the protein concentration. For high MW GAGs, multiple binding sites are on the GAG and not on the protein, and therefore the n value could be <1. For more details, please refer to ref. [9].

6. The first injection is usually not accurate due to the diffusion of the ligand into the cell during the temperature equilibration. So, inject a small volume first (2 µL) and then inject 4 µL or more for the remaining titrations. One could also inject 2 µL for the first 4–5 injections and then increase the injection volume if one wants to switch between high data coverage at larger noise to lower data coverage at less noise.

7. We suggest that the protein is taken in the cell and the GAG in the syringe for two reasons. As discussed above, the concentration in the syringe must be higher (preferably $n \times 30$) than the concentration in the cell. We observe that titrating protein into GAGs in the cell results in precipitation, and not if we titrate GAG into the protein. Secondly, more GAG will be required if taken in the cell compared to taken in the syringe. Though this may not be limiting for crude and cost-effective heparins, it could be limiting due to the high cost of the smaller and homogeneous GAGs. It is possible that precipitation may not be an issue for 1:1 stoichiometry, and in which case, the relative cost of the GAG and protein could determine what goes in the syringe and cell (*see* **Note 10**).

8. The affinity of protein–GAG interaction could increase with increasing GAG length. Therefore, if the c value is low for a particular chain length, longer GAGs could result in a better binding isotherm especially if the protein concentration cannot be increased (*see* **Note 11**).

3.2 Data Analysis

1. Ensure that the concentrations of the protein and ligand are correct.

2. The peak integration baseline of each run should be inspected and adjusted manually if required.

3. If necessary, subtract the heats obtained from the protein and GAG dilution experiments.

4. Neglect the first data point, which may have a systematic error due to the diffusion effect.

5. Choose an appropriate model using the fitting software provided by Microcal (Origin). The most common model is 1:1 binding. For a protein containing single set of n identical binding sites, the following equations are used:

$$Q = \frac{nP_t \Delta H V_0}{2} \left[1 + \frac{X_t}{nP_t} + \frac{1}{nK_aP_t} - \sqrt{\left(1 + \frac{X_t}{nP_t} + \frac{1}{n_aP_t}\right)^2 - \frac{4X_t}{nP_t}} \right] \quad (2)$$

$$\Delta Q_{(i)} = Q_{(i)} + \frac{dV_{(i)}}{V_0}\left[\frac{Q_{(i)} + Q_{(i-1)}}{2}\right] + Q_{(i-1)} \quad (3)$$

where, P_t is the concentration of protein in the active volume, $dV_{(i)}$ is the displaced volume, V_0 is the volume of the cell, X_t is the concentration of the ligand in the active volume, $Q_{(i)}$ and $Q_{(i-1)}$ are the total heat evolved or absorbed for ith and $(i-1)$th injections, respectively. Detailed derivation of the equations can be found in the VP ITC manual. Provide an initial estimate of n, K_a and ΔH, which calculates $\Delta Q_{(i)}$ for each injection. Then the program calculates the values of n, K_a, and ΔH that provide the best fit to the calculated and experimental $\Delta Q_{(i)}$. Note that good initial estimates are essential to prevent the algorithm from terminating at a sub-optimum.

4   Notes

1. ITC can be used to describe the thermodynamics of how individual amino acids contribute to the binding process by characterizing a panel of mutants. ITC can be carried out at different conditions including as a function of pH, ionic strength, buffers, and temperature to gain insights into the role of electrostatic interactions and counter ions, ionization state, and binding mechanisms.

2. The protein and GAG concentrations must be accurate. If GAG is taken in the syringe, an error in the GAG concentration will result in systematic errors in all of the thermodynamic parameters as shown in Fig. 2b. For instance, if the measured GAG concentration is higher than the actual concentration, the experimental value of n and ΔH will be less than the actual n and ΔH, whereas the experimental K_d will be higher than the actual K_d. Therefore preparing stock solutions of high GAG concentration (>1 mM) could minimize the weighing error and hence the concentration error. If protein is taken in the cell, an error in the protein concentration will result in errors in

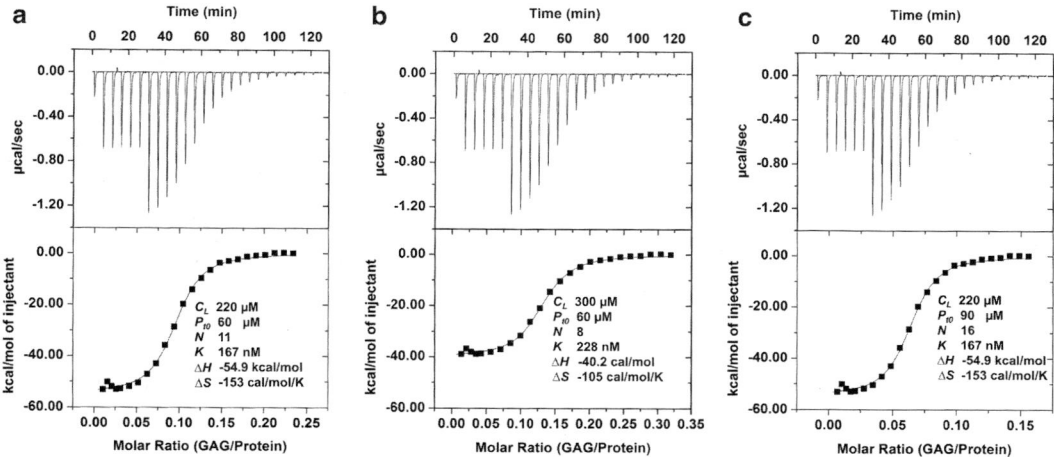

Fig. 2 Relationship between concentration errors and thermodynamic parameters. (**a**) The protein (P_{t0}) and GAG (C_L) concentrations were 60 and 220 μM, respectively. The first six injections were 2 μL and the rest were 4 μL. The stoichiometry (n), equilibrium dissociation constant (K_d), enthalpy and entropy data are shown. (**b**) An error of increased GAG concentration affected the values of all thermodynamic parameters. The stoichiometry and enthalpy decreased proportionally, and ΔS value also changed due to changes in ΔH and K_d. Note that the stoichiometry is protein/GAG. (**c**) An error of increased protein concentration affected the value of stoichiometry alone and all other values remained unchanged

stoichiometry alone and not in any other parameter as shown in Fig. 2c. The error in protein concentration will linearly change the estimated stoichiometry. Thus a 10 % inaccuracy in protein concentration will cause a 10 % error in stoichiometry. For instance, if the measured protein concentration is higher than the actual concentration (90 vs. 60 μM), then the measured stoichiometry will be higher ($n = 16$ vs. 11; the calculated numbers have been rounded to the nearest integer).

3. Errors in protein and GAG concentrations may not be limiting if the objective of the study is to compare different variants of a protein. For instance, any errors will propagate in the same manner when comparing thermodynamic parameters of a panel of protein mutants using the same GAG stock solution. However, the resulting dissociation constant values will have an unknown offset that will depend on the real concentration of the GAG stock solution. We recommend the following for those who use their own fitting procedures rather than the manufacturer-provided analysis package. In Eq. 2, the stoichiometry factor, n, is associated with the cell concentration of the protein, $n \times P_t$. Alternatively, one could associate this factor with the syringe concentration of the ligand, $n \times X_t$. The best practice is to have the factor "n" associated with the more uncertain of the two concentrations, P_t or X_t. In this way, the binding constant is least affected by concentration errors.

4. Meticulous cleaning of the cell and the syringe are very important. Please refer to the ITC manual for the instructions, and ensure that both the cell and syringe are as dry as possible to minimize dilution errors.

5. Take utmost care while loading samples in the cell and the syringe. There should be no bubbles in the cell or the syringe. The syringe should not bend. Even a small bending of the syringe will lead to a poor baseline.

6. Choose the time interval in such a way that heat released due to one injection does not overlap with that of another injection. The most common spacing is 300 s.

7. If protein is taken in the cell, 1.45 mL of sample is required for each experiment for the VP-ITC calorimeter. The protein concentration required depends on the affinity of protein–GAG complex; if unknown, we suggest a concentration between 10 and 100 μM.

8. A thorough knowledge of the protein of interest (or of related proteins) from literature including any prior studies on GAG binding, and knowledge of the structures and solution properties is essential. If the protein exists as monomers and dimers, knowledge of the dimerization constants and whether GAG binding influences dimerization would be useful. If GAG binding and dimerization are coupled, it is necessary to use appropriate equations which will be different from Eq. 2.

9. For a protein for which nothing is known regarding its GAG binding properties, initial experiments will determine whether ITC experiments are feasible and can provide meaningful results. Ideally, the protein must not be limiting, and so availability of milligram (mg) and preferably tens of mgs of protein would be useful. The apparent binding affinities could vary by orders of magnitude from mM to nM. In principle, ITC can measure binding affinities over this range, but multiple parameters must be optimized to obtain reliable thermodynamic data.

10. Stoichiometry of binding can vary among different proteins, and can also vary for a given protein depending on the size of the heparin used. The MW of GAG-binding proteins can also vary by orders of magnitude from a few kDa to 100 s of kDa. Further, GAG-binding residues could be highly clustered or span a large surface for any given protein. It has been proposed that chemokines and growth factors bind GAGs like beads on a string on the basis that the stoichiometry increases with increasing heparin length [6, 9]. As many as ten or more protein molecules have been reported to bind a single GAG. From an experimental perspective, this would mean that much less GAG is required for a titration on a mol:mol basis. Some studies also suggest that GAG-binding and chemokine dimerization are coupled. The measured binding affinities also vary for

different heparin sizes and are also sensitive to solution pH and buffer conditions. These factors are also critical for deciding whether the protein or GAG should be in the syringe and the amount of protein or GAG that will be required.

11. The major challenges for characterizing protein–GAG complexes are large sample requirements, binding-induced protein dimerization/oligomerization, precipitation, and intrinsic heterogeneity of the GAGs. We suggest that the initial experiments use crude high MW GAGs, as these GAGs are less expensive and are also more likely bind with tighter affinity compared to homogeneous shorter oligosaccharides. Tighter binding enables experiments at lower protein concentrations thus minimizing binding-induced aggregation and precipitation issues. However, the thermodynamic parameters obtained using crude heterogeneous GAGs are average of all GAGs present in the sample. It is also possible that precipitation might occur with these GAGs due to high MW, in which case, we suggest using size-fractionated smaller GAGS. Size-fractionated GAGs are more homogeneous and also provide more defined thermodynamic parameters. Therefore, once feasibility is established and binding parameters more or less optimized, one could use size-fractionated GAGs whose results can be interpreted with greater confidence. However, the experimental design must take into consideration the fact that smaller GAGs could lead to reduced heat release and so would require larger quantities.

Acknowledgements

This work was supported in part by grants P01HL1071521 and R21AI097975 to K.R. and R01GM049760 to J.R. from the National Institutes of Health. The authors would like to thank Dr. Luis Holthauzen for technical support.

References

1. Lindahl U, Kjellén L (2013) Pathophysiology of heparan sulphate: many diseases, few drugs. J Intern Med 273:555–571

2. Bishop JR, Schuksz M, Esko JD (2007) Heparan sulphate proteoglycans fine-tune mammalian physiology. Nature 446:1030–1037

3. Ladbury JE, Klebe G, Freire E (2010) Adding calorimetric data to decision making in lead discovery: a hot tip. Nat Rev Drug Discov 9:23–27

4. Salanga CL, Handel TM (2011) Chemokine oligomerization and interactions with receptors and glycosaminoglycans: the role of structural dynamics in function. Exp Cell Res 317: 590–601

5. Gangavarapu P, Rajagopalan L, Kolli D, Guerrero-Plata A, Garofalo RP, Rajarathnam K (2012) The monomer-dimer equilibrium and glycosaminoglycan interactions of chemokine CXCL8 regulate tissue-specific neutrophil recruitment. J Leukoc Biol 91:259–265

6. Kuschert GSV, Coulin F, Power CA, Proudfoot AEI, Hubbard RE, Hoogewerf AJ, Wells TNC (1999) Glycosaminoglycans interact selectively

with chemokines and modulate receptor binding and cellular responses. Biochemistry 38: 12959–12968

7. Pace CN, Vajdos F, Fee L, Grimsley G, Gray T (1995) How to measure and predict the molar absorption coefficient of a protein. Protein Sci 4:2411–2423

8. Rajarathnam K, Rösgen J (2013) Isothermal titration calorimetry of membrane proteins – progress and challenges. Biochim Biophys Acta (in press). doi: pii: S0005-2736(13)00169-7. 10.1016/j.bbamem.2013.05.023

9. Venkataraman G, Sasisekharan V, Herr AB, Ornitz DM, Waksman G, Cooney CL, Langer R, Sasisekharan R (1996) Preferential self-association of basic fibroblast growth factor is stabilized by heparin during receptor dimerization and activation. Proc Natl Acad Sci 93: 845–850

10. Wiseman T, Williston S, Brandts JF, Lin L-N (1989) Rapid measurement of binding constants and heats of binding using a new titration calorimeter. Anal Biochem 179: 131–137

Chapter 26

Characterizing Protein–Glycosaminoglycan Interactions Using Solution NMR Spectroscopy

Prem Raj B. Joseph, Krishna Mohan Poluri, Krishna Mohan Sepuru, and Krishna Rajarathnam

Abstract

Solution nuclear magnetic resonance (NMR) spectroscopy and, in particular, chemical shift perturbation (CSP) titration experiments are ideally suited for characterizing the binding interface of macromolecular complexes. ^{1}H-^{15}N-HSQC-based CSP studies have become the method of choice due to their simplicity, short time requirements, and not requiring high-level NMR expertise. Nevertheless, CSP studies for characterizing protein–glycosaminoglycan (GAG) interactions have been challenging due to binding-induced aggregation/precipitation and/or poor quality data. In this chapter, we discuss how optimizing experimental variables such as protein concentration, GAG size, and sensitivity of NMR instrumentation can overcome these roadblocks to obtain meaningful structural insights into protein–GAG interactions.

Key words Nuclear magnetic resonance (NMR), Chemical shift perturbation, Protein–ligand interactions, Glycosaminoglycan, Dissociation constant, Heparan sulfate, Heparin

1 Introduction

Crosstalk between macromolecules in both the intracellular and extracellular milieu orchestrates all cellular processes, and understanding the structural basis and molecular mechanisms that dictate the specificity and affinity requires a detailed atomistic description of the complexes. Nuclear magnetic resonance (NMR) spectroscopy and X-ray crystallography are both routinely used for structure determination of macromolecular complexes. However, NMR is the only avenue available when complexes fail to crystallize, which could be the case if one or both of the partners are dynamic and/or if the interactions are weak in nature.

Glycosaminoglycans, such as heparin and heparan sulfate, are highly negatively charged linear polysaccharides that are widely expressed by most cell types. They mediate a wide variety of crucial functions due to their ability to bind and regulate the activities of large classes of proteins [1, 2]. We refer the reader to other

Kuberan Balagurunathan et al. (eds.), *Glycosaminoglycans: Chemistry and Biology*, Methods in Molecular Biology, vol. 1229, DOI 10.1007/978-1-4939-1714-3_26, © Springer Science+Business Media New York 2015

chapters for a more detailed description of GAG structures and properties. Remarkably very little is known regarding the structural basis of how GAGs interact with proteins and mediate function. The challenges that have plagued both NMR and X-ray studies can be attributed to multiple interrelated factors that include binding-induced protein precipitation, especially at the high concentrations used in structural studies, the flexibility and high negative charge of GAGs, intrinsic heterogeneity and GAG size, and the dynamic nature of the protein–GAG interface.

In this chapter, we outline how NMR chemical shift perturbation (CSP) methods can be effectively used to study protein–GAG interactions. The popularity of this method is due to the high sensitivity and robustness of 2D ^{1}H-^{15}N HSQC experiments. Chemical shifts are exquisitely sensitive to binding-induced changes in the local electronic environment, and can provide residue-level details of the binding interface, binding affinities, and timescales of protein–ligand interactions, and together with molecular docking tools like HADDOCK can also provide structural models of the protein–ligand complexes [3].

CSP-based NMR experiments have been routinely used for studying protein binding to other proteins, peptides, DNA, and small molecule ligands, but have been challenging for characterizing protein–GAG interactions. In addition to the other challenges described above, one other reason could be working under nonoptimal conditions. In this chapter, we discuss different factors that must be taken into consideration in the experimental design, and in particular, the importance of protein concentration, GAG size, sensitivity of NMR instruments, and do's and don'ts for obtaining quality NMR data. Our guidelines are based on our experience of characterizing the binding of heparin oligosaccharides to neutrophil-activating chemokines ([4]; unpublished results). Structure-function and animal model studies have shown that gradients formed by chemokines bound to the cell surface and soluble GAGs direct and regulate neutrophil trafficking from the blood to the tissue [5].

An HSQC spectrum provides a fingerprint of the backbone amides of a protein. Each backbone amide is represented by a peak corresponding to the chemical shifts of the ^{1}H (x-axis) and ^{15}N (y-axis) nuclei. The total number of peaks corresponds to the number of amino acids in the protein chain excluding prolines (which do not have a backbone amide NH) plus those corresponding to side chain NH_2 of asparagine and glutamine residues. In general, chemical shifts are available from literature or the BMRB data bank (http://www.bmrb.wisc.edu); if not, the chemical shifts must be assigned using established NMR procedures.

An overlay of a series of HSQC spectra of a heparin oligosaccharide titration to a ^{15}N-labeled chemokine, and the 2D histogram plot of the weighted average of ^{1}H and ^{15}N chemical shift changes

$((\Delta\delta = [\Delta\delta_H{}^2 + (\Delta\delta_N/5)^2]^{1/2})$ as a function of individual residues, are shown in Fig. 1a and c, respectively. The data show selective perturbation of a subset of protein peaks indicating specific binding. It is generally assumed that residues showing the largest CSP mediate the binding process. However, CSP changes are correlative and not necessarily causative of the binding process. Therefore, it is possible that not all of the interfacial residues will show significant CSP upon binding. Considering that binding is predominantly mediated via lysine NH_2 and arginine guanidinium groups, NMR chemical shifts of some of the Lys/Arg amide backbone may not be sensitive enough due to the long intervening side chain. Similarly, changes in chemical shift may also occur due to indirect binding, arising from structural rearrangements or changes in packing interactions remote from the binding interface. Therefore it is important that the CSP data be interpreted cautiously.

If the structure of the protein is known, a residue-level description of the binding site can be obtained (Fig. 1d). Analysis of the protein structure also helps in teasing out direct interactions from indirect interactions. Binding affinities can be obtained from the CSP of the individual residues as described previously (Fig. 1b) [6, 7].

2 Materials

2.1 Protein Concentration and NMR Titrations

NMR experimental conditions that enable titrations at low protein concentrations would substantially increase the probability of acquiring quality data, considering high concentration could lead to aggregation/precipitation. Availability of high field NMR instruments and advances in cryoprobes and gradient accessories have significantly improved the scope and complexity of experiments including the ability to work at low protein concentrations. For instance, we have acquired an HSQC spectrum of a ~100 µM chemokine sample (MW ~16 kDa) in 10 min on an 800 MHz spectrometer equipped with cryoprobe and field gradient accessories. The amount of time required can be reduced by a factor of ~2 when using higher field instruments (800 vs. 600 or 500 MHz), and by a factor of ~8 when using an NMR instrument with and without a cryoprobe. However, the sensitivity of a cryoprobe is dependent on ionic strength, with low salt conditions providing the best sensitivity. In terms of protein concentration, reducing the concentration by a factor of 2 will increase the time requirement by a factor of 4. Therefore, time requirement for a 50 µM compared to a 200 µM sample will be 16 times higher to achieve a similar signal to noise (s/n).

For some proteins, binding studies may be feasible only at concentrations as low as ~10 µM. It is still possible to collect an HSQC spectrum of a ~10 µM sample in less than 12 hrs on an 800 MHz

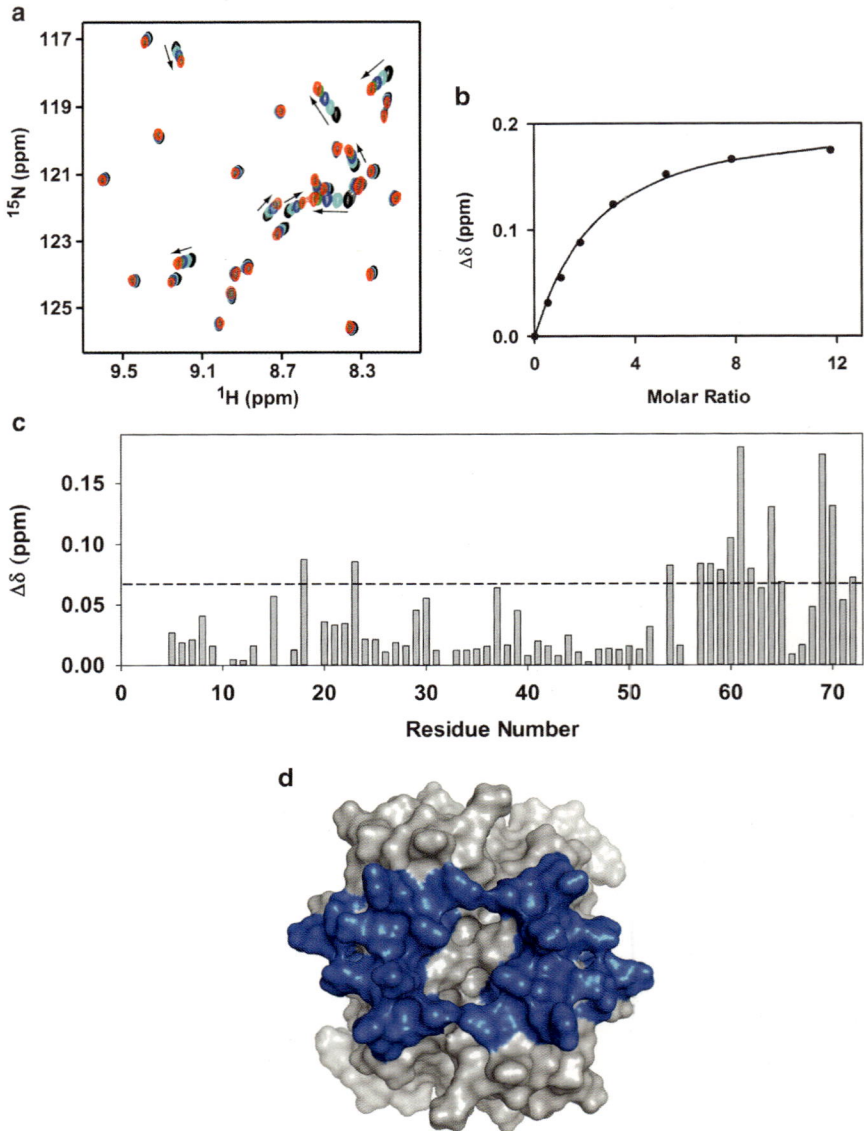

Fig. 1 Mapping of chemokine-heparin oligosaccharide binding interface using NMR chemical shift perturbation data. (**a**) A section of a ^{1}H-^{15}N HSQC spectrum showing heparin oligosaccharide binding-induced chemical shift changes in a chemokine. The unbound and final bound peaks are in *black* and *red* and the intermediate peaks are shown in *cyan, blue*, and *green*. Note the selective perturbation of a subset of amino acids indicated by an *arrow*. (**b**) A representative plot of CSP vs. GAG/chemokine concentration (in molar ratio), which allows calculation of the dissociation constant (K$_d$). (**c**) A histogram of the CSP as a function of chemokine sequence. *Dotted line* indicates the cutoff for residues to be considered perturbed. (**d**) A surface plot of the chemokine showing residues (*blue*) that are significantly perturbed on GAG binding

spectrometer with a cryoprobe. As titration experiments may not be practical and cost effective, we suggest collecting the spectrum of the unbound protein at high concentrations and collecting one or two spectra of the GAG-bound protein at the low concentrations. Ideally, these data could identify which residues mediate binding and allow describing the geometry of the binding interface. However, chemical shift assignments of all the residues in the GAG-bound form may not be possible, especially if the interface residues undergo significant binding-induced chemical shift changes. In this case, additional experiments must be performed to better define the binding-interface residues. We also suggest, if possible, characterizing the protein–GAG complexes using other biophysical techniques such as dynamic light scattering (DLS) and sedimentation velocity to gain insights into the relationship between protein concentration and aggregation state of the complexes before carrying out the NMR experiments.

2.2 Choice of GAG for NMR Titrations

NMR studies require homogeneous GAG samples, and as only heparin is commercially available in different sizes, we discuss our experimental strategy on the basis of our experience using heparin oligosaccharides. Nevertheless, our discussion is most likely applicable to heparan sulfate and all other GAGs. Various studies have shown that the structural and functional properties of heparin mimic heparan sulfate, especially protein binding to the highly sulfated regions of the heparan sulfate. Most biophysical and structural studies reported in literature have also used unfractionated or size-fractionated heparin oligosaccharides. Heparin oligosaccharides are available in various sizes from a disaccharide to a 26 mer from specialized vendors like Neoparin, Iduron, and V-labs.

2.3 NMR Sample Preparation

The protein must be isotopically labeled with ^{15}N for characterizing binding interactions using ^{1}H-^{15}N HSQC titrations. Isotopically labeled proteins are overexpressed in E. coli grown in minimal media using $^{15}NH_4Cl$ as the sole nitrogen source. Since the growth characteristics of E. coli in minimal media are somewhat compromised compared to growth in rich media (LB media), growing larger cultures is necessary to produce sufficient quantities of protein. In recent years, adding labeled growth supplements (available from vendors such as Cambridge Isotopes and Spectra Isotopes) circumvents this problem by increasing cell growth rates and overall protein expression. Once the protein is purified, the purity and molecular weight are confirmed using mass spectrometry. Sample purity is important because contaminations could complicate interpreting real signals from the background spurious noise peaks.

Protein samples for NMR studies are typically ~500 μl in volume, containing ~5–10 % D_2O for spectrometer frequency lock, 1 mM 2,2-dimethyl-2-silapentanesulfonic acid (DSS) for spectral referencing, and 1 mM sodium azide (NaN_3) to prevent microbial growth. Sample temperature for NMR data collection can vary depending on the behavior of the protein, and typically is between 20 and 40 °C.

The choice of buffer could influence the quality of the GAG-bound spectra. If the initial choice of buffer results in poor quality spectra, we suggest collecting spectra at different buffers, pH, and ionic strength. In our experience, significant line broadening could occur under nonoptimal pH conditions (*see* **Note 1**).

It is also important to ensure that the protein and GAG samples are prepared in the same buffer to minimize chemical shift changes due to pH changes which could complicate and in worst case scenario lead to wrong interpretation on the binding interactions. If necessary, dialyze the protein and ligand in the same buffer. Alternatively, the lyophilized powders can be dissolved in the NMR buffer of interest, but the pH of the samples needs to be checked and adjusted before proceeding with the experiments.

3   Methods

3.1 Experimental Design

Prior knowledge of the protein, including behavior of the protein in solution, dimerization and oligomerization properties (including K_d), its GAG binding properties including binding-induced oligomerization and precipitation issues would be useful. Using the right protein concentration is a critical parameter for successful titration. A major problem is binding-induced precipitation, especially at high protein concentrations typically used in NMR studies. We propose an initial concentration of ~200 μM, and in the event of precipitation, continuing to reduce the concentration until there is no evidence of precipitation. For chemokines, we carried out titrations on samples anywhere between 50 and 150 μM. In principle, lower concentrations will require longer data acquisition times, and therefore whenever possible, we strongly recommend using spectrometers with cryoprobes.

We propose that the initial experiments are carried out in low ionic strength buffers, which could lead to stronger binding, resulting in lower GAG requirements. If the binding data suggest nonspecific interactions, spectra collected at varying ionic strengths could allow teasing out specific vs. nonspecific interactions (*see* **Note 2**). If starting protein concentrations is dependent on the oligosaccharide length, then titrations have to be carried out with shorter oligosaccharides or the protein concentration must be reduced sufficiently where the complex does not aggregate or precipitate (*see* **Notes 3** and **4**).

3.2 NMR Titrations and Data Analysis

CSP experiments involve collecting a series of HSQC spectra by adding GAG aliquots until essentially there are no binding-induced changes in protein backbone amide shifts. We suggest collecting a minimum of 6 to 8 spectra, which includes those of the free protein, around 50 % fraction bound population, and at saturation (*see* **Note 5**). More data points collected around 50 % bound population help in better defining the binding isotherms for accurate calculation of the dissociation constants. We advise using stock solutions of ~10–15 mM heparin oligosaccharides in order to minimize errors in protein concentration due to dilution. Prior knowledge of an estimate of binding affinities could be useful in selecting the starting protein concentration and amount of GAG to be added.

NMR data processing, analysis, and spectra viewing can be carried out using NMRpipe, NMRview, Sparky, or instrument-specific Bruker and Varian software [8–10]. Most processing and analysis script and programs for data fitting are available in the respective software websites. Binding affinities of protein–GAG interactions can vary by orders of magnitude (nM to mM), and accordingly kinetics (especially off constant, k_{off}) can vary by many orders. The kinetics of binding are classified as slow, intermediate, and fast exchange on the NMR time scale, which can influence the nature and quality of the spectra [7]. If binding occurs in the slow exchange regime, spectra will contain separate peaks at the chemical shifts of the free (δ_P) and GAG-bound (δ_{PL}) protein. As the ligand is titrated into the protein, the peak intensity at δ_P will decrease and of the peak at δ_{PL} will increase. If binding occurs in the fast exchange regime, spectra will contain a single set of peaks at the population-weighted average chemical shifts. As the ligand is titrated into the protein, the peak position will move from δ_P to δ_{PL}. If binding is in the intermediate exchange regime, peaks are exchange-broadened resulting in poor quality spectra (*see* **Note 6**).

The CSP experiments are best performed under the fast exchange regime as assigning chemical shifts of the GAG-bound form is straightforward as shown in Fig. 1a, b. However, in the intermediate and slow exchange regimes, assigning chemical shifts of all the residues in the GAG-bound form may not be possible as the information on the direction and magnitude of the individual peak movements is missing (*see* **Note 6**). In particular, this will be the case if the binding residues are in the crowded region of the spectrum and/or undergo large chemical shift changes. Therefore, it may be necessary to change experimental conditions so that the binding occurs in the fast exchange regime by using smaller GAGs, increasing ionic strength, and/or other experimental parameters.

In the fast exchange regime, the CSP follows a hyperbolic dependence as a function of ligand concentration. Proper choice of protein and ligand concentrations combined with nonlinear least squares analyses using two independent variables (ligand and protein concentrations) and two parameters (K_d and maximum CSP) can increase the accuracy of measured dissociation constant as shown in Fig. 1b [6].

4 Notes

1. Choice of pH could be a critical factor in obtaining quality NMR CSP data. One can collect trial HSQC spectra of a protein–GAG sample by varying pH conditions (say pH 5.0–8.0). This can help in arriving at the right pH condition for the titration experiments. In the case of some chemokines, we observe severe line broadening below pH 7.0 for some heparin oligosaccharides but not all.

2. Initial experiments must be carried out in low ionic strength buffers. In our experience, we observed no significant differences in the perturbation profile or the binding affinities between low and high salt conditions for various oligosaccharides, though the overall extent of perturbation was higher in low salt buffers. Using low ionic strength buffers can result in non-native interactions leading to aggregation and precipitation. On the other hand, use of high salt buffers can lead to screening of electrostatic interactions and weak binding.

3. Binding-induced oligomerization and precipitation effects are highly sensitive to heparin oligosaccharide chain length and the protein of interest. We suggest starting with shorter oligosaccharides such as a disaccharide and then progressing to higher oligosaccharides. For some proteins, titrations even with tetrasaccharide have led to precipitation.

4. Working with shorter oligosaccharides would be a compromise but nonetheless can provide useful information on binding. Some limitations include smaller chemical shift perturbations, lower binding affinity, multiple binding modes, and nonspecific interactions. Therefore it may not be possible to come up with a unique binding model, and one needs to be cautious and not over interpret the data. On the other hand, selective perturbations would suggest specific binding and can provide the binding geometry. Further, studying shorter oligosaccharides can also be exploited for designing GAG decoys that could function as therapeutics in a clinical setting.

5. When performing titration experiments, it is important that the sample is thoroughly mixed. After collecting the first HSQC spectrum of the free protein, the sample is transferred back to an Eppendorf tube using a glass pipette before addition of the ligand. After adding an aliquot of GAG (~2–10 μl), the sample is mixed well by pipetting the solution a few times. Look for any cloudiness or precipitation. Precipitate sticking to the sides of the NMR tube will interfere with the shimming of the sample. Switch sample to a new NMR tube if necessary. Once the sample is in the NMR machine it is shimmed before starting the next experiment. It is not uncommon for the initial cloudiness to disappear on subsequent addition of GAG

aliquots. We advise not to add GAG directly into the NMR tube, which can cause uneven mixing leading to errors in the estimation of the protein–GAG molar ratios.

6. Line broadening could occur due to aggregation or due to intermediate exchange binding kinetics. In the latter case, the peaks would reappear on excess GAG titration (fractional bound population >90 %). Therefore, we suggest not to abort the experiment, but to collect a few more spectra in the presence of excess GAG.

References

1. Lindahl U, Kjellén L (2013) Pathophysiology of heparan sulphate: many diseases, few drugs. J Intern Med 273:555–571

2. Kamhi E, Joo EJ, Dordick JS, Linhardt RJ (2013) Glycosaminoglycans in infectious disease. Biol Rev Camb Philos Soc 88:928–943

3. Dominguez C, Boelens R, Bonvin AM (2003) HADDOCK: a protein-protein docking approach based on biochemical or biophysical information. J Am Chem Soc 125:1731–1737

4. Poluri KM, Joseph PR, Sawant KV, Rajarathnam K (2013) Molecular basis of glycosaminoglycan heparin binding to the chemokine CXCL1 dimer. J Biol Chem 288:25143–25153

5. Gangavarapu P, Rajagopalan L, Kolli D, Guerrero-Plata A, Garofalo RP, Rajarathnam K (2011) The monomer-dimer equilibrium and glycosaminoglycan interactions of chemokine

CXCL8 regulate tissue-specific neutrophil recruitment. J Leukoc Biol 91:259–265

6. Markin CJ, Spyracopoulos L (2012) Increased precision for analysis of protein-ligand dissociation constants determined from chemical shift titrations. J Biomol NMR 53:125–138

7. Mittermaier A, Meneses E (2013) Analyzing protein-ligand interactions by dynamic NMR spectroscopy. Methods Mol Biol 1008:243–266

8. Delaglio F, Grzesiek S, Vuister GW, Zhu G, Pfeifer J, Bax A (1995) NMRPipe: a multidimensional spectral processing system based on UNIX pipes. J Biomol NMR 6:277–293

9. Johnson BA, Blevins RA (1994) NMR View: a computer program for the visualization and analysis of NMR data. J Biomol NMR 4:603–614

10. Goddard TD, Kneller DG. SPARKY 3, University of California, San Francisco

Chapter 27

Glycosaminoglycan–Protein Interaction Studies Using Fluorescence Spectroscopy

Rio S. Boothello, Rami A. Al-Horani, and Umesh R. Desai

Abstract

Fluorescence spectroscopy is a quantitative analytical tool that has been extensively used to provide structural and dynamical information on GAG–protein complexes. It possesses major advantages including high sensitivity, relative ease of applicability, and wide range of available fluorescence labels and probes. It has been applied to practically every protein–GAG system through the use of either intrinsic (e.g., Trp) or extrinsic (e.g., a non-covalent fluorophore) probe. For studies involving GAGs, it forms the basis for measurement of dissociation constant of complexes and the stoichiometry of binding, which helps elucidate many other thermodynamic and/or mechanistic parameters. We describe the step-by-step procedure to measure the affinity of GAG–protein complexes, parse the ionic and nonionic components of the free energy of binding, and identify the site of GAG binding through competitive binding experiments.

Key words Activation, Binding affinity, Competitive binding, Fluorescence spectroscopy, GAG–protein interactions, Inhibition, Serpins

1 Introduction

1.1 Structures and Functions of Glycosaminoglycans

Glycosaminoglycans (GAGs) are heterogeneous, polydisperse, variably sulfated (heparin/heparin sulfate (H/HS), chondroitin sulfate (CS), dermatan sulfate (DS), and keratan sulfate (KS)) or unsulfated (hyaluronic acid (HA)), linear polysaccharides (*see* Chaps. 2 and 3 in this volume). At the physiological pH, the carboxylic acid and sulfate groups of GAGs are fully ionized, which induce number of interactions with a wide range of proteins [1]. For example, a recent report identified 435 human proteins as the possible interactome of H/HS [2]. These multifarious interactions ensure that GAGs play important roles in many physiological and pathological processes such as coagulation, growth and development, angiogenesis, inflammation, and microbial infections among others [1, 3].

One of the most important targets of GAGs is the family of serpins (*ser*ine *p*rotease *in*hibitors), which includes antithrombin (AT), heparin cofactor II (HCII), protein C inhibitor (PCI), and others.

Kuberan Balagurunathan et al. (eds.), *Glycosaminoglycans: Chemistry and Biology*, Methods in Molecular Biology, vol. 1229, DOI 10.1007/978-1-4939-1714-3_27, © Springer Science+Business Media New York 2015

Serpins typically utilize GAGs to enhance inhibition of target proteases. For example, H binds to AT, HCII, and PCI, and accelerates the inhibition of factor Xa, factor IXa, thrombin, acrosin, and activated protein C. Likewise, HCII is able to utilize DS as a cofactor in its accelerated inhibition of thrombin.

Therapeutically, the most useful GAGs are heparins, which include unfractionated heparin (UFH), low molecular weight heparins (LMWHs), and fondaparinux. Although UFH interacts with several proteins in the coagulation cascade, its interaction with AT forms the basis for its clinical use as anticoagulant [4]. AT, a glycoprotein present in plasma at substantial levels (2–3 μM), regulates coagulation through the *serpin* "mousetrap" mechanism. The main physiological targets of AT include thrombin, factor Xa, and factor IXa, although by itself it is a poor inhibitor of these enzymes. The slow rates of inhibition are dramatically increased ($\sim10^2$–10^6-fold) in the presence of UFH. Two mechanisms are involved in UFH-dependent activation of AT inhibition of the procoagulant enzymes. The first is allosteric activation of AT induced by a specific five-residue sequence called DEFGH and the second is the sulfated polysaccharide initiated bridging of AT and the target enzyme. The DEFGH-driven allosteric activation of antithrombin, also called the conformational activation mechanism, is an important mechanism that enhances the rate of factor Xa and factor IXa inhibition some 300-fold. Likewise, the bridging mechanism can bring about a 1,000-fold increase in the rate of enzyme inhibition under physiological conditions, as observed for thrombin and factor IXa [5].

The study of sulfated GAGs binding to proteins is challenging primarily because of their heterogeneous and polydisperse properties, which engineer interaction with multiple sites on a target. Yet, a wide range of techniques have been exploited including fluorescence spectroscopy [6–22], nuclear magnetic resonance (NMR) spectroscopy, surface plasmon resonance (SPR), isothermal titration calorimetry (ITC), circular dichroism, and crystallography, affinity chromatography, electromobility shift assays, immunochemical methods, and in silico computational modeling [5]. The quantitative methods among these, e.g., fluorescence spectroscopy, are particularly useful because of their ability to provide structural and/or dynamical information on GAG–protein complexes. Fluorescence spectroscopy possesses major advantages including high sensitivity, relative ease of applicability, and wide range of available fluorescence labels and probes (Table 1). Fluorescence spectroscopy has been used to study the interaction of several AT variants including site-directed mutants and glycoforms with different heparin variants including UFH, high affinity heparin (HAH), low affinity heparin (LAH), LMWHs, and other pentasaccharides and oligosaccharides (Table 2). Fluorescence spectroscopy has also been most often used in the study of GAG or GAG mimetics binding to coagulation

Table 1
Excitation and emission wavelengths of a small group of typically used fluorophores[a]

Fluorophore	Excitation wavelength (λ_{EX}) (nm)	Emission wavelength (λ_{EM}) (nm)
Phenylalanine	260	282
Tyrosine	270	304
Tryptophan	280	340
ANS	385	485
TNS	330	432
NBD	466	539
DANS-Cl	335	500
AMCA	350	450
Cascade Blue	360	450
Fluorescein	495	519
FITC	488	525
PE	488	575
APC	630	650
PerCP™	488	680
Texas Red™	610	630
Tetramethylrhodamine-amines	550	575
Coumarin-phalloidin	350	450
CY3(Indotrimethinecyanines)	540	575
CY5(Indopentamethinecyanines)	640	670

[a]Wavelengths may vary according to solvent and/or pH

proteins such as AT, HCII, PCI, and thrombin, chemokines such as interleukin-8 [15–18], growth factors such as FGF-1 and FGF-2 [19], lipid-binding proteins such as apolipoprotein E [20], neutrophil elastase inhibitor [21], squamous cell carcinoma antigens (SCCA-1 and SCCA-2) [22], parathyroid hormone [23], and met-myoglobin [24] (Table 3).

1.2 Principle and Application of Fluorescence Spectroscopy

Fluorescence arises from electronic transitions in molecules when absorption of a photon results in the formation of an excited singlet state, which emits light in the form of fluorescence to return to the ground state. Non-radiative processes, including vibrational loss (heat), formation of triplet state, and collision with a quencher (iodide, oxygen), may compete with fluorescence to deactivate the excited state and reduce its efficiency. The emission rate of most

Table 2
Heparin–antithrombin interactions studied by intrinsic fluorescence spectroscopy

Heparin–AT interaction	K_D (µM)	Comments
Plasma AT and heparin or LMWH [25, 26]		
UFH–AT	0.019 ± 0.009	Trp enhancement 32 %
Enoxaparin–AT	0.058 ± 0.15	Trp enhancement 32 %
Fragmin–AT	0.0146 ± 0.0066	Trp enhancement 32 %
Ardeparin–AT	0.022 ± 0.00032	Trp enhancement 32 %
H–Enoxaparin–AT	0.019 ± 0.0009	Trp enhancement 32 %
H–Fragmin–AT	0.093 ± 0.02	Trp enhancement 32 %
LAH–AT	19 ± 6	8 ± 3 % Trp enhancement
Mutants of AT and heparins [7, 27–29]		
HAH–AT$_{N135A}$	0.017 ± 0.002	Trp enhancement, pH 7.4, 25 °C, I 0.4
HAH–AT$_{F121A/N135A}$	0.21 ± 0.07	Trp enhancement, pH 7.4, 25 °C, I 0.4
HAH–AT$_{F121A/N135A}$	0.038 ± 0.008	Trp enhancement, pH 7.4, 25 °C, I 0.3
HAH–AT$_{F122L/N135A}$	6.7 ± 0.51	Trp enhancement, pH 7.4, 25 °C, I 0.3
HAH–AT$_{F122L/N135A}$	0.28 ± 0.025	Trp enhancement, pH 7.4, 25 °C, I 0.15
H5–AT$_{F122L/N135A}$	0.068 ± 0.005	Trp enhancement, pH 6, 25 °C, I 0.075
HAH–AT$_{W49K}$	0.021 ± 0.004	Trp enhancement 36 ± 4 %, pH 7.4, 25 °C, I 0.15
H5–AT$_{W49K}$	0.257 ± 0.044	Trp enhancement 31 ± 4 %, pH 7.4, 25 °C, I 0.15
H5–AT$_{N135Q}$	0.002 ± 0.0002	Trp enhancement 35 ± 2 %, pH 7.4, 25 °C, I 0.15
H5–AT$_{W49K}$	0.017 ± 0.005	Trp enhancement 26 ± 2 %, pH 6, 25 °C, I 0.02
H5–AT$_{K114R/N135A}$	0.029 ± 0.004	Trp enhancement, pH 7.4, 25 °C, I 0.15
H5–AT$_{K125R/N135A}$	0.006 ± 0.003	Trp enhancement, pH 7.4, 25 °C, I 0.15
H5–AT$_{R129K/N135A}$	0.19 ± 0.02	Trp enhancement, pH 7.4, 25 °C, I 0.15
H5–AT$_{K114R/N135A}$	0.28 ± 0.02	Trp enhancement, pH 7.4, 25 °C, I 0.3
H5–AT$_{K125R/N135A}$	0.16 ± 0.01	Trp enhancement, pH 7.4, 25 °C, I 0.3
H5–AT$_{R129K/N135A}$	2 ± 0.9	Trp enhancement, pH 7.4, 25 °C, I 0.3
H26–AT$_{N135A}$	0.007 ± 0.001	Trp enhancement, pH 7.4, 25 °C, I 0.3
H26–AT$_{K114R/N135A}$	0.049 ± 0.004	Trp enhancement, pH 7.4, 25 °C, I 0.3
H26–AT$_{K125R/N135A}$	0.043 ± 0.002	Trp enhancement, pH 7.4, 25 °C, I 0.3
H26–AT$_{R129K/N135A}$	0.57 ± 0.02	Trp enhancement, pH 7.4, 25 °C, I 0.3
H–AT$_{S380W}$	0.009 ± 0.001	~29 % Trp enhancement, red shift: 337–354 nm

(continued)

Table 2
(continued)

Heparin–AT interaction	K_D (µM)	Comments
Glycoforms of AT and heparin [30]		
H–AT Plasma	0.021 ± 0.001	Trp enhancement, pH 7.4, 25 °C, 100 mM NaCl
H–AT WT	0.022 ± 0.002	Trp enhancement, pH 7.4, 25 °C, 100 mM NaCl
H–AT Gln96	0.009 ± 0.0013	Trp enhancement, pH 7.4, 25 °C, 100 mM NaCl
H–AT Gln135	0.0031 ± 0.0004	Trp enhancement, pH 7.4, 25 °C, 100 mM NaCl
H–AT Gln155	0.0098 ± 0.0007	Trp enhancement, pH 7.4, 25 °C, 100 mM NaCl
H–AT Gln192	0.0058 ± 0.0009	Trp enhancement, pH 7.4, 25 °C, 100 mM NaCl
AT and oligosaccharides [9, 11]		
DEFGH–AT	0.05 ± 0.006	32 ± 3 % Trp enhancement, pH 7.4, 25 °C, I 0.15
DEFG*–AT	17 ± 2	28 ± 1 % Trp enhancement, pH 7.4, 25 °C, I 0.15
EFGH″–AT	61 ± 3	49 ± 1 % Trp enhancement, pH 7.4, 25 °C, I 0.15
DEFGH–AT$_{N135Q}$	0.011 ± 0.002	32 ± 1 % Trp enhancement, pH 6, 25 °C, I 0.35
DEFGH–AT$_{N135Q}$	0.032 ± 0.01	36 ± 3 % Trp enhancement, pH 6, 25 °C, I 0.45
DEFGH–AT$_{N135Q}$	0.114 ± 0.024	36 ± 3 % Trp enhancement, pH 6, 25 °C, I 0.55
DEFGH–AT$_{N135Q}$	0.174 ± 0.017	35 ± 2 % Trp enhancement, pH 6, 25 °C, I 0.65
3-*O*-Desulfonated DEFGH–AT$_{N135Q}$	2.8 ± 0.2	36 ± 2 % Trp enhancement, pH 6, 25 °C, I 0.025
3-*O*-Desulfonated DEFGH–AT$_{N135Q}$	6.2 ± 0.7	40 ± 2 % Trp enhancement, pH 6, 25 °C, I 0.037
3-*O*-Desulfonated DEFGH–AT$_{N135Q}$	11 ± 2	35 ± 4 % Trp enhancement, pH 6, 25 °C, I 0.062
3-*O*-Desulfonated DEFGH–AT$_{N135Q}$	27 ± 3	38 ± 1 % Trp enhancement, pH 6, 25 °C, I 0.087

Table 3
Other GAG–protein interactions studied using extrinsic or intrinsic fluorescence spectroscopy

GAG–protein interaction	K_D (μM)	Comments
Coagulation proteins [8, 10, 12–14, 35]		
Heparin–AT-TNS (H5–AT-TNS)	~0.002 (~0.006)	60 % decrease of TNS fluorescence with a blue shift of ~3 nm; H binds to different AT mutants of residues 134–137 resulting in K_D's of 10 ± 1 to 79 ± 15 nM; H5 binds to same mutants with a range of affinity corresponds to 37 ± 1 to 283 ± 94 nM
Heparin–Thrombin-PABAM	2.8 ± 0.3 to 10 ± 1	Analysis by specific and nonspecific models based on ~16 % quench of *p*-aminobenzamidine fluorescence
Heparin disaccharides–Thrombin-PABZ	2.8 ± 0.3 to 71 ± 6	Analysis based on >90 % quench of *p*-aminobenzamidine fluorescence
Heparin–PCI-TNS	0.03 ± 0.006 to 8.0 ± 1.0	K_D is highly dependent on the form of PCI and heparin size; ~39 % quench in the extrinsic fluorescence of TNS with a 2–3 nm blue shift; cleaved PCI shows only a 2 % fluorescence change
Heparin–HCII-TNS	26 ± 6	49 % quench in the extrinsic fluorescence of TNS; Affinity, fluorescence quenching, and λ_{EM} are highly dependent on heparin size
DS–HCII	~25.3	13 % decrease of Trp intrinsic fluorescence; 6 % decrease observed with dansylated HCII (K_D ~5.1 μM)
Heparin–rPN-1	<0.001	~25 % increase of Trp fluorescence at 341 nm
Chemokines/regulatory proteins [15–18]		
DS–IL-8	5.5 ± 1.0	H6 results in ~30 % decrease of Trp fluorescence
CS4–IL-8	4.8 ± 0.7	H6 results in ~15 % decrease of Trp fluorescence
CS6–IL-8	1.4 ± 0.4	H6 results in ~35 % decrease of Trp fluorescence
HA–IL-8	5.5 ± 1.3	H6 results in ~25 % decrease of Trp intrinsic fluorescence
Heparin–Platelet factor 4	<0.5	~75 % enhancement of Tyr60 intrinsic fluorescence at 307 nm

Heparin–RANTES	0.32 ± 0.006	H18 results in an increase of Trp57 intrinsic fluorescence
Heparin–RANTES	0.43 ± 0.009	H14 results in an increase of Trp57 intrinsic fluorescence
Heparin–Arrestin	0.26 ± 0.02 7.33 ± 0.05	Two binding events detected by ~45 % diminution of Tyr fluorescence
Growth factors [19]		
Heparin–Fibroblast growth factor 1 and 2	0.001	20–25 % quench of Trp30 intrinsic fluorescence
Lipid binding proteins [20]		
Heparin–ApoE4	~1	H8 results in an increase of intrinsic fluorescence at 353 nm
Miscellaneous		
Heparin–Mucous proteinase inhibitor (neutrophil elastase inhibitor) [21]	0.05	~90 % enhancement of Trp30 intrinsic fluorescence and resulting in 6 nm blue shift of λ_{EM} from 336 to 330 nm
Heparin–SCCA-1 [22]	4.20 ± 0.46	~80 % decrease of Trp fluorescence at 340 nm
Heparin–SCCA-2 [22]	2.03 ± 0.15	~80 % decrease of Trp fluorescence at 340 nm
Heparin–Parathyroid hormone [23]	–	Significant decreases of Trp and ANS fluorescence as well as blue shift in λ_{EM} (Trp: 7 nm; ANS: 10–14 nm); shifts were highly temperature dependent
Heparin–Metmyoglobin [24]	–	Increase of Trp7 and Trp14 fluorescence as well as 10 nm red shift in λ_{EM}; profile dependent on pH and ionic strength

fluorophores is typically 10^8 s^{-1} so that a typical fluorescence lifetime is ~10 ns. This lifetime falls in the range of conformational transitions, which means that fluorescence spectroscopy is particularly suited for sensing conformational changes that may occur upon ligand binding. A fluorometer used to these changes minimally consists of five key components including the light source, an excitation wavelength monochromator, a sample compartment, an emission wavelength monochromator, and a photon detector/counter. Other components such as polarizers can also be present. Fluorescence spectroscopy is useful as long as ligand binding induces a change in the fluorophore, which may be an intrinsic part of a protein (e.g., Trp) or may be introduced extrinsically (e.g., a non-covalent probe). In a vast majority of studies on GAG binding to proteins, changes in fluorescence intensity of either an intrinsic or extrinsic probe are monitored; however, polarization, anisotropy, and energy transfer approaches could be also exploited.

Several GAG–protein interaction parameters can be readily deduced using fluorescence spectroscopy including the equilibrium dissociation constant (K_D) and the stoichiometry (n) of binding. Other experiments could be developed based on the demonstration that either intrinsic or extrinsic fluorescence changes follow ligand binding. These include measurement of ionic and nonionic components of the free energy of binding, assessing the sites of ligand binding using a competitive binding study, and measurement of heat capacity and other thermodynamic parameters. Finally, mechanistic insights into the nature of interactions can also be deduced using stopped-flow fluorescence spectroscopy.

2   Materials

Antithrombin (AT), a serpin, is considered the prototypical protein for studying GAG–protein interaction considering that UFH, LMWHs, and the DEFGH pentasaccharide bind this protein with high affinity. Another group of special interest for studying GAG interaction is the coagulation serine proteases including thrombin and factor Xa. Buffers of appropriate concentrations (20–50 mM) of sodium phosphate or Tris–HCl at pH 7.4 or 6.0 containing NaCl (25–150 mM), 0.1 % PEG 8000, and 0.1 mM EDTA (with or without 2.5 mM $CaCl_2$) are typically used in these studies. Probes used include TNS (2-(p-toluidinyl)-naphthalene-6-sulfonic acid), which is an external reversible fluorophore, and NBD (N,N'-dimethyl-N-acetyl-N'-(7-nitrobenz-2-oxa-1,3-diazol-4-yl) ethylenediamine) and dansyl (5-dimethylaminonaphthalene-1-sulfonyl-) groups, which are covalently attached to the protein of interest. Steady-state fluorescence experiments are typically performed in a ratiometric mode on a spectrofluorometer.

3 Methods

3.1 Excitation and Emission Scans

To obtain an appropriate interaction profile of GAG–protein complex, a sufficient change in fluorescence signal (>10 %) upon complexation is necessary. The approximate excitation and emission maxima for most proteins and/or external fluorophores of interest are documented in literature and/or provided by the manufacturer (Table 1). However, these may change a little due to experimental variations (buffers, excipients, etc.). Thus, the excitation and emission wavelengths (λ_{EX} and λ_{EM}, respectively) for the system being studied should be measured. These wavelengths are measured by preparing a solution of the sample in an appropriate buffer (0.22 µm filtered and degassed) in a quartz cuvette at an optimal temperature. The concentration of the sample should be sufficient so as to give neither a very low nor high intensity spectra. The emission spectrum is first measured by recording the fluorescence of the sample at wavelengths higher than the published λ_{EX} with a resolution of 0.25–0.50 nm and averaging over at least three experiments. The maximal λ_{EM} is then used to record the excitation spectrum by varying the λ_{EX}. The two experiments together help determine the optimal λ_{EX} and λ_{EM} to be used for titration (or other experiments). It is possible that a wavelength other than the maximal λ_{EX} be used to avoid inter-filter effects arising from either too high extinction coefficient or concentration of fluorophore.

3.2 Affinity of GAG–Protein Complexes

3.2.1 Intrinsic Fluorescence of Serpins

Intrinsic protein fluorescence typically relies on the ability of protein's tryptophan to fluoresce with adequate intensity and may provide a reasonable change in signal following GAG binding, as observed for a group of serpins. Tryptophan's fluorescence (λ_{EX} ~280 nm; λ_{EM} ~340 nm) depends on the environment, which typically changes upon binding of the highly electronegative GAG sequence. The literature is replete with a number of examples of the use of intrinsic fluorescence in studying GAG–protein interactions (Tables 1 and 2). In a majority of cases, a simple titration is used to measure the affinity of GAG for its target. In this experiment, the serpin (at fixed concentration) is titrated with increasing levels of GAG until no significant change in fluorescence is noticeable. The raw data is plotted as a function of the concentration of the GAG and fitted using the quadratic equation to calculate the binding affinity. Following step-by-step procedure has been used to measure heparin (H) affinity for AT.

1. Warm up a spectrofluorometer for about 30 min and set λ_{EX} and λ_{EM} at 280 nm and 340 nm, respectively. Set excitation and emission slits to 1–2 mm each (*see* **Note 1**).

2. Prepare a 20 mM sodium phosphate buffer, pH 7.4, containing 100 mM NaCl, 0.1 % PEG 8000, and 0.1 mM EDTA in an acrylic disposal cuvette and allow it to equilibrate to the desired temperature (25 °C) (*see* **Note 2**).

3. Measure the fluorescence intensity of the buffer, which is the background fluorescence, and ensure that this intensity is not high (*see* **Note 3**).

4. Add AT from a stock solution to give an appropriate working concentration (~50–500 nM) and mix gently. Record the fluorescence at 340 nm multiple times to ensure stability and uniformity of protein solution. The basal fluorescence of AT is the increase in fluorescence over the background (F_{AT}). This intensity should be at least 10–20 times higher than the background of **step 3** (*see* **Note 4**).

5. Prepare a 10 μM solution of H in the same buffer used for titration (**step 2**) (*see* **Notes 5 and 6**).

6. Add 1 μl of the H stock to the titration cuvette, mix gently without generating air bubbles, and record the fluorescence of the sample ($F_{H:AT}$).

7. Repeat **step 6** multiple times with gradually increasing aliquot volumes until further addition results in either a decrease or minimal change in fluorescence emission intensity (*see* **Note 6**). Measure the fluorescence intensity after the system reaches equilibrium following each addition.

8. To analyze the data, transform the raw data into percent change in intensity following addition of each aliquot (Eq. 1). In this equation, $\Delta F/\Delta F_0$ is the change in fluorescence intensity, $F_{H:AT}$ is the intensity after addition of each aliquot, and F_{AT} is the basal intensity of the protein (AT).

$$\frac{\Delta F}{\Delta F_0}(\%) = \frac{F_{H:AT} - F_{AT}}{F_{AT}} \times 100 \tag{1}$$

9. Plot $\Delta F/\Delta F_0$ versus H concentration and fit the data using quadratic equation 2 to calculate the K_D (affinity) and ΔF_{MAX} (maximal change in fluorescence upon saturation of AT) (*see* **Notes 7–10**).

$$\frac{\Delta F}{\Delta F_0} = \frac{\Delta F_{MAX}}{[AT]_0} \times \frac{\left([AT]_0 + [H]_0 + K_D\right) - \sqrt{\left(([AT]_0 + [H]_0 + K_D)^2 - 4[AT]_0[H]_0\right)}}{2} \tag{2}$$

An example of H–AT titration is shown in Fig. 1. In this system, the intrinsic tryptophan fluorescence of AT increases ~30–40 % at saturation. The affinities of various forms of H for AT and other proteins measured using this technique are listed in Table 1. Heparin oligosaccharides induce varying fluorescence enhancements in AT depending on the microscopic structure of the variants. Low-affinity heparin induces a ΔF_{MAX} of only 8 % and under appropriate conditions 3-*O*-desulfated heparin pentasaccharide shows 28–49 % fluorescence enhancement [11, 25]. Other examples include low molecular weight heparins binding to plasma

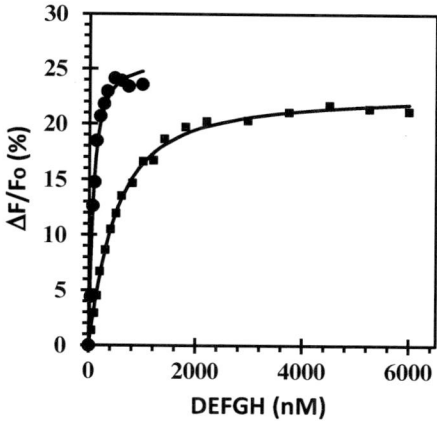

Fig. 1 A typical fluorescence titration for K_D measurement. The increase in intrinsic antithrombin fluorescence change ($\Delta F/F_0$, in %) as a function of increasing concentration of pentasaccharide DEFGH is fitted using the quadratic equation 2 (*solid line*), described in the text. The nonlinear regression analysis gives the maximal fluorescence change (ΔF_{MAX}) and affinity K_D under the conditions of the experiment. Experiments were performed using 50 nM AT in 20 mM sodium phosphate buffer, pH 7.4, in the presence of 100 mM (*filled circle*) and 300 mM (*filled square*) NaCl at 25 °C. The λ_{EX} was 280 nm and λ_{EM} was 340 nm and the excitation and emission slits were 2 mm each

AT [26], full-length heparin and heparin pentasaccharide binding to AT mutants [7, 9, 11, 27–29], and AT glycoforms [30]. These systems behave in a similar fashion to the plasma AT–heparin system. Some variances are possible as displayed by AT containing Ser380Trp mutation, which exhibits a 17 nm red shift in λ_{EM} (354 nm) following heparin binding [7] as opposed to heparin binding to plasma wild-type AT, which induces a blue shift of 15 nm in Trp49 fluorescence and a red shift of 5 nm in Trp225 fluorescence [31].

Tryptophan fluorescence-based affinity measurement has also been used for proteins other than AT. For example, SCCAs interact with heparin leading to more than 80 % quench in the fluorescence of four tryptophans from which affinities of 2.0–4.2 μM were calculated [22]. The change in intrinsic fluorescence was also exploited to study heparin binding to protease nexin-1, which results in a saturable increase of tryptophan fluorescence of ~25 % [14].

3.2.2 Extrinsic Fluorophores

For interactions where GAG binding to a target does not affect the intrinsic fluorescence, extrinsic probes may be used. There are two types of external probes, reversible and covalently attached fluorophores. For these types of probes too, the literature is replete with a number of successful examples. Experimentally, the process is similar to the above intrinsic fluorescence method, except for the presence of an extrinsic probe.

An example of very useful fluorophore is TNS (2-(p-toluidinyl)-naphthalene-6-sulfonic acid), which is a fairly hydrophobic, small probe that binds to many proteins to form a highly fluorescent complex. When AT–H interaction is studied using TNS (λ_{EM} ~432 nm; λ_{EX} ~330 nm), a 50–90 % decrease in fluorescence is observed with a blue shift of ~3 nm [10]. A concentration of about 5–40 µM of TNS (K_D of AT–TNS complex ~125 µM) is used and the GAG is titrated in a manner similar to that described above. The fluorescence change profile as a function of ligand concentration is analyzed using the quadratic equation above (Eq. 2) to determine the affinity of interaction. The affinities of cleaved and latent AT for H and DEFGH have been also measured using TNS as extrinsic probe [30]. Likewise, this protocol was also exploited to study H binding to AT mutants devoid of amino acids 134–137 [10].

The TNS-based affinity measurement is especially useful when the intrinsic tryptophan fluorescence does not change upon GAG binding (*see* **Notes 11** and **12**). An example of this is the HCII–H interaction for which TNS was used [12]. The λ_{EM} of HCII–TNS complex was found to be 448 nm, rather than 432 nm for AT–TNS complex. While small heparin-derived oligosaccharides (<8 units) did not affect the HCII–TNS fluorescence, heparin chains of 8–12 monosaccharides displayed a 40–70 % increase in TNS fluorescence and about 1.5 nm blue shift. Likewise, chains with ≥14 units caused a 30–70 % fluorescence quench and 7.5 nm red shift [12]. Other examples of the use of TNS include PCI, which displayed a 39 % quench at λ_{EM} of 448 nm with a 2–3 nm blue shift upon heparin binding. On the other hand, cleaved PCI showed only a 2 % decrease in TNS fluorescence following heparin binding [13].

Extrinsic fluorophores could also be covalently linked to the protein of interest. The fluorophores are typically loaded onto a protein target by a chemical reaction. An example of this protocol is NBD (N,N'-dimethyl-N-acetyl-N'-(7-nitrobenz-2-oxa-1,3-diazol-4-yl) ethylenediamine), which is a small hydrophobic probe for site-specific attachment to reactive cysteine residues on proteins. When introduced at the P1 position of the mutant ATArg393Cys, the fluorescence of NBD at 542 nm ($\lambda_{EX} = 465$ nm) shows dramatic increases of 32–105 % upon interaction with dextran sulfate, low-affinity heparin, and full-length heparin corresponding to significant differences in the reactive center loop conformation/electrostatics following GAG binding [32]. Another fluorophore that has been used often is the dansyl group (5-dimethyl-aminonapthalene-1-sulfonyl-). The interaction of dansylated heparin with AT results in an approximately 80 % increase in fluorescence at 510 nm ($\lambda_{EX} = 335$ nm), which serves as an excellent quantitative measure of interaction [33, 34]. Alternatively, dansylated HCII has been used to study its differential interaction with H and dermatan sulfate [35].

3.3   Nature of Forces Involved in GAG–Protein Interactions

Although GAGs tend to predominantly utilize ionic interactions in recognizing their protein targets, they may utilize nonionic forces, such as hydrogen bonds. Thus, while many GAG–protein systems are nonspecific and involve many ionic sulfate–Arg/Lys interactions, a few are highly specific and form hydrogen bonds between the same types of groups. The thrombin–heparin system is an example of a nonspecific system, while the H–AT system is an example of a specific system. It is important to understand and quantify the different forces involved in these interactions. Note that the sulfate group introduces both ionic and nonionic forces in the interaction.

To decipher the ionic and nonionic components of the binding energy, the dissociation constant (K_D) should be measured at several ionic strengths. Fluorescence spectroscopy is particularly useful for this purpose as NaCl levels typically do not affect fluorescence quantum yields. The steps for this experiment are described below using the prototypical example of DEFGH binding to AT.

1. Measure the intrinsic AT fluorescence as a function of DEFGH concentration in the manner of titration described above in 20 mM sodium phosphate buffer, pH 7.4, containing 100 mM NaCl, 0.1 mM EDTA, and 0.1 % PEG 8000 at 25 °C. Ideally, reach a high level of AT saturation with DEFGH (approximately $[\text{DEFGH}] = 20 \times K_D$).

2. Perform multiple titrations in the same buffer except for different levels of salt (e.g., 25, 50, 150, and 250 mM).

3. Calculate the K_D at each NaCl concentration using Eq. 2. Typically, five K_Ds should be calculated so as to span a reasonable range of pentasaccharide affinity (2–3 orders of magnitude). For most GAG–protein systems, the affinity of the polymer decreases as the NaCl concentration increases.

4. Plot log $K_{D,obs}$ versus log $[\text{Na}^+]$ (*see* **Note 13**) and perform a least squares analysis using Eq. 3, in which $K_{D,obs}$ is the observed affinity of the GAG for its protein target; $K_{D,nonionic}$ is the nonionic affinity of the GAG for its protein; Z is the number of ionic interactions; and Ψ is the number of Na$^+$ ions released upon formation of one electrostatic interaction and was found to be 0.8 for heparin. This equation is derived on the basis of protein–polyelectrolyte interaction, as described in the literature [36, 37]. The number of the ionic interactions between H and AT, or a GAG and a protein, can be calculated from the slope of the regressional line. The intercept provides the affinity of GAG for protein at a high [NaCl] of 1 mM, which is assumed to fully screen ionic forces and thereby reflect all nonionic forces involved in the interaction. This salt dependence study indicates the presence of four to five ionic interactions

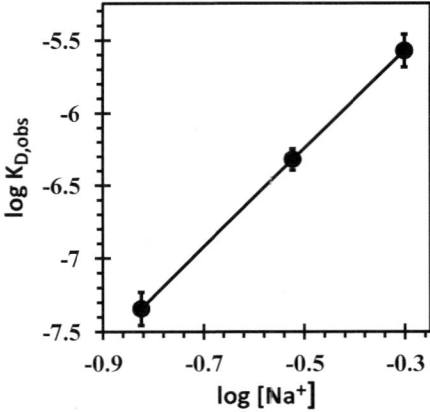

Fig. 2 Resolution of ionic and nonionic forces contributing to GAG–protein interaction. Shown is a plot of the observed affinity ($K_{D,obs}$) of DEFGH binding to antithrombin at 25 °C in 20 mM sodium phosphate buffer, pH 7.4, at varying salt concentrations (150–500 mM). The affinity measurements were performed as described in Fig. 1 through spectrofluorometric titrations. *Solid line* represents linear regression fits to the data using Eq. 3. Error bars in symbols represent ±S.E

for AT–DEFGH system with a nonionic interaction affinity of approximately 25 μM (Fig. 2) [9, 28].

$$\log K_{D,obs} = \log K_{D,nonionic} + Z\Psi \log\left[Na^+\right] \qquad (3)$$

5. The ionic and nonionic binding energies can be easily calculated from the above derivations. Thus, $\Delta G_{nonionic}$ is equal to $-2.303RT \times \log K_{D,nonionic}$ and ΔG_{ionic} at a defined NaCl concentration, e.g., physiologic, can be calculated from the difference $\Delta G_{obs} - \Delta G_{nonionic}$, where $\Delta G_{obs} = -2.303RT \times \log K_{D,obs}$ at [NaCl] = 100 mM.

3.4 Competitive Binding Studies with Glycosaminoglycans

Competitive binding studies form a powerful means of establishing the nature and extent of GAG-binding site(s). Two relatively similar molecules may induce a slightly different fluorescence property in the GAG-binding protein to enable a competitive binding experiment. Several examples of such studies have been reported in the literature [9], e.g., DEFGH variants, newly discovered GAGs, or GAGs mimetic binding to GAG-binding proteins. The prototypical example of this study is the study of a trisaccharide variant FGH‴ of DEFGH to assess its interaction with AT. Tetrasaccharide DEFG* ($K_D = 34 \pm 3$ nM at pH 6.0 I 0.025) was used as the competitor because DEFGH binds so potently to AT (K_D ~0.001 nM at pH 6.0 I 0.025) that extremely high levels of FGH‴ ($K_D = 11 \pm 1$ μM at pH 6.0 I 0.025) would be needed to displace DEFGH [9]. Briefly, the K_D of DEFG* and FGH‴ for AT was first measured using the increase in fluorescence, as described above.

Using fixed concentrations of FGH‴ (~1 and ~4 times the K_D of AT–FGH‴ complex), the K_D of DEFG* for AT was measured using the residual change in tryptophan fluorescence of AT. If the competition between DEFG* and FGH‴ was ideal, then K_D of DEFG*–AT complex at different fixed concentrations of FGH‴ can be calculated using the Dixon–Webb equation (Eq. 4). If the observed K_Ds for DEFG* are similar to the ones predicted by Eq. 4, then the two ligands possibly compete for the same binding site on AT. However, if the measured K_Ds did not change in parallel to the predicted ones, the two ligands do not compete in an ideal manner (*see* **Notes 14** and **15**).

$$K_{D(DEFG^*)predicted} = K_{D(DEFG^*),no-competitor}\left(1 + \frac{[FGH''']_0}{K_{D,FGH^-}}\right) \qquad (4)$$

In this study, the $K_{D,app}$ for DEFG* binding to AT increased from 34 ± 3 nM in the absence of FGH‴ to values of 89 ± 6 and 181 ± 13 nM in the presence of 13.3 and 43.3 μM FGH‴, respectively. These measured K_Ds were indistinguishable from the calculated K_Ds of 75 ± 10 and 167 ± 20 nM based on Eq. 4. Thus both oligosaccharide variants were found to bind to the same site on AT [9].

4   Notes

1. The slit widths should be such that provides optimal resolution and/or signal to noise ratio, while also not interfering with possible wavelength overlap. Typically, 1 mm width corresponds to about 8 nm wavelength dispersion, which means that λ_{EX} and λ_{EM} should be more than 8 nm apart.

2. Degas and filter the buffer to reduce quenching effects of oxygen. Likewise, it is important to use cuvettes that do not absorb either the protein or the GAG. This is especially true for experiments performed at low protein concentrations (~10 nM). It may be advantageous to explore coating of cuvette walls with either BSA or PEG 20000.

3. It may be helpful to record photon counts in dark to ensure that stray light does not enter the monochromators and distort results.

4. Mixing gently is important because of the multiple additions of ligand to be performed over the course of the experiment. Many proteins denature because of repeated shear forces and this may be a significant cause of irreproducibility.

5. The H stock should be prepared such that the volume of the titrant added through the end should not exceed more than

10 % of the total volume in the cuvette. For affinity titrations, the final concentration of the titrant should reach 20–25-fold the affinity of protein–ligand complex so as to observe good saturation of the protein.

6. To define the characteristic hyperbolic profile expected of an affinity titration, a minimum of nine fluorescence readings ($F_{H:AT}$) should be recorded. These would include three each for defining the initial straight line, middle curved section, and the final straight line. Additional readings may be necessary to accurately define each section of the hyperbolic plot. Appropriate volume of the stock titrant should be prepared.

7. The generic quadratic equation for analyzing the interaction of a ligand for its protein target is Eq. 5, which takes into consideration the stoichiometry of binding (n) also.

$$\frac{\Delta F}{F_0} = \frac{\Delta F_{MAX}}{F_0} \times \frac{\left(n[P]_0 + [L]_0 + K_D\right) - \sqrt{\left(n[P]_0 + [L]_0 + K_D\right)^2 - 4n[P]_0[L]_0}}{2[P]_0}. \tag{5}$$

This means that the K_D of protein–GAG interaction is most accurately measured in fluorescence titrations when the protein concentration is approximately in the range of the affinity, while stoichiometry of binding "n" is better determined when the protein concentration is approximately 10–100-fold above the K_D. For example, concentrations of ~20 and ~500 nM are better suited to measure the K_D and n, respectively, of AT–full length heparin interaction under physiological conditions.

8. Some GAG–protein interactions result in a decrease (quench) in fluorescence. Such decrease could be utilized to deduce binding parameters. However, fluorescence can be also lost by collisional quenchers, static quenching, resonance energy transfer, and inner filter effect. It is important to ensure that fluorescence quenching relates to the ligand binding interaction before calculation of affinities.

9. Turbidity because of interfering substance(s) is a serious issue in fluorescence experiments. If the interfering substance absorbs light, fluorescence emission will be reduced. If the interfering substance does not absorb light, the fluorescence readings will not be affected unless there is so much turbidity that the emitted light cannot penetrate through. In fact, if the interfering substance is reflective, turbidity can create light scatter and fluorescence readings may increase. Samples should be filtered before titration to minimize the effect of turbidity. Turbidity effects can be easily accounted for by using the same experiment medium for standards, blanks, and samples, and handling them exactly in the same manner.

10. Fluorescence is also affected by changes in temperature. As temperature increases, molecular collisions increase causing a subsequent loss of energy and hence a fluorescence decrease. Although temperature effect may vary because the temperature coefficient varies depending upon the compound being measured, potential temperature related inconsistencies should be avoided. Other factors that could affect the fluorescence profiles and should be considered are the possible effect of pH as well as photochemical decay of the fluorescent molecules.

11. Affinity measurement using intrinsic fluorescence spectroscopy relies on changes introduced on one or more tryptophans of the protein by GAG binding. In case of AT, Trp49, Trp189, Trp225, and Trp307 contribute 8 %, 10 %, 19 %, and 63 %, respectively, of the total fluorescence [31]. Particularly, Trp225 and Trp307 account for most of fluorescence enhancement induced by heparin binding [4]. Thus, if a tryptophan in a protein is located at a distant site from the ligand binding site, it may not report on ligand binding.

12. Some GAG sequences may possess poor affinity for the target protein, necessitating higher concentration of protein in titration for good signal-to-noise ratio. For example, titrations of low-affinity heparin were performed using 10 μM AT, which is high enough to introduce inner filter effect. To reduce this interference, a λ_{EX} of 295 nm, rather 280 nm, was used [25]. Likewise, a smaller path length on the excitation side (2 mm instead of 10 cm) could also be used to reduce inner filter effects.

13. Note that the buffer will have Na^+ counter ion and its concentration should be added to the level of NaCl. This can be calculated using the Henderson-Hasselbalch equation.

14. It is important to choose competitors that are relatively close in affinity to the ligand being analyzed. For example, if a GAG displays an affinity of 10 μM for a target protein and a competing GAG derivative displays an affinity of 0.1 nM for the same protein, the Dixon-Webb competition experiment may not yield an "ideal" competitive behavior because of the widely different affinities.

15. It is important to use competitor concentration approximately one to four times its K_D to observe substantial changes in the apparent K_d of the ligand.

Acknowledgements

This work was supported by the grants HL090586 and HL107152 from the National Institutes of Health.

References

1. Gandhi NS, Mancera RL (2008) The structure of glycosaminoglycans and their interactions with proteins. Chem Biol Drug Des 72: 455–482

2. Ori A, Wilkinson MC, Fernig DG (2011) A systems biology approach for the investigation of the heparin/heparan sulfate interactome. J Biol Chem 286:19892–19904

3. Capila I, Linhardt RJ (2002) Heparin–protein interactions. Angew Chem Int Ed 41:391–412

4. Henry BL, Desai UR (2010) Anticoagulants: drug discovery and development. In: Rotella D, Abraham DJ (eds) Burger's medicinal chemistry, 7th edn. Wiley, New York, pp 365–408

5. Rein CM, Desai UR, Church FC (2011) Serpin–glycosaminoglycan interaction. Methods Enzymol 501:105–137

6. Olson ST, Shore JD (1981) Binding of high affinity heparin to antithrombin. III. Characterization of the protein fluorescence enhancement. J Biol Chem 256:11065–11072

7. Huntington JA, Olson ST, Fan B, Gettins PGW (1996) Mechanism of heparin activation of antithrombin. Evidence for reactive center loop preinsertion with expulsion upon heparin binding. Biochemistry 35:8495–8503

8. Olson ST, Halvorson HR, Björk I (1991) Quantitative characterization of the thrombin-heparin interaction. Discrimination between specific and nonspecific binding models. J Biol Chem 266:6342–6352

9. Desai UR, Petitou M, Björk I, Olson ST (1998) Mechanism of heparin activation of antithrombin: role of individual residues of the pentasaccharide activating sequence in the recognition of native and activated states of antithrombin. J Biol Chem 273:7478–7487

10. Meagher JL, Olson ST, Gettins PG (2000) Critical role of the linker region between helix D and strand 2A in heparin activation of antithrombin. J Biol Chem 275:2698–2704

11. Richard B, Swanson R, Olson ST (2009) The signature 3-O-sulfo group of the anticoagulant heparin sequence is critical for heparin binding to antithrombin but is not required for allosteric activation. J Biol Chem 284:27054–27064

12. O'Keeffe D, Olson ST, Gasiunas N, Gallagher J, Baglin TP, Huntington JA (2004) The heparin binding properties of heparin cofactor II suggest an antithrombin-like activation mechanism. J Biol Chem 279:50267–50273

13. Li W, Adams TE, Kjellberg M, Stenflo J, Huntington JA (2007) Structure of native protein c inhibitor provides insight into its multiple functions. J Biol Chem 282:13759–13768

14. Arcone R, Chinali A, Pozzi N, Parafati M, Maset F, Pietropaolo C, Filippis VD (2009) Conformational and biochemical characterization of a biologically active rat recombinant protease nexin-1 expressed in E. coli. Biochim Biophys Acta 1794:602–614

15. Pichert A, Samsonov SA, Theisgen S, Thomas L, Baumann L, Schiller J, Beck-Sickinger AG, Huster D, Pisabarro MT (2012) Characterization of the interaction of interleukin-8 with hyaluronan, chondroitin sulfate, dermatan sulfate and their sulfated derivatives by spectroscopy and molecular modeling. Glycobiology 22:134–145

16. Rek A, Brandner B, Geretti E, Kungl AJ (2009) A biophysical insight into the RANTES glycosaminoglycan interaction. Biochim Biophys Acta 1794:577–582

17. Wilson CJ, Copeland RA (1997) Spectroscopic characterization of arrestin interactions with competitive ligands: study of heparin and phytic acid binding. J Protein Chem 16:755–763

18. Loscalzo J, Melnick B, Handin RI (1985) The interaction of platelet factor four and glycosaminoglycans. Arch Biochem Biophys 240: 446–455

19. Li LY, Seddon AP (1994) Fluorospectrometric analysis of heparin interaction with fibroblast growth factors. Growth Factors 11:1–7

20. Dong J, Peters-Libeu CA, Weisgraber KH, Segelke BW, Rupp B, Capila I, Hernáiz MJ, LeBrun LA, Linhardt RJ (2001) Interaction of the N-terminal domain of apolipoprotein E4 with heparin. Biochemistry 40:2826–2834

21. Faller B, Mely Y, Gerard D, Bieth JG (1992) Heparin-induced conformational change and activation of mucus proteinase inhibitor. Biochemistry 31:8285–8290

22. Higgins WJ, Fox DM, Kowalski PS, Nielsen JE, Worrall DM (2010) Heparin enhances serpin inhibition of the cysteine protease cathepsin L. J Biol Chem 285:3722–3729

23. Kamerzell TJ, Joshi SB, McClean D, Peplinskie L, Toney K, Papac D, Li M, Middaugh CR (2007) Parathyroid hormone is a heparin/polyanion binding protein: binding energetics and structure modification. Protein Sci 16: 1193–1203

24. Fedunová D, Antalík M (1998) Studies on interactions between metmyoglobin and heparin. Gen Physiol Biophys 17:117–131

25. Streusand VJ, Björk I, Gettins PGW, Petitou M, Olson ST (1995) Mechanism of acceleration of antithrombin-proteinase reactions by low affinity heparin. J Biol Chem 270: 9043–9051

26. Lin P, Sinha U, Betz A (2001) Antithrombin binding of low molecular weight heparins and inhibition of factor Xa. Biochim Biophys Acta 1526:105–113

27. Jairajpuri MA, Lu A, Desai U, Olson ST, Bjork I, Bock SC (2003) Antithrombin III phenylalanines 122 and 121 contribute to its high affinity for heparin and its conformational activation. J Biol Chem 278:15941–15950

28. Monien BH, Krishnasamy C, Olson ST, Desai UR (2005) Importance of tryptophan 49 of antithrombin in heparin binding and conformational activation. Biochemistry 44:11660–11668

29. Schedin-Weiss S, Arocas V, Bock SC, Olson ST, Björk I (2002) Specificity of the basic side chains of Lys114, Lys125, and Arg129 of antithrombin in heparin binding. Biochemistry 41:12369–12376

30. Olson ST, Frances-Chmura AM, Swanson R, Björk I, Zettlmeissl G (1997) Effect of individual carbohydrate chains of recombinant antithrombin on heparin affinity and on the generation of glycoforms differing in heparin affinity. Arch Biochem Biophys 341:212–221

31. Meagher JL, Beechem JM, Olson ST, Gettins PGW (1998) Deconvolution of the fluorescence emission spectrum of human antithrombin and identification of the tryptophan residues that are responsive to heparin binding. J Biol Chem 273:23283–23289

32. Futamura A, Beechem JM, Gettins PGW (2001) Conformational equilibrium of the reactive center loop of antithrombin examined by steady state and time-resolved fluorescence measurements: consequences for the mechanism of factor Xa inhibition by antithrombin-heparin complexes. Biochemistry 40:6680–6687

33. Piepkorn MW (1981) Dansyl (5-dimethyl-aminonaphthalene-1-sulphonyl)-heparin binds antithrombin III and platelet factor 4 at separate sites. Biochem J 196:649–651

34. Piepkorn MW, Lagunoff D, Schmer G (1980) Binding of heparin to antithrombin III: the use of dansyl and rhodamine labels. Arch Biochem Biophys 205:315

35. Liaw PCY, Austin RC, Fredenburgh JC, Stafford AR, Weitz JI (1999) Comparison of heparin- and dermatan sulfate-mediated catalysis of thrombin inactivation by heparin cofactor II. J Biol Chem 274:27597–27604

36. Mascotti DP, Lohman TM (1995) Thermodynamics of charged oligopeptide-heparin interactions. Biochemistry 34:2908–2915

37. Olson ST, Björk I, Sheffer R, Craig PA, Shore JD, Choay J (1992) Role of the antithrombin-binding pentasaccharide in heparin acceleration of antithrombin-proteinase reactions. Resolution of the antithrombin conformational change contribution to heparin rate enhancement. J Biol Chem 267:12528–12538

Chapter 28

Studying Glycosaminoglycan–Protein Interactions Using Capillary Electrophoresis

Aiye Liang and Umesh R. Desai

Abstract

Methods for studying interactions between glycosaminoglycans (GAGs) and proteins have assumed considerable significance as their biological importance increases. Capillary electrophoresis (CE) is a powerful method to study these interactions due to its speed, high efficiency, and low sample/reagent consumption. In addition, CE works effectively under a wide range of physiologically relevant conditions. This chapter presents state-of-the-art on CE methods for studying GAG–protein interactions including affinity capillary electrophoresis (ACE), capillary zone electrophoresis (CZE), frontal analysis (FA)/frontal analysis continuous capillary electrophoresis (FACCE), and capillary electrokinetic chromatography (CEC) with detailed experimental protocols for ACE and CZE methods.

Key words Affinity capillary electrophoresis, Biophysical technique, Capillary electrophoresis, Capillary zone electrophoresis, GAG–protein interactions, Glycosaminoglycans, Heparin

1 Introduction

The interaction between glycosaminoglycans (GAGs) with proteins is gaining increasing significance as new functions of GAGs are discovered. GAGs have been found to interact with a range of proteins including growth factors, extracellular matrix proteins, proteinases, and proteinase inhibitors [1–3]. GAG interaction with a protein may modify the function of the protein. For example, antithrombin (AT) is a poor inhibitor of thrombin, factor Xa, and factor IXa. However, heparin induces AT to become a better inhibitor (300–600-fold). It is assumed that majority of GAG interactions with proteins are induced by specific sequences present in the GAG chain. For example, a specific pentasaccharide sequence in heparin is known to bind to AT (*see* Chapters 2 and 3 in this volume). Likewise, basic fibroblast growth factor (bFGF) and acidic fibroblast growth factor (aFGF) bind to a distinct hexasaccharide in heparin, thereby aiding

Kuberan Balagurunathan et al. (eds.), *Glycosaminoglycans: Chemistry and Biology*, Methods in Molecular Biology, vol. 1229, DOI 10.1007/978-1-4939-1714-3_28, © Springer Science+Business Media New York 2015

the interaction between the growth factors and their receptors [4]. More than 400 proteins have been suggested to bind to heparin [5]. GAGs may stabilize the target protein and prolong its half-life in vivo. Likewise, these interactions may also play important roles in cellular metabolism [6]. These varied biological functions imply that it is crucial to study the interactions between GAGs and proteins to elucidate the fundamentals underlying the mechanism of disorders and to understand structure–function relationships of GAGs.

Many techniques have been used to investigate GAG–protein interactions, of which capillary electrophoresis (CE) [7–9] is particularly useful for studying GAG mixtures that are typically produced from biological systems. CE is a robust analytical technique owing to its many advantages such as short analysis time, low sample size requirement, high efficiency, and flexible applications. CE can be performed in buffers that are physiologically relevant. CE displays an extremely high resolving power, which aids analysis of complex species present in the sample [9]. CE affords measurement of the binding constants and stoichiometry, which significantly enhances the utility of CE in studying GAG–protein interactions. Several modes of CE have been developed to study GAG–protein interactions including affinity capillary electrophoresis (ACE) [10–30], capillary zone electrophoresis (CZE) [31–44], capillary electrokinetic chromatography [CEC] [45–47], frontal analysis (FA) and frontal analysis continue capillary electrophoresis (FACCE) [48–52]. Of these, ACE and CZE are most often used.

2   Basic Theory

The quantitative parameter most useful to derive for a receptor (protein)–ligand interaction is the binding constant. Assuming that a receptor (protein) has n_i independent sites for binding, then r, the ratio of the bound form of the receptor (or ligand) to the total ligand (or receptor), can be calculated using the Law of Mass Action by Eq. 1, in which K_i is the binding constant for a defined site, C_f is the concentration of free receptor (or ligand) in the equilibrium system, and m is the number of different types of sites that have n_i independent binding sites [53].

$$r = \sum_{i=1}^{m} \frac{n_i K_i C_f}{1 + K_i C_f} \tag{1}$$

If there is only one type of the binding site on a protein (and n independent sites of this type), which is typically the case for majority of systems, then Eq. 1 reduces to Eq. 2, which can be linearized to the famous Scatchard equation 3. Under appropriate conditions, Eq. 3 can afford the binding constant K and the stoichiometry of binding n [54].

$$r = \frac{nKC_f}{1 + KC_f} \tag{2}$$

$$\frac{r}{C_f} = nK - rK \tag{3}$$

Scatchard analysis applies well to medium- or high-affinity interactions [55], for which a single apparent binding constant is a good representation of the interaction between the two species. It can be rearranged to Klotz equation 4 [56], which has also been used in the literature.

$$\frac{1}{r} = \frac{1}{n} + \frac{1}{nKC_f} \tag{4}$$

If there are two types of binding sites, Eq. 1 will yield Eq. 5 with two binding constants and two stoichiometries of binding. Assuming a good collection of experimental data, Eq. 5 can be used for nonlinear regression to derive the four parameters.

$$r = \frac{n_1 K_1 C_f}{1 + K_1 C_f} + \frac{n_2 K_2 C_f}{1 + K_2 C_f} \tag{5}$$

For use in ACE, Scatchard equation 3 was transformed to Eq. 6 by Whitesides and coworkers [57]. Briefly, if μ_0^{cp} and μ_t^{cp} are the protein's electrophoretic mobilities in the buffer without and with GAG, respectively, then $\Delta\mu^{cp} = \mu_t^{cp} - \mu_0^{cp}$ represents the change in the mobility of the protein due to the presence of GAG, which directly correlates with the concentration of the protein in the bound form. When the protein is fully saturated with the GAG, the change in the electrophoretic mobility will reach a maximum ($\Delta\mu_{max}^{cp}$). Thus, the ratio of the protein in the bound form to the total protein (r) will be equal to the fraction $\Delta\mu^{cp}/\Delta\mu_{max}^{cp}$. Under rapid equilibrium conditions, Eq. 6 can then be derived as the electrophoretic version of the traditional Scatchard equation 3, assuming $n = 1$ [54].

$$\frac{\Delta\mu^{ep}}{C_f} = -K_A \Delta\mu^{ep} + K_A \Delta\mu_{max}^{ep} \tag{6}$$

In this equation K_A is apparent equilibrium association constant, $\Delta\mu_{max}^{cp}$ is the maximal change in mobility of protein when saturated with ligand L ($C_f >> [\text{protein}]$). Thus, a plot of $\Delta\mu^{cp}/C_f$ versus $\Delta\mu^{cp}$ will provide a value of K_A from its slope. The electrophoretic mobility of protein is to be calculated from Eq. 7 [54]. In this equation, L_e is the effective separation length of the capillary, L_t is the total length of the capillary, t_p is the migration time of protein, t_{nm} is the migration time of the neutral marker, and V is the applied voltage during the separation process.

$$\mu^{ep} = \frac{\left(\dfrac{L_e}{t_p} - \dfrac{L_e}{t_{nm}}\right)}{\dfrac{V}{L_t}} \qquad (7)$$

3   Applications of CE to Study of GAG–Protein Interactions

3.1   Affinity Capillary Electrophoresis (ACE) of GAG–Protein Interactions

In ACE, the electrophoretic mobility of the protein changes as the ligand concentration increases due to the formation of the complex with an altered mobility. ACE is normally used in intermediate to fast dissociation systems, which corresponds to binding constants in the range of µM to M. ACE is not particularly useful for measuring the stoichiometry of binding (n). An internal marker is required to correct for background mobility changes arising from altered current as the concentration of ligand increases. A specific advantage of ACE is that protein concentration is not strictly required to measure the binding constant as long as the concentration of the ligand is well established. This is important for GAGs for which it may be difficult to ascertain their exact concentration owing to their origin from bodily fluids.

In most GAG–protein interaction studies, GAGs are added to the run buffer and the protein is injected as an analyte. Since most GAGs are sulfated and carboxylated, the negative charge of the protein–GAG complex increases resulting in an increase in the electrophoretic mobility under normal polarity conditions. This decreases the migration rate of the protein because electroosmotic flow (EOF) is the major force under normal polarity, which pulls the analyte to the detector, and opposes the normal electrophoretic mobility. In contrast, the protein's migration time will decrease under reverse polarity conditions because EOF is negligible and resolution relies only on its electrophoretic mobility.

Heegaard's group has contributed extensively to the investigation of heparin with different peptides using ACE [10]. The group studied the interactions between peptide mixture and heparin using bare silica capillary under normal polarity. The signal corresponding to a distinct peptide lagged behind that of other peptide peaks following addition of heparin to the run buffer helping identify the peptide sequence that preferentially bound to heparin [10]. Likewise, the group identified two proteolytic fragments obtained from the degradation of human serum amyloid-P as the reason behind the protein's heparin-binding activity [12, 22]. A new peak corresponding to a complex was observed for one peptide (fragment 3), which bound to heparin with high affinity. The best condition for studying those interactions was 10 mM HEPES buffer at pH 8.17. Under these conditions, dissociation constants between serum amyloid-P and bovine lung heparin in the presence and absence of Ca^{2+} were measured to be 40 nM and 120 nM, respectively, which were lower than those measured using other methods under physiological ionic strength at neutral pH (200–500 nM K_D) [23, 24]. It is possible that lower K_Ds may result from the low ionic strength buffer used by the group. Heparin binding to folate-binding protein (FBP) was studied by ACE at neutral pH in the presence of nonionic detergents. Heparin resolved FBP into several peaks and although the affinity of interaction was not measured, the method quantified free FBP in the sample pre-incubated with folate [25].

An interesting application of CE technique is to assess whether different protein conformations exhibit altered heparin affinity. The interaction of two conformational states f and s of Lys^{58}-β_2m microglobulin with heparin has been studied by Heegaard et al. The mobility of the two forms, especially form s, changed upon heparin addition to the run buffer resulting in their baseline separation. The resolution allowed calculation of the heparin affinities of both the components (0.6 mM (s) and 2.2 mM (f)) [30]. Although the study supported the idea that pure analytes are not necessary for ACE analysis, a critical requirement is that the association and dissociation rates must be fast. More specifically, the dissociation half-life of the complex should be equal to or less than 1 % of the time to separate the free analyte and complex. This will ensure that peak tailing, broadening, splitting, or disappearance do not complicate the results [17]. Other systems studied using ACE include low-affinity heparin binding to AT [15] and low molecular weight heparin, porcine mucosal heparin, and heparan sulfate binding to amyloid precursor protein (APP) [18]. The latter study also measured the dissociation constants between different GAGs and APP using linear and nonlinear regression analysis, both of which gave essentially same results.

A particularly difficult problem with some proteins is their adsorption to bare capillary wall, which reduces reproducibility significantly. An example of such proteins is human beta2-glycoprotein I ($\beta_2$gpI, pI = 8), which was studied for heparin binding using bare silica capillary by Bohlin et al. [14]. The group used pH hysteresis behavior of fused silica surfaces to reduce the interaction between the alkaline protein and capillary wall resulting in good repeatability of electrophoretic runs.

ACE can quickly assess whether the target protein possess two significantly different binding sites. Liu et al. have studied the interaction of fibronectin with heparin using coated capillary under reverse polarity conditions [19]. The fibronectin migration time was found to consistently decrease, but not that of the internal marker, with the increase in heparin concentration. Two linear Scatchard plots were observed, from which two widely different binding constants were calculated. Likewise, ACE has been used to study other sulfated polysaccharides, e.g., fucoidan, binding to AT and proteins of the complement classical pathway [20, 21]. Binding constants were calculated from the change of migration time and the results indicated that the interaction significantly dependent on the molecular weight of fucoidan.

Dependence on chain length was also the focus of an ACE-based study by Anderot et al. The group studied the interaction of 17 and 3 kDa heparin fractions with human serum albumin (HSA) [27]. In this method, the capillary was injected with different length of heparin plugs, followed by an injection of a constant amount of HSA. During separation, HSA was expected to migrate faster than heparin resulting in mixing during the electrophoretic run. The interaction of heparin with HSA decreased the mobility of HSA and increased its migration time. The K_Ds of the two HSA–heparin complexes (17 and 3 kDa) were found to be 33 μM and 504 μM, respectively. Likewise, the interaction between aminoacridone (AMAC)-labeled hyaluronan (HA) and hyaluronan-binding proteins (HABPs) was investigated by ACE with laser-induced fluorescent detection (LIF) [28]. Different sizes of AMAC-labeled HA oligomers, e.g., HA_4 to HA_{22}, were found to separate well due to their interaction with HABP present in the run buffer. Whereas the HA–fibrinogen interaction led to a retardation of HA migration time arising from the less negatively charged HA–fibrinogen complex, the interaction between HA and HABP from bovine nasal cartilage resulted in loss of HA peak. Inhibition studies were carried out by injecting AMAC-labeled HA_{20} with HABP and unlabeled HA oligosaccharides of different chain lengths in the run buffer. The unlabeled HA competed with the AMAC-labeled HA_{20} for binding to HABP. Results showed that high molecular weight HA oligomers were good inhibitors and the decasaccharide was the minimum size necessary for complete recognition. ACE provided a more precise assessment of HA binding by HABPs.

Besides measuring affinities of different GAGs, ACE can be exploited in mechanistic studies too, as exemplified by the Desai group. To devise an alternative anticoagulation approach based on AT, the Desai group initiated a synthetic medicinal chemistry program and designed several small, nonsugar, highly sulfated, aromatic molecules [58–63]. An interesting feature of these highly sulfated, aromatic molecules is that they simultaneously possess both hydrophobic and anionic characters [64]. The group investigated the interaction of four sulfated tetrahydroisoquinolines with AT by ACE [26] and demonstrated the ability of ACE to differentiate different modes of binding for structurally different antithrombin activators. The binding affinities between three tetrahydroisoquinoline-based polysulfated molecules and AT were found in the range of 40–60 µM in 20 mM sodium phosphate buffer, pH 7.4. For a pentasulfated molecule, a biphasic profile with affinities of 4.7 and 30 µM was observed. An interesting component of the study was the elucidation of the nature of forces that play a role in AT recognition. When the affinities were measured at varying salt concentrations, nearly 44 % of binding energy was found to arise from nonionic forces (Fig. 1). Competitive binding studies were also carried out and showed that the sulfated tetrahydroisoquinolines do not compete with high affinity heparin pentasaccharide. In contrast, the affinity of these tetrahydroisoquinoline derivatives decreased dramatically in the presence of an extended heparin-binding site ligand. ACE has been exploited a lot for antithrombin–heparin studies including the study on AT binding to heparin from different sources (porcine, bovine, and ovine mucosa) [29]. The study measured affinities between 16 heparin samples and AT (K_Ds 14.2–56.1 nM) and found good correlation with in vitro anticoagulant activity.

Overall, ACE is a fairly valuable and convenient approach for studying GAG–protein interactions. It affords faster analysis and small sample consumption and is highly suitable for medium throughput screening of interactions.

3.2 Capillary Zone Electrophoresis (CZE) of GAG–Protein Interactions

In CZE, different ratios of the protein and ligand are pre-equilibrated and then separated by CE. The expectation is that the complex and free analyte resolve from each other due to their different electrophoretic mobilities. CZE is normally used for slow interaction and stable complex systems. The Scatchard equation 3 is used to calculate the binding constant and stoichiometry of binding "n" through the measurement of peak height or peak area.

Heegaard et al. investigated the interaction between heparin and synthetic peptides using CZE conditions and measured dissociation constants [32]. It was an early demonstration of the suitability of CZE to study the interactions involving complex acidic polysaccharides. In this study, peptides were held at constant concentration and the concentration of heparin was varied. The peak height and peak area of peptide decreased, which indicated the formation of a stable complex, although it could not migrate through the detection window under normal polarity electrophoresis at pH 2.5. If the free protein signal is found to be sharp and symmetric, it indicates little dissociation of the complex. This implies that interaction parameters can also be calculated from the free protein peak. Although this initial work was performed using a pH 2.5 buffer, which may not be considered physiological, the group's later work using a phosphate buffer at pH 7.5 gave essen-

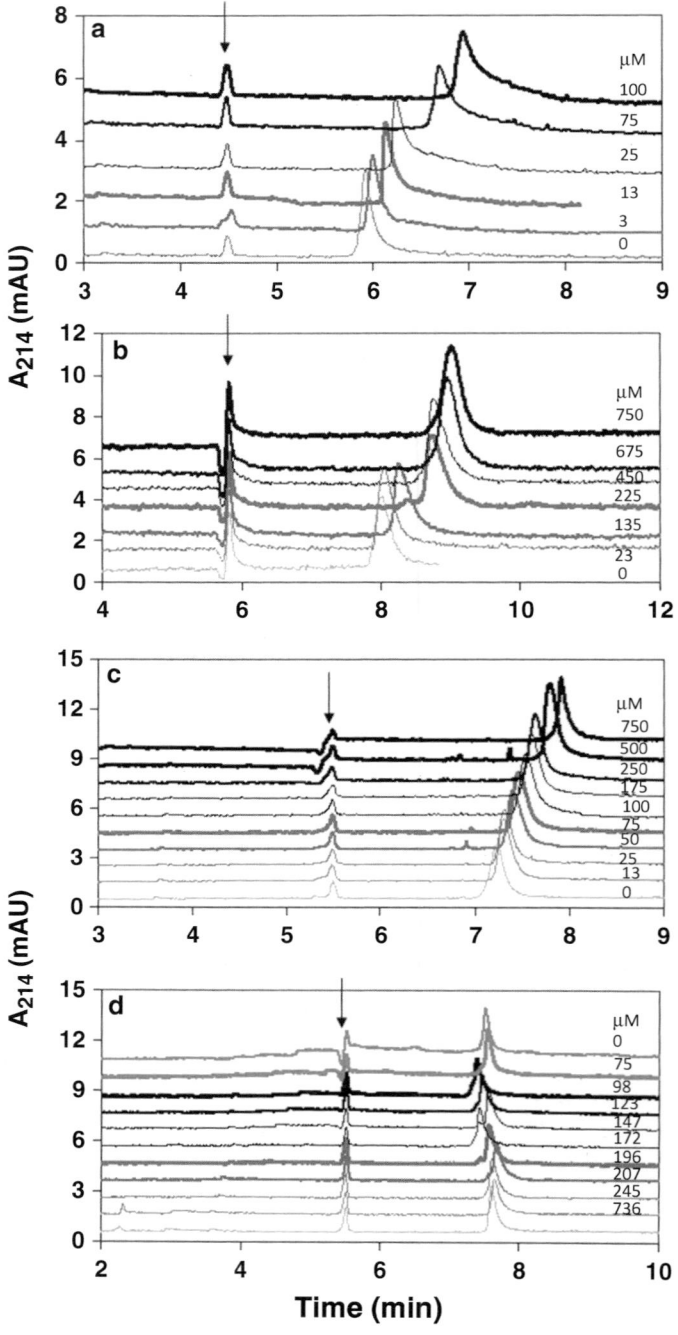

Fig. 1 Electropherograms of 3.4 μM AT in 20 mM sodium phosphate buffer, pH 7.4, containing 0 (Panel **a**), 10 (Panel **b**), 20 (Panel **c**) or 50 mM (Panel **d**) NaCl in the presence of varying concentrations of IAS$_5$, which are displayed on the *right hand side*. Naphthol (*arrow*) was used as a neutral marker. Experimental conditions: Capillary: 40.2 cm total; 30 cm effective length; 50 μm I.D.; Applied voltage: 10 kV; Injection: 4 s at 0.5 psi pressure; Detection: 214 nm (Reproduced with permission from ref. 26)

tially similar results [33]. The Heegaard group has demonstrated the general applicability of CZE in studying heparin–peptide interactions, especially of high affinity, under a variety of conditions [16]. No peaks corresponding to the complex(es) was(were) observed in these studies. Most probably the complex migrated much slower than the electrophoretic rate resulting in lack of observation in the detection window. Another reason could be that the highly heterogeneous heparin gives multiple complexes present in small proportions.

Dimitrellos et al. developed a new method to investigate the interactions of GAG–protein interactions [31]. Heparin–BSA conjugate was synthesized by reductive amination and its interaction with bFGF was investigated to calculate the binding constant. The heparin–BSA conjugate displayed good UV absorbance, which was used to investigate the purity of heparin as well as identify bFGF-interacting heparin preparations of biopharmaceutical importance.

Using CZE, Liang et al. [38, 39] have demonstrated that heparin, carrageenan, and dextran sulfate interact with a hematopoietic growth factor, granulocyte colony-stimulating factor (G-CSF), and have potential therapeutic effect on cancers through inhibition of growth and induction of differentiation of leukemia cells. These interactions are dependent on the molecular weight and location of the sulfate groups on the chain. Interestingly, the number of carrageenan binding sites increased with sulfate content, but exactly opposite was found for dextran sulfate. The latter was suggested to arise from steric effects. G-CSF was found to minimally require a chain with a molecular weight of 1,000 Da [40]. Likewise, N-desulfated and 2,3-O-desulfated heparin were studied for G-CSF binding. The results showed that the former had an affinity similar to heparin for G-CSF (Fig. 2), but 2,3-O-desulfation decreased the affinity ~1,000-fold [42]. Likewise, Liang et al. used CZE in the study of heparin/low molecular weight heparin (LMWH) with interleukin-2 [39] and granulocyte-macrophage colony-stimulating factor [41]. The results demonstrated that the interaction between IL-2 and LMWH falls under the "fast on–off" kinetics regime since the peak height decreases, but the peak area essentially keeps the same (Fig. 3). This study demonstrated that CZE might be used to study not only slow on-and-off rate interactions, but also fast ones. The binding constant can be calculated easily and the method appears to be useful for a wide range of heparin–protein interactions.

Militsopoulou et al. developed a simple CZE method for identification of HS oligosaccharides interacting with bFGF [35]. Variably sulfated unsaturated HS di- and oligosaccharides were incubated with bFGF and then resolved using 50 mM phosphate, pH 3.5, as the run buffer under reverse polarity conditions (30 kV) with detection at 232 nm. In a similar manner, Ling et al. studied the interactions of heparin with programmed cell death 5 (PDCD5) and

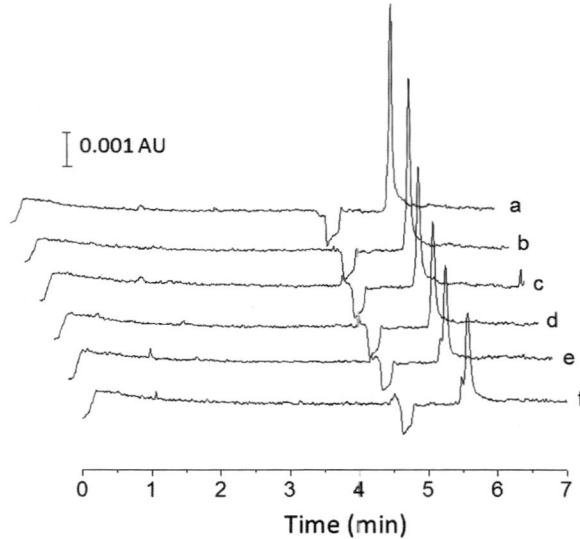

Fig. 2 Electropherograms of 0.083 g/L G-CSF mixed with various concentrations of N-desulfated heparin. (a) 0 g/L, (b) 4.17 g/L, (c) 7.50 g/L, (d) 10.42 g/L, (e) 12.50 g/L and (f) 18.75 g/L. CE conditions: injection, 0.5 psi for 4 s; detection wavelength, 210 nm; applied voltage, 8 kV; capillary, uncoated capillary of 31.2 cm (effective length 21 cm) with an ID of 50 μm (Reproduced with permission from ref. 42)

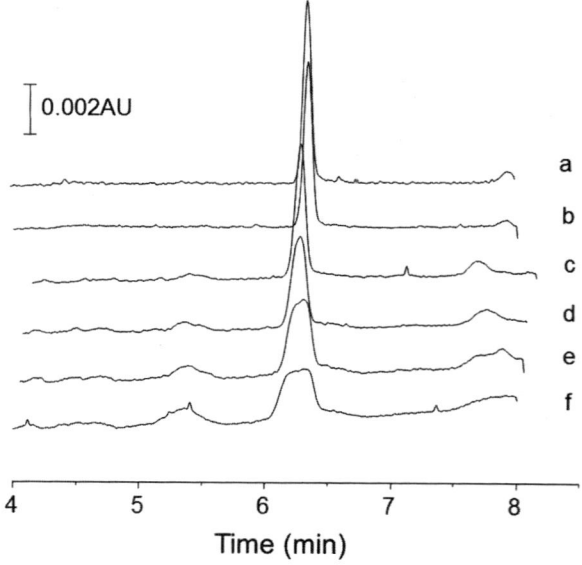

Fig. 3 Electropherograms of 0.1 g/L IL-2 mixed with various concentrations of LMWH. (a) 0 g/L. (b) 2.0 g/L. (c) 10.0 g/L. (d) 16.7 g/L. (e) 25.0 g/L. (f) 33.3 g/L. Experimental conditions: Capillary: 31.2 cm total length; 21 cm effective length; 50 μm I.D.; Applied voltage: 8 kV; Injection: 4 s at 0.5 psi pressure; Detection: 201 nm; the temperatures of the cartridge and sample room were 25 °C and 20 °C, respectively. (Reproduced with permission from ref. 39)

its related peptides using CZE [36]. Heparin-PDCD5 interaction was quantitatively studied, which gave a binding constant of 4.2×10^4 M^{-1}. CZE was further exploited to identify the heparin-binding site on PDCD5 by using PDCD5-related peptides. A related study explored the interaction of peptides isolated from secreted human CKLF1 and heparin using CZE [37], which led to identification of a critical Lys in this interaction. Likewise, a Pro was also found to play an important role. The work highlights the advantages of CZE as a simple and flexible method to study the interactions of heparin with peptides, in particular, and GAGs with proteins, in general.

In summary, CZE has the advantages of small sample consumption, fast analysis, no need for an internal marker, and applicability of affinity analysis to one or more components of a mixture. The binding constants and binding stoichiometry can also be measured simultaneously, as long as the system exhibits tight binding.

3.3 Frontal Analysis (FA)/Frontal Analysis Continuous Capillary Electrophoresis (FACCE) of GAG–Protein Interactions

In frontal analysis (FA), and its derivative frontal analysis continuous capillary electrophoresis (FACCE), a large plug of pre-equilibrated mixtures of different ratios of receptor and ligand are injected and electrophoresed. Samples are in equilibrium in the large injection plug (~10-fold more than that in a CZE experiment) during CE separation. The free species form plateau peaks and cannot be baseline separated from other peaks because of the high levels injected. However, the plateau heights are proportional to their concentrations in the original sample. As the concentration of the ligand changes in the plug, the concentration of free species will change in the manner related to affinity. In this method, only the protein and the ligand should generate a response and all other species should be silent to give good interaction profile. FA is applicable to weak interactions. Although there are several GAG–protein interactions with high μM to mM affinities, no report has been published on the use of FA probably because the method requires significant amount of samples. FACCE, a derivative method of FA, also quantifies the free protein through the height of the plateau peaks, which is used to calculate the binding constant and stoichiometry. FACCE is often used for interactions associated with polymers. Since the GAG–protein complex is more negatively charged than the free protein, the first plateau is free protein and the second plateau is that of the complex. Compared to FA, the pre-equilibrated mixture of GAG and protein is continually injected throughout the separation [34]. The free protein concentration is constant in the mixing zone (the first plateau region) as the loss of the free protein due to interaction with the GAG is made up by supplying free protein in the continuous injection. Therefore, the complex is always in a steady equilibrium [48]. FACCE is applicable to fast on–off binding interactions.

Hattori et al. investigated the interaction between bovine serum albumin (BSA) and heparin [48] using FACCE. Two plateaus were detected in the electropherogram, one corresponding to free BSA and the other corresponding to the complex. The heights of the two plateaus increased proportionally with BSA concentration, but heparin concentration was fixed suggesting interaction between the two species. The authors studied the binding constants and stoichiometry of binding between BSA and heparin at pH 6.5, 6.8, and 7.0. The affinity increased significantly as the ionic strength of the buffer was reduced suggesting a primarily electrostatic interaction between heparin and BSA. In another study, the group studied the interaction between heparin and β-lactoglobulin (and BSA), and found that the binding constant depends on the flexibility of the polymeric chain and not on the density of the charge [49]. Saux et al. used both ACE and FACCE to investigate the interaction between heparin fragments and AT [50]. ACE was used for study tetrasaccharide–AT interaction, while FACCE was used for studying interactions of higher oligosaccharides with AT. The results indicated that both methods are complementary. Unfractionated (14 kDa) and low molecular weight heparin (5 kDa)–AT interactions were also studied by Dubin et al. [51] using FACCE at various ionic strengths. The work indicated a significant role for long-range, nonspecific, protein–polyelectrolyte electrostatic interactions. Further, a new strategy based on the tandem FACCE–electrospray ionization mass spectrometry (ESI-MS) has been developed by the Daniel group [52] using AT–sulfated pentasaccharide interaction. The free and bound forms of AT were well separated affording a fine method of detecting the complex by ESI-MS in positive ionization mode under nondenaturing conditions. The measurement of the complex mass allowed accurate determination of the stoichiometry of the complex. The FACCE–ESI-MS method appears to be more sensitive than the traditional protocol based on zone analysis. This strategy is likely to open a more effective pathway to ligand fishing experiments with heterogeneous GAG mixtures, which will benefit greatly from accurate characterization of bound GAG sequences.

3.4 Capillary Electrokinetic Chromatography (CEC) of GAG–Protein Interactions

CEC is an affinity-based method similar to the affinity column chromatography. The affinity ligand is immobilized or covalently bonded onto the inner surface of the capillary, or printed on the column material or gel. The analyte, which binds to the affinity ligand migrates slower and elutes later, while most other species with no interaction with the ligand elute much faster. In CEC, the migration rate of the affinity ligand is zero, which maximizes the interaction. Despite its simplicity, CEC has some disadvantages including (1) the nature of the tether between the ligand and the capillary surface interfering with the recognition process, (2) the difficulty of measuring binding constants because of inaccuracy of knowing the concentration of the active ligand on capillary surface, and (3) the difficulty of regenerating the active surface after completion of the binding phenomenon.

The Linhardt group covalently immobilized heparin on the surface of an etched capillary through a silane spacer [46]. The injected protein, either AT or secretory leukocyte proteinase inhibitor, bound to the immobilized heparin and the authors used it to measure its affinity for heparin. This approach is similar to traditional affinity chromatography, which generally consumes a lot more sample then CEC (mg versus ng). Linhardt's group also immobilized heparin on the inner surface of capillary through biotin-avidin [47]. The EOF decreased after the immobilization of heparin. The system was used to study the interaction between heparin and aFGF. It was found that aFGF peptides that differ only in the stereochemistry of the proline residue can be readily separated using this capillary. Lipponen et al. studied the interaction between heparin and lipoprotein using CEC [45]. Both noncovalent and covalent heparin coatings for APTES-modified silica capillaries were developed and studies on the interaction of heparin with selected peptide fragments of apoB-100, apoE, and low- and high-density lipoproteins (LDL and HDL) with and without apoE were performed. Although the study did not measure the binding constants, the affinities were expressed in terms of retention factors and reduced mobilities. The work showed that heparin interacts strongly with apoB-100 peptide than with apoE peptide and confirmed that the sulfate groups in heparin play an important role in these interactions.

In the following sections, we provide details on studying GAG–protein interactions using two different CE protocols. Detailed step-by-step methods for ACE and CZE, two most common methods used in such studies, are described. Both methods consume very small amount of an analyte compared to other methods.

4   Materials

Prepare solutions in high-purity water (18.2 megohms) and use analytical grade reagents.

1. Sodium hydroxide solution: 0.1–1 M.

2. Hydrochloric acid solution: 0.1–1 M.

3. CE run buffer: 20 or 50 mM sodium phosphate, pH 7.4 (or 6.5, or 5.5) (*see* **Note 13**).

4. GAGs such as heparin, low molecular weight heparin, chondroitin sulfate, and heparin oligosaccharides.

5. 0.22 μm filter.

6. 20 mL syringes or vacuum system for filtering buffers.

7. NaCl.

8. CE Instrument: Beckman (Fullerton, CA) P/ACE MDQ system with PDA detector (*see* **Notes 14** and **15**).

9. Capillary: 50 or 75 µm ID fused silica capillaries with a total length of 31.2 or 40.2 cm and a 5 mm detection window at 21.0 or 30.0 cm length, respectively, from the injection point (*see* **Notes 11, 12** and **16**).

10. Ethanol.

11. Dry ice.

12. Naphthol.

5 Methods

5.1 ACE and CZE

1. Voltage: Typically electrophoresis is performed at a constant voltage of 8 or 10 kV for 31.2 cm and 40.2 cm capillary, respectively (*see* **Notes 1–3**). Temperature: During electrophoresis, the capillary and sample trays are to be held at a constant temperature of 25 °C (*see* **Note 4**).

2. Activation of a new capillary: A sequential wash of 1 M HCl for 20 min, high-purity water for 3 min, 1 M NaOH for 20 min, high-purity water for 3 min, and run buffer for 20 min at 20 psi should be used to activate a new capillary.

3. Activation of capillary for daily use: A sequential wash of 1 M HCl for 10 min, high-purity water for 3 min, 1 M NaOH for 10 min, and high-purity water for 3 min, and run buffer for 10 min at 20 psi is used to activate a capillary every day before the start of analysis.

4. Rinses between runs: Before every electrophoretic run, the capillary is rinsed with the run buffer for 3 min at 20 psi (*see* **Note 5**).

5. Sample preparation: Stock proteins are prepared in the run buffer and diluted to appropriate concentration with the run buffer. Stock GAG solutions are prepared in high-purity water and diluted ($\geq$10-fold dilution) with run buffer as run electrolytes. A saturated stock solution of neutral marker, naphthol, is prepared using high-purity water and diluted 100–300-fold with run buffer (*see* **Notes 6–8**).

6. Experimental Procedure: After turning on the instrument, set up the methods protocol according to the instrument manufacturer's instructions and **step 1** above. Activate the capillary as described in **step 2** or **3** above. Allow sufficient time (10–20 min) for the capillary to equilibrate to the desired temperature and then inject the sample (*see* **Notes 9** and **17**).

7. Sample injection for ACE: Samples containing protein and naphthol (300-fold dilutions from the stock solution) are injected at the anodic end using 0.5 psi pressure for 4 s and detected at the cathodic end at a wavelength of 214 or 280 nm.

8. Sample injection for CZE: Samples containing a constant protein concentration and series of different ligand concentration are injected at the anodic end using 0.5 psi pressure for 4 s and detected at the cathodic end at a wavelength of 214 or 280 nm (*see* **Note 10**).

9. Perform ACE or CZE electrophoresis runs using the experimental settings described in **step 1** above (*see* **Note 18**). Note that for ACE experiments, the signal will arise from the protein alone, although its interaction with the protein will change its migration time. This change in migration time will be used for quantitative analysis of binding parameters (*see* analysis below). For CZE experiments, the peak height and/or peak area of protein will change based on the interaction between the protein and ligand, which will be used in the quantitative analysis of binding parameters (*see* **Note 19**).

10. Analysis of ACE profiles:

    (a) Find the migration time of the protein and neutral marker at different concentrations of the GAG and calculate the electrophoretic mobility of protein at each nonzero concentration of GAG in the run buffer, μ_t^{cp}, using Eq. 7. Likewise, calculate the electrophoretic mobility μ_0^{cp} in the absence of GAG using the same equation.

    (b) Subtract μ_t^{cp} by μ_0^{cp} to get $\Delta\mu^{\text{cp}}$, $\Delta\mu^{\text{cp}} = \mu_t^{\text{cp}} - \mu_0^{\text{cp}}$.

    (c) Plot $\dfrac{\Delta\mu^{\text{cp}}}{C_f}$ versus $\Delta\mu^{\text{cp}}$, perform linear regression and deduce the slope of the line, which will be the negative value of K_A.

11. Analysis of CZE profiles:

    (a) Use at least six concentration of the protein and obtain a calibration curve by plotting peak height or peak area versus protein concentration. Fit it using a regression equation for subsequent use in calculation of free and bound protein concentrations (*see* below).

    (b) After finishing **step 9** above, measure the peak height or peak area for each protein peak at different concentrations of GAG. Use the above calibration equation to calculate the free concentration of protein.

    (c) Subtract the free concentration of the protein from the total concentration of the protein to obtain the bound concentration of protein at each concentration of GAG.

(d) Divide the bound concentration of protein by the total concentration of GAG to obtain r and plot r/C_f versus r. Linear regressional fit to the data will give the slope, which will be the negative value of K_A.

(e) Divide the intercept by K_A will give the value of binding site n.

6   Notes

1. Voltage is determined by the length of capillary, diameter of the capillary, ionic strength of buffer, and purpose of separation. A 100–600 V/cm can be used and 250 V/cm is commonly used. High ionic strength buffer may limit the use of high voltage because of Joule Heating. Use of a smaller diameter capillary, lower ionic strength buffer, and cooling of the capillary may allow the use of high voltage.

2. If constant voltage does not give good reproducibility, constant current can be used and may give better reproducibility.

3. For choosing the capillary: the length of the capillary is based on the best separation of the sample. If only one protein is going to be analyzed, use the shortest capillary possible for best reproducibility and fastest analysis. If more analytes need to be analyzed, use the capillary that can achieve the best separation. Do not use a capillary longer than you need because longer the capillary, longer the migration time, lower the reproducibility, and more adsorption of the analyte to the capillary wall. This may make the system nonworkable. The diameter of the capillary is also important. A 50 or 75 µm I.D. is commonly used. The smaller the capillary, better the resolution. However, this may require higher concentration of analyst.

4. The temperatures of capillary and sample trays can be different based on the needs of the analysis. The temperature can be typically held constant from 4 to 37 °C. If protein is very sensitive to temperature, 4 °C is typically used for the sample tray.

5. If reproducibility is not good when washing only with buffer in between of runs, a sequence washing with 0.1 M HCl for 3 min, high-purity water for 3 min, 0.1 M NaOH for 3 min, and high-purity water for 3 min, and run buffer for 3 min should be used.

6. Make stock solutions of protein aliquot into microtubes, flash freeze aliquots using dry ice and ethanol, and store at −80 °C. Remove aliquots and discard after use. Do not re-freeze aliquots.

7. All buffers should be filtered through 0.22 µm filter.

8. Other neutral markers, such as dimethylformamide, dimethyl-sulfoxide, and mesityl oxide are also often used to indicate the EOF.

9. Methods for new capillary activation, daily capillary activation, and analysis can be developed, saved, and easily used later. A sequence can also be set for most of the CE instruments so to automate the collection of data at different concentrations of the protein and/or GAGs using the same method. Different methods can also be used in a sequence for different samples. Remember to set up the inlet and outlet vials for each run in a sequence. An important thing to remember is the use of sample volumes more than the minimum (usually 10 μL). Otherwise, sample evaporation may affect the analysis.

10. Protein concentrations in a binding study are normally range from 0.1 to 10 μM depending on the extinction coefficient of each protein at the detection wavelength. The concentration of the ligand depends on the binding affinity between the protein and the ligand. The tighter the binding the lower the ligand concentration and vice versa. Normally, the ligand concentration should range from 0 to 10 times the dissociation constant K_D of the protein–ligand complex.

11. Coatings on capillary windows should be removed completely and the window should not be too big, otherwise, the capillary is very easy to break. A window maker will help do a better job on making the detection windows on capillaries.

12. Do not use scissors to cut capillary. Use fixed blade or ceramic cleaning tool to cut the capillary for even cutting.

13. Run buffer should be refilled for every three runs for better reproducibility.

14. Electrodes and areas around electrodes should be cleaned periodically.

15. Pay attention to electrodes bent or capillary bent between runs and correct them immediately. Otherwise, the electrodes and capillary may be broken in the next run.

16. If a capillary is not going to be used for a while, wash the capillary with 1 M HCl, deionized water, 1 M NaOH, deionized water, respectively, followed by pressure drying. This can be obtained by putting a clean empty vial for both the inlet and outlet, and then apply 20 psi pressure from inlet to outlet for 5–10 min. The water in the capillary should be pushed out and dried.

17. Turn the CE on for at least half an hour before running your sample for better reproducibility. This time period can be used to activate the capillary by sequential washing with 1 M HCl, deionized water, 1 M NaOH, deionized water, buffer, respectively.

18. It is better to finish an entire interaction experiment in 1 day (both standards and different concentrations of samples).

19. The protein–GAG interaction can be normally followed by holding the protein concentration constant and varying the GAG concentration or vice versa. It also can be achieved by changing the ratio of protein and GAG concentration. Since most GAGs lack UV absorptivity, the protein concentration is usually held constant and the GAG concentrations varied.

Acknowledgments

This work was supported by grants SC EPSCoR/IDeA and SCICU to AL and grants HL090586 and HL107152 from the National Institutes of Health to URD. We thank Ms. Yingzi Jin of VCU for helping with the preparation of this chapter.

References

1. Mulloy B, Linhardt RJ (2001) Order out of complexity—protein structures that interact with heparin. Curr Opin Struct Biol 11: 623–628

2. Yamada S, Sakamoto K, Tsuda H, Yoshida K, Sugiura M, Sugahara K (1999) Structural studies of octasaccharides derived from the low-sulfated repeating disaccharide region and octasaccharide serines derived from the protein linkage region of porcine intestinal heparin. Biochemistry 38:838–847

3. Jiao QC, Liu Q, Sun C, He H (1999) Investigation on the binding site in heparin by spectrophotometry. Talanta 48:1095–1101

4. Gallagher JT, Lyon M (2000) Heparan sulfate (molecular structure and interactions with growth factors and morphogens). In: Lozzo RV (ed) Proteoglycans (structure, biology, and molecular interactions). Marcel Dekker, Inc., New York, NY, pp 27–60

5. Whitelock JM, Iozzo RV (2005) Heparan sulfate: a complex polymer charged with biological activity. Chem Rev 105:2745–2764

6. Dong J, Peters-Libeu CA, Weisgraber KH, Segelke BW, Rupp B, Capila I, Hernaiz MJ, LeBrun LA, Linhardt RJ (2001) Interaction of the N-terminal domain of apolipoprotein E4 with heparin. Biochemistry 40:2826–2834

7. Fromm JR, Hileman RE, Caldwell EEO, Weiler JM, Linhardt RJ (1995) Differences in the interaction of heparin with arginine and lysine and the importance of these basic-amino-acids

in the binding of heparin to acidic fibroblast growth-factor. Arch Biochem Biophys 323: 279–287

8. Heegaard NHH, De Lorenzi E (2005) Interactions of charged ligands with beta (2)-microglobulin conformers in affinity capillary electrophoresis. Biochim Biophys Acta 1753:131–140

9. Heegaard NHH (1998) Capillary electrophoresis for the study of affinity interactions. J Mol Recognit 11:141–148

10. Heegaard NHH, Mortensen HD, Roepstorff P (1995) Demonstration of a heparin-binding site in serum amyloid-P component using affinity capillary electrophoresis as an adjunct technique. J Chromatogr A 717:83–90

11. Heegaard NHH (1999) Microscale characterization of the structure-activity relationship of a heparin-binding glycopeptide using affinity capillary electrophoresis and immobilized enzymes. J Chromatogr A 853:189–195

12. Heegaard NHH, Heegaard PMH, Roepstorff P, Robey FA (1996) Ligand-binding sites in human serum amyloid P component. Eur J Biochem 239:850–856

13. Heegaard NHH (1998) A heparin-binding peptide from human serum amyloid P component characterized by affinity capillary electrophoresis. Electrophoresis 19:442–447

14. Bohlin ME, Kogutowska E, Blomberg LG, Heegaard NHH (2004) Capillary electrophoresis- based analysis of phospholipid and

glycosaminoglycan binding by human beta(2)-glycoprotein. J Chromatogr A 1059:215–222

15. Gunnarsson K, Valtcheva L, Hjerten S (1997) Capillary zone electrophoresis for the study of the binding of antithrombin to low-affinity heparin. Glycoconj J 14:859–862

16. Heegaard NHH, Nilsson S, Guzman NA (1998) Affinity capillary electrophoresis: important application areas and some recent developments. J Chromatogr B 715:29–54

17. Heegaard NHH, Nissen MH, Chen DDY (2002) Applications of on-line weak affinity interactions in free solution capillary electrophoresis. Electrophoresis 23:815–822

18. McKeon J, Holland LA (2004) Determination of dissociation constants for a heparin-binding domain of amyloid precursor protein and heparins or heparan sulfate by affinity capillary electrophoresis. Electrophoresis 25:1243–1248

19. Liu JP, Abid S, Hail ME, Lee MS, Hangeland J, Zein N (1998) Use of affinity capillary electrophoresis for the study of protein and drug interactions. Analyst 123:1455–1459

20. Varenne A, Gareil P, Colliec-Jouault S, Daniel R (2003) Capillary electrophoresis determination of the binding affinity of bioactive sulfated polysaccharides to proteins: study of the binding properties of fucoidan to antithrombin. Anal Biochem 315:152–159

21. Tissot B, Montdargent B, Chevolot L, Varenne A, Descroix S, Gareil P, Daniel R (2003) Interaction of fucoidan with the proteins of the complement classical pathway. Biochim Biophys Acta 1651:5–16

22. Heegaard NHH, He X, Blomberg LG (2006) Binding of Ca^{2+}, Mg^{2+}, and heparin by human serum amyloid P component in affinity capillary electrophoresis. Electrophoresis 27:2609–2615

23. Hamazaki H (1987) Ca^{2+} mediated association of human-serum amyloid-P component with heparan-sulfate and dermatan-sulfate. J Biol Chem 262:1456–1460

24. Li XA, Hatanaka K, Guo L, Kitamura Y, Yamamoto A (1994) Binding of serum amyloid-P component to heparin in human serum. Biochim Biophys Acta 1201:143–148

25. Heegaard NHH, Hansen SI, Holm J (2006) A novel specific heparin-binding activity of bovine folate-binding protein characterized by capillary electrophoresis. Electrophoresis 27: 1122–1127

26. Liang A, Raghuraman A, Desai UR (2009) Capillary electrophoretic study of small, highly sulfated, non-sugar molecules interacting with antithrombin. Electrophoresis 30:1544–1551

27. Anderot M, Nilsson M, Vegvari A, Moeller EH, Weert M, Isaksson R (2009) Determination of dissociation constants between polyelectrolytes and proteins by affinity capillary electrophoresis. J Chromatogr B 877:892–896

28. Kinoshita M, Kakehi K (2005) Analysis of the interaction between hyaluronan and hyaluronan-binding proteins by capillary affinity electrophoresis: significance of hyaluronan molecular size on binding reaction. J Chromatogr B 816: 289–295

29. Gotti R, Parma B, Spelta F, Liverani L (2013) Affinity capillary electrophoresis in binding study of antithrombin to heparin from different sources. Talanta 105:366–371

30. Heegaard NH, Roepstorff P, Melberg SG, Nissen MH (2002) Cleaved beta 2- microglobulin partially attains a conformation that has amyloidogenic features. J Biol Chem 277: 11184–11189

31. Dimitrellos V, Lamari FN, Militsopoulou M, Kanakis I, Karamanos NK (2003) Capillary electrophoresis and enzyme solid phase assay for examining the purity of a synthetic heparin proteoglycan-like conjugate and identifying binding to basic fibroblast growth factor. Biomed Chromatogr 17:42–47

32. Heegaard NHH, Robey FA (1992) Use of capillary zone electrophoresis to evaluate the binding of anionic carbohydrates to synthetic peptides derived from human serum amyloid-P component. Anal Chem 64:2479–2482

33. Hernaiz MJ, LeBrun LA, Wu Y, Sen JW, Linhardt RJ, Heegaard NHH (2002) Characterization of heparin binding by a peptide from amyloid P component using capillary electrophoresis, surface plasmon resonance and isothermal titration calorimetry. Eur J Biochem 269:2860–2867

34. Guijt-van Duijn RM, Frank J, van Dedem GWK, Baltussen E (2000) Recent advances in affinity capillary electrophoresis. Electrophoresis 21:3905–3918

35. Militsopoulou M, Lamari F, Karamanos NK (2003) Capillary electrophoresis: a tool for studying interactions of glycans/proteoglycans with growth factors. J Pharm Biomed Anal 32:823–828

36. Ling X, Liu Y, Fan H, Zhong Y, Li D, Wang Y (2007) Studies on interactions f programmed cell death 5 (PDCD5) and its related peptides with heparin by capillary zone electrophoresis. Anal Bioanal Chem 387:909–916

37. Liu Y, Zhang S, Ling X, Li Y, Zhang Y, Han W, Wang Y (2008) Analysis of the interactions

between the peptides from secreted human CKLF1 and heparin using capillary zone electrophoresis. J Pept Sci 14:984–988

38. Liang A, He X, Du Y, Wang K, Fung Y, Lin B (2004) Capillary zone electrophoresis investigation of the interaction between heparin and granulocyte-colony stimulating factor. Electrophoresis 25:870–875

39. Liang A, He X, Du Y, Wang K, Fung Y, Lin B (2005) Capillary zone electrophoresis characterization of low molecular weight heparin binding to interleukin 2. J Pharm Biomed Anal 38:408–413

40. Liang A, Chao Y, Liu X, Du Y, Wang K, Qian S, Lin B (2005) Separation, identification, and interaction of heparin oligosaccharides with granulocyte-colony stimulating factor using capillary electrophoresis and mass spectrometry. Electrophoresis 26:3460–3467

41. Liang A, Du Y, Wang K, Lin B (2006) Quantitative investigation of interaction between granulocyte-macrophage colony-stimulating factor and heparin by capillary zone electrophoresis. J Sep Sci 29:1637–1641

42. Liang A, Liu X, Du Y, Wang K, Lin B (2008) Further characterization of the binding of heparin to granulocyte colony-stimulating factor: importance of sulfate groups. Electrophoresis 29:1286–1290

43. Liang A, Zhou X, Wang Q, Liu X, Qin J, Du Y, Wang K, Lin B (2006) Interactions of dextran sulfates with granulocyte colony-stimulating factor and their effects on leukemia cells. Electrophoresis 27:3195–3201

44. Liang A, Zhou X, Wang Q, Liu X, Liu X, Du Y, Wang K, Lin B (2006) Structural features in carrageenan that interact with a heparin-binding hematopoietic growth factor and modulate its biological activity. J Chromatogr B 843:114–119

45. Lipponen K, Liu Y, Patricia WS, Oorni K, Kovanen PT, Riekkola M (2012) Capillary electrochromatography and quartz crystal microbalance, valuable techniques in the study of heparin-lipoprotein interactions. Anal Biochem 424:71–78

46. Wu XJ, Linhardt RJ (1998) Capillary affinity chromatography and affinity capillary electrophoresis of heparin binding proteins. Electrophoresis 19:2650–2653

47. VanderNoot VA, Hileman RE, Dordick JS, Linhardt RJ (1998) Affinity capillary electrophoresis employing immobilized glycosaminoglycan to resolve heparin-binding peptides. Electrophoresis 19:437–441

48. Hattori T, Kimura K, Seyrek E, Dubin PL (2001) Binding of bovine serum albumin to heparin determined by turbidimetric titration and frontal analysis continuous capillary electrophoresis. Anal Biochem 295:158–167

49. Hattori T, Kimura K, Seyrek E, Dubin PL (2001) The use of frontal analysis continuous capillary electrophoresis to compare protein binding by natural and synthetic polyelectrolyte. Anal Sci 17:93–95

50. Saux TL, Varenne V, Perreau F, Siret L, Duteil S, Duhau L, Gareil P (2006) Determination of the binding parameters for antithrombin-heparin fragment systems by affinity and frontal analysis continuous capillary electrophoresis. J Chromatogr A 1132:289–296

51. Seyrek E, Dubin PL, Henriksen J (2007) Nonspecific electrostatic binding characteristics of the heparin-antithrombin interaction. Biopolymers 86:249–259

52. Fermas S, Gonnet F, Varenne A, Gareil P, Daniel R (2007) Frontal analysis capillary electrophoresis hyphenated to electrospray ionization mass spectrometry for the characterization of the antithrombin/heparin pentasaccharide complex. Anal Chem 79: 4987–4993

53. He X, Ding Y, Li D, Lin B (2004) Recent advances in the study of biomolecular interactions by capillary electrophoresis. Electrophoresis 25:697–711

54. Scatchard G (1949) The attraction of proteins for small molecules and ions. Ann N Y Acad Sci 51:660–672

55. Keyes RS, Bobst AM (1993) A comparative study of Scatchard-type and linear lattice models for the analysis of EPR competition experiments with spin-labeled nucleic acids and sin. Biophys Chem 45:281–303

56. Klotz IM, Hunston DL (1971) Properties of graphical representations of multiple classes of binding sites. Biochemistry 10:3065–3069

57. Colton IJ, Carbeck JD, Rao J, Whitesides GM (1998) Affinity capillary electrophoresis: a physical-organic tool for studying interactions in biomolecular recognition. Electrophoresis 19:367–382

58. Olson ST, Björk I, Sheffer R, Craig PA, Shore JD, Choay J (1992) Role of the antithrombin-binding pentasaccharide in heparin acceleration of antithrombin-proteinase reactions. Resolution of the antithrombin conformational change contribution to heparin rate enhancement. J Biol Chem 267:12528–12538

59. Desai UR, Petitou M, Björk I, Olson ST (1998) Mechanism of heparin activation of antithrombin. Role of individual residues of the pentasaccharide activating sequence in the recognition of native and activated states of antithrombin. J Biol Chem 273: 7478–7487

60. Gunnarsson GT, Desai UR (2002) Interaction of designed sulfated flavanoids with antithrombin: lessons on the design of organic activators. J Med Chem 45:4460–4470

61. Gunnarsson GT, Desai UR (2002) Designing small, nonsugar activators of antithrombin using hydropathic interaction analyses. J Med Chem 45:1233–1243

62. Gunnarsson GT, Riaz M, Adams J, Desai UR (2005) Synthesis of per-sulfated flavonoids using 2,2,2-trichloro ethyl protecting group and their factor Xa inhibition potential. Bioorg Med Chem 13:1783–1789

63. Gunnarsson GT, Desai UR (2003) Exploring new non-sugar sulfated molecules as activators of antithrombin. Bioorg Med Chem Lett 13: 579–583

64. Raghuraman A, Riaz M, Hindle M, Desai UR (2007) Rapid and efficient microwave-assisted synthesis of highly sulfated organic scaffolds. Tetrahedron Lett 48: 6754–6758

Chapter 29

Histochemical Analysis of Heparan Sulfate 3-*O*-Sulfotransferase Expression in Mouse Brain

Tomio Yabe and Nobuaki Maeda

Abstract

In situ hybridization provides information for understanding the localization of gene expression in various tissues. The relative expression levels of mRNAs in a single cell can be sensitively visualized by this technique. Furthermore, since in situ hybridization is a histological technique, tissue structure is maintained after fixation, and it is possible to accurately identify cell types. We have examined the expression of heparan sulfate sulfotransferases by in situ hybridization to better understand the functions of heparan sulfate in the development of mouse nervous system. This chapter describes methods of in situ hybridization analyses using cRNA probes labeled with nonradioactive nucleotides.

Key words In situ hybridization, cRNA probe, Digoxigenin-11-UTP, Paraffin-embedded section, Heparan sulfate 3-*O*-sulfotransferase

1 Introduction

In situ hybridization as well as Northern blotting is a technique using the hybridization of a labeled nucleic acid probe to a complementary sequence of mRNA. Unlike Northern blotting, however, in situ hybridization is extremely sensitive, and can detect small amounts of mRNA expressed in a single cell. Furthermore, since in situ hybridization is a histological technique, tissue structure is maintained after fixation, and it is possible to accurately identify the cell types expressing specific mRNAs. On the other hand, Northern blotting would only provide information concerning the presence or absence of mRNAs in the tissue samples, because the starting material for Northern blotting is total RNA extracted from tissue homogenate. Here, we describe our current in situ hybridization protocol to detect the expression of heparan sulfate (HS) sulfotransferase genes.

HS is ubiquitously distributed on the cell surface and in the extracellular matrix. HS plays important roles in a number of biological phenomena such as blood coagulation, viral infection,

Kuberan Balagurunathan et al. (eds.), *Glycosaminoglycans: Chemistry and Biology*, Methods in Molecular Biology, vol. 1229, DOI 10.1007/978-1-4939-1714-3_29, © Springer Science+Business Media New York 2015

tumor metastasis, and various developmental processes (for review, *see* refs. 1–5). HS chains show enormous structural diversity, and structurally different HS variants exhibit different affinities for a variety of proteins such as growth factors, enzymes, and extracellular matrix components, thus regulating the biological activities of these proteins in a structure-dependent manner [6].

HS is synthesized in the Golgi apparatus by a sequential modification involving over 20 different enzymes acting in a concerted fashion after the polymerization of repeating disaccharide units consisting of a *N*-acetylglucosamine (GlcNAc) and a glucuronic acid (GlcA) by HS copolymerases, EXTs. In this process, the polymerized chains are partly *N*-deacetylated and *N*-sulfated by glucosaminyl *N*-deacetylase/*N*-sulfotransferases (NDSTs), leading to the production of regions that remain unmodified (NA domain), domains with contiguous *N*-sulfated sequences (NS domain), and mixed sequences of *N*-acetylated and *N*-sulfated disaccharides (NA/NS domain). Subsequently, some GlcAs are C5 epimerized to iduronic acid (IdoA) by glucuronyl C5-epimerase. After that, *O*-sulfation reactions occur at various positions: 2-*O*-sulfation of GlcA/IdoA by HS 2-*O*-sulfotransferase (2-OST), and 6- and 3-*O*-sulfation of glucosamine units by 6- and 3-OSTs, respectively (for review, *see* ref. 7). Since C5 epimerization and *O*-sulfation reactions occur in close vicinity of *N*-sulfated units, HS chains produced in this manner have segments that remain relatively unmodified and heavily modified domains.

Most HS sulfotransferases are composed of several isoforms with distinct enzymatic properties, each of which is supposed to be involved in the generation of structurally and functionally distinct HS domains. Therefore, it is considered that specific combinations of HS sulfotransferase isoforms expressed in a cell would lead to the generation of specific functional HS domains in that cell. Recently, we examined expression profiles of the isoforms of 3-OST (3-OST-1, –2, –3B, and –4), NDST (NDST-1, –2, –3, and –4), and 6-OST (6-OST-1, –2, and –3) in the developing mouse cerebral cortex and cerebellum using in situ hybridization [8]. We observed that the expression of these HS sulfotransferase genes was dynamically regulated during brain development. Furthermore, the various types of neurons expressed distinct but partially overlapping combinations of these genes. These findings suggested that different types of neurons synthesize HS chains with different structures, which are also dynamically altered during development. Unexpectedly, we found that multiple isoforms of each sulfotransferase were often expressed by the same neurons at the same time. This suggests that neurons synthesize complex HS chains with many different functional domains generated by multiple HS sulfotransferase isoforms. However, it should be noted that neurons have highly polarized structure with functionally distinct regions, such as dendrites, axons, cell bodies, spines, and axon terminals.

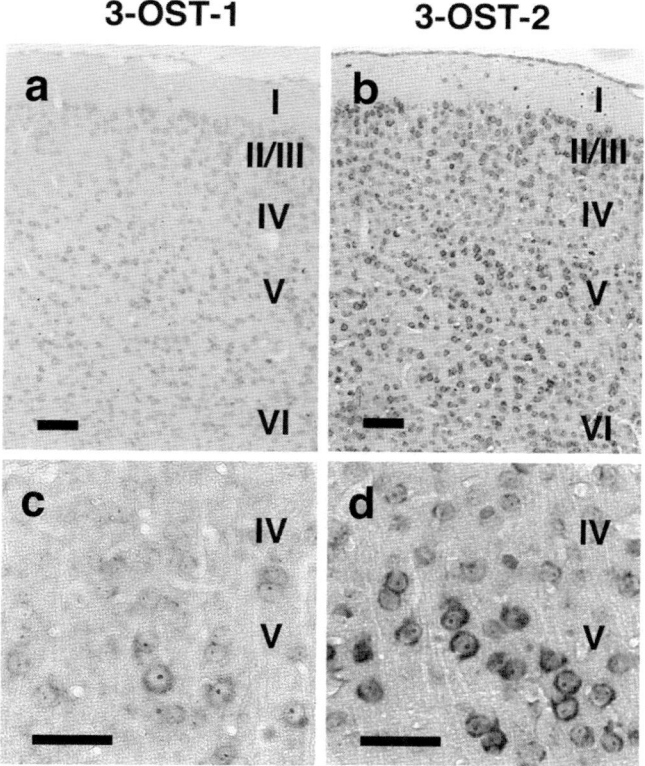

Fig. 1 In situ hybridization analysis of the expression of 3-OST-1 and 3-OST-2 mRNA in the adult cerebrum. 3-OST-1 (**a**) was weakly expressed in the cerebral cortex. 3-OST-2 (**b**) was strongly expressed in the cells in layer V. High magnification of the cerebral cortex showed that 3-OST-1 (**c**) and 3-OST-2 (**d**) were clearly expressed by the pyramidal neurons in layer V. *I–VI* cortical layers I-VI; scale bars, 100 μm (**a** and **b**) and 50 μm (**a** and **d**)

It is considered that development of these distinct parts is regulated by different signaling pathways, which might require structurally different HS chains. Thus, there is a possibility that structurally different HS chains generated by multiple isoforms in one neuron might be delivered to these distinct neuronal parts. Future fine structural analyses are necessary to evaluate whether distinct regions in one neuron express structurally different HS chains.

Numerous studies on the critical roles of HS in various biological events have stimulated interests in elucidating their detailed structures. In situ hybridization analyses of HS sulfotransferase genes provide useful insights about the structural and functional aspects of HS chains. However, there is a methodological limitation as described above. Combination of in situ hybridization and immunohistochemical analyses using specific anti-HS antibodies would be very useful. In this chapter, we present detailed protocol for in situ hybridization analysis of the expression of HS 3-OST in mouse brain (Fig.1.).

2 Materials

2.1 Animals

1. BALB/c mice.
2. The mating day is considered embryonic day 0 (E0), and the day of birth is considered postnatal day 0 (P0).

2.2 Stock Solutions

1. 50× TAE buffer: 2 M Tris, 1 M acetic acid, 50 mM EDTA, pH 8.0.
2. Ethidium bromide: 0.44 mg/mL.
3. 5 M NaCl.
4. DEPC-water: water treated with 1 % diethyl pyrocarbonate (DEPC).
5. 0.5 M ethylenediaminetetraacetic acid (EDTA).
6. 3 M sodium acetate, pH 5.2.
7. Phosphate buffer (PB): 0.4 M sodium phosphate, pH 7.4.
8. 1 M HCl.
9. 1 M Tris–HCl, pH 7.5 and pH 8.0.
10. 20× standard saline citrate (SSC): 3 M NaCl and 0.3 M sodium citrate, pH 7.0.
11. Maleic acid buffer: 0.1 M maleic acid, 0.15 M NaCl; adjust with NaOH to pH 7.5.

2.3 Preparation of Hybridization RNA Probes

1. Mouse brain 5′-stretch plus cDNA library (Clontech, BD Biosciences Clontech, Mountain View, CA).
2. DNA agarose gel electrophoresis: 1.5 % agarose gel in 0.5× TAE buffer.
3. NucleoTrap (Takara Bio Inc., Otsu, Japan).
4. PCR Polishing Kit (Stratagene, La Jolla, CA).
5. *Pfu* DNA polymerase.
6. pPCR-Script Amp SK(+) cloning vector (Stratagene).
7. PCR-Script Amp Cloning Kit (Stratagene).
8. *Escherichia coli* XL10-Gold Kan^r ultracompetent cell (Stratagene).
9. Restriction enzyme: *Bam* HI, *Not* I.
10. Phenol/chloroform/isoamyl alcohol (PCI, 25:24:1).
11. Digoxigenin-11-UTP (DIG) RNA Labeling Kit (Roche Diagnostics GmbH, MannHeim, Germany).
12. T7 RNA polymerase.
13. T3 RNA polymerase.
14. DNase I (RNase-free).
15. RNA sizing solution: 40 mM $NaHCO_3$, 60 mM Na_2CO_3.

16. 1 M dithiothreitol (DTT).

17. Yeast RNA (10 mg/mL).

18. RNase inhibitor.

2.4 Paraffin Processing of Tissue

1. 4 % Paraformaldehyde (PFA): prepare as follows (for 250 mL) just before use. Dissolve 10 g of PFA in 62.5 mL of 0.4 M PB and 170 mL of distilled water with heat and stir for 2 h. After cooling down on ice, adjust the volume to 250 mL with distilled water and then filtrate.

2. Ethanol/toluene (1:1).

3. Paraffin: melting point 51–53 °C.

4. Cassette for paraffin embedding.

5. FINE FROST glass slide (Matsunami Glass Ind., Ltd., Osaka, Japan).

2.5 In Situ hybridization

1. Proteinase K: 20 µg/mL of proteinase K in 0.1 M Tris–HCl, pH 7.5 containing 50 mM EDTA.

2. Salmon sperm DNA (10 mg/mL).

3. Yeast tRNA (10 mg/mL).

4. Denhardt's solution (50×): 1 % ficoll, 1 % polyvinylpyrrolidone, and 1 % bovine serum albumin in sterile distilled water.

5. Hybridization solution: 10 ng RNA probe (1 µg/mL), 500 µg/mL yeast tRNA, 1× Denhardt's solution, 50 % formamide, 10 % dextran sulfate, 125 µg/mL salmon sperm DNA, and 4× SSC.

6. RNase A.

2.6 Detection of Signals

1. DIG Nucleic Acid Detection Kit (Roche Diagnostics).

2. Washing buffer: Maleic acid buffer with 0.3 % (v/v) Tween 20.

3. Blocking reagent: dissolve Blocking reagent (in DIG Nucleic Acid Detection Kit) in Maleic acid buffer to a final concentration of 10 % (w/v) with shaking and heating in a microwave oven. Autoclave stock solution.

4. Blocking solution: prepare a 1× working solution by diluting 10× Blocking reagent 1:10 with Maleic acid buffer.

5. Antibody solution: dilute anti-digoxigenin-AP conjugate (in DIG Nucleic Acid Detection Kit) 1:1,000 in Blocking solution.

6. Detection buffer: 0.1 M Tris–HCl, 0.1 M NaCl, pH 9.5.

7. Color substrate solution: add 200 µL of 5-bromo-4-chloro-3-indolyl phosphate and *p*-nitroblue tetrazolium (BCIP/NBT) stock solution (in DIG Nucleic Acid Detection Kit) to 10 mL of Detection buffer.

8. Crystal/Mount aqueous mounting medium (Biomeda, Foster City, CA).

3 Methods

3.1 Primer Design

Primers to make RNA probes for in situ hybridization are designed by the open source program "Primer3" (URL = "http://primer3.sourceforge.net/"). The general strategy is to choose primers within the coding sequence with an optimal length between 20 and 30 nucleotides, a melting temperature between 50 and 60 °C, and a limited number of G and C in the 3′-end of the oligonucleotides. The amplicon length varies between 200 and 250 bp. Table 1 shows genes chosen with their GenBank accession numbers and references. Care should be taken in design to ensure the specificity of the primers (*see* **Note 1**).

3.2 Making Plasmids Containing Target Genes

1. Amplification of gene fragments: heparan sulfate sulfotransferase gene fragments are amplified by polymerase chain reaction (PCR) from mouse brain 5′-stretch plus cDNA library using the primers listed in Table 1.

2. Purification of the amplicon: the amplicon is applied to a DNA agarose gel electrophoresis. After staining a DNA agarose gel with ethidium bromide, DNA band of the amplicon is excised from gel and purified by appropriate purification kit such as NucleoTrap.

3. Generation of blunt-ended DNA fragments from the purified amplicon: the purified amplicon is polished by PCR polishing kit including *Pfu* DNA polymerase (*see* **Note 2**) according to the instruction of the manufacturer.

4. Inserting the amplicon into a cloning vector: the blunt-ended amplicon is inserted into *Srf* I site of pPCR-Script Amp SK(+) cloning vector (10 ng/μL). To prepare the ligation reaction,

Table 1
Primers for HS 3-*O*-sulfotransferase genes

Gene	GenBank	Primers	Ref.
HS 3-OST-1	NM_010474	Fwd: 5′-GTGCGCAAGGGTGGTACCCG-3′ Rev: 5′-GCACTTTGGGCGAAGTGAAATAGG-3′	[11]
HS 3-OST-2	NM_001081327	Fwd: 5′-CCTGATGCCGAGGACCTTGGAG-3′ Rev: 5′-GCATGATCCCGTGCATGCTGAACATG-3′	[12]
HS 3-OST-3B	NM_018805	Fwd: 5′-GTCCCGCGCTACTCACGCTC-3′ Rev: 5′-CGATGATGATGGCCTGCGGCAG-3′	[13]
HS 3-OST-4	NM_001252072	Fwd: 5′-AGAGCTTGCTTGTTAGTTGTTCACAGACAC-3′ Rev: 5′-CAGGTTGCATTAAGCTTTATAACAACA CAGTTTCC-3′	[8, 14]

add the components into a 0.5-mL microcentrifuge tube according to PCR-Script Amp Cloning Kit instruction manual (*see* **Note 3**). Mix the ligation reaction gently and incubate this reaction for 1 h at room temperature. Heat the ligation reaction for 10 min at 65 °C. Store the ligation reaction on ice until ready to use for transformation into a competent cells.

5. Amplifying the constructed plasmid: *E. coli* XL10-Gold Kan[r] ultracompetent cell is transformed by the constructed plasmid for amplifying the gene according to the instruction of the manufacturer. The inserted gene into obtained plasmid is confirmed by restriction enzyme digestion.

3.3 Linearization of the Constructed Plasmids

1. To prepare the antisense and the sense probes for the target genes, the constructed plasmids are linearized by restriction enzymes; *Bam* HI for antisense probe (T7 polymerase) and *Not* I for sense probe (T3 polymerase).

2. PCI extraction: add the same volume of PCI as the digested samples, mix, and centrifuge at $10,000 \times g$ for 5 min. Transfer the aqueous layer to another tube, add equal volume of PCI, mix, and centrifuge at $10,000 \times g$ for 5 min. Tranfer the aqueous layer to another tube, add 0.05 volume of 5 M NaCl and 2.5 volume of ice-cold ethanol, mix, and keep at −80 °C for 1 h. Thaw rapidly to room temperature and centrifuge at $10,000 \times g$ at 4 °C for 30 min. Dissolve precipitates in 10 μL of DEPC-water.

3.4 Transcription and Labeling of RNA Probes

Linearized DNA (1 μg/μL) is applied to DIG RNA Labeling Kit for transcription by T7 or T3 RNA polymerase and labeling with digoxigenin-11-UTP according to the instruction of the manufacturer.

1. Incubate the reaction tube for 2 h at 37 °C.

2. Add 2 μL of DNase I (RNase-free) into the sample and incubate for 15 min at 37 °C.

3. Add 2 μL of 0.2 M EDTA (pH 8.0) to quit degradation of DNA.

4. Store DIG-labeled RNA at −20 °C.

3.5 Alkali Hydrolysis

1. Dissolve DIG-labeled RNA in 95 μL of RNA sizing solution and 5 μL of 1 M DTT, mix, and centrifuge briefly.

2. Incubate at 60 °C for the appropriate time ([9], *see* **Note 4**).

3. Cool in ice and add 10 μL of 3 M sodium acetate (pH 5.2), 5 μL of yeast RNA (10 mg/mL) and 300 μL of ice-cold ethanol, mix, and keep at −80 °C for 1 h.

4. Thaw rapidly to room temperature and centrifuge at $10,000 \times g$ for 30 min at 4 °C.

5. Wash precipitates with 100 μL of ice-cold 70 % ethanol and dry.

6. Dissolve precipitates in 50 μL of DEPC-water, add 1 μL of RNase inhibitor and keep at −20 °C.

3.6 Fixation of Tissues

1. Anesthetize mouse with diethyl ether (*see* **Note 5**).

2. Subject to cardiac perfusion with 4 % PFA in 0.1 M PB (volume is approximately 2/3 of body weight) at room temperature.

3. Dissect out the brain and cut it in the sagittal plane.

4. Postfix in 4 % PFA in 0.1 M PB for 2 h at 4 °C.

5. Transfer fixed tissues to 70 % ethanol and store them at 4 °C overnight.

3.7 Paraffin Infiltration and Embedding Tissues in Paraffin Blocks

Tissue is dehydrated through a series of graded ethanol baths to displace the water and then infiltrated with wax. The infiltrated tissues are then embedded into paraffin wax blocks.

1. Put tissues in a labeled cassette.

2. Soak the tissues in 70 % ethanol for 1 h.

3. Soak the tissues in 95 % ethanol for 1 h.

4. Soak the tissues in first absolute ethanol for 30 min.

5. Soak the tissues in second absolute ethanol for 1 h.

6. Soak the tissues in third absolute ethanol for 1 h.

7. Rinse the tissues in ethanol/toluene (1:1) for 10 min.

8. Soak the tissues three times in toluene for 10 min each.

9. Soak the tissues in first paraffin wax at 55 °C for 30 min.

10. Soak the tissues twice in paraffin wax at 55 °C for 1 h each.

11. Put suitable amount of molten paraffin in embedding tray from paraffin reservoir heated at 58 °C.

12. Using warm forceps, transfer tissue into the tray, placing cut side down.

13. Float the tray on iced water. When the paraffin wax is completely cooled and hardened, sink the tray into iced water. Completed paraffin block will float on iced water within 1 h (*see* **Note 6**).

3.8 Sectioning Tissues

Tissues are sectioned using a microtome. Turn on the paraffin stretching plate and check that the temperature is 37 °C. Set the dial to cut 7 μm sections.

1. Pour 100 μL of DEPC-water onto the surface of clean glass slides such as FINE FROST glass slide.

2. Cut 7-μm-thick sections by a microtome, pick them up with a fine paint brush, and float them on the surface of the DEPC-water on the glass slides.

3. After the sections extend well, take away excessive water from the glass slides.

4. Keep the slides at 37 °C overnight.

5. Keep the slides at 4 °C until use (*see* **Note 7**).

3.9 In Situ Hybridization: Deparaffinization

The slides with paraffin sections must be deparaffinized and rehydrated. Incomplete removal of paraffin can cause poor staining of the section.

1. Place the slides in a rack.

2. Rinse the slides three times in xylene for 5 min each.

3. Rinse the slides twice in absolute ethanol for 5 min each.

4. Rinse the slides in 95 % ethanol for 5 min.

5. Rinse the slides in 70 % ethanol for 5 min.

6. Wash the slides three times in DEPC-water for 3 min each.

7. Rinse the slides in 0.2 M HCl for 20 min at room temperature.

8. Wash the slides three times in DEPC-water for 3 min each.

3.10 In Situ Hybridization: Prehybridization

1. Treat slides with 20 μg/mL of proteinase K in 0.1 M Tris–HCl for 10 min at room temperature (*see* **Notes 8** and **9**).

2. Wash the slides three times in 0.1 M Tris–HCl (pH 7.5) for 3 min each.

3. Wash the slides three times in DEPC-water for 3 min each.

4. Rinse the slides in 70 % ethanol for 3 min.

5. Rinse the slides in 95 % ethanol for 1 min then air dry.

3.11 In Situ Hybridization: Hybridization

Before making the hybridization solution, RNA probe (10 μL) is heated at 80 °C for 5 min and cool on ice. Both the antisense and sense probes are diluted to a final concentration of 1 μg/mL. Salmon sperm DNA is denatured by boiling just before use. Dextran sulfate is heated at 70 °C and add it in the end.

1. Mix the hybridization solution well with pippetter and drop 10 μL of the solution onto each section.

2. Put coverslip on the section and incubate overnight at 53 °C in a moist chamber.

3.12 In Situ Hybridization: Stringency Washing

1. Remove coverslip from the section in 2× SSC for 15 min at 53 °C.

2. Wash the section in 2× SSC for 30 min at 53 °C.

3. Rinse the section twice in 10 mM Tris–HCl (pH 8.0) containing 0.5 M NaCl for 5 min each.

4. Treat the section with 20 μg/mL RNase A in 10 mM Tris–HCl (pH 8.0) containing 0.5 M NaCl for 30 min at 37 °C.

5. Rinse the section twice in 2× SSC.

6. Wash the section in 2× SSC for 1 h at 57 °C (*see* **Note 10**).

7. Wash the section in 1× SSC for 1 h at 57 °C.

8. Wash the section in 0.2× SSC for 1 h at 57 °C.

3.13 Detection of Signals

1. Wash the section in Washing buffer for 5 min.

2. Incubate the section with Blocking solution for 30 min at room temperature.

3. Incubate the section with Antibody solution at 4 °C overnight.

4. Wash the section twice with Washing buffer for 15 min each.

5. Equilibrate in Detection buffer for 5 min at room temperature.

6. Drop freshly prepared Color substrate solution onto the section and incubate for 6 h at room temperature in the dark.

7. Stop the reaction by washing twice with sterile distilled water for 5 min each.

3.14 Mount

1. Mount tissue sections with three drops of Crystal/Mount.

2. Dry the glass slides for 30 min at 50 °C.

---

4 Notes

1. Primer specificity should be tested by searching the complementary sequence in the whole mouse genome. Only one sequence corresponding to the gene of interest should be obtained. A single band of the expected size should be obtained after resolving the PCR products in a DNA agarose gel electrophoresis. The primers described in this chapter are all tested in this manner.

2. The PCR Polishing Kit is designed to polish the ends of the 3′-overhang extensions of polymerase-generated DNA fragments directly from a PCR amplification reaction. The PCR Polishing Kit can also be used to perform complete fill-in of 5′ overhangs to generate blunt ends. Dramatic increase in the population of blunt-ended DNA fragments following polishing treatment results in a drastic increase in overall experimental efficiency associated with procedures utilizing blunt-ended ligation.

3. For ligation, this kit requires an insert-to-vector molar ratio, that is higher than the molar ratios used in many other cloning procedures. For the sample DNA, a range from 40:1 to 100:1 is recommended.

4. Sizing time in min $= (L_0 - L_f)/(0.11 \times L_0 \times L_f)$
   L_0: original transcript length in kb. L_f: final length in kb (0.075 is currently used).

5. The heart should not be weakened before perfusion. Therefore, anesthetization of mouse with diethyl ether should be carefully performed (not too much, not too little).

6. Once the tissue is embedded, it is stable for many years.

7. The slides with paraffin sections should be used within 1 week.

8. Proteinase K treatment using 20 μg/mL for 10 min at room temperature results in much higher signals than that using 1 μg/mL for 30 min at 37 °C.

9. This is important as permeabilization with proteinase K has been shown to improve the signal intensity after in situ hybridization presumably because they either loosen up the membranes allowing greater penetration of the probes or because they partly digest proteins associated with the RNA allowing longer regions to access to the probes [10].

10. Temperature of stringency wash is 2–5 °C above the hybridization temperature.

References

1. Bernfield M, Gotte M, Park PW, Reizes O, Fitzgerald ML, Lincecum J, Zako M (1999) Functions of cell surface heparan sulfate proteoglycans. Annu Rev Biochem 68:729–777

2. Kreuger J, Kjellén L (2012) Heparan sulfate biosynthesis: regulation and variability. J Histochem Cytochem 60:898–907

3. Lindahl U, Kjellén L (2013) Pathophysiology of heparan sulphate: many diseases, few drugs. J Intern Med 273:555–571

4. Mizumoto S, Sugahara K (2013) Glycosaminoglycans are functional ligands for receptor for advanced glycation end-products in tumors. FEBS J 280:2462–2470

5. Tiwari V, Maus E, Sigar IM, Ramsey KH, Shukla D (2012) Role of heparan sulfate in sexually transmitted infections. Glycobiology 22:1402–1412

6. Kinnunen T, Huang Z, Townsend J, Gatdula MM, Brown JR, Esko JD, Turnbull JE (2005) Heparan 2-O-sulfotransferase, hst-2, is essential for normal cell migration in Caenorhabditis elegans. Proc Natl Acad Sci U S A 102:1507–1512

7. Esko JD, Selleck SB (2002) Order out of chaos: assembly of ligand binding sites in heparan sulfate. Annu Rev Biochem 71:435–471

8. Yabe T, Hata T, He J, Maeda N (2005) Developmental and regional expression of heparan sulfate sulfotransferase genes in the mouse brain. Glycobiology 15:982–993

9. Cox KH, DeLeon DV, Angerer LM, Angerer RC (1984) Detection of mRNAs in sea urchin embryos by in situ hybridization using asymmetric RNA probes. Dev Biol 101:485–502

10. Wilcox JN, Gee CE, Roberts JL (1986) In situ cDNA:mRNA hybridization: development of a technique to measure mRNA levels in individual cells. Methods Enzymol 124:510–533

11. Shworak NW, Liu J, Fritze LM, Schwartz JJ, Zhang L, Logeart D, Rosenberg RD (1997) Molecular cloning and expression of mouse and human cDNAs encoding heparan sulfate D-glucosaminyl 3-O-sulfotransferase. J Biol Chem 272:28008–28019

12. Shworak NW, Liu J, Petros LM, Zhang L, Kobayashi M, Copeland NG, Jenkins NA, Rosenberg RD (1999) Multiple isoforms of heparan sulfate D-glucosaminyl 3-O-sulfotransferase. Isolation, characterization, and expression of human cDNAs and identification of distinct genomic loci. J Biol Chem 274:5170–5184

13. Shukla D, Liu J, Blaiklock P, Shworak NW, Bai X, Esko JD, Cohen GH, Eisenberg RJ, Rosenberg RD, Spear PG (1999) A novel role for 3-O-sulfated heparan sulfate in herpes simplex virus 1 entry. Cell 99:13–22

14. Lawrence R, Yabe T, Hajmohammadi S, Rhodes J, McNeely M, Liu J, Lamperti ED, Toselli PA, Lech M, Spear PG, Rosenberg RD, Shworak NW (2007) The principal neuronal gD-type 3-O-sulfotransferases and their products in central and peripheral nervous system tissues. Matrix Biol 26:442–455

Chapter 30

Keratan Sulfate: Biosynthesis, Structures, and Biological Functions

Kenji Uchimura

Abstract

Keratan sulfate is a glycosaminoglycan that has been investigated in the cornea and skeletal tissues for decades. Endoglycosidases and monoclonal antibodies specific for keratan sulfate have been developed. These materials have facilitated the analysis of keratan sulfate biosynthesis and structures. Likewise, they have expedited study of the biological roles of keratan sulfate in vitro and in vivo. It has been shown that keratan sulfate is also expressed in the central nervous system and functions as a regulator of neuronal regeneration/sprouting. Here, we describe methods to determine the enzymatic activity of GlcNAc6ST, which is involved in keratan sulfate biosynthesis, and to extract and prepare ocular keratan sulfate for a disaccharide composition analysis. Immunohistochemistry for an anti-keratan sulfate epitope in the brain is also described.

Key words Keratan sulfate, Sulfotransferase, *N*-acetylglucosamine, Galactose, Immunohistochemistry, Disaccharide analysis, Enzymatic specificity

1 Introduction

Keratan sulfate (KS) is one of the glycosaminoglycans, occurring as keratan sulfate proteoglycans (KSPGs) on the cell surface and in the extracellular matrix. KS is found in the cornea [1] and in skeletal and nervous tissues [2]. Three classes of KS, namely, KS-I, KS-II, and KS-III, have been designated on the basis of structural differences in the linkage oligosaccharides that connect KS to the protein core [3]. KS-I is *N*-linked KS chains abundant in the cornea. KS-II is KS chains *O*-linked through *N*-acetylgalactosamine (GalNAc) found at high levels in cartilage. KS extended from *O*-linked mannose was shown in proteoglycans expressed in the brain [4] and defined later as KS-III [3]. Phosphacan is a major proteoglycan that carries KS-III in the central nervous tissue [5, 6]. The building blocks of KS are repeating disaccharides of galactose (Gal) and *N*-acetylglucosamine (GlcNAc). KS can be capped at their nonreducing termini with various types of monosaccharides

Kuberan Balagurunathan et al. (eds.), *Glycosaminoglycans: Chemistry and Biology*, Methods in Molecular Biology, vol. 1229, DOI 10.1007/978-1-4939-1714-3_30, © Springer Science+Business Media New York 2015

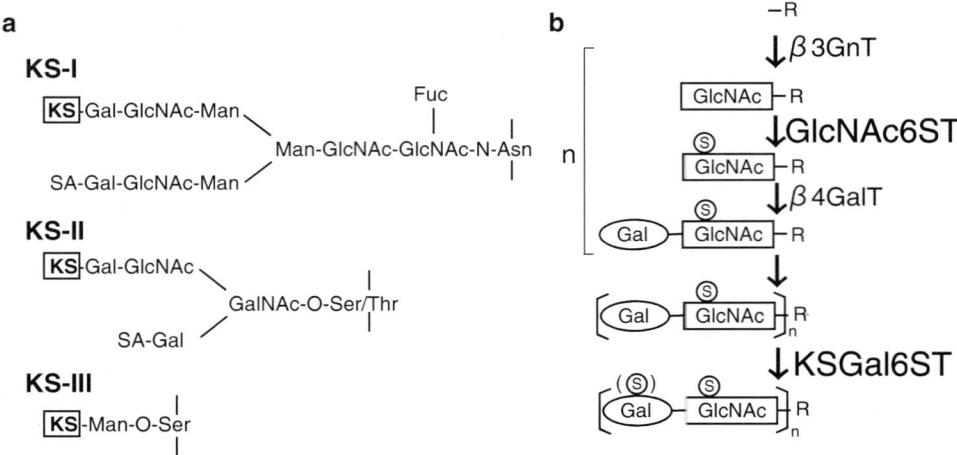

Fig. 1 Keratan sulfate linkage oligosaccharides and biosynthesis pathway. (**a**), Structural differences in the oligosaccharides linking keratan sulfate (KS) to protein define the three classes of KS. (**b**), KS chain is extended by the alternative actions of a ß-1,3-*N*-acetylglucosaminyltransferase (ß3GnT) and a ß-1,4-galactosyltransferase (ß4GalT). The sequence of biosynthesis is *N*-acetylglucosaminylation, C-6 sulfation of a GlcNAc residue exposed at the nonreducing end by a GlcNAc6ST and galactosylation. KSGal6ST sulfates some Gal residues after formation of the polysaccharide chain. *SA* sialic acid, *Fuc* fucose, *Gal* galactose, *GlcNAc N*-acetylglucosamine, *Man* mannose, *GalNAc N*-acetylgalactosamine, *S* sulfate

including α(2, 3)- or α(2–6)-linked sialic acid, α(1–3)-linked Gal and β(1–3)-linked sulfated GalNAc [7]. Golgi-localized enzymes catalyze the transfer of sulfates as well as elongation of the KS chains. The KS polysaccharide is extended by the alternative actions of a ß1,4-galactosyltransferase (ß4GalT) and a ß1,3-*N*-acetylglucosaminyltransferase (ß3GnT). C-6 sulfation modifications of the majority of GlcNAc residues and a significant proportion of adjacent Gal residues are observed within KS (Fig. 1).

GlcNAc-6-*O*-sulfotransferase-1 (GlcNAc6ST1, encoded by the gene *CHST2*) transfers a sulfate group to C-6 of GlcNAc exposed at the nonreducing end [8–11]. Thus far, five members of the GlcNAc6ST family have been identified in humans, four of which have mouse orthologs [12]. It has been shown that GlcNAc6ST1 is an enzyme responsible for the synthesis of KS [6, 13] that is induced in the brain and spinal cord of adult mice after injury, and that the loss of KS facilitates axonal regeneration/sprouting [14, 15]. GlcNAc6ST5 (encoded by the gene *CHST6*) is essential for the synthesis of corneal KS. Mutation in the *CHST6* gene was identified as a cause of macular corneal dystrophy [16]. Keratan sulfate galactose 6-*O*-sulfotransferase (KSGal6ST, encoded by the gene *CHST1*) and chondroitin sulfotransferase-1 (C6ST-1, encoded by the gene *CHST3*) have been shown to generate C6-sulfated Gal in vitro [17–19]. Genetic disruptions in KSGal6ST are associated with deficiency in the biosynthesis of KS in the eye [20].

Several monoclonal antibodies against KS have been developed. These include 5D4, BCD4, TRA-1, I22, 373E1, and R10G [21–26]. Each of these antibodies recognizes sulfated epitopes within the KS. 5D4 has been used extensively to elucidate KS expression. As determined by immunocytochemistry, 5D4 reacts with cornea [16, 27], articular cartilage [22, 28] and N-linked KS in aggrecan [29]. In the central nervous system (CNS), the epitope is constitutively expressed in a subpopulation of microglia of the adult brain [30, 31]. The 5D4 epitope is upregulated in the adult CNS after injury [32, 33] and in neurodegenerative diseases [34–37]. In the developing brain, expression of the 5D4 epitope is spatiotemporally regulated as demonstrated in rat and mouse [6, 38, 39]. Methods to determine the activities of KS sulfotransferases and to analyze the expression and structures of KS have been established and have facilitated their study. Here, we describe methods of an assay for GlcNAc6ST, KS disaccharide analysis, and immunohistochemical detection of the 5D4 KS epitope.

2 Materials

2.1 Determination of Enzymatic Specificity of GlcNAc6STs

1. Expression plasmids of N-terminal-IgM signal peptide and Protein A-fused versions of GlcNAc6ST1, GlcNAc6ST2, and GlcNAc6ST3 [9] (see **Note 1**).

2. Oligosaccharide substrates: GlcNAcß1-6Man-O-methyl (Sigma), GlcNAcß1-2Man (Dextra Laboratories, Reading, UK), GlcNAcß1-6[Galß1-3]GalNAc-p-nitrophenyl (core 2-p-nitrophenyl), and GlcNAcß1-3GalNAc-p-nitrophenyl (core 3-p-nitrophenyl) (Toronto Research Chemicals, Ontario, Canada); GlcNAcß1-3Galß1-4GlcNAcß1-3Galß1-4GlcNAc was prepared from Galß1-4GlcNAcß1-3Galß1-4GlcNAcß1-3Galß1-4GlcNAc as described previously [8].

3. COS-7 cells.

4. Dulbecco's Modified Eagle's Medium (DMEM) supplemented with 10 % fetal bovine serum.

5. DMEM supplemented with 2 % IgG-free fetal bovine serum.

6. IgG-Sepharose.

7. LipofectAMINE PLUS transfection reagent.

8. 1 M Tris–HCl, pH 7.5.

9. 1 M $MgCl_2$.

10. Phosphate-buffered saline (PBS).

11. 50 mM adenosine monophosphate (AMP).

12. 1 M NaF.

13. 0.2 M $MnCl_2$.

14. [^{35}S] 3'-phosphoadenosine-5'-phosphosulfate (PAPS, 1.9 Ci/mmol, PerkinElmer, Waltham, MA) (*see* **Note 2**).

15. Thin layer chromatography (TLC) aluminum sheet coated with cellulose (0.1 mm thick).

16. TLC-developing buffer: ethanol/pyridine/n-butyl alcohol/water/acetate (100:10:10:30:3, by volume).

17. BAS2000 bioimaging analyzer (Fuji Film, Tokyo, Japan).

2.2 Structural Analysis of Keratan Sulfate

1. Adult (>8-week-old) C57BL/6 mice.

2. 0.2 N NaOH.

3. 4 N HCl.

4. 1 M Tris–HCl, pH 7.2.

5. DNase I (10 mg/ml).

6. RNase A (10 mg/ml).

7. 1 M Tris–HCl, pH 8.0.

8. Actinase E (10 mg/ml).

9. Diethylaminoethanol (DEAE) Sepharose.

10. 99.5 % ethanol.

11. DEAE-wash buffer: 50 mM Tris–HCl, pH 7.2, and 0.1 M NaCl (*see* **Note 3**).

12. DEAE-elution buffer: 50 mM Tris–HCl, pH 7.2, and 2 M NaCl.

13. 50 mM Tris–acetate, pH 6.0.

14. 2 mg/ml glycogen.

15. 99.5 % ethanol/1.3 % potassium acetate.

16. Keratanase (*Pseudomonas* sp., 50 U/ml), keratanase II (*Bacillus* sp. Ks 36, 0.5 U/ml) (Seikagaku, Tokyo, Japan), α-1,3/4-fucosidase (*Streptomyces* sp. 142, 1 mU/ml, Takara Bio, Shiga, Japan) and neuraminidase (*Arthrobacter ureafaciens*, 10 U/ml) (*see* **Note 3**).

17. Nanosep 3 K centrifugal device (Pall, Port Washington, NY).

2.3 Immunohisto-chemical Analysis for a Keratan Sulfate Epitope in the Brain

1. C57BL/6 mouse embryos (embryonic day 15.5).

2. 5D4 monoclonal antibody (1 mg/ml, Seikagaku).

3. Cy3-conjugated goat anti-mouse IgG1 antibody (1.5 mg/ml, Jackson Immuno Research Laboratories, West Grove, PA).

4. 0.1 % BSA in PBS.

5. O.C.T. compound and plastic molds (Sakura Finetek, Torrance, CA).

6. Ice-cold acetone.

7. Harris' hematoxylin.

8. Blocking reagent: 3 % BSA, 0.01 % NaN$_3$ in PBS (*see* **Note 4**).

9. FluorSaver™ Reagent.

10. MAS-coated glass slide with three 15 mm wells (Catalog #SF17293, Matsunami Glass, Osaka, Japan).

11. Cryostat (model CM1950, Leica Microsystems, Wetzlar, Germany).

3 Methods

3.1 Determination of Enzymatic Specificity of GlcNAc6STs

Enzymatic assays utilizing various oligosaccharides derived from *N*-linked or *O*-linked glycoprotein-bound glycans and agalacto-*N*-acetyllactosamine repeats are applicable to compare the substrate specificity of GlcNAc6STs and to elucidate their contribution to extend KS chains and/or initiate the addition of KS to different core oligosaccharides linking to scaffold proteins. Here, TLC is employed to assay a number of samples.

1. Culture COS-7 cells on 75 cm^2 flasks and passage them until they reach ~70 % confluence with DMEM 10 % fetal bovine serum. Rinse the cells with warm PBS, and then transfect transiently with an expression plasmid of IgM signal peptide/ Protein A-fused GlcNAc6ST using LipofectAMINE PLUS transfection reagent according to the manufacturer's instructions. Replace the medium to DMEM 2 % IgG-free fetal bovine serum after the transfection. Incubate the cells for 48 h.

2. Collect 10 mL of conditioned medium (CM) and mix with 10 μL of an IgG-Sepharose resin in a tube. Rotate the tube for 3 h at 4 °C.

3. Collect the resin by centrifugation at 12,000 × *g* and then wash with PBS three times. Suspend the resin in 30 μL of 50 mM Tris–HCl, pH 7.5, and use as an enzyme.

4. Mix 5 μL of the enzyme suspension in a tube with 1 μmol Tris–HCl, pH7.5, 0.2 μmol MnCl$_2$, 0.04 μmol AMP, 2 μmol NaF, 20 μmol oligosaccharide, 150 pmol [^{35}S]PAPS (1.5 × 10^6 cpm), 0.05 % Triton-X in a final volume of 20 μL. Incubate the reaction mixture at 30 °C for 1 h (*see* **Note 5**).

5. Centrifuge the tube at 12,000 × *g* for 1 min at room temperature.

6. Apply aliquots of 2 μL of the supernatant to TLC plates and then develop them with TLC-developing buffer. Stop development of the plates when the solvent front reaches the top.

7. The radioactivity of the ^{35}S-labeled products is visualized and measured with a BAS2000 bioimaging analyzer [9] (Table 1).

Table 1
Comparison of the substrate specificities of GlcNAc6ST-1, GlcNAc6ST-2, and GlcNAc6ST-3 secreted into the culture medium by transfected COS-7 cells. Table taken from [9] with permission

	Enzyme activity[a] [pmol/h/ml of medium (%)[b]]		
Acceptor	GlcNAc6ST-1	GlcNAc6ST-2	GlcNAc6ST-3
GlcNAcß1-6ManOMe	20.6 (100)	6.8 (100)	N.D.
GlcNAcß1-2Man	26.3 (128)	10.3 (151)	N.D.
GlcNAcß1-6[Galß1-3]GalNAc-pNP (core 2)	39.3 (191)	9.9 (145)	5.7[c]
GlcNAcß1-3GalNAc-pNP (core 3)	N.D.[d]	12.5 (184)	N.D.
GlcNAcß1-3Galß1-4GlcNAcß1-3 Galß1-4GlcNAc	4.3[c] (21)	5.0[c] (73)	N.D.

[a]The values represent the averages of two independent experiments.
[b]The percentage of the activity compared with that of GlcNAcß1-6ManOMe is also shown
[c]The actually observed radioactivities were around 30,000 cpm, while the assay without the enzyme or with IgG-Sepharose exposed to culture supernatant of mock-transfected cells gave values less than 500 cpm
[d]N.D., less than 0.1 pmol/h/ml of medium

3.2 Structural Analysis of Keratan Sulfate

The disaccharide compositions of KS are determined by reversed-phase ion-pair chromatography with post-column fluorescent labeling adapted from a method in a previous report [40]. Here, we describe extraction of ocular KS and preparation of disaccharide samples for the reversed-phase ion-pair chromatography [6, 20, 41].

1. Suspend adult mouse eyes (~50 mg) in 2 mL of 0.2 N NaOH and incubate at room temperature overnight. Neutralize the samples with 4 N HCl.

2. Add 1 M Tris–HCl, pH 8.0 (1/20 volume of the sample), and 1 M MgCl$_2$ (1/100 volume of the sample) to the tube.

3. Add 10 μL of DNase I and 10 μL of RNase A. Incubate the tube at 37 °C for 3 h.

4. Add 100 μL of 10 mg/mL Actinase E to the tube. Incubate the tube at 50 °C overnight on a rocking platform (*see* **Note 6**).

5. Heat the tube in boiling water for 10 min. Centrifuge the tube at 800×g for 10 min at room temperature. Transfer the supernatant to a new tube and then mix with the same volume of 50 mM Tris–HCl, pH 7.2.

6. Apply the mixture to a DEAE Sepharose column (*see* **Note 7**).

7. Wash the Sepharose with 2 mL of 50 mM Tris–HCl, pH 7.2, 0.1 M NaCl.

8. Elute the bound keratan sulfate with 0.6 mL of 50 mM Tris–HCl, pH 7.2, 2 M NaCl. Collect the eluent in a 5 mL tube.

9. Add 1.8 mL (triple volume of the eluent) of 99.5 % ethanol and then divide into 1.5 mL centrifuge tubes. Place the tubes at −80 °C for 30 min (*see* **Note 8**).

10. Centrifuge the tubes at $20,000 \times g$ in a microcentrifuge for 15 min at 4 °C.

11. Remove the supernatants and then rinse the pellets with 70 % ethanol. Place the tubes at −80 °C for 30 min and then centrifuge as above.

12. Dissolve the pellets in 100 μL of distilled water (*see* **Note 9**).

13. For pretreatment, mix 25 μL of the sample in a tube with 5 mM Tris–acetate, pH 6, 1 U/ml neuraminidase, 0.04 mU/ml α-1,3/4 fucosidase in a total volume of 50 μL. Incubate the reaction mixture at 37 °C for 2 h. Terminate the reaction by placing the tube at 95 °C for 5 min. Add 5 μL of 2 mg/ml glycogen and 110 μL of 99.5 % ethanol/1.3 % potassium acetate. Place the tubes at −80 °C for 30 min and then centrifuge as above. Discard the supernatants.

14. Dissolve the pellets in 44 μL of distilled water and then mix with 5 mM Tris–acetate, pH 6, and 1 mU/μl keratanase or 10 μU/μl keratanase II in a total volume of 50 μL. Incubate the reaction mixture at 37 °C for 24 h. Terminate the reaction by placing the tube at 95 °C for 5 min. Add 50 μL of distilled water to the tube.

15. Apply the mixture on a Nanosep 3 K filter and then spin on the Nanosep 3 K device at $16,000 \times g$ for 15 min at room temperature (*see* **Note 10**). Subject the filtered materials to reversed-phase ion-pair chromatography (Fig. 2).

3.3 Immunohisto-chemical Analysis for a Keratan Sulfate Epitope in the Brain

The 5D4 anti-KS antibody recognizes GlcNAc-6- and Gal-6-sulfated poly-*N*-acetyllactosamine structures within KS. We have shown that the 5D4 epitope expressed in developing and early postnatal brains was abolished in mice deficient in GlcNAc6ST1 and KSGal6ST [6, 13]. Here, we describe immunohistochemical staining for the 5D4 KS epitope in developing mouse brain sections.

1. Dissect the head of 15.5-day mouse embryos, embed in O.C.T. compound-containing plastic molds and then freeze on dry ice.

2. Cut sections of the embedded tissue block using a cryostat at a thickness of 10 μm. Mount sections on warm MAS-coated glass slides.

3. After air drying for 30–60 min, fix sections in ice-cold acetone for 15 min.

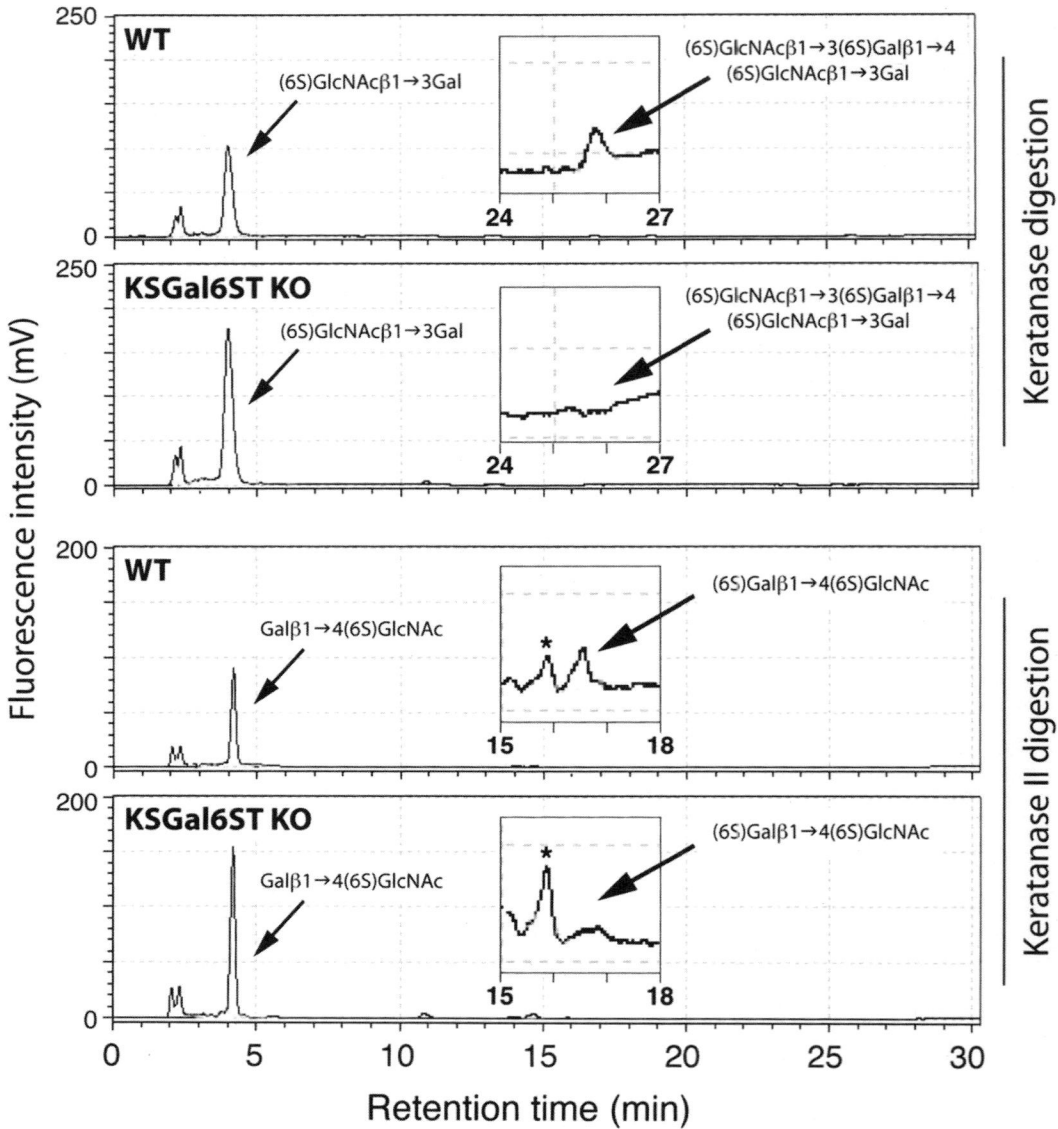

Fig. 2 Structural analysis of keratan sulfate in KSGal6ST-deficient mice. Ocular keratan sulfates (KS) from eyes of wild-type (WT) and KSGal6ST KO mice were digested with keratanase or keratanase II. The oligosaccharide compositions of the KS were determined by reversed-phase ion-pair chromatography. Standard substances were eluted at the peak positions indicated by *arrows*. Elution profiles around the peak positions of (6S) GlcNAcβ1→3(6S)Galβ1→4(6S)GlcNAcβ1→3Gal and (6S)Galβ1→4(6S)GlcNAc are magnified in insets. Peaks indicated by *asterisks* were not identified. Figure taken from [20] with permission

4. After air drying for 30 min, wash sections with PBS and then incubate with blocking reagent for 15 min at room temperature.

5. Incubate sections with a mixture of 5D4 (5 μg/mL in 0.1 % BSA) at 4 °C overnight.

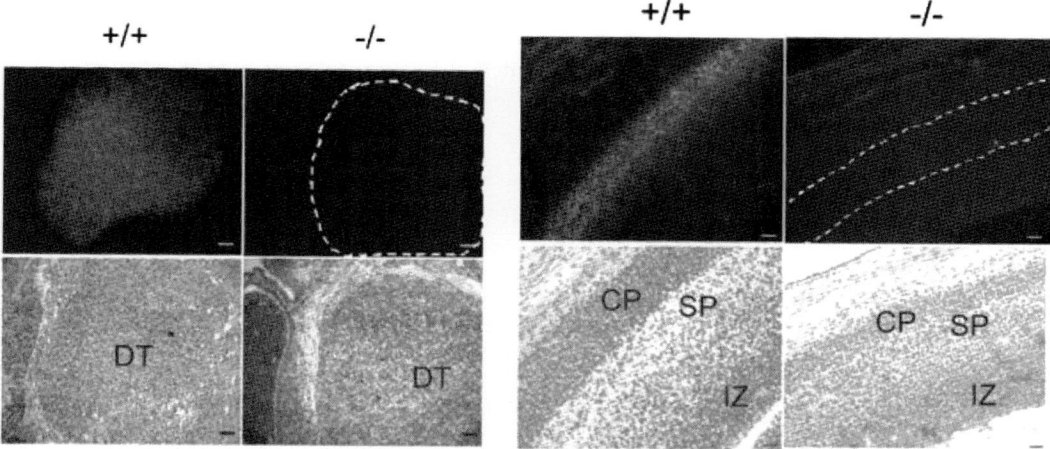

Fig. 3 Immunohistochemical staining of an anti-ketratan sulfate epitope in GlcNAc6ST1-deficient mice. Brain specimens obtained from wild-type (+/+) embryonic-day-15.5 mice were stained with the 5D4 antibody. The epitope of 5D4 anti-keratan sulfate (KS) antibody is expressed in the dorsal thalamus (DT) and subplate (SP) during development. These signals are absent in the brain of GlcNAc6ST1-deficient (−/−) mice [6, 13]. *Dark field* photos show KS expression (*white*). *Bright field* photos show hematoxylin staining. *CP* cortical plate, *IZ* intermediate zone. Bars: 25 μm. Figure taken from [13] with permission

6. Wash sections twice with PBS and then detect the primary antibody with Cy3-conjugated goat anti-mouse IgG$_1$ (6 μg/mL in 0.1 % BSA). Counterstain sections with hematoxylin. Wash sections twice with PBS. Add one drop of FluorSaver™ Reagent on each section. Cover sections with coverslips.

7. Capture digital images by fluorescence microscopy (Fig. 3).

4 Notes

1. Microsomal fractions of the cells transfected with an expression plasmid encoding full-length GlcNAc6ST1 are also useful as enzymes.

2. The reagent should be diluted with water and aliquots should be stored frozen.

3. For the analysis of heparan sulfate and chondroitin sulfate, 0.2 M NaCl is used. Keratanase and keratanase II are made as small aliquots and stored at −20 °C. Repetition of freezing and thawing should be restricted to less than five times.

4. The reagent should be filtered using a 0.22 μm filter and stored at 4 °C.

5. AMP is thought to increase the affinity of [^{35}S]PAPS to Glc NAc6ST protein. NaF is added to inhibit dephosphorylation

of [^{35}S]PAPS. C-6 sulfation modifications on the GlcNAc residues alter the properties of these oligosaccharides to be unsusceptible to Jack bean ß-N-acetylhexosaminidase.

6. For the analysis of heparan sulfate and chondroitin sulfate, 10 μL of Actinase E is added to 2 mL of the sample. The mixture is incubated at 37 °C overnight.

7. Pack 0.2 mL of DEAE Sepharose in an empty plastic column. Prewash the Sepharose with 2 mL of 50 mM Tris–HCl, pH 7.2.

8. For the analysis of heparan sulfate and chondroitin sulfate, the volume of ethanol is double that of the eluent.

9. The samples can be placed at –80 °C for long-term storage.

10. Nanosep 3K device should be prewashed with 200 μL of 70 % methanol and 200 μL of distilled water to avoid nonspecific fluorescent peak signals in the reversed-phase ion-pair chromatography.

Acknowledgements

Supported by Japanese Health and Labour Sciences Research Grants [H19-001 and H22-007], Grants-in-Aid from the Ministry of Education, Science, Sports and Culture [22790303 and 24590349, and for Scientific Research on Innovative Areas] and in part by the Takeda Science Foundation.

References

1. Meyer K, Linker A, Davidson EA et al (1953) The mucopolysaccharides of bovine cornea. J Biol Chem 205:611–616

2. Funderburgh JL (2000) Keratan sulfate: structure, biosynthesis, and function. Glycobiology 10:951–958

3. Funderburgh JL (2002) Keratan sulfate biosynthesis. IUBMB life 54:187–194

4. Krusius T, Finne J, Margolis RK et al (1986) Identification of an O-glycosidic mannose-linked sialylated tetrasaccharide and keratan sulfate oligosaccharides in the chondroitin sulfate proteoglycan of brain. J Biol Chem 261:8237–8242

5. Margolis RK, Rauch U, Maurel P et al (1996) Neurocan and phosphacan: two major nervous tissue-specific chondroitin sulfate proteoglycans. Perspect Dev Neurobiol 3:273–290

6. Hoshino H, Foyez T, Ohtake-Niimi S et al (2014) KSGal6ST is essential for the 6-sulfation of galactose within keratan sulfate in early postnatal brain. J Histochem Cytochem 62(2):145–56. doi:10.1369/0022155413511619

7. Tai GH, Huckerby TN, Nieduszynski IA (1996) Multiple non-reducing chain termini isolated from bovine corneal keratan sulfates. J Biol Chem 271:23535–23546

8. Uchimura K, Muramatsu H, Kadomatsu K et al (1998) Molecular cloning and characterization of an N-acetylglucosamine-6-O-sulfotransferase. J Biol Chem 273:22577–22583

9. Uchimura K, El-Fasakhany FM, Hori M et al (2002) Specificities of N-acetylglucosamine-6-O-sulfotransferases in relation to L-selectin ligand synthesis and tumor-associated enzyme expression. J Biol Chem 277:3979–3984

10. Uchimura K, Gauguet JM, Singer MS et al (2005) A major class of L-selectin ligands is

eliminated in mice deficient in two sulfotransferases expressed in high endothelial venules. Nat Immunol 6:1105–1113

11. Fujiwara M, Kobayashi M, Hoshino H et al (2012) Expression of long-form N-acetylglucosamine-6-O-sulfotransferase 1 in human high endothelial venules. J Histochem Cytochem 60:397–407

12. Uchimura K, Rosen SD (2006) Sulfated L-selectin ligands as a therapeutic target in chronic inflammation. Trends Immunol 27:559–565

13. Zhang H, Muramatsu T, Murase A et al (2006) N-Acetylglucosamine 6-O-sulfotransferase-1 is required for brain keratan sulfate biosynthesis and glial scar formation after brain injury. Glycobiology 16:702–710

14. Ito Z, Sakamoto K, Imagama S et al (2010) N-acetylglucosamine 6-O-sulfotransferase-1-deficient mice show better functional recovery after spinal cord injury. J Neurosci 30:5937–5947

15. Imagama S, Sakamoto K, Tauchi R et al (2011) Keratan sulfate restricts neural plasticity after spinal cord injury. J Neurosci 31:17091–17102

16. Akama TO, Nishida K, Nakayama J et al (2000) Macular corneal dystrophy type I and type II are caused by distinct mutations in a new sulphotransferase gene. Nat Genet 26:237–241

17. Habuchi O, Hirahara Y, Uchimura K et al (1996) Enzymatic sulfation of galactose residue of keratan sulfate by chondroitin 6-sulfotransferase. Glycobiology 6:51–57

18. Fukuta M, Inazawa J, Torii T et al (1997) Molecular cloning and characterization of human keratan sulfate Gal-6-sulfotransferase. J Biol Chem 272:32321–32328

19. Torii T, Fukuta M, Habuchi O (2000) Sulfation of sialyl N-acetyllactosamine oligosaccharides and fetuin oligosaccharides by keratan sulfate Gal-6-sulfotransferase. Glycobiology 10:203–211

20. Patnode ML, Yu SY, Cheng CW et al (2013) KSGal6ST generates galactose-6-O-sulfate in high endothelial venules but does not contribute to L-selectin-dependent lymphocyte homing. Glycobiology 23:381–394

21. Andrews PW, Banting G, Damjanov I et al (1984) Three monoclonal antibodies defining distinct differentiation antigens associated with different high molecular weight polypeptides on the surface of human embryonal carcinoma cells. Hybridoma 3:347–361

22. Caterson B, Christner JE, Baker JR (1983) Identification of a monoclonal antibody that specifically recognizes corneal and skeletal keratan sulfate. Monoclonal antibodies to cartilage proteoglycan. J Biol Chem 258(14):8848–8854

23. Funderburgh JL, Funderburgh ML, Rodrigues MM et al (1990) Altered antigenicity of keratan sulfate proteoglycan in selected corneal diseases. Invest Ophthalmol Vis Sci 31:419–428

24. Glant TT, Mikecz K, Roughley PJ et al (1986) Age-related changes in protein-related epitopes of human articular-cartilage proteoglycans. Biochem J 236:71–75

25. Magro G, Perissinotto D, Schiappacassi M et al (2003) Proteomic and postproteomic characterization of keratan sulfate-glycanated isoforms of thyroglobulin and transferrin uniquely elaborated by papillary thyroid carcinomas. Am J Pathol 163:183–196

26. Kawabe K, Tateyama D, Toyoda H et al (2013) A novel antibody for human induced pluripotent stem cells and embryonic stem cells recognizes a type of keratan sulfate lacking oversulfated structures. Glycobiology 23:322–336

27. Funderburgh JL, Caterson B, Conrad GW (1987) Distribution of proteoglycans antigenically related to corneal keratan sulfate proteoglycan. J Biol Chem 262:11634–11640

28. Poole CA, Glant TT, Schofield JR (1991) Chondrons from articular cartilage. (IV). Immunolocalization of proteoglycan epitopes in isolated canine tibial chondrons. J Histochem Cytochem 39(9):1175–1187

29. Poon CJ, Plaas AH, Keene DR et al (2005) N-linked keratan sulfate in the aggrecan interglobular domain potentiates aggrecanase activity. J Biol Chem 280:23615–23621

30. Bertolotto A, Caterson B, Canavese G et al (1993) Monoclonal antibodies to keratan sulfate immunolocalize ramified microglia in paraffin and cryostat sections of rat brain. J Histochem Cytochem 41:481–487

31. Jander S, Schroeter M, Fischer J et al (2000) Differential regulation of microglial keratan sulfate immunoreactivity by proinflammatory cytokines and colony-stimulating factors. Glia 30:401–410

32. Jones LL, Tuszynski MH (2002) Spinal cord injury elicits expression of keratan sulfate proteoglycans by macrophages, reactive microglia, and oligodendrocyte progenitors. J Neurosci 22:4611–4624

33. Zhang H, Uchimura K, Kadomatsu K (2006) Brain keratan sulfate and glial scar formation. Ann N Y Acad Sci 1086:81–90

34. Manuelidis L, Fritch W, Xi YG (1997) Evolution of a strain of CJD that induces BSE-like plaques. Science 277:94–98

35. Miao J, Vitek MP, Xu F et al (2005) Reducing cerebral microvascular amyloid-beta protein deposition diminishes regional neuroinflammation in vasculotropic mutant amyloid precursor protein transgenic mice. J Neurosci 25:6271–6277

36. Vidal R, Barbeito AG, Miravalle L et al (2009) Cerebral amyloid angiopathy and parenchymal amyloid deposition in transgenic mice expressing the Danish mutant form of human BRI2. Brain Pathol 19:58–68

37. Hirano K, Ohgomori T, Kobayashi K et al (2013) Ablation of keratan sulfate accelerates early phase pathogenesis of ALS. PLoS One 8:e66969

38. Meyer-Puttlitz B, Milev P, Junker E et al (1995) Chondroitin sulfate and chondroitin/keratan sulfate proteoglycans of nervous tissue: developmental changes of neurocan and phosphacan. J Neurochem 65:2327–2337

39. Miller B, Sheppard AM, Pearlman AL (1997) Developmental expression of keratan sulfate-like immunoreactivity distinguishes thalamic nuclei and cortical domains. J Comp Neurol 380:533–552

40. Toyoda H, Kinoshita-Toyoda A, Fox B et al (2000) Structural analysis of glycosaminoglycans in animals bearing mutations in sugarless, sulfateless, and tout-velu. Drosophila homologues of vertebrate genes encoding glycosaminoglycan biosynthetic enzymes. J Biol Chem 275:21856–21861

41. Patnode ML, Cheng CW, Chou CC et al (2013) Galactose 6-o-sulfotransferases are not required for the generation of siglec-f ligands in leukocytes or lung tissue. J Biol Chem 288:26533–26545

Chapter 31

The Sulfs: Expression, Purification, and Substrate Specificity

Kenji Uchimura

Abstract

Sulf-1 and Sulf-2 are endo-acting extracellular sulfatases. The Sulfs liberate 6-*O* sulfate groups, mainly from *N*, 6-*O*, and 2-*O* trisulfated disaccharides of heparan sulfate (HS)/heparin chains. The Sulfs have been shown to modulate the interaction of a number of protein ligands including growth factors and morphogens with HS/heparin and thus regulate the signaling of these ligands. They also play important roles in development and are dysregulated in many cancers. The establishment of the expression of the Sulfs and methods of assaying them has been desirable to investigate these enzymes. In this chapter, methods to express and purify recombinant Sulfs and to analyze HS structures in an extracellular fraction of HSulf-transfected HEK293 cells are described. The application of these enzymes for ex vivo degradation of an anti-HS epitope accumulated in the brain of a neurodegenerative disease model mouse is also described.

Key words Heparan sulfate, Sulf-1, Sulf-2, Sulfatase, Endo-acting enzyme, Protein ligands, Immunoblot, Transfection, FLAG-tag, His-tag

1 Introduction

The Sulfs are members of the sulfatase family. Sulfatases hydrolyze sulfate esters of macromolecules such as glycosaminoglycans and sulfatides [1]. The majority of eukaryotic sulfatases are localized in lysosomes where they decompose sulfated molecules as exo-acting enzymes. In contrast, the Sulfs are endo-acting enzymes that are secreted and catabolize heparin/heparin sulfate (HS) glycosaminoglycans. Human Sulf-1 (HSulf-1), an ortholog of QSulf-1 in quail [2], was identified and characterized as an extracellular heparin/HS endosulfatase [3]. The closely related human Sulf-2 (HSulf-2) was also cloned [3]. Each Sulf contains a signal peptide and two sulfatase-related domains, which are separated by a large hydrophilic domain [3–7]. Sulf-1 and Sulf-2 are synthesized as pre-proproteins [8], posttranslationally modified with formylglycine [9, 10] and *N*-linked glycans [11], processed by furin-like endoproteases [3, 12, 13] and secreted into the extracellular space

Kuberan Balagurunathan et al. (eds.), *Glycosaminoglycans: Chemistry and Biology*, Methods in Molecular Biology, vol. 1229, DOI 10.1007/978-1-4939-1714-3_31, © Springer Science+Business Media New York 2015

or anchored on the cell surface. Both Sulf-1 and Sulf-2 remove sulfate groups at the C-6 position of glucosamines in the *N*-, 6-*O*-, and 2-*O*-trisulfated disaccharides of heparin/HS chains [3, 14–18], as well as having arylsulfatase activity against 4-methylumbelliferyl sulfate (4-MUS) [3, 12]. Analysis of HS chains in tissues of Sulf-deficient mice confirmed the substrate specificity of the Sulfs [19–21]. The Sulfs demonstrate optimal enzymatic activity in the neutral pH range, in contrast to lysosomal sulfatases that function in the acidic pH range [3].

Heparan sulfate proteoglycans (HSPGs) are secreted or cell-associated extracellular matrix proteins that are covalently modified with HS glycosaminoglycans. They regulate the localization and biological activities of growth factors, morphogens and cytokines, by interacting with these ligands via the HS [22–24]. HS sequences are variable, but are spatiotemporally controlled [25]. The Sulfs are novel postsynthetic regulators that remodel HS sequences and thereby modulate the signaling of heparin/HS binding ligands. It has been demonstrated that Sulfs regulate the signaling of Wnts [2, 12, 26, 27], PDGF [28], BMP [16, 29], FGF-2 [30–33], HGF [34, 35], TGF-β [36] and GDNF [37].

The Sulfs are dysregulated in many cancers [8]. Sulf-2 is pro-oncogenic in several cases [26–28, 38]. Sulf-2 has also been shown to be proangiogenic, presumably through its ability to reverse the association between angiogenic factors and heparin/HS [39, 40]. The Sulfs play important roles in embryonic development [2, 41, 42]. Sulf-1/Sulf-2 doubly deficient mice are more severely affected than single null mice, which is consistent with overlapping and redundant functions for the enzymes in development [19, 37, 43–46]. The Sulfs also regulate corneal wound healing by promoting the migration of corneal epithelial cells [47]. Establishment of the expression of the Sulfs and methods of assaying them has been desirable to investigate the enzymes [17]. Here, we describe methods to express and purify the Sulfs and to analyze the substrate specificity of these enzymes. We also describe application of these enzymes for the degradation of an anti-HS epitope expressed in the brain of a neurodegenerative disease model mouse.

2 Materials

2.1 Expression, Purification, and Immunoblotting of FLAG-His Sulfs

1. N-Terminal-FLAG, C-terminal-His-tagged versions of HSulf1, HSulf2, HSulf1ΔCC, or HSulf2ΔCC expression plasmids [12, 40] (*see* **Note 1**).

2. Human embryonic kidney (HEK) 293 cells.

3. Dulbecco's Modified Eagle's Medium (DMEM) supplemented with 10 % fetal bovine serum.

4. OptiMEM I reduced serum medium.

5. FuGENE6 transfection reagent.

6. Ni-NTA agarose resin (Pro-Bond™ resin, Invitrogen).

7. Poly-Prep chromatography column (Bio-Rad, Hercules, CA).

8. Resin-washing buffer: 50 mM HEPES, pH 7.5, 10 mM MgCl$_2$ and 0.05 % Tween-20.

9. Elution buffer: 50 mM HEPES, pH 7.5, 10 mM MgCl$_2$ and 100 mM imidazole (*see* **Note 2**).

10. Reducing SDS-7.5 % polyacrylamide gel.

11. Polyvinylidene fluoride (PVDF) membrane.

12. 1 M Tris–HCl, pH 8.0.

13. Phosphate buffered saline containing 0.1 % Tween-20 (PBS-T).

14. Membrane blocking reagent: 5 % skim milk in PBS-T.

15. Mouse monoclonal anti-FLAG antibody (Sigma, St. Louis, MO) and horseradish peroxidase (HRP)-conjugated goat anti-mouse IgG1 antibody.

16. Super Signal West Pico Chemiluminescent reagent (Thermo Fisher Scientific, Waltham, MA).

17. LAS-3000 mini-luminescent image analyzer (Fujifilm, Tokyo, Japan).

2.2 Structural Analysis of Heparan Sulfate in HSulf-Transfected HEK 293 Cells

1. DNase I (10 mg/mL).

2. RNase A (10 mg/mL).

3. 1 M Tris–HCl, pH 7.2.

4. 1 M Tris–HCl, pH 8.0.

5. 1 M MgCl$_2$.

6. Actinase E (10 mg/mL).

7. Diethylaminoethanol (DEAE) Sepharose.

8. 99.5 % ethanol.

9. DEAE-wash buffer: 50 mM Tris–HCl, pH 7.2, and 0.2 M NaCl.

10. DEAE-elution buffer: 50 mM Tris–HCl, pH 7.2, and 2 M NaCl.

11. 0.5 M HEPES-NaOH, pH 7.5.

12. Heparinase I (2 mU/μL) (EC 4.2.2.7), heparinase II (0.5 mU/μL) and heparinase III (0.2 mU/μL) (EC 4.2.2.8) (Sigma) (*see* **Note 3**).

13. Nanosep 3K centrifugal device (Pall, Port Washington, NY).

2.3 Ex Vivo Assay for the HSulfs

1. Heterozygotic transgenic mice that expressed the human amyloid precursor protein (hAPP) bearing the Swedish (K670N, M671L) mutation (Tg2576 strain, Taconic Farms, Inc., NY).

2. Phage display-derived RB4CD12 anti-HS antibody [17].

3. Cy3-conjugated monoclonal mouse anti-VSV-G antibody.

4. Bead-bound or purified FLAG/His-tagged HSulfs [17, 54] (*see* Subheadings 2.1 and 3.1).

5. 0.5 M HEPES-NaOH, pH 7.5.

6. 1 M $MgCl_2$.

7. O.C.T. compound and plastic molds (Sakura Finetek, Torrance, CA).

8. Phosphate buffered saline (PBS).

9. Ice-cold acetone.

10. Blocking reagent: 3 % BSA and 0.01 % NaN_3 in PBS (*see* **Note 4**).

11. FluorSaver™ Reagent.

12. MAS-coated glass slide with three 15 mm wells (Catalog #SF17293, Matsunami Glass, Osaka, Japan).

13. Cryostat (model CM1950, Leica Microsystems, Wetzlar, Germany).

14. Fluorescent microscope (Model BX50; Olympus, Tokyo, Japan).

3 Methods

3.1 Expression, Purification, and Immunoblotting of FLAG-His Sulfs

1. Culture HEK293 cells on 75 cm^2 flasks and passage them until they reach ~50 % confluence. Rinse the cells with warm PBS and OptiMEM I, and then transfect transiently with a FLAG-His-tagged HSulf expression plasmid using FuGENE6 transfection reagent according to the manufacturer's instructions. Replace the medium to OptiMEM I after the transfection. Incubate the cells for 48 h.

2. Collect 10 mL of conditioned medium (CM) and mix with 100 μL of a Ni-NTA agarose resin in a tube. Rotate the tube for 4 h at 4 °C.

3. Incorporate the mixture into a chromatography column. Wash the resin with 2 mL of resin-washing buffer (20 times the resin volume).

4. Elute the resin-bound FLAG-His-tagged HSulf proteins with 200 μL of elution buffer (double the volume of the resin) (*see* **Note 5**).

5. Separate the eluted HSulf proteins in 16 μL of the eluent or the Ni-NTA agarose-bound FLAG-His-tagged HSulf proteins prepared from 0.8 mL of HEK293 transfectant CM by electrophoresis on a reducing SDS-7.5 % polyacrylamide gel.

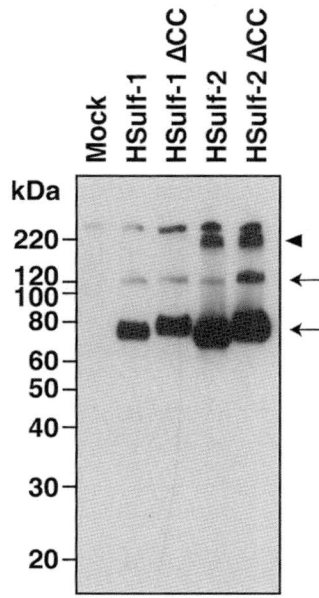

Fig. 1 Expression of recombinant HSulf-1 and HSulf-2. Recombinant HSulfs (HSulf-1 and HSulf-2) and their enzymatically inactive forms (HSulf-1ΔCC and HSulf-2ΔCC) tagged with N-terminal FLAG and C-terminal His were produced in HEK293 cells. The proteins were purified on Ni-NTA beads and subjected to immunoblotting for the FLAG tag. The "Mock" control was based on testing bead-bound materials from vector control-transfected HEK293 cells. Specific bands detected for HSulf-1 and HSulf-1ΔCC (~130 and ~75 kDa, *arrows*) and for HSulf-2 and HSulf-2ΔCC (~240, ~130, and ~75 kDa, arrows and an arrowhead) are indicated. See review [8] for processing of the Sulfs. Figure taken from [17] with permission

6. Blot the proteins onto a PVDF membrane.

7. Block the membrane with membrane blocking reagent for 1 h and then incubate overnight with an anti-FLAG tag antibody (0.2 μg/mL) in membrane blocking reagent at 4 °C.

8. Wash the membrane six times in PBS-T for 10 min and then incubate with HRP-conjugated goat anti-mouse IgG1 antibody (0.016 μg/mL in membrane blocking reagent) for 1 h.

9. Wash the membrane six times in PBS-T for 10 min and then visualize bound antibodies with chemiluminescent reagent (*see* **Note 5**) (Fig. 1).

3.2 Structural Analysis of Heparan Sulfate in HSulf-Transfected HEK 293 Cells

The disaccharide compositions of heparan sulfate are determined by reversed-phase ion-pair chromatography with postcolumn fluorescent labeling adapted from a method described previously [48]. The level of total heparan sulfate is determined by summing the amounts of all disaccharides detected in each sample [49].

Here, we describe the extraction of heparan sulfate in a pericellular/extracellular fraction and preparation of disaccharide samples for the reversed-phase ion-pair chromatography.

1. Culture HEK293 cells and then transiently transfect them with a FLAG-His-tagged HSulf expression plasmid as described above. Incubate the cells for 48 h with OptiMEM I (*see* **Note 6**).

2. Trypsinize HSulf-transfected adherent cells (2×10^6 cells) for 15 min. Collect the supernatants, which contain cell surface/extracellular matrix heparan sulfate (*see* **Note 7**).

3. Heat the tube in boiling water for 15 min. Add 1 M Tris–HCl, pH 8.0 (1/20 volume of the supernatants), and 1 M $MgCl_2$ (1/100 volume of the supernatants) to the tube.

4. Add 10 μL of DNase I and 10 μL of RNase A. Incubate the tube at 37 °C for 2 h.

5. Add 10 mg/mL Actinase E (10 μL for 2 mL of the supernatants). Incubate the tube at 37 °C overnight.

6. Heat the tube in boiling water for 10 min. Centrifuge the tube at $800 \times g$ for 10 min at room temperature. Transfer the supernatant to a new tube and then mix with the same volume of 50 mM Tris–HCl, pH 7.2.

7. Apply the mixture to a DEAE Sepharose column (*see* **Note 8**).

8. Wash the Sepharose with 2 mL of 50 mM Tris–HCl, pH 7.2, and 0.2 N NaCl.

9. Elute the bound heparan sulfate with 0.6 mL of 50 mM Tris–HCl, pH 7.2, and 2 N NaCl. Collect the eluent in a 5 mL tube.

10. Add 1.2 mL (double the volume of the eluent) of 99.5 % ethanol and then divide into 1.5 mL centrifuge tubes. Place the tubes at –80 °C over 30 min.

11. Centrifuge the tubes at $20,000 \times g$ in a microcentrifuge for 15 min at 4 °C.

12. Remove the supernatants and then rinse the pellets with 70 % ethanol. Place the tubes at –80 °C for 30 min and then centrifuge as above.

13. Dissolve the pellets in 100 μL of distilled water (*see* **Note 9**).

14. Mix 25 μL of the sample in a tube with 50 mM HEPES, pH 7.5, 10 mM $MgCl_2$, 2 mU heparinase I, 0.5 mU heparinase II, and 0.2 mU heparinase III in a total volume of 50 μL. Incubate the reaction mixture at 37 °C overnight. Terminate the reaction by placing the tube at 95 °C for 5 min. Add 50 μL of distilled water.

15. Apply the mixture to a Nanosep 3K filter and then spin the device at 13,000 RPM for 15 min at room temperature (*see* **Note 10**). Apply filtered materials to the reversed-phase ion-pair chromatography (Table 1) or store the materials at –20 °C.

Table 1
Disaccharide composition of pericellular/extracellular heparan sulfate of HSulf-transfected HEK293 cells

Transfectant	Unsaturated disaccharide[a]						Total 6-O-sulfation[b]
	ΔDi-0S	ΔDi-NS	ΔDi-6S	ΔDi-(N,6)diS	ΔDi-(N,2)diS	ΔDi-(N,6,2)triS	
Mock	49.7 (100)	14.5 (100)	13.5 (100)	9.1 (100)	9.2 (100)	3.9 (100)	26.5 (100)
HSulf-1	55.1 (110.7)	15.8 (109.1)	14.3 (105.7)	4.8 (52.8)	9.2 (100.3)	0.8 (19.2)	19.8 (74.7)
HSulf-1 ΔCC	50.9 (102.3)	14.5 (100)	13.4 (99.5)	9.3 (102.3)	7.8 (84.2)	4.1 (104.2)	26.8 (101.1)
Hsulf-2	51.6 (103.6)	15.7 (108.2)	14.0 (103.4)	6.9 (75.4)	10.2 (111.1)	1.6 (42.1)	22.5 (84.9)
HSulf-2 ΔCC	49.5 (99.6)	15.2 (104.9)	13.5 (99.9)	9.2 (101.4)	7.7 (83.5)	4.8 (123.2)	27.5 (103.8)

The values are representative of two independent experiments. Table taken from [17] with permission

[a]% of total disaccharides (% of "Mock")

[b]% of total 6-O-sulfation (% of "Mock") determined by the sum of the percentages of 6-O-sulfated disaccharide products

3.3 Ex Vivo Assay for the HSulfs

RB4CD12 phage display antibody recognizes *N*- and *O*-sulfated saccharides of HS/heparin [50, 51]. Its recognition epitope is proposed to be [-GlcNSO$_3$(6-OSO$_3$)-IdoA(2-OSO$_3$)-GlcNSO$_3$(6-OSO3)-], a trisulfated disaccharide-containing HS oligosaccharide [52]. RB4CD12-based assays, which measure the enzymatic activity of the Sulfs, have been established [27, 53]. We have shown that the RB4CD12 epitope accumulates in cerebral amyloid ß plaques of Alzheimer's disease [54]. Here, we describe ex vivo degradation of the RB4CD12 epitope in brain sections of Tg2576 Alzheimer model mice with HSulfs.

1. Dissect out fresh brains of aged (>1-year-old) Tg2576 mice, embed in the O.C.T. compound-containing plastic molds and then freeze on dry ice.

2. Cut sections of the embedded brain block with a cryostat at a thickness of 10 μm. Mount sections on warm MAS-coated glass slides.

3. After air drying for 30–60 min, fix sections in ice-cold acetone for 15 min.

4. After air drying for 30 min, wash sections one time in PBS for 5 min and then incubate them with blocking reagent for 15 min at room temperature.

5. Pretreat sections with bead-bound or purified FLAG/His-tagged HSulfs (0.4 μg of HSulf-1 or HSulf-2) in 100 μL of a reaction mixture containing 50 mM HEPES-NaOH, pH 7.5, and 10 mM MgCl$_2$ at 37 °C overnight (*see* **Note 11**).

6. Wash sections twice with PBS and then incubate with a mixture of RB4CD12 (1:100 dilution) for 1 h at RT.

7. Wash sections twice with PBS and then detect primary antibodies with Cy3-conjugated monoclonal anti-VSV-G (4 μg/mL). Mount sections in FluorSaver™ Reagent.

8. Capture digital images by fluorescence microscopy at the same setting for all images (Fig. 2).

4 Notes

1. An empty vector is used as a control ("Mock" control).

2. The concentration of imidazole for elution of the bound proteins can be increased up to 250 mM.

3. All three enzymes are made as small aliquots and stored at −20 °C. The number of repetitions of freezing and thawing should be restricted to less than 5.

4. The reagent should be filtered using a 0.22 μm filter and stored at 4 °C.

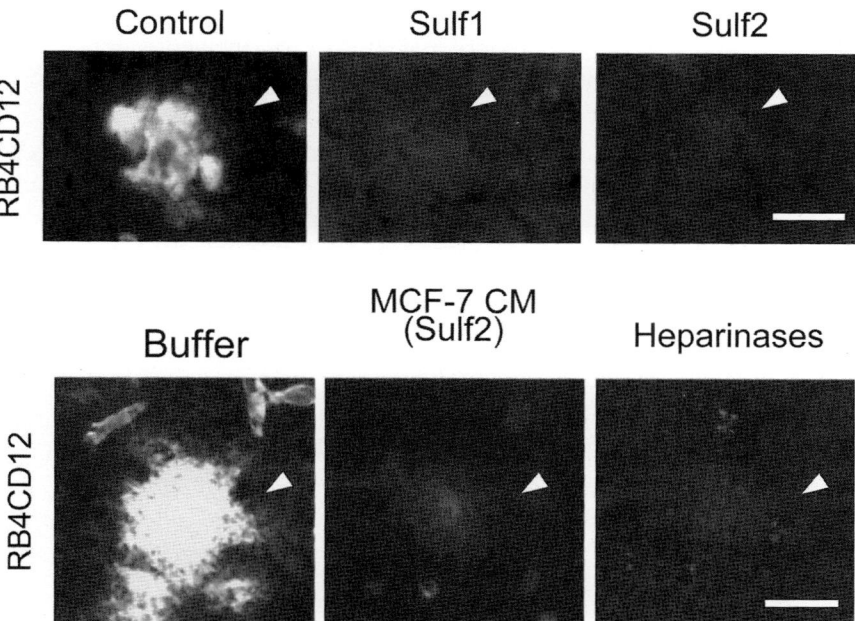

Fig. 2 Ex vivo degradation of the RB4CD12 epitope in amyloid plaques of Tg2576 mouse brains by HSulf-1, HSulf-2, and conditioned medium of HSulf-2-expressing cells. Consecutive cryostat-cut sections of 18-month-old Tg2576 mouse brains were incubated overnight with recombinant HSulf-1 and HSulf-2 prepared from CM of transfected HEK293 cells (Sulf1, Sulf2), buffer only (Buffer) or CM of MCF-7 human breast cancer cells (MCF-7 CM). The Ni-NTA resin-bound materials that were prepared from HEK293 cells transfected with the empty vector were eluted and used (Control). A mix of bacterial heparinases (Heparinases) served as a positive control (*see* **Note 11**). RB4CD12 binding was visualized using a Cy™3-conjugated anti-VSV tag antibody (*white*). The data are representative of two independent experiments. Bar denotes 20 μm. Figure adapted from [54] with permission

5. The eluted proteins are stored at 4 °C until use. Avoid freezing and thawing of the eluate. A slight shift of bands for eluted HSulf-1ΔCC and HSulf-2ΔCC is seen in immunoblotting.

6. The CM contains secreted heparan sulfate. The collected CM can be subjected to the disaccharide analysis described above if necessary.

7. The pellets of trypsinized cells contain intracellular heparan sulfate. The pellets can also be subjected to the disaccharide analysis. Suspend the pellets in 2 mL of 0.2 N NaOH and place the mix at room temperature overnight. Add 1 μL of phenol red solution and then neutralize the mix with 4 N HCl by monitoring the color.

8. Pack 0.2 mL of DEAE Sepharose in an empty plastic column. Prewash the Sepharose with 2 mL of 50 mM Tris–HCl, pH 7.2.

9. The samples are placed at –80 °C for long-term storage.

10. The Nanosep 3K device should be prewashed with 200 µL of 70 % methanol and 200 µL of distilled water to avoid having nonspecific fluorescent peak signals in the reversed-phase ion-pair chromatography.

11. One hundred-fold concentrated CM of MCF-7 cells (10 µL) exhibits Sulf activity. The CM is used as native HSulf-2. A mix of heparinase I (2 mU), heparinase II (0.5 mU) and heparinase III (0.2 mU) serves as a positive control.

Acknowledgements

Supported by Japanese Health and Labour Sciences Research Grants [H19-001 and H22-007], Grants-in-Aid from the Ministry of Education, Science, Sports and Culture [22790303 and 24590349, and Scientific Research on Innovative Areas] and in part by the Takeda Science Foundation.

References

1. Diez-Roux G, Ballabio A (2005) Sulfatases and human disease. Annu Rev Genomics Hum Genet 6:355–379

2. Dhoot GK, Gustafsson MK, Ai X et al (2001) Regulation of Wnt signaling and embryo patterning by an extracellular sulfatase. Science 293:1663–1666

3. Morimoto-Tomita M, Uchimura K, Werb Z et al (2002) Cloning and characterization of two extracellular heparin-degrading endosulfatases in mice and humans. J Biol Chem 277: 49175–49185

4. Ai X, Do AT, Kusche-Gullberg M et al (2006) Substrate specificity and domain functions of extracellular heparan sulfate 6-O-endosulfatases, QSulf1 and QSulf2. J Biol Chem 281: 4969–4976

5. Frese MA, Milz F, Dick M et al (2009) Characterization of the human sulfatase Sulf1 and its high affinity heparin/heparan sulfate interaction domain. J Biol Chem 284:28033–28044

6. Ohto T, Uchida H, Yamazaki H et al (2002) Identification of a novel nonlysosomal sulphatase expressed in the floor plate, choroid plexus and cartilage. Genes Cells 7:173–185

7. Nagamine S, Koike S, Keino-Masu K et al (2005) Expression of a heparan sulfate remodeling enzyme, heparan sulfate 6-O-endosulfatase sulfatase FP2, in the rat nervous system. Brain Res Dev Brain Res 159:135–143

8. Rosen SD, Lemjabbar-Alaoui H (2010) Sulf-2: an extracellular modulator of cell signaling and a cancer target candidate. Expert Opin Ther Targets 14:935–949

9. Dierks T, Schmidt B, Borissenko LV et al (2003) Multiple sulfatase deficiency is caused by mutations in the gene encoding the human C(alpha)-formylglycine generating enzyme. Cell 113:435–444

10. Cosma MP, Pepe S, Annunziata I et al (2003) The multiple sulfatase deficiency gene encodes an essential and limiting factor for the activity of sulfatases. Cell 113:445–456

11. Ambasta RK, Ai X, Emerson CP Jr (2007) Quail Sulf1 function requires asparagine-linked glycosylation. J Biol Chem 282:34492–34499

12. Tang R, Rosen SD (2009) Functional consequences of the subdomain organization of the sulfs. J Biol Chem 284:21505–21514

13. Nagamine S, Keino-Masu K, Shiomi K et al (2010) Proteolytic cleavage of the rat heparan sulfate 6-O-endosulfatase SulfFP2 by furin-type proprotein convertases. Biochem Biophys Res Commun 391:107–112

14. Saad OM, Ebel H, Uchimura K et al (2005) Compositional profiling of heparin/heparan sulfate using mass spectrometry: assay for specificity of a novel extracellular human endosulfatase. Glycobiology 15:818–826

15. Ai X, Do AT, Lozynska O et al (2003) QSulf1 remodels the 6-O sulfation states of cell surface heparan sulfate proteoglycans to promote Wnt signaling. J Cell Biol 162:341–351

16. Viviano BL, Paine-Saunders S, Gasiunas N et al (2004) Domain-specific modification of heparan sulfate by Qsulf1 modulates the binding of the bone morphogenetic protein antagonist Noggin. J Biol Chem 279:5604–5611

17. Hossain MM, Hosono-Fukao T, Tang R et al (2010) Direct detection of HSulf-1 and HSulf-2 activities on extracellular heparan sulfate and their inhibition by PI-88. Glycobiology 20:175–186

18. Dai Y, Yang Y, MacLeod V et al (2005) HSulf-1 and HSulf-2 are potent inhibitors of myeloma tumor growth in vivo. J Biol Chem 280:40066–40073

19. Lamanna WC, Baldwin RJ, Padva M et al (2006) Heparan sulfate 6-O-endosulfatases: discrete in vivo activities and functional co-operativity. Biochem J 400:63–73

20. Lamanna WC, Frese MA, Balleininger M et al (2008) Sulf loss influences N-, 2-O-, and 6-O-sulfation of multiple heparan sulfate proteoglycans and modulates fibroblast growth factor signaling. J Biol Chem 283:27724–27735

21. Nagamine S, Tamba M, Ishimine H et al (2012) Organ-specific sulfation patterns of heparan sulfate generated by extracellular sulfatases Sulf1 and Sulf2 in mice. J Biol Chem 287:9579–9590

22. Bishop JR, Schuksz M, Esko JD (2007) Heparan sulphate proteoglycans fine-tune mammalian physiology. Nature 446:1030–1037

23. Bernfield M, Gotte M, Park PW et al (1999) Functions of cell surface heparan sulfate proteoglycans. Annu Rev Biochem 68:729–777

24. Gallagher JT (2001) Heparan sulfate: growth control with a restricted sequence menu. J Clin Invest 108:357–361

25. Nakato H, Kimata K (2002) Heparan sulfate fine structure and specificity of proteoglycan functions. Biochim Biophys Acta 1573: 312–318

26. Nawroth R, van Zante A, Cervantes S et al (2007) Extracellular sulfatases, elements of the Wnt signaling pathway, positively regulate growth and tumorigenicity of human pancreatic cancer cells. PLoS One 2:e392

27. Lemjabbar-Alaoui H, van Zante A, Singer MS et al (2010) Sulf-2, a heparan sulfate endosulfatase, promotes human lung carcinogenesis. Oncogene 29:635–646

28. Phillips JJ, Huillard E, Robinson AE et al (2012) Heparan sulfate sulfatase SULF2 regulates PDGFRalpha signaling and growth in human and mouse malignant glioma. J Clin Invest 122:911–922

29. Otsuki S, Hanson SR, Miyaki S et al (2010) Extracellular sulfatases support cartilage homeostasis by regulating BMP and FGF signaling pathways. Proc Natl Acad Sci U S A 107:10202–10207

30. Lai J, Chien J, Staub J et al (2003) Loss of HSulf-1 up-regulates heparin-binding growth factor signaling in cancer. J Biol Chem 278:23107–23117

31. Wang S, Ai X, Freeman SD et al (2004) QSulf1, a heparan sulfate 6-O-endosulfatase, inhibits fibroblast growth factor signaling in mesoderm induction and angiogenesis. Proc Natl Acad Sci U S A 101:4833–4838

32. Li J, Kleeff J, Abiatari I et al (2005) Enhanced levels of Hsulf-1 interfere with heparin-binding growth factor signaling in pancreatic cancer. Mol Cancer 4:14

33. Narita K, Staub J, Chien J et al (2006) HSulf-1 inhibits angiogenesis and tumorigenesis in vivo. Cancer Res 66:6025–6032

34. Lai JP, Chien JR, Moser DR et al (2004) hSulf1 Sulfatase promotes apoptosis of hepatocellular cancer cells by decreasing heparin-binding growth factor signaling. Gastroenterology 126:231–248

35. Lai JP, Chien J, Strome SE et al (2004) HSulf-1 modulates HGF-mediated tumor cell invasion and signaling in head and neck squamous carcinoma. Oncogene 23:1439–1447

36. Yue X, Li X, Nguyen HT et al (2008) Transforming growth factor-beta1 induces heparan sulfate 6-O-endosulfatase 1 expression in vitro and in vivo. J Biol Chem 283:20397–20407

37. Ai X, Kitazawa T, Do AT et al (2007) SULF1 and SULF2 regulate heparan sulfate-mediated GDNF signaling for esophageal innervation. Development 134:3327–3338

38. Lai JP, Sandhu DS, Yu C et al (2008) Sulfatase 2 up-regulates glypican 3, promotes fibroblast growth factor signaling, and decreases survival in hepatocellular carcinoma. Hepatology 47:1211–1222

39. Uchimura K, Morimoto-Tomita M, Bistrup A et al (2006) HSulf-2, an extracellular endoglucosamine-6-sulfatase, selectively mobilizes heparin-bound growth factors and chemokines: effects on VEGF, FGF-1, and SDF-1. BMC Biochem 7:2

40. Morimoto-Tomita M, Uchimura K, Bistrup A et al (2005) Sulf-2, a proangiogenic heparan sulfate endosulfatase, is upregulated in breast cancer. Neoplasia 7:1001–1010

41. Fujita K, Takechi E, Sakamoto N et al (2010) HpSulf, a heparan sulfate 6-O-endosulfatase, is

involved in the regulation of VEGF signaling during sea urchin development. Mech Dev 127:235–245

42. Freeman SD, Moore WM, Guiral EC et al (2008) Extracellular regulation of developmental cell signaling by XtSulf1. Dev Biol 320:436–445

43. Ratzka A, Kalus I, Moser M et al (2008) Redundant function of the heparan sulfate 6-O-endosulfatases Sulf1 and Sulf2 during skeletal development. Dev Dyn 237:339–353

44. Lum DH, Tan J, Rosen SD et al (2007) Gene trap disruption of the mouse heparan sulfate 6-O-endosulfatase gene, Sulf2. Mol Cell Biol 27:678–688

45. Holst CR, Bou-Reslan H, Gore BB et al (2007) Secreted sulfatases Sulf1 and Sulf2 have overlapping yet essential roles in mouse neonatal survival. PLoS One 2:e575

46. Kalus I, Salmen B, Viebahn C et al (2009) Differential involvement of the extracellular 6-O-endosulfatases Sulf1 and Sulf2 in brain development and neuronal and behavioral plasticity. J Cell Mol Med 13:4505–4521

47. Maltseva I, Chan M, Kalus I et al (2013) The SULFs, extracellular sulfatases for heparan sulfate, promote the migration of corneal epithelial cells during wound repair. PLoS One 8: e69642

48. Toyoda H, Kinoshita-Toyoda A, Fox B et al (2000) Structural analysis of glycosaminoglycans in animals bearing mutations in sugarless, sulfateless, and tout-velu. Drosophila homologues of vertebrate genes encoding glycosaminoglycan biosynthetic enzymes. J Biol Chem 275:21856–21861

49. Hosono-Fukao T, Ohtake-Niimi S, Nishitsuji K et al (2011) RB4CD12 epitope expression and heparan sulfate disaccharide composition in brain vasculature. J Neurosci Res 89: 1840–1848

50. Jenniskens GJ, Oosterhof A, Brandwijk R et al (2000) Heparan sulfate heterogeneity in skeletal muscle basal lamina: demonstration by phage display-derived antibodies. J Neurosci 20:4099–4111

51. Dennissen MA, Jenniskens GJ, Pieffers M et al (2002) Large, tissue-regulated domain diversity of heparan sulfates demonstrated by phage display antibodies. J Biol Chem 277: 10982–10986

52. Jenniskens GJ, Hafmans T, Veerkamp JH et al (2002) Spatiotemporal distribution of heparan sulfate epitopes during myogenesis and synaptogenesis: a study in developing mouse intercostal muscle. Dev Dyn 225:70–79

53. Uchimura K, Lemjabbar-Alaoui H, van Kuppevelt TH et al (2010) Use of a phage display antibody to measure the enzymatic activity of the Sulfs. Methods Enzymol 480:51–64

54. Hosono-Fukao T, Ohtake-Niimi S, Hoshino H et al (2012) Heparan sulfate subdomains that are degraded by Sulf accumulate in cerebral amyloid ss plaques of Alzheimer's disease: evidence from mouse models and patients. Am J Pathol 180:2056–2067

Chapter 32

The Detection of Glycosaminoglycans in Pancreatic Islets and Lymphoid Tissues

Marika Bogdani, Charmaine Simeonovic, Nadine Nagy,
Pamela Y. Johnson, Christina K. Chan, and Thomas N. Wight

Abstract

In this chapter, we describe the detection of the glycosaminoglycans hyaluronan and heparan sulfate in pancreatic islets and lymphoid tissues. The identification of hyaluronan in tissues is achieved by utilizing a highly specific hyaluronan binding protein (HABP) probe that interacts with hyaluronan in tissue sections. The HABP probe is prepared by enzymatic digestion of the chondroitin sulfate proteoglycan aggrecan which is present in bovine nasal cartilage, and is then biotinylated in the presence of bound hyaluronan and the link protein. Hyaluronan is then removed by gel filtration chromatography. The biotinylated HABP–link protein complex is applied to tissue sections and binding of the complex to tissue hyaluronan is visualized by enzymatic precipitation of chromogenic substrates.

To determine hyaluronan content in tissues, tissues are first proteolytically digested to release hyaluronan from the macromolecular complexes that this molecule forms with other extracellular matrix constituents. Digested tissue is then incubated with HABP. The hyaluronan–HABP complexes are extracted and the hyaluronan concentration in the tissue is determined using an ELISA-like assay.

Heparan sulfate is identified in mouse tissues by Alcian blue histochemistry and indirect immunohistochemistry. In human tissues, heparan sulfate is best detected by indirect immunohistochemistry using a specific anti-heparan sulfate monoclonal antibody. A biotinylated secondary antibody is then applied in conjunction with streptavidin-peroxidase and its binding to the anti-heparan sulfate antibody is visualized by enzymatic precipitation of chromogenic substrates.

Key words Hyaluronan, Heparan sulfate, Pancreatic islets, Lymphoid tissue, Hyaluronan binding protein, Immunohistochemistry

1 Introduction

Hyaluronan and heparan sulfate are ubiquitous glycosaminoglycans present on cell surfaces and in the extracellular matrix that have been increasingly implicated in various biological processes including cell growth, differentiation and migration, angiogenesis, tissue regeneration, and inflammation [1–6]. The distribution and mass of hyaluronan and the cell-associated and extracellular levels of heparan sulfate are crucial for their biological functions [1–3, 6].

Kuberan Balagurunathan et al. (eds.), Glycosaminoglycans: Chemistry and Biology, Methods in Molecular Biology, vol. 1229, DOI 10.1007/978-1-4939-1714-3_32, © Springer Science+Business Media New York 2015

Therefore identifying the hyaluronan and heparan sulfate morphologic patterns and determining hyaluronan size and abundance are important in addressing questions relating to the role these molecules play in physiologic and pathologic processes affecting different tissues, including pancreatic islets and secondary lymphoid organs. We have found that hyaluronan is located in the extracellular matrix in pancreatic islets and that heparan sulfate is localized at extraordinarily high levels intracellularly in normal insulin-producing islet β cells, as well as in the peri-islet basement membrane [7–10]. We and others have also identified hyaluronan and heparan sulfate as components of the extracellular matrix in specialized regions of immune cell activation in the spleen and lymph nodes. The morphologic patterns and abundance of hyaluronan and heparan sulfate are altered in islets and lymphoid tissue in type 1 diabetes, suggesting a potential role for these molecules in the pathogenesis of this disease [8, 10–13]. During the course of our studies, we have modified previously developed techniques for hyaluronan and heparan sulfate detection by light microscopy and for determination of hyaluronan content and size by biochemistry, and these modified procedures are described herein.

2 Materials

2.1 Histochemistry of Hyaluronan

The identification of hyaluronan in tissues is achieved by utilizing a highly specific hyaluronan binding protein (HABP) probe that interacts with hyaluronan in tissue sections [14–16]. The HABP probe is prepared by enzymatic digestion of the chondroitin sulfate proteoglycan aggrecan present in bovine nasal cartilage [15, 17–22]. The HABP is applied to tissue sections and its binding to tissue hyaluronan is visualized by enzymatic precipitation of chromogenic substrates.

2.1.1 Preparation of Biotinylated Hyaluronan Binding Protein (bHABP)

1. Bovine nasal septum cartilage.
2. Surform pocket plane, cheesecloth, Whatman no. 1 filter paper.
3. Guanidine buffer: 4.0 M Guanidine HCl, 0.5 M sodium acetate, pH 5.8.
4. HEPES buffer: 0.1 M HEPES, 0.1 M sodium acetate, pH 7.3.
5. Trypsin (type III).
6. Soybean trypsin inhibitor (type I-S).
7. Coomassie blue staining reagent.
8. Sulfo-NHS-LC biotin.
9. Hyaluronan-Sepharose [14, 18, 23]: Digest hyaluronan (1 g) (see Note 1) with testicular hyaluronidase (1 mg/mL in 0.02 M phosphate buffer, 0.01 % BSA, pH 7.0) in 500 mL of

0.15 M NaCl/0.15 M sodium acetate, pH 5.0 for 3 h at room temperature; boil for 20 min and then centrifuge at 10,000×g for 15 min; discard supernatant and wash the precipitate in 75 % EtOH. Mix the digested hyaluronan with 100 mL of EAH sepharose 4B and 2 g of 1-ethyl-3-(3-dimethylamino-propy)-L carbodiimide. Adjust the mixture pH to 5.0 and incubate for 24 h at room temperature. Add 10 mL of acetic acid to the mixture and incubate for 6 h. Wash the gel with 1 L of 1 M NaCl, 1 L of 0.05 M formic acid and 1 L of distilled water followed by a wash with 0.5 M sodium acetate, pH 5.7, 0.02 % sodium azide. Store in Corning glass bottles at 4 °C.

10. Fraction collector FC-203B (Gilson).

11. Dialysis membranes 12–14,000 MWCO.

12. Econo column (2.5×20 cm, Bio-Rad).

13. Glycerol.

2.1.2 Tissue Preparation for Histochemistry and Immunohistochemistry

1. Human pancreas, spleen, and pancreatic lymph nodes are collected from brain-dead organ donors and procured by the Network of Pancreatic Organ Donors with Diabetes (nPOD) [24, 25] (*see* **Notes 2** and **3**).

2. Mouse pancreas, spleen, and lymph nodes are collected in animals euthanized with carbon dioxide [26, 27].

3. Dissecting instruments: stille straight and tissue serrated forceps, stille and iris straight scissors, dissecting boards.

4. 10 % neutral buffered formalin.

5. Methyl Carnoy's fixative: 10 % glacial acetic acid, 60 % methanol, 30 % chloroform.

6. Methyl Carnoy's post-fixative: 42 % isopropanol 28 % methanol, 30 % distilled water.

7. Tissue cassettes.

8. Liquid nitrogen, aluminum foil for tissue snap freezing.

9. Automatic paraffin processor.

10. Paraffin.

11. Paraffin Embedding Station.

12. Superfrost Plus slides.

2.1.3 Histochemical Localization of Hyaluronan using bHABP

1. 5 μm paraffin-embedded tissue sections mounted on Superfrost Plus slides, baked 1 h at 50 °C.

2. Graded EtOH series for rehydration: 100, 95, 70, and 50 %.

3. Xylene.

4. Phosphate buffered saline (PBS): 8 g NaCl, 0.22 g KCl, 1.15 g Na_2HPO_4, 0.2 g KH_2PO_4, 800 mL distilled H_2O. Adjust pH to 7.4 with HCl and bring volume to 1 L with distilled H_2O.

5. Calcium and magnesium-free phosphate buffer saline (PBS-A).

6. Tris-HCl buffer (TB): 0.05 M, pH 7.6 with 1 N HCl.

7. Acetate buffer: 50 mM NaOAc, 0.15 M NaCl, pH 5.2.

8. 0.7 % H_2O_2 in absolute methanol.

9. Blocking solution: 10 % normal goat serum (NGS) in PBS.

10. PBS/0.1 % BSA: Add 1 mg of bovine serum albumin, globulin-free (BSA)/mL of PBS.

11. Biotinylated hyaluronan binding protein (bHABP) 100 µg/mL in PBS-A; bHABP stock solution 5 mg/mL in distilled water, stored at −20 °C.

12. Vectastain Elite ABC kit (Vector Laboratories).

13. DAB substrate kit (Vector Laboratories).

14. 2 % methyl green in sodium acetate buffer, pH 4.2.

15. *Streptomyces* hyaluronidase: 1 U/µL dissolved in 10 % calf serum in PBS-A.

16. High molecular weight hyaluronan (>1,000 kDa).

17. Humidified chamber with lid.

2.2 Biochemical Determination of Hyaluronan Content in Tissues

The determination of hyaluronan content in tissues requires the release of hyaluronan from its complexes with other extracellular matrix molecules. Tissues are first proteolytically digested. The digested tissue is then incubated with the HABP; hyaluronan–HABP complexes are extracted and the hyaluronan concentration in the tissue is determined using an ELISA-like assay [28].

1. Proteinase K.

2. 100 mM ammonium acetate pH 7.0.

3. Lyophilizer.

4. Calcium and magnesium-free phosphate buffer saline (PBS-A).

5. 10 % calf serum in PBS.

6. Hyaluronan-BSA: Dissolve 100 mg hyaluronan in 500 mL of 0.2 M NaCl. Adjust the pH to 4.7. Add 100 mg of BSA followed by 20 mg of 1-ethyl-3(3-dimethylaminopropyl) carbodiimide. Dialyze extensively against PBS with 0.02 % sodium azide. Aliquot and store at −20 °C.

7. Hyaluronan standards at the concentrations of 0, 50, 100, 200, 400, 600, 800, 1,000 ng/mL in PBS.

8. 96-well plate.

9. Peroxidase-labeled streptavidin.

10. Peroxidase substrate: 0.03 % H_2O_2 in 3-ethylbenzthiazoline-6-sulfonic acid; Sigma.

11. 0.1 M sodium citrate, pH 4.2.

12. 2 mM sodium azide.

13. OPTImax microplate reader.

2.3 Determination of Hyaluronan Size Distribution in Tissues

To determine size distribution of hyaluronan in tissue samples, the tissue extract obtained from proteolytic digestion is first enriched for hyaluronan using anion-exchange chromatography. The enriched product is then fractionated using gel-filtration chromatography. The hyaluronan concentration in each fraction is then determined by an ELISA-like assay.

1. Diethylaminoethyl-Sephacel (DEAE) equilibrated with 8 M urea buffer.

2. Poly-prep chromatography columns, 0.8×4 cm.

3. 8 M urea buffer: 480.48 g urea, 0.59 g EDTA, 6.06 g Tris Base, 800 mL distilled H_2O. Adjust the pH to 7.5 and bring the volume to 1 L with distilled H_2O.

4. 0.7×30 cm chromatography column packed with Sephacryl S-1000 in PBS with 0.02 % sodium azide. Fill column with a thick slurry of gel suspension (*see* **Note 4**) and equilibrate with PBS containing 0.02 % sodium azide. Connect the inflow tubing to buffer reservoir and the outflow tubing to a fraction collector.

5. Hyaluronan standards at the following molecular weights: 30, 200, 1,500 kDa.

2.4 Histochemistry and Immunohisto-chemistry of Heparan Sulfate

Heparan sulfate can be localized in mouse pancreas by Alcian blue histochemistry. Heparan sulfate in human pancreas sections is routinely detected by immunohistochemistry, using anti-heparan sulfate monoclonal antibodies.

2.4.1 Histochemical Localization of Heparan Sulfate in Mouse Pancreatic Islets

The selective staining of heparan sulfate by Alcian blue, a cationic dye, requires the stringent conditions of 0.65 M $MgCl_2$ at pH 5.8, as specified by the Critical Electrolyte Concentration (CEC) principle of differential staining of glycosaminoglycans using salt solutions [29]. The staining procedure below is a modification of the method published by Calvitti et al. [30].

2.4.2 Preparation of Alcian Blue Stain

1. 1 % Alcian blue 8GX in deionized H_2O.

2. 1 M $MgCl_2$ (in 100 mL deionized H_2O, *see* **Note 5**).

3. 1 M acetate buffer: 1 M glacial acetic acid (19 mL), 1 M sodium acetate trihydrate (181 mL). Adjust the pH to 5.8 with glacial acetic acid (*see* **Note 5**).

4. Alcian blue working solution: 1 % Alcian blue (0.5 mL), 1 M acetate buffer (5 mL), 1 M $MgCl_2$ (3.25 mL), deionized water (41.25 mL) i.e., total volume of 50 mL (*see* **Note 6**).

2.4.3 Histochemical Localization of Heparan Sulfate

1. 4-μm thick formalin-fixed paraffin-embedded unstained mouse pancreas sections (*see* Subheading 2.1.2) mounted on uncoated Superfrost Plus slides, baked for approximately 1 h at 70 °C.

2. 0.1 M acetate buffer.

3. Xylene.

4. Graded ethanol series for rehydration: 2×100 %, 2×90 %, 70 % and tap water.

5. Coplin or Heyerdahl glass staining jars (50 mL).

6. 0.01 % Safranin O in deionized H_2O as counterstain.

7. Micromount mounting medium.

2.4.4 Immunohisto-chemical Localization of Heparan Sulfate in Human Pancreatic Islets

1. 5-μm thick formalin-fixed paraffin-embedded human pancreas sections provided by nPOD (*see* Subheading 2.1.2).

2. Xylene.

3. Graded EtOH series for rehydration: 2×100 %, 1×90 %, 1×70 %, tap water.

4. 30 % H_2O_2, 3 % H_2O_2 in methanol, 3 % H_2O_2 in deionized H_2O.

5. 0.5 mg/mL (0.05 %) Pronase in deionized H_2O.

6. Animal Free Block diluted to 20 % in deionized H_2O.

7. Phosphate-buffered saline (PBS): 8 g NaCl/L, 1.25 g Na_2 $HPO_4.2H_2O/L$, 0.35 g $NaH_2PO_4.H_2O/L$ in deionized H_2O.

8. Protein concentrate from M.O.M.-peroxidase kit (Vector Laboratories) diluted 1/13.5 in PBS.

9. Mouse anti-human anti-heparan sulfate monoclonal antibody 10E4 mAb, 1 mg/mL (*see* **Notes 7** and **8**).

10. Isotype control mouse IgM (0.25 mg/mL).

11. Polyclonal rabbit anti-mouse Ig-horseradish peroxidase (HRP) secondary antibody.

12. 3-amino-9-ethylcarbazole (AEC chromagen, Sigma): 8 mg/mL in *N*-*N*-dimethyl formamide.

13. 0.2 μm chemically resistant filter.

14. Acetate buffer: 0.1 M, pH 5.2: 0.1 N acetic acid (10.5 mL), 0.1 M sodium acetate (39.5 mL).

15. Gill's hematoxylin.

16. Ammonium H_2O: 100 μL ammonia in 250 mL deionized H_2O.

17. Glycergel mounting medium.

18. 37 °C incubator.

19. Custom-made covered large staining tray with lid (humidified) and a small immunostaining tray for incubation at 37 °C.

20. Diamond pen.

3 Methods

3.1 Histochemistry of Hyaluronan

3.1.1 Preparation of Biotinylated Hyaluronan Binding Protein (bHABP)

1. Shred the bovine nasal cartilage into fine pieces using a Surform pocket plane. Add 10 mL of guanidine buffer for each gram of cartilage and incubate overnight at 4 °C (*see* **Note 9**).

2. Pour mixture through several layers of cheese cloth, then centrifuge at $10,000 \times g$ for 45 min at 4 °C. Filter supernatant through Whatman filter paper.

3. Dialyze supernatant against distilled water (water volume is 200 times the sample volume, *see* **Note 10**). Lyophilize the extract, aliquot, and store at –20 °C.

4. Dissolve 3 g of lyophilized extract in 100 mL of HEPES buffer, incubate overnight at 4 °C.

5. Add 1.6 mg of trypsin and incubate for 2 h at 37 °C.

6. Add 3.2 mg of soybean trypsin inhibitor and adjust the pH to 8.0.

7. Determine the protein content using Coomassie blue staining reagent.

8. Add 0.1 mg sulfo-NHS-LC biotin per 1 mg protein. Allow the coupling reaction to proceed for 1–2 h at room temperature to form bHABP.

9. Dialyze the mixture against three changes of guanidine buffer.

10. Wash 100 mL of hyaluronan-sepharose in a Buchner funnel with fritted disc with 4 M guanidine buffer. Transfer hyaluronan-sepharose and the bHABP mixture into a large dialysis bag and dialyze against 10 volumes of distilled water, overnight at 4 °C (*see* **Note 11**).

11. Pour the mixture into a glass column.

12. Wash the column with 1 M NaCl and then with 3 M NaCl.

13. Connect the column to a fraction collector and elute bHABP with guanidine buffer. Each fraction is assayed for protein concentration. Pool the bHABP-containing fractions and dialyze against 0.15 M NaCl (*see* **Note 12**).

14. Mix bHABP with glycerol (1:1 vol/vol), aliquot, and store at –20 °C (*see* **Notes 13–15**).

3.1.2 Tissue Preparation for Histochemistry and Immunohistochemistry

Human Tissues

1. Tissue slicing [24, 25].

   Pancreas: Divide the pancreas into three regions of head, body, and tail. Slice each pancreas region in a transverse "bread loaf" manner and prepare slices that are ~1.5 × 1.5 × 0.5 cm.

   Spleen: Slice splenic tissue in pieces of ~1.5 × 1.5 × 0.5 cm.

   Pancreatic lymph node (*PLN*): Isolate PLN from fat and trim tissue surrounding the PLN capsule. Divide PLN in half.

2. Fix tissue pieces in neutral buffered formalin for 16–24 h (*see* **Notes 16–18**). Fixed tissues are then paraffin-embedded and sectioned for histochemistry and immunohistochemistry (*see* **Note 19**).

3. Wrap tissue pieces in aluminum foil, snap freeze in liquid nitrogen and store at –80 °C for biochemical analysis of hyaluronan.

Mouse Tissues

1. Anesthetize and euthanize the mouse according to the institution's requirements for animal care and use. Pin the mouse down on a dissection board with the abdomen facing up and wipe down fur with 70 % EtOH. Open the abdomen cutting first along the ventral midline and continue through the sternum to open the thorax, and then laterally and down on both sides to create two flaps of skin. Pin the two skin flaps down on the dissection board.

2. Dissect out pancreas first and then the spleen and lymph nodes (*see* **Notes 20 and 21**).

3. Fix tissues in methyl Carnoy's solution for 1–2 h at 4 °C (*see* **Note 22**). Store tissues in methyl Carnoy's post-fixation solution at 4 °C until processed for paraffin embedding for hyaluronan histochemistry (*see* **Note 23**).

4. For heparan sulfate histochemistry and immunohistochemistry, fix tissues in 10 % neutral buffered formalin for at least 2 days at room temperature until processing and paraffin embedding.

5. Wrap tissues in aluminum foil, snap freeze in liquid nitrogen and store at –80 °C for biochemical analysis of hyaluronan.

3.1.3 Histochemical Localization of Hyaluronan using bHABP

1. Cut 5-μm thick tissue sections and mount sections on Superfrost Plus slides.

2. Deparaffinize tissue sections in three changes of xylene, 5 min each (*see* **Note 24**).

3. Rinse tissue in two changes of 100 % EtOH, 2 min each.

4. Quench endogenous tissue peroxidase by incubating tissue in 0.7 % H_2O_2 in absolute methanol for 20 min.

5. Hydrate tissue sections in graded EtOH series.

6. Rinse for 10 min in PBS.

7. Incubate sections in 10 % NGS in PBS for 30 min to block nonspecific binding (*see* **Notes 25 and 26**).

8. Apply bHABP (5 μg/mL in PBS-A) diluted in PBS with 0.1 % BSA. Incubate overnight at 4 °C (*see* **Note 27**).

9. Rinse tissue sections in three changes of PBS, 10 min each.

10. Prepare Vectastain Elite avidin biotin complex (ABC): add two drops of reagent A to 5 mL of buffer in the ABC reagent

mixing bottle; add two drops of reagent B to the same mixing bottle, mix immediately. Allow the ABC reagent to incubate for 30 min before applying to tissue.

11. Apply the ABC reagent to sections for 1 h at room temperature.

12. Rinse sections in three changes of PBS, 5 min each

13. Prepare the DAB substrate solution: Add two drops of buffer stock solution to 5.0 mL of distilled water; add four drops of the DAB stock solution; add two drops of the H_2O_2 solution. If a gray-black reaction product is desired add two drops of the nickel solution. Mix well before use (*see* **Note 28**).

14. Incubate sections with the DAB substrate solution for 10 min at 37 °C. Stop reaction by washing sections in PBS.

15. Rinse sections with H_2O for 20 min.

16. Counterstain with methyl green for 5 min (*see* **Note 29**). Dehydrate through 95 % and 100 % EtOH (two changes, 1 min each), clear in xylene (three times, 5 min each) and cover slip.

17. Examine slides under a light microscope (*see* **Notes 30** and **31**).

18. Controls for hyaluronan staining and specificity include digestion with hyaluronidase and preincubation of the bHABP with excess hyaluronan prior to bHABP application to the tissue section. Sections are digested with *Streptomyces* hyaluronidase (20 U/mL in sodium acetate buffer) at 37 °C for 1 h. Undigested control sections are also stained in parallel and incubated in sodium acetate buffer only. For the preincubation experiments, bHABP (at working concentration of 5 μg/mL) is mixed with 100 μg/mL hyaluronan (>1,000 kDa) in order to block the hyaluronan-binding sites. Slides are then processed as described in **steps 6–16**.

3.2 Biochemical Determination of Hyaluronan Content in Tissues

3.2.1 Extraction of Hyaluronan from Pancreas and Lymphoid Tissue

1. Lyophilize frozen tissue and measure dry weight.

2. Digest tissue with proteinase K (250 μg/mL) in 100 mM ammonium acetate pH 7.0 overnight at 60 °C.

3. Stop the reaction by heating the tissue digest to 100 °C for 20 min.

3.2.2 Quantitative Evaluation of Extracted Hyaluronan from Pancreas and Lymphoid Tissue [20]

1. Coat each well (96-well plate) with 100 μL hyaluronan-BSA and incubate for 1 h at room temperature.

2. Wash wells with PBS (three times, 200 μL each).

3. Block with 100 μL of 10 % bovine calf serum in PBS for 1 h at room temperature.

4. Add an equal volume of 3 mg/mL bHABP in 10 % calf serum in PBS to proteinase K-digested tissue samples and to the hyaluronan standards. Incubate for 1 h at room temperature.

5. Rinse the plate with 100 µL PBS after blocking. Add 70 µL aliquot of sample to each well and incubate for 1 h at room temperature.

6. Wash thoroughly with distilled water (four times, 200 µL each).

7. Incubate for 20 min with 100 µL/well of peroxidase labeled streptavidin (1 mg/mL in 50 % glycerol, diluted 1:500 in 10 % calf serum in PBS).

8. Wash plates with distilled water (four times, 200 µL each). Add 100 µL of peroxidase substrate consisting of 0.03 % H_2O_2, 0.5 mg/mL 3-ethylbenzthiazoline-6-sulfonic acid in 0.1 M $C_2H_3NaO_2$, pH 4.2.

9. Terminate the reaction after 30 min by adding 25 µL/well of 2 mM sodium azide.

10. Use an ELISA plate reader to determine the OD_{405} reading (*see* **Note 32**).

11. Plot standard curve on semi-log graph. The curve is linear over the hyaluronan concentration range of 50–1,000 ng/mL.

3.3 Determination of Hyaluronan Size in Pancreas and Lymphoid Tissues

3.3.1 Concentration and Purification of Proteinase K-Digested Samples by DEAE Micro-chromatography.

1. Digest tissues as described under Subheading 3.2.1.

2. Equilibrate DEAE-Sephacrel resins with 8 M urea buffer. Pack 300 µL matrix bed by adding 600 µL of DEAE Sephacrel slurry to an Econo column. Wash off any excess resins on the side of the column with 5–10 mL of 8 M urea buffer.

3. Spin down the proteinase K-digested sample. Collect supernatant and pour it onto the column. Wait until the sample goes completely into the column.

4. Wash the column with 8 M urea buffer (four times, 10 mL each).

5. Elute hyaluronan with urea buffer with 0.25 M NaCl (three times, 300 µL each).

6. Store eluents at –20 °C.

3.3.2 Gel Filtration Chromatography [31]

1. Set the fraction collector to collect fractions every 1.5 min (fraction volume is 0.3 mL).

2. Apply an aliquot (200 µL containing about 6–7 µg of hyaluronan) of DEAE purified sample onto an analytical Sephacryl S-1000 column. Add 10 µL of Vitamin B12 (10 mg/mL, red color) to mark the end of the column.

3. Run the hyaluronan standards to calibrate the column.

4. Elute column with PBS at a flow rate of 12–15 mL/h. The column is completed when the red color is eluted off the column.

5. Subject an aliquot of each fraction to ELISA (*see* Subheading 3.2.2) to generate a hyaluronan profile.

3.4 Histochemical Localization of Heparan Sulfate

1. Cut paraffin sections of mouse pancreas at 4 μm and mount sections on untreated Superfrost slides.

2. Dewax in xylene, 2 × 2 min.

3. Rehydrate sections in graded EtOH series (2 × 100 %, 1 × 90 %, 1 × 70 % to tap water).

4. Treat sections with working buffer: 1 M acetate buffer (5 mL), 1 M MgCl$_2$ (3.25 mL), deionized H$_2$O (41.75 mL) for 10 min (use 50 mL staining jar).

5. Transfer sections to Alcian blue working solution for 40 min.

6. Wash sections in running tap water for 2 min and flick off excess water.

7. Counterstain in 0.01 % safranin for 5 min.

8. Blot sections and air dry completely.

9. Mount in micromount mounting medium and coverslip.

10. Examine slides under a light microscope (*see* **Notes 33–38**).

3.5 Immunohisto-chemical Localization of Heparan Sulfate

1. Paraffin sections of formalin-fixed human pancreas provided by nPOD (*see* Subheading 2.1.2).

2. Prepare 0.1 N acetic acid and 0.1 M sodium acetate, store at 4 °C (*see* **Note 39**).

3. Deparaffinize tissue sections in xylene, 2 × 1 min.

4. Rehydrate tissue sections in graded EtOH series (2 × 100 %, 1 × 90 %, 1 × 70 %, tap water).

5. Block endogenous peroxidase activity by incubating sections in 3 % H$_2$O$_2$ for 10 min.

6. Rinse in PBS, 2 × 2 min, and then in running tap H$_2$O, 2 × 2 min.

7. Prepare 0.5 mg/mL (0.05 %) pronase for antigen retrieval.

8. Transfer the slides to fresh tap H$_2$O.

9. Remove excess H$_2$O from sections.

10. Incubate sections with 0.05 % pronase in a small humidified immunostaining tray placed in a 37 °C incubator for 10 min.

11. Rinse sections in PBS, 2 × 2 min.

12. Prepare 20 % Animal Free Block (*see* Subheading 2.4.4, **item 6**).

13. Prepare 1/13.5 mL dilution of Protein Concentrate (M.O.M. diluent, *see* Subheading 2.4.4, **item 8**).

14. Dilute the anti-heparan sulfate 10E4 mAb to 0.2 mg/mL in M.O.M. diluent.

15. Dilute isotype control IgM to 0.2 mg/mL in M.O.M. diluent.

16. Remove excess PBS from sections and incubate with 20 % Animal Free Block for 5 min at room temperature.

17. Tip off Animal Free Block and remove excess from the sections.

18. Incubate sections with 0.2 mg/mL 10E4 mAb or mouse IgM for 30 min at room temperature.

19. Prepare 8 mg/mL AEC stock solution (*see* **Note 40**).

20. Rinse sections with PBS and wash 2×2 min in PBS.

21. Dilute secondary rabbit anti-mouse Ig-HRP antibody to 0.03 mg/mL in M.O.M. diluent.

22. Remove excess PBS from sections and incubate in secondary antibody for 30 min at room temperature.

23. Prepare 0.1 M acetate buffer (*see* Subheading 2.4.4, **item 14** and **Note 41**).

24. Prepare AEC working solution: 0.1 M acetate buffer (4.75 mL), 8 mg/mL AEC stock solution (0.25 mL), 3 % H_2O_2 in deionized H_2O (25 μL). Sterile filter using a CR filter (*see* **Notes 42–44**).

25. Rinse sections with PBS and then wash 2×2 min in PBS.

26. Remove excess PBS from sections and incubate with AEC working solution for 30 min at room temperature.

27. Rinse sections with deionized H_2O and then wash 3× in deionized H_2O over 10 min (total wash time).

28. Counterstain in Gill's hematoxylin.

29. Wash 2× in deionized H_2O, 2× brief immersion in diluted ammonium H_2O, wash 2× in deionized H_2O.

30. Mount in liquid Glycergel mounting medium and coverslip.

31. Examine slides under a light microscope.

4   Notes

1. It is important that highly purified hyaluronan be used for preparation of biotinylated hyaluronan.

2. Donor recovery and tissue procurement are performed by organ procurement organizations in the USA, through subsequent referral to the National Disease Research Interchange or the International Institute for the Advancement of Medicine. The nPOD program provides access to high quality biospecimens from donors selected on the basis of inclusion and exclusion criteria. Donor demographics, laboratory assays, and histopathological characterizations of the tissues are available

online at the nPOD website http://www.jdrfnpod.org. Pancreas, spleen, and nonpancreatic lymph nodes are recovered from cadaveric organ donors, placed in a sterile container with media, submerged in ice, and shipped to The Organ Processing and Pathology Core at the University of Florida.

3. Tissues recovered several hours after the donor has been pronounced dead may be prone to autolytic changes; therefore the level of preservation of tissue integrity should be examined and considered when interpreting histochemical and immunohistochemical staining patterns.

4. Allow gel to pack and make sure it settles without visible interfaces. Equilibrate gel with 2 or 3 column volumes of eluent buffer.

5. 1 M MgCl$_2$ and 1 M acetate buffer can be stored at 4 °C for up to 2 months.

6. Alcian blue working solution is made freshly on the day of staining.

7. 10E4 mAb recognizes N-sulfated/N-acetylated glucosamine in disaccharides in heparan sulfate [32, 33].

8. Heparatinase treatment of tissue sections abolishes 10E4 immunohistochemical detection of heparan sulfate, confirming the specificity of the 10E4 mAb for heparan sulfate [32].

9. The mixture is poured into a large beaker placed on a shaking table. The solution is too thick to use a stirring bar.

10. Dialyze against several changes of distilled H$_2$O to ensure complete removal of guanidine HCl buffer.

11. For the first 4 h it is important to resuspend the gel by turning the dialysis bag upside down every 30–40 min.

12. SDS-PAGE analysis of the final product shows two bands, one at ~70–80 kDa (the hyaluronan-binding domain of aggrecan) and one at ~43 kDa (link protein).

13. The HABP concentration is about 100 µg/mL.

14. The probe is stable and can be stored for about 5 years.

15. Purified bHABP is also commercially available at EMD Millipore (Billerica, MA).

16. Tissues should be immersed immediately into fixative.

17. The choice of a particular tissue fixative is determined by the nature of the epitopes of interest to be preserved while ensuring adequate tissue integrity for evaluation of staining patterns. Neutral buffered formalin is routinely used for fixation of human tissue specimens since it permits the successful application of a wide range of special stains. Hyaluronan histochemistry of neutral buffered formalin fixed human tissues generates reproducible patters of intense staining.

18. Formalin fixed tissues are stored in 70 % EtOH if not processed immediately for paraffin-embedding for analysis of hyaluronan only. Formalin-fixed tissues for heparan sulfate histochemistry are not to be stored in 70 % ethanol prior to processing, as this method of storage appears to interfere with heparan sulfate detection by Alcian blue staining.

19. Other methods of tissue embedding (cryostat, plastic) may be used.

20. Pancreas should be fixed whole.

21. The spleen is cut lengthwise with a scalpel prior to fixation. The cut surface is embedded face down prior to sectioning.

22. Mouse pancreas fixation in methyl Carnoy's solution should not exceed 2 h.

23. Application of the same concentration of bHABP on mouse tissues fixed in neutral buffered formalin generates a less intense hyaluronan staining than in tissues fixed in Carnoy's solution without altering the hyaluronan staining pattern.

24. Never allow sections to dry out.

25. All incubation steps are performed in humidified chambers.

26. Bovine serum albumin should be globulin-free (immunohisto-chemical grade) when used to block nonspecific protein interactions.

27. To detect hyaluronan by fluorescent microscopy, fluorescent HABP [16] can be applied instead of bHABP. Sections are rinsed in PBS, cover slipped, and examined under a fluorescent microscope.

28. Sodium azide is an inhibitor of peroxidase activity and should not be included in buffers used to make peroxidase substrate or the ABC reagent.

29. Counterstaining with Harris haematoxylin can also be used.

30. Intense hyaluronan staining has been observed in different human tissues fixed in neutral buffered formalin [34–47]. The hyaluronan staining pattern of normal human pancreatic islets does not change with aging. Hyaluronan distribution and abundance are altered in human pancreatic islets in type 1 diabetes, the degree of alteration varying with disease duration and severity of islet inflammation. Increased hyaluronan deposits also occur in human spleen and pancreatic lymph nodes in type 1 diabetes.

31. The intensity of hyaluronan staining in tissue sections may vary as a function of fixation techniques [48]. The bHABP probe generates a more intense hyaluronan staining in mouse tissues fixed in Carnoy's solution than in tissues fixed in neutral

buffered formalin. However, the patterns of hyaluronan staining in the mouse tissues are not influenced by the type of the fixative used.

32. Hyaluronan ELISA-like assay kits are commercially available for purchase at Echelon Bioscience (Salt Lake City, UT), Corgenix (Broomfield, CO) and R&D Systems (Minneapolis, MN) [28].

33. Islets in normal mouse pancreas are distinguished from surrounding exocrine pancreas tissue by their heparan sulfate positive staining. Heparan sulfate is localized in insulin-producing islet beta cells [10].

34. The specificity of the Alcian blue staining (0.65 M MgCl$_2$/ pH 5.8) of heparan sulfate in mouse islets has been confirmed by pretreatment of sections with nitrous acid (pH 4.0) which cleaves N-sulfated glucosamines that are present in heparan sulfate but not in other glycosaminoglycans [10].

35. Routinely, heparan sulfate in small- and medium-size islets is stained with a very intense blue color, whereas the intensity of blue staining in very large islets can be more variable.

36. Although heparan sulfate is also localized in the peri-islet basement membrane [9], light microscopic examination of Alcian blue-stained pancreas sections clearly identifies intra-islet heparan sulfate in islet beta cells.

37. Heparan sulfate is selectively lost from islet beta cells in the T1D mouse pancreas [10].

38. Histochemical staining of heparan sulfate using Alcian blue has also been successfully demonstrated on surgically resected 10 % formalin-fixed human pancreas specimens, using a staining time of 45 min with Alcian blue working solution.

39. Diluted acetic acid and sodium acetate are prepared immediately prior to the assay.

40. Stock AEC solution is prepared during incubation of sections with primary antibody (or isotype control Ig), and then stored at 4 °C until the preparation of the AEC working solution.

41. 0.1 M acetate buffer is prepared during the secondary antibody incubation.

42. AEC working solution is protected from light and used within 2 h of preparation.

43. Test AEC working solution with residual secondary antibody for coloration of chromagen.

44. Prepare liquid Glycergel by warming stock bottle in a beaker of hot tap water.

Acknowledgements

This research was performed with the support of the Network for Pancreatic Organ Donors with Diabetes (nPOD), a collaborative type 1 diabetes research project sponsored by JDRF, Grant 25-2010-648, National Institutes of Health grants U01 AI101984, CSGADP Innovative Project (under AI101984), and P01 HL098067 (T.N.W.). Organ Procurement Organizations (OPO) partnering with nPOD to provide research resources are listed at www.jdrfnpod.org/our-partners.php. This work was also supported by a National Health and Medical Research Council of Australia (NH&MRC)/Juvenile Diabetes Research Foundation (JDRF) Special Program Grant in Type 1 Diabetes (#418138; C.S.), a NHMRC Project Grant (#1043284), JDRF nPOD Research Grant 25-2010-716 (C.S.), a research grant from the Roche Organ Transplantation Research Foundation (ROTRF)/JDRF (#477554991; C.S.), and a Deutsche Forschungsgemeinschaft (DFG) Research Grant NA 965/2-1 (N.N.). We thank Anne Prins for assistance with the Alcian blue histochemical methodology and Lora Jensen and Sarah Popp for optimizing the heparan sulfate immunohistochemistry on nPOD human pancreas sections.

References

1. Laurent TC, Laurent UB, Fraser JR (1996) The structure and function of hyaluronan: an overview. Immunol Cell Biol 74:A1–A7

2. Stern R, Asari AA, Sugahara KN (2006) Hyaluronan fragments: an information-rich system. Eur J Cell Biol 85:699–715

3. Jiang D, Liang J, Noble PW (2011) Hyaluronan as an immune regulator in human diseases. Physiol Rev 91:221–264

4. Perrimon N, Bernfield M (2000) Specificities of heparan sulphate proteoglycans in developmental processes. Nature 404:725–728

5. Iozzo RV (2001) Heparan sulfate proteoglycans: intricate molecules with intriguing functions. J Clin Invest 108:165–167

6. Parish CR (2006) The role of heparan sulphate in inflammation. Nat Rev Immunol 6:633–643

7. Hull RL, Johnson PY, Braun KR, Day AJ, Wight TN (2012) Hyaluronan and hyaluronan binding proteins are normal components of mouse pancreatic islets and are differentially expressed by islet endocrine cell types. J Histochem Cytochem 60:749–760

8. Bogdani M, Johnson PY, Potter-Perigo S, Nagy N, Day AJ, Bollyky PL, Wight TN (2014) Hyaluronan and hyaluronan binding proteins accumulate in both human type 1 diabetic islets and lymphoid tissues and associate with inflammatory cells in insulitis. Diabetes 63:2727–2743

9. Irving-Rodgers HF, Ziolkowski AF, Parish CR, Sado Y, Ninomiya Y, Simeonovic CJ, Rodgers RJ (2008) Molecular composition of the peri-islet basement membrane in NOD mice: a barrier against destructive insulitis. Diabetologia 51:1680–1688

10. Ziolkowski AF, Popp SK, Freeman C, Parish CR, Simeonovic CJ (2012) Heparan sulfate and heparanase play key roles in mouse β cell survival and autoimmune diabetes. J Clin Invest 122:132–141

11. Brown TJ, Kimpton WG, Fraser JR (2000) Biosynthesis of glycosaminoglycans and proteoglycans by the lymph node. Glycoconj J 17:795–805

12. Kramer RH, Rosen SD, McDonald KA (1988) Basement-membrane components associated with the extracellular matrix of the lymph node. Cell Tissue Res 252:367–375

13. Kaldjian EP, Gretz JE, Anderson AO, Shi Y, Shaw S (2001) Spatial and molecular organization of lymph node T cell cortex: a labyrinthine

cavity bounded by an epithelium-like mono-layer of fibroblastic reticular cells anchored to basement membrane-like extracellular matrix. Int Immunol 13:1243–1253

14. Tengblad A (1979) Affinity chromatography on immobilized hyaluronate and its application to the isolation of hyaluronate binding proteins from cartilage. Biochim Biophys Acta 578: 281–289

15. Ripellino JA, Klinger MM, Margolis RU, Margolis RK (1985) The hyaluronic acid binding region as a specific probe for the localization of hyaluronic acid in tissue sections. J Histochem Cytochem 33: 1060–1066

16. Knudson CB, Toole BP (1985) Fluorescent morphological probe for hyaluronate. J Cell Biol 100:1753–1758

17. Ripellino JA, Bailo M, Margolis RU, Margolis RK (1988) Light and electron microscopic studies on the localization of hyaluronic acid in developing rat cerebellum. J Cell Biol 106: 845–855

18. Banerjee SD, Toole BP (1991) Monoclonal antibody to chick embryo hyaluronan-binding protein: changes in distribution of binding protein during early brain development. Dev Biol 146:186–197

19. Azumi N, Underhill CB, Kagan E, Sheibani K (1992) A novel biotinylated probe specific for hyaluronate. Its diagnostic value in diffuse malignant mesothelioma. Am J Surg Pathol 16:116–121

20. Underhill CB, Nguyen HA, Shizari M, Culty M (1993) CD44 positive macrophages take up hyaluronan during lung development. Dev Biol 155:324–336

21. Toole BP, Yu Q, Underhill CB (2001) Hyaluronan and hyaluronan-binding proteins. Probes for specific detection. Methods Mol Biol 171:479–485

22. Toole BP (2004) Hyaluronan: from extracellular glue to pericellular cue. Nat Rev Cancer 4:528–539

23. Underhill CB, Zhang L (2000) Analysis of hyaluronan using biotinylated hyaluronan-binding proteins. Methods Mol Biol 137:441–447

24. Campbell-Thompson ML, Montgomery EL, Foss RM, Kolheffer KM, Phipps G, Schneider L, Atkinson MA (2012) Collection protocol for human pancreas. J Vis Exp 63:e4039

25. Campbell-Thompson M, Wasserfall C, Kaddis J, Albanese-O'Neill A, Staeva T, Nierras C, Moraski J, Rowe P, Gianani R, Eisenbarth G, Crawford J, Schatz D, Pugliese A, Atkinson M (2012) Network for Pancreatic Organ Donors

with Diabetes (nPOD): developing a tissue biobank for type 1 diabetes. Diabetes Metab Res Rev 28:608–617

26. AVMA (American Veterinary Medical Association), Guidelines for the Euthanasia of Animals: 2013 Edition, Schaumburg, IL

27. Artwohl J, Brown P, Corning B, Stein S (2006) Report of the ACLAM Task Force on rodent euthanasia. J Am Assoc Lab Anim Sci 45: 98–105

28. Haserodt S, Aytekin M, Dweik RA (2011) A comparison of the sensitivity, specificity, and molecular weight accuracy of three different commercially available Hyaluronan ELISA-like assays. Glycobiology 21:175–183

29. Scott JE, Dorling J (1965) Differential staining of acid glycosaminoglycans (mucopolysaccharides) by alcian blue in salt solutions. Histochemie 5:221–233

30. Calvitti M, Baroni T, Calastrini C, Lilli C, Caramelli E, Becchetti E, Carinci P, Vizzotto L, Stabellini G (2004) Bronchial branching correlates with specific glycosidase activity, extracellular glycosaminoglycan accumulation, TGF beta(2), and IL-1 localization during chick embryo lung development. J Histochem Cytochem 52:325–334

31. Yingsung W, Zhuo L, Morgelin M, Yoneda M, Kida D, Watanabe H, Ishiguro N, Iwata H, Kimata K (2003) Molecular heterogeneity of the SHAP-hyaluronan complex. Isolation and characterization of the complex in synovial fluid from patients with rheumatoid arthritis. J Biol Chem 278:32710–32718

32. David G, Bai X, Van der Schueren B, Cassiman JJ, Van den Berghe H (1992) Developmental changes in heparan sulfate expression: in situ detection with monoclonal antibodies. J Cell Biol 119:961–975

33. van den Born J, Salmivirta K, Henttinen T, Ostman N, Ishimaru T, Miyaura S, Yoshida K, Salmivirta M (2005) Novel heparan sulfate structures revealed by monoclonal antibodies. J Biol Chem 280:20516–20523

34. Tammi R, Ripellino JA, Margolis RU, Tammi M (1988) Localization of epidermal hyaluronic acid using the hyaluronate binding region of cartilage proteoglycan as a specific probe. J Invest Dermatol 90:412–414

35. Tammi R, Tammi M, Hakkinen L, Larjava H (1990) Histochemical localization of hyaluronate in human oral epithelium using a specific hyaluronate-binding probe. Arch Oral Biol 35:219–224

36. Parkkinen JJ, Hakkinen TP, Savolainen S, Wang C, Tammi R, Agren UM, Lammi MJ, Arokoski

J, Helminen HJ, Tammi MI (1996) Distribution of hyaluronan in articular cartilage as probed by a biotinylated binding region of aggrecan. Histochem Cell Biol 105:187–194

37. Wang C, Tammi M, Guo H, Tammi R (1996) Hyaluronan distribution in the normal epithelium of esophagus, stomach, and colon and their cancers. Am J Pathol 148:1861–1869

38. Anttila MA, Tammi RH, Tammi MI, Syrjanen KJ, Saarikoski SV, Kosma VM (2000) High levels of stromal hyaluronan predict poor disease outcome in epithelial ovarian cancer. Cancer Res 60:150–155

39. Bohm J, Niskanen L, Tammi R, Tammi M, Eskelinen M, Pirinen R, Hollmen S, Alhava E, Kosma VM (2002) Hyaluronan expression in differentiated thyroid carcinoma. J Pathol 196:180–185

40. Auvinen P, Tammi R, Kosma VM, Sironen R, Soini Y, Mannermaa A, Tumelius R, Uljas E, Tammi M (2013) Increased hyaluronan content and stromal cell CD44 associate with HER2 positivity and poor prognosis in human breast cancer. Int J Cancer 132:531–539

41. de la Motte CA, Hascall VC, Drazba J, Bandyopadhyay SK, Strong SA (2003) Mononuclear leukocytes bind to specific hyaluronan structures on colon mucosal smooth muscle cells treated with polyinosinic acid:polycytidylic acid: inter-a-trypsin inhibitor is crucial to structure and function. Am J Pathol 163:121–133

42. Selbi W, de la Motte CA, Hascall VC, Day AJ, Bowen T, Phillips AO (2006) Characterization of hyaluronan cable structure and function in renal proximal tubular epithelial cells. Kidney Int 70:1287–1295

43. Aytekin M, Comhair SA, de la Motte C, Bandyopadhyay SK, Farver CF, Hascall VC, Erzurum SC, Dweik RA (2008) High levels of hyaluronan in idiopathic pulmonary arterial hypertension. Am J Physiol Lung Cell Mol Physiol 295:L789–L799

44. Lewis A, Steadman R, Manley P, Craig K, de la Motte C, Hascall V, Phillips AO (2008) Diabetic nephropathy, inflammation, hyaluronan and interstitial fibrosis. Histol Histopathol 23:731–739

45. de la Motte CA, Drazba JA (2011) Viewing hyaluronan: imaging contributes to imagining new roles for this amazing matrix polymer. J Histochem Cytochem 59:252–257

46. Boregowda RK, Appaiah HN, Siddaiah M, Kumarswamy SB, Sunila S, Thimmaiah KN, Mortha K, Toole B, Banerjee S (2006) Expression of hyaluronan in human tumor progression. J Carcinog 5:2

47. Tan KT, McGrouther DA, Day AJ, Milner CM, Bayat A (2011) Characterization of hyaluronan and TSG-6 in skin scarring: differential distribution in keloid scars, normal scars and unscarred skin. J Eur Acad Dermatol Venereol 25:317–327

48. Lin W, Shuster S, Maibach HI, Stern R (1997) Patterns of hyaluronan staining are modified by fixation techniques. J Histochem Cytochem 45:1157–1163

Chapter 33

Nonradioactive Glycosyltransferase and Sulfotransferase Assay to Study Glycosaminoglycan Biosynthesis

Cheryl M. Ethen, Miranda Machacek, Brittany Prather, Timothy Tatge, Haixiao Yu, and Zhengliang L. Wu

Abstract

Glycosaminoglycans (GAGs) are linear polysaccharides with repeating disaccharide units. GAGs include heparin, heparan sulfate, chondroitin sulfate, dermatan sulfate, keratan sulfate, and hyaluronan. All GAGs, except for hyaluronan, are usually sulfated. GAGs are polymerized by mono- or dual-specific glycosyltransferases and sulfated by various sulfotransferases. To further our understanding of GAG chain length regulation and synthesis of specific sulfation motifs on GAG chains, it is imperative to understand the kinetics of GAG synthetic enzymes. Here, nonradioactive colorimetric enzymatic assays are described for these glycosyltransferases and sulfotransferases. In both cases, the leaving nucleotides or nucleosides are hydrolyzed using specific phosphatases, and the released phosphate is subsequently detected using malachite reagents.

Key words Glycosyltransferase assay, Sulfotransferase assay, Glycosaminoglycans, Enzyme kinetics, Heparin, Heparan sulfate, Chondroitin sulfate, Keratan sulfate, Hyaluronan

1 Introduction

Glycosaminoglycans (GAGs) are linear polysaccharides with repeating disaccharide units. GAGs include heparin, heparan sulfate (HS), chondroitin sulfate (CS), dermatan sulfate (DS), keratan sulfate (KS), and hyaluronan (HA).

Heparin and HS are glucosaminoglycan polymers containing N-acetylglucosamine alternating with glucuronic acid. Heparin and HS are synthesized by the EXT family of dual glycosyltransferases that have both glucuronyltransferase and N-acetylglucosaminyltransferase activities and may be subsequently sulfated by 14 specific sulfotransferases (Table 1) [1]. CS and DS are galactosaminoglycan polymers containing N-acetylgalactosamine alternating with glucuronic acid. CS and DS are polymerized by the CHSY family of dual glycosyltransferases that have both N-acetylgalactosyltransferase and glucuronyltransferase activities and may be subsequently sulfated by 8 DS/CS-specific sulfotransferases [2].

Kuberan Balagurunathan et al. (eds.), *Glycosaminoglycans: Chemistry and Biology*, Methods in Molecular Biology, vol. 1229, DOI 10.1007/978-1-4939-1714-3_33, © Springer Science+Business Media New York 2015

Table 1
GAG synthesis enzymes

Enzymes	Protein	Possible acceptor	Donor
Core tetrasaccharide synthesis enzymes	XYLT1	GAG core protein	UDP-Xyl
	XYLT2	GAG core protein	UDP-Xyl
	B4GALT7	D-xylose	UDP-Gal
	B3GALT6	Galβ1,4Xylβ-O-benzyl	UDP-Gal
	FAM20B[a]	D-xylose	ATP
	B3GAT1[b]	alpha-lactose/Galβ1,3Galβ1,4Xyl	UDP-GlcA
	B3GAT3	Galβ1,3Galβ1,4Xyl	UDP-GlcA
HS/heparin polymerases	EXT1	HS/heparosan	UDP-GlcA, UDP-GlcNAc
	EXT2	HS/heparosan	UDP-GlcA, UDP-GlcNAc
	EXTL1[b]	GlcAβ1,4GlcNAc	UDP-GlcNAc
	EXTL2	GlcA	UDP-Gal, UDP-GlcNAc
	EXTL3[b]	GlcAβ1,3Galβ1,3Galβ1,4Xyl	UDP-GlcNAc
HS/heparin sulfotransferases	NDST1[b]	HS/heparosan	PAPS
	NDST2	HS/heparosan	PAPS
	NDST3	HS/heparosan	PAPS
	NDST4[b]	HS/heparosan	PAPS
	HS2ST1[b]	HS/heparin	PAPS
	HS3ST1	HS/heparin	PAPS
	HS3ST2	HS/heparin	PAPS
	HS3ST3A[b]	HS/heparin	PAPS
	HS3ST3B[b]	HS/heparin	PAPS
	HS3ST4	HS/heparin	PAPS
	HS3ST5[b]	HS/heparin	PAPS
	HS6ST1	HS/heparin/heparosan	PAPS
	HS6ST2[b]	HS/heparin	PAPS
	HS6ST3	HS/heparin	PAPS
CS/DS polymerases	CHSY1	Chondroitin	UDP-GlcA, UDP-GalNAc
	CHSY2	Chondroitin	UDP-GlcA, UDP-GalNAc
	CHSY3	Chondroitin	UDP-GlcA, UDP-GalNAc
	CSGalNAcT-1	Chondroitin/(GlcAβ1,3GalNAc)₃	UDP-GalNAc
	CSGalNAcT-2	Chondroitin	UDP-GalNAc
	CSGlcAT	Chondroitin	UDP-GlcA
CS/DS sulfotransferases	UST	Dermatan sulfate	PAPS
	CHST7[b]	Chondroitin	PAPS
	CHST11	Chondroitin	PAPS
	CHST12	Chondroitin	PAPS
	CHST13[b]	Chondroitin	PAPS
	CHST14[b]	Dermatan sulfate	PAPS
	CHST3[b]	Chondroitin	PAPS
	CHST15[b]	Chondroitin sulfate A	PAPS
KS[c] polymerases	B4GALT1[b]	GlcNAc	UDP-Gal
	B3GNT2	N-acetyllactosamine	UDP-GlcNAc
	B3GNT3[b]	Lactose	UDP-GlcNAc
	B3GNT4	Beta-lactose	UDP-GlcNAc

(continued)

Table 1
(continued)

Enzymes	Protein	Possible acceptor	Donor
KS sulfotransferases	CHST6[b]	GlcNAc	PAPS
	CHST1[b]	Lactose	PAPS
HA polymerases	HAS1	Hyaluronan	UDP-GlcNAc, UDP-GlcA
	HAS2	Hyaluronan	UDP-GlcNAc, UDP-GlcA
	HAS3	Hyaluronan	UDP-GlcNAc, UDP-GlcA

[a]This is a kinase and can be assayed using a similar methods [19]
[b]Commercial sources tested using assay methodology
[c]KS can be O-linked and N-linked. The enzymes for generating the core structure of N-glycan and O-glycan are not listed, but the current assay is applicable to enzymes involved in both O- and N-glycan synthesis

Both glucosaminoglycans and galactosaminoglycans are extended at the nonreducing ends of a common tetrasaccharide sequence [3]. The core tetrasaccharide has the sequence of GlcUAβ1–3Galβ1–3Galβ1–4Xylβ1 where the xylose is covalently attached to the serine residues on core proteins by XYLT1 and XYLT2 [4]. Addition of the two galactose residues occurs via two distinct galactosyltransferases, B4GALT7 and B3GALT6, respectively [5, 6]. The linkage tetrasaccharide is completed by the addition of the first glucuronic acid residue by a specific β-1,3 glucuronyltransferase B3GAT3 [7]. Further addition of an *N*-acetylgalactosamine to the tetrasaccharide by CSGalNAcT1 and CSGalNAcT2 initiates the formation of a galactosaminoglycan [8], whereas addition of an *N*-acetylglucosamine to the tetrasaccharide by EXTL initiates the formation of a glucosaminoglycan [9].

KS does not link to the core tetrasaccharide like HS, CS, and DS, but is attached to either O-glycan or N-glycan [10]. KS polymers contain galactose alternating with *N*-acetylglucosamine and are polymerized by B4GALT [11] and the B3GNT family of proteins [12]. KS may be subsequently sulfated by two specific sulfotransferases, CHST1 and CHST6 [13].

HA is not covalently attached to protein and is a constituent of the extracellular matrix. HA polymers contain *N*-acetylglucosamine alternating with glucuronic acid and are polymerized by membrane-bound dual enzymes HAS1, HAS2, and HAS3 that have both β1-3 *N*-acetylglucosaminyltransferase and β1-4 glucuronyltransferase activity [14]. HA is usually not sulfated like other GAGs.

Given that GAG synthesis is not template driven, the enzymes involved in modulating the chain length and sulfation pattern of GAGs specifically respond to biological conditions [1]. Therefore, it is imperative to understand the kinetics of these biosynthetic enzymes. Enzymatic assays in the past have been complicated, generally relying on the use of radiolabeled sugars followed by purification or separation methodology. Here, nonradioactive colorimetric enzymatic assays are described for glycosyltransferases [15] and sulfotransferases [16].

In each glycosyltransferase reaction, a sugar monomer is transferred from an activated donor substrate, such as UDP-Gal, UDP-GlcNAc, UDP-GlcA, UDP-GalNAc, UDP-Xyl, and GDP-Fuc, to an acceptor substrate, generating a free nucleotide UDP or GDP [15, 17]. In each sulfotransferase reaction, a sulfonate group is transferred from the donor substrate, PAPS, to an acceptor substrate, generating a leaving nucleotide, PAP [16, 18]. In both cases, the leaving nucleotides are hydrolyzed using specific phosphatases and the released phosphate is subsequently detected using malachite reagents (Fig. 1).

a Glycosyltransferase assay

b Sulfotransferase assay

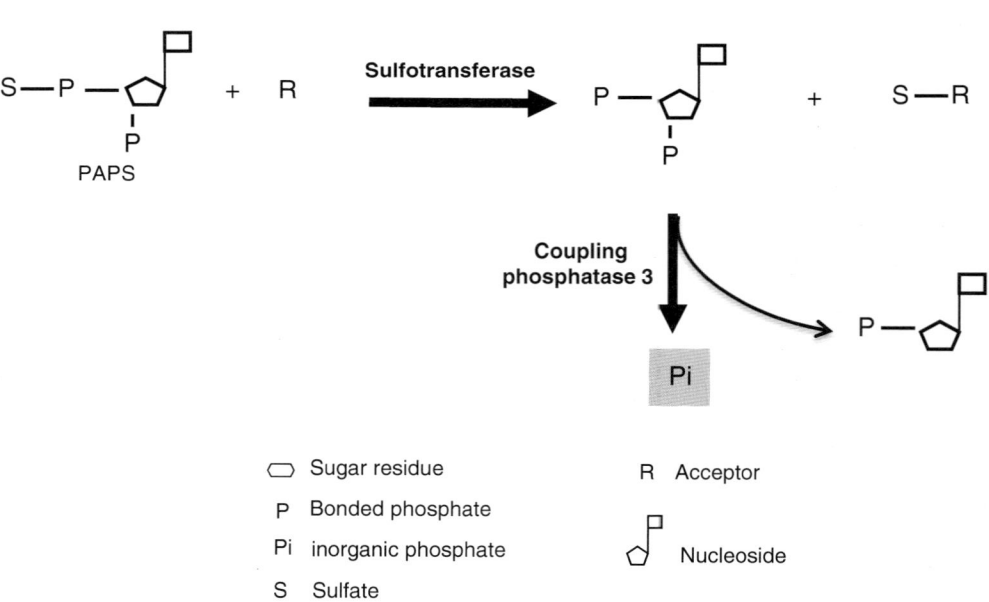

Fig. 1 Scheme for glycosyltransferase assay and sulfotransferase assay. (**a**) Coupling phosphatase 1 releases phosphate from the leaving nucleotide of a glycosyltransferase reaction. (**b**) Coupling phosphatase 3 releases phosphate from the leaving nucleotide of a sulfotransferase reaction. In both cases, the released phosphate is detected using Malachite Green Reagents

2   Materials

Prepare all solutions using ultrapure water (prepared by purifying deionized and distilled water with resistivity ≥ 18 MΩ/cm at 25 °C). All reagents require minimal phosphate content to reduce background signals (*see* **Note 1**). Malachite reagents contain concentrated sulfuric acid and should be handled with appropriate personal protection equipment. Diligently follow all waste disposal regulations when disposing waste materials.

2.1 Glycosyltransferase Assay Components

Dilution of reagents should be performed in the assay buffer indicated (*see* **Notes 2** and **3**).

1. Malachite Green Phosphate detection reagent A (ammonium molybdate in 3 M sulfuric acid) and Malachite Green Phosphate detection reagent B (malachite green oxalate and polyvinyl alchohol).

2. Assay buffer for glycosyltransferase assay : 25 mM Tris, 10 mM CaCl$_2$, 10 mM MnCl$_2$, pH 7.5.

3. Phosphate standard: 1 mM KH$_2$PO$_4$.

4. Donor substrate for glycosyltransferases: 0.5–5 mM of UDP-GlcNAc, UDP-Gal, UDP-Xyl, UDP-GlcA, UDP-GalNAc (Table 1).

5. Acceptor substrate for glycosyltransferases (0.5–10 mM) (*see* Table 1).

6. Coupling Phosphatase 1: recombinant humanCD39L3 at 10 ng/µL (*see* **Note 4**).

2.2 Sulfotransferase Assay Components

Dilution of reagents should be performed in the assay buffer (*see* **Notes 3** and **5**). A fixed concentration of high-purity PAPS is recommended to minimize the background signal from the hydrolysis of PAPS by Coupling Phosphatase 3 (*see* **Note 6**). If varied concentrations of PAPS are tested, a negative control for every concentration of PAPS is necessary.

1. Malachite Green Phosphate detection reagent A (ammonium molybdate in 3 M sulfuric acid) and Malachite Green Phosphate detection reagent B (malachite green oxalate and polyvinyl alchohol).

2. Assay buffer for sulfotransferase assay : 50 mM Tris, 15 mM MgCl$_2$, pH 7.5.

3. Phosphate standard: 1 mM KH$_2$PO$_4$.

4. Donor substrate for sulfotransferase assay: 1 mM PAPS.

5. Acceptor substrate for sulfotransferase assay: HS, heparin, CS, DS, KS (5–50 mg/mL) (Table 1).

6. Coupling Phosphatase 3: Recombinant mouse IMPAD1 at 50 ng/µL (*see* **Note 7**).

2.3 Equipment

96-Well microplates.

Plate reader or spectrometer capable of detection at 620 nm.

Pipettes.

3 Methods

3.1 Phosphate
Standard Curve
Determination

1. Prepare 8 twofold serial dilutions of phosphate standard starting with 100 µM in assay buffer leaving the 8th void of phosphate to serve as a blank.

2. Add 50 µL of each dilution into a well of a microplate.

3. Add 30 µL of Malachite Green Reagent A to each well (*see* **Note 8**).

4. Add 100 µL of water to each well.

5. Add 30 µL of Malachite Green Reagent B to each well.

6. Incubate the plate at room temperature for 20 min (*see* **Note 9**).

7. Read the plate in a spectrometer at 620 nm (*see* **Note 10**).

8. Subtract the blank OD from the OD of each dilution to obtain ΔOD.

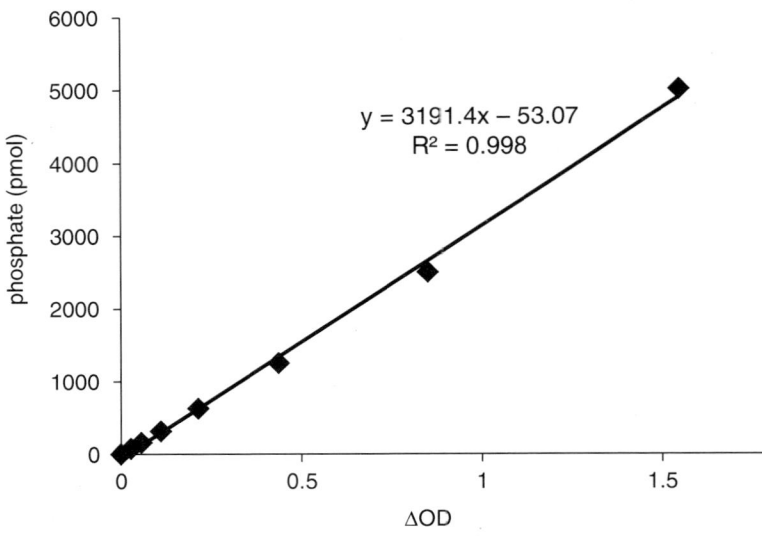

Fig. 2 Phosphate standard curve. The slope of the linear regression line represents the amount of phosphate corresponding to a unit of absorbance at 620 nm and is defined as the phosphate conversion factor (CF). In this example, it is calculated to be 3191 pmol/OD

9. After background subtraction, plot the phosphate content versus the ΔOD to obtain a linear slope, which equals the amount of phosphate corresponding to 1 unit of the absorbance, defined as the conversion factor (CF) of phosphate detection (*see* **Note 11**) (Fig. 2).

3.2 Enzymatic Assay   This protocol may be used for both glycosyltransferase (Fig. 3a) and sulfotransferase assays (Fig. 3b). For the glycosyltransferase assay, the reaction time may be prolonged to several hours (*see* **Note 12**); however, for the sulfotransferase assay, we recommend using a reaction time of 20 min (*see* **Note 13**).

1. Prepare substrate mix according to the following list for each reaction. Multiply the volumes by the number of reactions desired. Donor substrate (10 μL, volume per reaction), Acceptor substrate (10 μL, volume per reaction), Assay buffer (5 μL per reaction), and with a total of 25 μL volume per reaction.

2. Prepare enzyme mix according to the following list for each reaction. Multiply the volumes by the number of reactions desired. Test enzyme (10 μL per reaction), Coupling phosphatase (10 μL per reaction), Assay buffer (5 μL, per reaction), with 25 μL (total volume per reaction).

3. Pipette 25 μL of the enzyme mix into a well of a 96-well plate.

4. Pipette 15 μL of assay buffer and 10 μL of coupling phosphatase into another well to serve as a negative control.

5. Initiate the reaction by adding 25 μL of the substrate mix to each well.

6. Incubate the reaction at room temperature or 37 °C for 20 min.

7. Add 30 μL of Malachite Green Reagent A to each well.

8. Add 100 μL of water to the each well.

9. Add 30 μL of Malachite Green Reagent B to each well.

10. Incubate the plate at room temperature for 20 min.

11. Read the absorbance at 620 nm with a spectrometer.

12. Subtract the OD of the negative control from the OD of each dilution to obtain ΔOD.

13. Calculate the specific activity (SA) of the enzyme using the conversion factor (CF) determined from the phosphate standard curve in Subheading 3.1.

$$SA = \frac{\Delta OD \times CF}{\text{reaction time} \times \text{amount of the enzyme}}$$

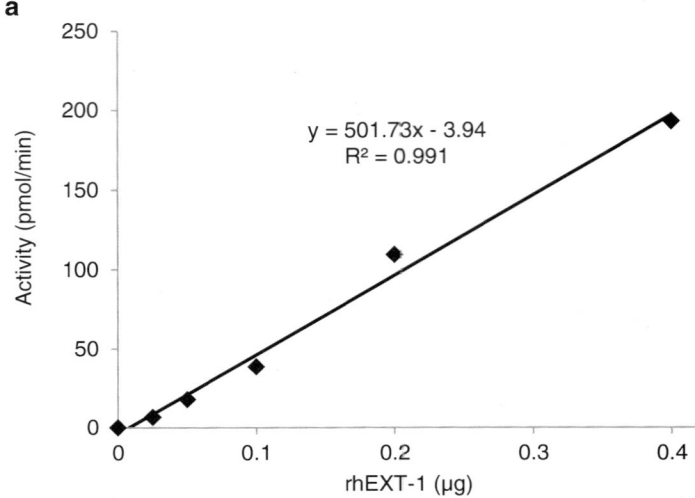

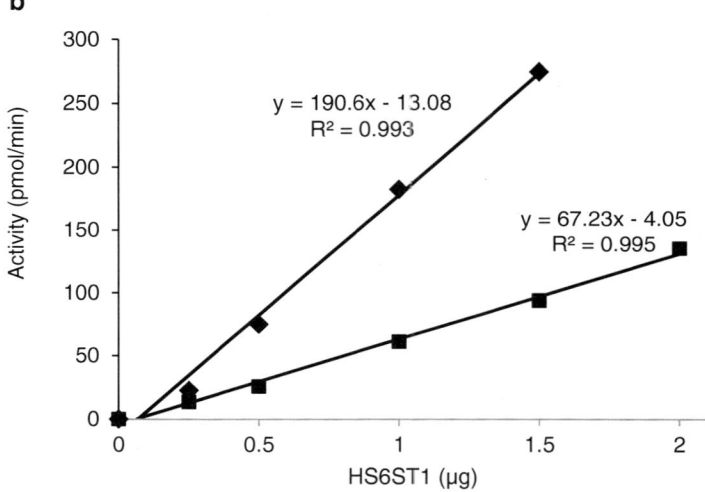

Fig. 3 Enzymatic activity assay examples. (**a**) Recombinant human EXT-1 (rhEXT-1) glycosyltransferase assay. Increasing amounts of rhEXT-1 were assayed using 0.5 mM UDP-GlcA and 0.5 mM UDP-GlcNAc as donors and 4 mg/mL heparan sulfate as the acceptor for 20 min at 37 °C in buffer containing Tris, $MnCl_2$, $CaCl_2$, pH 7.5. The specific activity was determined to be 501.7 pmol/min/μg. (**b**) Recombinant human HS6ST1 sulfotransferase assay. Increasing amounts of rhHS6ST1 were assayed using 0.2 mM PAPS as the donor and either 2 mg/mL heparan sulfate (diamonds) or 2 mg/mL heparosan (squares) as the acceptor for 20 min at 37 °C in buffer containing Tris, $MgCl_2$, pH 7.5. The specific activity was determined to be 190.6 pmol/min/μg using heparan sulfate or 67.2 pmol/min/μg using heparosan as the acceptor

4 Notes

1. These methods are based on phosphate detection. Any phosphate from the reagents will cause background. Removal of phosphate is required to obtain a higher signal to background ratio.

2. The Coupling Phosphatase 1 requires Ca^{2+} for activity and shows optimal activity from pH 7.0 to 8.5. It loses about 30 % of its activity at pH 5.5 and approximately 50 % activity when NaCl increases to 0.3 M.

3. If the reaction conditions for a sulfotransferase or glycosyltransferase and the respective coupling phosphatase are not compatible, a decoupled assay may be performed in which the phosphatase reaction can be performed following the main reaction. In this case, the strength of the phosphatase buffer is recommended to be 4× higher than that of the primary reaction buffer.

4. The Coupling Phosphatase 1 at 100 ng/well is sufficient to release phosphate from 10 nmol/well of UDP in the recommended assay conditions. Should different assay conditions be used, the amount of Coupling Phosphatase 1 may need to be adjusted to reach the same rate.

5. Coupling Phosphatase 3 shows activity across a wide pH range with the optimal activity around pH 7.5 and requires Mg^{2+} for activity. Mn^{2+} is not suggested in all buffers as it causes high background in the assays due to elevated activity on the donor PAPS. Na^+ and Li^+ are inhibitors of the enzyme, with a K_i of 17.2 mM and 0.122 mM, respectively.

6. In a sulfotransferase assay, Coupling Phosphatase 3 shows about 0.5 % activity on the donor substrate PAPS than on the leaving nucleotide PAP. To minimize the background signal from the hydrolysis of PAPS, a fixed PAPS concentration is suggested here.

7. The exact relationship between inorganic phosphate produced and PAPS consumed is reflected in the coupling rate [19], a parameter that defines the product–signal conversion. For accurate enzyme kinetic determination, the final assay data may be adjusted using the determined coupling rate. The Coupling Phosphatase 3 at 500 ng/well will achieve a coupling rate of approximately 0.96 using the recommended assay conditions. Should different assay conditions be used, the amount of Coupling Phosphatase 3 may need to be adjusted to reach the same coupling rate.

8. Malachite Green Reagent A contains sulfuric acid and addition to the reaction wells effectively stops the enzymatic reaction.

9. After addition of Malachite Green Reagent B, the yellow color will fade with time as the reaction stabilizes. Twenty minutes is required for this to occur.

10. Pipetting concentrated proteins, polypeptides, or substrates may cause foaming or surface bubbles in the plate wells. These should be eliminated prior to absorbance measurement.

11. The dynamic range for phosphate detection using malachite reagents is from 100 to 4,000 pmol with the current format. The amount of phosphate produced in a coupled reaction should fall into this range for accurate enzyme kinetic determination. This can be achieved by changing the amount of glycosyltransferase or sulfotransferase or reaction time.

12. For the glycosyltransferase assay, the reaction time can be prolonged to several hours without adverse effect as the enzyme remains active over long periods of time. The enzyme loses about 20 % of activity at 37 °C after overnight incubation.

13. For the sulfotransferase assay, the reaction time is not recommended to go beyond 20 min, as the coupling phosphatase may hydrolyze the donor substrate PAPS and thus increase background. In addition, long incubation times may cause PAPS degradation and contribute to increased background.

Acknowledgement

This work was supported by R&D Systems, and we would like to thank all our colleagues who made contributions through product development.

References

1. Esko JD, Lindahl U (2001) Molecular diversity of heparan sulfate. J Clin Invest 108:169–173

2. Silbert JE, Sugumaran G (2002) A starting place for the road to function. Glycoconjugate J 19:227–237

3. Uyama T, Kitagawa H, Tamura Ji J, Sugahara K (2002) Molecular cloning and expression of human chondroitin N-acetylgalactosaminyltransferase: the key enzyme for chain initiation and elongation of chondroitin/dermatan sulfate on the protein linkage region tetrasaccharide shared by heparin/heparan sulfate. J Biol Chem 277:8841–8846

4. Goetting C, Kuhn J, Zahn R, Brinkmann T, Kleesiek K (2000) Molecular cloning and expression of human UDP-D-xylose:proteoglycan core protein beta-d-xylosyltransferase and its first isoform XT-II. J Mol Biol 304:517–528

5. Bai X, Zhou D, Brown JR, Crawford BE, Hennet T, Esko JD (2001) Biosynthesis of the linkage region of glycosaminoglycans: cloning and activity of galactosyltransferase II, the sixth member of the beta 1,3-galactosyltransferase family (beta 3GalT6). J Biol Chem 276: 48189–48195

6. Almeida R, Levery SB, Mandel U, Kresse H, Schwientek T, Bennett EP, Clausen H (1999) Cloning and expression of a proteoglycan UDP-galactose:beta-xylose beta1,4-galactosyltransferase I. A seventh member of the human beta4-galactosyltransferase gene family. J Biol Chem 274:26165–26171

7. Kitagawa H, Tone Y, Tamura J, Neumann KW, Ogawa T, Oka S, Kawasaki T, Sugahara K (1998) Molecular cloning and expression of glucuronyltransferase I involved in the

biosynthesis of the glycosaminoglycan-protein linkage region of proteoglycans. J Biol Chem 273:6615–6618

8. Sato T, Gotoh M, Kiyohara K, Akashima T, Iwasaki H, Kameyama A, Mochizuki H, Yada T, Inaba N, Togayachi A, Kudo T, Asada M, Watanabe H, Imamura T, Kimata K, Narimatsu H (2003) Differential roles of two N-acetylgalactosaminyltransferases, CSGalNAcT-1, and a novel enzyme, CSGalNAcT-2. Initiation and elongation in synthesis of chondroitin sulfate. J Biol Chem 278:3063–3071

9. van Hul W, Wuyts W, Hendrickx J, Speleman F, Wauters J, de Boulle K, van Roy N, Bossuyt P, Willems PJ (1998) Identification of a third EXT-like gene (EXTL3) belonging to the EXT gene family. Genomics 47:230–237

10. Funderburgh JL (2000) KS: structure, biosynthesis and function. Glycobiology 10:951–958

11. Amado M, Almeida R, Schwientek T, Clausen H (1999) Identification and characterization of large galactosyltransferase gene families: galactosyltransferases for all functions. Biochim Biophys Acta 1473:35–53

12. Lee PL, Kohler JJ, Pfeffer SR (2009) Association of β-1,3-N-acetylglucosaminyltransferase 1 and β-1,4-galactosyltransferase 1, trans-Golgi

enzymes involved in coupled poly-N-acetyllactosamine synthesis. Glycobiology 19:655–664

13. Torii T, Fukuta M, Habuchi O (2000) Sulfation of sialyl N-acetyllactosamine oligosaccharides and fetuin oligosaccharides by keratan sulfate Gal-6-sulfotransferase. Glycobiology 10:203–211

14. Weigel PH, DeAngelis PL (2007) Hyaluronan synthases: a decade-plus of novel glycosyltransferases. J Biol Chem 282:36777–36781

15. Wu ZL, Ethen CM, Prather B, Machacek M, Jiang W (2011) Universal phosphatase-coupled glycosyltransferase assay. Glycobiology 21:727–733

16. Prather B, Ethen CM, Machacek M, Wu ZL (2012) Golgi-resident PAP-specific 3'-phosphatase-coupled sulfotransferase assays. Anal Biochem 423:86–92

17. Breton C, Snajdrova L, Jeanneau C, Koca J, Imberty A (2006) Structures and mechanisms of glycosyltransferases. Glycobiology 16:29R–37R

18. Paul P, Suwan J, Liu J, Dordick JS, Linhardt RJ (2012) Recent advances in sulfotransferase enzyme activity assays. Anal Bioanal Chem 403:1491–1500

19. Wu ZL (2011) Phosphatase-coupled universal kinase assay and kinetics for first-order-rate coupling reaction. PLos One 6:8

Chapter 34

Mapping Proteoglycan Functions with Glycosidases

Mauricio Cortes, Leslie K. Cortes, and Nancy B. Schwartz

Abstract

The intrinsic and extrinsic factors that contribute to stem and neuronal precursor cell maintenance and/or differentiation remain poorly understood. Proteoglycans, major residents of the stem cell microenvironment, modulate key signaling cues and are of particular importance. We have taken a loss-of-function approach, by developing a library of bacterial lyases and sulfatases to specifically remodel the ECM and test the functional role of glycosaminoglycans (GAGs) in cell self-renewal, maintenance, and differentiation.

Key words Proteoglycan, Glycosaminoglycan, Chondroitin sulfate, Heparan sulfate, Glycosidase, ES cells, Cloning, Cortical slices

1 Introduction

1.1 Proteoglycans in the Stem Cell Niche

The extracellular matrix (ECM) is a complex environment where cell–cell and cell–growth factor interactions take place, but the functional contribution of specific molecules of the ECM to cell signaling pathways that regulate maintenance and/or differentiation is poorly understood. The ability of the ECM to regulate cell function is of particular interest in the area of stem cell and neuronal precursor biology (Fig. 1). The cell niche or microenvironment plays an important role in cell homeostasis by providing and modulating cues for self-renewal and differentiation into a particular cell lineage. Despite recent advances in stem cell biology and glycobiology, the functional contribution of specific glycosaminoglycan (GAG) modifications in stem cell maintenance and the guided differentiation of a pluripotent embryonic stem cell to a neuronal precursor remains to be elucidated.

The major ECM constituents involved in stem cell and central nervous system biology are the heparan sulfate and chondroitin sulfate proteoglycans (Fig. 2). Their respective GAG chains are covalently attached to core proteins through a common linkage region and consist of linear chains of distinct repeating disaccharide units that are uniquely decorated with sulfate substitutions (Fig. 2).

Kuberan Balagurunathan et al. (eds.), *Glycosaminoglycans: Chemistry and Biology*, Methods in Molecular Biology, vol. 1229, DOI 10.1007/978-1-4939-1714-3_34, © Springer Science+Business Media New York 2015

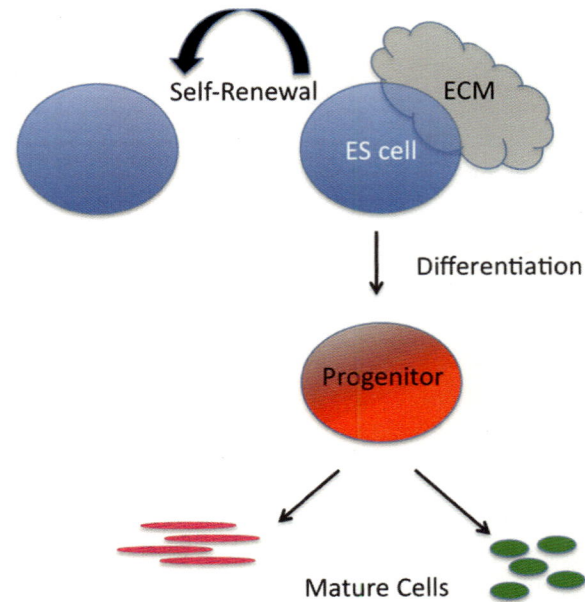

Fig. 1 Schematic diagram depicting stem cell self-renewal and differentiation of progenitor cells

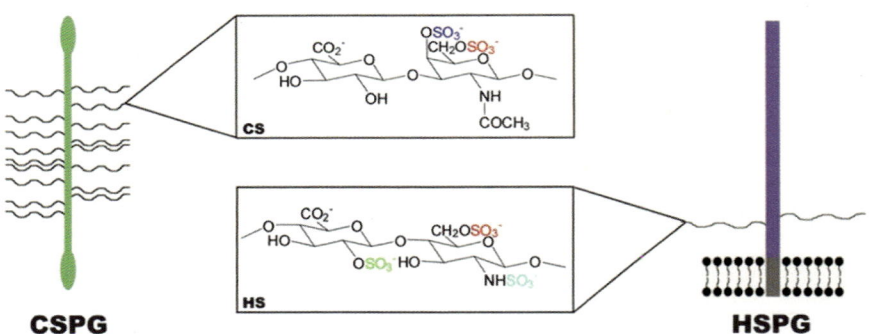

Fig. 2 Schematic representation depicting chondroitin sulfate proteoglycans (CSPGs) and heparan sulfate proteoglycans (HSPGs). The core proteins of CSPGs (*green*) tend to have multiple glycosaminoglycan (GAG) chains (*black* wavy lines) per core protein composed of repeating GalNAc units sulfated either at the 4 (*blue*, CS-A), or 6 (*red*, CS-C) positions. HSPGs' core proteins (*purple*) tend to have fewer GAG chains per core protein. They are composed of repeating GlcNAc units sulfated at 6 (*red*) position, and 2-N-sulfated (teal), and uronic acid units sulfated in the 2 position (*green*)

The GAG chains and/or their SO_4 substitutions can be targets for specific degrading enzymes.

Studies focused on heparan sulfate composition in stem cells showed that increased N-sulfation and O-sulfation is important for the differentiation of ES cells into neuronal progenitors [1]. Further studies on neuronal stem cells and in the chick and mouse

Table 1
Anti-CS and anti-HS antibodies

CS56	Mouse IgM
Di-0S (1-B-5)	Mouse IgG
Di-4S (2-B-6)	Mouse IgG
Di-6S (3-B-3)	Mouse IgM
HS (F58-10E4)	Mouse IgM

See **Note 2**

ventricular zone showed that chondroitin sulfate proteoglycans (CSPGs) are important in controlling neuronal progenitor self-renewal and differentiation into the glial lineage [2, 3]. These studies highlight the importance of the niche and its extracellular components, especially CSPGs. Characterization of the extracellular environment and how changes in the microenvironment influence signaling cues is key to understanding control of stem/precursor cell proliferation and differentiation. Therefore, we describe here methodologies to characterize the proteoglycan content in ES cells and brain tissue using immunohistochemistry and fluorophore-assisted carbohydrate electrophoresis (FACE). In addition, we describe the cloning of a representative bacterial glycosidase as an example of an approach to alter the ECM matrix in vivo to elucidate the role of chondroitin sulfate (CS) and heparan sulfate (HS) in mouse cortical development.

To determine which sulfated proteoglycans are produced by ES cells, immunohistochemistry with a battery of antibodies directed against the CS and HS GAG chain epitopes was performed on AB1.1 ES cells, and revealed that ES cells express both HS and CSPGs, albeit with different distribution patterns. Immunofluorescence using the CS antibody CS-56 which recognizes intact CS chains, revealed weak expression of CS in ES cell colonies. Pretreatment of cultured ES cells with chondroitinase ABC prior to immunostaining revealed strong cellular expression of CS-4 between the cells within an ES cell colony using the 2-B-6 antibody that specifically recognizes chondroitin-4-sulfate moieties. Similarly, immunostaining using antibody 10E4, which recognizes HS, revealed weak, but specific, staining surrounding the ES cell colonies as well as feeder-cell membranes (Table 1, Fig. 3).

To complement the immunohistochemistry experiments and to obtain quantitative analysis of the types and amounts of sulfated GAG chains in the ECM of ES cells, FACE was performed. The FACE profile of CS species confirmed that ES cells contain mostly CS-4 and the double-sulfated disaccharide CS-4,6 also known as

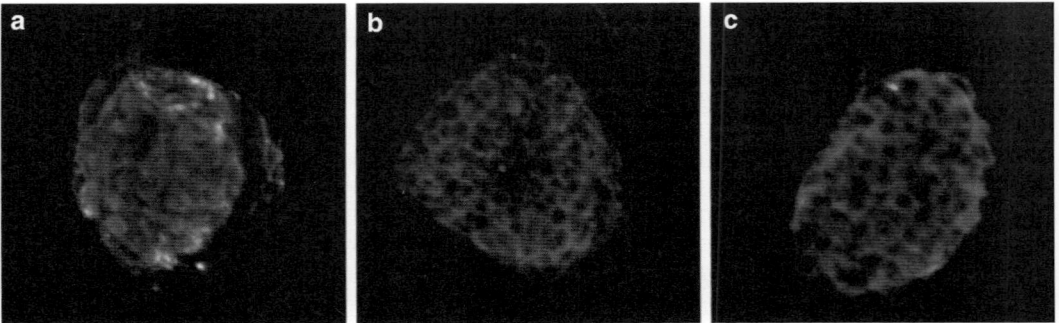

Fig. 3 Sulfated epitope analysis in ES cells. AB1.1 ES cells were cultured in complete ES cell media with LIF, and immunostained with antibodies which recognize specific sulfated epitopes. Immunofluorescence experiments revealed the presence of the Δdi-6S (**a**) and Δdi-4S (**b**), and CS-D (**c**)

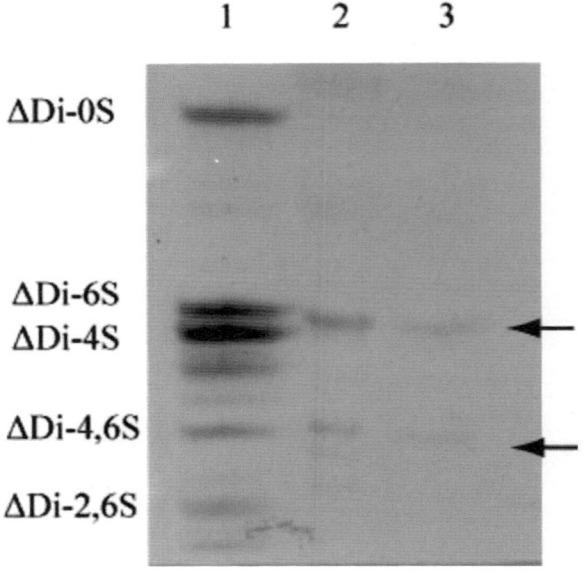

Fig. 4 Sulfated glycosaminoglycan content in ES cells. Biochemical analysis of the GAG content of the AB1.1 ES cell colonies using FACE revealed the presence of CS-4 and the double sulfated CS-E (*lanes 2, 3*), as determined by their mobility compared to known CS standards (*lane 1*)

CS-E, as determined by their mobility compared to known standards (Fig. 4). Similarly, the FACE profile of HS revealed the presence of multiple sulfated HS species, but they did not correspond to known standards, and therefore could not be identified. The presence of sulfated GAG chains, the high amount of CS-4 and the unique expression of the rare double-sulfated CS-E suggest that sulfation may be required for some aspect of ES cell function and/or maintenance. Based on the existence of commercially available antibodies that recognize distinct HS and CS moieties, changes in ECM composition can be probed and visualized as ES cells undergo differentiation to specific cell types. In addition, FACE permits the quantification of ECM composition during ES cell maintenance and differentiation.

1.2 Use of Proteoglycan Lyases to Remodel the Microenvironment

Based on our recent work on the role of CSPGs in modulating hedgehog signaling in the developing growth plate [4] and on the prevalence of these sulfated moieties in ES cells, we took a loss-of-function approach using a set of bacterial lyases and sulfatases to specifically remodel the ECM. A chondroitin sulfate lyase that degrades all types of CS, chondroitinase ABC (ChABC), was cloned from the bacterium *Proteus vulgaris*. A second chondroitin lyase, chondroitinase AC (ChAC), and heparanase III were cloned from the bacterium *Pedobacter heparinus*. All three genes were engineered for both bacterial and mammalian expression. Unlike ChABC, the ChAC lyase only degrades chondroitin-4-sulfate and chondroitin-6-sulfate structures, leaving dermatan sulfate structures intact, while heparanase degrades HS. Expression in cells and live tissue demonstrated that CS can be degraded both in culture and in vivo (Fig. 5). To address the functional importance of sulfate modifications on the CS chains, we also cloned two putative bacterial sulfatases from the bacterium *Yersinia pseudoturberculosis*, and characterization of these candidate sulfatases is currently underway.

All of these recombinant enzymes have been expressed in *E. coli*, purified and assayed. Expression constructs were electroporated into ES cells or cortical slice cultures in which degradation of the respective GAG moieties was subsequently observed. To determine whether the ECM of living brain tissue can be modified by over-expressing a bacterial lyase which digests CS chains (ChABC), the *P. vulgaris* ChABC gene was cloned into a mammalian plasmid together with a signal peptide coding sequence for efficient expression and secretion. Upon electroporation of the resulting pcDNA3.1 ChABC *myc* His-SP plasmid into cortical slice cultures, the activity of expressed ChABC was assessed by measuring CS degradation by immunohistochemistry, FACE, and [35-S]-sulfate labeling experiments [5]. As shown in Fig. 5, there was strong expression of both the control (anti-β-galactosidase) and ChABC-expressing plasmids (anti-myc), only the cortical slices electroporated with the ChABC construct were stained by anti-CS-6 antibodies. Strong anti-CS-6 staining throughout the slice indicates that ChABC was secreted and enzymatically active in digesting CS chains. In sum, we have begun to characterize the GAG content within mouse ES cells and brain cortical slices using a library of ECM-remodeling enzymes to test the functional role of GAGs in self-renewal and differentiation of ES cells and neuronal precursors.

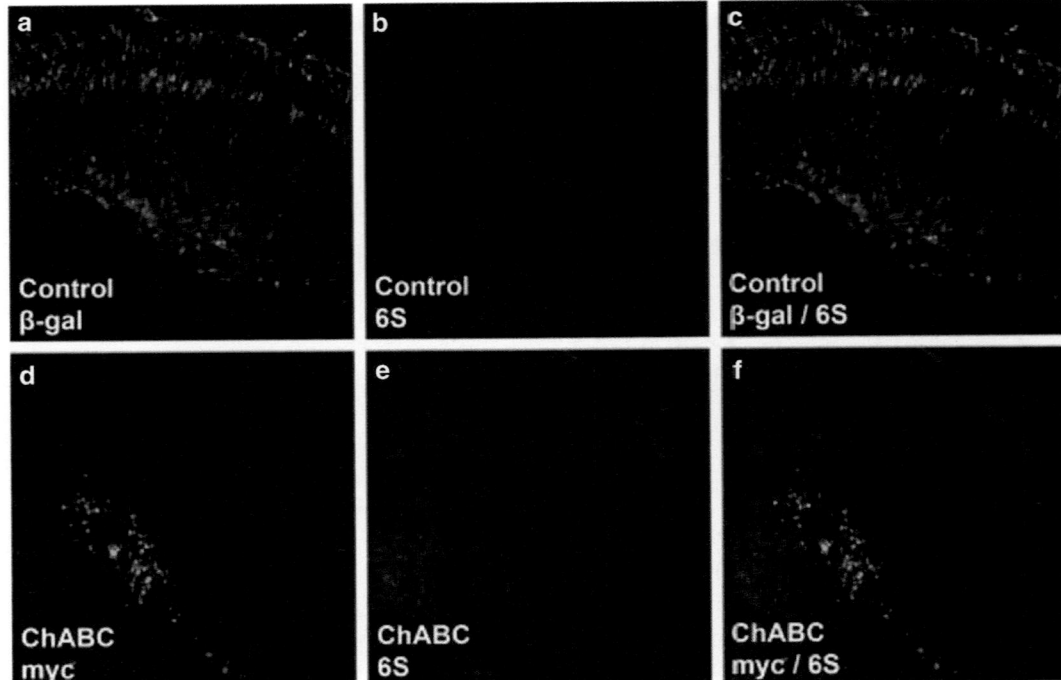

Fig. 5 Digestion of CSPGs in cortical slice cultures. E15.5 brains were electroporated *ex utero* with a plasmid encoding a secreted form of ChABC or a control plasmid encoding LacZ. Coronal slices were cultured for 4 days in vitro and then costained for either β-galactosidase (control plasmid) or myc (ChABC plasmid) to demarcate plasmid expression as well as the 6S antibody, which only recognizes CS chains sulfated at the 6-position after ChABC digestion. Slice electroporated with the control plasmid stained for β-galactosidase (**a**), 6S (**b**), merge (**c**). Note the lack of 6S staining due to a lack of ChABC digestion despite strong expression of the control plasmid. Slice electroporated with ChABC stained for myc (**d**), 6S (**e**), and merge (**f**). Note the strong 6S staining throughout the slice, due to digestion of CPSGs by the expression of secreted ChABC from the electroporated plasmid

2 Materials (*See* Note 1)

2.1 General

1. Sterile plastic Petri dishes (Falcon 1029) and standard sterile plastic tissue culture dishes (Falcon 3003 and 3002) from Fisher. For gelatin-coated dishes, treat with 0.1 % gelatin solution, remove excess.

2. Sterile 17×100-mm polypropylene snap-cap tubes Sarstedt or Fisher and 10-mL sterile, individually wrapped pipets (Fisher) (both presumed to be RNase-free).

3. Sterile 15-mL conical-bottom screw-cap centrifuge tubes from Sarstedt.

4. Microcentrifuge tubes: 1.5-mL, and 1–200, and 100–1,000 μL pipettor tips (USA Scientific); all are RNase/DNase/pyrogen-free as supplied, autoclave if sterility is desired.

5. Glass 4-well chamber slides (Lab-Tek): Gelatin-treat as for gelatin-coated tissue culture dishes.

6. Poly-L-lysine- and laminin-coated 1.0 μm track-etched polyethylene terephthalate (PET) cell culture inserts

7. ChemiDoc XRS Imaging system with Quantity One software (Bio-Rad).

8. Cell lines/bacterial strains: AB1.1 ES cells; STO-Neo-LIF (SNL) feeder cells; *E. coli* TOP10 competent cells (Invitrogen).

9. Nucleic acids: *P. vulgaris* genomic DNA (ATCC); pcDNA3.1/myc-His plasmid (Invitrogen); pcDNA3.1-Shh.

10. PCR primers: custom-synthesized by commercial source (i.e., Integrated DNA Technologies). Bold type denotes restriction enzyme sites.

ChABCFwd-5′-**CGGGATCCGCCACCAGCAATCCTGC ATTTGATCCT-3**′; ChABCRev-5′-**CCGCTCGAGCG AGGGAGTGGCGAGAGTTTGATTTCTT-3**′; pcDNA3.1Fwd-5′-CAAGTGTAT**CATATG**CCAAGTAC GCCCCCTATT-3′; ShhRevSP-5′-C**GGGATCC**GGGCC CACAGGCCAGCCCGGGGCACACCAG-3′.

2.2 Reagents

1. 0.1 % Triton X-100 in phosphate-buffered saline (PBS).

2. 0.1 % Triton X-100 in PBS supplemented with 3 % normal lamb serum (NLS, Gibco).

3. 0.1 % Triton X-100 in PBS supplemented with 1.5 % NLS.

4. 100 mU/mL heparatinase (Glyko).

5. 100 mU/mL chondroitinase (Seikagaku).

6. 12.5 mM 2-aminoacridone (AMAC, Molecular Probes).

7. Ethanol.

8. 1.25 M sodium cyanoborohydride.

9. Glycerol.

10. MONO gels (Glyko).

11. Mitomycin C (Roche).

12. Fluoromount-G (Southern Biotech).

13. Chondroitin sulfate (CS) and heparin sulfate (HS) (Seikagaku).

14. PfuUltra High Fidelity (HF) DNA polymerase (Agilent/Stratagene, supplied with 10× reaction buffer).

15. T4 DNA ligase (New England Biolabs).

16. BamHI, XhoI and NdeI restriction enzymes (New England Biolabs).

17. 12.5 M solution in 85 % DMSO/15 % acetic acid.

18. QIAquick Gel Extraction Kit (Qiagen).

19. QIAquick Nucleotide Removal Kit (Qiagen).

20. QIAprep Spin Miniprep Kit (Qiagen).

21. TOP10 chemically competent *E. coli* (Life Technologies).

22. 4 % Parafomaldehyde (PFA): Make a 4 g per 100 mL solution in PBS.

23. Antibodies: Anti-CS-4, anti-CS-6 (3-B-3), and anti-HS antibodies (Seikagaku); fluorophore-conjugated secondary antibodies (Invitrogen); anti-β-galactosidase (control plasmid slices) (Chemicon, anti-rabbit) or anti-myc (ChABC plasmid slices) (Upstate, anti-rabbit); anti-6S antibody (Seikagaku, anti-mouse IgM) (*see* **Note 2**).

24. DAPI (4′,6-diamidino-2-phenylindole, Sigma), a fluorescent stain that binds strongly to A-T rich regions in DNA: Make a 1 mg/mL stock solution, dilute 1:10,000 (0.1 mL/L) in PBS for use).

25. Proteinase K (Roche): Make 1 mL of a 2.5 mg/mL solution in 100 mM ammonium acetate buffer, pH 7.0, 0.005 % phenol red.

2.3 Cell Growth Media

1. M15 medium: Dulbecco's modified Eagle medium (DMEM) (high glucose, no sodium pyruvate) supplemented with 15 % fetal bovine serum (FBS), 100 units/mL penicillin, 0.1 mg/mL streptomycin, 1 mM L-glutamine (all from Life Technologies), and 0.1 % β-mercaptoethanol (Sigma).

2. LB liquid medium: 10 g/L tryptone, 5 g/L yeast extract, 10 g/L NaCl, adjust pH to 7.5 with 5 N NaOH, autoclave.

3. LB agar plates: Make up LB medium containing 1.5 % agar (15 g/L), autoclave, then cool to ~45 °C. For 100 μg/mL ampicillin-agar plates, add 1 mL/L of a 1,000× ampicillin sodium (Sigma) stock solution (100 mg/mL in ddH$_2$O, filter-sterilized) to the molten sterilized agar. Pour the agar into 100 mm sterile plastic Petri dishes, approximately 25 mL per dish (~half-full) and let cool to room temperature.

4. M15 medium supplemented with 10^3 U/mL ESGRO leukemia inhibitory factor (LIF, Millipore).

3 Methods

3.1 ES Cell Culture

1. Grow AB1.1 ES cells on mitomycin C-treated SNL feeder cells, in M15 medium at 37 °C.

2. Feed cells daily with fresh M15 media, and passage upon 80 % confluency.

3. Grow feeder-free AB1.1 ES cells in 0.1 % gelatin-coated dishes in M15 medium supplemented with ESGRO LIF at a concentration of 10^3 U/mL.

3.2 Immunofluo-rescence Detection of CS/HS Species in Cultured Cells

1. Plate AB1.1 ES cells on gelatin-treated glass chamber slides at low density (50–100 cells per well in 4-well chamber slides) and incubate for 3–4 days until colonies are visible.

2. Fix cells with 4 % PFA for 10 min, then permeabilize by soaking in 0.1 % Triton X-100 in PBS for 15 min.

3. Wash and block cells with 3 % NLS, 0.1 % Triton X-100 in PBS for 2 h at room temperature.

4. Incubate cells in 1.5 % NLS, 0.1 % Triton X-100 in PBS with primary antibodies at the following dilutions: anti-CS-4, 1:150; anti-CS-6 3-B-3, 1:400 and anti-HS, 1:100.

5. Rinse cells with 0.1 % Triton X-100 in PBS and incubate with fluorophore-conjugated secondary antibodies diluted 1:500 in 1.5 % NLS, 0.1 % Triton X-100 in PBS for 2 h at room temperature.

6. Incubate cells with 100 ng/mL DAPI in PBS for 5 min, then wash several times with PBS before mounting with Fluoromount-G.

3.3 Fluorophore-Assisted Carbohydrate Electrophoresis (FACE)

1. Perform FACE as previously described [6–8] with minor modifications.

2. Treat cells with 50 μL of 2.5 mg/mL proteinase K solution for 2 h at 60 °C, followed by a second 2 h incubation with a fresh addition of proteinase K, as before.

3. Boil samples for 15 min to heat-inactivate the proteinase K and then centrifuge at $10,000 \times g$ for 15 min (in a SA-600 Sorvall rotor).

4. Dilute 100-μL supernatant aliquots tenfold in 100 mM ammonium acetate buffer, pH 7.2, then digest with either 100 mU/mL heparatinase or chondroitinase for 3 h at 37 °C.

5. Remove undigested macromolecules by ethanol precipitation overnight at –20 °C, followed by centrifugation at $10,000 \times g$ for 15 min at 4 °C.

6. Supernatants of digested samples were transferred to a new tube and vacuum-dried for 2 h.

7. Label the digested samples by adding 40 μL of 12.5 mM AMAC to each dried sample, incubate for 15 min at room temperature, then add 40 μL of 1.25 M sodium cyanoborohydride and incubate for 16 h at 37 °C in the dark. Add 20 μL of glycerol to each sample in preparation for gel loading.

8. Prepare disaccharide standards for HS and CS by digestion and labeling in the same manner. Load 5 μL of each sample onto a MONO gel and perform electrophoresis at 4 °C with constant current of 60 mA for 40 min.

9. Visualize and quantify the resulting bands using an image capture and analysis system.

3.4 Cloning of Chondrointinase ABC into pcDNA3.1/ myc-His

1. Amplify the chondrointinase ABC lyase coding sequence from *Proteus vulgaris*, lacking its bacterial leader sequence (amino acids 25–1021), by PCR from *P. vulgaris* genomic DNA. Use a forward primer designed to contain a BamHI restriction enzyme site to allow ligation of the 5′-end of the ChABC coding sequence in-frame with the mammalian signal peptide coding sequence and a reverse primer including an XhoI restriction site and lacking a stop codon, permitting cloning of the 3′-end of the ChABC coding fragment in-frame with the *myc*-His tag coding sequence of pcDNA3.1/myc-His. The primer pair ChABCFwd and ChABCRev fulfill these specifications (*see* Subheading 2).

2. Perform PCR with the above-mentioned template and primers using PfuUltra HF polymerase and the following cycling parameters: denaturation at 95 °C for 4 min, 35 cycles of 95 °C for 30 s, 58 °C for 30 s, and 72 °C for 3 min; and an additional 2 min at 72° followed by cooling to 4 °C.

3. Purify the resulting PCR fragment by agarose gel electrophoresis followed by DNA extraction using a QIAquick Gel Extraction Kit (Qiagen).

4. Digest the pcDNA3.1/myc-His plasmid and ChABC PCR fragment separately with both BamHI and XhoI at 37 °C for 2 h. Purify the digested DNA samples with the QIAquick Nucleotide Removal Kit (Qiagen).

5. Ligate the ChABC fragment into the linearized pcDNA3.1/ myc-His plasmid at 16 °C overnight using T4 DNA ligase at a vector-to-insert ratio of 1:3 based on DNA molar ends.

6. Use 1/10th of the ligation reaction to transform TOP10 competent cells; then plate the cells on LB + ampicillin agar plates and incubate overnight at 37 °C. Pick well-separated colonies and transfer to a fresh LB + ampicillin plate, arranging in a grid pattern and assigning isolate labels.

7. Inoculate 5 mL LB + ampicillin cultures in 13-mL tubes from single isolate colonies and grow overnight at 37 °C, then isolate plasmid DNA using a QIAprep Spin Miniprep Kit (Qiagen).

8. Identify insert-positive clones by restriction enzyme digestion using BamHI and XhoI followed by DNA agarose gel electrophoresis.

9. Confirm positive clones via DNA sequencing of the inserts and the flanking plasmid sequences.

3.5 Engineering of N-Terminal Mammalian Signal Peptide

1. The signal peptide from murine sonic hedgehog (MLLLARCFLVILASSLLVCPG) is added to the N-terminal end of ChABC by amplifying a PCR fragment encompassing part of the pcDNA3.1/myc-His vector sequence containing the CMV promoter and the adjacent, downstream Shh signal

peptide coding sequence from a pcDNA3.1-Shh construct. Use a forward primer complementary to the nucleotide sequence in pcDNA3.1/myc-His from 484 to 505 that includes the unique NdeI site from pcDNA3.1/myc-His together with a reverse primer complementary to the Shh signal peptide coding sequence and having an engineered BamHI restriction site at its 5′-end in order to generate an open reading frame ligation with the ChABC coding sequence. The primer pair pcDNA3.1Fwd and ShhRevSP (*see* Subheading 2) satisfies those conditions.

2. Amplify a 500-base-pair PCR fragment using PfuUltra HF DNA polymerase, the primers given above and pcDNA3.1-Shh as template, with the following PCR conditions: denaturation at 95 °C for 4 min, 35 cycles of 95 °C for 30 s, 65 °C for 30 s, and 72 °C for 40 s; and an additional 2 min at 72° followed by cooling to 4 °C.

3. Purify the resulting PCR product via gel electrophoresis followed by DNA gel extraction using a QIAquick Gel Extraction Kit.

4. Separately digest the pcDNA3.1ChABCmycHis vector and signal peptide coding fragment PCR product with both NdeI and BamHI at 37 °C for 2 h.

5. Purify the restriction-enzyme-digested linearized vector and signal peptide coding fragment using the QIAquick Nucleotide Removal Kit.

6. Ligate the signal peptide coding fragment into pcDNA3.1ChABCmycHis vector overnight at 16 °C, using T4 DNA ligase at a ratio of 1:3 vector to insert based on molar quantity of DNA ends.

7. Use 1/10th of the ligation reaction to transform one vial of TOP10 chemically competent *E. coli*. Following transformation, plate the cells on LB agar + 100 μg/mL ampicillin plates, and incubate at 37 °C overnight, as previously described [9].

8. Pick well-separated colonies to fresh plates, and use isolate colonies to inoculate 5-mL LB + ampicillin tube cultures as before. Incubate overnight at 37 °C with shaking and purify plasmid DNA from individual clone cultures using the QIAprep Spin Miniprep Kit.

9. Identify positive clones by double restriction digestion with NdeI and HindIII followed by DNA gel electrophoresis. Positive clones will yield an ~500-bp band corresponding to the signal peptide coding fragment. Confirm all putative constructs by sequencing the full inserts and flanking plasmid regions.

10. The resulting plasmid is pcDNA3.1-ChABCmycHis-SP.

*3.6 Chondroitinase
ABC Digestion
of CSPGs in Cortical
Slice Cultures*

1. Reagents and equipment are as in Polleux and Ghosh [10].

2. Separately electroporate a plasmid encoding a secreted form of ChABC (pcDNA3.1-ChABCmycHis-SP) or a control plasmid encoding LacZ (pcDNA3.1-LacZ) at concentrations of 1 μg/μL into E14.5 mouse brains *ex utero* with 3×70 V square wave pulses (100 ms intervals).

3. Prepare coronal slices (250 μm) and culture for 4 days in vitro on poly-L-lysine- and laminin-coated 1.0 μm track-etched polyethylene terephthalate inserts, as previously described [10].

4. Fix and costain the slices as described [10] for either β-galactosidase (control plasmid slices) or myc (ChABC plasmid slices) to demarcate plasmid expression, and with the anti-6S antibody, which only recognizes CS chains sulfated at the 6-position after ChABC digestion. Therefore, a positive signal from the 6S antibody only in the slices electroporated with the ChABC-expressing plasmid indicates ChABC activity in those tissues.

4   Notes

1. The suppliers indicated for materials are those used by our laboratory. Other suppliers' products may work as well, but without experience we cannot verify.

2. Seikagaku antibodies were distributed by Associates of Cape Cod, Inc. and Amsbio (http://www.amsbio.com/Glycobiology-Glycosaminoglycans-monoclonal-antibodies.aspx) until March 2012. Seikagaku Inc. no longer offers antibodies or other research products.

Acknowledgements

This work was supported by NIH grants HD-009402, HD 017332, HD 054275, HD 06976, and 10035 from the Mizutani Foundation for Glycoscience. We thank James Mensch for help in assembling and proofing this manuscript.

References

1. Johnson CE, Crawford BE et al (2007) Essential alterations of heparan sulfate during the differentiation of embryonic stem cells to sox1-enhanced green fluorescent protein-expressing neural progenitor cells. Stem Cells 25(8):1913–1923

2. Domowicz MS, Sanders TA et al (2008) Aggrecan is expressed by embryonic brain glia and regulates astrocyte development. Dev Biol 315(1):114–124

3. Sirko S, von Holst A et al (2007) Chondroitin sulfate glycosaminoglycans control proliferation, radial glia cell differentiation and neurogenesis in neural stem/progenitor cells. Development 134(15):2727–2738

4. Cortes M, Baria AT et al (2009) Sulfation of chondroitin sulfate proteoglycans is necessary for proper Indian hedgehog signaling in the developing growth plate. Development 136(10):1697–1706

5. Domowicz M, Mangoura D et al (2000) Cell specific-chondroitin sulfate proteoglycan expression during CNS morphogenesis in the chick embryo. Int J Dev Neurosci 18(7): 629–641

6. Calabro A, Benavides M et al (2000) Microanalysis of enzyme digests of hyaluronan and chondroitin/dermatan sulfate by fluorophore-assisted carbohydrate electrophoresis (FACE). Glycobiology 10(3):273–281

7. Calabro A, Hascall VC et al (2000) Adaptation of FACE methodology for microanalysis of total hyaluronan and chondroitin sulfate composition from cartilage. Glycobiology 10(3):283–293

8. Calabro A, Midura R et al (2001) Fluorophore-assisted carbohydrate electrophoresis (FACE) of glycosaminoglycans. Osteoarthritis Cartilage 9(Suppl A):S16–S22

9. Cortes M, Mensch JR et al (2012) Proteoglycans: gene cloning. Methods Mol Biol 836:3–21

10. Polleux F, Ghosh A (2002) The slice overlay assay: a versatile tool to study the influence of extracellular signals on neuronal development. Sci STKE 2002(136):pl9

Cell Substrate Patterning with Glycosaminoglycans to Study Their Biological Roles in the Central Nervous System

Tony W. Hsiao, Vimal P. Swarup, Colin D. Eichinger, and Vladimir Hlady

Abstract

Microcontact printing (μCP) based techniques have been developed for creating cell culture substrates with discrete placement of CNS-expressed molecules. These substrates can be used to study various components of the complex molecular environment in the central nervous system (CNS) and related cellular responses. Macromolecules such as glycosaminoglycans (GAGs), proteoglycans (PGs), or proteins are amenable to printing. Detailed protocols for both adsorption based as well as covalent reaction printing of cell culture substrates are provided. By utilizing a modified light microscope, precise placement of two or more types of macromolecules by sequential μCP can be used to create desired spatial arrangements containing multicomponent PG, GAG, and protein surface patterns for studying CNS cell behavior. Examples of GAG stripe assays for neuronal pathfinding and directed outgrowth, and dot gradients of PG + laminin for astrocyte migration studies are provided.

Key words Microcontact printing, Chondroitin sulfate proteoglycan, Glycosaminoglycans, Cell choice assay, Random dot gradient, Neuronal pathfinding, Astrocyte migration

1 Introduction

Proteoglycans (PGs) have been implicated in various roles of central nervous system (CNS) development, maintenance, and injury [1]. Along with the diverse types of PG molecules in the CNS, there is also an inherent heterogeneity in the glycosaminoglycan (GAG) structures on the polysaccharides themselves. Since PGs integrate into the ECM and influence neuronal growth and differentiation, developing in vitro models to study such interactions can help to better understand the effect of PGs or GAGs on neurons. Since its inception by Whitesides et al. [2] microcontact printing (μCP) based on soft lithography has become a powerful tool for investigating cell-surface interactions. μCP is based on

Kuberan Balagurunathan et al. (eds.), *Glycosaminoglycans: Chemistry and Biology*, Methods in Molecular Biology, vol. 1229, DOI 10.1007/978-1-4939-1714-3_35, © Springer Science+Business Media New York 2015

creating elastomeric stamps that replicate features made via photo-lithography, and using the stamps to transfer macromolecules such as proteins or polysaccharides to suitable substrates. In its simplest form, μCP is used with macromolecules, such as proteins, that will adsorb to solid substrates via nonspecific interactions. When attaching proteins, such as NGF, neurotrophins, or other growth factors, to solid substrates, different regions of the molecules may be targeted for covalent surface attachment. In addition to uniform surface density of stamped macromolecules, gradients of multiple types have been created by controlling stamp feature density or utilizing diffusion [3–6]. Complex surface patterns of proteoglycans can also be produced by microcontact printing and are only constrained by the limits of soft lithography [7]. Here, we report on μCP-based methods that have been developed to precisely deposit PGs or GAGs in specific spatial patterns onto substrates to investigate their roles in directing neuronal outgrowth and axonal pathfinding.

The patterns described below include basic lanes or stripes [8] as well as dot gradients that use random distributions of μm-scale features to avoid the appearance of periodicity that could affect neurite outgrowth. Different surface conjugation chemistry can be implemented to create covalently attached macromolecules that will not detach or leach away from the substrates. We also report on a method to pattern binary or ternary combinations of macromolecules in a desired spatial arrangement using sequential stamping and/or solution adsorption steps. One challenge in multiple μCP steps is the ability to place sequential patterns in registration with previously printed patterns. Other methods, such as dip-pen nanolithography [9] and inkjet printing [10] have automated systems to produce in-register sequential patterns, but have limitations such as patterning over large-scale areas and creating micrometer-scale patterns, respectively. In the protocols described in this chapter, we describe the methods to prepare multipatterns of chondroitin sulfate proteoglycans (CSPGs) or chondroitin sulfate GAG chains for studying neuronal outgrowth and guidance [11, 12]. These methods will allow for the study of particular ECM components chosen from the complex environment of the CNS. In addition to neuronal outgrowth and guidance, other cells from the CNS, such as astrocytes, can be studied as well.

2   Materials

Use pure water (double distilled and deionized) and reagent grade materials for all procedures. Surfaces for cell culture should be sterilized prior to any experiments.

**2.1 PDMS Stamp
Fabrication
and Adsorption
Stamping**

1. *Patterned photolithographic masks made via standard techniques.* Transfer given patterns to silicon based substrates with appropriately exposed and developed photoresist to provide 3-D structure. The provided examples use AZ 1500 photoresist on chromium plated soda lime glass masks (Telic) that were exposed using either a micropattern generator (Electromask) or a μPG 101 tabletop laser pattern generator (Heidlberg Instruments).

2. *Polydimethylsiloxane (PDMS) mix to create elastomeric stamps.* Use a 10:1 mass ratio of polymer base to crosslinking agent such as Dow Corning Sylgard 184.

3. *Phosphate buffered saline (PBS)*: Combine 0.01 M NaHPO$_4$, 0.0027 M KCl and 0.137 M NaCl, pH 7.4 in water.

4. *Proteoglycan solutions.* Dissolve desired amounts into PBS. Here, commercial aggrecan was used at a concentration of 25 μg/mL PBS. Proteins, like laminin (5–500 μg/mL PBS) can also be applied in the same manner.

**2.2 Thiolation
and Attachment
of GAGs**

1. *Stock GAGs thiolation*: To chemically conjugate GAGs to surfaces via thiol linkages [13], react GAGs with 3,3′-dithiobis(propionic hydrazide) (DTP) to create GAGs with a low density of thiols. 1-ethyl-3-(3-dimethylaminopropyl)carbodiimide (EDC) and dithiothreitol (DTT) are also required for the reaction. To visualize these GAGs, fluorescent dyes (for example Alexa Fluor 594) can be combined in solution with the GAGs so that they are colocalized after printing.

2. *GAG purification*: Following the reaction, purify modified GAGs by performing dialysis with a 1 kDa cut-off membrane. A 100 mM NaCl solution in water will be required. Lyophilize and store at 4 °C.

3. *Substrate activation*: Poly-L-Lysine (pLL), is used to coat surfaces and bind the crosslinking molecule, sulfosuccinimidyl 4-[N-maleimidomethyl]cyclohexane-1-carboxylate (sSMCC), to the surfaces. Prepare sSMCC activation buffer: 0.1 M NaHPO$_4$, 0.15 M NaCl, pH 7.2.

**2.3 Microscope
Adaptation for Stamp
Manipulation**

1. *Modified light microscope*: The precise machining of the optics for a light microscope can be utilized to facilitate accurate registration of multiple stamps for sequential μCP. A schematic of the necessary modifications and an example of the produced cell substrates is given in Fig. 1. The key components are numbered according to the figure. Modify a microscope objective (Fig. 1a-5) with an attached auxiliary stage (Fig. 1a-4) that will hold the substrate to be patterned (Fig. 1a-3) directly over the objective lens. This auxiliary stage can be a simple ring or cylinder with the desired height attached to the objective in such

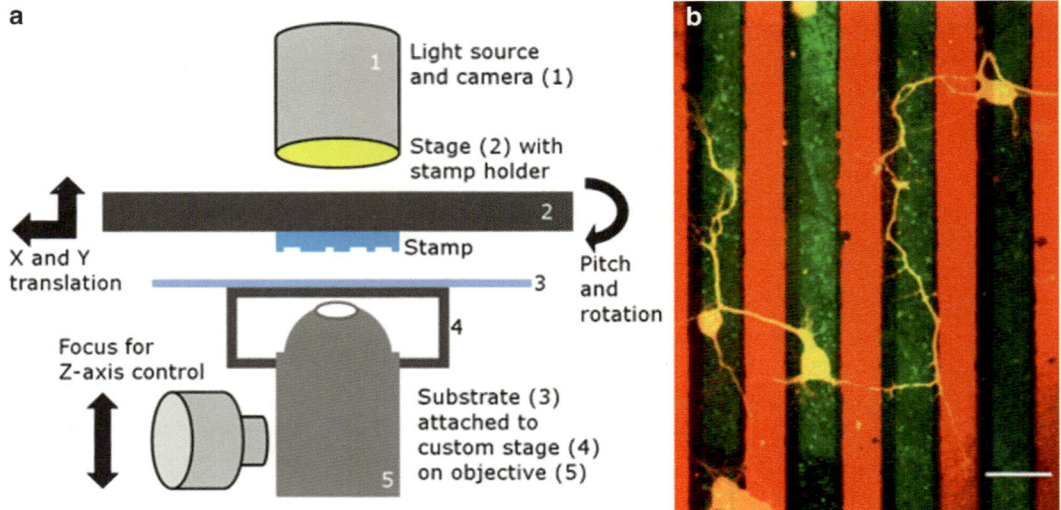

Fig. 1 Multicomponent microcontact printing via microscope. (**a**) Schematic of customized microscope to allow real-time registration and printing of multiple stamps. (**b**) Neuronal guidance on a binary-CS patterned surface. Neurons (*orange*) cultured for 48 h preferred CS-E (*green*) and avoided CS-C (*red*) lanes. Scale bar = 25 μm, *orange*: anti-rat TAU/MAPT, *red*: AlexaFluor 594 colocalized with CS-C, *green*: AlexaFluor 488 colocalized with CS-E

way that the substrate which is to be printed is in focal range of the objective. In the case shown in Fig. 1a, a plastic ring was custom fit to a Fluor 10× phase objective. A custom solution will need to be built for any given microscope objective, but clamping, glues, or 2-sided tape can be used to secure items in place. The substrate must be transparent, e.g. glass coverslips, to allow for real time leveling and alignment of multiple stamps. Design a stamp holder that attaches to the main microscope stage (Fig. 1a-2). A simple option is a standard glass microscope slide clamped to the stage. If the stage does not have any control for tilt or rotation, this should be added to this holder so that the stamp can be leveled with respect to the substrate. For example, a kinematic mount was mounted to the stage to provide motion control. PDMS stamps will adhere to many smooth surfaces like glass and metals via capillary forces alone, so a specific tool for stamp attachment is usually not necessary.

2. *CCD camera capable of real-time imaging and display*: A CCD camera (Fig. 1a-1) will be necessary for registering subsequent stamps and stamp leveling.

3. *Computer equipped with live imaging processing software*: Multiple stamp registration is achieved by overlaying a static image of a previous stamp in contact with the substrate with a live image of the current stamp in close proximity to the substrate. The open source imaging software with live overlay features, Micro-Manager (www.micro-manger.org), is recommended.

2.4   Cell Cultures and Immunocytochemistry

1. *Primary cells*: Numerous cell types are available for specific tailoring to the desired study. Here, hippocampal neurons (HNs) are used to investigate the behavior of neurons found in the brain, isolated dorsal root ganglia neurons (DRGs) [14] for neurons in the spinal cord, and cortical astrocytes [15] for glia in the CNS.

2. *Culture media*: Nbactiv1 media is used for HNs. Dulbecco's modified eagle medium supplemented with F12 nutrient mixture (DMEM/F12) and 10 % fetal bovine serum (FBS) is used for astrocytes, and SATO serum free media [16] is used for DRG cultures.

3. *Cell fixatives*: 4 % Paraformaldehyde (PFA) or 3.7 % formalin in PBS solutions.

4. *Antibodies for immunolabeling*: Label HN with chicken anti-rat TAU/MAPT, followed by goat anti-rat IgG secondary antibody conjugated to fluorescent markers such as Alexa Fluor 488 from Molecular Probes. DRGs can be labeled using antibodies for anti-rat neurofilament such as mouse anti-NF160. Astrocytes can be visualized using anti-GFAP such as rabbit anti-GFAP. DAPI can be used to visualize nuclei of all cells.

5. *Blocking Solution*: 4 % goat serum in PBS. This solution may be supplemented with 0.05 % Triton X-100 to access antigens within the cell.

6. *A fluorescence microscope*: An inverted microscope with the appropriate filters to excite fluorescence and capture the emitted light of the fluorescent markers used is required to observe and record cell images.

3   Methods

3.1   Printing Proteoglycan Gradients via Nonspecific Adsorption

1. Generate random pixel distributions using Mathematica (Wolfram) or similar software. In Mathematica, the commands to create a continuous dot gradient are as follow:

```
Cell 1 - gradientPercolation[n_, m_]:=
    Module[{p=0},
    createRow=(p+=1/(n-1);
    h[#, Table[Floor[Random[]+p],{m}]])&
Cell 2 - Apply[List,
    Flatten[Nest[createRow, Table[0,{m}], n-1], Infinity, h]]]
Show[Graphics[RasterArray[gradientPercolation[1000, 50]/.
    {0 ->RGBColor[1, 1, 1],1 ->RGBColor[0,0,0]}]]],
    AspectRatio->Automatic]
```

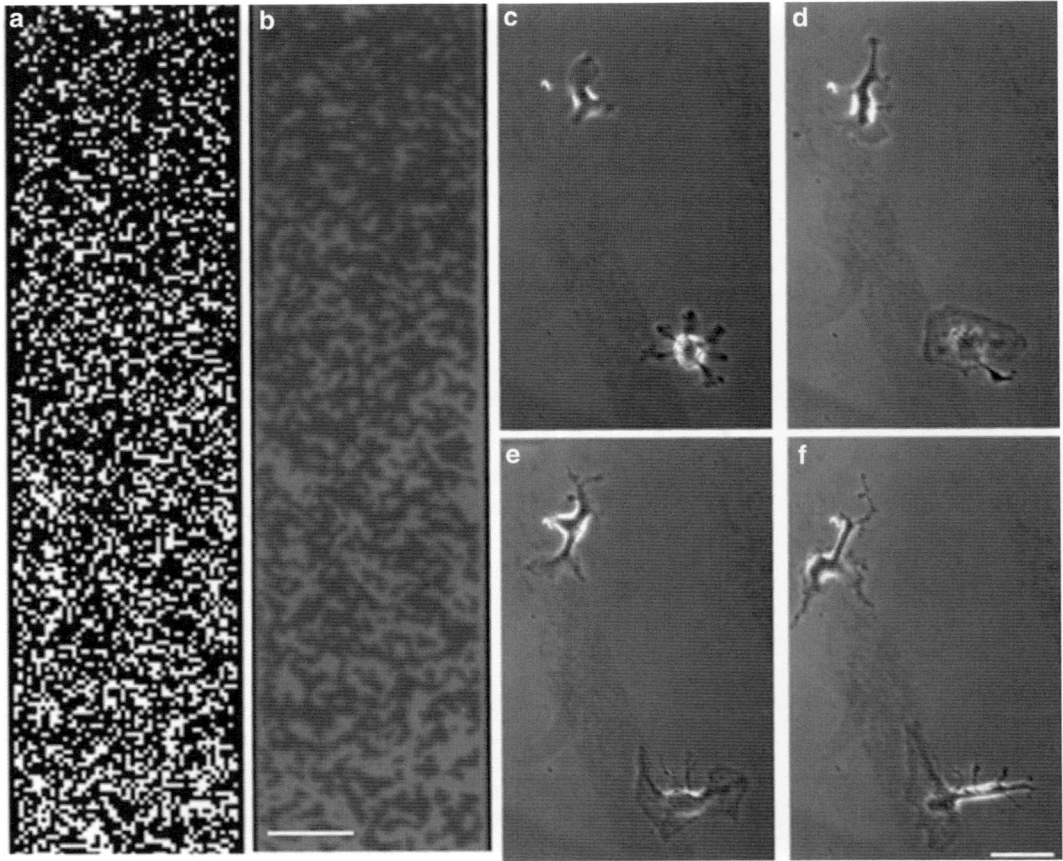

Fig. 2 Astrocyte spreading on a CSPG/Laminin dot gradient. (**a**) Region of computer generated random dot gradient template and (**b**) subsequent pattern of fluorescent protein created by printing given pattern onto glass (scale bar = 20 μm). (**c–f**) Culture of astrocytes on aggrecan (*lighter gray*) gradients printed onto uniform laminin (*darker gray*). At 1 h post seeding (**c**), astrocytes attached and began to spread. At 4 (**d**), 12 (**e**), and 24 (**f**) h after seeding, astrocytes preferentially spread and shifted their nucleus to regions of higher laminin and lower aggrecan concentrations. For **c–f**, scale bar = 50 μm

The graphics output can then be copied as a bitmap to lithography patterning software. This command created a 50 unit wide 1,000 unit long gradient. An example of the middle portion of the gradient graphics, its printed pattern, and use in astrocyte migration experiments are shown in Fig. 2.

2. Create a photolithographic template by spin-coating photoresist on a silicon or glass substrate and then exposing the generated pattern. Develop the resist to provide 3-D features in the template (*see* **Note 1**).

3. Cast PDMS stamps by curing the PDMS mix on top of the photolithographic template (*see* **Note 2**).

4. Adsorb PDMS stamps with proteoglycan solutions for 30 min at room temperature (r.t.) (*see* **Note 3**).

5. Rinse the stamp with water and dry with a nitrogen stream.

6. Print the proteoglycans by placing the coated stamp in conformal contact with a coverslip. Allow the stamp to remain in contact with the coverslip for 1–2 min. Carefully remove the stamp from the coverslip. Wash the stamp with detergent and water and store for reuse (*see* **Note 4**).

7. To create a pattern made of two macromolecules there are two approaches: precoating the coverslip with the first macromolecule prior to the printing of the second macromolecule, or coating (so-called back-filling) the coverslip with the second macromolecule *after* it was stamped with the first macromolecule. The approach chosen depends on the desired combination of molecules. For example, coverslips coated with laminin solution (100–500 μg/mL PBS) for 1 h at r.t. can be rinsed with water, dried in nitrogen, and printed with aggrecan solution to create surfaces with laminin and aggrecan-on-laminin features. Substrates with laminin and laminin-on-aggrecan features can be similarly produced by reversing the order of the two macromolecules.

3.2 Printing GAGs via sSMCC Reactions

1. Thiolate 500 mg of GAGs in 50 mL water by adding 42 mg of DTP and 16.8 mg of EDC. Adjust the pH to 4.75 and let the reaction continue for 4 h at r.t.

2. Add 125 mg of DTT to reaction mixture and increase the pH to 8.5. Continue the reaction for 24 h.

3. Dialyze the mixture with a 1 kDa cut-off dialysis membrane in 100 mM NaCl, pH 3.0 for 3 days. Continue the dialysis for another 3 days in water, pH 3.0. Lyophilize and store at 4 °C until ready to use (*see* **Note 5**).

4. Coat coverslips with poly-L-Lysine solution (0.05 mg/mL in water) for 1 h at r.t. Remove the pLL solution and rinse with water.

5. Add sSMCC solution (0.8 mg/mL in sSMCC activation buffer) to the coated surfaces and incubate for 1 h at r.t. Remove sSMCC solution and rinse with water.

6. Prepare stamps during sSMCC incubation by coating the stamp features with GAG solutions. Allow GAG to adsorb to the stamp for 20 min at r.t., rinse stamps with water, and then dry with nitrogen.

7. Print GAGs by placing the stamp onto the sSMCC-activated surface. Allow it to stand for 2 min and then carefully remove the stamp.

8. Incubate the patterned substrates for 8 h at r.t. to allow the thiolated GAG to sSMCC reaction to complete (*see* **Note 6**).

9. Use the substrate for cell culture. A schematic of this process and the resulting substrates are shown in Fig. 3.

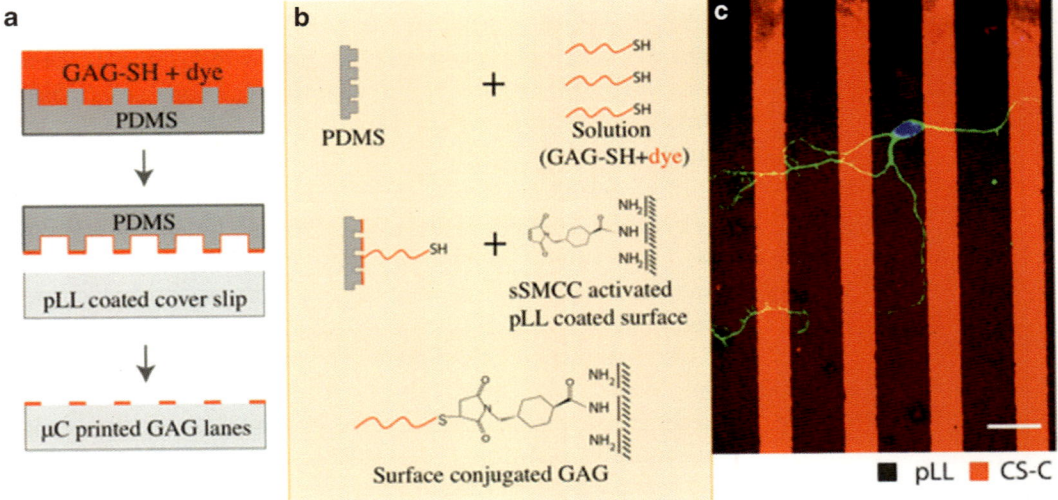

Fig. 3 Using μCP to deposit GAGs and studying the directional response of hippocampal neurons on CS patterned substrate. (**a**) Scheme for creating GAG lanes on a pLL-coated coverslip. (**b**) Surface chemistry involved in conjugating thiolated GAGs onto preactivated pLL. The reaction steps shown correspond to the μC printing step presented in (**a**). (**c**) Representative image of a hippocampal neuron (*green*) grown on CS-C lanes (*red*) for 48 h. As suggested by the image, hippocampal neurons avoided growing on regions containing CS-C in comparison to the flanking pLL lanes (*black*). Scale bar = 25 μm, *green*: anti-rat TAU/MAPT, *blue*: DAPI, *red*: AlexaFluor 594 colocalized with CS-C

3.3 Multicomponent Microcontact Printing

1. Prepare stamps for printing by coating them with the desired macromolecules.

2. Attach nonfeatured back of stamp to the stamp holder on the adapted microscope stage (Fig. 1).

3. Attach the coverslip substrate to the auxiliary stage on the adapted microscope objective and rotate the objective into viewing position.

4. Slowly bring the stamp surface into the view of the objective. Bring the stamp nearly into focus, but stop prior to contacting the substrate.

5. Level the stamp by checking each of the four corners of the stamp by adjusting the tilt of the stamp holder until all four corners are equally in focus.

6. Print the surface by using the main focus knob of the microscope to bring the stamp into contact with the substrate. Take an image of the stamp in contact with the substrate as a reference for alignment of subsequent stamps.

7. Carefully retract the stamp from the patterned surface (*see* **Note 7**).

8. Place the next prepared stamp onto the stage stamp holder.

9. Slowly bring this stamp into view of the objective and level as in **step 4**.

10. Compare the live image of the leveled stamp with an overlay of the previous reference image taken in **step 6** to determine and adjust registration of the patterns, both translationally and rotationally.

11. Once the desired alignment is attained, proceed to print the surface with the main focus knob of the microscope. Retract this stamp from the surface.

12. Repeat **steps 8–11** for as many components as desired.

3.4 CNS Cell Culture

1. *Cell Plating:* For HNs, add HNs suspended in Nbactiv1 media (80,000 cells/mL) to each well containing patterned coverslips (2 mL/well for a 6-well plate, 1 mL/well for a 12 well plate, and 0.5 mL/well for 24 well plate). For DRGs, similar neuron suspensions can be done in DMEM/F12 or SATO media. Astrocytes suspended in DMEM/F12 should be added to wells containing patterned coverslips at sparse densities (25,000 cells/mL) to minimize intercellular interactions if the objective is to study individual cells.

2. Maintain CNS cell cultures at 37 °C and 5 % CO_2 for desired time periods (e.g. 24, 48, and 72 h)

3. Fix the cells. Exchange media with equal volume of 4 % PFA and allow for 15 min of incubation. For HNs, apply 3.7 % formalin solution for 1 min.

3.5 Immuno-cytocheistry

1. Incubate fixed samples in Blocking Solution for 1 h at r.t.

2. Prepare primary and secondary (containing fluorescent dye) antibody solutions according to manufacturer guidelines. Dilutions ranged from 1:100 to 1:1,000 of antibody in blocking solution.

3. Rinse the fixed and blocked samples with PBS thrice.

4. Apply primary antibody solutions to the cells for 1 h at r.t. and then rinse thrice with PBS.

5. Add the secondary antibody solution to the samples for 1 h at r.t. and rinse thrice with PBS (*see* **Note 8**).

6. Store samples in PBS or mount to slides for imaging.

3.6 Imaging and Analysis

1. Choose isolated neurons that are not contacting other neurons or neurites for outgrowth analysis. To build statistically relevant populations, sample forty to fifty neurons. Individual astrocytes should also be chosen to determine their interactions with the substrate rather than with neighboring cells.

2. Measure neuron outgrowth lengths using the segmented line tool and outgrowth alignment using the angle tool in ImageJ.

4   Notes

1. Be sure to double check which regions of your pattern will be exposed by your photolithography device. If using a negative resist, the exposed areas will be developed away therefore those will be the features stamped. The opposite will happen if a positive resist is used.

2. For optimal PDMS removal from mask templates, it is best to remove the PDMS from the mask before it is fully cured and rigid. This prevents fracturing of the PDMS. Partial curing is accomplished by placing samples in a 100 °C oven for 30 min, after which the PDMS can be removed from the photoresist surface. The PDMS alone, without the mold, is then placed back into the oven and allowed to fully cure. To control for stamp thickness, the volume of PDMS mix can be adjusted.

3. To facilitate the spreading of solutions and to minimize reagent use, a sterile coverslip can be placed on top of a small volume of solution that has been applied to the top of the PDMS stamp to provide full stamp coverage. If the hydrophobicity of the PDMS is causing difficulties with adsorption of inking proteoglycans, a brief plasma treatment [17], either in oxygen or corona discharge in air, immediately before adsorption can render the surface hydrophilic, although the surface tends to revert to being hydrophobic over time.

4. After use, PDMS stamps should be sonicated in a mild detergent and rinsed with water. The stamps can be stored submerged in water or be dried and sealed in petri-dishes to avoid dust particles and contaminants from adhering to the features.

5. Thiolated GAGs should be always stored under nitrogen or argon. Storing the molecules in air can result in reduction of thiolated moieties due to the formation of disulfide bonds.

6. Antibody staining can be done to ensure uniform deposition of printed PGs or GAGs.

7. When retracting the stamp from contact with the substrate, occasionally the stamp will release from its holder and remain on the substrate instead. This should not pose a problem so long as the substrate does not translate with respect to the objective to which it is attached. The stamp can be removed from the mounted substrate with forceps, taking care not to move the substrate.

8. The incubation steps for immunocytochemistry can be extended without adverse effects with the exception of the secondary antibody application. Prolonged incubation with the dye containing molecules may lead to excessive background fluorescence. The PBS solutions may be supplemented with 0.1 % sodium azide to prevent contamination of samples.

Acknowledgment

This work was supported by NIH grant R01 NS57144.

References

1. Swarup VP, Mencio CP, Hlady V, Kuberan B (2013) Sugar glues for broken neurons. Biomol Concepts 4:233–257

2. Xia Y, Whitesides GM (1998) Soft lithography. Annu Rev Mater Sci 28:153–184

3. Fricke R, Zentis PD, Rajappa LT et al (2011) Axon guidance of rat cortical neurons by microcontact printed gradients. Biomaterials 32:2070–2076. doi:10.1016/j.biomaterials.2010.11.036

4. Von Philipsborn AC, Lang S, Bernard A et al (2006) Microcontact printing of axon guidance molecules for generation of graded patterns. Nat Protoc 1:1322–1328. doi:10.1038/nprot.2006.251

5. Von Philipsborn AC, Lang S, Loeschinger J et al (2006) Growth cone navigation in substrate-bound ephrin gradients. Development 133:2487–2495. doi:10.1242/dev.02412

6. Mai J, Fok L, Gao H et al (2009) Axon initiation and growth cone turning on bound protein gradients. J Neurosci 29:7450–7458. doi:10.1523/JNEUROSCI.1121-09.2009

7. Qin D, Xia Y, Whitesides GM (2010) Soft lithography for micro- and nanoscale patterning. Nat Protoc 5:491–502. doi:10.1038/nprot.2009.234

8. Knoll B, Weinl C, Nordheim A, Bonhoeffer F (2007) Stripe assay to examine axonal guidance and cell migration. Nat Protoc 2:1216–1224. doi:10.1038/nprot.2007.157

9. Piner RD, Zhu J, Xu F et al (1999) "Dip-Pen" nanolithography. Science 283:661–663. doi:10.1126/science.283.5402.661

10. Calvert P (2001) Inkjet printing for materials and devices. Chem Mater 13:3299–3305. doi:10.1021/cm0101632

11. Eichinger CD, Hsiao TW, Hlady V (2011) Multiprotein microcontact printing with micrometer resolution. Langmuir 28:2238–2243. doi:10.1021/la2039202

12. Swarup VP, Hsiao TW, Zhang J et al (2013) Exploiting differential surface display of chondroitin sulfate variants for directing neuronal outgrowth. J Am Chem Soc 135:13488–13494. doi:10.1021/ja4056728

13. Liu Y, Cai S, Shu XZ et al (2007) Release of basic fibroblast growth factor from a crosslinked glycosaminoglycan hydrogel promotes wound healing. Wound Repair Regen 15:245–251. doi:10.1111/j.1524-475X.2007.00211.x

14. Meng F, Hlady V, Tresco PA (2012) Inducing alignment in astrocyte tissue constructs by surface ligands patterned on biomaterials. Biomaterials 33:1323–1335. doi:10.1016/j.biomaterials.2011.10.034

15. McCarthy KD, de Vellis J (1980) Preparation of separate astroglial and oligodendroglial cell cultures from rat cerebral tissue. J Cell Biol 85:890–902. doi:10.1083/jcb.85.3.890

16. Bottenstein JE, Sato GH (1979) Growth of a rat neuroblastoma cell line in serum-free supplemented medium. Proc Natl Acad Sci 76:514–517

17. Anderson JR, Chiu DT, Wu H et al (2000) Fabrication of microfluidic systems in poly (dimethylsiloxane). Electrophoresis 21:27–40

Chapter 36

Analyzing the Role of Heparan Sulfate Proteoglycans in Axon Guidance In Vivo in Zebrafish

Fabienne E. Poulain

Abstract

One of the most fascinating questions in the field of neurobiology is to understand how neuronal connections are properly formed. During development, neurons extend axons that are guided along defined paths by attractive and repulsive cues to reach their brain target. Most of these guidance factors are regulated by heparan sulfate proteoglycans (HSPGs), a family of cell-surface and extracellular core proteins with attached heparan sulfate (HS) glycosaminoglycans. The unique diversity and structural complexity of HS sugar chains, as well as the variety of core proteins, have been proposed to generate a complex "sugar code" essential for brain wiring. While the functions of HSPGs have been well characterized in *C. elegans* or *Drosophila*, relatively little is known about their roles in nervous system development in vertebrates. In this chapter, we describe the advantages and the different methods available to study the roles of HSPGs in axon guidance directly in vivo in zebrafish. We provide protocols for visualizing axons in vivo, including precise dye labeling and time-lapse imaging, and for disturbing the functions of HS-modifying enzymes and core proteins, including morpholino, DNA, or RNA injections.

Key words Axon pathfinding, Sugar code, Syndecan, Glypican, Enzyme, Dye labeling, Injection, Mutant

1 Introduction

Brain connectivity and function depend on the proper development of long-range neuronal projections, which in turn relies on the guidance of individual axons as they elongate and grow. When they navigate to their targets, axons respond to diverse attractive and repulsive cues acting at a distance or locally by contact. Concomitantly or subsequently to this guidance process, refinement mechanisms involving pruning or degeneration correct axons that have deviated from the right path, thereby ensuring the formation of accurate neuronal circuits. While many growth factors and guidance cues regulate axon pathfinding, the combined information they provide does not seem sufficient to sculpt the entire neuronal network. Heparan sulfate proteoglycans (HSPGs) are cell-surface and extracellular core proteins with attached heparan

Kuberan Balagurunathan et al. (eds.), *Glycosaminoglycans: Chemistry and Biology*, Methods in Molecular Biology, vol. 1229, DOI 10.1007/978-1-4939-1714-3_36, © Springer Science+Business Media New York 2015

sulfate (HS) glycans that are thought to play crucial roles in axon guidance. The diversity and structural complexity of their HS chains allow them to interact with many factors and orchestrate most if not all guidance pathways essential for neuronal wiring. In addition, accumulative observations indicate that core proteins also have functional specificities, and that cooperation between them and their HS chains appears essential for some HSPG functions in nervous system development.

1.1 Roles of HS Chains in Axon Guidance

Several biochemical experiments have shown that HS interacts with guidance molecules and is important for their functions. For instance, netrins were originally purified using heparin affinity columns [1], and their receptor DCC was shown to bind to HS chains in vitro [2]. Similarly, HS critically regulates the function of the guidance cue Slit and its receptor Robo by forming a ternary signaling complex at the surface of axons [3–8]. Additional morphogens such as Wnt, FGF, BMP, or Shh, whose role as guidance molecules has been later identified, also bind to HS with a high affinity [9]. The importance of HS in axon guidance has further been demonstrated in animal models by chemically or genetically modifying HS levels. In *Xenopus*, adding HS to the developing retinotectal pathway or removing HS with heparitinase prevents retinal axons from entering their brain target, the tectum [10, 11]. In mice, conditional depletion of HS in the nervous system induces severe guidance errors in major commissural tracts, revealing an essential role of HS in midline axon pathfinding [12]. Similarly in zebrafish, drastic reduction in HS induces many retinal axon guidance defects including projections into the forebrain, the hindbrain, and the opposite eye, as well as missorting of axons along the optic tract [13, 14]. Pathfinding of peripheral sensory neurons is also altered in mutants lacking HS [15]. Overall, these different interactions and functions designate HS as a "master-regulator" of axon guidance in vivo.

HS disaccharides are subject to a large number of modifications responsible for their high diversity. These modifications give HS its specific binding affinities and have thus been proposed to generate a "sugar code" for the recognition of guidance factors during development. At least 14 biochemical steps occur in HS chain synthesis. HS are synthesized in the Golgi, where HS polymerases generate a nonsulfated sugar backbone consisting of alternating *N*-acetylglucosamine and D-glucoronic acid repeats. Initiation and polymerization of this precursor disaccharide are catalyzed by glycosyltransferases of the exostosin family (Ext). Then several modifications occur nonuniformly along the HS chain, creating distinct and specific domains. The first step involves deacetylation and sulfation in *N*-acetylglucosamine catalyzed by the *N*-deacetylase-*N*-sulfotransferase (NDST) class of enzymes. Next, epimerases convert some glucuronic acid units to the isomeric iduronic acid. Then, sulfotransferases add sulfate to specific residues, creating in this way a unique HS fine structure. 2-O-sulfotransferases (Hs2st or 2-OST) attach sulfate to uronic acid residues, whereas 3-O- and 6-O-sulfotransferases (Hs3st and Hs6st or 3-OST and 6-OST) add it to glucosamine residues. Finally, 6-O-endosulfatases (sulf) "edit" HS chains at the plasma membrane by removing sulfate from defined domains. Consequently, these numerous modifications confer HS chains an exceptional diversity (up to 10^{36} types of HS isoforms) and thus, the potential to provide a large amount of information required for axon pathfinding.

The "sugar code" hypothesis has first emerged from the observation in *C. elegans* that mutants with different impaired HS modifying enzymes show distinct axon development phenotypes. *C. elegans* offers the great advantage of having only single orthologs of the various HS-modifying enzymes, making genetic manipulation and analysis of the resulting phenotype easier: deacetylation and epimerization are catalyzed by hst-1 and hse-5, respectively, while 2-O, 3-O, and 6-O sulfations are performed by hst-2, hst3.1 and hst3.2, and hst-6. Interestingly, distinct classes of neurons require the activity of hse-5, hst-2, and hst-6 in different combinations for their axon to be guided properly, while hst-3.1 and hst3.2 appear to regulate more refined steps of later differentiation, controlling branching in a context-dependent manner [16–19]. These results, together with biochemical studies dissecting the structural requirements for HS interaction with different factors, suggest that specific modifications of HS regulate the response of axons to different cues in an instructive manner [17, 20, 21]. It should be noted however that some enzymes can partially compensate each other, suggesting that the presumptive HS code may be degenerate [17, 22, 23]. While the contribution of specific HS motifs has been determined in *C. elegans* and *Drosophila*, it remains largely unknown in vertebrates. Many mouse mutants indeed have early embryonic patterning defects or die perinatally. Determining the roles of specific modifications is further complicated by the large number of isoforms for each class of enzymes. A few studies have nonetheless shown that retinal axons in mice lacking Hs2st or Hs6st1 make distinct errors at the chiasm, confirmed that different sulfations regulate specific aspects of axon pathfinding [24, 25].

1.2   Roles of HS Core Proteins

Syndecans (SDCs) and Glypicans (GPCs) are the two major families of cell surface core proteins highly expressed in the nervous system. Tetrapod genomes typically contain four SDC and six GPC genes, each expressed in a specific spatiotemporal expression pattern. In contrast, only one SDC and two GPCs have been identified in *Drosophila* and *C. elegans*. Functional studies in these two invertebrates revealed essential roles of SDC and GPCs in nervous system development. Mutation in the single *sdc* gene induces defects in midline axon guidance through impaired Slit/Robo signaling [5, 26–28]. Both SDC and the GPC dally-like are required for proper axon guidance and visual system function in Drosophila [29]. Finally, the GPC *lon2* controls motor axon guidance in *C. elegans* [17]. To date, very little is known about similar roles in vertebrates. This lack of information is even more surprising considering that core proteins may influence HS levels or composition [30]. Mice lacking SDC3 show neural migration defects that may have made the detection of misguided axons difficult [31, 32]. Only one recent study in chick has demonstrated a role for GPC1 in mediating the repulsive response of postcrossing axons to Shh at the floorplate [33].

1.3 Studying HSPGs In Vivo in Zebrafish

Studying axon guidance in vivo in classical vertebrate models like the mouse presents several difficulties: the generation of knockout models is long and fastidious, and importantly, analysis of axonal trajectories is done a posteriori by fixing and labeling tissues. Observing axon turning, retracting, or degenerating during the course of their navigation thus proves to be very challenging. In contrast, the zebrafish offers several advantages [34, 35]. The optical transparency of zebrafish embryos allows a direct visualization of axons and is particularly suited for high-resolution imaging, especially time-lapse analysis. External fertilization and large clutches provide many embryos that can be observed at different stages. The recent characterization of the zebrafish genome is particularly suited for genetic analysis and allows the fast generation of mutants. Finally, chimeric individuals can easily be obtained by cell transplants. The zebrafish is thus a model of choice to study the roles of HSPGs in axon guidance in vivo, and test the "sugar code" hypothesis.

Several mutants of the HS synthetic pathways have already been identified in a screen for retinal axon guidance defects. *Dackel* (*dak*), *Boxer* (*box*), and *Pinscher* (*pin*) lack functional Ext2, Extl3, and Papst1 (a sulfate transporter), respectively, and can be compared to wild-type (WT) animals for testing the role of HS in axonal development [13, 14, 36–38]. The other HS-modifying enzymes, as well as SDC and GPC core proteins, have been cloned and can be tested for their functions in vivo during development [39–46]. In this chapter, we describe methods for (1) visualizing axon pathfinding directly in the embryo, and (2) down-regulating the expression of genes of interest to test HSPGs' functions. Directly imaging transgenic embryos that express fluorescent proteins in the neurons of interest constitutes the easiest way to visualize axons as they develop (Fig. 1b). Some transgenes like Tg[*elavl3:EGFP*] [47] label most neurons and their projections, whereas others are more specific of a class of neurons. Available transgenic lines can be found in the Zfin database at http://zfin.org/action/fish/search. Alternatively, DNA encoding the transgene of interest can be injected at one cell stage in the embryo to either generate

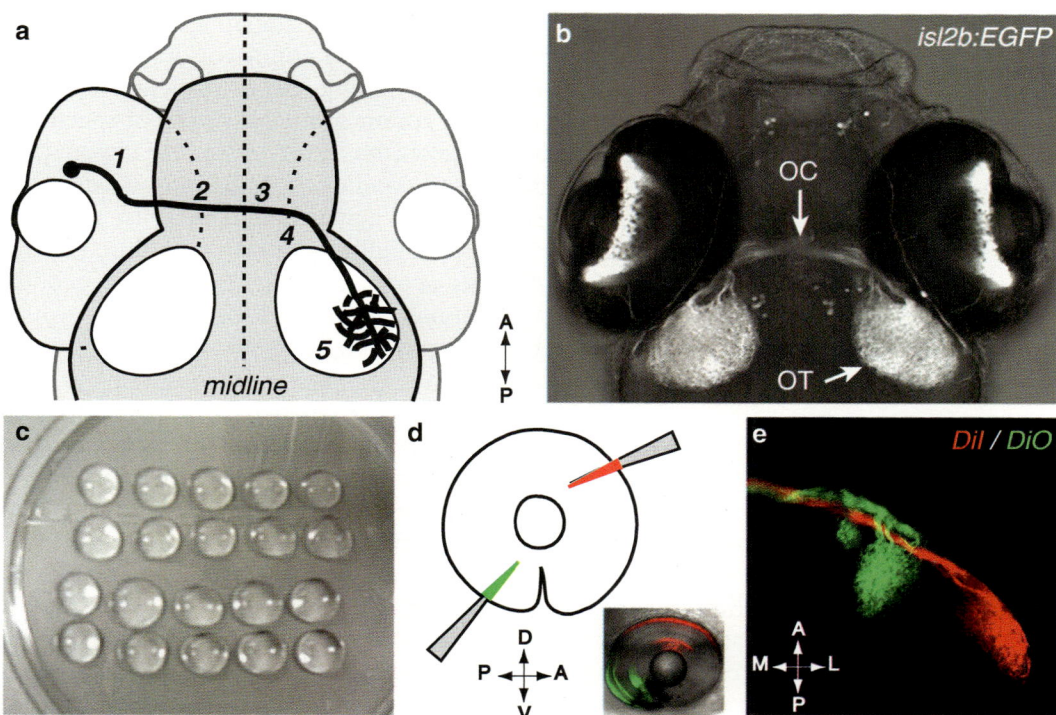

Fig. 1 Methods for visualizing retinal axons. (**a**) Diagram of the retinal axon pathway. Retinal axons navigate to the optic nerve head (*1*), pass through the optic nerve and exit the eye (*2*), cross the midline at the chiasm (*3*), and grow dorsally along the optic tract (*4*) to reach the tectum (*5*). (**b**) Dorsal view of a Tg[*isl2b:EGFP*]^{zc7} transgenic embryo, in which EGFP is specifically expressed in all RGCs, allowing a direct visualization of retinal projections. Courtesy of A. Pittman. *OC* optic chiasm, *OT* optic tectum. (**a**, **b**) dorsal views, anterior up. *Maximum intensity projection, confocal microscopy*. (**c–e**) Focal injection of dyes in the retina allows visualization of retinal axons making topographic connections in the tectum. (**c**) Embryos are mounted laterally in low-melt agarose drops placed on a Petri dish lid. (**d**) DiI- (*red*) and DiO-(*green*) coated glass micropipettes are briefly inserted in a peripheral direction into the retina to label dorsonasal (DN in *red*) and ventrotemporal (VT in *green*) retinal neurons (method described in detail in Subheading 3.2). *A* anterior, *P* posterior, *D* dorsal, *V* ventral. (**e**) Dorsal view of the corresponding retinal axon projections in the brain target, the tectum. *A* anterior, *P* posterior, *M* medial, *L* lateral. *Maximum intensity projection, confocal microscopy*

transgenic lines, or label transiently a subpopulation of neurons. This last approach is particularly useful to visualize a single axon, and is described in the method section. Finally, lipophilic carbocyanine dyes like DiI, DiA, DiD or DiO can be injected in fixed or live embryos and are particularly suited to label specific subpopulations of neurons (Fig. 1c–e). A detailed protocol for dye injection is provided in the method section. All these different approaches can be performed on the available mutants described above to test the role of HS in axon pathfinding. Heparinase can also be injected at specific places in the embryos to assess the effects of locally removing HS [15]. To further investigate the roles of specific enzymes or core proteins, one needs to experimentally manipulate the expression of corresponding genes. A first common approach is to inject stable antisense morpholino oligonucleotides (MOs) into one-cell stage embryos. MOs inhibit either protein translation when targeted near the start codon of mRNAs [48] or splicing of the pre-mRNAs when targeted to exon–intron or intron–exon boundaries [49]. Under good conditions, MOs can quickly reveal required functions for a targeted gene, though their use is subject to several caveats, such as loss of efficacy as they are diluted during development [50]. Another approach is to inject RNAs encoding sequence-specific synthetic nucleases called TALENs (Transcription Activator-Like Effector Nucleases) to generate targeted knockouts [51, 52]. TALENs combine a TAL effector DNA binding domain with a DNA cleavage domain to target specific sequences in the genome and induce mutations at any locus. Finally, DNA constructs encoding dominant negative forms of the protein of interest can be transiently or stably expressed. Spatial or temporal control can be provided by cell-specific promoters or the *hsp70l* heat shock promoter, respectively. Gain-of-function experiments can also be performed by overexpressing genes of interest at specific times or locations. At last, cell-autonomy of HSPGs' function can be tested by expressing a WT gene in a specific tissue in the corresponding mutant embryo, and test whether the mutant phenotype is rescued. All these different approaches rely on the injection of MOs, RNA, or DNA in the embryo at one-cell stage [53], for which we provide a detailed protocol below.

2   Materials

2.1 Zebrafish Embryos

Wild-type (WT) and mutant embryos are obtained from natural matings, raised at 28.5 °C in E3 medium in the presence of 150 mM of 1-phenyl-2-thiourea (PTU) to prevent pigment formation, and staged by age and morphology [54]. They are dechorionated before dye injection experiments.

– E3 medium: 5 mM NaCl, 0.17 mM KCl, 0.33 mM $CaCl_2$, and 0.33 mM $MgSO_4$

2.2 Material Needed for Dye Injection

– DiI or DiO crystals (Molecular Probes)

– Glass capillary (World Precision Instruments, Inc.) with an outer diameter of 1.0 mm and an inner diameter of 0.58 mm to prepare the microneedle for injections. Pull the capillary to make a microneedle with a final taper length of 9.0 mm and a tip size of 2 μm.

– 4 % PFA: 4 % paraformaldehyde in 0.1 M phosphate buffer, pH 7.4, if the experiment is performed on fixed embryos (*see* **Note 1**).

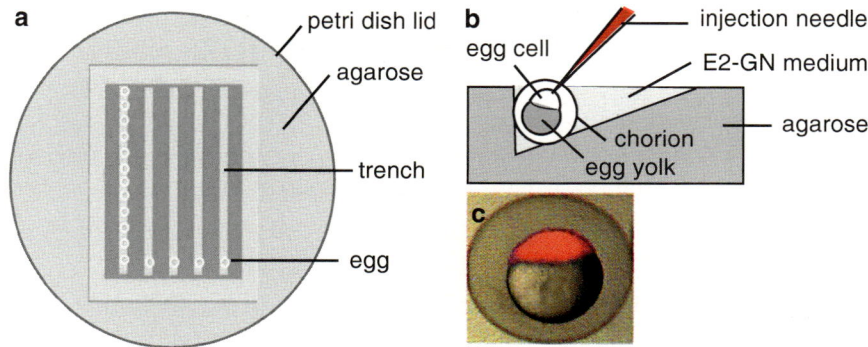

Fig. 2 Injecting one-cell stage embryos. (**a**) Diagram of the injection plate. Agarose is poured in a Petri dish, and trenches are made with a plastic mold set into the agarose. Eggs collected at one-cell stage are aligned in the trenches, and covered with E3-GN medium. (**b**) Egg positioning for injection. The egg in its chorion is positioned so that its cell is oriented up. Using the micromanipulator (not shown), the needle is brought next to the egg. It is then moved in a smooth movement to pierce the surface of the chorion and enter the cell. (**c**) Picture showing an egg successfully injected in the cell with a solution labeled with phenol red

- Tricaine stock: 0.4 % tricaine, 10 mM HEPES, pH 7.4, if the experiment is performed on live embryos (*see* **Note 2**).

- 1 % low-melt agarose in water (for fixed embryos) or PBS (phosphate-buffered saline) for live embryos.

- 50 % and 80 % glycerol in water.

- 1 % low-melt agarose in E3/GN/tricaine if the experiment is performed on live embryos.

- E3/GN/tricaine: 10 µg/ml gentamicin in E3 medium, 0.02 % tricaine.

- A three-axis micromanipulator, a needle holder, and a microscope.

2.3 Material Needed for Injections at One-Cell-Stage

- Glass capillary (World Precision Instruments, Inc.) with an outer diameter of 1.0 mm; the glass capillary is pulled into two needles that are stored on playdough lines in a 150 mm Petri dish.

- Injection plate (Fig. 2a): pour approximately 40 mL of 2 % agarose in E3 medium in a 150 mm Petri dish on a level surface. Set the plastic mold for making slots (Adaptive Science Tools) teeth down into the agarose, and tap gently to eliminate bubbles. After the agarose sets, add a small amount of E3 medium, remove the mold, wrap the Petri dish in parafilm, and store at 4 °C.

- A microinjector, a three-axis micromanipulator, a needle holder, and a microscope.

- MOs: MOs are designed and made by Gene Tools, LLC. They are lyophilized when delivered and are then resuspended in Danieau's solution. MOs in solution can be stored at −20 °C.

- Danieau's solution: 1.5 mM HEPES, 0.18 mM Ca(NO3)$_2$, 0.12 mM MgSO$_4$, 0.21 mM KCl, 17.4 mM NaCl.
- DNA and mRNA are diluted in water. Plasmid DNA is purified with plasmid miniprep purification kits. Pure mRNA is made by in vitro transcription and purified on micro Bio-Spin chromatography columns (Bio-Rad).
- Phenol red is used at a final concentration of 0.5 % as a marker dye for all solutions to be injected.
- E3/GN: 10 µg/ml gentamicin in E3 medium.

3   Methods

3.1 Precise Labeling with Injection of Lipophilic Dyes

Originally developed by Torsten Trowe [55], this method uses glass microneedles coated with lipophilic carbocyanine dyes to focally deposit dye into a region of interest. It can be adapted to label different population of neurons such as retinal neurons, as shown in Fig. 1 [14, 35, 56]. We provide a detailed protocol for labeling fixed embryos, but a similar approach can be used in live embryos (*see* **Notes 3** and **4**).

1. Fix zebrafish embryos at required stage in 4 % PFA at room temperature for the first 2 h, then at 4 °C for 10 h. Embryos can be kept at 4 °C as long as wanted for future experiments.

2. To coat the microneedle with dye, place a few dye crystals on a cover glass and melt them at 100 °C on a hot plate. Dip the tip of the microneedle horizontally into the dye paste and roll it to cover the tip equally on all sides. Wipe off as much dye from the tip as possible onto the cover glass.

3. Prepare 30 ml of 1 % low-melt agarose in water and keep on heating block at 37 °C to prevent from solidifying. Use a Petri dish lid to embed embryos for dye injection. Rinse embryos in water, transfer them onto the lid, and cover them with a drop of 1 % low-melt agarose. Orient them in an appropriate position, so that the region to be injected is easily accessible to the micropipette.

4. Use a standard pipette holder and three-axis micromanipulator to hold the dye-coated microneedle. Insert the microneedle into the region of interest by advancing in a peripheral direction at a roughly 45° angle (this angle usually allows good visualization and penetration of the tissue). The time during which the needle needs to be left in the tissue depends on how big the region of interest is or how many neurons need to be labeled. Leaving the needle for 5–20 s ensures a small injection site and labeling of few axons. The coated microneedle can be reused for several injections before it has to be coated with fresh dye again.

5. After finishing the injections, cover embedded embryos with water to avoid drying. This step also washes off excessive dye. Store the embryos for a few hours at room temperature for fast diffusion of the dye, or keep them at 4 °C overnight if slower diffusion is desired. Long incubation times can result in nonspecific diffusion of the dye, which can prevent clear imaging results later on.

6. Recover embryos from agarose drops using forceps, and rinse them with water. Transfer embryos to 50 % glycerol/H_2O and incubate them under agitation for 3 h at 4 °C. Change the medium to 80 % glycerol/H_2O, and store embryos at 4 °C overnight. Now that they are cleared, embryos can be mounted for confocal imaging in 80 % glycerol between two coverslips.

3.2 Injection in Embryos at One-Cell Stage

As mentioned earlier, this approach is used to inject MOs, RNA or/and DNA. Optimal concentrations for these different compounds are discussed in Subheading 4 (*see* **Note 5**).

1. The night before the experiment, set up fish in breeding tanks with dividers in place. The next morning, remove the dividers, wait about 20 min, and collect eggs using a strainer.

2. Align the eggs in the trenches of the injection plate using a transfer pipette (Fig. 2a). Cover the eggs with E3/GN, and position them such as the cell is oriented up (Fig. 2b) (*see* **Note 6**).

3. Load the needle with 3 μL of the solution to be injected (the media used to dilute MOs, RNA, and DNA are described in Subheading 2) (*see* **Note 7**). Insert the needle into the pipette holder and micromanipulator connected to the microinjector. Check that the micromanipulator is in a proper position to allow movement and adjustment of the needle. The needle should be positioned such as it is directed towards the egg and especially, its cell part (Fig. 2b). Bring the needle tip into the plane of view of the microscope and focus on the thinnest part of its tip.

4. Cut the needle at its tip with a pair of forceps, so that it is narrow enough to pierce the chorion but still capable of delivering a consistent volume. Press the foot pedal of the injector and monitor the size of the drop released in the medium. Volume can be adjusted by trimming the needle, adjusting the injection pressure, or the duration of injection. Injection volumes of 500 pL or 1 nL are typically used.

5. Ensure that the embryos are still at one-cell stage before pursuing the injections.

6. Using the micromanipulator, bring the needle next to the egg, pierce the surface of the chorion and enter the cell in one smooth movement. Inject the solution in the cell while being careful not to move to avoid tearing the membrane.

After injecting, remove the needle, and slowly move the injection plate with your hand to proceed with the next egg. Always keep some eggs uninjected as a control.

7. After finishing, use E3 medium to move the injected eggs into a clean Petri dish. At the end of the day, remove dead or damaged embryos, and count the number of embryos you have injected (*see* **Note 8**).

4   Notes

1. Do not inject lipophilic dyes if embryos have been permealized with triton or tween for other experiments such as immunolabeling. Permeabilization would affect the diffusion of lipophilic dyes along plasma membranes, resulting in a blurry, nonspecific staining.

2. Injection of lipophilic dyes can also be performed on live embryos to visualize axons as they develop. In this case, anesthetize embryos in 0.015 % tricaine, and mount them in 1 % agarose, 0.015 % tricaine in E3/GN on a Petri dish lid (first pipet the embryo in the agarose tube, and then pipet the agarose containing the embryo on the lid, so that manipulation of embryos is kept at a minimum).

3. Using dyes diluted at a concentration of 3 or 4 % in DMSO can also be used and is often more effective for labeling live embryos. Keep the tube containing the diluted dye at 37 °C, so that the dye remains liquid, and fill an injecting needle connected to a microinjector with the dye solution. Embryos are mounted in the same way as described previously, but they are submerged with water only. It is important to limit the use of saline solutions like PBS, as salts are known to facilitate dye precipitation. Before inserting the microneedle into the region of interest and injecting the dye, use an agarose drop as a test to determine the parameters of injection corresponding to the volume that needs to be injected.

4. Injecting dyes in live embryos allows a direct visualization of elongating axons in vivo by time-lapse microscopy. To perform time-lapse imaging, mount embryos in 1 % low-melt agarose in E3 medium with 150 mM PTU and 0.015 % tricaine in a glass-bottomed Petri dish. Chamber temperature should be maintained at 28.7 °C using a heated stage. Using a confocal microscope, z-series can be acquired a regular intervals. It is important to keep the power of the lasers at a minimum, so that embryos survive the procedure. Maximal intensity projections for each time point can be compiled and aligned using ImageJ software and StackReg plugin [57].

5. Concentration of DNA, mRNA, and MO should be determined by the user. Usually, between 20 and 60 pg of DNA can be injected. Higher concentrations are often toxic, leading to embryos' death. MOs are usually injected at a final concentration of 0.5–1 mM (which corresponds to around 4–8 ng, depending on the MO sequence). Finally, mRNA appears less toxic than DNA or MO and can be injected at higher concentrations ranging from 25 pg to 1 ng.

6. For good efficiency, DNA needs to be injected as early as possible at one-cell stage. When performing DNA injections, it is recommended to load the injection needle and prepare the injection set up before collecting eggs, so that DNA can be injected right away as soon as eggs are transferred to the injection plate. In contrast to DNA, MOs, and mRNAs can be injected in the egg yolk right below the cell for a similar efficiency.

7. The Tol2kit system is a powerful tool widely used to generate transgenic lines. It uses site-specific recombination-based cloning (multisite Gateway technology) to allow modular and quick assembly of constructs in a Tol2 transposon backbone. Plasmid DNA generated with this system is coinjected with *transposase* mRNA at one-cell stage [58].

8. As MOs can have off-target effects or can be diluted during development [59], performing several controls is required before drawing any conclusion about a gene's function. At least two different MOs targeting distinct regions of the gene of interest should induce the phenotype. Alternatively or in complement, the MO phenotype should be rescued by coinjecting a synthetic mRNA encoding the protein from the targeted locus but not the MO target sequence (MO and mRNA are injected separately with different needles). Embryos injected with the MO alone should be compared to embryos injected with the mRNA alone or with both the mRNA and the MO.

Acknowledgments

F.E. Poulain is supported by a grant from the NINDS (K99-1K99NS083714-01).

References

1. Serafini T, Kennedy TE, Galko MJ, Mirzayan C, Jessell TM, Tessier-Lavigne M (1994) The netrins define a family of axon outgrowth-promoting proteins homologous to C. elegans UNC-6. Cell 78:409–424
2. Bennett KL, Bradshaw J, Youngman T, Rodgers J, Greenfield B, Aruffo A, Linsley PS (1997) Deleted in colorectal carcinoma (DCC) binds heparin via its fifth fibronectin type III domain. J Biol Chem 272:26940–26946
3. Hu H (2001) Cell-surface heparan sulfate is involved in the repulsive guidance activities of Slit2 protein. Nat Neurosci 7:695–701

4. Ronca F, Andersen JS, Paech V, Margolis RU (2001) Characterization of Slit protein interactions with glypican-1. J Biol Chem 276: 29141–29147

5. Johnson KG, Ghose A, Epstein E, Lincecum J, O'Connor MB, Van Vactor D (2004) Axonal heparan sulfate proteoglycans regulate the distribution and efficiency of the repellent slit during midline axon guidance. Curr Biol 14:499–504

6. Zhang F, Ronca F, Linhardt RJ, Margolis RU (2004) Structural determinants of heparan sulfate interactions with slit proteins. Biochem Biophys Res Commun 317:352–357

7. Hussain SA, Piper M, Fukuhara N, Strochlic L, Cho G, Howitt JA, Ahmed Y, Powell AK, Turnbull JE, Holt CE, Hohenester E (2006) A molecular mechanism for the heparan sulfate dependence of slit-robo signaling. J Biol Chem 281:39693–39698

8. Fukuhara N, Howitt JA, Hussain SA, Hohenester E (2008) Structural and functional analysis of slit and heparin binding to immunoglobulin-like domains 1 and 2 of Drosophila Robo. J Biol Chem 283:16226–16234

9. Strigini M (2005) Mechanisms of morphogen movement. J Neurobiol 64:324–333

10. Walz A, McFarlane S, Brickman YG, Nurcombe V, Bartlett PF, Holt CE (1997) Essential role of heparan sulfates in axon navigation and targeting in the developing visual system. Development 124:2421–2430

11. Irie A, Yates EA, Turnbull JE, Holt CE (2002) Specific heparan sulfate structures involved in retinal axon targeting. Development 129:61–70

12. Inatani M, Irie F, Plump AS, Tessier-Lavigne M, Yamaguchi Y (2003) Mammalian brain morphogenesis and midline axon guidance require heparan sulfate. Science 302:1044–1046

13. Lee JS, von der Hardt S, Rusch MA, Stringer SE, Stickney HL, Talbot WS, Geisler R, Nusslein-Volhard C, Selleck SB, Chien CB, Roehl H (2004) Axon sorting in the optic tract requires HSPG synthesis by ext2 (dackel) and extl3 (boxer). Neuron 44:947–960

14. Poulain FE, Chien CB (2013) Proteoglycan-mediated axon degeneration corrects pretarget topographic sorting errors. Neuron 78:49–56

15. Wang F, Wolfson SN, Gharib A, Sagasti A (2012) LAR receptor tyrosine phosphatases and HSPGs guide peripheral sensory axons to the skin. Curr Biol 22:373–382

16. Bülow HE, Hobert O (2004) Differential sulfations and epimerization define heparan sulfate specificity in nervous system development. Neuron 41:723–736

17. Bülow HE, Tjoe N, Townley RA, Didiano D, van Kuppevelt TH, Hobert O (2008) Extracellular sugar modifications provide instructive and cell-specific information for axon-guidance choices. Curr Biol 18:1978–1985

18. Tecle E, Diaz-Balzac CA, Bülow HE (2013) Distinct 3-O-sulfated heparan sulfate modification patterns are required for kal-1-dependent neurite branching in a context-dependent manner in Caenorhabditis elegans. G3 (Bethesda) 3:541–552

19. Gysi S, Rhiner C, Flibotte S, Moerman DG, Hengartner MO (2013) A network of HSPG core proteins and HS modifying enzymes regulates netrin-dependent guidance of D-type motor neurons in Caenorhabditis elegans. PLoS One 8:e74908

20. Shipp EL, Hsieh-Wilson LC (2007) Profiling the sulfation specificities of glycosaminoglycan interactions with growth factors and chemotactic proteins using microarrays. Chem Biol 14: 195–208

21. Zhang F, Moniz HA, Walcott B, Moremen KW, Linhardt RJ, Wang L (2013) Characterization of the interaction between Robo1 and heparin and other glycosaminoglycans. Biochimie pii:S0300-9084(13)00290-3

22. Kamimura K, Koyama T, Habuchi H, Ueda R, Masu M, Kimata K, Nakato H (2006) Specific and flexible roles of heparan sulfate modifications in Drosophila FGF signaling. J Cell Biol 174:773–778

23. Dejima K, Takemura M, Nakato E, Peterson J, Hayashi Y, Kinoshita-Toyoda A, Toyoda H, Nakato H (2013) Analysis of Drosophila glucuronyl C-5 epimerase: implications for developmental roles of heparan sulfate sulfation compensation and 2-O sulfated glucuronic acid. J Biol Chem 288:34384–34393

24. Pratt T, Conway CD, Tian NM, Price DJ, Mason JO (2006) Heparan sulphation patterns generated by specific heparan sulfotransferase enzymes direct distinct aspects of retinal axon guidance at the optic chiasm. J Neurosci 26:6911–6923

25. Conway CD, Howe KM, Nettleton NK, Price DJ, Mason JO, Pratt T (2011) Heparan sulfate sugar modifications mediate the functions of slits and other factors needed for mouse forebrain commissure development. J Neurosci 31:1955–1970

26. Steigemann P, Molitor A, Fellert S, Jackle H, Vorbruggen G (2004) Heparan sulfate proteoglycan syndecan promotes axonal and myotube guidance by slit/robo signaling. Curr Biol 14:225–230

27. Rhiner C, Gysi S, Frohli E, Hengartner MO, Hajnal A (2005) Syndecan regulates cell migration and axon guidance in C. elegans. Development 132:4621–4633

28. Smart AD, Course MM, Rawson J, Selleck S, Van Vactor D, Johnson KG (2011) Heparan sulfate proteoglycan specificity during axon pathway formation in the Drosophila embryo. Dev Neurobiol 71:608–618

29. Rawson JM, Dimitroff B, Johnson KG, Rawson JM, Ge X, Van Vactor D, Selleck SB (2005) The heparan sulfate proteoglycans Dally-like and Syndecan have distinct functions in axon guidance and visual-system assembly in Drosophila. Curr Biol 15:833–838

30. Chen RL, Lander AD (2001) Mechanisms underlying preferential assembly of heparan sulfate on glypican-1. J Biol Chem 276:7507–7517

31. Hienola A, Tumova S, Kulesskiy E, Rauvala H (2006) N-Syndecan deficiency impairs neural migration in brain. J Cell Biol 174:569–580

32. Bespalov MM, Sidorova YA, Tumova S, Ahonen-Bishopp A, Magalhães AC, Kulesskiy E, Paveliev M, Rivera C, Rauvala H, Saarma M (2011) Heparan sulfate proteoglycan syndecan-3 is a novel receptor for GDNF, neurturin, and artemin. J Cell Biol 192:153–169

33. Wilson NH, Stoeckli ET (2013) Sonic hedgehog regulates its own receptor on postcrossing commissural axons in a glypican1-dependent manner. Neuron 79:478–491

34. Hutson LD, Campbell DS, Chien CB (2004) Analyzing axon guidance in the zebrafish retinotectal system. Methods Cell Biol 76:13–35

35. Poulain FE, Gaynes JA, Hörndli C, Law MY, Chien CB (2010) Analyzing retinal axon guidance in zebrafish. Methods Cell Biol 100:3–26

36. Karlstrom RO, Trowe T, Klostermann S, Baier H, Brand M, Crawford AD, Grunewald B, Haffter P, Hoffmann H, Meyer SU, Muller BK, Richter S, van Eeden FJ, Nusslein-Volhard C, Bonhoeffer F (1996) Zebrafish mutations affecting retinotectal axon pathfinding. Development 123:427–438

37. Trowe T, Klostermann S, Baier H, Granato M, Crawford AD, Grunewald B, Hoffmann H, Karlstrom RO, Meyer SU, Muller B, Richter S, Nusslein-Volhard C, Bonhoeffer F (1996) Mutations disrupting the ordering and topographic mapping of axons in the retinotectal projection of the zebrafish, Danio rerio. Development 123:439–450

38. Clément A, Wiweger M, von der Hardt S, Rusch MA, Selleck SB, Chien CB, Roehl HH (2008) Regulation of zebrafish skeletogenesis by ext2/dackel and papst1/pinscher. PLoS Genet 4(7):e1000136

39. Cadwallader AB, Yost HJ (2006) Combinatorial expression patterns of heparan sulfate sulfotransferases in zebrafish: II. The 6-O-sulfotransferase family. Dev Dyn 235:3432–3437

40. Cadwallader AB, Yost HJ (2006) Combinatorial expression patterns of heparan sulfate sulfotransferases in zebrafish: I. The 3-O-sulfotransferase family. Dev Dyn 235:3423–3431

41. Cadwallader AB, Yost HJ (2007) Combinatorial expression patterns of heparan sulfate sulfotransferases in zebrafish: III. 2-O-sulfotransferase and C5-epimerases. Dev Dyn 236:581–586

42. Kramer KL, Barnette JE, Yost HJ (2002) PKCgamma regulates syndecan-2 inside-out signaling during Xenopus left-right development. Cell 111:981–990

43. Arrington CB, Yost HJ (2009) Extra-embryonic syndecan 2 regulates organ primordia migration and fibrillogenesis throughout the zebrafish embryo. Development 136:3143–3152

44. Hofmeister W, Devine CA, Key B (2013) Distinct expression patterns of syndecans in the embryonic zebrafish brain. Gene Expr Patterns 13:126–133

45. Topczewski J, Sepich DS, Myers DC, Walker C, Amores A, Lele Z, Hammerschmidt M, Postlethwait J, Solnica-Krezel L (2001) The zebrafish glypican knypek controls cell polarity during gastrulation movements of convergent extension. Dev Cell 1:251–264

46. Gorsi B, Whelan S, Stringer SE (2010) Dynamic expression patterns of 6-O endosulfatases during zebrafish development suggest a subfunctionalisation event for sulf2. Dev Dyn 239:3312–3323

47. Park HC, Kim CH, Bae YK, Yeo SY, Kim SH, Hong SK, Shin J, Yoo KW, Hibi M, Hirano T, Miki N, Chitnis AB, Huh TL (2000) Analysis of upstream elements in the HuC promoter leads to the establishment of transgenic zebrafish with fluorescent neurons. Dev Biol 227:279–293

48. Nasevicius A, Ekker SC (2000) Effective targeted gene 'knockdown' in zebrafish. Nat Genet 26:216–220

49. Draper BW, Morcos PA, Kimmel CB (2001) Inhibition of zebrafish fgf8 pre-mRNA splicing with morpholino oligos: a quantifiable method for gene knockdown. Genesis 30:154–156

50. Eisen JS, Smith JC (2008) Controlling morpholino experiments: don't stop making antisense. Development 135:1735–1743

51. Dahlem TJ, Hoshijima K, Jurynec MJ, Gunther D, Starker CG, Locke AS, Weis AM, Voytas DF, Grunwald DJ (2012) Simple methods for generating and detecting locus-specific mutations induced with TALENs in the zebrafish genome. PLoS Genet 8:e1002861

52. Bedell VM, Wang Y, Campbell JM, Poshusta TL, Starker CG, Krug Ii RG, Tan W, Penheiter

SG, Ma AC, Leung AY, Fahrenkrug SC, Carlson DF, Voytas DF, Clark KJ, Essner JJ, Ekker SC (2012) In vivo genome editing using a high-efficiency TALEN system. Nature 491:114–118

53. Rosen JN, Sweeney MF, Mably JD (2009) Microinjection of zebrafish embryos to analyze gene function. J Vis Exp pii:1115. doi:10.3791/1115

54. Kimmel CB, Ballard WW, Kimmel SR, Ullmann B, Schilling TF (1995) Stages of embryonic development of the zebrafish. Dev Dyn 203:253–310

55. Trowe T (2000) Analyse von Mutationen mit Einfluss aud die topographische Ordnung von Axonen im retinotektalen System des Zebrabärblings, Danio rerio. Ph.D. thesis, Eberhard-Karls-Universität Töbingen

56. Stacher Hörndli C, Chien CB (2012) Sonic hedgehog is indirectly required for intraretinal axon pathfinding by regulating chemokine expression in the optic stalk. Development 139:2604–2613

57. Thevenaz P, Ruttimann UE, Unser M (1998) A pyramid approach to subpixel registration based on intensity. IEEE Trans Image Process 7:27–41

58. Kwan KM, Fujimoto E, Grabher C, Mangum BD, Hardy ME, Campbell DS, Parant JM, Yost HJ, Kanki JP, Chien CB (2007) The Tol2kit: a multisite gateway-based construction kit for Tol2 transposon transgenesis constructs. Dev Dyn 236:3088–3099

59. Bill BR, Petzold AM, Clark KJ, Schimmenti LA, Ekker SC (2009) A primer for morpholino use in zebrafish. Zebrafish 6:69–77

Chapter 37

Murine Models in the Evaluation of Heparan Sulfate-Based Anticoagulants

David Gailani, Qiufang Cheng, and Ivan S. Ivanov

Abstract

Evaluating anticoagulants in animal thrombosis models is a standard component of preclinical drug testing. Mice are frequently used for these initial evaluations because a variety of thrombosis models have been developed and are well characterized in this species, and the animals are relatively inexpensive to maintain. Because mice have a natural resistance to forming intravascular thrombi, vessel injury is required to induce intravascular clot formation. Several methods have been established for inducing arterial or venous thrombosis in mice. For the purpose of testing heparin-based drugs, we adapted a well-established model in which thrombus formation in the carotid artery is induced by exposing the vessel to ferric chloride. For studying anticoagulant effects on venous thrombosis, we use a model in which the inferior vena cava is ligated and the size of the resulting clots is measured. The most common adverse effect of anticoagulation therapy is bleeding. The effect of heparin-based anticoagulants can be tested in mice in a simple tail bleeding assay.

Key words Mouse, Heparin, Anticoagulant, Arterial thrombosis, Venous thrombosis, Tail bleeding time

1 Introduction

Animal models of hemostasis and thrombosis have contributed substantially to our understanding of normal and pathologic blood coagulation in humans [1–5]. Mice, because they are small in size, have a short gestation period, have high reproductive capacity, and are relatively inexpensive to maintain, are commonly used as test subjects for initial in vivo analyses of antithrombotic compounds intended for therapeutic use in humans. These animals have a number of features that make them suitable surrogates for testing anticoagulant drugs. Mice and humans have similar complements of plasma coagulation factors and regulatory proteins, and coagulation proteins from one species usually demonstrate reasonable activity (with a few exceptions) in the plasma of the other [6–8]. Heparan sulfate, heparin, and dermatan sulfate produce their anticoagulant effects through inhibition of the key coagulation

Kuberan Balagurunathan et al. (eds.), *Glycosaminoglycans: Chemistry and Biology*, Methods in Molecular Biology, vol. 1229, DOI 10.1007/978-1-4939-1714-3_37, © Springer Science+Business Media New York 2015

protease thrombin, and through inhibition of proteases responsible for thrombin generation [9]. These glycosaminoglycans primarily inhibit coagulation proteases indirectly by enhancing the activities of plasma serine protease inhibitors (serpins) such as antithrombin through allosteric- and template-based mechanisms (serpin-dependent effects). In some cases they directly inhibit proteases by binding to them (serpin-independent effect). The anion binding sites on coagulation proteases and serpins required for productive interactions with glycosaminoglycans are largely conserved between mice and humans, suggesting the two species have fundamentally similar mechanisms for regulating thrombin generation.

Similarities between human and murine plasma coagulation factors can be useful for evaluating a drug candidate, even when the compound interacts only with the human version of a protein target. The genomes of mice can be manipulated to produce constitutive or conditional deficiencies of a protein of interest [6–8]. To generate a model for testing a drug that interacts exclusively with a human plasma protein, a mouse deficient in that protein can be "reconstituted" by intravenous infusion of its human counterpart. For coagulation proteases, this usually restores the wild-type phenotype in hemostasis and thrombosis models [10–12].

There are important differences in normal (hemostasis) and pathologic (bleeding or thrombosis) coagulation between mice and humans that must be considered when interpreting results from mouse models. The size discrepancy alone between a large biped and a several thousand-fold smaller quadruped results in different forces on tissues, presenting different challenges for the respective hemostatic systems. Humans lacking coagulation factor IX have a condition (hemophilia B) characterized by a propensity to develop recurrent hemorrhage into large joints (hemarthrosis) such as knees, ankles, and elbows that can be crippling [13]. While factor IX deficient mice also have a hemorrhagic disorder, hemarthrosis is not a prominent feature, perhaps due to the relatively smaller mechanical forces on their joints [14]. The atherosclerotic changes in large arteries that are a major contributor to arterial thrombosis in humans may develop over decades. Mice, with their shorter life-spans and high plasma levels of high density lipoprotein, are resistant to atherosclerosis, and require dietary or genetic manipulation to produce atherosclerotic plaque [15]. Even mice genetically altered to develop atherosclerosis (for example, mice lacking apolipoprotein E) tend not to develop occlusive thrombi at sites of plaque rupture in the same manner as humans [16, 17]. In humans, venous thrombi form preferentially in the deep veins of the lower extremities and pelvis. Blood stasis aggravated by upright posture is a major contributor to formation of this type of clot [18]. The long-term effects of hydrostatic forces on leg veins are unlikely to be important in mice, which rarely develop spontaneous venous thrombi in their extremities.

The natural resistance of mice to formation of occlusive thrombi requires that a vessel be injured in some manner to induce local thrombosis, and almost all mouse thrombosis models involve formation of a clot in a vessel that was healthy immediately prior to injury. This contrasts with thrombosis in humans, which typically occurs in a diseased vessel. A variety of techniques are employed to induce venous and arterial thrombosis in mice. In this chapter ferric chloride-induced injury of the carotid artery and ligature-induced venous stasis in the inferior vena cava are described. These techniques were chosen because they have been widely used, and because they are sensitive to heparin. The major adverse side effect of anticoagulants is bleeding. The last section of this chapter describes a simple heparin-sensitive tail bleeding time assay that can be used to evaluate the anti-hemostatic effects of a compound of interest.

1.1 Ferric Chloride-Induced Arterial Thrombosis Model

A variety of approaches are used to induce acute thrombus formation in large (e.g. carotid), medium (e.g. mesenteric), and small (e.g. cremaster) arteries in mice. Injury to the vascular endothelium to promote thrombus formation can be induced by chemical exposure (e.g. ferric chloride) [1–3, 19–21], photochemical techniques [1–3], direct laser-based techniques [1–4], or mechanical methods [2]. Thrombus formation is detected by monitoring changes in blood flow through the vessel with a Doppler flow-probe, by directly observing thrombus formation by intravital microscopy, or by observing histologic changed to the injured vessel. It has not been established which approach most closely reflects processes that occur during arterial thrombosis in humans. Many groups have found ferric chloride-induced injury to be a reproducible method for generating a clot that histologically resembles platelet-rich arterial thrombi in humans [19–23]. Typically one or more small pieces of paper saturated with $FeCl_3$ solution are applied to a vessels adventitial surface. $FeCl_3$ defuses through the vessel wall to the luminal surface. Initially, it was thought that this treatment resulted in denudation of the vessel endothelium, exposing thrombogenic subendothelial matrix to flowing blood. While this may occur at high $FeCl_3$ concentrations, recent work from Barr et al. suggests the endothelium remains largely intact after $FeCl_3$ application [21]. These investigators observed that erythrocytes initially bind to the altered endothelium, and ferric ions localize primarily to adherent erythrocytes and erythrocyte-derived structures and not to the endothelial cells themselves. Subsequently, platelets bind to adherent erythrocytes in a manner dependent on the platelet receptor glycoprotein 1b-α, with platelet aggregates eventually filling the lumen of the vessel.

We use ferric chloride-induced carotid artery occlusion to study the effects of heparin [23] and heparan-based anticoagulants on thrombus formation. Thrombus formation in mice induced by

FeCl$_3$ requires contributions from tissue factor-initiated coagulation (extrinsic pathway) [24] and from factors XII and XI (intrinsic pathway) [12, 23, 25]. The method has been particularly useful for studying the effects of heparan-based compounds that are designed to inhibit specific protease components of the coagulation mechanism. Our approach involves establishing the lowest FeCl$_3$ concentration that reproducibly induces vessel occlusion in untreated wild-type mice, and then demonstrating that an anticoagulant compound prevents vessel occlusion at this FeCl$_3$ concentration [22, 23]. If an antithrombotic effect is observed, the drug can then tested at progressively higher FeCl$_3$ concentrations until the effect of the drug is overcome.

1.2 Inferior Vena Cava Stasis-Induced Venous Thrombosis Model

Venous thrombi that cause symptoms in humans form primarily in the deep veins of the lower extremities and the pelvis [18]. Embolization of clot to the pulmonary circulation is a major cause of mortality in patients with deep vein thrombosis. While platelets contribute to venous thrombus formation, the clots are predominantly comprised of fibrin and erythrocytes, and differ significantly in their histology from the platelet-rich thrombi that form in the arterial circulation. Blood has a tendency to pool in the deep leg veins in humans, due to hydrostatic forces created by our upright posture. The resulting stasis is a major contributor to venous thrombus formation in humans. Predisposition to venous thrombosis in humans also has a strong genetic component, and a number of inherited conditions have been identified that enhance basal thrombin generation and, consequently, risk for venous thromboembolism. The pathophysiologic processes that cause venous thrombosis in humans are, therefore, difficult to reproduce in mice. Injury to a vein with a chemical such as FeCl$_3$ will cause formation of a venous thrombus, but the platelet-rich clots tend to look like those produced in the arterial circulation [5]. Venous stasis models involving partial or complete ligation of the inferior vena cava in the abdomen have been developed to study the effects of anticoagulants on thrombus formation [1–5, 26]. Here we present a method that involves incomplete ligation of the inferior vena cava [26].

1.3 Tail Bleeding Model

Many antithrombotic drugs, including heparin, produce therapeutic effects by inhibiting processes required for normal hemostasic responses to injury. Therefore, the major trade-off for the beneficial drug effect is a significantly increased risk of bleeding [27]. Some newer oral agents may have better safety profiles than older drugs such as heparin and warfarin, but still target the key hemostatic proteases thrombin and factor Xa and, therefore, increase bleeding risk [27]. Antithrombotic compounds are now under development that target plasma proteases such as factor XIa and factor XIIa that do not play major roles in hemostasis but that contribute to thrombus growth [28–31]. The anticipated advantage of such drugs is not necessarily better efficacy than currently available agents, but an improved safety profile, permitting anticoagulation therapy to be applied to a wider range of patients. Mice deficient in factor XI or factor XII (the zymogen precursors of factor XIa and factor XIIa) exhibit resistance to injury-induced thrombosis in a number of models, including the $FeCl_3$-induced carotid artery thrombosis and inferior vena cava ligation models described above, but do not have obvious hemostatic abnormalities [12, 23, 25]. Drugs specifically targeting factor XIa and factor XIIa should, therefore, prevent thrombus formation but not increase bleeding. Drugs based on heparan-like structures have been reported that inhibit the activity of factor XIa [31]. As heparans have a tendency to bind to multiple targets, it is important to test these compounds for off-target effects that could cause undesirable consequences, particularly bleeding.

The tail bleeding time has been used extensively to study hemostasis in mice [32–34]. This assay is simple to perform, and is sensitive to the effects of heparin [23]. Removal of the tip of the tail with a scalpel transects several blood vessels including two large lateral veins and the ventral artery. The tail tip is usually immersed in warm normal saline and time to cessation of bleeding and/or total blood loss can be determined. While wild-type mice usually bleed for 1–3 min, mice with certain types of bleeding disorders or mice who have received anticoagulation therapy may have prolonged bleeding that is punctuated by periods of 1–2 min in which no bleeding occurs. For example mice lacking factor VIII or factor IX (hemophilia A or B, respectively) have similar initial tail bleeding times as wild-type mice. However, shortly (30–60 s) after the initial cessation of bleeding, hemorrhage recurs [34]. We observe mice for up to 30 min after tail transection to account for "rebleeding", with the bleeding time recorded as the time it takes for all bleeding to stop. It must be recognized that any bleeding model only provides information about hemostasis in response to a specific-type of injury in a specific vascular bed, and will not necessarily reflect the propensity to bleed when injury involves other parts of the body. Nevertheless, the ease with which the tail bleeding time is performed has made it a mainstay of evaluating hemostasis in mice.

2   Materials

2.1 Ferric Chloride-Induced Arterial Thrombosis Model

1. A variety of inbred and mixed breeds of mice have been used to study hemostasis and thrombosis. C57Bl/6 mice have been used extensively in this regard and our work with the $FeCl_3$ arterial thrombosis model is standardized with this readily

available mouse line [22, 23]. Assay reproducibility can be enhanced by attention to a number of animal related factors (*see* **Note 1**).

2. Pentobarbital. Mice are placed under general anesthesia for the $FeCl_3$ model with pentobarbital. This drug is a controlled substance and will require DEA licensure to obtain and use. Other general anesthetics may be used, but the sensitivity of the assay to vessel injury may be different than with pentobarbital (*see* **Note 2**).

3. $FeCl_3$. A 20 % stock solution is prepared by bringing 200 mg of $FeCl_3$ up to 1 ml with deionized water. Subsequent dilutions of the stock are prepared with deionized water. We prepare fresh stock solution every 1–2 weeks.

4. Filter paper for application of $FeCl_3$. Whatman 3MM Chromatography paper (catalog # 3030-917) is cut into small rectangles measuring ~1×1.5 mm.

5. Flow probe. We use a Transonic (Ithaca, New York) TS4020 transit time perivascular flow meter fitted with a 0.5VB504 Doppler flow probe (Catalog Mao-5VB). The flow meter is connected to an ML866 PowerLab 4/30 data acquisition system (AD Instruments, Dunedin, New Zealand) interfaced with a Macintosh computer.

6. Phosphate buffered saline (PBS). PBS is used to keep tissues moist during the procedure and as a vehicle for diluting heparan-based anticoagulants for intravenous administration. We use 1× Cellgro Dulbecco's phosphate buffered saline without calcium or magnesium (Mediatech, Manassas, VA), but any source of sterile PBS should work.

2.2 Inferior Vena Cava Stasis-Induced Venous Thrombosis Model

1. C57Bl/6 mice. Same as in the $FeCl_3$ thrombosis model (*see* **Note 1**).

2. Pentobarbital. Same as in the $FeCl_3$ thrombosis model (*see* **Note 2**).

3. Phosphate buffered saline (PBS). Same as in the $FeCl_3$ thrombosis model.

4. Surgical supplies. 4-0 Vicryl suture is used to ligate the inferior vena cava. 4-0 Steelex metal monofilament suture (Braun Catalog #F1614037).

2.3 Tail Bleeding Model

1. C57Bl/6 mice. Same as in the $FeCl_3$ thrombosis model (*see* **Note 1**).

2. Pentobarbital. Same as in the $FeCl_3$ thrombosis model (*see* **Note 2**).

3. Phosphate buffered saline (PBS). Same as in the $FeCl_3$ thrombosis model.

3 Methods

3.1 Ferric Chloride-Induced Arterial Thrombosis Model

1. Mice are anesthetized by administration of pentobarbital (50 mg/kg) through an intraperitoneal injection. The injection is best given into the right side of the abdomen to avoid injuring the cecum or spleen.

2. Once the animals are under anesthesia, they are placed on their backs on a 37 °C warming pad, and the extremities are immobilized with pieces of tape. The neck is opened along the midline and the carotid artery and jugular vein are exposed on one side of the neck. The carotid artery is separated from surrounding tissues by blunt dissection (Fig. 1a) (*see* **Note 3**).

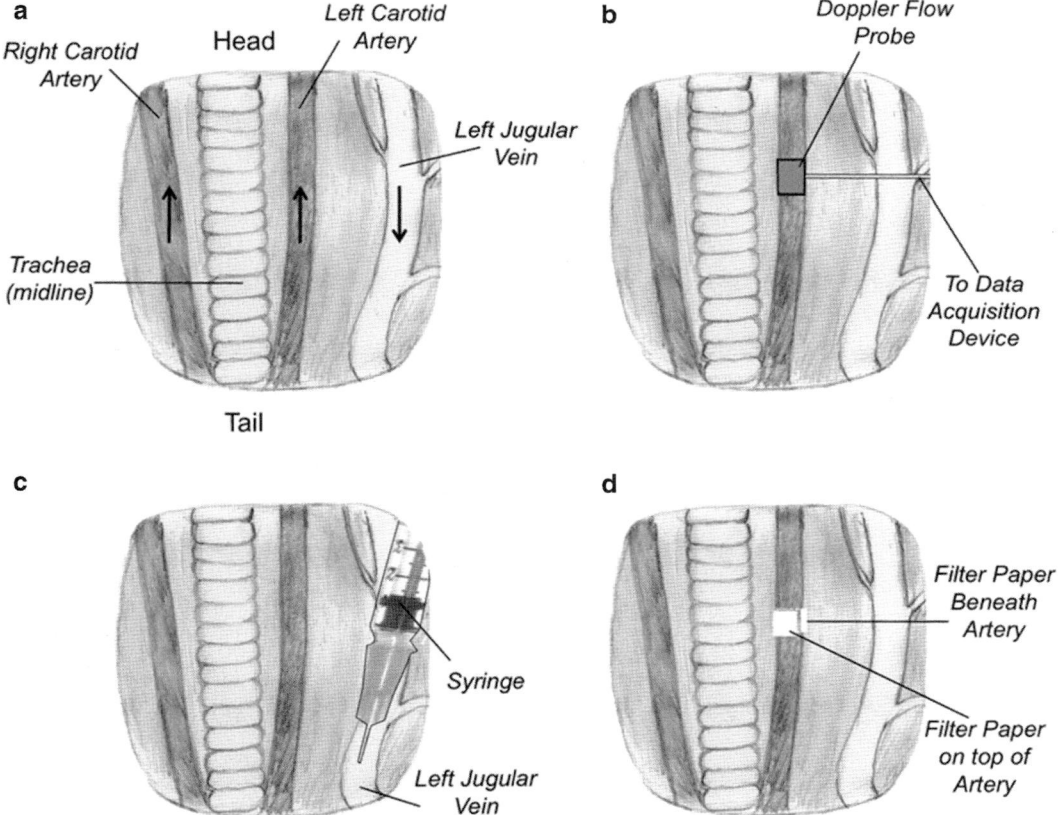

Fig. 1 Ferric chloride carotid artery thrombosis model. (**a**) Anatomy of ventral surface of the mouse neck showing the relative positions of the carotid arteries and jugular vein to the trachea, which runs along the midline. The positions of the animal's head and tail relative to the drawing are indicated. The *black arrows* indicate the direction of blood flow in the major vessels. (**b**) A Doppler flow probe is placed on the carotid artery to monitor blood flow. (**c**) Anticoagulant compounds to be tested are administered through an intravenous injection into the jugular vein in the direction of blood flow (toward the heart). (**d**) Pieces of filter paper (two total) saturated with ferric chloride solution are applied beneath and on top of the carotid artery

3. The flow probe is attached to the artery, and the area is bathed with PBS to insure proper signal transduction from the vessel to the probe (Fig. 1b). A baseline flow rate is established (typically 0.5–0.8 ml/min in an adult mouse). The flow probe is then removed.

4. Anticoagulant compounds to be tested are diluted up to 100 μl with PBS. The drug (or vehicle control) is infused into the jugular vein using a 300 μl tuberculin syringe fitted with a 30-G needle, with the needle tip pointing toward the heart (the direction of blood flow) (Fig. 1c) (*see* **Note 4**).

5. Five minutes after drug infusion, the area around the carotid artery is dried with cotton Q-tips. Two Whatman chromatography paper pads are saturated with 50 μl each of $FeCl_3$ solution (Fig. 1d). Initial studies are typically performed at 3.5 %. Fifty microliters is more solution than the pad can hold, and nonabsorbed solution is discarded. The pads are applied to the surface of the carotid artery, on opposite sides of the artery from each other. After 3 min, the pads are removed, the area is washed with PBS, the Doppler flow probe is replaced (Fig. 1b) and flow is monitored for up to 30 min. Changes in flow over time, and time to vessel occlusion are determined. Mice are sacrificed prior to recovering from anesthesia (*see* **Note 5**).

3.2 Inferior Vena Cava Stasis-Induced Venous Thrombosis Model

1. Mice are anesthetized by administration of pentobarbital (50 mg/kg) through an intraperitoneal injection as described in the section on the $FeCl_3$ thrombosis model.

2. The mouse is placed on its back on a 37 °C warming pad, and the extremities are immobilized with pieces of tape. Heparan-based drug or vehicle in 100 μl of PBS is infused into a lateral tail vein using a 1 ml tuberculin syringe fitted with a 27-G needle. If the drug appears to compromise subsequent surgery, it can be administered shortly after the surgical procedure described below is complete (*see* **Note 6**).

3. A midline vertical incision is made through the skin and abdominal wall with a scalpel. The inferior vena cava is exposed between the iliac bifurcation and the renal veins by pushing the abdominal contents (bowel) to the left side of the animal (Fig. 2a). The bowel is kept moist by covering it with cotton gauze soaked in PBS. The vena cava is gently separated from the aorta by blunt dissection (*see* **Note 7**).

4. A 4-0 coated Vicryl suture is placed underneath the vena cava immediately below the renal veins, and a 4-0 Steelex metal monofilament suture is placed longitudinally over the IVC (Fig. 2b, **Step 1**) (*see* **Note 8**). The Vicryl suture is tied over the IVC and the metal suture to stop blood flow through the vessel (Fig. 2b, **Step 2**). Then the metal suture is gently removed by

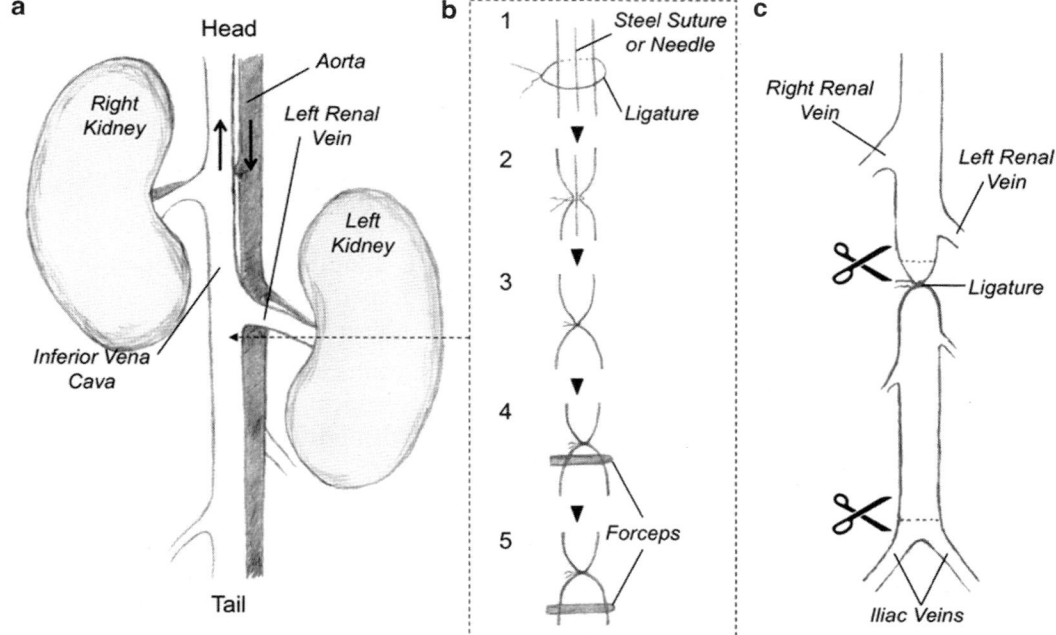

Fig. 2 Inferior vena cava venous thrombosis model. (**a**) Anatomy of the retroperitoneum of the mouse as viewed from a ventral abdominal incision. The positions of the animal's head and tail relative to the drawing are indicated. The *black arrows* indicate the direction of blood flow in the major vessels. (**b**) Thrombosis model. *Step 1:* A ligature is loosely placed around the inferior vena cava caudal to the left renal vein. A steel suture or other linear object such as a needle is also placed within the ligature. *Step 2:* The ligature is tightened around the vessel and steel suture. *Step 3:* The steel suture is gently removed from the ligature. This results in the ligature restricting flow through the vena cava without completely blocking it. *Step 4:* Forceps are used to crimp the vena cava immediately below the suture, and (*Step 5*) 5 mm caudal to the suture to injure the vessel endothelium. (**c**) Twenty-four hours after surgery, the abdomen is reopened, and the vena cava is excised for processing by cutting the vessel immediately above (cranial) to the ligature, and just above the iliac bifurcation

sliding it out from ligature (Fig. 2b, **Step 3**). Removal of the steel suture restores a small amount of blood flow through the vena cava (*see* **Note 9**).

5. A sterile forceps with serrated tip is used to compress (crimp) the vena cava immediately below the suture (toward the tail) for 20 s (Fig. 2b, **Step 4**). The procedure is repeated at a location 5 mm below the first compression site (Fig. 2b, **Step 5**). This "crush" injury will cause endothelial cell damage and serve as a nidus for thrombus formation within the vessel lumen (*see* **Note 9**).

6. Bowel is returned to the abdominal cavity, and the abdominal wall is closed with sutures. The overlying skin is closed with surgical clips. The animal is observed until it recovers from anesthesia.

7. 24 h post surgery, the mouse is sacrificed with an intracardiac infusion of pentobarbital (100 mg/kg) and the abdomen is

reopened. The vena cava is cut above the ligature and at the distal end above the iliac bifurcation (Fig. 2c). Vessel contents are pushed out of the distal end of the vena cava by running forceps along the length of the vessel starting at the ligature. Expressed clot is placed in 10 % formalin for 24 h. After fixing, the clot is dried on a piece of filter paper and weighed (*see* **Note 10**).

3.3 Tail Bleeding Model

1. Mice are anesthetized by administration of pentobarbital (50 mg/kg) through an intraperitoneal injection as described in the section on the FeCl$_3$ thrombosis model.

2. Once under anesthesia, the animal is placed on a 37 °C heating pad. Heparan based drug, or vehicle, in 100 μl of PBS is infused into a lateral tail vein using a 1 ml tuberculin syringe fitted with a 27-G needle (*see* **Note 6**).

3. The tail is transected with a scalpel 2 mm above the tip (*see* **Note 11**).

4. The bleeding tail is immediately immersed in a 1.7 ml microfuge tube filled with 1.2 ml of PBS kept at 37 °C with a heating block.

5. The animal is observed for up to 30 min. The time to cessation of bleeding (including rebleeding) is noted (*see* **Note 12**). Mice are sacrificed prior to recovering from anesthesia.

6. Volume of PBS plus blood is recorded to establish the amount of blood lost (*see* **Note 13**).

4 Notes

1. A number of animal-related factors can produce variability in this model. Different mouse strains vary in their propensity to form thrombi in response to FeCl$_3$. Each laboratory should determine the sensitivity of their mouse line (inbred or mixed) to different concentrations of FeCl$_3$ before testing anticoagulants. Older mice are larger than younger animals and have thicker vessel walls. This can alter the response to injury. We prefer to use mice that are in a relatively narrow age range (12–20 weeks) to limit this effect. Many investigators confine analysis to male mice, to avoid effects of variation in the levels of coagulation proteins during the estrus cycle in females. We have not noticed a significant inter-gender difference with the FeCl$_3$ carotid artery injury model, but it is possible that certain anticoagulants may be sensitive to changes due to the estrus cycle.

2. The response of the animal to FeCl$_3$ injury (i.e. the lowest concentration required to reproducibly induce thrombus

formation) will vary with different anesthetics. For example, exposure of the carotid artery to 3.5 % $FeCl_3$ will reproducibly cause thrombus formation in C57Bl/6 mice anesthetized with pentobarbital [23], while the same mice anesthetized by isoflurane inhalation will occlude with 2.5 % $FeCl_3$. For each anesthetic, a range of $FeCl_3$ concentrations should be tested to identify the lowest concentration that reproducibly causes vessel occlusion.

3. An obvious concern with using a surgery-based model to test anticoagulants is peri-operative bleeding. We find that if care is taken not to injure structures underlying the skin, the ferric chloride arterial injury model and the venous stasis model can be performed with minimal blood loss even on mice with severe hemophilia (factor VIII or IX deficiency) or in mice who have received heparin.

4. In our experience, $FeCl_3$ arterial injury models are less sensitive to anti-platelet agents than to inhibitors of thrombin generation. However, when using relatively low $FeCl_3$ concentrations, anti-platelet agents can influence results (produce an antithrombotic effect). Given this, we avoid using nonsteroidal anti-inflammatory analgesics to treat pain because they can produce an anti-platelet effect that will alter results in the thrombosis model.

5. Changing the size or number of the $FeCl_3$-soaked filter papers, or the duration of exposure of the vessel to $FeCl_3$, will change the extent of vessel injury and influence the result. If these factors are standardized, the assay should have a high degree of reproducibility.

6. Administering drugs by tail vein injection is a skill that requires practice. The diameter of the target vessel is small and the skin and underlying tissue of the tail is tough and can be difficult to penetrate with a small-gauge needle. Warming the tail for a few minutes with a warm cloth to dilate the vessels can make injection easier.

7. While the anatomy of the carotid artery varies relatively little between mice, there can be considerable variation in the anatomy of the inferior vena cava, even among animals of the same strain. The number and size of collateral branches, and the positions of branch points of important vessels such as the renal veins can vary. Some investigators will ligate large collaterals so that flow to the vessel comes largely from the lower extremities, but we have not found that this affects results appreciably. We have observed that an occasional animal may develop paralysis of the hind limbs a few hours after the procedure. This might be caused by trauma to the aorta or to the small arteries that branch off of it at numerous points to supply the spinal chord.

For beginners, it is often difficult to distinguish the vena cava from the aorta, leading to inadvertent ligation of both vessels. Practice is required to obtain proficiency in isolating and manipulate the inferior vena cava without injuring the aorta.

8. The 4-0 Steelex metal monofilament suture used in **step 4** under Subheading 3 can be replaced with another object of comparable diameter, such as a 33-G needle. The extent of the partial obstruction of blood flow can be adjusted by using needles of different gauge.

9. This model requires a significant amount of practice to perform reproducibly compared to the $FeCl_3$-induced arterial injury model, as the inferior vena cava is considerably more delicate than the carotid artery, the vessel is closely associated with the descending aorta in the abdominal cavity, and the required manipulations of the vessel are relatively complex. When crimping the vessel with forceps, excessive force can result in laceration.

10. A relatively large range of thrombus sizes should be expected, even in a control group of mice of the same strain. On occasion, a control mouse may even fail to develop a detectable thrombus. For this reason, it is usually necessary to test a larger number of animals than would be used in a more reproducible assay such as the $FeCl_3$-induced arterial injury model.

11. While a common strategy is to transect the animal's tail ~2 mm from the tip, some published models use injuries at different points on the tail. It is important to note that tail anatomy varies between animals, resulting in different amounts of tissue injury if a fixed distance from the tail tip is used as the point of injury. Some investigators chose to injure tails at a point where the cross-sectional areas are the same. This can be achieved by creating a template (typically a piece of plastic) with an aperture of desired cross-sectional area. The animal's tail is drawn through the aperture and transected using the surface of the template to guide the scalpel.

12. Bleeding time should always be time to cessation of all bleeding, including rebleeding. Some investigators will determine when bleeding has stopped for at least 60 s as an indication that all bleeding has stopped. We prefer to observe the animal for a full 30 min to insure that bleeding does not recur.

13. We feel that the bleeding time alone is not sufficient to accurately reflect the extent of bleeding. A mouse may have a prolonged bleeding time, but lose relatively little blood during that time, compared to another animal that bleeds more briskly for a short period of time. We recommend measuring total blood loss as well as time to cessation of bleeding.

Acknowledgment

The work described in this manuscript was supported by awards HL81326, HL58837, and HL107152 from the National Heart, Lung and Blood Institute.

References

1. Day SM, Reeve JL, Myers DD, Fay WP (2004) Murine thrombosis models. Thromb Haemost 92:486–494

2. Westrick RJ, Winn ME, Eitzman DT (2007) Murine models of vascular thrombosis. Arterioscler Thromb Vasc Biol 27:2079–2093

3. Sachs UJ, Nieswandt B (2007) In vivo thrombus formation in murine models. Circ Res 100:979–991

4. Furie B (2009) Pathogenesis of thrombosis. Hematology 2009:255–258

5. Diaz JA, Obi AT, Myers DD Jr, Wrobleski SK, Henke PK, Mackman N, Wakefield TW (2012) Critical review of mouse models of venous thrombosis. Arterioscler Thromb Vasc Biol 32:556–562

6. Hogan KA, Weiler H, Lord ST (2002) Mouse models of coagulation. Thromb Haemost 87:563–574

7. Emeis JJ, Jirouskova M, Muchitsch EM, Shet AS, Smyth SS, Johnson GJ (2007) A guide to murine coagulation factor structure, function, assays, and genetic alterations. J Thromb Haemost 5:670–679

8. McManus MP, Gailani D (2012) Mouse models of coagulation factor deficiencies. In: Wang X (ed) Animal models of diseases: translational medicine perspective for drug discovery and development. Bentham Scientific Publishers, Sharjah, pp 67–121

9. Tollefsen DM, Zhang L (2013) Heparin and vascular proteoglycans. In: Marder VJ, Aird WC, Bennett JS, Schulman S, White GC (eds) Hemostasis and thrombosis, basic principles and clinical practice, 6th edn. Lippincott, Williams and Wilkins, Philadelphia, PA, pp 585–597

10. Kung SH, Hagstrom JN, Cass D, Tai SJ, Lin HF, Stafford DW, High KA (1998) Human factor IX corrects the bleeding diathesis of mice with hemophilia B. Blood 91:784–790

11. Geng Y, Verhamme IM, Smith SB, Sun MF, Matafonov A, Cheng Q, Smith SA, Morrissey JH, Gailani D (2013) The dimeric structure of factor XI and zymogen activation. Blood 121:3962–3969

12. Renné T, Pozgajová M, Grüner S, Schuh K, Pauer HU, Burfeind P, Gailani D, Nieswandt B (2005) Defective thrombus formation in mice lacking coagulation factor XII. J Exp Med 202:271–281

13. Carcao M, Moorehead P, Lillicrap D (2013) Hemophilia A and B. In: Hoffman RH, Benz EJ, Silberstein LE, Heslop H, Weitz JI, Snastasi J (eds) Hematology, basic principles and practice, 6th edn. Saunders-Elsevier, Philadelphia, PA, pp 1940–1960

14. Lin HF, Maeda N, Smithies O, Straight DL, Stafford DW (1997) A coagulation factor IX-deficient mouse model for human hemophilia B. Blood 90:3962–3966

15. Zadelaar S, Kleemann R, Verschuren L, de Vries-Van der Weij J, van der Hoorn J, Princen HM, Kooistra T (2007) Mouse models for atherosclerosis and pharmaceutical modifiers. Arterioscler Thromb Vasc Biol 27:1706–1721

16. Zhang SH, Reddick RL, Burkey B, Maeda N (1994) Diet-induced atherosclerosis in mice heterozygous and homozygous for apolipoprotein E gene disruption. J Clin Invest 94:937–945

17. Nakashima Y, Plump AS, Raines EW, Breslow JL, Ross R (1994) ApoE-deficient mice develop lesions of all phases of atherosclerosis throughout the arterial tree. Arterioscler Thromb Vasc Biol 14:133–140

18. Lim W (2013) Venous thromboembolism. In: Hoffman RH, Benz EJ, Silberstein LE, Heslop H, Weitz JI, Snastasi J (eds) Hematology, basic principles and practice, 6th edn. Saunders-Elsevier, Philadelphia, PA, pp 2039–2047

19. Eckly A, Hechler B, Freund M, Zerr M, Cazenave JP, Lanza F, Mangin PH, Gachet C (2011) Mechanisms underlying FeCl₃-induced arterial thrombosis. J Thromb Haemost 9:779–789

20. Owens AP 3rd, Lu Y, Whinna HC, Gachet C, Fay WP, Mackman N (2011) Towards a standardization of the murine ferric chloride-induced carotid arterial thrombosis model. J Thromb Haemost 9:1862–1863

21. Barr JD, Chauhan AK, Schaeffer GV, Hansen JK, Motto DG (2013) Red blood cells mediate

the onset of thrombosis in the ferric chloride murine model. Blood 121:3733–3741

22. Wang X, Xu L (2005) An optimized murine model of ferric chloride-induced arterial thrombosis for thrombosis research. Thromb Res 115:95–100

23. Wang X, Cheng Q, Xu L, Feuerstein GZ, Hsu MY, Smith PL, Seiffert DA, Schumacher WA, Ogletree ML, Gailani D (2005) Effects of factor IX or factor XI deficiency on ferric chloride-induced carotid artery occlusion in mice. J Thromb Haemost 3:695–702

24. Wang L, Miller C, Swarthout RF, Rao M, Mackman N, Taubman MB (2009) Vascular smooth muscle-derived tissue factor is critical for arterial thrombosis after ferric chloride-induced injury. Blood 113:705–713

25. Cheng Q, Tucker EI, Pine MS, Sisler I, Matafonov A, Sun MF, White-Adams TC, Smith SA, Hanson SR, McCarty OJ, Renné T, Gruber A, Gailani D (2010) A role for factor XIIa-mediated factor XI activation in thrombus formation in vivo. Blood 116:3981–3989

26. Revenko AS, Gao D, Crosby JR, Bhattacharjee G, Zhao C, May C, Gailani D, Monia BP, MacLeod AR (2011) Selective depletion of plasma prekallikrein or coagulation factor XII inhibits thrombosis in mice without increased risk of bleeding. Blood 118:5302–5311

27. Weitz JI (2013) Antithrombotic drugs. In: Hoffman RH, Benz EJ, Silberstein LE, Heslop H, Weitz JI, Snastasi J (eds) Hematology, basic principles and practice, 6th edn. Saunders-Elsevier, Philadelphia, PA, pp 2102–2119

28. Schumacher WA, Luettgen JM, Quan ML, Seiffert DA (2010) Inhibition of factor XIa as a new approach to anticoagulation. Arterioscler Thromb Vasc Biol 30:388–392

29. Löwenberg EC, Meijers JC, Monia BP, Levi M (2010) Coagulation factor XI as a novel target for antithrombotic treatment. J Thromb Haemost 8:2349–2357

30. Woodruff RS, Sullenger B, Becker RC (2011) The many faces of the contact pathway and their role in thrombosis. J Thromb Thromb 3:9–20

31. Al-Horani RA, Ponnusamy P, Mehta AY, Gailani D, Desai UR (2013) Sulfated Pentagalloylglucoside is a potent, allosteric, and selective inhibitor of factor XIa. J Med Chem 56:867–878

32. Greene TK, Schiviz A, Hoellriegl W, Poncz M, Muchitsch EM, Animal Models Subcommittee of the Scientific And Standardization Committee Of The Isth (2010) Towards a standardization of the murine tail bleeding model. J Thromb Haemost 8:2820–2822

33. Liu Y, Jennings NL, Dart AM, Du X-J (2012) Standardizing a simpler, more sensitive and accurate tail bleeding assay in mice. World J Exp Med 2:30–36

34. Broze GJ Jr, Yin ZF, Lasky N (2001) A tail vein bleeding time model and delayed bleeding in hemophiliac mice. Thromb Haemost 85:747–748

Chapter 38

Genetic Approaches in the Study of Heparan Sulfate Functions in *Drosophila*

Masahiko Takemura and Hiroshi Nakato

Abstract

Several classes of heparan sulfate proteoglycan (HSPG) core proteins and all HS biosynthetic/modifying enzymes are evolutionarily conserved from human to *Drosophila melanogaster*. This genetically tractable model offers highly sophisticated techniques to manipulate gene function in a spatially and temporally controlled manner. Thus, *Drosophila* has been a powerful system to explore the functions of HSPGs in vivo. In this chapter, we will introduce two genetic techniques available in *Drosophila*: TARGET (temporal and regional gene expression targeting) and MARCM (mosaic analysis with a repressible cell marker).

Key words *Drosophila melanogaster*, HSPG, GAL4/UAS, TARGET, MARCM

1 Introduction

1.1 HSPGs in Drosophila

There are several evolutionarily conserved classes of HSPGs. Syndecans have a single transmembrane domain, while glypicans are linked to the cell surface by a GPI anchor. There are also secreted proteoglycans, including perlecan, a major constituent of the basement membrane. *Drosophila* has a single Syndecan (Sdc) [1], two glypicans, Dally and Dally-like protein (Dlp) [2–4], and the *Drosophila* perlecan (called Trol) [5–7]. *Drosophila* also has a complete set of HS biosynthetic/modifying enzymes (Table 1) and produces complex HS structures that are equivalent to mammalian HS. HSPGs play important roles in *Drosophila* development by regulating signaling and distribution of BMPs, Wnts, Hedgehog, FGFs [8, 9], and cytokines [10, 11].

Mutations for all these conserved HSPG core proteins are available in *Drosophila*. In addition, mutations for virtually all HS biosynthetic/modifying enzymes have been generated. *Drosophila* has only one gene (or two for Hs3st) for each class of these enzymes (Table 1). Therefore, there is minimal genetic redundancy, which could hamper the genetic analysis of these molecules in mammalian systems. Transgenic fly strains bearing overexpression and

Kuberan Balagurunathan et al. (eds.), *Glycosaminoglycans: Chemistry and Biology*, Methods in Molecular Biology, vol. 1229, DOI 10.1007/978-1-4939-1714-3_38, © Springer Science+Business Media New York 2015

Table 1
Evolutionarily conserved HSPG core proteins and HS biosynthetic/modifying enzymes

Core proteins/enzymes	Mammal	Drosophila
Glypican	Glypican1–6	Dally, Dally-like
Syndecan	Syndecan1–4	Syndecan
Perlecan	Perlecan	Trol
HS copolymerases	Ext1, Ext2, Extl3	Ttv, Sotv, Botv
N-deacetylase/N-sulfotransferase	NDST1–4	Sulfateless
C_5-epimerase	Hsepi	Hsepi
2-O-sulfotransferase	Hs2st	Hs2st
6-O-sulfotransferase	Hs6st1–3	Hs6st
3-O-sulfotransferase	Hs3st1, 2, 3A, 3B, 4, 5, 6	Hs3st-A, B
6-O-endosulfatases	Sulf1–2	Sulf1

RNAi-mediated knockdown constructs for these genes are also available. These genetic tools enable us to manipulate HSPGs in vivo in a temporally and spatially controlled manner. Although there is no technique specifically designed for HSPG research, *Drosophila* provides a powerful system to address unanswered questions regarding in vivo functions of this class of molecules. In the following sections, we describe two genetic techniques: TARGET and MARCM in *Drosophila*.

1.2   TARGET System

The GAL4/UAS system is the most popular genetic technique to express specific genes in specific tissues [12]. The GAL4/UAS system consists of the yeast GAL4 transcriptional activator and a transgene under the control of a UAS (upstream activation sequence). In the absence of GAL4, transgene expression is not induced. However, when GAL4 is present, it binds the UAS and activates expression of the downstream transgene. Thousands of GAL4 lines, each of which shows a specific expression pattern, have been generated by fusing defined promoter sequences upstream of *GAL4* or by randomly inserting *GAL4* into the genome to "trap" nearby enhancer elements. Thus, genetic manipulation, such as overexpression and RNAi-mediated knockdown, in a tissue-specific manner can be performed by examining the progeny produced from crossing a fly carrying a GAL4 under the control of specific regulatory elements with a fly carrying a UAS-transgene.

The GAL4/UAS system does not provide an independent temporal control over transgene expression. This is critical in some cases. For example, temporal control of GAL4 activity is

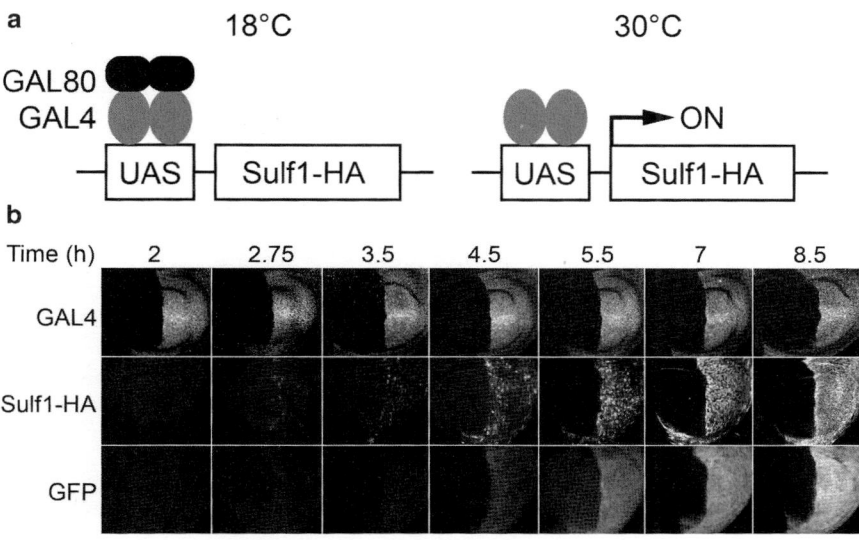

Fig. 1 Induction of Sulf1 using the TARGET system. (**a**) Graphic depiction of the TARGET system. In the conventional GAL4/UAS system, GAL4 protein (*gray oval*) binds to UAS and activates the downstream transgene (Sulf1-HA and UAS-GFP, in this case). In the TARGET system, a temperature-sensitive version of GAL80 protein (GAL80ts, *black oval*) driven ubiquitously by the *tubulin* promoter represses GAL4-mediated transgene expression at the permissive temperature (18 °C). At the restrictive temperature (30 °C), GAL4 is relieved from repression by GAL80ts and activates the HA-tagged Sulf1 and GFP expressions. (**b**) Immunostaining of mid-third instar larval wing discs showing the gradual induction of Sulf1-HA and GFP after shifting larvae to the restrictive temperature. Expression of GAL4 protein and GFP at each time point is also shown. GAL4 protein was detected at a constant level throughout the time course. In this case, GAL4 is expressed only in the posterior compartment of the wing disc by *hh-GAL4* and the anterior compartment serves as an internal control. This figure is modified from [15]

desired to avoid developmental effects caused by constitutive expression. The TARGET system was developed to add greater temporal control to the conventional GAL4/UAS system by incorporating a temperature-sensitive variant of a GAL4 repressor, GAL80 (GAL80ts) [13, 14]. Figure 1a illustrates the TARGET system. At 18 °C (permissive temperature), GAL80ts supresses GAL4 activity. At 30 °C (restrictive temperature), however, GAL4 is relieved from the repression of GAL80ts and activates UAS-transgene expression. Therefore, it is possible to express transgenes at any stage by simply changing the culture temperature of the flies. Figure 1b shows the time-course of expression of GAL4 protein (which is constant) and two target genes, GFP and Sulf1-HA in the wing disc [15]. In this experiment, fly larvae carrying *hedgehog* (*hh*)-*GAL4* (which is expressed specifically in the posterior compartment of the wing disc), *UAS-GFP*, *UAS-Sulf1-HA*, and *tub-GAL80^s* (which expresses GAL80ts ubiquitously driven by a *tubulin* 1α promoter) are transferred from 18 to 30 °C and incubated for 2–8.5 h. Posterior compartment-specific expression of GFP and Sulf1-HA became detectable about 2.5 h after the temperature shift and gradually increased over time.

1.3 MARCM System

Mosaic analysis is a genetic technique that allows scientists to generate patches of homozygous mutant cells in an otherwise heterozygous background. This is very useful for analyzing a homozygous lethal mutation of a gene and for determining cell autonomy of gene activity. Mosaic analysis using flippase (FLP)/flippase recognition target (FRT) system-mediated mitotic recombination has been widely used in *Drosophila* for studying gene function [16]. The FLP/FRT system is illustrated in Fig. 2a. FLP under the control of a heat-shock promoter (*hs-FLP*) is commonly used for inducing FLP ubiquitously at the desired time. FLP mediates mitotic recombination between two FRT sites present on homologous chromosomes. Mosaic animals are thus created by recombination (after chromosomal duplication) in somatic cells. If a fly has a mutation distal to a FRT site on one chromosome and a marker (usually ubiquitously-expressed GFP) distal to the FRT site on the homologous chromosome, segregation of recombinant chromosomes at mitosis can produce two daughter cells: a homozygous mutant cell, which is negatively marked by the loss of marker, and a homozygous wild-type cell, which is brightly labeled by the two copies of marker. This "twin-spot" wild-type cell serves as a control, allowing for direct comparison of wild-type and mutant cells in one tissue sample.

In some complex tissues such as nervous system, the negatively marked mutant cells induced by the above mentioned method can be difficult to identify. To overcome this problem, Lee and Luo [17] developed the MARCM system by combining the GAL4/UAS system and the GAL80 repressor protein with the FLP/FRT system to positively label mutant clones [17, 18]. In this system, GAL4-dependent expression of a UAS-marker (GFP etc.) is repressed by the presence of the *tub-GAL80* transgene, which is placed *in trans* to a mutant gene of interest (Fig. 2b). After FLP/FRT-mediated mitotic recombination, only homozygous mutant cells can express a marker gene via loss of GAL80. Furthermore, this system allows for the addition of another UAS-transgene for gene manipulation in labeled cells, which can, for example, be useful in a rescue experiment. Thus, the advantage of the MARCM system is that mutant clones are positively marked in an unlabeled background, which is extremely useful for visualizing the morphology of mutant cells, such as neuronal axons, and for lineage tracing of multipotent stem cells [19].

2 Materials

2.1 A Sample Set for Fig. 1

1. tub-GAL80^{ts}; hh-GAL4 UAS-GFP/TM6B.

2. UAS-Sulf1-HA.

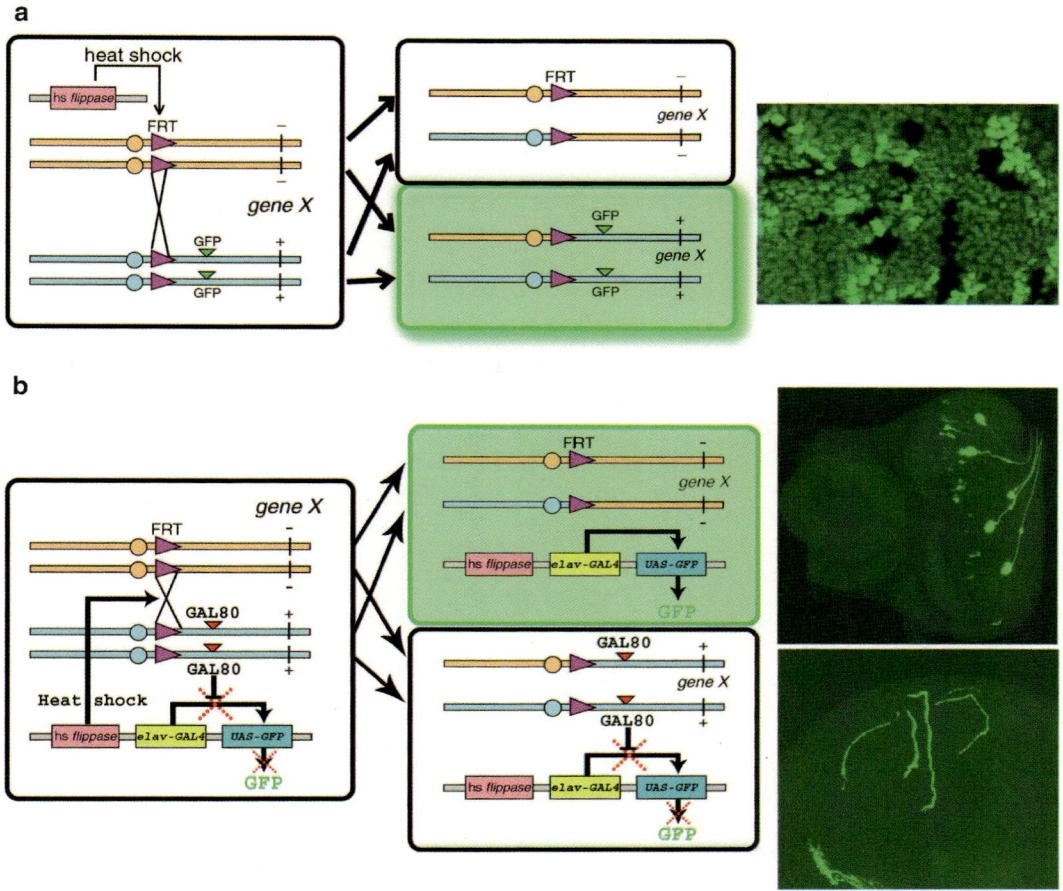

Fig. 2 Mosaic analysis in *Drosophila*. (**a**) The conventional "twin-spot" mosaic analysis. In a dividing heterozygous mother cell, FLP expression induced by heat shock mediates mitotic recombination between FRT sites. Segregation of recombinant chromosomes can produce two homozygous mutant cells (−/−) and homozygous wild-type cells (+/+). Subsequent cell divisions result in clones from each of the original daughter cells. Only the heterozygous and wild-type cells bear one copy or two copies of GFP, respectively, which are distinguishable by strength of GFP signal. Homozygous mutant cells can be identified by the absence of GFP signal. *Right panel* shows a confocal image of a wing disc bearing somatic clones induced by the FLP/FRT system. (**b**) MARCM system for positively labeling mutant cells. Mosaic tissues are generated in a manner identical to the method described in (**a**). The wild-type chromosome, instead of bearing a marker, bears the *tub-GAL80* transgene which blocks GAL4-mediated transcription. Cells which are heterozygous or homozygous for this *tub-GAL80-bearing* chromosome do not express GFP under the control of GAL4/UAS system (present on another chromosome) due to the repression by GAL80. Only cells homozygous for the mutant chromosome will lack a copy of the *tub-GAL80* transgene and hence express GFP. In the *right panel* are shown an eye disc (*top*) and the optic lobe (*bottom*) from the same third instar larva post heat shock. Pan-neuronal GAL4 driver *elav-GAL4* is used for marking neurons in this case. This figure is modified from [21]

2.2 A Sample Set for Fig. 2

To make MARCM clones, the fly must carry *FLP*, *FRT*, a ubiquitous or tissue-specific *GAL4*, *UAS-marker*, and *tub-GAL80*. It is useful to make MARCM-ready flies, which have all the required transgenes (*see* **Note 1**). These flies are ready to cross to a fly carrying the corresponding *FRT* and mutation. Further information about designing MARCM experiments is available in [20].

1. *elav-GAL4 UAS-mCD8::GFP hs-FLP; FRTG13 tub-GAL80* (MARCM-ready fly).

    - *elav-GAL4* is a neuron-specific GAL line.

    - *UAS-mCD8::GFP* is used for expressing a membrane-targeted GFP, which is useful for marking neuronal morphology.

2. *FRTG13* (Use the same FRT insertion with the MARCM-ready fly).

    - To generate a mutant clone, a mutation of the gene of interest has to be placed distal to the FRT on the same chromosome arm. This can be achieved by meiotic recombination.

2.3 Reagents

1. PBS (phosphate-buffered saline): 137 mM NaCl, 2.7 mM KCl, 10 mM Na_2HPO_4, 1.8 mM KH_2PO_4.

2. 0.1 % PBST: 0.1 % w/v Triton X-100 in PBS.

3. 37 % Formaldehyde.

4. Blocking solution: 10 % normal goat serum in 0.1 % PBST.

5. Appropriate primary antibody solution.

6. Appropriate secondary antibody solution.

2.4 Equipment

1. Vials and Standard fly food.

2. Incubator set at 18, 25, and 30 °C.

3. Water bath set at 37 °C.

4. Dissecting microscope.

5. 9-Well glass depression plate.

6. Forceps.

7. Dissecting scissors.

8. 1.5-ml microcentrifuge tube.

9. Microscope slides.

10. Coverslips, 22×22 mm.

11. Mounting medium.

12. Nail polish.

13. Fluorescence microscope or Confocal microscope.

3 Methods

3.1 Cross and Heat Shock (for the TARGET System)

1. Cross ten virgin females of *tub-GAL80^ts^; hh-GAL4 UAS-GFP* flies with five to ten males of *UAS-Sulf1-HA* flies in a fly food vial kept in an incubator at 18 °C, to repress GAL4-mediated transcriptional activation of the UAS transgene. Flip them into a fresh fly food vial every 2 days to avoid overcrowding and to repeat experiments.

2. Transfer the vial containing mid-third instar larvae to a 30 °C incubator at desired timing (*see* **Note 2**).

3. After desired time, dissect and immunostain the wing discs according to Subheading 3.3.

3.2 Cross and Heat Shock (for the MARCM System)

1. Collect ten virgin females of *FRT42B* flies and five to ten males of MARCM-ready flies to make the cross.

2. Culture the parental flies in a fly food vial for 2 days and let them lay eggs. Keep transferring the parental flies to new vials to expand the progeny (*see* **Note 2**).

3. Allow the offspring to develop until a specific developmental stage. Then, apply heat shock by incubating the vials in a water bath set at 37 °C for 30–60 min (*see* **Note 3**).

4. After heat shock, return the vial to the incubator set at 25 °C. Wait until the desired developmental stage for analysis and proceed to Subheading 3.3.

3.3 Dissection and Immunohisto-chemistry of Wing Discs

All steps are done at room temperature unless otherwise noted. When you remove the supernatant solution from a 1.5-ml tube, take care not to damage samples.

1. Mix 900 μl of PBS and 100 μl of formaldehyde (final concentration is 3.7 %) in a 1.5-ml tube and place on ice.

2. Collect third-instar wandering larvae with the right genotype and wash them with PBS in a 9-well glass plate.

3. Make drops of PBS on a soft dissection pad.

4. Pick one larva and place into the drop.

5. With a pair of fine forceps in one hand, hold the larva; with scissors in the other hand, cut the larva in half.

6. Invert the anterior half of the larva, exposing imaginal discs.

7. Remove the fat body, midgut, brain, and other imaginal discs.

8. Immediately transfer the inverted anterior half of the larva into a 1.5-ml tube containing the 3.7 % formaldehyde solution placed on ice.

9. Repeat the dissection.

10. Fix with 3.7 % formaldehyde solution on a nutator for 15 min at room temperature (*see* **Note 4**).

11. Wash 3 times with 0.1 % PBST, 10 min each.

12. Incubate in blocking solution for 30 min.

13. Incubate in appropriate primary antibody solution at 4 °C overnight.

14. Wash 3 times with 0.1 % PBST, 10 min each.

15. Incubate in secondary antibody solution for 2 h.

16. Wash 3 times with 0.1 % PBST, 10 min each.

17. Cut off a pipet tip (for widening the opening) and transfer the samples onto a soft dissection plate.

18. Put a drop of mounting medium on a microscope slide.

19. Separate the wing discs from larval cuticle using forceps and transfer into the mounting medium drop.

20. Cover with a 22 × 22 mm coverslip.

21. Seal the edges with nail polish.

22. Analyze the samples by fluorescence or confocal microscopy.

4   Notes

1. The MARCM-ready flies may be weak due to the presence of multiple transgenes. If this is the case, relocation of transgenes may be required.

2. If enough offspring are not obtained, increase the number of virgin females.

3. Timing and duration of heat shock need to be optimized depending on each experiment and *hs-FLP* used. For example, to increase the efficiency of mitotic recombination, increase the duration of heat shock. If flies are sensitive to longer heat shock, multiple rounds of shorter heat shock with hour-long intervals is an option.

4. Optimize the way and duration of fixing depending on the tissue dissected and antibodies used.

Acknowledgment

We are grateful to Daniel Levings and Pui Choi for helpful comments to the manuscript. The authors are supported by Japan Society for the Promotion of Science (to M. T.) and National Institutes of Health Grant R01 HD042769 (to H. N.).

References

1. Spring J, Paine-Saunders SE, Hynes RO et al (1994) *Drosophila* syndecan: conservation of a cell-surface heparan sulfate proteoglycan. Proc Natl Acad Sci U S A 91:3334–3338

2. Nakato H, Futch TA, Selleck SB (1995) The *division abnormally delayed* (*dally*) gene: a putative integral membrane proteoglycan required for cell division patterning during postembryonic development of the nervous system in *Drosophila*. Development 121:3687–3702

3. Khare N, Baumgartner S (2000) Dally-like protein, a new *Drosophila* glypican with expression overlapping with *wingless*. Mech Dev 99:199–202

4. Baeg GH, Lin X, Khare N et al (2001) Heparan sulfate proteoglycans are critical for the organization of the extracellular distribution of Wingless. Development 128:87–94

5. Datta S, Kankel DR (1992) *l(1)trol* and *l(1)devl*, loci affecting the development of the adult central nervous system in *Drosophila melanogaster*. Genetics 130:523–537

6. Voigt A, Pflanz R, Schäfer U et al (2002) Perlecan participates in proliferation activation of quiescent *Drosophila* neuroblasts. Dev Dyn 224:403–412

7. Park Y, Rangel C, Reynolds MM et al (2003) *Drosophila* Perlecan modulates FGF and Hedgehog signals to activate neural stem cell division. Dev Biol 253:247–257

8. Kirkpatrick CA, Selleck SB (2007) Heparan sulfate proteoglycans at a glance. J Cell Sci 120:1829–1832

9. Yan D, Lin X (2009) Shaping morphogen gradients by proteoglycans. Cold Spring Harb Perspect Biol. doi: 10.1101/cshperspect.a002493

10. Hayashi Y, Sexton TR, Dejima K et al (2012) Glypicans regulate JAK/STAT signaling and distribution of the unpaired morphogen. Development 139:4162–4171

11. Zhang Y, You J, Ren W et al (2013) *Drosophila* glypicans Dally and Dally-like are essential regulators for JAK/STAT signaling and unpaired distribution in eye development. Dev Biol 375:23–32

12. Brand AH, Perrimon N (1993) Targeted gene expression as a means of altering cell fates and generating dominant phenotypes. Development 118:401–415

13. McGuire SE, Le PT, Osborn AJ et al (2003) Spatiotemporal rescue of memory dysfunction in *Drosophila*. Science 302:1765–1768

14. McGuire SE, Mao Z, Davis RL (2004) Spatiotemporal gene expression targeting with the TARGET and gene-switch systems in *Drosophila*. Sci STKE. doi:10.1126/stke.2202004pl6

15. Kleinschmit A, Takemura M, Dejima K et al (2013) *Drosophila* heparan sulfate 6-*O*-endosulfatase Sulf1 facilitates Wingless (Wg) protein degradation. J Biol Chem 288:5081–5089

16. Xu T, Rubin GM (1993) Analysis of genetic mosaics in developing and adult *Drosophila* tissues. Development 117:1223–1237

17. Lee T, Luo L (1999) Mosaic analysis with a repressible cell marker for studies of gene function in neuronal morphogenesis. Neuron 22:451–461

18. Lee T, Luo L (2001) Mosaic analysis with a repressible cell marker (MARCM) for *Drosophila* neural development. Trends Neurosci 24:251–254

19. Biteau B, Hochmuth CE, Jasper H (2011) Maintaining tissue homeostasis: dynamic control of somatic stem cell activity. Cell Stem Cell 9:402–411

20. Wu JS, Luo L (2006) A protocol for mosaic analysis with a repressible cell marker (MARCM) in Drosophila. Nat Protoc 1:2583–2589

21. Selleck SB, Nakato H (2004) Functional dissection of glycoconjugates during development: lessons from the fruitfly. Trends Glycosci Glycotechnol 16:95–108

Chapter 39

Measuring Sulfatase Expression and Invasion in Glioblastoma

Anna Wade, Jane R. Engler, Vy M. Tran, and Joanna J. Phillips

Abstract

Extracellular sulfatases (SULF1 and SULF2) selectively remove 6-O-sulfate groups from heparan sulfate proteoglycans (HSPGs) and by this process control important interactions of HSPGs with extracellular factors including morphogens, growth factors, and extracellular matrix components. The expression of SULF1 and SULF2 is dynamically regulated during development and is altered in pathological states such as glioblastoma (GBM), a highly malignant and highly invasive brain cancer. SULF2 protein is increased in an important subset of human GBM and it helps regulate receptor tyrosine kinase signaling and tumor growth in a murine model of the disease. By altering ligand binding to HSPGs, SULF2 has the potential to modify the extracellular availability of factors important in a number of cell processes including proliferation, chemotaxis, and migration. Diffuse invasion of malignant tumor cells into surrounding healthy brain is a characteristic feature of GBM that makes therapy challenging. Here, we describe methods to assess SULF2 expression in human tumor tissue and cell lines and how to relate this to tumor cell invasion.

Key words GBM, Glioma, Invasion, Sulfatase, SULF2, HSPG, Tumor microenvironment, NPC

1 Introduction

Heparan Sulfate Proteoglycans (HSPGs) comprise a protein core attached to one or more glycosaminoglycan chains. Plasma membrane-associated HSPGs or those secreted into the extracellular environment participate in a diverse array of interactions mediating many important functions, such as adhesion, migration, and morphogen/growth factor signaling [1, 2]. An important determinant of these interactions is the extent of HSPG sulfation, particularly 6-O-sulfation, which can be regulated during HSPG biosynthesis or extracellularly by the sulfatases, SULF1 and SULF2 [3]. The SULFs specifically remove sulfates from the 6-O position of glucosamine [4, 5]. This modification changes the ability of HSPGs to bind specific growth factors and morphogens, modulating HSPG-dependent signaling in a temporally and spatially

Kuberan Balagurunathan et al. (eds.), *Glycosaminoglycans: Chemistry and Biology*, Methods in Molecular Biology, vol. 1229, DOI 10.1007/978-1-4939-1714-3_39, © Springer Science+Business Media New York 2015

controlled manner, e.g., Wnts, fibroblast growth factor (FGF2), glial cell line-derived neurotrophic factor (GDNF), and vascular endothelial growth factor (VEGF) [4, 6–9].

Consistent with the important roles for extracellular sulfatases in signaling, SULF2 expression is increased in a variety of human cancers including brain, lung, breast, head and neck, and pancreatic cancer [9–16]. In human and murine models for GBM, loss of SULF2 results in decreased signaling through the platelet-derived growth factor receptor-alpha (PDGFRα) and decreased cell proliferation [11]. Furthermore, ablation of Sulf2 expression in tumor cells prolongs survival in vivo, demonstrating a functional role for Sulf2 in malignant glioma [11]. Interestingly, SULF2 loss is also associated with decreased activation of EPHA2 and IGF1Rβ, demonstrating SULF2 can modulate multiple signaling pathways simultaneously.

Invasion of tumor cells into the adjacent brain is a hallmark of GBM and severely complicates therapy [17]. Mechanisms of tumor cell invasion are complex, involving multiple extracellular and intracellular cues [18]. As demonstrated in development, wound repair, and cancer, the SULFs can help to regulate extracellular migratory cues and alter cell migration [7, 19, 20]. SULF2 can promote glioma development [11, 21] but its role in glioma invasion is unclear. Similar to the role of SULF2 in tumor development, the specific molecular alterations present in a tumor may be an important determinant of its function in tumor invasion.

Post-synthetic modifications of HSPGs are increasingly being recognized for their importance in tumor biology and as potential therapeutic targets. Here we describe methods to assess SULF2 mRNA and protein expression and to quantify tumor cell invasion of human and murine glioma cells in vitro.

2   Materials

2.1   SULF2 mRNA Expression Analysis

1. Reagents to isolate mRNA from cultured cells, such as the Qiagen RNeasy Mini Kit which includes RLT lysis buffer.

2. DEPC-treated water.

3. Superscript III First-Strand Synthesis System for RT-PCR (Life Technologies, Grand Island, NY).

4. FastStart Universal SYBR Green Master.

5. Primers:

   Human SULF2 forward primer: ACAGGGATGTCCTCAA CCAG

   Human SULF2 reverse primer: CTTCCCACAGTTGTCC CAGT

Expected size of amplified product is 204 bp [Homo sapiens sulfatase 2 (SULF2), mRNA transcript variant 1 (NM_018837.3); and transcript variant 3, (NM_001161841.1)]; and 195 bp [mRNA transcript variant 2 (NM_198596.2)].

Mouse SULF2 forward primer: GAAAGACCACAAGCTGCACA

Mouse SULF2 reverse primer: GGAGCCTTTGTGCTTGAGAC

Expected size of amplified product is 169 bp [Mus musculus sulfatase 2 (Sulf2), mRNA transcript variant 3 (NM_001252579.1); transcript variant 2 (NM_028072.5); transcript variant 1 (NM_001252578.1)].

6. Control primers are essential and can include primers to GAPDH, ß-tubulin, cyclophilin, actin, elongation factor 1, adenine phosphoribosyl transferase (aprt), and cytoplasmic ribosomal protein L2. We commonly normalize expression of SULF2 to the expression of GAPDH.

Human GAPDH forward primer: CGACAGTCAGCCGCATCTT

Human GAPDH reverse primer: CCGTTGACTCCGACCTTCA

Mouse GAPDH forward primer: AGGTCGGTGTGAACGGATTTG

Mouse GAPDH reverse primer: TGTAGACCATGTAGTTGAGGTCA

7. 7900 HT Fast Real Time PCR machine.

2.2 SULF2 Protein Expression Analysis

1. Protein lysis buffer: 1× of 10× Lysis buffer (Cell Signaling Technology, Boston, MA), 1× of 100× Protease Inhibitor Cocktail (Sigma Chemical Company, St. Louis, MO, USA), and 1× of 100× Halt™ Phosphatase Inhibitor Single-Use Cocktail (Thermo Fisher Scientific Inc, Waltham, MA) in milli-Q water.

2. Cell scraper (Falcon, Corning Life Sciences, Corning, NY).

3. Dulbecco's PBS (D-PBS), ice cold.

4. 22½ gauge needle or sonicator Diagenode Bioruptor®.

5. 1× Sample Buffer plus reducing agents: 1× of 4× LDS Sample Buffer, Tris(2-carboxyethyl)phosphine (TCEP), and Dithiothreitol (DTT).

6. NuPAGE® 4–12 % Bis-Tris Gels (Life Technologies, Grand Island, NY).

7. NuPAGE® MOPS SDS Running Buffer (Life Technologies, Grand Island, NY).

8. Invitrogen XCELL II Blot Module system (Life Technologies, Grand Island, NY, #EI9051).

9. Polyvinylidene fluoride (PVDF) 0.45 μm pore size membrane.

10. Transfer buffer 10× stock (1 L): 75 g Tris-base, 187.5 g Glycine, milli-Q water.

11. Transfer buffer 1×: Transfer buffer 10× stock, 20 % Methanol, milli-Q water.

12. BIO-RAD Mini-PROTEAN Tetra-System (Bio-Rad Laboratories Inc, Hercules, CA, #165-8000).

13. Tris-buffered saline with tween (TBST): Tris-buffered saline plus 0.05 % Tween.

14. Blocking buffer: 5 % Bovine serum albumin (BSA) (w/v) in TBST.

15. Mouse anti-SULF2 antibody 2B4 (Novus Biologicals, Littleton, CO).

16. Loading control antibody: mouse anti-GAPDH antibody.

17. Anti-mouse IgG horseradish peroxidase linked F(ab′)2 fragment from sheep.

18. Electrochemical luminescence (ECL) detection system.

2.3   Method to Assess Invasion in SULF2 Expressing Systems: 3D Invasion Assay

1. 96 well round bottom ultra-low attachment plate.

2. 96 well flat bottom tissue culture plate.

3. BD Growth Factor Reduced Matrigel™ (BD Biosciences, San Jose, CA).

4. ImageJ (http://rsbweb.nih.gov/ij/).

2.4   Method to Assess Invasion in SULF2 Expressing Systems: Transwell Migration Assay

1. BD BioCoat™ Growth Factor Reduced Matrigel™ Invasion chamber (BD Biosciences, San Jose, CA #354483).

2. 24 well tissue culture treated plate.

3. Growth factor-free/serum-free media.

4. Chemoattractant media.

5. 4 % paraformaldehyde (PFA) in PBS.

6. Crystal Violet.

7. Methanol.

3   Methods

3.1   RNA Extraction, cDNA Synthesis, and Quantitative Real Time PCR (qRTPCR)

1. Perform all steps on ice.

2. Cell culture media is removed, RLT lysis buffer is added, and cell lysates are scraped and transferred to a 1.5 mL eppendorf on ice (*see* **Note 1**).

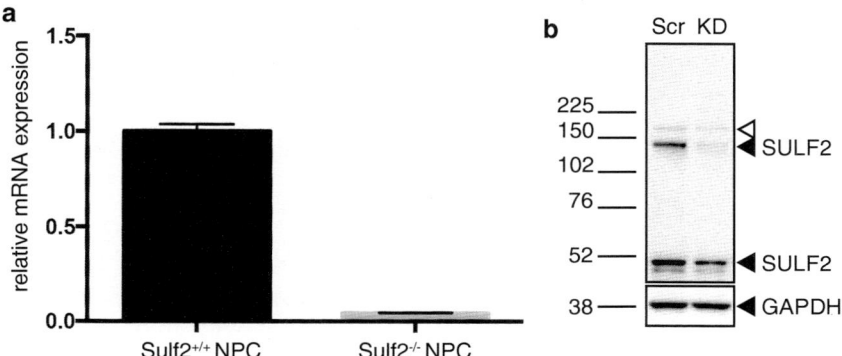

Fig. 1 Assessment of SULF2 mRNA and protein expression. There is an absence of Sulf2 mRNA expression in NPC isolated from Sulf2$^{-/-}$ mice demonstrating the specificity of the primers for Sulf2 (**a**). *Bars* represent the mean $\pm$ SEM of triplicate samples. For each sample the GAPDH Ct was subtracted from the Sulf2 Ct to generate a Delta Ct, this was averaged across technical triplicates, normalized to Sulf2 mRNA expression in Sulf2 wild-type NPC (Sulf2$^{+/+}$), and expressed as relative quantification [2$^{-(\text{normalized DeltaCt})}$]. In (**b**) specificity of the 2B4 mouse anti-Sulf2 antibody is demonstrated by the decrease in full-length 140 kDa and C-terminal fragment 50 kDa SULF2 (*black arrowheads*) in U251 cells which have shRNA knockdown of SULF2 (kDa) compared to a scrambled shRNA control (Scr) U251 cells. Equivalent quantities of protein are demonstrated with the use of GAPDH as a loading control. A high molecular weight nonspecific band is indicated by the *white arrowheads*. The shRNA construct has been reported previously [16] (*see* **Note 10**)

3. Samples in RLT buffer can be immediately processed for RNA extraction or transferred to −80 °C for long-term storage. The Qiagen RNeasy Mini Kit is used to extract RNA from cells following the manufacturer's protocol and RNA is resuspended in DEPC-treated water.

4. cDNA synthesis with Superscript III First-Strand Synthesis System following manufacturer's protocol and 50 μM oligo(dT) primer. Use the same amount of RNA (200–300 ng) across all samples. Perform the additional RNase H step from the manufacturer's protocol after cDNA synthesis to remove RNA template from the reaction mix.

5. Prepare mastermix for PCR: 450 nM of forward and reverse primers, 7.5 μL of FastStart Universal SYBR Green Master, and DEPC-treated water to a final reaction volume of 10 μL. Add 5 μL of cDNA from **step 4** to each well. Run each sample in triplicate (Fig. 1a).

6. Use 7900 HT Fast Real Time PCR machine standard cycling conditions 10 min at 95 °C for cDNA denaturation, 40 cycles at 95 °C for 15 s and 60 °C for 1 min, for primer annealing and amplicon elongation respectively.

3.2 SULF2 Protein Expression Analysis

1. Protein lysate buffer prepared immediately before use and stored on ice. Perform all steps on ice unless otherwise noted.

2. Adherent cells: Remove cell culture media, gently wash cells with ice-cold D-PBS, add lysis buffer directly to cells, collect

cells using a cell scraper with plate on ice, and transfer lysate to an ice-cold eppendorf tube (*see* **Note 2**).

3. Non-adherent cells: Collect cells by centrifugation, wash cells once by resuspending in ice-cold D-PBS, repeat the centrifugation at 4 °C, remove D-PBS, and resuspend cells in lysis buffer on ice.

4. Extract proteins on ice for 20 min.

5. To aid cell lysis and shred chromosomal DNA, either pass samples 12 times through 22½ gauge needle or sonicate sample using the Diagenode Bioruptor® with an on/off interval of 30 s with high intensity ultrasonic waves for 10 min.

6. Spin homogenized protein lysate at $21,000 \times g$ to pellet cell debris.

7. Transfer protein lysate supernatant to a fresh eppendorf and store at –80 °C.

8. Measure protein concentration per standard protocol.

9. Prepare 20 μg of protein with 1× Loading Buffer plus reducing agents (*see* **Note 3**).

10. Boil samples for 5 min, place immediately on ice for 3 min, pulse spin, and leave on ice.

11. Run the samples into the stacking gel at 110 mV and then increase the voltage to 125 mV for 3–4 h.

12. Activate PVDF membrane with sequential 5 min washes in methanol, milli-Q water, and transfer buffer.

13. Transfer protein from PAGE gel to PVDF membrane using BIO-RAD Mini-PROTEAN Tetra-System. Transfer overnight at 4 °C at constant current 20 mA.

14. Wash membranes with TBST and clip bottom right corner of membrane for orientation.

15. Block nonspecific binding to the membrane with Blocking Buffer for 1 h at room temperature.

16. Incubate membrane with primary SULF2 antibody at 1:500 diluted in Blocking Buffer overnight at 4 °C.

17. Wash the membrane 5 times for 5 min with 25 mL TBST.

18. Incubate membrane with secondary anti-mouse antibody conjugated to horseradish peroxidase (HRP) at 1:5,000 diluted in Blocking Buffer for 1 h at room temperature.

19. Wash the membrane 5 times for 5 min with 25 mL TBST.

20. Electrochemical luminescence (ECL) detection system to visualize SULF2 protein (Fig. 1b, *see* **Note 4**).

3.3 Method to Assess Invasion in SULF2 Expressing Systems: 3D Spheroid Invasion Assay

1. Prepare single cell suspension of cells of interest in media (*see* **Note 5**).

2. To form spheroids plate 500–1,000 cells in 100 μL of standard growth media per well of 96 well round bottom plate. Centrifuge at $200 \times g$ for 3 min at room temperature.

3. Allow spheres to form over 3 days under standard growing conditions. Plate a minimum of 3 wells per condition to be tested.

4. Take care while handling matrigel. Thaw matrigel on ice, use cooled tips (stored at –20 °C) to transfer and mix matrigel, and perform all steps on ice to prevent spontaneous polymerization and in tissue culture hood to keep sterile.

5. Coat 96 well flat bottom plate with a matrigel plug to prevent migration of cells along the plastic of the dish. Mix matrigel with pre-cooled full growth media at a 1:1 ratio. Add 50 μL matrigel:media mix to each well and transfer to 37 °C for 30 min to promote matrigel polymerization.

6. Cool plates on ice. Transfer spheroids in 50 μL of media from the round bottom plate to the matrigel plug-coated plate using cooled and blunted p200 pipette tips.

7. Gently add 50 μL of matrigel to each well and incubate for 1 h at 37 °C for matrigel polymerization.

8. Add 50 μL of warmed full growth media to the top of each well containing a spheroid in matrigel.

9. Use inverted tissue culture microscope with low magnification objective (2.5×) to take picture of spheroid size at 0 h (Fig. 2a) and at specific time points afterward such as 16 h, 24 h, 48 h (Fig. 2b; *see* **Note 6**).

10. Quantify cell invasion using ImageJ. Overall spheroid outgrowth can be determined and the fold change in area can be compared across wells after normalizing spheroid outgrowth to the spheroid size at 0 h. To do this: In "Analyze," "Set Measurements" choose "Area" and "Limit to Threshold." Adjust the threshold so the spheroid is highlighted, and select "Analyze," "Measure." If there are issues with uneven illumination across the field of view, the polygon tool can be used to limit the area analyzed to the spheroid area of interest.

3.4 Method to Assess Invasion in SULF2 Expressing Systems: Transwell Migration Assay

1. Transfer the appropriate number of chambers into a 24 well tissue culture plate in tissue culture hood and thaw for 5 min. Each condition should be tested in triplicate.

2. To rehydrate the chamber inserts, add 500 μL of warm media and incubate for at least 2 h at 37 °C.

3. Prepare cells as a single cell suspension in a growth factor-free/serum-free media (*see* **Note 7**).

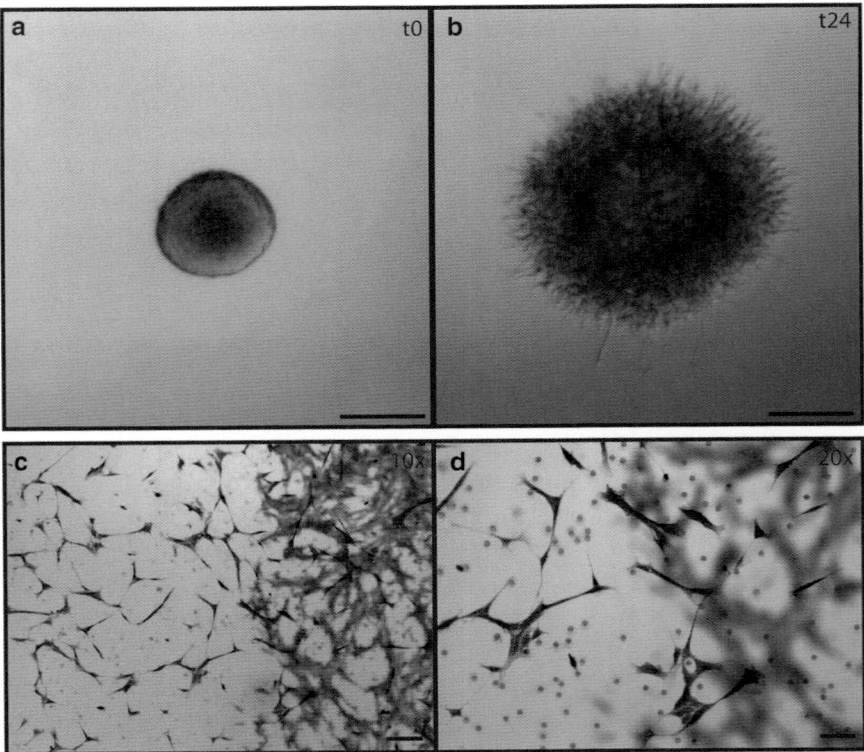

Fig. 2 Examples of assays used to measure invasion in vitro. NPC spheroids imaged immediately after plating in matrigel are compact spheres (*t*0) but 24 h later (*t*24) cells have invaded into the surrounding matrix (**a, b**). NPCs migrating across the matrigel membrane toward the chemoattractant-containing lower chamber are stained purple in (**c**) and (**d**) at both ×100 and ×200 magnification. The non-wiped area of the membrane is evident on the right-hand side and is not in the same focal plane as the cells that have migrated through the membrane. Scale bars are 50 µm (**a**) and (**b**), 100 µm (**c**), and 50 µm (**d**)

4. In a new 24 well place add 750 µL of media containing a chemoattractant of interest to each well (*see* **Note 8**).

5. Remove the media used to rehydrate the inserts, and transfer inserts into wells containing chemoattractant media. Be careful not to trap air bubbles beneath the membrane.

6. Immediately add 500 µL of cell suspension into the inserts.

7. Incubate the Invasion Chambers for 16 h at 37 °C 5 % CO_2 (*see* **Note 9**).

8. Carefully remove media from each insert and transfer inserts into wells containing 500 µL of 4 % PFA for 10 min on ice.

9. Transfer inserts into wells containing 500 µL of 0.1 % crystal violet diluted in methanol at room temperature for 10 min.

10. Wash inserts by immersing in three changes of milli-Q water, 2× dips in each.

11. Wick off excess water by touching edge of insert with Kimwipe.

12. Gently rinse interior of chamber with 100 μL of milli-Q water.

13. Wipe the inside of one-half of the membrane with one swipe of a cotton swab moistened with water. The area of the membrane left un-wiped is later used to determine the depth of focus (Fig. 2c, d).

14. Let inserts dry for at least 1 h.

15. Invert insert and remove membrane using a scalpel blade cutting along the membranes edge.

16. Place membrane upper chamber side down on a drop of immersion oil on a microscope slide.

17. Place another drop of immersion oil on top of the membrane and mount a coverslip on top.

18. Identify the cells that have invaded through the membrane. Take six photos at 200× magnification per insert (Fig. 2d). Count the number of invading cells and average the results across replicate wells.

4 Notes

1. Volume of RLT lysis buffer added is dependent on cell number. For example, one confluent 10 cm^2 plate of adherent U251 cells is lysed with 600 μL of RLT buffer. For murine neural progenitor cell (NPC) cultures, 2–4 confluent wells of a 6 well plate are pooled, centrifuged at $100 \times g$ for 4 min, and resuspended in 350 μL of RLT buffer.

2. Adjust cell lysate buffer volume according to the number of cells being processed. Approximately 500 μL per 10 cm dish of confluent cells and 150 μL per 6 well plate.

3. The use of fresh reducing agents is important for detecting the 75 kDa SULF2 band.

4. The 2B4 antibody recognizes both human and mouse SULF2, binding to an epitope in the C-terminus to resolve the C-terminal 50 kDa fragment and the full-length/unprocessed, SULF2 at 135–140 kDa [22]. In some human and murine cells, the antibody detects a nonspecific band at ~155 kDa, which is not altered by ablation of SULF2 expression.

5. The 3D spheroid assay can be used with adherent cells such as U251 or with non-adherent cells such as NPCs.

6. The optimum time points to assess invasion will vary between different cell types and their ability to invade.

7. The plating density needs to be optimized according to the cell types used. For murine NPC cultures, dilute cells to 2×10^5 cells/mL in growth factor-free/serum-free media and use 1×10^5 cells per insert.

8. Epidermal growth factor (EGF)/Fibroblast Growth factor (FGF) for NPC or serum-containing media for U251 cells.

9. The invasion chamber incubation time will vary across cells lines but must be minimized to reduce confounding cell proliferation.

10. We would like to acknowledge and thank Drs. Steven D. Rosen and Mark S. Singer for their helpful advice and for providing the SULF2 knockdown constructs.

References

1. Bernfield M, Gotte M, Park PW et al (1999) Functions of cell surface heparan sulfate proteoglycans. Annu Rev Biochem 68:729–777

2. Esko JD, Lindahl U (2001) Molecular diversity of heparan sulfate. J Clin Invest 108(2):169–173

3. Habuchi H, Habuchi O, Kimata K (2004) Sulfation pattern in glycosaminoglycan: does it have a code? Glycoconj J 21(1–2):47–52

4. Dhoot GK, Gustafsson MK, Ai X et al (2001) Regulation of Wnt signaling and embryo patterning by an extracellular sulfatase. Science 293(5535):1663–1666

5. Ai X, Do AT, Lozynska O et al (2003) QSulf1 remodels the 6-O-sulfation states of cell surface heparan sulfate proteoglycans to promote Wnt signaling. J Cell Biol 162(2):341–351

6. Freeman SD, Moore WM, Guiral EC et al (2008) Extracellular regulation of developmental cell signaling by XtSulf1. Dev Biol 320(2):436–445

7. Ai X, Kitazawa T, Do AT et al (2007) SULF1 and SULF2 regulate heparan sulfate-mediated GDNF signaling for esophageal innervation. Development 134(18):3327–3338

8. Fujita K, Takechi E, Sakamoto N et al (2010) HpSulf, a heparan sulfate 6-O-endosulfatase, is involved in the regulation of VEGF signaling during sea urchin development. Mech Dev 127(3–4):235–245

9. Lui NS, van Zante A, Rosen SD et al (2012) SULF2 expression by immunohistochemistry and overall survival in oesophageal cancer: a cohort study. BMJ Open 2(6):pii:e001624

10. Wade A, Robinson AE, Engler JR et al (2013) Proteoglycans and their roles in brain cancer. FEBS J 280(10):2399–2417

11. Phillips JJ, Huillard E, Robinson AE et al (2012) Heparan sulfate sulfatase SULF2 regulates PDGFRalpha signaling and growth in human and mouse malignant glioma. J Clin Invest 122(3):911–922

12. Dai Y, Yang Y, MacLeod V et al (2005) HSulf-1 and HSulf-2 are potent inhibitors of myeloma tumor growth in vivo. J Biol Chem 280(48):40066–40073

13. Rosen SD, Lemjabbar-Alaoui H (2010) Sulf-2: an extracellular modulator of cell signaling and a cancer target candidate. Expert Opin Ther Targets 14(9):935–949

14. Morimoto-Tomita M, Uchimura K, Bistrup A et al (2005) Sulf-2, a proangiogenic heparan sulfate endosulfatase, is upregulated in breast cancer. Neoplasia 7(11):1001–1010

15. Lemjabbar-Alaoui H, van Zante A, Singer MS et al (2010) Sulf-2, a heparan sulfate endosulfatase, promotes human lung carcinogenesis. Oncogene 29(5):635–646

16. Nawroth R, van Zante A, Cervantes S et al (2007) Extracellular sulfatases, elements of the Wnt signaling pathway, positively regulate growth and tumorigenicity of human pancreatic cancer cells. PLoS One 2(4):e392

17. Preusser M, de Ribaupierre S, Wohrer A et al (2011) Current concepts and management of glioblastoma. Ann Neurol 70(1):9–21

18. Calvo F, Sahai E (2011) Cell communication networks in cancer invasion. Curr Opin Cell Biol 23(5):621–629

19. Maltseva I, Chan M, Kalus I et al (2013) The SULFs, extracellular sulfatases for heparan sulfate, promote the migration of corneal epithelial cells during wound repair. PLoS One 8(8):e69642

20. Peterson SM, Iskenderian A, Cook L et al (2010) Human sulfatase 2 inhibits in vivo tumor growth of MDA-MB-231 human breast cancer xenografts. BMC Cancer 10:427

21. Johansson FK, Goransson H, Westermark B (2005) Expression analysis of genes involved in brain tumor progression driven by retroviral insertional mutagenesis in mice. Oncogene 24(24):3896–3905

22. Tang R, Rosen SD (2009) Functional consequences of the subdomain organization of the sulfs. J Biol Chem 284(32):21505–21514

Chapter 40

Synthesis and Biomedical Applications of Xylosides

Mausam Kalita, Maritza V. Quintero, Karthik Raman, Vy M. Tran, and Balagurunathan Kuberan

Abstract

Xylosides modulate the biosynthesis of sulfated glycosaminoglycans (GAGs) in various cell types. A new class of xylosides called "click-xylosides" has been synthesized for their biostability, ease of chemical synthesis, and tunable sulfated GAG biogenesis in vitro and in vivo. These click-xylosides have several therapeutic and biomedical applications in the regulation of angiogenesis, tumor inhibition, and regeneration. This protocol focuses on the synthesis of click-xylosides, their cellular priming activities, and biomedical applications.

Key words Xylosides, Proteoglycans, Glycosaminoglycans, Heparan sulfate, Chondroitin sulfate, Glycosaminoglycan priming, Anti-angiogenesis, Anti-invasion

1 Introduction

Proteoglycans are expressed on several cell types and participate in various physiological functions such as blood clotting, angiogenesis, neural growth and guidance, growth factor signaling etc. [1–4]. Glycosaminoglycans (GAGs) contribute more than 50 % to the total weight of proteoglycans [5, 6]. Sulfated GAGs are decorated with sulfate ($-SO_3$) groups on disaccharide repeat units of hexosamine and hexuronic acid. The anionic sulfated GAGs are comprised of three major types of glycans: heparin/heparan sulfate (HS), chondroitin sulfate (CS)/dermatan sulfate (DS), and keratan sulfate (KS) (Fig. 1). These classifications are based on (a) the type of hexosamine residues: glucosamine or galactosamine, (b) the percentage of uronic acid epimer content: glucuronic acid and iduronic acid, (c) the position of sulfate groups, and (d) the stereochemistry of glycosidic linkage: α- and β-.

The first step in the biosynthesis of GAGs is the attachment of an activated xylose sugar to the serine residues of the core protein [7, 8]. This xylosylation step takes place in the late endoplasmic reticulum and/or *cis* Golgi apparatus [9, 10]. Subsequent steps involve consecutive attachment of β1-4 galactose-, β1-3 galactose-,

Kuberan Balagurunathan et al. (eds.), *Glycosaminoglycans: Chemistry and Biology*, Methods in Molecular Biology, vol. 1229, DOI 10.1007/978-1-4939-1714-3_40, © Springer Science+Business Media New York 2015

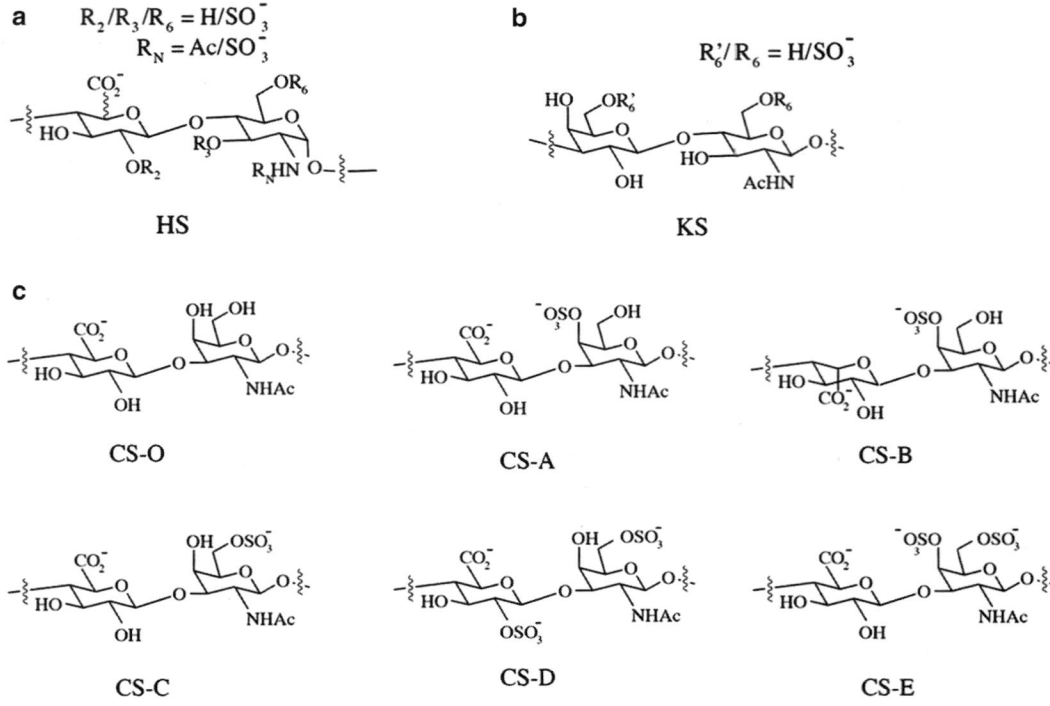

Fig. 1 Sulfated GAGs can contain sulfate groups at various positions of the repeated disaccharide units. (**a**) HS is composed of glucuronic acid/iduronic acid and *N*-sulfo/*N*-acetylglucosamine, (**b**) KS is composed of galactose and N-acetylglucosamine, (**c**) CS is composed of glucuronic acid/iduronic acid and *N*-acetylgalactosamine repeating units. CS has six structural variants—0, A, B, C, D, and E based on sulfation pattern

and β1-3 glucuronic acid residues to the xylose sugar yielding a tetrasaccharide linkage assembly (Fig. 2a). The fifth sugar added to the nonreducing end of the chain dictates the type of GAG chain to be made; addition of a β-1-4 *N*-acetylgalactosamine results in CS/DS whereas α-1-4 *N*-acetylglucosamine yields heparin/HS.

The xyloside primed GAG chains are often secreted outside the cell and compete with endogenous GAG chains found as proteoglycans on the cell surface and thereby alter the interaction between proteoglycans and other bio-macromolecules such as growth factors, cell adhesion molecules, chemokines, cytokines, etc. These altered interactions can have significant patho-physiological consequences such as elongated zebra fish phenotype, altered angiogenesis, antiproliferative effect etc. [11, 12]. Thus, secreted xyloside primed GAG chains, not attached to a core protein, can be used as molecular tools to investigate the role of proteoglycans in pathophysiological events. GAGs can be primed by xylosides in the Golgi by incubating cells with xyloside molecules in cell culture. A new class of click-xylosides has been designed for the biostability of triazole rings, ease of synthesis, and tunable aglycone

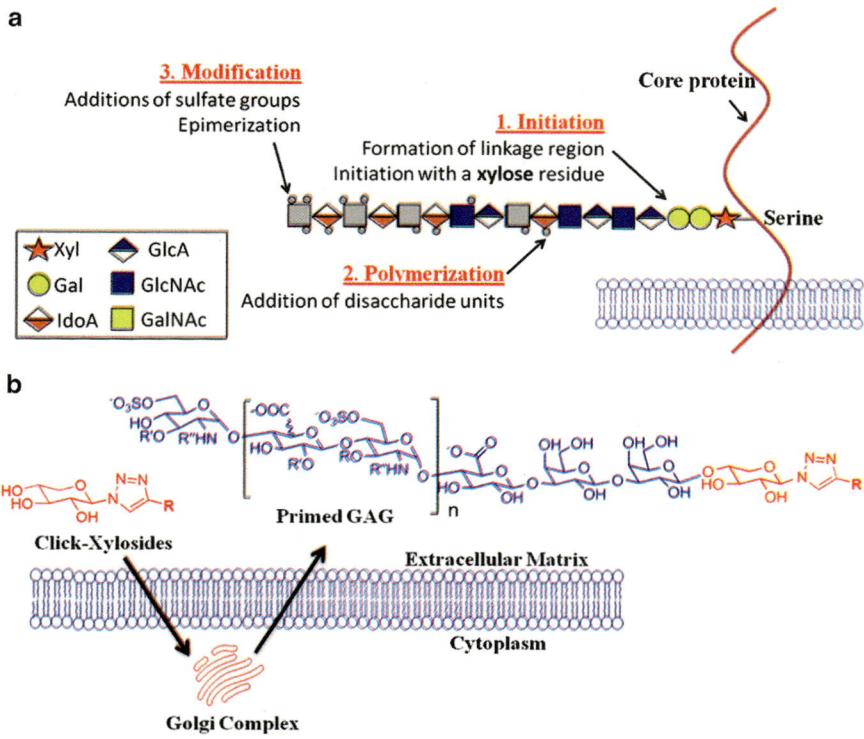

Fig. 2 (**a**) Proteoglycans (PGs) are transmembrane proteins containing one or more GAG chains. Biosynthesis of PGs starts with the attachment of an activated xylose sugar to the serine amino acid of the core protein followed by sequential transfer of three monosaccharides moieties. (**b**) Click-xylosides compete with endogenous xylosides for GAG biosynthesis pathway and get incorporated in the GAG chains, which exit through the cell membrane and modulate the ECM structures

structures [13, 14]. Aglycone groups impart hydrophobicity to click-xylosides for cellular uptake and can modulate the amount of HS and CS produced by modifying the endogenous biochemical pathways (Fig. 2b). There are several biomedical applications of the xylosides: controlling angiogenesis, modulating invasion during progression of cancer, and stimulating neuronal regeneration following CNS injury.

Inhibition of tumor-associated angiogenesis using small molecules has been one of the primary methods of controlling cancers [15]. Heparan sulfate proteoglycans (HSPGs), which are composed of sulfated GAG chains attached to a protein core via serine residues, are key players in modulating tumor-associated angiogenesis. HSPGs are involved in cell–cell signaling and act as co-receptors for a variety of pro- and anti-angiogenic factors such as FGF, VEGF, and endostatin [16]. Additionally, HSPGs are known to bind several chemokines and are involved in cell–cell and cell–extracellular matrix communication [17]. The effects of "click" xylosides on angiogenesis and tumor cell invasion can easily be assessed using in vitro assays before studying their efficacy in vivo.

2   Materials

Sodium acetate, acetic acid, acetic anhydride, xylosyl azide, sodium chloride, sodium methoxide, copper sulfate (II) pentahydrate, sodium ascorbate, DMSO (cell culture grade). Organic solvents: acetonitrile, acetone, ethyl acetate, hexanes, and anhydrous dichloromethane are available commercially.

2.1   Synthesis of Click-Xylosides

1. TLC plates 60 F_{254}.
2. TLC chamber.
3. Flash silica (230–400 mesh).
4. Chromatography column.
5. Milli-Q water.
6. Ceric Ammonium Molybdate TLC stain.
7. Rotary Evaporator.

2.2   TLC Staining Agent: Hanessian's Stain

1. Ceric Ammonium Molybdate stain (visualizes compounds containing hydroxyl groups).
2. 40 mL concentrated H_2SO_4; 10 g ammonium molybdate; 4 g ceric ammonium sulfate; 360 mL H_2O.

2.3   Priming of GAG Chains by Click-Xylosides

1. Chinese Hamster Ovary Cell line—CHO pgsA-745.
2. Radiolabeled sodium sulfate, ^{35}S-Na_2SO_4; 10 mCi/mL solution.
3. Hams F-12 Growth Medium: Media requires the addition of 50 mL of cell culture grade Fetal Bovine Serum and 5 mL of cell culture grade penicillin/streptomycin antibiotic per 500 mL of commercially purchased Hams F-12 media.
4. GAG Priming Media: Dialyze 100 mL of cell culture grade fetal bovine serum in dialysis membrane (1,500 MWCO). Add 50 mL of freshly dialyzed and filter-sterilized serum and 5 mL of cell culture grade penicillin/streptomycin to 500 mL of Hams-F12 media.
5. Sterile Phosphate Buffered Saline (PBS).
6. Xylosides stock solution: Solutions of click-xyloside to be tested should be prepared at 10 mM concentration by weighing the appropriate amount of click-xyloside based on its individual molecular weight. The solid click-xyloside should be dissolved in cell culture grade DMSO and/or water based on their solubility. These solutions can be filter sterilized to prevent contamination of cells used in the screen.
7. DEAE Wash Buffer: In a clean 1 L flask add 1.36 g KH_2PO_4, 1 g CHAPS, and add milli-Q water for a total volume of 1 L. Adjust to pH 6.0 (*see* **Note 1**).

8. DEAE Elution Buffer: Add 1.36 g KH_2PO_4, 1 g CHAPS, and 58.44 g NaCl into a clean 1 L flask. Add milli-Q water for a total volume of 1 L. Adjust the pH to 6.0 (*see* **Note 1**).

2.4 In Vitro Angiogenesis Tube Formation Assay

1. Reduced Growth Factor Matrigel.
2. BLMVEC (bovine lung microvascular endothelial cells).
3. Complete endothelial media.
4. Sulforaphane.

2.5 In Vitro Cell Invasion Assay

1. 6-well Matrigel coated invasion chambers with 8 μm pores.
2. U87 Mg Glioma cells.
3. Tweezers.
4. Razor blade.

3 Methods

3.1 Synthesis of Peracetylated Click-Xylosides

3.1.1 Copper Catalyzed Click Chemistry

1. Equip a 100 mL round bottom flask with a magnetic stir bar.
2. Charge peracetylated xylosyl azide (**1**, Scheme 1) (1 g, 3.31 mmol) to the round bottom flask.
3. Add 10 mL of 1:1 acetone:water mixture to dissolve protected xylosyl azide.
4. Add the alkyne-functionalized molecule (3.972 mmol, 1.2 equivalent) to the above reaction mixture.
5. Transfer 1 M sodium ascorbate (0.7 μL) followed by 1 M $CuSO_4 \cdot 5H_2O$ (1.3 μL) to the above reaction (*see* **Note 2**).
6. Allow the heterogeneous reaction mixture to stir vigorously at room temperature for 16 h.
7. Run the reaction mixture in TLC using hexane (H):ethyl acetate (E) mobile phase (*see* **Note 3**).
8. Visualize the click product on the TLC plate under UV light (short wave, 254 nm).

3.1.2 Purification of Click-Xylosides

1. Evaporate off the solvents of the reaction mixture by using a rotary evaporation.
2. Redissolve the crude reaction mixture in dichloromethane (10 mL).

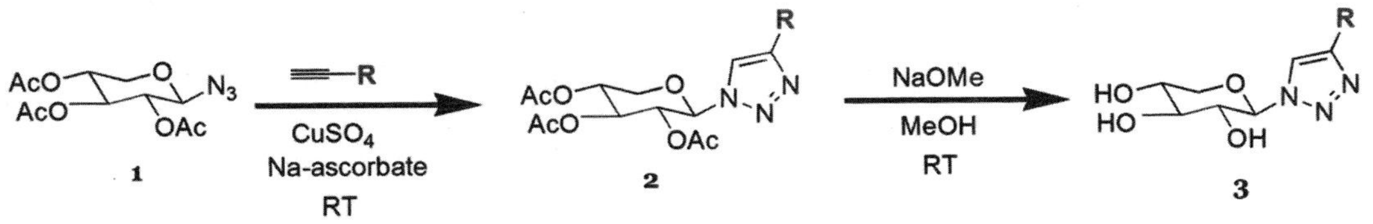

Scheme 1 Synthesis of xylosides using click chemistry

3. Pack a column with silica gel (75 g) immersed in hexane (500 mL).

4. Load the crude reaction mixture dissolved in dichloromethane on the top of the silica gel column.

5. Purify the desired compound by running a solvent gradient of H and E (*see* **Note 4**) and collecting the eluted compound in 30 mL fractions.

6. Analyze TLC of each eluted fraction and combine the fractions containing the desired click-xyloside.

7. Remove the solvents by rotary evaporation and dry the resulting compound (**2**, Scheme 1) under vacuum.

3.2 Deprotection of Peracetylated Click-Xylosides

1. Transfer protected click-xyloside (**2**, Scheme 1) (0.5 g) to a 100 mL round bottom flask equipped with magnetic stir bar.

2. Dissolve this compound in 10 mL dry methanol.

3. Add approximately 10 mL of freshly prepared sodium methoxide solution (*see* **Note 5**) to the above reaction mixture until the solution reaches pH 9.

4. Stir the above reaction overnight.

5. Spot the reaction mixture in a silica gel TLC plate using H:E (1:2) as TLC solvent and then observe the polar product under UV light (short wave, 254 nm).

6. Neutralize the pH of the reaction mixture by adding Amberlite IR 120 hydrogen form (~500 mg) (*see* **Note 6**).

7. Filter the reaction immediately using a glass frit filtration unit.

8. Concentrate the solution by rotary evaporation and bring the volume to ~3 mL.

9. Load the concentrated compound on a silica gel (50 g) column and separate the deprotected click-xyloside by using H:E solvent mixture (*see* **Note 4**).

10. Analyze TLC of all the eluted fractions and combine similar fractions.

11. Remove the solvent by rotary evaporation and dry the resulting compound (**3**, Scheme 1) under vacuum.

3.3 Priming of GAG Chains by Click-Xylosides (Schematically Shown in Fig. 3)

1. Plate approximately 1×10^5 cells per well in a 6-well cell culture plate.

2. Add 1.0 mL of prepared Hams F-12 Growth Media per well. Incubate the plate for 24 h at 37 °C with 5 % CO_2 to reach a confluence of about 50 %.

3. Remove the media and wash the cells with sterile PBS three times and replace it with 995 μL of GAG priming media.

4. Add 5 μL of click-xyloside stock solution for a final xyloside concentration of 100 μM and 100 μCi of ^{35}S-Na_2SO_4.

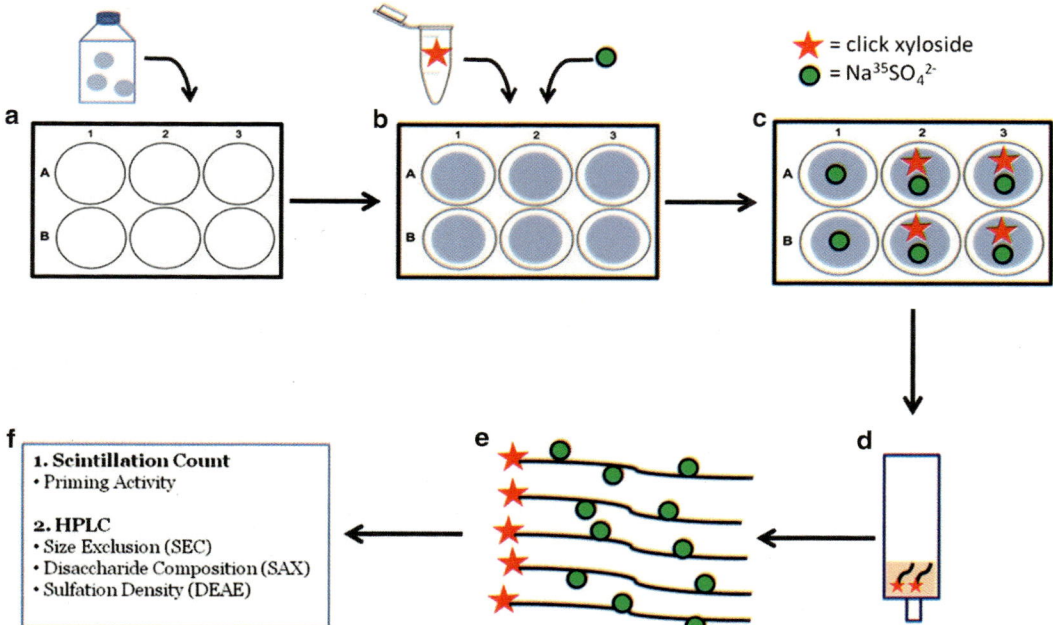

Fig. 3 (**a**) pgsA-745 CHO cells are seeded into a 6-well plate (1×10^5/well) with growth media. Cells are grown to ~50 % confluence, (**b**) Used growth media is removed and each well is washed with PBS. GAG priming media, $Na_2{}^{35}SO_4$, and click-xyloside stock solution are added to cells and incubated for 24 h, (**c**) Click-xylosides prime glycosaminoglycan biosynthesis and $^{35}SO_4{}^{2-}$ is incorporated into the chain, (**d**) The contents of each well is diluted and loaded on DEAE resin for GAG isolation, (**e**) Elution buffer removes bound GAG from the column, (**f**) Isolated GAG can be analyzed by scintillation count to confirm and compare $^{35}SO_4{}^{2-}$ incorporation. Further GAG analysis can be performed using HPLC

The radioactive sulfate will label the GAG chains that are produced by the cells. The experiment should be performed in triplicates to quantitatively assess the priming activity.

5. Use 5 μL of PBS instead of xyloside for the negative control.

6. Incubate the plate for 24 h at 37 °C with 5 % CO_2.

3.4 Purification of Click-Xyloside Primed GAGs in CHO Cells

1. Transfer the entire content of each well to a 2 mL microcentrifuge tube and add 0.8 mL of milli-Q water.

2. Centrifuge the above condition media at $10,000 \times g$ for 10 min to sediment insoluble cell extracts leaving the supernatant for purification.

3. Prepare a DEAE-sepharose column by adding 0.6 mL sepharose resin to the column (*see* **Note 7**).

4. Allow the resin to settle but ensure it does not dry out.

5. Wash the gel with 2 mL milli-Q water to remove ethanol.

6. Equilibrate the gel with 10 column volumes (3 mL) of wash buffer.

7. Load the entire diluted supernatant from **step 2** onto the DEAE-sepharose column.

8. Wash the column with 30 column volumes (9 mL) of wash buffer.

9. Elute the bound GAG chains using 6 column volumes (1.8 mL) into a clean collection tube (*see* **Note 8**).

3.5 Quantification of Click-Xyloside Primed GAGs

1. Place 5 mL of scintillation fluid into each scintillation vial.

2. Add 10 μL of purified GAGs to the vial and shake to mix well. Place the vial in the scintillation counter and compare the scintillation counts between the control and the treated samples. Measurements need to be performed in triplicates.

3. To remove salt from the elution, filter through a 3,000 MWCO spin filtration unit at $10,000 \times g$ for 10 min each spin, leaving at least 100 μL of solution above the filter. Discard flow-through in appropriate radioactive waste containers. Repeat this process until the entire sample has been loaded onto the filter unit. Desalt the sample by adding milli-Q water (350 μL) and spinning down using the filtration unit. Repeat this step at least five more times.

4. Flip the filter unit over a new microcentrifuge tube and spin again to transfer entire liquid from the filter unit to the microcentrifuge tube.

5. Dilute the sample by adding milli-Q water to make up a total final volume of 500 μL.

6. Inject an appropriate volume corresponding to one million counts into an analytical HPLC equipped with radio-detector for GAG structural analysis using anion-exchange or size exclusion chromatography columns.

3.6 Angiogenesis Assay

1. Culture BLMVEC cells to confluence in endothelial media.

2. 1 day before the angiogenesis experiment, trypsinize and split cells into two containers (50 % confluence) to bring them into the log stage of growth.

3. Thaw 50 μL per well of Matrigel overnight at 4 °C.

4. Add 50 μL per well of matrigel in a 96-well plate. Triplicate experimental wells are recommended.

5. Leave the matrigel, undisturbed, in a 37 °C incubator for 2 h to allow it to solidify (*see* **Note 9**).

6. Dilute Sulforaphane to a 100 μM solution in endothelial medium.

7. Add 200 μL of endothelial media per well containing matrigel.

8. Trypsinize log-phase BLMVEC and add 8×10^4 cells into each well (remove appropriate amounts of media to maintain the total volume at 200 μL per well) (*see* **Note 10**).

9. Add 10 μM of sulforaphane or any other test xylosides in each well (*see* **Note 11**).

10. Leave cells undisturbed for 12 h at 37 °C in a humidified incubator.

11. After this incubation time, prepare a solution of 2 μM Calcein AM in PBS.

12. Use a paper towel or a pipette to slowly siphon off media from the side of each well. Be careful not to disturb the tubes (*see* **Note 12**).

13. Gently add PBS to wash cells once and remove PBS using a paper towel or pipette.

14. Add 100 μl of Calcein AM solution per well and incubate cells for 30–45 min at 37 °C.

15. Image cells in the green channel using a fluorescence microscope.

3.7 Invasion Assay

1. Remove matrigel coated chambers from packaging and add 2 mL of DMEM per well above (in the chamber) and below (in the well). Duplicate or triplicate experiments are recommended (*see* **Note 13**).

2. Leave chambers in a 37 °C incubator for 2 h.

3. After the incubation period, remove the media and wash once with PBS. Be careful not to disturb the matrigel in the chamber.

4. Add 2 mL of HAMS F-12 media in each chamber and well.

5. Trypsinize U87 Mg Glioma cells and add ~5×10^4 cells into each chamber. (Minimize the amount of trypsin added. Remove appropriate amount of media when cells are added.) (*see* **Note 14**)

6. Add any experimental compounds to appropriate chambers and wells while maintaining the volume at 2 mL total in each, the chamber and well. Compounds are added to both the chamber and well (*see* **Note 15**).

7. After 3 days, invasive cells will migrate to the bottom of each chamber.

8. Use cotton swabs to gently disturb the matrigel in the top of each chamber (do not disturb the cells that are on the bottom of each chamber).

9. Wash the top of the chamber with 2 mL of PBS three times to ensure that all matrigel and cells are removed from the top.

10. Remove media from each well, wash once with PBS, and add 2 mL PBS into each.

11. Invert the chambers and cut out the PET membrane using a razor but slicing along the walls of the membrane in a circle.

Use tweezers to put the PET membrane into the PBS in each well, with the invaded cells facing up.

12. Remove the PBS and add 1 mL of trypsin into each well.

13. Count cells in a hemocytometer after trypsinization (*see* **Note 16**).

---

4   Notes

1. CHAPS may damage the pH meter. Be sure to wash the probe thoroughly with milli-Q water before and after it touches CHAPS.

2. 1 M Sodium ascorbate solution (yellow color) and 1 M $CuSO_4 \cdot 5H_2O$ solution (blue color) must be prepared in milli-Q water immediately before for each reaction.

3. The TLC solvents are prepared by mixing hexane (H) with ethyl acetate (E) at different ratios

4. Column chromatography purification is started with 100 % hexane and then with gradual increase in ethyl acetate amounts.

5. Sodium methoxide solution was prepared by slowly dissolving 2.3 g of sodium metal in 100 mL of dry methanol.

6. Amberlite IR-120 hydrogen form should be added into the stirring reaction mixture in 50 mg every 5 min. pH of the solution should be checked after every addition. Be extra careful as pH drops below 8 since the pH will change faster after this point.

7. A syringe plugged with wet cotton or a purchased column with frit can be used. Plug the column and add water. Tap the column gently to ensure all air bubbles have been removed from frit or cotton plug.

8. For best results, elute the primed GAGs with four aliquots of 0.45 mL elution buffer.

9. Tubes generally form in approximately 12 h. If cells are disturbed during tube formation, the tubes may break or not form at all.

10. Adding more than 80,000 cells results in very large tubes, which are hard to discern and quantify. Ideally cell count should be between 60,000 and 80,000 cells.

11. For added compounds, minimize the amount of DMSO added to the media. At more than 3 % DMSO, tube formation is hindered.

12. Cells can also be fixed after the 12 h incubation. Soon after tube formation, cells begin to undergo apoptosis. By 18 h, most tubes are broken and cells will begin to appear as clumps.

13. The invasion assay uses a well within a well. In this protocol the top well is referred to as the chamber. It consists of a PET membrane with a matrigel coating and 8 μm pores. The bottom well is referred to as a well.

14. U87 Mg cells seem to invade more effectively in HAMS F-12 media compared to DMEM.

15. Compounds are added to both the chamber and the well in order to minimize diffusion of the compound from one into the other. If compounds are not added into both the well and the chamber, the effective concentration is half of what is added.

16. Some cells, such as SF 295, are not fit for this assay due to their low invasive capability in vitro.

Acknowledgements

This work was supported in part by NIH grants (P01HL107152 and R01GM075168) to B.K. and by the NIH fellowship F31CA168198 to K.R.

References

1. Sasisekharan R, Shriver Z, Venkataraman G, Narayanasami U (2002) Roles of heparan sulphate glycosaminoglycans in cancer. Nat Rev Cancer 2:521–528

2. Swarup V, Mencio CP, Hlady V, Kuberan B (2013) Proteoglycans in the nervous system. Sugar Glues for Broken Neurons. Biomol Concepts 4: 233–257

3. Capila I, Linhardt RJ (2002) Heparin-protein interactions. Angew Chem Int Ed Engl 41:391–412

4. Powell AK, Yates EA, Fernig DG, Turnbull JE (2004) Interactions of heparin/haparan sulfate with protein: appraisal of the structural factors and experimental approaches. Glycobiology 14:17R–30R

5. Salmivirta M, Lidholt K, Lindahl U (1996) Heparan sulfate: a piece of information. FASEB J 10:1270–1279

6. Esko JD, Selleck SB (2002) Order out of chaos: Assembly of ligand binding sites in heparan sulfate. Annu Rev Biochem 71:435–471

7. Lindahl U, Cifonelli JA, Lindahl B, Roden L (1965) The role of serine in the linkage of heparin to protein. J Biol Chem 240:2817–2820

8. Bourdon MA, Krusius T, Campbell S, Schwartz NB, Ruoslahti E (1987) Identification and synthesis of a recognition signal for the attachment of glycosaminoglycans to proteins. Proc Natl Acad Sci U S A 84:3194–3198

9. Vertel BM, Walters LM, Flay N, Kearns AE, Schwartz NB (1993) Xylosylation is an endoplasmic reticulum to Golgi event. J Biol Chem 268:11105–11112

10. Lohmander LS, Shinomura T, Hascall VC, Kimura JH (1989) Xylosyl transfer to the core protein precursor of the rat chondrosarcoma proteoglycan. J Biol Chem 264:18775–18780

11. Nguyen TKN, Tran VM, Sorna V, Eriksson I, Kojima A, Koketsu M, Loganathan D, Kjellén L, Dorsky RI, Chien C-B, Kuberan B (2013) Dimerized glycosaminoglycan chains increase FGF signaling during Zebrafish development. ACS Chem Biol 8(5):939–948

12. Nilsson U, Jacobsson M, Johnsson R, Mani K, Ellervik U (2009) Antiproliferative effects of peracetylated naphthoxylosides. Bioorg Med Chem Lett 19:1763–1766

13. Victor XV, Nguyen TKN, Ethirajan M, Tran VM, Nguyen KV, Kuberan B (2009) Investigating the elusive mechanism of glycosaminoglycan biosynthesis. J Biol Chem 284(38):25842–25853

14. Tran VM, Nguyen TKN, Sorna V, Loganathan D, Kuberan B (2013) Synthesis and assessment of glycosaminoglycan priming activity of

cluster-xylosides for potential use as proteoglycan mimetic. ACS Chem Biol 8(5):949–957

15. Folkman J (1971) Tumor angiogenesis: therapeutic implications. N Engl J Med 285(21): 1182–1186

16. Iozzo RV, Zoeller JJ, Nystrom A (2009) Basement membrane proteoglycans: modula-tors par excellence of cancer growth and angiogenesis. Mol Cells 27(5):503–513

17. Casu B, Naggi A, Torri G (2010) Heparin-derived heparan sulfate mimetics to modulate heparan sulfate-protein interaction in inflammation and cancer. Matrix Biol 29(6): 442–452

Chapter 41

A Strategic Approach to Identification of Selective Inhibitors of Cancer Stem Cells

Nirmita Patel, Somesh Baranwal, and Bhaumik B. Patel

Abstract

Cancer stem-like cells (CSC) have been implicated in resistance to conventional chemotherapy as well as invasion and metastasis resulting in tumor relapse in majority of epithelial cancers including colorectal cancer. Hence, targeting CSC by small molecules is likely to improve therapeutic outcomes. Glycosaminoglycans (GAGs) are long linear polysaccharide molecules with varying degrees of sulfation that allows specific GAG-protein interaction which plays a key role in regulating cancer hallmarks such as cellular growth, angiogenesis, and immune modulation. However, identifying selective CSC-targeting GAG mimetic has been marred by difficulties associated with isolating and enriching CSC in vitro. Herein, we discuss two distinct methods, spheroid growth and EMT-transformed cells, to enrich CSC and set up medium- and high-throughput screen to identify selective CSC-targeting agents.

Key words Cancer stem cells, MTT, Colonosphere assay, EMT, Selective anti-CSC agents

1   Introduction

It is estimated that cancer will claim over half a million lives in the USA in the year 2013 alone [1]. Majority of cancer-related fatalities are attributed to metastasis as well as primary and acquired resistance to cytotoxic therapies resulting in disease recurrence [2]. These two deadly attributes of cancer can be reconciled under a unifying cancer stem cell (CSC) hypothesis which posits a paradigm-shifting direction for discovery of novel anticancer therapies [3, 4]. These so-called selective anti-CSC agents would target a small population of tumor cells, which possess the ability to self-renew and reconstitute the entire tumor; invade and metastasize; as well as resist killing by conventional cytotoxic agents [3, 4]. Understandably, it is a tall order to achieve for any of the known anticancer agents currently in clinical use or on the horizon. Yet, a specific group of natural polymers hold major promise in simultaneously modulating and regulating several key growth factors and morphogens that mediate CSC phenotype [5–11]. This group is

Kuberan Balagurunathan et al. (eds.), *Glycosaminoglycans: Chemistry and Biology*, Methods in Molecular Biology, vol. 1229, DOI 10.1007/978-1-4939-1714-3_41, © Springer Science+Business Media New York 2015

the family of glycosaminoglycans (GAGs), which are long polysaccharides, composed of repeating disaccharide units of alternating uronic acids and glycosamines in either the free form in extracellular matrix (ECM) or in the form of proteoglycans that are cell membrane bound. Only in recent time select group of GAG mimetics have been reported to possess anticancer properties [12, 13]. Although no single GAG mimetic has been shown to selectively target CSC, it will not be long before one is discovered.

CSC hypothesis has been around for over a decade; however only a handful of synthetic or natural agents that fit the bill of selective anti-CSC drug have been identified to date [14, 15]. Although attractive, targeting CSCs is challenging. CSCs comprise only small percent of tumor cell population, which implies that high-throughput screening (HTS) using bulk cancer cells is likely not to identify CSC-specific agents [16]. Most popular method of growing cancer cells in adherent fashion on a plastic as a monolayer does not allow enrichment of CSCs as it promotes differentiation of these cells and hence CSCs cannot be stably grown in this fashion [17]. In this chapter we describe two strategies that are used by our laboratory to enrich CSCs and apply medium- or high-throughput system to screen a GAG mimetic library to identify selective anti-CSC agents.

The first method of enrichment of CSC has been in vogue to grow cancer as well as a few adult stem cells in ex vivo cultures [18]. When certain cancer cells are cultured in serum-free media supplanted with B27 supplements and growth factors (CSC media), they grow as spheroids which are significantly enriched in CSCs and early progenitor cells in undifferentiated state compared to same cells grown as monolayer in serum-containing media [17] (Fig. 1). As regards these spheroids, when disintegrated into single-cell suspension and passaged in CSC media, only self-renewing stem cells but not progenitor cells make sizable spheroids in subsequent cultures (secondary/tertiary spheroids) [17, 19, 20]. Herein, we describe a strategy based on the above concept to identify selective anti-CSC agents. In this tandem screening strategy we initially identify compounds that inhibit growth only in spheroid but not in monolayer growth condition which are likely to be selective against stem/progenitor cells (Fig. 2a). In the subsequent screen, we identify compounds that show growth inhibition not only in primary but also several subsequent passages in spheroid condition without further treatment, suggesting inhibition of self-renewal, a property that only self-renewing stem cells but not progenitor cells possess (Fig. 2b). This tandem screening method is attractive due to its ease of use, familiarity in CSC research circles, suitability for moderate-throughput screening, and relatively modest cost. However, on the downside it is laborious and not suitable for very-high-throughput screening of GAG mimetics.

Epithelial-to-mesenchymal transition (EMT) in epithelial cancer is a common phenomenon and is associated with invasion,

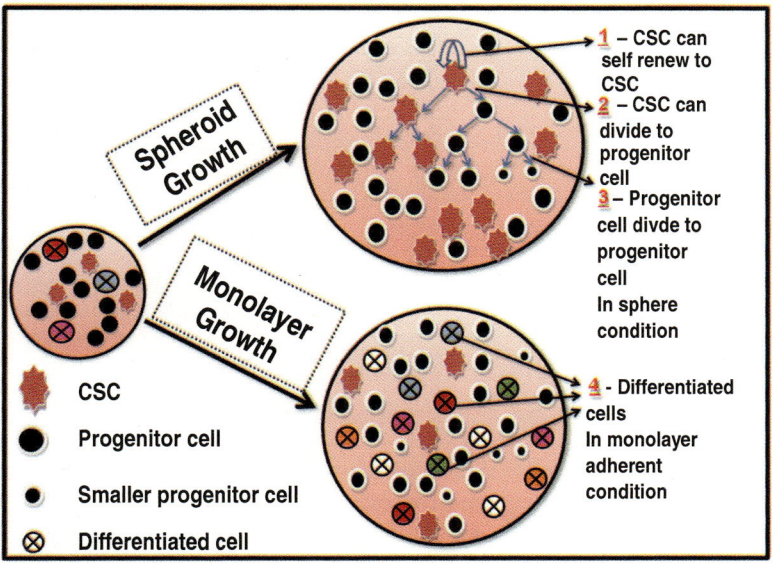

Fig. 1 Pictorial representation of cancer cell growth and its fate in monolayer (adherent to plastic) vs. spheroid condition (non-adherent and serum free)

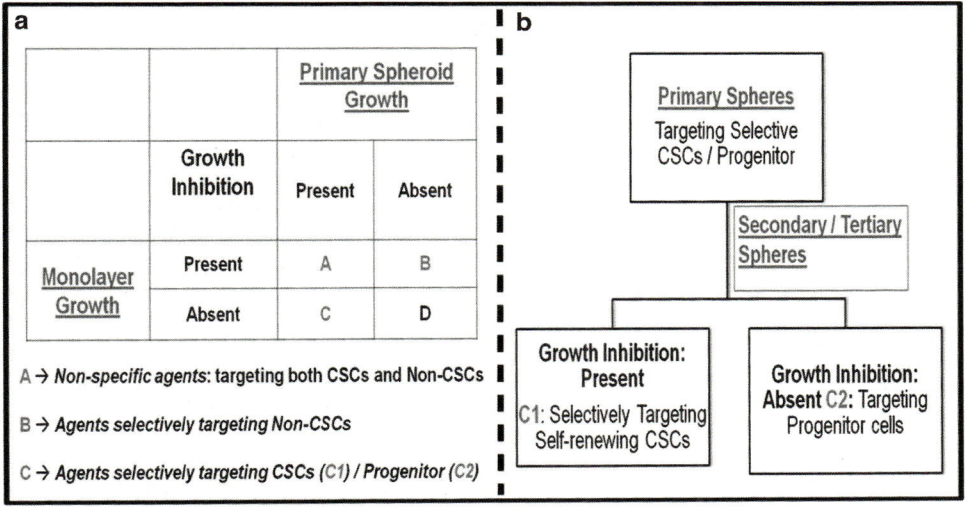

Fig. 2 Schematic representation of screening strategy # 1—tandem approach involving differential cell growth inhibition in monolayer vs. primary spheroid (CSCs/progenitor enriched) condition (**a**) which can be further refined by differential growth in primary vs. subsequent spheroid cultures to identify selective CSC-targeting agents in a medium-throughput fashion (**b**)

metastasis, and chemotherapy resistance akin to CSC phenotype [21–23]. This observation was exploited by Gupta et al. to generate stable stem cell line by inducting EMT in telomerase-immortalized human mammary epithelial (HMLE) by knockdown of E-cadherin [14]. Indeed, they observed a significant upregulation of stem cell markers, gain of capacity to form mammosphere, and increased resistance to established chemotherapeutics, suggesting acquisition of stable CSC phenotype [14]. Using this method, they identified salinomycin as a selective anti-CSC agent that showed >20-fold higher growth inhibition towards E-cadherin shRNA-transfected HMLE cells compared to their scramble-transfected counterparts. This particular method is less labor intensive and suitable for high-throughput screening [14]. However, this particular strategy is somewhat complex to be widely utilized in several small-scale laboratories and even though E-cadherin knockdown HMLE cells possess stem-like properties, they are essentially normal epithelial cells lacking complex cancer genetic changes that drive CSC growth. We propose a modified version of this technique wherein HT-29 (as well as Caco-2) colon cancer cell line(s) that shows mostly epithelial phenotype can be induced to express mesenchymal phenotype by stable knockdown of E-cadherin and resultant enrichment of CSC (Fig. 3). The appealing aspect to this approach is that both scramble-transfected and E-cadherin shRNA-transfected cells possess capacity to form spheroids albeit to a different degree, allowing valid comparison between low vs. high CSC-expressing cancer cells. This method is suitable for both moderate- and high-throughput screening of GAG mimetic library to identify selective anti-CSC agents.

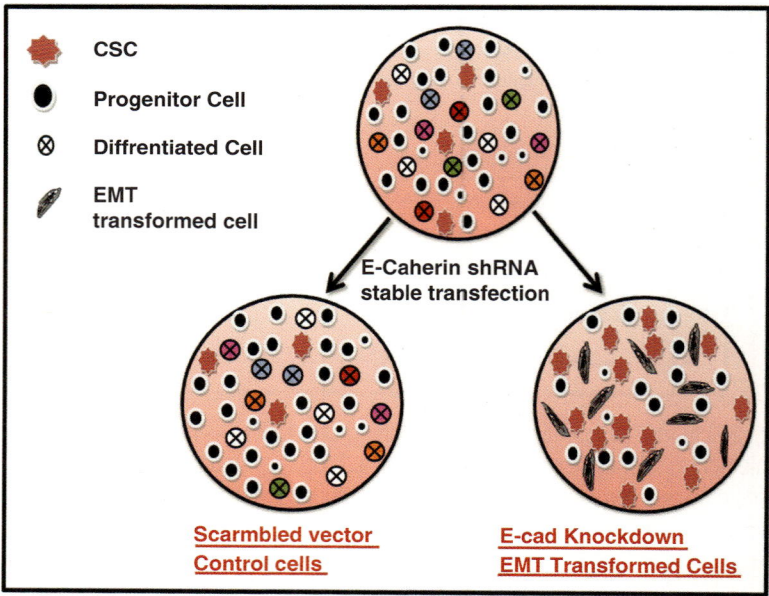

Fig. 3 Pictorial representation of screening strategy # 2—differential cell growth inhibition (MTT assay or equivalent) in scrambled control vs. E-cadherin knockdown (EMT/CSC enriched) cells in monolayer condition can identify selective CSC-targeting agents in a high-throughput fashion

2 Materials

Prepare all solutions using ultrapure cell culture-grade water and store solutions as indicated below (*see* **Note 1**).

2.1 Reagents for Culture Media

1. 1× DMEM-F12 (1:1).
2. 1× DMEM.
3. Phosphate-buffered saline (PBS).
4. Fetal calf serum (FBS).
5. Penicillin/streptomycin (AA).
6. Trypsin/EDTA.
7. B27 supplements.
8. EGF, human recombinant.
9. FGF2, human recombinant.

2.2 Reagents for Generation for Concentrated E-Cadherin shRNA-Containing Lentiviral Particles and Viral Transduction

1. Plasmid 12259: pMD2.G—Second-generation envelope plasmid.
2. Plasmid 12260: psPAX2—Second-generation packaging plasmid.
3. Plasmid 18801: pLKO.1 puro shRNA E-cadherin.
4. Plasmid 1864: pLKO.1 puro shRNA scramble.
5. Producer cell line—293T human embryonic kidney cells (ATCC, cat. no. CRL-11268).
6. HD transfection reagent—FuGENE®.

2.3 Media and Solutions

1. Cell culture media: Mix 50 ml of heat-inactivated FBS and 5 ml of 100× penicillin/streptomycin (AA) to 450 ml of DMEM/F12 and store at 4 °C.
2. Colonosphere formation media/stem cell media (SCM): In 50 ml of DMEM/F12, add 1 ml of 50× B27, 5.1 μl of EGF, and 51 μl of FGF (final concentration of 20 ng/ml EGF and 10 ng/ml of FGF).
3. Media and solution for viral transduction—OptiMEM® media.

2.4 Other Reagents

1. MTT 3-(4,5-dimethylthiazol-2-yl)-2,5-diphenyltetrazolium bromide solution.
2. HCl in isopropanol.
3. Trypan blue.
4. Puromycin.
5. Hexadimethrine bromide dissolved in 8 mg/ml water (filter sterilize using 0.22 μm filter).
6. 500 ml filter.

3 Methods

3.1 Cell Proliferation Assay/MTT Assay (See Notes 2 and 3)

The MTT assay allows quantitative measurement of cell viability and, as such, the proliferation rate of cells upon drug administration. MTT (3(4, 5-dimethylthiazol-2-yl)-2, 5-diphenyltetrazolium bromide) is processed by metabolically active cells resulting in the generation of a purple formazan. Release of insoluble formazan by viable cells is achieved through cell lysis using solubilization agent as dimethyl sulfoxide (DMSO). The absorbance of the resulting solution is measured at 550 nm wavelength by a spectrophotometer, and it corresponds to cell viability/proliferation (Fig. 4a).

1. Plate 2.5×10^3 cells in 100 µl of growth media/well of 96-well tissue culture-treated plate.

2. Incubate overnight at 37 °C in 5 % CO_2.

3. Add vehicle/treatment of interest at desired concentration and again incubate for 60–72 h (*see* **Note 4**).

4. At the end of incubation, add 10 µl of 5 mg/ml MTT solution (*see* **Note 5**) suspended in PBS to each well and incubate at 37 °C for a minimum of 1.5–3 h until the crystals form.

5. After 2 h, add 150 µl of 4 mM HCl in isopropanol solution to each well and triturate the mixture until the crystal dissolves completely (*see* **Note 6**).

6. Read the plate on the whole-plate spectrophotometer reader at 590 nm wavelength.

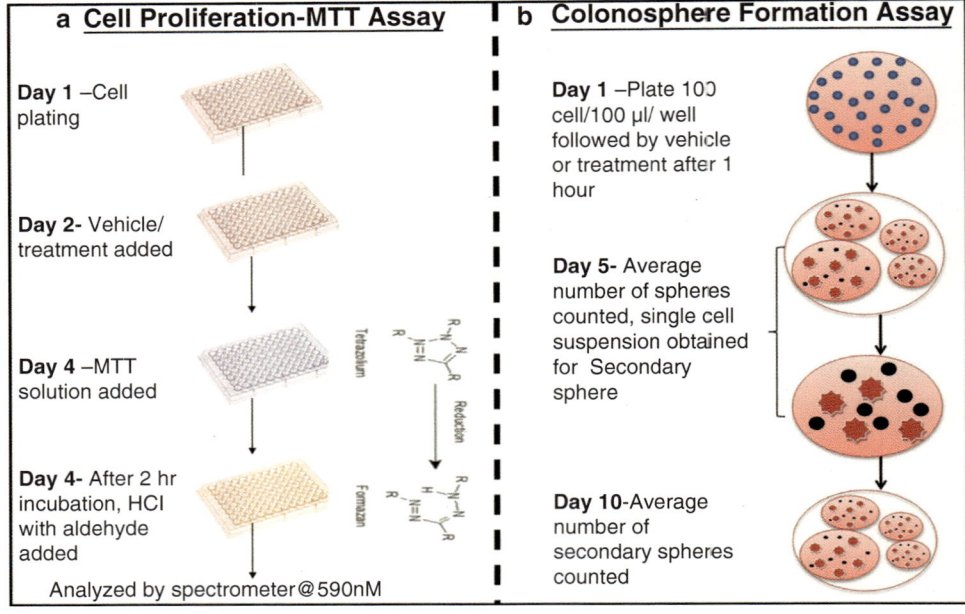

Fig. 4 Pictorial representation of assays used to determine cell growth in monolayer condition (MTT assay) (**a**) as well as spheroid condition (sphere formation assay) (**b**)

3.2 Colonosphere Formation Assay

The standard colonosphere culture selectively enhances the growth of cancer stem cells which possess a capacity to self-renew like normal stem cell. These cells when grown in serum-free condition in the presence of additives such as B27 supplements and growth factors (especially EGF and FGFs) in low-adhesion plate form a floating cluster of cells that are sphere shaped known as colonosphere. It has been demonstrated that each colonosphere is derived from a single colonosphere-forming cell (i.e., CSC). Hence, it is possible to determine the number of these CSCs in a culture by limiting dilution assay. Additionally, this forms the basis for propagating single cells derived from primary colonosphere cells into subsequent CSC cultures, i.e., secondary/tertiary spheres to allow true self-renewing CSC to enrich. This concept as illustrated in Fig. 4b is used to identify the compounds that selectively target CSC.

3.2.1 Primary Sphere Formation

1. On day 1, collect healthy cells with less than ten passages which are grown to about 70–80 % confluence by detaching them with 1 ml of 0.05 % trypsin/EDTA. Incubate at 37 °C for 2 min and check for detached cells. Add 6 ml of serum-containing media to neutralize trypsin effect. Collect cells in a 15 ml Falcon sterile conical centrifuge tube. After centrifugation at 1,000 rpm for 1.25 min, resuspend directly in SCM media and count in hemocytometer with 1:1 ratio of trypan blue. 100 cells/100 µl/well are plated in serum-free SCM media on low-adhesion 96-well plate.

2. After 6 h of incubation, add vehicle/treatment of interest at desired concentrations.

3. On day 5, numbers of spheres ranging from 50 to 150 µm in diameter are counted using phase-contrast microscope. Data is presented as percent of control.

3.2.2 Secondary Sphere Formation

1. On day 5 of primary sphere formation, centrifuge the 96-well plate at speed of 1,000 rpm for 1 min and remove the supernatant. Spheres are settled at the base of the plate. Add 20 µl of trypsin/well and triturate to dissociate them into single-cell suspension (*see* **Notes** 7 and **8**).

2. Count the number of cell with 1:5 ratio of cell:trypan blue and plate 100 cells/100 µl/well in SCM media in a low-adhesion plate.

3. No further treatment will be added.

4. Count an average number of spheres on day 5 as described above.

3.2.3 Tertiary Sphere Formation

1. For tertiary sphere formation from secondary spheres, follow the same method as secondary sphere formation.

2. Count the number of cell with trypan blue after single cell and plate 100 cells/100 µl/well in SCM media on low-adhesion plate.

3. No further treatment will be added.

4. Count an average number of spheres on day 5 as described above.

3.2.4 Applying Medium-Throughput Screening Using Monolayer vs. Spheroid Growth Conditions to Identify Selective Anti-CSC Agents

In order to identify compounds that selectively target CSCs, they are initially screened in monolayer vs. spheroid growth conditions. Growth inhibition in monolayer condition is identified by MTT assay as less than 50 % viable cells. Compounds that have growth inhibition in monolayer condition are screened in sphere formation assay, where growth inhibition is identified as less than 50 % sphere-forming cells compared to control. Compounds that have growth inhibition in monolayer condition but have no significant inhibition in sphere formation assay are not selective for CSCs (Fig. 2a; category B). On the other hand, if compounds have growth inhibition in both monolayer and sphere formation assay, they are also not selective for CSCs (Fig. 2a; category A). However, compounds that did not have significant growth inhibition in monolayer but have growth inhibition in sphere formation assay may be selective for CSCs (Fig. 2a; category C). Growth inhibition in sphere formation assay can be due to compound's effect on true CSCs (Fig. 2b; C1) or progenitor cells (Fig. 2b; C2). These compounds that have selective growth inhibition in primary sphere formation assay can be analyzed further by secondary sphere formation assay. If growth inhibition is observed in secondary sphere, those compounds are targeting self-renewing CSCs (Fig. 2b; C1). If growth inhibition is absent in secondary sphere formation assay, those compounds are likely targeting progenitor cells (Fig. 2b; C2).

3.3 Generation and Transduction of Lentivirus Particle for E-Cadherin Knockdown

This protocol is suitable for making lentivirus from five 10 cm tissue culture dishes which is sufficient for transducing more than ten cell lines (*see* **Note 9**).

3.3.1 Protocol for Generating Concentrated Lentiviral Particles

1. Plate 1.2×10^6 293T/17 cell into 10 ml of DMEM media with 10 % FBS without antibiotic.

2. Next day make a transfection mix of plasmids—9 µg pLKO.1 puro shRNA scramble or pLKO.1 puro shRNA E-cadherin plasmid, 6.25 µg packaging plasmid psPAX2, and 2 µg envelope plasmid pMD2.G in 300 µl OptiMEM media.

3. In another tube make a master mix of 50 µl FuGENE® HD transfection reagent in 450 µl optimum media. Centrifuge briefly and incubate for 5 min at room temperature.

4. Add HD transfection reagent + OptiMEM master mix into the plasmid mix. Briefly centrifuge and incubate for 20 min at room temperature.

5. Dropwise add 160 µl plasmid + HD transfection reagent and OptiMEM mix into each tissue culture dish, swirl gently, and incubate overnight at 37 °C in 5 % CO₂ incubator.

6. Next morning replace media with 10 ml of 10 % FBS DMEM with antibiotic and incubate for 60 h at 37 °C in 5 % CO_2 incubator.

7. Collect supernatant containing lentivirus particles, and centrifuge at $430 \times g$ for 15 min (*see* **Note 10**).

8. Transfer virus supernatant into the thin-walled ultracentrifuge tube and spin it for 28,000 rpm ($100,000 \times g$) for 90 min in SW28 swinging bucket ultracentrifuge rotor.

9. After centrifugation, decant the supernatant and resuspend the virus pellet into 400 µl of ca^{++} Mg^{++} free PBS pH 7.4 (*see* **Note 11**).

3.3.2 Protocol for Transducing Colorectal Cancer Cell Lines

1. Plate 5×10^5 HT29 or Caco-2 cells in six well of tissue culture plate in 10 % FBS + DMEM with antibiotic. Incubate overnight at 37 °C with 5 % CO_2.

2. Cells should be 60 % confluent at the time of transduction with the lentiviral particles.

3. Prepare an 8 µg/ml polybrene solution in complete media.

4. For transducing two six-well plate, add 40 µl of concentrated scramble or E-cadherin lentivirus particle in 4 ml of polybrene-containing media.

5. Infect target cells with scramble and E-cadherin shRNA after adding 2 ml of virus-containing media into six-well plates. Use two wells as MOCK controls with no viral transduction.

6. After allowing for an overnight transduction, replace with normal media the next morning (*see* **Note 12**).

7. On day 3 post-infection, trypsinize cells from the six-well plates and transfer them into a 10 cm tissue culture dish with 2 µg/ml puromycin in complete media.

8. Replace media with fresh 2 µg/ml puromycin every other day. Pooled clones will be generated within 1 week and can be used for further experiments.

9. Confirm knockdown of E-cadherin using western blotting with mouse monoclonal antibody.

3.3.3 Characterization of Stably Transduced E-Cadherin-Depleted Cells for Stem Cell Marker Expression

1. Plate 3×10^4 HT29 scramble and HT29 E-cadherin shRNA cells in DMEM 10 % FBS media in six-well tissue culture plates.

2. Trypsinize cells and make a single-cell suspension with 1×10^6 cells/ml in PBS (pH 7.4).

3. Add 1:30 diluted CD133-APC conjugated to the 100 µl of cell suspension.

4. Incubate on ice for 30 min, wash once with 500 µl of PBS, and centrifuge at 1,500 rpm for 5 min.

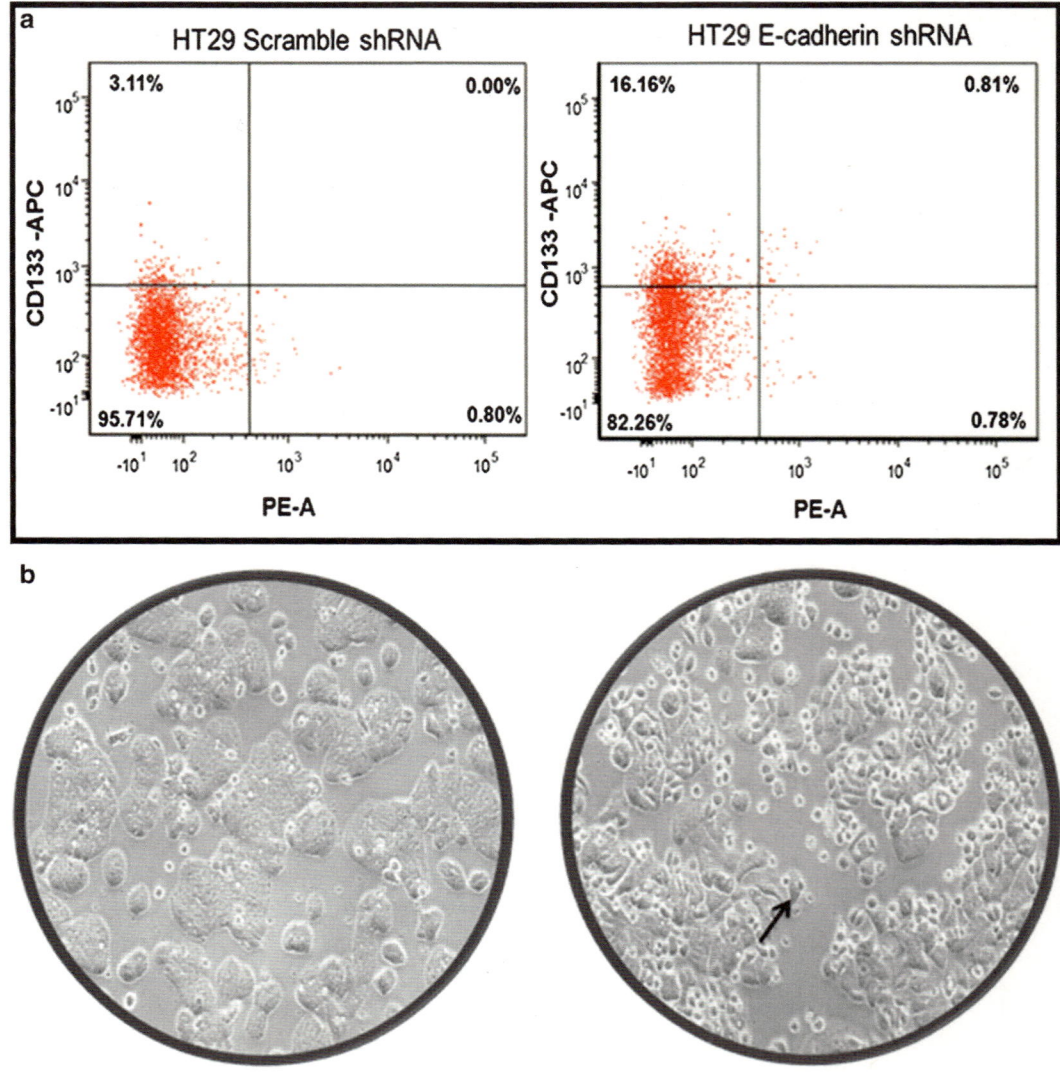

Fig. 5 Flow cytometry analysis of E-cadherin-depleted cells showing the fivefold increase in the expression of CD133-positive cells (**a**) and photomicrograph showing morphology of HT-29 cells transfected with scrambled or E-cadherin shRNA grown in monolayer. The latter cells demonstrate spindle cell morphology (*black arrow*) which is characteristic of EMT transformation (**b**)

5. Resuspend cells in 300 µl PBS buffer and analyze for surface expression of CD133-positive cells using flow cytometry (*see* **Note 13**) (Fig. 5).

3.3.4 Applying Medium- or High-Throughput Screening Using E-Cadherin Knockdown Cells (Enriched in CSC) to Identify Selective Anti-CSC Agents

To identify the compounds selectively targeting cancer stem cells, dose–response profile (0.1 nM→1 mM) will be generated for E-cadherin knockdown and scrambled cells using abovementioned MTT-growth assay and IC50 will be calculated [14]. The compounds exhibiting >5-fold lower IC_{50} for E-cadherin knockdown cells compared to scrambled control will be selected as compounds of interest should be tested for their effect on CSC surface markers and self-renewal factors using a combination of western blot, quantitative reverse transcriptase polymerase chain reaction and flow cytometry [14, 17, 19]. Detail discussion of these methods is beyond the scope of this chapter.

4   Notes

1. All tissue culture procedures should be carrying out in sterile hood. All media and stock solutions should be prepared in sterile disposable tubes and/or bottles to minimize the issue of contamination.

2. Before any tissue culture experiment, check for cell confluency which ideally should be between 70 and 80 %. We also recommend using cells with low passage number (<10) to reduce experimental variations.

3. Cell concentration to be plated varies according to cell type and length of the proposed experiment. For example, HCT 116 and HT 29 cell lines give approximately 20–24 spheres with plating 100 cells/100 μl while Panc1 and Mia Paca cell lines give approximately 20–24 spheres with plating 200 cells/100 μl.

4. We also recommend using appropriate vehicle treatment in the control sample as modest amount of solvent, e.g., DMSO, can significantly affect spheroid growth.

5. For cell proliferation assay, we strongly recommend making a fresh MTT solution on the day of assay.

6. For cell proliferation assay, if crystal does not dissolve after initial trituration, keep the plate in CO_2 incubator for 5–10 min and re-triturate.

7. For secondary colonosphere formation assay, any cell dissociation solution can be used instead of trypsin for single-cell suspension especially if you encounter significant cell death during mechanical and enzymatic digestion process.

8. For secondary/tertiary colonosphere formation, dissociation of sphere to single cell can be achieved by mechanical pipetting for approximately 80–90 times in conjunction with enzymatic digestion with trypsin.

9. All protocol must be approved by your institute biosafety office and recommended biosafety precautions should be observed while working with lentiviruses. In general, one must follow standard operating procedure for BSL2+ laboratory practices as outlined by American Biological Safety Association.

10. As an alternative to centrifugation, virus particle may be filtered through 0.45 μm syringe filter to remove the dead cells.

11. We recommend aliquoting concentrated lentivirus since repeated freeze/thaw significantly reduces viral titer.

12. In case of significant cytotoxicity due to viral transduction, infection time can be reduced to 4 h. Puromycin should be added only after 36 h post-transduction. To achieve optimal degree of knockdown, we highly recommend testing different puromycin-resistant clones for target gene knockdown as lentivirus integrates randomly into the host genome.

13. We also recommend that samples should be filtered before flow cytometry analysis.

Acknowledgements

This work was supported by VA Merit Award to B.B.P.

References

1. Siegel R, Naishadham D, Jemal A (2013) Cancer statistics, 2013. CA Cancer J Clin 63(1):11–30

2. Donnenberg VS, Donnenberg AD (2005) Multiple drug resistance in cancer revisited: the cancer stem cell hypothesis. J Clin Pharmacol 45(8):872–877

3. Clarke MF et al (2006) Cancer stem cells—perspectives on current status and future directions: AACR Workshop on cancer stem cells. Cancer Res 66(19):9339–9344

4. Ricci-Vitiani L et al (2007) Identification and expansion of human colon-cancer-initiating cells. Nature 445(7123):111–115

5. Reya T, Clevers H (2005) Wnt signalling in stem cells and cancer. Nature 434(7035): 843–850

6. Zhao C et al (2009) Hedgehog signalling is essential for maintenance of cancer stem cells in myeloid leukaemia. Nature 458(7239): 776–779

7. Reya T et al (2001) Stem cells, cancer, and cancer stem cells. Nature 414(6859): 105–111

8. Piccirillo SG et al (2006) Bone morphogenetic proteins inhibit the tumorigenic potential of human brain tumour-initiating cells. Nature 444(7120):761–765

9. Mishra L, Derynck R, Mishra B (2005) Transforming growth factor-beta signaling in stem cells and cancer. Science 310(5745):68–71

10. Esko JD, Kimata K, Lindahl U (2009) Proteoglycans and sulfated glycosaminoglycans. In: Varki A et al (eds) Essentials of glyco-biology. Cold Spring Harbor Laboratory Press, NY

11. Tumova S, Woods A, Couchman JR (2000) Heparan sulfate proteoglycans on the cell surface: versatile coordinators of cellular functions. Int J Biochem Cell Biol 32(3):269–288

12. Yip GW, Smollich M, Gotte M (2006) Therapeutic value of glycosaminoglycans in cancer. Mol Cancer Ther 5(9):2139–2148

13. Kim SH, Turnbull J, Guimond S (2011) Extracellular matrix and cell signalling: the dynamic cooperation of integrin, proteoglycan and growth factor receptor. J Endocrinol 209(2):139–151

14. Gupta PB et al (2009) Identification of selective inhibitors of cancer stem cells by high-throughput screening. Cell 138(4):645–659

15. Yi SY et al (2013) Cancer stem cells niche: a target for novel cancer therapeutics. Cancer Treat Rev 39(3):290–296

16. Korkaya H, Wicha MS (2007) Selective targeting of cancer stem cells: a new concept in cancer therapeutics. BioDrugs 21(5):299–310

17. Kanwar SS et al (2010) The Wnt/beta-catenin pathway regulates growth and maintenance of colonospheres. Mol Cancer 9:212

18. Dontu G et al (2003) In vitro propagation and transcriptional profiling of human mammary stem/progenitor cells. Genes Dev 17(10): 1253–1270

19. Kanwar SS et al (2011) Difluorinated-curcumin (CDF): a novel curcumin analog is a potent inhibitor of colon cancer stem-like cells. Pharm Res 28(4):827–838

20. Yu Y et al (2009) Elimination of colon cancer stem-like cells by the combination of curcumin and FOLFOX. Transl Oncol 2(4): 321–328

21. Kong D et al (2011) Cancer stem cells and epithelial-to-mesenchymal transition (EMT)-phenotypic cells: are they cousins or twins? Cancers (Basel) 3(1):716–729

22. Mani SA et al (2008) The epithelial-mesenchymal transition generates cells with properties of stem cells. Cell 133(4):704–715

23. Xia H et al (2010) miR-200a regulates epithelial-mesenchymal to stem-like transition via ZEB2 and beta-catenin signaling. J Biol Chem 285(47):36995–37004

Chapter 42

Analysis of the Heavy-Chain Modification and TSG-6 Activity in Pathological Hyaluronan Matrices

Mark E. Lauer, Jacqueline Loftis, Carol de la Motte, and Vincent C. Hascall

Abstract

During inflammation and developmental processes, heavy chains (HCs) from inter-α-inhibitor (IαI) are covalently transferred to hyaluronan (HA) via the enzyme tumor-necrosis-factor-stimulated-gene 6 (TSG-6) to form a HC-HA complex. In this manuscript, we describe a gel-based assay to detect HC-HA and TSG-6 activity in tissues.

Key words Hyaluronan, TSG-6, Heavy chain, Inter-α-inhibitor

1 Introduction

Under normal conditions, hyaluronan (HA) is synthesized as a large (>1,500 kDa) glycosaminoglycan, lacking protein modifications of any kind. During inflammation, a pathological change occurs to HA in which the enzyme, tumor-necrosis-factor-stimulated-gene 6 (TSG-6), transfers heavy chains (HCs) from the serum-derived proteoglycan inter-α-inhibitor (IαI) to the 6-OH of N-acetylglucosamine residues in HA to form a covalent HC-HA complex [1, 2]. While co-localization of biotinylated HA-binding protein (HABP) with HC or IαI antibodies in pathological HA matrices, using fluorescent microscopy, implies the presence of HCs on HA, it does not rule out the possibility that intact IαI is simply bathing the tissue from serum exudates (*see* **Note 1**). Furthermore, measurement of TSG-6 protein levels from diseased tissues has been difficult due to the low expression levels of this enzyme. Thus, a method to detect TSG-6 activity is likely to improve the sensitivity of TSG-6 detection since a single TSG-6 molecule can produce many measurable HC-HA products. We now describe a gel-based method to document the presence of HC-HA, and to measure TSG-6 activity, in a variety of tissues (*see* **Note 2**).

Kuberan Balagurunathan et al. (eds.), *Glycosaminoglycans: Chemistry and Biology*, Methods in Molecular Biology, vol. 1229, DOI 10.1007/978-1-4939-1714-3_42, © Springer Science+Business Media New York 2015

The principle behind the detection of HC-HA in tissues is that HC-HA is too large to enter a standard SDS-PAGE gel. Upon digestion with a hyaluronidase that specifically digests only HA (*Streptomyces* hyaluronidase), HCs are released from HC-HA, permitting them to migrate as a single ~85 kDa band that can be detected with an antibody against IαI. The principle behind the detection of TSG-6 activity in tissues is that when TSG-6 transfers HCs to a small HA oligosaccharide (HA14 for example), a single HC band appears on an SDS-PAGE gel at ~85 kDa plus the MW of the oligo. Thus, in the presence of endogenous IαI and an exogenous HA oligo added to the reaction mixture, TSG-6 activity can be detected by the appearance of a HC band on a Western blot probed with an antibody against IαI (Fig. 1). The reader is also

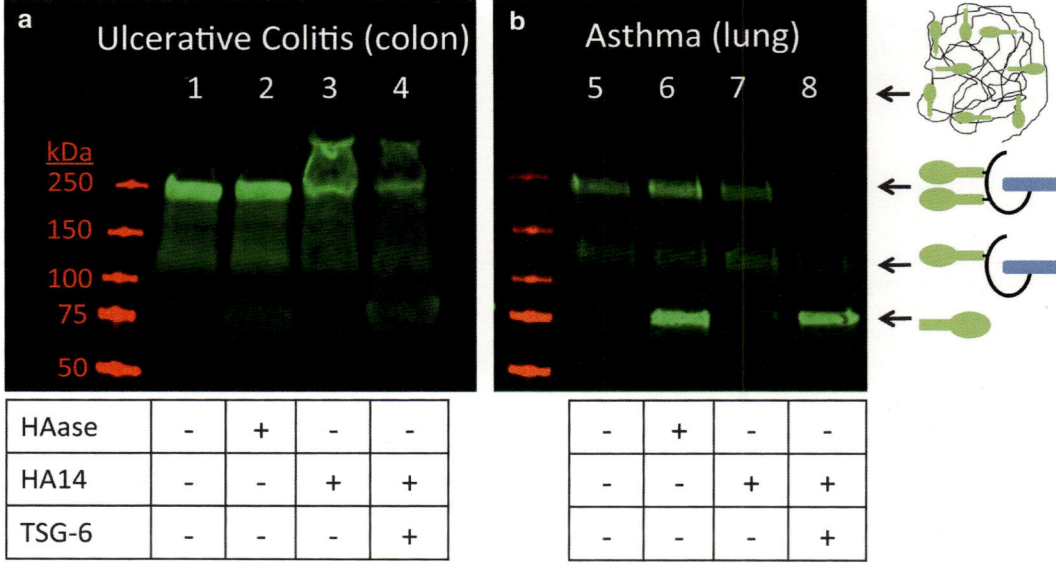

Fig. 1 Western blot demonstrating the presence of the pathological form of hyaluronan (i.e., HC-HA) in inflamed colon and lung tissue from human patients with ulcerative colitis and asthma (panels **a** and **b**, respectively). This blot was probed with the Dako IαI antibody (*green*). Molecular weight standards are shown in *red*. The *green lobules* in the cartoon models represent HCs attached to bikunin (*blue rectangle*) via a single chondroitin sulfate chain (*black line*). The cartoon with two HCs represents IαI. The cartoon with one HC represents pre-IαI. When hyaluronidase (HAase) was added to the minced tissue, the HC-HA complex, which was too large to enter the gel, was digested, releasing HCs that migrate as a single band at ~85 kDa (compare lanes 1 and 2 for the gut tissue, and lanes 5 and 6 for the lung tissue). When an HA oligo, 14 monosaccharides in length, is added to the minced tissue, endogenous TSG-6 activity has an HA substrate on which to transfer HCs from endogenous IαI, which now appear as a single ~85 kDa band on the gel (lanes 3 and 7). Recombinant TSG-6 was added as a positive control for TSG-6 activity (lanes 4 and 8), proving that the reaction conditions were suitable for measuring TSG-6 activity if endogenous TSG-6 was present. In both the colon and lung tissues, no detectable endogenous TSG-6 activity was present. The broader HC band toward lower molecular weights in the colon tissue (lanes 2 and 4) may reflect endogenous protease activity in the original tissue. Furthermore, the smearing of the IαI band in the colon tissue (lanes 3 and 4) could also be caused by endogenous proteolytic activity during the 2-h incubation at 37 °C. Notice that this smearing was not observed in the lung tissue (lanes 7 and 8) which was treated the same as the colon tissue

advised to consult a recently published, alternate, ELISA based method for measuring TSG-6 activity [3].

2 Materials

1. Approximately 75 mg (wet weight) of frozen tissue is needed for this assay (*see* **Note 3**).

2. *Streptomyces* hyaluronidase (EMD Millipore; 389561): Add 0.5 ml of PBS to one vial containing 100 U of enzyme. Gently swirl and incubate at RT for 10 min. Freeze 25 µl aliquots at −80 °C.

3. Any hyaluronan oligosaccharide containing 8–20 monosaccharides can be used. Oligosaccharides containing 8 and 10 monosaccharides can be purchased from Hyalose, LLC (HYA-OLIOG8EF-1 and HYA-OLIGO10EF-1, respectively). Solubilize the oligosaccharide at 1 mg/ml in water.

4. Recombinant TSG-6 (R&D Systems; 2104-TS-050): Add 200 µl PBS to 50 µg of the dried protein to make a 5 µg/ml stock solution, which can be stored at −80 °C.

5. Polyclonal IαI antibody (*see* **Note 1**) (A0301, Dako North America, Inc. Carpinteria, CA): This antibody is also available through the Program of Excellence in Glycosciences (http://pegnac.sdsc.edu/cleveland-clinic/files/PEGNAC-DAKO-Inter-alpha-Inhibitor-Antibody-v1-1.pdf).

6. SDS-PAGE gels (4–15 % acrylamide) and appropriate reagents and equipment for Western blotting.

7. Additional miscellaneous items: Analytical balance (1 mg sensitivity), dry ice, forceps, No. 15 scalpel, eye protection, fine scissors, 1.5 ml tube Pestle Mixer, 10 cm Petri dishes.

3 Methods

1. Label 1.5 ml centrifuge tubes. Weigh the tubes and record their weights to within 1 mg in a lab notebook. Chill on dry ice. Leave them on dry ice unless otherwise instructed. It is essential to keep the tissue frozen until it is minced (**step 6**).

2. Transfer frozen tissue to dry ice and bring it to a lab bench. Using forceps, transfer a tissue chunk to a 10 cm Petri dish pre-chilled on dry ice. With appropriate eye protection, use a No. 15 scalpel to cut off a chunk of tissue approximately the volume of 100 µl. Transfer the cut tissue into a pre-weighed 1.5 ml tube.

3. Repeat **step 2** for all of the other tissues.

4. Weigh the tubes, containing the cut tissues, and record the weights to within 1 mg in a lab notebook. Note that 75 mg of

tissue is the minimum amount of tissue needed for this assay. Calculate the amount of pre-chilled PBS to add to each tissue such that there is 0.33 mg tissue for each 1 µl of PBS.

5. Label four new tubes for each tissue (tube #s described below).

6. Mince one of the tissues in the following manner: Transfer one of the tissues from dry ice to wet ice and let it thaw. Keep the tissue on wet ice during the mincing procedure. Using a pair of fine scissors, cut the tissue into pieces as small as possible (about 1 min, *see* **Note 4**). If available, we recommend the use of a bead homogenizer instead of scissors, such as the Beadbug Microtube Homogenizer (Benchmark Scientific).

7. Cut the tip off of a 200 µl pipet tip to widen the orifice. Transfer 50 µl aliquots of the minced tissue from **step 6** into the four new tubes labeled in **step 5**. Keep these on wet ice. Treat them as follows:

    (a) Tube #1 (the "– hyaluronidase" control): Add 5 µl of PBS to 50 µl of minced tissue and incubate on ice for 30 min. Centrifuge at $13,000 \times g$ for 5 min at 4 °C and transfer the supernatant to a new tube. Incubate the supernatant at 37 °C for 30 min. After the 30-min incubation, store the supernatant in the freezer until ready to analyze by Western blot. Discard the pellet.

    (b) Tube #2 (the "+ hyaluronidase" control): Add 5 µl of Streptomyces hyaluronidase (at 100 TRU per 500 µl) to the 50 µl extract and incubate on ice for 30 min. Centrifuge at $13,000 \times g$ for 5 min at 4 °C and transfer the supernatant to a new tube. Incubate the supernatant at 37 °C for 30 min. After the 30-min incubation, store the supernatant in the freezer until ready to analyze by Western blot. Discard the pellet.

    (c) Tube #3 (measurement of endogenous TSG-6 activity): If the sample does not contain serum, or it is expected to have low levels of serum, add 2 µl of human serum to each 50 µl of human serum to each. Add 2 µl of an HA oligosaccharide (from HA 8-14) (stock concentration at 1 mg/ml) and 3 µl of PBS to 50 µl of minced tissue and incubate at 37 °C for 2 h. Centrifuge at $13,000 \times g$ for 5 min at 4 °C, transfer the supernatant to a new tube, and store the supernatant in the freezer until ready to analyze by Western blot. Discard the pellet.

    (d) Tube #4 (positive control for TSG-6 activity): If the sample does not contain serum, or it is expected to have low levels of serum, add 2 µl of human serum to each 50 µl aliquot of sample to provide a heavy chain donor (i.e. inter-alpha-inhibitor). Add 2 µl of an HA oligosaccharide (from HA 8-14) (stock concentration at 1 mg/ml), and

2 µl of recombinant TSG-6, and 1 µl of PBS to 50 µl of minced tissue and incubate at 37 °C for 2 h. Centrifuge at $13,000 \times g$ for 5 min at 4 °C, transfer the supernatant to a new tube, and store the supernatant in the freezer until ready to analyze by Western blot. Discard the pellet (*see* **Notes 5** and **6**).

8. Repeat **steps 6** and **7** for the remaining tissues.

---

4   Notes

1. We have tested HC antibodies by several other manufacturers and individuals and found that only the Dako IαI antibody has the ability to detect HCs in this assay. Our data suggest that the Dako antibody has an unusually high titer and sensitivity compared to other antibodies. The Dako antibody can be used for both human and mouse tissues.

2. We have found that it is difficult to process more than four samples at a time in this assay. The individual mincing of each tissue, followed by relatively short incubation times (i.e., 30 min), makes it difficult to process more than four samples while maintaining the schedule outlined in this protocol. Deviation from the suggested incubation times may result in lower sensitivity (for shorter incubation times) or proteolytic degradation of HCs (for longer incubation times). If more than four samples need to be processed, we recommend processing them in separate batches.

3. It is important not to let the tissue thaw until ready to begin the mincing step (**step 6**). The tissues should be kept on dry ice until this step to prevent proteolytic degradation. We recommend pre-chilling the tubes to the temperature described in each step to keep the temperature constant from step to step.

4. Effective mincing of the tissue is essential to the highest yield possible for each sample. Mincing of different types of tissues may require more or less mincing time for effective mincing. Once mincing with the scissors, or homogenizer, no longer has any further effect, additional mincing should be stopped.

5. We have noticed that endogenous TSG-6 activity is rarely detectable in end-stage diseased tissues in which inflammation has reached a chronic and advanced state. For example, the colon tissue analyzed in this manuscript came from a portion of colon that was surgically removed from a patient with ulcerative colitis because of the advanced stage of disease in that region. Furthermore, the lung tissue presented in this manuscript was postmortem tissue that came from a patient

who died of asthma. Although we are unable, at this time, to publish existing data demonstrating endogenous TSG-6 activity, we are able to propose that tissues from less advanced stages of disease may demonstrate endogenous TSG-6 activity.

6. The manuscript by Swaidani et al. [4] was the first to utilize the gel-based method of HC-HA detection as outlined in this protocol. It demonstrated that HC-HA was not detectable in the lungs of naïve wild-type or TSG-6$^{-/-}$ asthmatic mice, but was readily present in wild-type asthmatic mice. This emphasizes the importance of TSG-6 in the formation of HC-HA and shows that, at least in the airways, TSG-6 is essential for the formation of HC-HA.

References

1. Milner CM, Higman VA, Day AJ (2006) TSG-6: a pluripotent inflammatory mediator? Biochem Soc Trans 34(Pt 3):446–450

2. Milner CM, Tongsoongnoen W, Rugg MS et al (2007) The molecular basis of inter-alpha-inhibitor heavy chain transfer on to hyaluronan. Biochem Soc Trans 35(Pt 4):672–676

3. Wisniewski HG, Colón E, Liublinska V et al (2013) TSG-6 activity as a novel biomarker of progression in knee osteoarthritis. Osteoarth Cart 22(2):235–241

4. Swaidani S, Cheng G, Lauer ME et al (2013) TSG-6 protein is crucial for the development of pulmonary hyaluronan deposition, eosinophilia, and airway hyperresponsiveness in a murine model of asthma. J Biol Chem 288(1):412–422

Chapter 43

Heparan Sulfate Modulates Slit3-Induced Endothelial Cell Migration

Hong Qiu, Wenyuan Xiao, Jingwen Yue, and Lianchun Wang

Abstract

Heparan sulfate is a long, linear polysaccharide with sulfation modifications and belongs to the glycosaminoglycan family. Our recent studies elucidated that the axon guidance molecule Slit3 is a new heparan sulfate-binding protein and a novel angiogenic factor by interacting with its cognate receptor Robo4, which is specifically expressed in endothelial cells. Here we describe using heparan sulfate-deficient mouse endothelial cells to determine the co-reception function of heparan sulfate in Slit3-induced endothelial cell migration in a Boyden chamber trans-well migration assay.

Key words Heparan sulfate, Endothelial cell, Migration, Slit3, Robo4, Angiogenesis

1 Introduction

Heparan sulfate (HS) is a long, linear polysaccharide with four types of sulfation modification, including N-, 2-O, 6-O, and 3-O sulfation. HS presents abundantly on cell surfaces and in the extracellular matrix where it modulates various cell functions in both physiological and pathological processes by interacting with a wide range of protein ligands, including growth factors, morphogens, cytokines, adhesion molecules, proteinases, and proteinase inhibitors. Mechanism studies have suggested that cell surface HS may function as a co-receptor, endocytic receptor, chemokine presenter, or receptor for cell adhesion and mobility [1]. In our previous studies, we generated mouse endothelial cell lines that are deficient in various HS biosynthetic enzymes, such as *Ext1*, *Ndst1*, and *Ndst2*, to lack or express truncated HS [2–6]. These HS mutant cell lines have allowed to determine the functions of endothelial HS in inflammation, tumor angiogenesis, vascular development, and related angiogenic signaling.

Slit-Robo pathway is a conserved axon guidance signaling in mammals. There are three Slits (Slit1-3) and four Robos (Robo1-4) [7]. Slit proteins contain four tandem leucine-rich repeats, six

Kuberan Balagurunathan et al. (eds.), *Glycosaminoglycans: Chemistry and Biology*, Methods in Molecular Biology, vol. 1229, DOI 10.1007/978-1-4939-1714-3_43, © Springer Science+Business Media New York 2015

epidermal growth factor-like repeats, a laminin G-like domain, additional three epidermal growth factor-like repeats, and a C-terminal cysteine knot domain [7]. To date, Slit2 has been studied the most [8]. Slit2 can be cleaved into Slit2-N terminal (~140 kDa) and Slit2-C terminal (~55 kDa), and, interestingly, each Slit2 fragment contains an HS-binding site [9, 10]. Cell-based studies showed that Slit2-N is tightly associated with cell membranes in an HS-dependent manner [11]. Robo1–3 contain five immunoglobulin repeats, three fibronectin type III repeats, one transmembrane domain, and four conserved cytoplasmic motifs (CC0–CC3) [12]. Robo4 is structurally divergent from other Robos as it only has two immunoglobulin repeats, two fibronectin type III repeats, and two conserved cytoplasmic motifs (CC0 and CC2) [13]. The intracellular domain of Robo4 has been shown to interact with Mena and neural Wiskott-Aldrich syndrome proteins to activate Rho GTPases and to block small TGPases, ARf6, to critically modulate endothelial cell migration [14].

Cell migration is a highly integrated, multistep process that essentially orchestrates angiogenesis, the process to form new blood vessels. Cell migration can be evaluated through several different methods including scratch assays, cell-exclusion zone assays, microfluidic based assays, and Boyden chamber trans-well assays. The most widely applied cell migration technique is the Boyden chamber trans-well migration assay. The Boyden chamber system is based on a chamber of two medium-filled compartments separated by a microporous membrane (Fig. 1a). In general, cells are placed in the upper compartment and are allowed to migrate through the pores of the membrane into the lower compartment, following a chemoattractant gradient, which is generated by adding chemoattractant, such as chemokine and growth factor, in the lower compartment (Fig. 1b). Migratory cells are then stained and quantified to determine the migratory property of the cells induced by the added chemoattractant (Fig. 1c).

In vasculature Robo1 and Robo4 and their ligands Slit2 and Slit3 are expressed in both endothelial cells and vascular smooth muscle cells [15]. Our recent studies elucidated that Slit3 is a novel angiogenic factor and interacts with Robo4, not Robo1, to promote endothelial cell proliferation and migration [5, 15]. We also determined that Slit3 is a new HS-binding protein [5, 16]. Here we describe using the newly generated mouse diaphragmatic endothelial cell line, which contains conditionally targeted "floxed" *Ndst1* alleles (*Ndst1^{f/f}*, wild type), and its daughter cell line which is deficient in *Ndst1* (*Ndst1^{-/-}*, HS mutant), to determine the co-reception function of HS in its facilitation of Slit3-induced endothelial cell migration in the Boyden chamber trans-well migration assay [5].

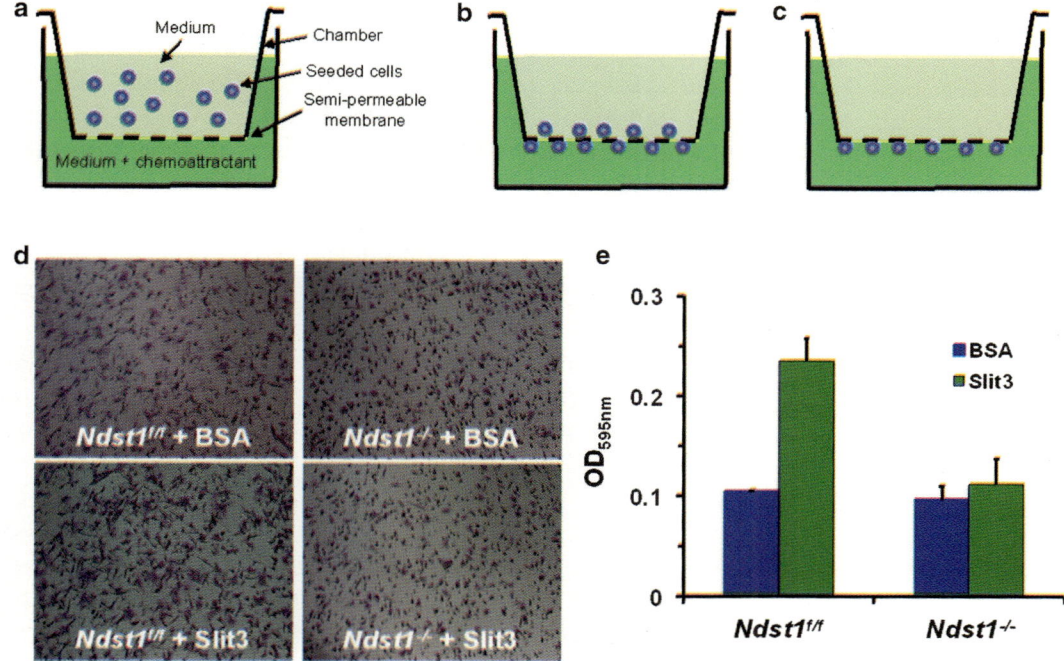

Fig. 1 Slit3-induced diaphragmatic endothelial cell migration in Boyden chamber assay. (**a–c**) Principle of the Boyden chamber assay. Cell suspension is placed in the upper chamber (**a**); migratory cells pass through a semipermeable membrane and adhere to the bottom side. Nonmigratory cells remain in the upper chamber (**b**). After removal of nonmigratory cells, the stained migrated cells are quantified (**c**). (**d**) Photo of migrated diaphragmatic endothelial cells that were induced by Slit3 with BSA as negative control. (**e**) OD_{595nm} measurement of migrated diaphragmatic endothelial cells after lysis. Data present as mean $\pm$ SEM

2   Materials

Prepare all solutions using ultrapure water (prepared by purifying deionized water to attain a sensitivity of 18 MΩ cm at 25 °C) and analytical grade reagents. Prepare all the reagents at room temperature and store at 4 °C. Strictly follow all waste disposal regulations.

2.1   Cells and Reagents

1. 1 % gelatin solution: 1 g tissue culture-grade gelatin power is added to 100 ml of water, shaking well, and then autoclaved at 121 °C for 1 h. Cool to room temperature and store at 4 °C.

2. Mouse diaphragmatic *Ndst1$^{f/f}$* and *Ndst1$^{-/-}$* endothelial cell lines were derived as reported [5] and are stored at 5×10^6 cells/vial in liquid nitrogen.

3. Endothelial cell culture medium: 500 ml of High-glucose DMEM is added with 56 ml of fetal bovine serum (FBS), 5.6 ml of penicillin stock at 10,000 U/ml, 5.6 ml of streptomycin

stock at 100 mg/ml, and 5.6 ml of nonessential amino acid stock at 100×. Store at 4 °C.

4. Endothelial cell growth supplement (ECGS, Sigma-Aldrich) stock solution: The vial containing 15 mg of ECGS is dissolved by adding 1.5 ml of endothelial cell culture medium to get the stock solution at 10 mg/ml. The solution is then aliquoted at 1 ml per vial and stored at −20 °C.

5. Porcine heparin stock solution: Power dissolved in water at 100 mg/ml, sterilized by filtering through a 0.2 μm Syringe Filter (Millipore), and stored at −20 °C until use.

6. 2 mM EDTA cell dissociation solution: 744.2 mg of EDTA-Na$_2$ and 1 g of bovine serum albumin are dissolved in 1,000 ml of water, sterilized by filtering through a 0.2 μm filter, and stored at 4 °C.

7. Crystal violet staining solution: 0.1 g of crystal violet and 0.618 g of boric acid are dissolved in 50 ml of water, 2 ml of ethanol (≥99.5 %, 200 proof) is added, and then water is added to reach a final volume of 100 ml. The solution is filtered with a 0.2 μm filter and stored at room temperature.

8. 10 % acetic acid solution: Add 10 ml of glacial acetic acid (>99.5 % acetic acid) drop by drop to 90 ml of water while the solution is mixed by stirring. Store at room temperature.

2.2 Instruments

1. Laminar flow hoods for tissue culture (Labconco).

2. Forma CO$_2$ incubator (37 °C, 5 % CO$_2$ atmosphere, Thermo Fisher Scientific).

3. Eclipse TE2000-S Inverted Microscope (Nikon).

4. OPTimax microplate reader (Tecan Group Ltd.).

3 Methods

3.1 Preparing the Endothelial Cells

1. Mouse diaphragmatic $Ndst1^{f/f}$ and $Ndst1^{-/-}$ endothelial cell stock vials (1 ml volume) are thawed in 37 °C water. Transfer each to an individual 15 ml centrifuge tube supplied with 4 ml of pre-warmed (37 °C) endothelial cell culture medium and spin down at $400 \times g \times 5$ min. Suspend the cell pellet with 1 ml of endothelial cell culture medium, and repeat the wash step once.

2. Suspend the cells in 10 ml of endothelial cell culture medium, add 100 μl of ECGS stock solution, and seed the cell solution in a 10 cm cell culture dish.

3. Change the culture medium every 3 days until the cells reach 70–80 % confluency. If necessary, overnight starvation

may be performed prior to using the cells for migration experiments.

4. Wash the cells twice with blank DMEM medium at 2 ml/well, detach the cells by adding 1 ml/ml 0.25 % trypsin-EDTA solution, and digest for 15 min at 37 °C.

5. Harvest the cells by centrifugation at $400 \times g \times 5$ min. Remove the digestion solution, resuspend the cells with 5 ml of blank DMEM medium, and centrifuge again.

6. Remove the medium and resuspend the cells in 1 ml of blank DMEM. Count and adjust the cells to 1×10^6 cells/ml in DMEM medium.

3.2 Loading and Assembling the Boyden Chamber

1. Under sterile conditions, allow the 24-well 8 μm pore size migration plate to warm up at room temperature for 10 min (*see* **Note 1**).

2. Add 600 μl of DMEM containing 1 % FBS and Slit3 (200 ng/ml) or bovine serum albumin (BSA; 200 ng/ml) to the lower chamber of the Boyden chamber system (*see* **Note 2**).

3. Add 100 μl of the endothelial cell suspension solution that contains 10^5 cells to the inside of each insert (*see* **Notes 2** and **3**).

4. Incubate for 3–6 h in the cell culture incubator.

5. Carefully aspirate the media in the inserts and fix cells at room temperature with 600 μl of 4 % paraformaldehyde in a new 24-well plate for 1 h.

6. Transfer the insert to a clean well containing 600 μl of crystal violet staining solution and incubate for 30 min at room temperature.

7. Wet the ends of cotton-tipped swabs with water, flatten the ends of the swabs by pressing against a clean hard surface, and gently swab the interior of the inserts to remove nonmigratory cells (*see* **Note 4**).

8. Gently wash the stained inserts five times with water and allow the inserts to air-dry.

3.3 Data Documentation and Analysis

1. Photograph and count migratory cells with a Nikon Eclipse TE-2000-S under high magnification, with at least five individual fields per insert (Fig. 1d) (*see* **Note 5**).

2. Alternatively, transfer each insert to an empty well, add 200 μl of 10 % acetic acid solution per well, and incubate for 20 min on an orbital shaker.

3. Transfer 100 μl from each sample to a 96-well microtiter plate and measure the OD_{595nm} using an OPTimax microplate reader (Fig. 1e).

4   Notes

1. It is recommended to plan for less than six experimental groups per 24-well chamber plate, which allows for at least four wells for an experimental group. This ensures that at least three representative wells can be selected for quantification for each experimental group. Using more experimental groups will take more time for preparation, which may cause cells in different experimental groups to remain in suspension under varied times, possibly affecting cell motility.

2. To avoid trapping bubbles in the wells of the chamber, the liquid should not be expelled completely from the pipette tip. In addition, for the lower chamber, it is important to load the correct volume of attractants, which should form a slightly positive meniscus when the well is filled. For loading the upper chamber, hold the pipette vertically so that the end of the pipette tip is against the side of the well just above the membrane and expel the liquid quickly from the pipette tip.

3. Choose a proper pore size. A 3 μm pore size is appropriate for leukocyte or lymphocyte migration; a 5 μm pore size is appropriate for fibroblast cells, monocytes, macrophages, and cancer cells such as NIH-3T3 and MDA-MAB 231 cells; and an 8 μm pore size is appropriate for most other cell types, such as the endothelial cells. The proper pore size supports optimal migration for most epithelial and fibroblast cells.

4. Take care not to puncture the polycarbonate membrane and be sure to completely remove cells on the inside perimeter of the insert.

5. With the aid of a digital camera and computer, the total number of stained cells in a well can be counted accurately and objectively. Directly counting the cells in selected fields from a well under a microscope at high magnification is not recommended.

Acknowledgement

This work was supported by grants from NIH R01HL093339 (L.W.) and RR005351/GM103390 (L.W.).

References

1. Sarrazin S, Lamanna WC, Esko JD (2011) Cold Spring Harb Perspect Biol 3(7)

2. Wang L, Fuster M, Sriramarao P, Esko JD (2005) Nat Immunol 6:902–910

3. Fuster MM, Wang L, Castagnola J, Sikora L, Reddi K, Lee PH, Radek KA, Schuksz M, Bishop JR, Gallo RL, Sriramarao P, Esko JD (2007) J Cell Biol 177:539–549

4. Wijelath E, Namekata M, Murray J, Furuyashiki M, Zhang S, Coan D, Wakao M, Harris RB, Suda Y, Wang L, Sobel M (2010) J Cell Biochem 111:461–468

5. Zhang B, Xiao WY, Qiu H, Zhang FM, Moniz HA, Condac E, Gutierrez-Sanchez G, Heiss C, Clugston RD, Azadi P, Greer JJ, Bergmann C, Moremen KW, Li D, Linhardt RJ, Esko JD, Wang L (2013) J Clin Invest 124:209–221

6. Qiu H, Jiang JL, Liu M, Huang X, Ding SJ, Wang L (2013) Mol Cell Proteomics 12:2160–2173

7. Carmeliet P, Tessier-Lavigne M (2005) Nature 436:193–200

8. Jaworski A, Tessier-Lavigne M (2012) Nat Neurosci 15:367–369

9. Wang KH, Brose K, Arnott D, Kidd T, Goodman CS, Henzel W, Tessier-Lavigne M (1999) Cell 96:771–784

10. Fukuhara N, Howitt JA, Hussain SA, Hohenester E (2008) J Biol Chem 283:16226–16234

11. Hu H (2001) Nat Neurosci 4:695–701

12. Wong K, Park HT, Wu JY, Rao Y (2002) Curr Opin Genet Dev 12:583–591

13. Park KW, Morrison CM, Sorensen LK, Jones CA, Rao Y, Chien CB, Wu JY, Urness LD, Li DY (2003) Dev Biol 261:251–267

14. Sheldon H, Andre M, Legg JA, Heal P, Herbert JM, Sainson R, Sharma AS, Kitajewski JK, Heath VL, Bicknell R (2009) FASEB J 23:513–522

15. Zhang B, Dietrich UM, Geng JG, Bicknell R, Esko JD, Wang L (2009) Blood 114:4300–4309

16. Condac E, Strachan H, Gutierrez-Sanchez G, Brainard B, Giese C, Heiss C, Johnson D, Azadi P, Bergmann C, Orlando R, Esmon CT, Harenberg J, Moremen K, Wang L (2012) Glycobiology 22:1183–1192

Chapter 44

Glycosaminoglycan Functionalized Nanoparticles Exploit Glycosaminoglycan Functions

James A. Vassie, John M. Whitelock, and Megan S. Lord

Abstract

Nanoparticles are being explored for a variety of applications including medical imaging, drug delivery, and biochemical detection. Surface functionalization of nanoparticles with glycosaminoglycans (GAGs) is an attractive strategy that is only starting to be investigated to improve their properties for biological and therapeutic applications. Herein, we describe a method to functionalize the surface of cerium oxide nanoparticles (nanoceria) with organosilane linkers, such as 3-(aminopropyl)triethoxysilane (APTES) and 3-(mercaptopropyl)trimethoxysilane (MPTMS), and GAGs, such as unfractionated and low molecular weight heparin. Examples of how the activity of these heparin functionalized nanoparticles are governed by the pendant GAGs are detailed. The activity of heparin covalently attached to the nanoceria was found to be unchanged when compared to unfractionated heparin using the activated partial clotting time (APTT) assay.

Key words Nanoparticle, Cerium oxide, Heparin, Glycosaminoglycan, Organosilane linker

1 Introduction

Glycosaminoglycan (GAG) functionalized materials have gained much research attention recently, in order to add biomimetic cues to materials and improve biofunctionality. Major classes of nanoparticles that have been functionalized with GAGs to date include polymeric micelles [1], polymer-drug conjugates [2], and metal oxide nanoparticles [3, 4]. GAGs, such as heparin and heparan sulfate, are attractive molecules to incorporate into nanoparticles as they possess diverse roles in biology and pathology through their ability to bind positively charged species, such as growth factors, proteases, chemokines, and anti-thrombin III [5], that may be exploited for therapeutic effect. Additionally, the functionalization of nanoparticles with heparin has been shown to reduce particle self-aggregation [4] and improve stability [6]. Nanoparticles themselves provide benefits over GAGs alone by increasing the intracellular uptake of GAGs that may be exploited for the delivery of GAG-bound molecules such as growth factors.

Kuberan Balagurunathan et al. (eds.), *Glycosaminoglycans: Chemistry and Biology*, Methods in Molecular Biology, vol. 1229, DOI 10.1007/978-1-4939-1714-3_44, © Springer Science+Business Media New York 2015

This chapter details a method to attach heparin to cerium oxide nanoparticles (nanoceria); however this method can be adapted for use with other metal oxide nanoparticles and GAGs. We have characterized the extent of functionalization of 14 nm nanoceria synthesized by flame spray pyrolysis [7] with APTES and either unfractionated or low molecular weight heparin. Examples of how the activity of these heparin functionalized nanoparticles are governed by the pendant GAGs are detailed in the activated partial clotting time assay (APTT). We have reported previously that nanoceria with a low level of unfractionated heparin functionalization were able to support the same level of fibroblast growth factor (FGF) 2 signaling in a BaF32 cell assay as unfractionated heparin [3], while at higher levels of heparin functionalization there was reduced FGF2 signaling [4]. Additionally, the level of heparin functionalization was found to direct the intracellular localization of the particles in endothelial cells. Nanoceria with a higher level of heparin functionalization were located in the lysosomes within 24 h of exposure, while particles with a lower level of heparin functionalization were predominantly located in the cytoplasm [4].

2 Materials

Prepare and store all reagents at room temperature, unless otherwise specified. Diligently follow all safety procedures outlined in this document and all waste disposal regulations relevant to your workplace when disposing of waste materials.

2.1 Conjugation of Organosilane Linker to Metal Oxide Nanoparticles

1. Nanoparticles: cerium oxide nanoparticles (nanoceria) synthesized via flame spray pyrolysis and characterized as reported previously [7]. Store in a container wrapped in aluminum foil due to their light sensitivity.

2. Aminosilane linker:

   (a) Anhydrous 3-(aminopropyl)triethoxysilane (APTES). Store at 4 °C. Allow the bottle to come to room temperature before use. Purge with N_2 after use and before resealing; or

   (b) Anhydrous 3-(mercaptopropyl)trimethoxysilane (MPTMS). Store at 4 °C. Allow the bottle to come to room temperature before use. Purge with N_2 after use.

3. Anhydrous dimethylformamide (DMF). Purge with N_2 after use.

4. Toluene, pure grade.

5. Acetone, pure grade.

6. N_2 gas.

7. Parafilm.

8. Aluminum foil.

9. 15 mL polypropylene tubes.

10. Sterile needles.

11. 40 mL flat-bottom, screw-top bottle (Schott).

12. 100 mL glass beakers.

13. 30 mL glass pipette.

14. Pipette gun.

15. Air gun.

16. 1 mL pipette.

17. Fume hood.

18. Sonicator bath.

19. Centrifuge ($7,445 \times g$ required).

20. Vegetable oil bath.

21. Magnetic stirring plate with heating element.

22. Solvent-compatible vacuum pump.

23. Vacuum desiccator chamber.

2.2 Conjugation of Heparin to Organosilane Functionalized Metal Oxide Nanoparticles

Items outlined in Subheading 2.1 and the following:

1. Organosilane functionalized metal oxide nanoparticles:

   (a) APTES-conjugated nanoceria (APTES-nanoceria). Store in a container wrapped in aluminum foil due to their light sensitivity; or

   (b) MPTMS-conjugated nanoceria (MPTMS-nanoceria). Store in a container wrapped in aluminum foil due to their light sensitivity.

2. 1-Ethyl-3-(3-dimethylaminopropyl)carbodiimide hydrochloride (EDC). Store at –20 °C. Allow the bottle to come to room temperature before use.

3. N-Hydroxysuccinimide (NHS).

4. Heparin:

   (a) Unfractionated heparin sodium salt from porcine intestinal mucosa (17–19 kDa).

   (b) Low molecular weight (LMW) Heparin sodium salt from porcine intestinal mucosa (4–6 kDa).

3 Methods

3.1 Conjugation of Organosilane Linker to Metal Oxide Nanoparticles

Carry out all procedures at room temperature unless otherwise specified.

1. Place a 100 mg of nanoceria into a 40 mL flat-bottom, screw-top bottle (*see* **Note 1**).

2. Transfer the bottle to a fume hood.

3. Add a clean magnetic stirrer and 25 mL DMF to the bottle (*see* **Note 2**). The concentration of nanoceria in DMF should not exceed 10 mg/mL.

4. Purge the bottle containing nanoceria in DMF with N_2 and replace the lid.

5. Purge the DMF bottle with N_2 and replace the lid.

6. Wrap 2–3 strips of parafilm around neck and lid of the bottle containing nanoceria in DMF to ensure a tight seal, then wrap in aluminum foil.

7. Sonicate the bottle containing nanoceria in DMF at 45 °C until the nanoceria are suspended (*see* **Note 3**).

8. Remove the APTES bottle from the fridge and allow it to come to room temperature (approximately 30 min) (*see* **Note 4**).

9. Place the bottle containing nanoceria in DMF on a magnetic stirring plate in a fume hood. Remove the lid. Set the magnetic stirring plate on a low speed.

10. Add 250 µL (1.1×10^{-3} mol) APTES to the bottle containing nanoceria in DMF. The concentration of APTES used should be in excess of the concentration of surface hydroxyl groups present on the nanoparticles.

11. Purge the bottle containing APTES and nanoceria in DMF with N_2 and replace the lid.

12. Purge the APTES bottle with N_2, replace the lid, and return to the fridge.

13. Transfer the bottle containing APTES and nanoceria in DMF to an oil bath and stir on a medium speed for 24 h at 45 °C.

14. Divide the APTES-nanoceria solution evenly into two 15 mL polypropylene tubes.

15. Centrifuge for 15 min at $7,445 \times g$ (*see* **Note 5**).

16. Pour the supernatant into a 100 mL glass beaker (*see* **Note 6**).

17. Add 10 mL toluene to each tube. Sonicate the tubes for 5 min to resuspend the nanoparticles.

18. Repeat **steps 15–17** twice.

19. Repeat **steps 15** and **16**.

20. Add 10 mL of acetone to each tube. Sonicate the tubes for 5 min to resuspend the nanoparticles.

21. Repeat **steps 15** and **16**.

22. Wrap the uncapped tubes (i.e., no lids) in aluminum foil and insert two small holes on the top with a small needle.

23. Air-dry the tubes in the fume hood for 24 h standing upright in a tube rack to remove the remaining acetone.

24. Break up the nanoparticle aggregates inside the tubes using a sterile needle.

25. Reseal the uncapped tubes with aluminum foil and insert one small hole on the top with a small needle.

26. Dry the tubes under vacuum at room temperature using solvent-compatible pump and desiccator for 24 h standing upright to facilitate the removal of trace amounts of acetone.

27. Release the vacuum slowly and replace the caps on the tubes.

28. Store the organosilane functionalized nanoceria away from light.

3.2 Conjugation of Heparin to Organosilane Functionalized Metal Oxide Nanoparticles

The level of organosilane surface functionalization needs to be determined before commencing this procedure (*see* **Note 7** and Table 1).

1. Calculate the amount of heparin required for a given amount of nanoparticles and a defined level of amine substitution using the following formula (*see* **Note 8**):

$$\text{mass heparin required} = \frac{A(m \times f)}{L} \times H$$

Table 1
Percent weight loss and number of molecules attached to 14 nm nanoceria after conjugation with APTES or APTES and heparin measured by TGA over the temperature range 20–1,000 °C heated at 5 °C/min

	LMW heparin			Unfractionated heparin		
	Weight loss (%)	Number of molecules per nanoceria		Weight loss (%)	Number of molecules per nanoceria	
		APTES	Heparin		APTES	Heparin
Nanoceria	2.5	–	–	2.5	–	–
APTES-nanoceria	15.4	413,900	–	15.4	413,900	–
Heparin-APTES-nanoceria	22.0	413,900	8,140	49.7	413,900	11,750

where A = proportion of amine functionalization required. m = mass of organosilane functionalized nanoparticles needed. f = proportion of nanoparticle mass attributed to organosilane linker determined by TGA. L = molecular weight of organosilane linker after functionalization. H = molecular weight of heparin.

2. Calculate the amounts of EDC and NHS required for a minimum molar ratio of heparin:EDC:NHS of 1:1.2:2. This minimum molar ratio is set to ensure that EDC is in excess with respect to heparin and NHS is in excess with respect to EDC (*see* **Note 9**).

3. Remove the EDC from the freezer and allow it to come to room temperature (approximately 30 min).

4. Place known mass of organosilane functionalized nanoceria into a 40 mL flat-bottom, screw-top bottle (*see* **Note 1**). Add a clean magnetic stirrer and close the lid.

5. Weigh required amounts of heparin, EDC, and NHS. Mix in a 40 mL flat-bottom, screw-top bottle. Add a clean magnetic stirrer and close the lid.

6. Transfer both bottles to a fume hood.

7. Add 10 mL DMF to the bottle containing heparin, EDC, and NHS (*see* **Note 2**).

8. Add enough DMF to the bottle containing organosilane functionalized nanoceria so that the concentration of particles does not exceed 10 mg/mL (*see* **Note 2**).

9. Purge each of the bottles with N_2 and replace the lid.

10. Purge the DMF bottle with N_2 and replace the lid.

11. Wrap 2–3 strips of parafilm around neck and lid of each bottle to ensure a tight seal, then wrap in aluminum foil.

12. Transfer the bottle containing heparin, EDC, and NHS to an oil bath and stir at 45 °C for 6 h.

13. After 3 h, sonicate the bottle containing organosilane functionalized nanoceria in DMF at 45 °C until the nanoceria are suspended (*see* **Note 3**).

14. Transfer the contents of the bottle containing heparin, EDC, and NHS to the bottle containing organosilane functionalized nanoceria and reseal. Stir for 3 days at room temperature.

15. Divide the heparin-APTES-nanoceria solution evenly into two 15 mL polypropylene tubes.

16. Centrifuge for 15 min at 7,445 × g (*see* **Note 5**).

17. Pour the supernatant into a 100 mL glass beaker (*see* **Note 6**).

18. Add 10 mL acetone to each tube. Sonicate the tubes for 5 min to resuspend the nanoparticles.

19. Repeats **steps 16–18** twice.

20. Repeat **steps 16** and **17**.

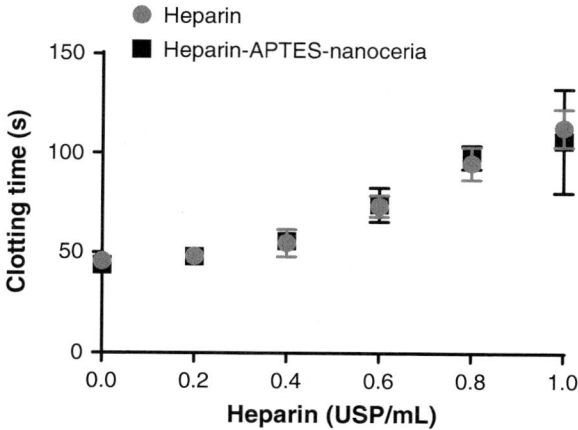

Fig. 1 Activated partial clotting times of unfractionated heparin and unfraction-ated heparin-APTES-nanoceria at different heparin concentrations. Data pre-sented as mean ± standard deviation ($n = 4$)

21. Wrap the uncapped tubes (i.e., no lids) in aluminum foil and insert two small holes on the top with a small needle.

22. Air-dry the tubes in the fume hood for 24 h standing upright in a tube rack to remove the remaining acetone.

23. Break up the nanoparticle aggregates inside the tubes using a sterile needle.

24. Reseal the uncapped tubes with aluminum foil and insert one small hole on the top with a small needle.

25. Dry the tubes under vacuum at room temperature using solvent-compatible pump and desiccator for 24 h standing upright to facilitate the removal of trace amounts of acetone.

26. Release the vacuum slowly and replace the caps on the tubes.

27. Store the heparin-APTES-nanoceria away from light until use. The heparin functionalized particles can be characterized for their GAG functions. For example, the activated partial throm-boplastin clotting time (APTT) of unfractionated heparin functionalized 14 nm nanoceria was the same as unfraction-ated heparin over the range of 0–1 USP/mL heparin using methods detailed in [8] (Fig. 1).

4   Notes

1. Weight nanoparticles in an enclosed balance outside the fume hood.

2. Pipette DMF out of container using a glass pipette. If using reusable glass pipettes, clean the glass pipette with acetone and dry with the air gun before use.

3. Sonication takes approximately 3 h and occasional stirring on a magnetic stirring plate may also facilitate suspension.

4. MPTMS may be used in place of APTES.

5. Program the centrifuge to accelerate rapidly and decelerate slowly so as not to reintroduce particles into the supernatant. Do not exceed $7,445 \times g$ as this may deform the polypropylene tubes.

6. Recover any particle aggregates in the supernatant in the beaker using the glass pipette. Allow any such aggregates to settle to the bottom of the pipette so as not to reintroduce the supernatant back into the tubes. After resuspending the particles following the first centrifugation step the particles may be combined into one tube, if desired. This may reduce the amount of particles lost during the drying step.

7. Determine the number of surface amine groups introduced by the covalent attachment of the organosilane linker by thermogravimetric analysis (TGA) [3, 4].

8. For example, for 50 mg of APTES-nanoceria (14 nm nanoceria) which were found to have a weight loss of 12.9 % by TGA attributed to APTES (Table 1) and a functionalization extent of 10 % was wanted using unfractionated heparin. The

$$\text{mass heparin required} = \frac{0.1(50\,\text{mg} \times 0.129)}{192.31\frac{\text{g}}{\text{mol}}} \times 18,000\frac{\text{g}}{\text{mol}} = 60.3\,\text{mg}.$$

9. For example, based on 60.3 mg of unfractionated heparin, use a minimum of 0.77 mg (4.0×10^{-6} mol) EDC and 0.77 mg (6.7×10^{-6} mol) NHS.

Acknowledgement

This work was supported by the Australian Research Council Discovery Project scheme.

References

1. Reyes-Ortega F, Rodríguez G, Aguilar MR, Lord M, Whitelock J, Stenzel MH, San Román J (2013) Encapsulation of low molecular heparin (bemiparin) into polymeric nanoparticles obtained from cationic block copolymers: properties and cell activity. J Mater Chem B 1:850–860

2. Zhao Y, Lord MS, Stenzel MH (2013) A polyion complex micelle with heparin for growth factor delivery and uptake into cells. J Mater Chem B 1:1635–1643

3. Ting SRS, Whitelock JM, Tomic R, Gunawan C, Teoh WY, Amal R et al (2013) Cellular uptake and activity of heparin functionalised cerium oxide nanoparticles in monocytes. Biomaterials 34:4377–44386

4. Lord MS, Tsoi B, Gunawan C, Teoh WY, Amal R, Whitelock JM (2013) Anti-angiogenic

activity of heparin functionalised cerium oxide nanoparticles. Biomaterials 28:8888–8999

5. Gandhi NS, Mancera RL (2008) The structure of glycosaminoglycans and their interactions with proteins. Chem Biol Drug Des 72:455–482

6. Yuk SH, Oh KS, Cho SH, Lee BS, Kim SY, Kwak BK et al (2011) Glycol chitosan/heparin immobilized iron oxide nanoparticles with a tumor-targeting characteristic for magnetic resonance imaging. Biomacromolecules 12:2335–2343

7. Lord MS, Jung M, Teoh WY, Gunawan C, Vassie JA, Amal R et al (2012) Cellular uptake and reactive oxygen species modulation of cerium oxide nanoparticles in human monocyte cell line U937. Biomaterials 33:7915–7924

8. Nilasaroya A, Poole-Warren LA, Whitelock JM, Martens PJ (2008) Structural and functional characterisation of poly(vinyl alcohol) and heparin hydrogels. Biomaterials 29:4658–4664

Chapter 45

Role of Glycosaminoglycans in Infectious Disease

Akiko Jinno and Pyong Woo Park

Abstract

Glycosaminoglycans (GAGs) have been shown to bind to a wide variety of microbial pathogens, including viruses, bacteria, parasites, and fungi in vitro. GAGs are thought to promote pathogenesis by facilitating pathogen attachment, invasion, or evasion of host defense mechanisms. However, the role of GAGs in infectious disease has not been extensively studied in vivo and therefore their pathophysiological significance and functions are largely unknown. Here we describe methods to directly investigate the role of GAGs in infections in vivo using mouse models of bacterial lung and corneal infection. The overall experimental strategy is to establish the importance and specificity of GAGs, define the essential structural features of GAGs, and identify a biological activity of GAGs that promotes pathogenesis.

Key words Heparan sulfate, Chondroitin sulfate, Proteoglycan, Syndecan, Pneumonia, Keratitis, Cathelicidin, Antimicrobial peptide, Host defense

1 Introduction

Despite significant improvements in hygienic conditions and prophylactic and therapeutic interventions, infectious diseases continue to be a major global health problem. Infections killed approximately 9.3 million people worldwide in 2010, accounting for 18 % of all global deaths [1]. Lower respiratory infections, in particular, are associated with high mortality (fourth leading cause of death) and killed approximately 2.8 million people worldwide in 2010 [1]. Several infectious diseases are also associated with significant morbidity. For example, corneal infections afflict approximately 500,000 patients globally [2] and can lead to reduced visual acuity, irreversible scarring, and blindness [3–5]. Furthermore, infections often exacerbate and dysregulate the host's inflammatory response, resulting in serious acute and chronic inflammatory complications [6–9]. In addition, infection by several pathogens can lead to malignant disease, such as gastric cancer by *Helicobacter pylori* [10], cervical cancer by human papillomavirus virus (HPV) [11], and liver cancer by hepatitis B and C viruses (HBV and HCV) [12].

Kuberan Balagurunathan et al. (eds.), *Glycosaminoglycans: Chemistry and Biology*, Methods in Molecular Biology, vol. 1229, DOI 10.1007/978-1-4939-1714-3_45, © Springer Science+Business Media New York 2015

A major gap in our scientific knowledge centers on how pathogens interact with host components and modulate or subvert their activities to promote pathogenesis in vivo.

Glycosaminoglycans (GAGs) have been shown to interact with a wide variety of pathogens, including viruses, bacteria, parasites, and fungi [13–15]. GAG–pathogen interactions have been implicated in many steps of pathogenesis, including host cell attachment and invasion, infection of neighboring cells, and dissemination and infection of distant tissues [14, 16]. In cell-based assays, many viruses, including HSV [17], HPV [18], HBV [19], HCV [20], and enterovirus [21], have been shown to bind to cell surface heparan sulfate (HS) and utilize HS as a receptor for their initial attachment to host cells. Several bacteria, such as *H. pylori* [22], *Pseudomonas aeruginosa* [23], and *Borrelia burgdorferi* [24], similarly bind to cell surface HS for their attachment. HS interactions have also been proposed to promote host cell invasion of intracellular pathogens, such as HSV [25], *Neisseria gonorrhoeae* [26], and *Listeria monocytogenes* [27], and to facilitate the dissemination of *Mycobacterium tuberculosis* [28] and replication of *Toxoplasma gondii* [29]. Furthermore, several bacterial pathogens have been shown to induce the release of dermatan sulfate (DS) from the extracellular matrix (ECM) [30] or HS from the cell surface [31–34] and exploit the ability of solubilized GAGs to counteract cationic antimicrobial factors or neutrophil-mediated host defense mechanisms. In addition, several pathogens have been shown to subvert HS to prevent detection by immune mechanisms [35, 36]. Altogether, these data suggest that GAG–pathogen interactions and the ability of pathogens to subvert GAG functions are important virulence mechanisms for a wide variety of microbes.

GAGs are unbranched polysaccharides composed of repeating disaccharide units. GAGs include HS, heparin, chondroitin sulfate (CS), DS, keratan sulfate (KS), and hyaluronan (HA), each with unique disaccharide units and chemical linkages. Except for HA, all GAGs in vivo are found covalently conjugated to specific core proteins as proteoglycans, and expressed ubiquitously on the cell surface, in the extracellular matrix (ECM), and in intracellular compartments. Biosynthesis of GAGs on proteoglycans is initiated with the assembly of a tetrasaccharide linkage region, which is attached to specific Ser residues in core proteins. An unmodified GAG precursor is polymerized and then extensively modified in the Golgi. For example, in HS biosynthesis, the unmodified HS precursor is sequentially modified by *N*-deacetylase *N*-sulfotransferases (NDSTs), C5 epimerase, 2-*O*-sulfotransferase (2OST), 6OSTs, and 3OSTs [37, 38]. These reactions do not go to completion, resulting in a highly heterogeneous mature HS chain. Importantly, the unique and complex sulfation patterns of GAGs enable them to bind specifically to many biomolecules and regulate diverse biological processes [39–41]. For example, efficient FGF-2 binding by HS requires N-sulfated glucosamine and 2-O-sulfated iduronic acid in a decasaccharide sequence, whereas antithrombin III binding by HS and heparin requires a central trisulfated (N-, 6-O-, and 3-O-sulfates) glucosamine residue in a minimal pentasaccharide sequence [42]. Several GAG–pathogen interactions are also dictated by GAG modifications. For instance, HCV envelope glycoproteins E1 and E2 require both and N- and 6-O-sulfate groups for efficient interaction with HS [43], whereas *Chlamydia trachomatis* OmcB interacts with 6-O-sulfated HS domains [44]. These observations suggest that microbes subvert specific GAG modifications to promote their pathogenesis, but whether GAG modifications are indeed important in vivo has yet to be determined. In fact, our knowledge of the role of GAGs in infections is mostly derived from cell-based experiments performed in vitro, and their physiological significance, relevance, and function in infectious diseases have yet to be determined.

There are diverse animal models to study the role of GAGs in infections in vivo, but each has its own advantages and disadvantages. Studies using invertebrates, such as *Caenorhabditis elegans* and *Drosophila melanogaster*, and lower vertebrates, such as *Danio rerio* (zebrafish), are simple and cost-effective, and have yielded valuable mechanistic information about host–pathogen interactions and innate immune responses to infections [45–47]. Mutant organisms lacking various GAGs, GAG modification enzymes, and proteoglycans have also been generated, and methods to specifically knockdown the expression of certain genes are established [48–50]. However, the lower organisms lack particular organs (e.g., lungs) and the structure and function of some organs do not closely resemble those of humans. The invertebrates also lack adaptive immunity.

Larger mammalian species, such as rabbits, dogs, and monkeys, have also been used and they too have generated much significant information about pathogenic and host defense mechanisms in vivo. Several drawbacks of these mammalian models include a relative slow rate of reproduction, high cost of maintenance, lack of specific experimental reagents to precisely determine molecular mechanisms, and ethical issues. Rodent models, in particular mouse models, are used frequently because of their small size, relative rapid reproduction cycle, relative cost-effectiveness, ease of handling, and abundant availability of specific experimental tools, including various transgenic mouse lines in which a particular gene is overexpressed or has been ablated globally or in a cell-specific manner. The availability of many inbred mouse strains (e.g., C57BL/6, BALB/c) also allows researchers to study genetically identical cohorts and reduces experimental variability from genetic variations. Furthermore, mice are readily amenable to experimental prophylactic and therapeutic approaches, and their immune system is well characterized. However, mice are not humans, and results from mouse studies should also be interpreted with caution when relating to human diseases. Regardless, for the above reasons, mice are currently the most frequently used animals to study mechanisms of various human diseases in vivo. Here we describe experimental approaches to study the role of GAGs in mouse models of bacterial lung and corneal infections. A method to investigate the role of GAGs in bacterial killing by innate antimicrobial factors is also described. The primary focus is on HS because a large number of microbes have been proposed to subvert HS and HSPGs for their pathogenesis, but the methods described below can be readily adapted to study the role of other GAGs and proteoglycans.

2 Materials

2.1 Intranasal Lung Infection Assay

1. Mice: Mice are used at the age of 5–10 weeks (*see* **Note 1**). Inbred wild-type (Wt) mice are available from several vendors (*see* **Note 2**).

2. Tryptic soy broth (TSB) and tryptic soy agar (TSA): Powder stocks, premade solutions, and premade plates are available from several vendors. When using powder stocks, sterilize resuspended broth by autoclaving.

3. Bacteria: *S. aureus* strain 8325-4 (*see* **Note 3**). Strain 8325-4 can be stored short term on TSA slants or plates at 4 °C, or long term in 40 % glycerol/TSB at –80 °C.

4. Inoculation materials: Micropipette and tips are from general supply vendors. Sterilize tips by autoclaving.

5. Surgical tools: Fine scissors and forceps. Sterilize surgical tools by autoclaving.

6. Phosphate-buffered saline (PBS), pH 7.4: Premade tablets and concentrated stock solutions are available from several commercial sources. Sterilize PBS by autoclaving or filtering through a sterile 0.22 μm filter.

7. Tissue straining medium: Dulbecco's modified Eagle's medium (DMEM) with 10 % fetal bovine serum (FBS). Do not add antibiotics to the medium (i.e., penicillin and streptomycin).

8. Cell strainer (70 μm mesh size), plunger from a 5 ml syringe, polystyrene petri dishes (i.e., ones for pouring bacterial plates, not for cell culture), and polypropylene microcentrifuge tubes are from general supply vendors. Sterilize microfuge tubes by autoclaving; others are sold as sterile supplies.

9. TSB containing 0.1 % (v/v) Triton X-100: Mix autoclaved TSB with Triton X-100 and filter sterilize.

10. Spectrophotometer: A conventional spectrophotometer and disposable plastic cuvettes are used to measure the turbidity of bacterial suspensions.

2.2 Effect of GAGs and GAG Antagonists in Bacterial Lung and Corneal Infection

1. GAGs: Purified GAGs are available from several commercial sources (*see* **Note 4**). Make concentrated stock solutions in autoclaved deionized water or neutral buffer (e.g., PBS) and store at 4 °C for short-term storage or at –80 °C for long-term storage.

2. GAG antagonists: Many general inhibitors of GAGs are available from commercial sources. These include the cationic compounds such as protamine and surfen, and polysulfated anionic compounds such as carrageenans and suramin, among others [14, 51]. Make concentrated stock solutions in autoclaved deionized water or neutral buffer and store at 4 °C for short-term storage or at –80 °C for long-term storage.

2.3 Effects of GAG Lyases and GAG Derivatives in Bacterial Lung and Corneal Infection

1. GAG lyases: Heparinase I, II, and III, chondroitinases A, B, C, AC, and ABC, hyaluronidase, and keratanase are available from commercial sources. Make concentrated stock solutions in autoclaved neutral buffer and store small aliquots at –80 °C. Do not repeat freeze-thaw.

2. GAG derivatives: Desulfated heparin compounds, oversulfated heparin, heparin oligosaccharides, CS oligosaccharides, oversulfated CS, DS oligosaccharides and oversulfated DS and HA oligosaccharides, among others are available from commercial sources (*see* **Note 5**). Make concentrated stock solutions in autoclaved deionized water or neutral buffer and store at 4 °C for short-term storage or at –80 °C for long-term storage.

2.4 Use of Transgenic Knockout (KO) Mice in Intranasal Lung and Corneal Infection

1. Global KO mouse lines: Several global KO mouse lines lacking genes for specific GAG modification enzymes or proteoglycan core proteins have been published. For HS modification enzymes, mice lacking *Ndst2* [52], *Ndst3* [53], *Hs6st2* [54], or *Hs3st1* [55] are viable. Global ablation of other HS modification enzymes results in either embryonic (e.g., *Ext1*, *HS6st1*) or perinatal (e.g., *Ndst1*, *Glce*, *Hs2st*) lethality. For HSPGs, global KO mice lacking syndecan-1 (*Sdc1*) [31, 56], *Sdc3* [57], *Sdc4* [58], glypican-1 (*Gpc1*) [59], *Gpc3* [60], *Gpc4* [61], serglycin (*Prg1*) [62], or collagen XVIII (*Col18a1*) [63] are viable. Global ablation of other HSPG core protein genes are either embryonic lethal (e.g., agrin, perlecan) or currently not available (e.g., syndecan-2). Contact the corresponding principal investigator for availability of these mice (*see* **Note 6**).

2. Conditional KO mouse lines: Several conditional KO lines for GAG modification enzymes have been published. For HS, mice harboring a floxed construct of *Ext1* [64], *Ndst1* [65], *Hs2st* [66], or *Hs6st1* [66] have been generated and ablated in various cell types by crossing with cell-specific Cre reporter lines. These floxed conditional mice can be crossed with other Cre reporter mice to ablate GAG modification enzymes in certain cells or tissues. Contact the corresponding principal investigator for availability of these mice.

2.5 Scarified Corneal Infection

1. Sterile 29 G syringe needles.
2. Stereomicroscope.
3. Fine scissors and forceps. Sterilize by autoclaving.
4. TSB containing 0.1 % (v/v) Triton X-100. Sterilize by filtering.
5. Microfuge tubes. Sterilize by autoclaving.
6. GAGs, GAG antagonists, GAG lyases, and GAG derivatives. Prepare as described in Subheadings 2.2 and 2.3.

2.6 Antimicrobial Peptide Killing Assay

1. Bacteria: *P. aeruginosa* strain PAO1. PAO1 can be stored short term in a TSA slant or plate at 4 °C or long term in 40 % glycerol/TSB at –80 °C.
2. LL-37. Make concentrated stock solutions in deionized water or neutral buffer and store at 4 °C for short-term storage or at –80 °C for long-term storage.
3. PBS, TSB, TSA plates.
4. Microcentrifuge tubes. Sterilize by autoclaving.
5. GAGs and GAG derivatives. Prepare as described in Subheadings 2.2 and 2.3.

3 Methods

We describe here procedures of mouse models of *S. aureus* lung and corneal infection. These general methods can also be used to study pathogenic mechanisms of other microbial pathogens proposed to subvert GAGs for their pathogenesis. We also describe approaches to adapt these procedures to determine the significance of GAGs and to identify essential GAG modifications, and to characterize how GAGs modulate antimicrobial factors. The role of GAGs in vivo is initially probed by exogenous administration of GAG antagonists or purified GAGs. If GAGs promote infection, addition of GAG antagonists will inhibit pathogenesis and result in reduced tissue bacterial burden and other parameters of infection. General GAG biosynthesis inhibitors (e.g., xyloside, chlorate) have been used in studies in vitro, but these are not recommended for use in vivo because of their strong toxicity. Exogenous administration of excess GAGs should also reduce the tissue bacterial burden by inhibiting bacterial attachment if the GAG under study binds to the pathogen and facilitates attachment. On the other hand, if the GAG under study promotes pathogenesis by inhibiting host defense, then administration of the particular GAG should enhance bacterial virulence by interfering with bacterial eradication. An example of the latter mechanism is shown where addition of HS or heparin, but not CS-A or heparosan, promotes *S. aureus* corneal infection (Fig. 1b).

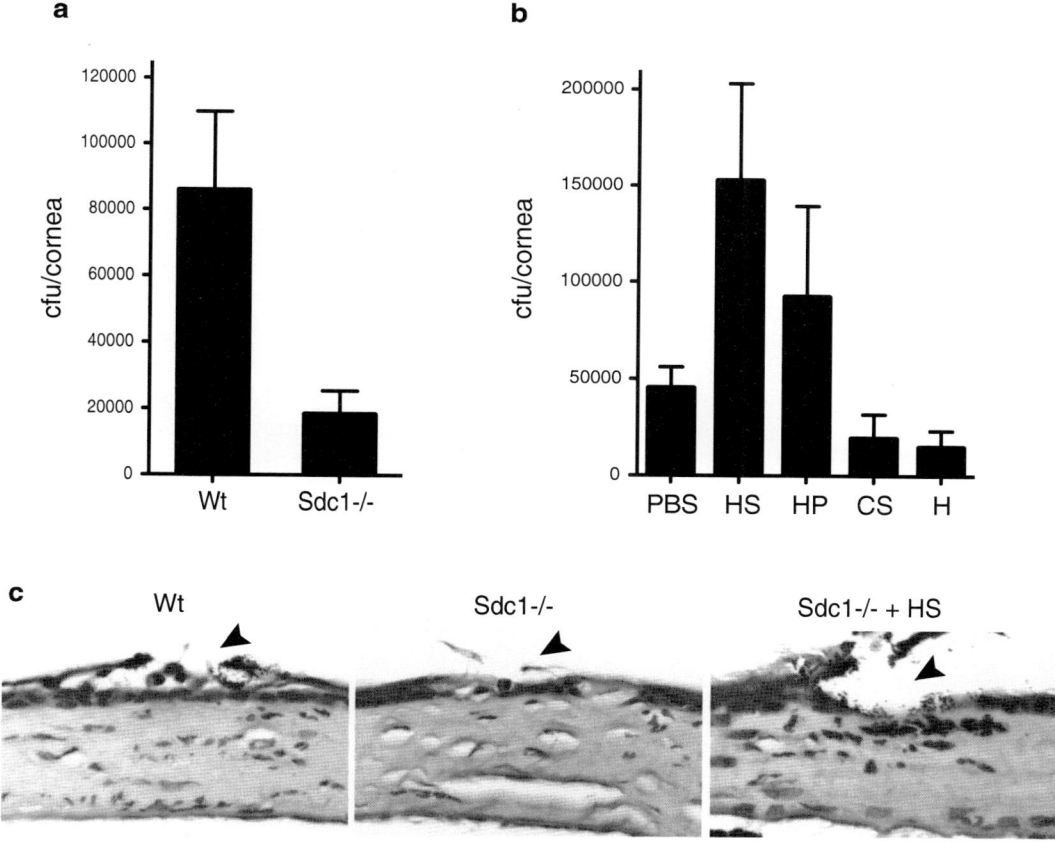

Fig. 1 Syndecan-1 promotes *S. aureus* corneal infection in an HS-dependent manner. (**a**) Corneas of anesthetized Wt and *Sdc1−/−* mice on the BALB/c background were scratched with a 29 G needle and infected topically with 1×10^9 cfu of *S. aureus* strain 8325-4. The corneal bacterial burden was quantified at 10 h postinfection. Data shown are mean ± S.E. (*n* = 9 in Wt and *n* = 6 in *Sdc1−/−* group). (**b**) Scarified Wt and *Sdc1−/−* corneas were infected with 1×10^9 cfu of 8325-4 with or without 200 ng of HS or heparin (HP), or 500 ng of CS-A (CS) or heparosan (H), and the corneal bacterial burden was quantified at 10 h postinfection. Data shown are mean ± S.E. (*n* = 11 in PBS, *n* = 10 in HS, *n* = 7 in HP, *n* = 4 in CS, and *n* = 5 in H group). (**c**) Paraffin-embedded eye sections of infected Wt and *Sdc1−/−* mice were Gram stained (*arrowhead* indicates injured areas). Note the increased number of Gram-positive cocci in Wt cornea infected with *S. aureus* only and *Sdc1−/−* cornea co-infected with *S. aureus* and HS compared to *Sdc1−/−* cornea infected with *S. aureus* only

GAG lyases and GAG derivatives are used to determine the essential structural features of GAGs that promote infection. Several GAG lyases selectively digest certain GAGs or regions in GAGs. For example, bacterial heparinase I and III digest sulfated and low sulfated regions of HS, respectively [67], thus allowing determination of whether sulfated or low sulfated HS domains are important in infection. Many selectively modified or size-defined GAG derivatives are also available, and these reagents are used to determine essential GAG modifications and minimum active size of

GAGs. For example, if *N*-sulfate groups of HS/heparin promote infection by inhibiting host defense mechanisms, bacterial virulence will be enhanced upon addition of intact heparin or 2-*O*- or 6-*O*-desulfated heparin, but not *N*-desulfated heparin.

The response of Wt and KO mice lacking genes for certain GAG biosynthesis and modification enzymes, or proteoglycan core proteins is compared to establish the physiological significance and relevance of certain GAGs, GAG modifications, and proteoglycans in infections. An example is shown where *Sdc1–/–* mice significantly resist *S. aureus* corneal infection relative to Wt mice (Fig. 1a), indicating that syndecan-1 is an important HSPG that promotes *S. aureus* pathogenesis in the cornea. Furthermore, *S. aureus* does not bind to syndecan-1 [34] and addition of exogenous HS or heparin markedly increases bacterial infection in the injured *Sdc1–/–* cornea (Fig. 1b, c), suggesting that HS chains of syndecan-1 promotes *S. aureus* corneal infection by interfering with host defense mechanisms.

Lastly, a method to explore the underlying biological mechanisms of how pathogens subvert GAGs is described. The in vivo studies should suggest whether GAGs are promoting infection by serving as an attachment site, facilitating dissemination, or inhibiting host defense mechanisms. Because many studies have examined the role of GAGs as attachment sites for viruses [68–70], bacteria [24, 71, 72], and parasites [73, 74], and microbial binding and attachment assays are established, these will not be discussed in this review. Instead, we describe a method to study whether GAGs interfere with the antibacterial activity of cationic antimicrobial peptides. Several studies suggest that pathogens not only subvert GAGs as attachment sites, but also as soluble effectors that counteract cationic antimicrobial peptides [31, 75–79]. Several potent antimicrobial peptides with broad activity towards many microbes have been identified, including the human cathelicidin LL-37 [80]. Here, we describe a method to measure the ability of HS to specifically inhibit the killing of *P. aeruginosa* by LL-37 (Fig. 2).

3.1 Intranasal Lung Infection Assay

1. Preparation of infectious inoculum. Grow 10 µl of *S. aureus* (strain 8325-4) from the glycerol stock overnight in 5 ml TSB at 37 °C with agitation. The next day, dilute the overnight culture and regrow 3–5 ml of the overnight culture in 30 ml TSB to mid-log growth phase ($OD_{600 \, nm}$: ~0.7) (*see* **Note 7**). Estimate the bacterial concentration by turbidity (i.e., based on the predetermined growth curve). Wash sufficient number of bacteria needed for the experiment by centrifuging at $10,000 \times g$ for 5 min, resuspending bacteria in PBS, and centrifuging at $10,000 \times g$ for 5 min. Discard supernatant and resuspend the bacterial pellet in PBS to the desired concentration. Plate out serial dilutions of the inoculum on TSA plates.

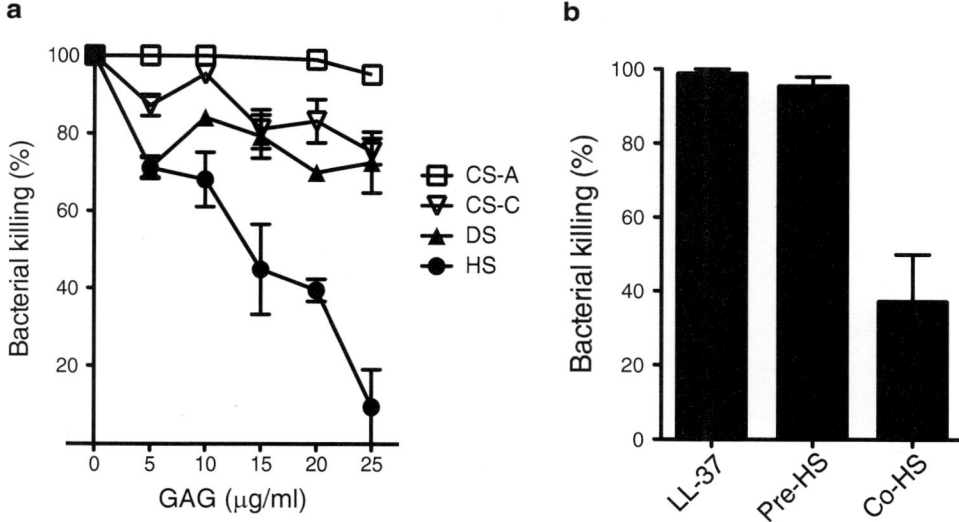

Fig. 2 HS specifically inhibits the killing of *P. aeruginosa* by LL-37. (**a**) *P. aeruginosa* (10^3 cfu) was incubated with LL-37 (3 μg/ml) in 30 μl PBS for 2 h at 37 °C in the absence or presence of increasing doses of CS-A, CS-C, DS, or HS. Bacterial killing was determined by plating out serial dilutions. Data shown are mean ± S.E. (*n* = 4 in each group). Note the significantly increased inhibitory activity of HS at doses ≥15 μg/ml compared to other GAGs. (**b**) *P. aeruginosa* was incubated with LL-37 (3 μg/ml) (LL-37 group), preincubated with HS (20 μg/ml) for 30 min, washed free of HS, and then incubated with LL-37 (Pre-HS group), or co-incubated with LL-37 and HS (Co-HS group) for 2 h at 37 °C in a microfuge tube. Bacterial killing was determined by plating out serial dilutions. Data shown are mean ± S.E. (*n* = 4 in each group). This experiment shows that HS does not inhibit LL-37 activity by binding to the bacteria, but rather by directly binding to LL-37 and inhibiting its antibacterial activity

Count number of colonies on the following day to determine the exact infectious dose. We generally infect with 10^7–10^9 cfu of *S. aureus* in 20 μl PBS per mouse.

2. Anesthetize mice with a mixture of ketamine (80–100 mg/kg) and xylazine (5–10 mg/kg) via intraperitoneal (i.p.) injection (*see* **Note 8**). Pinch a toe and see that there is no withdrawal reaction to confirm that a stable plane of anesthesia has been achieved to minimize variations in the inhalation of the infectious dose.

3. Slowly infect each nostril with 10 μl of the infectious inoculum using a micropipette (*see* **Note 9**). You are infecting too fast if you see expulsion of the inoculum from the nose (i.e., bubbles forming through the nostrils).

4. At various time points, euthanize mice by anesthesia followed by cervical dislocation. Carefully isolate the whole lung with fine scissors and forceps (*see* **Note 10**).

5. Weigh lungs and place in a 100 mm petri dish with 3 ml of DMEM with 10 % FBS (*see* **Note 11**). In the petri dish, strain lungs through a 70 μm filter using a plunger of a 5 ml polypropylene

syringe. Wash the strainer once with 1 ml DMEM with 10 % FBS to remove loosely attached strained tissues.

6. Transfer 1–2 ml of the strained tissue mixture to microcentrifuge tubes and centrifuge at $10,000 \times g$ for 10 min. Discard supernatant and resuspend pellet in 500 μl TSB containing 0.1 % Triton X-100.

7. Incubate for 30 min at room temperature with vigorous vortexing every 10 min to lyse host cells and to recover both intracellular and extracellular bacteria. *S. aureus* and most bacteria are not lysed by 0.1 % Triton X-100.

8. Prepare serial dilutions of the detergent extract in TSB and plate onto TSA plates.

9. Incubate overnight at 37 °C and count the colonies the following day. Based on the dilutions at **steps 6** and **8**, back calculate the bacterial burden per mg of lung tissue.

3.2 Effects of GAGs and GAG Antagonists in Intranasal Lung Infection (See Note 12)

1. To assess the effects of exogenous GAGs, dilute GAGs in PBS to the desired concentration. The dose of GAGs should be chosen based on preliminary titration experiments. For example, in studies examining the effects of heparin on intranasal *P. aeruginosa* lung infection [31], the dose of 35 ng per mouse was chosen based on preliminary experiments testing the ability of 10–300 ng of heparin to enhance *P. aeruginosa* lung virulence.

2. When co-administering GAGs with bacteria, resuspend washed bacteria in **step 1** of Subheading 3.1 in the GAG solution. Proceed to **steps 2–9** of Subheading 3.1 to determine the effects of exogenous GAGs on the lung bacterial burden. A PBS only control should be included in the assay as well as several other GAGs to serve as specificity controls.

3. The effects of treating mice with GAGs before or after infection can also be examined using this method. However, because intranasal infection of adult mice must be performed under anesthesia, any treatment before or after infection requires additional anesthesia. In these experiments, a control group treated identically with vehicle should be included.

4. To assess the effects of GAG antagonists, such as protamine, dilute antagonists in PBS to the desired concentration and resuspend the washed bacteria in **step 1** of Subheading 3.1 in the GAG antagonist solution. Proceed to **steps 2–9** of Subheading 3.1 to determine the effects of GAG antagonists on the lung bacterial burden (*see* **Note 12**).

3.3 Effects of GAGs Lyases and GAG Derivatives in Intranasal Lung Infection (See Note 12)

1. Dilute GAG lyases in PBS to the desired concentration. The effective dose should be selected based on preliminary titration experiments. In studies examining the effects of heparinase II and chondroitinase ABC on intranasal *P. aeruginosa* infection, we found that 0.3 mU per mouse of heparinase II and

chondroitinase ABC effectively removes HS and CS from the surface of airway epithelial cells. Moreover, we found that heparinase II significantly attenuates bacterial virulence in newborn lungs, whereas chondroitinase ABC had no effect, indicating that HS specifically promotes *P. aeruginosa* lung infection in newborn mice [31].

2. Resuspend the washed bacteria in **step 1** of Subheading 3.1 in the GAG lyase solution. Proceed to **steps 2–9** of Subheading 3.1 to determine the effects of GAG lyases on the lung bacterial burden.

3. To assess the effects of GAG derivatives, such as desulfated heparin compounds and HS oligosaccharides, dilute the derivatives in PBS to the desired concentration and resuspend the washed bacteria in **step 1** of Subheading 3.1 in this solution. Proceed to **steps 2–9** of Subheading 3.1 to determine the effects of exogenous GAG derivatives on the lung bacterial burden (*see* **Note 13**).

3.4 Use of Transgenic KO Mice in Intranasal Lung Infection

1. To compare the response of Wt and KO mice lacking certain GAGs, GAG modification enzymes, or proteoglycans to intranasal bacterial infection, proceed to **steps 1–9** of Subheading 3.1 using the control Wt mice and experimental KO mice (*see* **Note 10** to assess other parameters of infection).

3.5 Scarified Corneal Infection Assay

1. Prepare mid-log growth phase *S. aureus* (strain 8325-4) as described in Subheading 3.1, **step 1**. We generally infect an eye with 10^7–10^9 cfu in 5 μl PBS. Plate out serial dilutions of the inoculum on TSA plates. Count number of colonies on the following day to determine the exact infectious dose.

2. Anesthetize mice as described in Subheading 3.1, **step 2**.

3. A single vertical scratch is made with a 29 G needle in one of the corneas without penetrating beyond the superficial stroma under a dissecting microscope (*see* **Note 14**). The other eye serves as an uninjured control.

4. Carefully infect the scarified cornea topically with an inoculum of up to 5 μl using a micropipette. This volume will fill the entire ocular surface, but uninjured regions in the cornea and conjunctiva are not infected even with a high bacterial dose. Avoid volumes larger than 5 μl as they may not be retained by the surface tension of the ocular surface and spill over, resulting in variable results.

5. Treat mice with an analgesic, such as buprenorphine (0.1 mg/kg, subcutaneously, once). This is required by the IACUC to alleviate pain from the scratch injury.

6. At various time points, euthanize mice by anesthesia followed by cervical dislocation. Enucleate eyes with fine forceps.

7. Isolate the cornea from the sclera with fine scissors and forceps under a stereomicroscope (i.e., a basic dissecting microscope) and place the isolated corneas in microfuge tubes with 200 μl TSB containing 0.1 % Triton X-100.

8. Incubate for 30 min at room temperature with vigorous vortexing every 10 min.

9. Prepare serial dilutions of the detergent extract and plate onto TSA plates. Incubate overnight at 37 °C and count the colonies the following day to determine the bacterial burden per cornea (*see* **Note 15**).

10. To examine the effects of GAGs, GAG antagonists, GAG derivatives, or GAG lyases, dilute the test reagents in PBS to the desired concentration and resuspend the washed bacteria in this solution (*see* **Note 16**). The effective dose of the reagents should be determined in preliminary titration experiments. Proceed to **steps 2–9**, Subheading 3.5 (above) to determine the effects of the reagents on the corneal bacterial burden. Results from studies examining the effects of HS, heparin, CS-A, and heparosan on *S. aureus* corneal infection in *Sdc1−/−* mice are shown (Fig. 1b).

11. To compare the response of Wt and KO mice lacking certain GAGs, GAG modification enzymes, or proteoglycans, proceed to **steps 1–9**, Subheading 3.5 (above) using control Wt mice and experimental KO mice. Results from studies comparing the response of Wt and *Sdc1−/−* mice on the BALB/c background to *S. aureus* corneal infection are shown (Fig. 1a).

3.6 Antimicrobial Peptide Killing Assay

1. Grow 10 μl of *P. aeruginosa* (strain PAO1) from the glycerol stock in 5 ml TSB overnight at 37 °C with agitation. The next day, dilute the overnight culture and regrow 3–5 ml of the overnight culture in 30 ml TSB to mid-log growth phase. Estimate the bacterial concentration by turbidity (i.e., based on the predetermined growth curve), spin down sufficient number of bacteria needed for the assay at $10,000 \times g$ for 5 min, wash bacteria with PBS, and resuspend the washed bacteria to the desired concentration in PBS. Plate out serial dilutions of the resuspended bacteria onto TSA plates and count the number of colonies the following day to determine the exact bacterial concentration.

2. Dilute LL-37 (*see* **Note 17**) to the desired concentration in PBS. LL-37 has been reported to effectively kill *P. aeruginosa* at the dose range of 0.1–10 μg/ml [81, 82].

3. Dilute GAGs and GAG derivatives to the desired concentration in PBS.

4. Incubate bacteria and LL-37 in the absence or presence of GAGs and GAG derivatives for 2 h at 37 °C.

5. Make serial dilutions and plate out onto TSA plates. Incubate overnight and count the colonies the following day to determine the proportion of bacteria killed by LL-37. HS, but not CS-A, CS-C, or DS, potently inhibits the killing of *P. aeruginosa* by LL-37 at doses ≥15 µg/ml (Fig. 2a).

4   Notes

1. All animal experiments must be approved by the local Institutional Animal Care and Use Committee (IACUC) and comply with federal guidelines for research with experimental animals.

2. C57BL/6 and BALB/c are the most frequently used inbred strains in mouse models of infectious disease. However, there are strain-specific differences in the susceptibility to infection. For example, BALB/c mice are highly susceptible to *S. aureus* corneal infection, whereas C57BL/6 mice are relatively less susceptible [83, 84]. Another note of caution is that the BL/6 strains from JAX and Charles River have different origins, hence they are genetically different and the designation "BL/6J" is used for Wt mice on the C57BL/6 background from JAX. Also, when comparing the response of Wt and KO mice, one should use Wt littermates obtained from het crosses of the KO line unless the KO line has been backcrossed ≥10 times onto a particular background (i.e., congenic).

3. We frequently use *S. aureus* strain 8325-4 for our studies because methods to genetically manipulate this laboratory strain are established. However, other *S. aureus* strains, including clinical isolates, can also be genetically modified and they are available from ATCC. In fact, it is important to confirm key data with at least two different strains to exclude the possibility of strain-specific effects for any pathogen under study.

4. The tissue source of GAGs should be considered. For example, most commercial HS is isolated from bovine or porcine tissues. Although we found that porcine HS and heparin potently enhance bacterial infection in mouse corneas (Fig. 1 and [34]) and lung [31], they may not be active in other tissue compartments.

5. Methods to chemically desulfate heparin are established and can be performed in-house [85–87]. Alternatively, GAGs from mutant CHO cells lacking certain GAGs and GAG modifications [68, 88] can be isolated and tested in the infection assays.

6. When using global or conditional KO mice to establish the teoglycan, the off-target effects of the mutation must be carefully considered. For example, ablation of *Hs2st* in endothelial

cells increases N- and 6-O-sulfation of endothelial cell HS and results in enhanced neutrophil infiltration [89]. Increased or decreased innate immune responses may have profound effects on the outcome of infection.

7. The expression level of *S. aureus* virulence factors is regulated by growth phase [90]. In general, adhesin expression is high during early to mid-log growth, whereas exotoxin expression is high during late-log to stationary growth. The in vivo virulence of each bacterial species and strains at different growth phases should be determined in pilot studies.

8. An alternate anesthetic to ketamine/xylazine is isoflurane. Isoflurane anesthesia (2–4 %) is given by inhalation, with scavenging by either house vacuum or fume hood. Recovery from isoflurane anesthesia is faster than ketamine/xylazine, thus it is possible to increase the volume of the infectious inoculum up to ~50 μl in the intranasal lung infection assay.

9. Other routes of lung infection are inhalation, intratracheal, and peroral. Each route has its own advantages and disadvantages. Mice are obligate nose breathers; therefore, intranasal administration under anesthesia leads to lung infection and not gastric infection. However, deposition of bacteria can be variable because the inoculum is given in small volumes to avoid drowning of anesthetized mice. There is less experimental variability in intratracheal administration, but this method is invasive. The inhalation method is not invasive, does not require anesthesia, and uniform dosing can be achieved, but this method requires costly aerosol exposure systems and additional protective measures to protect personnel performing the assay and in the vicinity. The peroral approach mimics oropharyngeal aspiration, which is the route that causes aspiration pneumonia, and the procedure is simple. However, deposition of bacteria is asymmetric and nonuniform. We use the intranasal method because it is simple and cost-effective, and the drawbacks associated with variable dosing can be overcome with experience.

10. The procedure described in Subheading 3.1 measures the lung bacterial burden, but this method can be easily adapted to assess other key parameters of lung infection. For example, prior to isolation of lung lobes (Subheading 3.1, **step 4**), lungs can be lavaged with 1–3 ml of PBS to collect bronchoalveolar lavage (BAL) fluids [31]. BAL fluids can be used to assess inflammatory parameters, such as total protein (measure of lung injury, edema) by Bradford or BCA (kits available from Bio-Rad), cytokine levels by ELISA (Biolegend, Peprotech, R&D Systems), and leukocyte infiltration by differentially staining cytospun slides for leukocyte subsets by Giemsa (Fisher). Lungs can also be inflated, fixed, and processed for

histopathological analyses [31]. Lung sections can be stained with hematoxylin and eosin to assess inflammation or Gram's solution to visualize bacteria (*see* Fig. 1c), or immunostained to examine the expression of specific inflammatory factors (e.g., cytokines) and accumulation of leukocyte subsets (e.g., neutrophils, macrophages, lymphocytes). Lung lobes can also be homogenized to prepare total lung homogenates, which can be used to measure mRNA and protein levels of molecules relevant to infection.

11. Straining lung tissues in DMEM with 10 % FBS prevents non-specific adhesion of bacteria to plastic surfaces.

12. Off-target and adverse effects of test compounds (e.g., GAGs and derivatives, GAG antagonists, GAG lyases) on the bacteria and host should be determined. For adverse effects on the host, parameters such as weight loss, blood leukocyte counts (by CBC analysis), tissue injury (by histopathology or serum chemistry), and BAL total protein should be assessed in mice administered with test compounds only. The effects of test compounds on bacterial growth and viability should also be determined. For example, protamine has antibacterial activity at high concentrations in vitro (≥ 50 μg/ml) and it can also induce allergic inflammatory responses in vivo [91].

13. An alternate route to intranasal administration of test compounds is intravenous (i.v.) injection through the tail vein. However, because larger amounts of test compounds are required and i.v. administration is associated with unforeseen systemic effects, we recommend the local, intranasal route for the lung infection assay and local, topical route for the corneal infection assay.

14. Injury to the corneal epithelium is required to establish infection because intact corneas are highly resistant to infection. The degree of infection can be controlled by the number and size of scratches.

15. The procedure described in Subheading 3.5 measures the corneal bacterial burden. However, this method can be easily adapted to measure other parameters of corneal infection [34]. For example, ocular surface fluids from infected mice can be collected by consecutively incubating with 5 μl of 1 % N-acetylcysteine in PBS for 5 min (to break mucous layer of tear film) and 5 μl of PBS. The recovered ocular surface fluid can be used to measure levels of inflammatory (e.g., cytokines) and host defense (e.g., antimicrobials) factors. The infected corneas can also be processed for histopathological analyses. Eye sections can be stained with hematoxylin-eosin, Gram's solution (*see* Fig. 1c), or immunostained for inflammatory factors or leukocyte subsets. Corneal homogenates can also be

used to measure mRNA and protein levels of molecules relevant to infection. However, because of the small size, several corneas will have to be pooled to obtain sufficient amounts for these studies.

16. Mice can also be treated with GAGs or GAG-related reagents before or after infection. However, pre- or posttreatment requires additional anesthesia to prevent blinking during topical administration of these reagents.

17. The bacterial killing assay described here is for the human cathelicidin LL-37, but this method can be used to determine the effects of GAGs on other antimicrobial peptides, such as α- and β-defensins and other cathelicidins (CRAMP, PR-39) [31].

Acknowledgements

We would like to thank past and current members of the Park laboratory for developing essential reagents and constantly refining the described procedures. This work was supported by NIH grants R01 EY021765 and R01 HL107472.

References

1. Lozano R, Naghavi M, Foreman K et al (2012) Global and regional mortality from 235 causes of death for 20 age groups in 1990 and 2010: a systematic analysis for the Global Burden of Disease Study 2010. Lancet 380:2095–2128

2. Wilhelmus KR (2002) Indecision about corticosteroids for bacterial keratitis: an evidence-based update. Ophthalmology 109:835–842

3. Bourcier T, Thomas F, Borderie V, Chaumeil C, Laroche L (2003) Bacterial keratitis: predisposing factors, clinical and microbiological review of 300 cases. Br J Ophthalmol 87: 834–838

4. Limberg MB (1991) A review of bacterial keratitis and bacterial conjunctivitis. Am J Ophthalmol 112:2S–9S

5. Jett BD, Gilmore MS (2002) Host-parasite interactions in Staphylococcus aureus keratitis. DNA Cell Biol 21:397–404

6. Busse WW, Lemanske RF Jr, Gern JE (2010) Role of viral respiratory infections in asthma and asthma exacerbations. Lancet 376: 826–834

7. Abusriwil H, Stockley RA (2007) The interaction of host and pathogen factors in chronic obstructive pulmonary disease exacerbations and their role in tissue damage. Proc Am Thorac Soc 4:611–617

8. Folkesson A, Jelsbak L, Yang L et al (2012) Adaptation of Pseudomonas aeruginosa to the cystic fibrosis airway: an evolutionary perspective. Nat Rev Microbiol 10:841–851

9. Angus DC, van der Poll T (2013) Severe sepsis and septic shock. N Engl J Med 369:840–851

10. Cover TL, Blaser MJ (2009) Helicobacter pylori in health and disease. Gastroenterology 136:1863–1873

11. Bzhalava D, Guan P, Franceschi S, Dillner J, Clifford G (2013) A systematic review of the prevalence of mucosal and cutaneous human papillomavirus types. Virology 445:224–231

12. Rehermann B (2013) Pathogenesis of chronic viral hepatitis: differential roles of T cells and NK cells. Nat Med 19:859–868

13. Rostand KS, Esko JD (1997) Microbial adherence to and invasion through proteoglycans. Infect Immun 65:1–8

14. Bartlett AH, Park PW (2010) Proteoglycans in host-pathogen interactions: molecular mechanisms and therapeutic implications. Expert Rev Mol Med 12:e5

15. Spillmann D (2001) Heparan sulfate: anchor for viral intruders? Biochimie 83:811–817

16. Teng YH, Aquino RS, Park PW (2012) Molecular functions of syndecan-1 in disease. Matrix Biol 31:3–16

17. Shukla D, Liu J, Blaiklock P et al (1999) A novel role for 3-O-sulfated heparan sulfate in herpes simplex virus 1 entry. Cell 99:13–22

18. Johnson KM, Kines RC, Roberts JN, Lowy DR, Schiller JT, Day PM (2009) Role of heparan sulfate in attachment to and infection of the murine female genital tract by human papillomavirus. J Virol 83:2067–2074

19. Leistner CM, Gruen-Bernhard S, Glebe D (2008) Role of glycosaminoglycans for binding and infection of hepatitis B virus. Cell Microbiol 10:122–133

20. Shi Q, Jiang J, Luo G (2013) Syndecan-1 serves as the major receptor for attachment of hepatitis C virus to the surfaces of hepatocytes. J Virol 87:6866–6875

21. Tan CW, Poh CL, Sam IC, Chan YF (2013) Enterovirus 71 uses cell surface heparan sulfate glycosaminoglycan as an attachment receptor. J Virol 87:611–620

22. Guzman-Murillo MA, Ruiz-Bustos E, Ho B, Ascencio F (2001) Involvement of the heparan sulphate-binding proteins of Helicobacter pylori in its adherence to HeLa S3 and Kato III cell lines. J Med Microbiol 50:320–329

23. Bucior I, Mostov K, Engel JN (2010) Pseudomonas aeruginosa-mediated damage requires distinct receptors at the apical and basolateral surfaces of the polarized epithelium. Infect Immun 78:939–953

24. Isaacs RD (1994) Borrelia burgdorferi bind to epithelial proteoglycan. J Clin Invest 93:809–819

25. O'Donnell CD, Tiwari V, Oh MJ, Shukla D (2006) A role for heparan sulfate 3-O-sulfotransferase isoform 2 in herpes simplex virus type 1 entry and spread. Virology 346:452–459

26. Freissler E, Meyer auf der Heyde A, David G, Meyer TF, Dehio C (2000) Syndecan-1 and syndecan-4 can mediate the invasion of Opa_{HSPG}-expressing Neisseria gonorrhoeae into epithelial cells. Cell Microbiol 2:69–82

27. Alvarez-Dominguez C, Vasquez-Boland J, Carrasco-Marin E, Lopez-Mato P, Leyva-Cobian F (1997) Host cell heparan sulfate proteoglycans mediate attachment and entry of Listeria monocytogenes, and the listerial surface protein ActA is involved in heparan sulfate receptor recognition. Infect Immun 65:78–88

28. Pethe K, Alonso S, Biet F et al (2001) The heparin-binding haemagglutinin of M. tuberculosis is required for extrapulmonary dissemination. Nature 412:190–194

29. Bishop JR, Crawford BE, Esko JD (2005) Cell surface heparan sulfate promotes replication of Toxoplasma gondii. Infect Immun 73:5395–5401

30. Schmidtchen A, Frick I, Björck L (2001) Dermatan sulfate is released by proteinases of common pathogenic bacteria and inactivates antibacterial alpha-defensin. Mol Microbiol 39:708–713

31. Park PW, Pier GB, Hinkes MT, Bernfield M (2001) Exploitation of syndecan-1 shedding by Pseudomonas aeruginosa enhances virulence. Nature 411:98–102

32. Park PW, Foster TJ, Nishi E, Duncan SJ, Klagsbrun M, Chen Y (2004) Activation of syndecan-1 ectodomain shedding by Staphylococcus aureus alpha-toxin and beta-toxin. J Biol Chem 279:251–258

33. Chen Y, Bennett A, Hayashida A, Hollingshead S, Park PW (2005) Streptococcus pneumoniae sheds syndecan-1 ectodomains via ZmpC, a metalloproteinase virulence factor. J Biol Chem 282:159–167

34. Hayashida A, Amano S, Park PW (2011) Syndecan-1 promotes Staphylococcus aureus corneal infection by counteracting neutrophil-mediated host defense. J Biol Chem 285:3288–3297

35. Dubreuil JD, Giudice GD, Rappuoli R (2002) Helicobacter pylori interactions with host serum and extracellular matrix proteins: potential role in the infectious process. Microbiol Mol Biol Rev 66:617–629, table of contents

36. Duensing TD, Wing JS, van Putten JPM (1999) Sulfated polysaccharide-directed recruitment of mammalian host proteins: a novel strategy in microbial pathogenesis. Infect Immun 67:4463–4468

37. Esko JD, Selleck SB (2002) Order out of chaos: assembly of ligand binding sites in heparan sulfate. Annu Rev Biochem 71:435–471

38. Lindahl U, Kusche-Gullberg M, Kjellén L (1998) Regulated diversity of heparan sulfate. J Biol Chem 273:24979–24982

39. Perrimon N, Bernfield M (2000) Specificities of heparan sulphate proteoglycans in developmental processes. Nature 404:725–728

40. Funderburgh JL (2000) Keratan sulfate: structure, biosynthesis, and function. Glycobiology 10:951–958

41. Mikami T, Kitagawa H (2013) Biosynthesis and function of chondroitin sulfate. Biochim Biophys Acta 1830:4719–4733

42. Whitelock JM, Iozzo RV (2005) Heparan sulfate: a complex polymer charged with biological activity. Chem Rev 105:2745–2764

43. Kobayashi F, Yamada S, Taguwa S et al (2012) Specific interaction of the envelope glycoproteins E1 and E2 with liver heparan sulfate

involved in the tissue tropismatic infection by hepatitis C virus. Glycoconj J 29:211–220

44. Fechtner T, Stallmann S, Moelleken K, Meyer KL, Hegemann JH (2013) Characterization of the interaction between the chlamydial adhesin OmcB and the human host cell. J Bacteriol 195:5323–5333

45. O'Callaghan D, Vergunst A (2010) Non-mammalian animal models to study infectious disease: worms or fly fishing? Curr Opin Microbiol 13:79–85

46. Dorer MS, Isberg RR (2006) Non-vertebrate hosts in the analysis of host-pathogen interactions. Microbes Infect 8:1637–1646

47. Ferrandon D, Imler JL, Hetru C, Hoffmann JA (2007) The Drosophila systemic immune response: sensing and signalling during bacterial and fungal infections. Nat Rev Immunol 7:862–874

48. Lee JS, Chien CB (2004) When sugars guide axons: insights from heparan sulphate proteoglycan mutants. Nat Rev Genet 5:923–935

49. Nakato H, Kimata K (2002) Heparan sulfate fine structure and specificity of proteoglycan functions. Biochim Biophys Acta 1573: 312–318

50. Nishihara S (2010) Glycosyltransferases and transporters that contribute to proteoglycan synthesis in Drosophila: Identification and functional analyses using the heritable and inducible RNAi system. Methods Enzymol 480:323–351

51. Brown JR, Crawford BE, Esko JD (2007) Glycan antagonists and inhibitors: a fount for drug discovery. Crit Rev Biochem Mol Biol 42:481–515

52. Forsberg E, Pejler G, Ringvall M et al (1999) Abnormal mast cells in mice deficient in a heparin-synthesizing enzyme. Nature 400: 773–776

53. Pallerla SR, Lawrence R, Lewejohann L et al (2008) Altered heparan sulfate structure in mice with deleted NDST3 gene function. J Biol Chem 283:16885–16894

54. Sugaya N, Habuchi H, Nagai N, Ashikari-Hada S, Kimata K (2008) 6-O-sulfation of heparan sulfate differentially regulates various fibroblast growth factor-dependent signalings in culture. J Biol Chem 283:10366–10376

55. Shworak NW, HajMohammadi S, de Agostini AI, Rosenberg RD (2002) Mice deficient in heparan sulfate 3-O-sulfotransferase-1: normal hemostasis with unexpected perinatal phenotypes. Glycoconj J 19:355–361

56. Alexander CM, Reichsman F, Hinkes MT et al (2000) Syndecan-1 is required for Wnt-1-induced mammary tumorigenesis in mice. Nat Genet 25:329–332

57. Reizes O, Lincecum J, Wang Z et al (2001) Transgenic expression of syndecan-1 uncovers a physiological control of feeding behavior by syndecan-3. Cell 106:105–116

58. Echtermeyer F, Streit M, Wilcox-Adelman S et al (2001) Delayed wound repair and impaired angiogenesis in mice lacking syndecan-4. J Clin Invest 107:R9–R14

59. Jen YH, Musacchio M, Lander AD (2009) Glypican-1 controls brain size through regulation of fibroblast growth factor signaling in early neurogenesis. Neural Dev 4:33

60. Cano-Gauci DF, Song HH, Yang H et al (1999) Glypican-3-deficient mice exhibit developmental overgrowth and some of the abnormalities typical of Simpson-Golabi-Behmel syndrome. J Cell Biol 146:255–264

61. Allen NJ, Bennett ML, Foo LC et al (2012) Astrocyte glypicans 4 and 6 promote formation of excitatory synapses via GluA1 AMPA receptors. Nature 486:410–414

62. Abrink M, Grujic M, Pejler G (2004) Serglycin is essential for maturation of mast cell secretory granule. J Biol Chem 279:40897–40905

63. Li Q, Olsen BR (2004) Increased angiogenic response in aortic explants of collagen XVIII/endostatin-null mice. Am J Pathol 165: 415–424

64. Inatani M, Irie F, Plump AS, Tessier-Lavigne M, Yamaguchi Y (2003) Mammalian brain morphogenesis and midline axon guidance require heparan sulfate. Science 302:1044–1046

65. Wang L, Fuster M, Sriramarao P, Esko JD (2005) Endothelial heparan sulfate deficiency impairs L-selectin- and chemokine-mediated neutrophil trafficking during inflammatory responses. Nat Immunol 6:902–910

66. Stanford KI, Wang L, Castagnola J et al (2010) Heparan sulfate 2-O-sulfotransferase is required for triglyceride-rich lipoprotein clearance. J Biol Chem 285:286–294

67. Liu D, Shriver Z, Venkataraman G, El Shabrawi Y, Sasisekharan R (2002) Tumor cell surface heparan sulfate as cryptic promoters or inhibitors of tumor growth and metastasis. Proc Natl Acad Sci U S A 99:568–573

68. Avirutnan P, Zhang L, Punyadee N et al (2007) Secreted NS1 of dengue virus attaches to the surface of cells via interactions with heparan sulfate and chondroitin sulfate E. PLoS Pathog 3:1798–1812

69. Schowalter RM, Pastrana DV, Buck CB (2011) Glycosaminoglycans and sialylated glycans sequentially facilitate Merkel cell polyomavirus infectious entry. PLoS Pathog 7:e1002161

70. Hu YP, Lin SY, Huang CY et al (2011) Synthesis of 3-O-sulfonated heparan sulfate octasaccharides that inhibit the herpes simplex

virus type 1 host-cell interaction. Nat Chem 3:557–563

71. Bucior I, Pielage JF, Engel JN (2012) Pseudomonas aeruginosa pili and flagella mediate distinct binding and signaling events at the apical and basolateral surface of airway epithelium. PLoS Pathog 8:e1002616

72. Yabushita H, Noguchi Y, Habuchi H et al (2002) Effects of chemically modified heparin on *Chlamydia trachomatis* serovar L2 infection of eukaryotic cells in culture. Glycobiology 12:345–351

73. Love DC, Esko JD, Mosser DM (1993) A heparin-binding activity on Leishmania amastigotes which mediates adhesion to cellular proteoglycans. J Cell Biol 123:759–766

74. Oliveira FO Jr, Alves CR, Calvet CM et al (2008) *Trypanosoma cruzi* heparin-binding proteins and the nature of the host cell heparan sulfate-binding domain. Microb Pathog 44:329–338

75. Kaneider NC, Djanani A, Wiedermann CJ (2007) Heparan sulfate proteoglycan-involving immunomodulation by cathelicidin antimicrobial peptides LL-37 and PR-39. ScientificWorldJournal 7:1832–1838

76. Baranska-Rybak W, Sonesson A, Nowicki R, Schmidtchen A (2006) Glycosaminoglycans inhibit the antibacterial activity of LL-37 in biological fluids. J Antimicrob Chemother 57:260–265

77. Bergsson G, Reeves EP, McNally P et al (2009) LL-37 complexation with glycosaminoglycans in cystic fibrosis lungs inhibits antimicrobial activity, which can be restored by hypertonic saline. J Immunol 183:543–551

78. Wu H, Monroe DM, Church FC (1995) Characterization of the glycosaminoglycan-binding region of lactoferrin. Arch Biochem Biophys 317:85–92

79. Zou S, Magura CE, Hurley WL (1992) Heparin-binding properties of lactoferrin and lysozyme. Comp Biochem Physiol B 103:889–895

80. Zanetti M (2005) The role of cathelicidins in the innate host defenses of mammals. Curr Issues Mol Biol 7:179–196

81. Travis SM, Anderson NN, Forsyth WR et al (2000) Bactericidal activity of mammalian cathelicidin-derived peptides. Infect Immun 68:2748–2755

82. Schmidtchen A, Frick IM, Andersson E, Tapper H, Bjorck L (2002) Proteinases of common pathogenic bacteria degrade and inactivate the antibacterial peptide LL-37. Mol Microbiol 46:157–168

83. Hume EB, Cole N, Khan S et al (2005) A Staphylococcus aureus mouse keratitis topical infection model: cytokine balance in different strains of mice. Immunol Cell Biol 83:294–300

84. Girgis DO, Sloop GD, Reed JM, O'Callaghan RJ (2004) Susceptibility of aged mice to Staphylococcus aureus keratitis. Curr Eye Res 29:269–275

85. Inoue Y, Nagasawa K (1976) Selective N-desulfation of heparin with dimethyl sulfoxide containing water or methanol. Carbohydr Res 46:87–95

86. Ishihara M, Kariya Y, Kikuchi H, Minamisawa T, Yoshida K (1997) Importance of 2-O-sulfate groups of uronate residues in heparin for activation of FGF-1 and FGF-2. J Biochem 121:345–349

87. Kariya Y, Kyogashima M, Suzuki K et al (2000) Preparation of completely 6-O-desulfated heparin and its ability to enhance activity of basic fibroblast growth factor. J Biol Chem 275:25949–25958

88. Zhang L, Lawrence R, Frazier BA, Esko JD (2006) CHO glycosylation mutants: proteoglycans. Methods Enzymol 416:205–221

89. Axelsson J, Xu D, Kang BN et al (2012) Inactivation of heparan sulfate 2-O-sulfotransferase accentuates neutrophil infiltration during acute inflammation in mice. Blood 120:1742–1751

90. Novick RP (2003) Autoinduction and signal transduction in the regulation of staphylococcal virulence. Mol Microbiol 48:1429–1449

91. Porsche R, Brenner ZR (1999) Allergy to protamine sulfate. Heart Lung 28:418–428

Chapter 46

Isolation and Purification of Versican and Analysis of Versican Proteolysis

Simon J. Foulcer, Anthony J. Day, and Suneel S. Apte

Abstract

Versican is a widely distributed chondroitin sulfate proteoglycan that forms large complexes with the glycosaminoglycan hyaluronan (HA). As a consequence of HA binding to its receptor CD44 and interactions of the versican C-terminal globular (G3) domain with a variety of extracellular matrix proteins, versican is a key component of well-defined networks in pericellular matrix and extracellular matrix. It is crucial for several developmental processes in the embryo and there is increasing interest in its roles in cancer and inflammation. Versican proteolysis by ADAMTS proteases is highly regulated, occurs at specific peptide bonds, and is relevant to several physiological and disease mechanisms. In this chapter, methods are described for the isolation and detection of intact and cleaved versican in tissues using morphologic and biochemical techniques. These, together with the methodologies for purification and analysis of recombinant versican and a versican fragment provided here, are likely to facilitate further progress on the biology of versican and its proteolysis.

Key words Versican, Glycosaminoglycan, Extracellular matrix, Hyaluronan, Chondroitin sulfate, Affinity chromatography, A disintegrin-like and metalloprotease domain with thrombospondin type 1 motif, ADAMTS

Abbreviations

ADAMTS	A disintegrin-like and metalloprotease domain with thrombospondin type 1 motif
CS	Chondroitin sulfate
ECM	Extracellular matrix
GAG	Glycosaminoglycan
HA	Hyaluronan
PCM	Pericellular matrix

Kuberan Balagurunathan et al. (eds.), *Glycosaminoglycans: Chemistry and Biology*, Methods in Molecular Biology, vol. 1229, DOI 10.1007/978-1-4939-1714-3_46, © Springer Science+Business Media New York 2015

1   Introduction

1.1   Versican and the Extracellular Matrix

Versican, also known as CSPG2 or PG-M, is a large chondroitin sulfate proteoglycan with a widespread distribution [1]. It is prominently expressed during embryogenesis, and is found in adult brain, cardiovascular system, skin, and musculoskeletal tissues [2–4]. Versican exists primarily as a large aggregate with the glycosaminoglycan hyaluronan (HA) [5], a property it shares with other members of the hyalectan (or lectican) family, which include aggrecan, neurocan, and brevican. In contrast to versican, these family members have a restricted distribution, with aggrecan primarily confined to cartilage, and neurocan and brevican selectively present in the central nervous system.

Versican has a well-established significance in biology, especially during embryogenesis and in the pathogenesis of several human diseases. Versican is crucial for myocardial and valvular development, and indeed, *Vcan* deficient mice die at mid-gestation as a result of cardiac anomalies [6–8]. Versican is implicated in neural-crest cell migration, and is required for musculoskeletal development [9, 10]. *VCAN* mutations that alter normal *VCAN* exon splicing in humans give rise to a rare eye disorder called Wagner syndrome, and a related condition named erosive vitreoretinopathy [11].

Versican has been extensively investigated in the context of acquired human disorders, specifically in cardiovascular disorders such as atherosclerosis and arterial stenosis [12, 13], and more recently, in a wide variety of cancers, where its presence is typically associated with increased tumor malignancy and metastasis [14, 15]. In the context of these disorders, a variety of cellular effects have been attributed to specific versican domains in vitro [16–18], although the specific cellular mechanisms for the observed effects remain to be fully elucidated.

Like the other hyalectans, the versican core protein has amino- and carboxyl-terminal globular domains (G1 and G3 respectively, *see* Fig. 1) [19]. The structure of the intervening region bearing the CS chains, however, depends on the splice isoform [19, 20]. The *VCAN* gene encompasses 15 exons, of which exon 7 and exon 8 encode large chondroitin sulfate-binding domains named GAGα and GAGβ respectively. The splice isoform V2 contains only GAGα/exon 7, V1 contains only GAGβ/exon 8, and V0 contains both; the smallest versican isoform, V3, has only the globular domains without an intervening CS-attachment region (Fig. 1). The G1 domain contains two link protein modules that mediate its interaction with HA [21, 22] (Fig. 1). This interaction may be further stabilized by the inclusion of link protein to form a trimeric complex. The HA-binding property of versican is relevant to its purification, because dissociation from HA is required as a preliminary step for

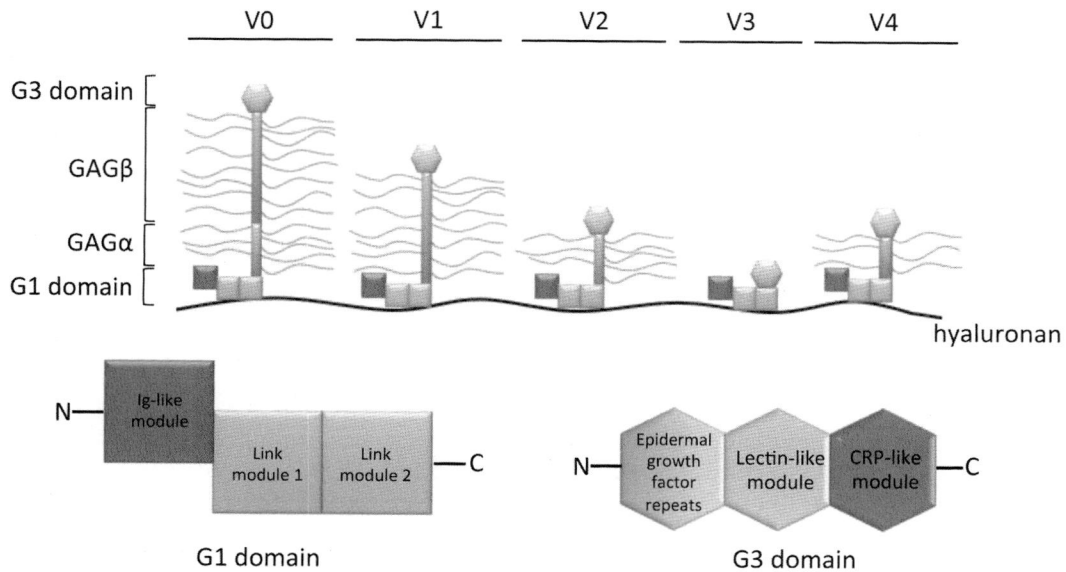

Fig. 1 A schematic showing the known versican isoforms and the modular structure of its N-terminal (G1) and C-terminal (G3) domains. Versican interacts with HA via its G1 domain and forms interactions with other ECM components via its G3 domain. There are two epidermal growth factor (EGF)-like repeats in this domain, the lectin-like module resembles C-type lectins, and the complement regulatory protein (CRP)-like module is also referred to as a complement control protein (CCP) module

isolation of the native proteoglycan. The CS-bearing region is a major contributor to the hydrodynamic properties of the HA–versican complex, and specific associations of the CS chains with cytokines (such as the C–C chemokine midkine, and TGFβ) and elastin-binding protein have been identified [9, 23]. The strongly anionic nature of the CS chains is a characteristic that is exploited for versican purification by ion-exchange chromatography. The G3 domain is known to bind to the extracellular matrix proteins fibrillin-1, tenascin-R, and fibulin-1 and -2 [24–26]. Through these N- and C-terminal interactions, versican participates in specific extracellular matrix networks. The connection of HA to cell surface receptors or HA synthases renders versican an important component of pericellular matrix (PCM, also referred to as the glycocalyx) in several cell types such as fibroblasts, neurons (where PCM is also known as the perineuronal net), myoblasts, and smooth muscle cells [27–30]. For readers interested in additional details, the diverse interactions of versican have been reviewed elsewhere [31].

Although many proteinase classes may have catalytic activity toward the versican core protein, there has been considerable recent interest in ADAMTS proteases [32], which have been shown to attack specific peptide bonds, i.e., Glu^{441}-Ala^{442} in the GAGβ domain of the V1 isoform (the corresponding cleavage site numbering in human versican V0 is Glu^{1428}-Ala^{1429}) and Glu^{405}-Gln^{406} in the

GAGα domain of the V2 and V0 isoforms [33]. The cleaved V1 N-terminal fragment (extending to Glu^{441}) is now referred to as versikine, and the corresponding V2 fragment (extending to Glu^{405}) was first isolated as GHAP (Glial hyaluronic acid binding protein) [34]. The generation of cleavage site-specific antibodies, termed neo-epitope antibodies, which react only with ADAMTS-cleaved versican, but not the intact form [33], was instrumental in revealing the biological significance of ADAMTS-mediated versican proteolysis [35–39]. Because versican is a component of the provisional matrix of the developing embryo, its clearance is required as development progresses and specialized matrices replace the embryonic provisional matrix. In addition, as a component of PCM, it stands at the interface between the cell membrane and its microenvironment, and consequently, versican proteolysis can influence cell behavior. For example, the amount of PCM versican and the activity of the versican-degrading protease ADAMTS5 in PCM strongly influence the fibroblast to myofibroblast transition [28], a pivotal step not only in wound healing, but also in its pathological counterpart, fibrosis. ADAMTS genes can be upregulated at sites of versican turnover in a coordinated fashion [38]. Single or compound ADAMTS gene deletions in mice have established that versican proteolysis is required during interdigital web regression, cardiovascular development (including valve and myocardial development), melanoblast colonization of skin, craniofacial development (palatogenesis), myogenesis, and ovulation [30, 35–43]. The embryologic roles of versican proteolysis were the subject of a recent review [44]).

In addition to the inherent complexity of ECM, the unique attributes of proteoglycans, i.e., their extensive, frequently variable glycosylation, large size and highly anionic nature, complicate the analysis of versican. However, these properties are useful for devising purification schemes. The majority of studies of versican to date have focused on the versican V1 isoform and the globular domains. Much less is known about the other isoforms. A recently identified novel isoform, V4, is generated from a cryptic splice site in exon 8 (Fig. 1) and is hitherto known only as an RNA species [14]. Here, we provide methods for isolation of versican from tissues and cultured cells, as well as versican characterization and tissue localization. The protocols provided encompass expression and purification of recombinant truncated forms of versican, including versikine.

1.2   Detection of Versican and Its Proteolytic Fragments in Tissue Sections

The different versican isoforms have distinct expression sites in tissues. Furthermore, it is becoming apparent that the G1 domain-containing versican V1 proteolytic fragment, versikine, may mediate processes that are different to those of unprocessed versican. These differences highlight the importance of correctly identifying the different versican isoforms using biochemical and immunohistochemical/immunofluorescent approaches as outlined here.

1.3 Analysis and Quantification of Versican Content in Tissues and Cell Culture

Often, analysis of versican in tissues by immunofluorescence is not sufficient to accurately characterize the versican content of tissue samples or cultures. Instead, an alternative approach that identifies the molecular species by Western blot or a dual approach may be appropriate. Here, we describe the extraction and purification of versican and versikine from both cells and tissue samples and provide an approach that can be used to quantify the extracted versican. Previously, methods have been described to isolate and purify versican from both animal and cell sources that used guanidine hydrochloride to disrupt versican interactions [8]. The present method uses urea as the chaotropic agent.

1.4 Expression and Purification of Recombinant Truncated Versican and Versikine

Methods for isolation of native versican from intact tissues and cell lines use tissue or ECM disruption with a chaotropic agent followed by ion-exchange chromatography, capitalizing on the highly anionic nature of the CS chains [21, 33, 45, 46]. Here, we describe methods for the expression and purification of recombinant versikine and V-5GAG (a truncated form of versican V1 that contains the G1 domain and the N-terminal-most five GAG chains, but lacks the G3 domain). These forms either do not exist in nature (V5-GAG) or may have low abundance in tissue (versikine) and yet, are useful for the analysis of versican processing and the biological impact of versican proteolysis respectively and are readily obtained in recombinant form. They are expressed and purified as recombinant proteins using Ni^{2+} NTA affinity for the $(His)_6$ tag as the first step, followed by HA-affinity purification.

2 Materials

2.1 Detection of Versican and Its Proteolytic Fragments in Tissue Sections

1. Citrate-EDTA antigen retrieval buffer: 10 mM citric acid, 2 mM EDTA, 0.05 % (v/v) Tween-20, adjust to pH 6.2.

2. PBS wash buffer: 137 mM NaCl, 2.7 mM KCl, 10 mM Na_2HPO_4, 1.8 mM KH_2PO_4, 0.1 % (v/v) Tween-20.

3. Vectashield or Vectashield containing 4',6-diamidino-2-phenylindole (DAPI) (Vector Labs, Burlingame, CA).

4. Polyclonal rabbit anti-mouse versican antibody 1360-1439 (Anti-GAGβ, EMD Millipore, Billerica, MA)—catalogue number AB1033.

5. Polyclonal rabbit anti-mouse versican antibody 535-598 (Anti-GAGβ, EMD Millipore, Billerica, MA)—catalog number AB1032.

6. Polyclonal rabbit anti-human/mouse versican V0, V1 neo-epitope antibody (anti-DPEAAE, ThermoFisher Scientific, Waltham, MA)—catalog number PA11748A. This antibody specifically detects ADAMTS cleaved versican isoforms V0 and

V1. The corresponding fragments are 1,428 and 441 amino acids in length.

7. Polyclonal rabbit anti-human versican V0, V2 neo-epitope antibody (anti-NIVSFE, ThermoFisher Scientific, Waltham, MA)—catalog number PA3-119. This antibody specifically detects the N-terminal fragment (GHAP) of ADAMTS cleaved versican isoforms V0 and V2.

8. Monoclonal mouse anti-human versican hyaluronate-binding region (Anti-G1, Developmental Studies Hybridoma Bank, Iowa City, IA)—catalog number 12C5.

9. Monoclonal mouse anti-human versican G3 domain (previously marketed by Seikagaku)—catalogue number 2B1.

10. Secondary antibodies are typically enzyme- or fluorophore-conjugated goat anti-rabbit IgG or goat anti-mouse IgG, as appropriate.

2.2 Analysis and Quantification of Versican Content in Tissues and Cell Culture

1. Extraction buffer 1: 8 M urea, 50 mM NaCl, 50 mM Tris–HCl, 1 mM EDTA, adjust to pH 7.5.

2. Extraction buffer 2: 8 M urea, 50 mM NaCl, 50 mM Tris–HCl, 1 mM EDTA, adjust to pH 8.0.

3. Wash buffer 1: 600 mM NaCl, 100 mM Na-acetate, 1 mM EDTA, adjust to pH 6.

4. Wash buffer 2: 50 mM NaCl, 50 mM Tris–HCl, 1 mM EDTA, adjust to pH 8.0.

5. Elution buffer 1: 1.5 mM NaCl, 100 mM Na-acetate, 1 mM EDTA, adjust to pH 6.0.

6. Elution buffer 2: 500 mM NaCl, 50 mM Tris–HCl, 1 mM EDTA, adjust to pH 8.0.

7. Q-Sepharose Fast Flow.

8. StrataClean Resin.

2.3 Expression and Purification of Recombinant Truncated Versican and Versikine

1. Polymeric hyaluronan (1.2 megaDalton HA. Hyalose, Oklahoma City, OK).

2. 1-ethyl-3-(3-dimethylaminot-propyl) carbodiimide.

3. Chondroitinase ABC lyase (C′ABC).

4. Dialysis Buffer 1: 4 M Guanidine-HCl (GuHCl), 100 mM NaH_2PO_4, 10 mM Tris–HCl, 10 mM imidazole, adjust to pH 8.

5. $(His)_6$ wash buffer 1: 100 mM NaH_2PO_4, 150 mM NaCl, 8 M urea, 20 mM imidazole, adjust to pH 8.

6. $(His)_6$ Wash buffer 2: 50 mM NaH_2PO_4, 500 mM NaCl, 20 mM imidazole, adjust to pH 8.

7. $(His)_6$ Elution buffer 2: 50 mM NaH_2PO_4, 500 mM NaCl, 250 mM imidazole, adjust to pH 8.

8. HA wash buffer 1: 150 mM NaCl, 50 mM Tris–HCl, 1 mM EDTA, adjust to pH 7.5.

9. HA elution buffer: 4 M GuHCl, 50 mM Tris–HCl, 1 mM EDTA, adjust to pH 7.5.

10. Ni^{2+}-NTA agarose.

11. EAH Sepharose-4B.

3 Methods

3.1 Detection of Versican and Its Proteolytic Fragments in Tissue Sections

3.1.1 Paraffin-Embedded Sections: Antigen Retrieval (See Note 1)

1. Remove paraffin by immersing slides in clearing agent (Histoclear or xylene) 2 × 5 min at room temperature (RT).

2. Rehydrate sections by successive immersion for 5 min at RT in 100 % EtOH (2×), 70 % EtOH, and dH_2O.

3. Place slides in a glass slide holder and fill with citrate/EDTA buffer. Place in microwave and heat at 500 W for 90 s, leave to stand for 30 s, and repeat heating step.

4. If necessary, top up the container with citrate/EDTA buffer to replace any lost through evaporation, leave to stand for 30 s, and repeat heating cycle (i.e., 90 s heat, 30 s stand, and 90 s heat).

5. Leave to stand in citrate buffer for 30 min to cool to room temperature.

3.1.2 Staining

If frozen sections are used, start the protocol from this step (*see* **Note 2**).

1. Block in 5 % (v/v in PBS) normal goat serum for 30 min at RT (*see* **Notes 3** and **4**).

2. Wash 3 × 5 min with wash buffer.

3. Add ~100 µl primary antibody (for concentrations *see* Table 1) for 1 h at RT or overnight (o/n) at 4 °C.

4. Wash 3 × 5 min with wash buffer.

5. Add secondary antibody, e.g., appropriately diluted enzyme/fluorophore-conjugated antibody and incubate at RT for 1 h (*see* **Note 5**).

6. Repeat **step 4**.

7. If using immunofluorescence, stain the nuclei by adding one drop of DAPI-containing mounting medium, seal and image (*see* Tables 2 and 3). Hematoxylin or another appropriate nuclear counterstain may be used for immunohistochemistry.

8. Staining is evident in extracellular matrix (*see* **Notes 6–8**).

Table 1
Antibody concentrations used for versican detection

Antibody[a]	Concentration (µg/ml) for IHC	Concentrations (µg/ml) for WB
AB1033	2	0.5
AB1023	2	0.5
PA11748A	1–2[b]	0.5
12C5	3	1
2B1	3	1

[a]Catalogue number (*see* Subheading 2)
[b]For paraffin-embedded sections use 2 µg/ml, for frozen sections use 1 µg/ml

Table 2
Versican domains detected by each antibody

Antibody[a]	Epitope position			
	G1[b]	GAGα[a]	GAGβ[b]	G3[b]
AB1033			x	
AB1023		x		
PA3-119		x		
PA11748A			x	
12C5	x			
2B1				x

[a]Catalogue number (*see* Subheading 2)
[b]Domain structure shown in Fig. 1

Table 3
Versican isoform recognized by each antibody

Antibody[a]	Isoform detected					
	V1[b]	V0[b]	V2[b]	V3[b]	Versikine[c]	GHAP[d]
AB1033	×	×				
AB1023			×			
PA3-119						×
PA11748A					×	
12C5	×	×	×	×	×	
2B1	×	×	×	×		

[a]Catalogue number (*see* Subheading 2)
[b]Isoform structure shown in Fig. 1
[c]Versikine is the G1 domain containing ADAMTS-cleaved versican V1 fragment (N-terminus to E^{441})
[d]GHAP (glial hyaluronic acid binding protein) is the ADAMTS-cleaved N-terminal fragment of versican V0 or V2 (N-terminus to E^{435} inclusive)

3.2 Analysis and Quantification of Versican Content in Tissues and Cell Culture

3.2.1 Extraction of Pericellular Versican from Cell Culture (See Note 9)

1. Culture cells of interest to ~80 % confluence, and wash with PBS (*see* **Note 10**).

2. Add serum-free medium and incubate for 24 h.

3. Remove serum-free medium and keep for further analysis.

4. Wash cells 3× with PBS.

5. Add 10 ml extraction buffer 1 for each 25 cm^2 cell culture plate, using a cell scraper to ensure that all cells are detached from the cell culture dish.

6. Adjust the volume to 30 ml with extraction buffer 1 and incubate on ice for 30 min, vortexing every 5 min.

7. Equilibrate 2 ml Q-Sepharose Fast Flow resin with 5 column volumes of extraction buffer 1.

8. Centrifuge extracted cells from **step 6** at $1,000 \times g$ to remove large particulates/cell debris.

9. Combine resin with supernatant and mix by rotating o/n at 4 °C.

10. Transfer resin slurry to a sintered gravity flow column and collect flow-through.

11. Wash with 5 column volumes extraction buffer 1.

12. Wash with 5 column volumes wash buffer 1.

13. Elute bound protein with 5 column volumes of elution buffer 1, collecting each eluted fraction.

14. Take media from **step 3** and add urea to a final concentration of 8 M.

15. Combine with 2 ml Q-Sepharose Fast Flow resin equilibrated with 5 column volumes of extraction buffer 1.

16. Extract versican by repeating **steps 10–13**.

3.2.2 Extraction of Pericellular Versikine from Cell Culture (See Note 9)

1. Culture cells of interest to ~80 % confluence, and wash with PBS (*see* **Note 10**).

2. Add serum-free medium and incubate for 24 h.

3. Remove serum-free medium and keep for further analysis.

4. Wash cells 3× with PBS.

5. Solubilize cells in 10 ml extraction buffer 2 for each 25 cm^2 cell culture plate, using a cell scraper to ensure all cells are detached from the cell culture dish.

6. Equilibrate 2 ml Q-Sepharose Fast Flow resin with 5 column volumes of extraction buffer 2.

7. Centrifuge extracted cells from **step 5** at $1,000 \times g$ to remove large particulates/cell debris.

8. Combine resin with supernatant and mix by rotating o/n at 4 °C.

9. Transfer resin slurry to a gravity flow column and collect flow-through.

10. Wash with 5 column volumes extraction buffer 2.

11. Wash with 5 column volumes wash buffer 2.

12. Elute bound protein with five column volumes of elution buffer 2 collecting each eluted fraction.

3.2.3 Extraction of Versican or Versikine from Tissue Samples

As described above for cell culture extraction, the extraction of versican or versikine from tissue is carried out using extraction, washing, and elution buffers 1 and 2 respectively.

1. Homogenize tissue sample in 1:10 (w/v) of tissue to the appropriate extraction buffer (e.g., 10 mg tissue added to 1 ml buffer) on ice using a tissue homogenizer. Ensure that the weights of the samples being compared are identical prior to homogenization.

2. Increase the volume of extraction buffer 1 fivefold and incubate on ice for 30 min, or overnight at 4 °C for tough tissues such as tendon, vortexing every 5 min for 30 s.

3. Centrifuge at $10,000 \times g$ for 10 min to remove large particulates.

4. Equilibrate 0.2 ml Q-Sepharose resin with either extraction buffer 1 or 2.

5. Combine equal volume of supernatant from **step 3** with the equilibrated Q-Sepharose resin.

6. Rotate o/n at 4 °C.

7. Transfer to a sintered gravity flow column and collect flow-through.

8. Wash resin with 5 column volumes of extraction buffer followed by 5 column volumes washing buffer.

9. Elute bound protein with 5 column volumes of elution buffer collecting all fractions.

3.2.4 Quantification of Versican/Versikine

1. For versican analysis incubate each elution fraction with C'ABC at a final concentration of 0.1 U/ml at 37 °C for 3 h. C'ABC digestion is not required for versikine analysis (*see* **Note 11**).

2. Add 15 μl StrataClean to each fraction and mix by rotating at RT for 30 min (*see* **Note 12**).

3. Centrifuge for 5 min at $10,000 \times g$.

4. Discard supernatant and resuspend beads in SDS-PAGE loading buffer (*see* **Note 12**).

5. Add 2-mercaptoethanol (10 % v/v) and/or dithiothreitol to a final concentration of 0.1 M. Heat at 100 °C for 5 min and analyze by either 8 % or 10 % SDS-PAGE (for versican and versikine respectively) alongside versican protein standards (either versikine (Fig. 2) or V-5GAG as appropriate).

6. Perform Western blot using an antibody specific to the desired region (G1 domain, neo-epitope, or GAGβ).

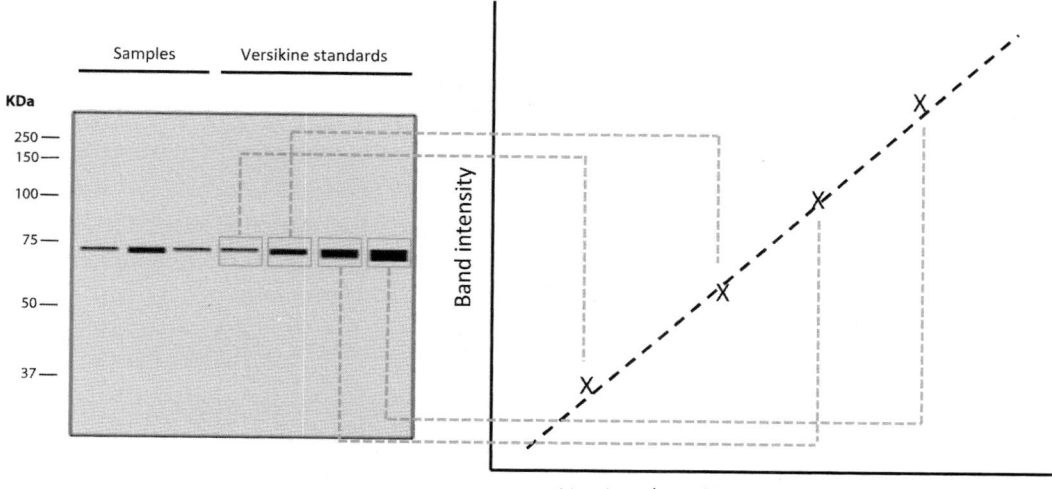

Fig. 2 Illustration of a method to quantify unknown versican (versikine) samples using known standards. Versikine standards are run alongside unknown samples (**a**). The standards are then used to construct a standard curve (**b**), which can be used to derive total protein content in the samples using the gradient of the line (i.e., $y = mx + c$). This technique is possible because the antibody epitopes are identical in both the samples

7. Quantify extracted protein by measuring the band intensity of the loaded standards and drawing a standard curve. Use this curve to quantify the extracted protein amount (*see* Fig. 2).

3.3 Expression and Purification of Recombinant Truncated Versican and Versikine

3.3.1 Expression (See Notes 13–15)

1. Stable cell lines expressing versikine (residues 1–441 of human versican) and V-5GAG (residues 1–695 of human versican) are obtained using zeocin selection and expanded to ~500 cm² cell culture area (*see* **Notes 15** and **16**).

2. At ~70 % confluence, remove cell culture medium, wash cells with PBS, and replace with serum-free medium.

3. Incubate cells for 48 h.

4. Remove serum-free medium and store it at –80 °C until used for protein purification, replace with medium containing 10 % (v/v) fetal calf serum, and incubate cells for 12 h (*see* **Note 17**).

5. Remove medium, wash cells with PBS, and replace with serum-free medium.

6. Incubate for 48 h and collect the medium, this constitutes a second collection.

3.3.2 Purification: Hyaluronan (HA) Affinity Column Preparation (to Make 40 ml HA-Affinity Resin)

1. Resuspend 10 mg polymeric hyaluronan (1.2 MDa HA) in 10 ml H_2O o/n at 4 °C.

2. Wash 40 ml EAH-Sepharose 4B resin with 250 ml H_2O.

3. Add the resin to the HA, bring the volume to 90 ml with H_2O, and adjust pH to 4.7.

4. Add 0.4 g 1-ethyl-3-(3-dimethylamino-propyl) carbodiimide to the HA resin slurry and stir at RT while readjusting the pH to 4.7 for 3 h (*see* **Note 18**).

5. Allow to stand o/n at RT.

6. Add 2 ml glacial acetic acid to block unsubstituted groups on the resin.

7. Wash sequentially with 500 ml 1 M NaCl, 500 ml 0.05 M formic acid, 500 ml H_2O, and 500 ml 0.5 M Na-acetate containing 0.02 % (w/v) Na-azide (*see* **Note 19**).

8. Store the HA-affinity resin in 0.5 M Na-acetate containing 0.02 % (w/v) Na-azide at 4 °C.

3.3.3 Purification: Stage 1—His$_6$-tag Affinity Purification

1. Concentrate the collected serum-free conditioned medium to ~50 ml using concentrators with a 10 kDa molecular weight cut-off (*see* **Note 20**).

2. Place into dialysis tubing (e.g., SnakeSkin, Pierce) and dialyze extensively against dialysis buffer 1 (*see* **Note 21**)

3. Equilibrate 5 ml Ni^{2+} NTA resin with 5 column volumes (i.e., 5×5 ml) of dialysis buffer 1.

4. Add the resin to the dialyzed conditioned medium and rotate at 4 °C o/n.

5. Transfer slurry to a sintered gravity flow column and collect flow-through.

6. Wash with 5 column volumes of (His)$_6$ wash buffer 1.

7. Wash with 5 column volumes of (His)$_6$ wash buffer 2 (*see* **Note 22**).

8. Elute with 5 column volumes of (His)$_6$ elution buffer 1, collecting 1 column volume fractions (Fig. 3a).

9. Dialyze washes and supernatant from **steps 5** and **6** into dialysis buffer 1 and repeat **steps 3–8**.

10. Digest elution fractions with 0.1U/ml chondroitinase ABC lyase (C'ABC) at 37 °C for 3 h, analyze by 10 % SDS-PAGE, and pool all fractions containing the purified proteins (Fig. 3) (*see* **Note 23**).

3.3.4 Purification: Stage 2—HA-Affinity Purification

1. Dialyze eluted fractions containing protein from Subheading 3.3.3 **steps 8** and **9** extensively against HA wash buffer 1.

2. Equilibrate 5 ml of HA resin with 5 column volumes HA wash buffer 1.

3. Combine dialyzed protein with HA-affinity resin and rotate at 4 °C o/n.

4. Transfer slurry to a sintered gravity flow column and collect flow-through.

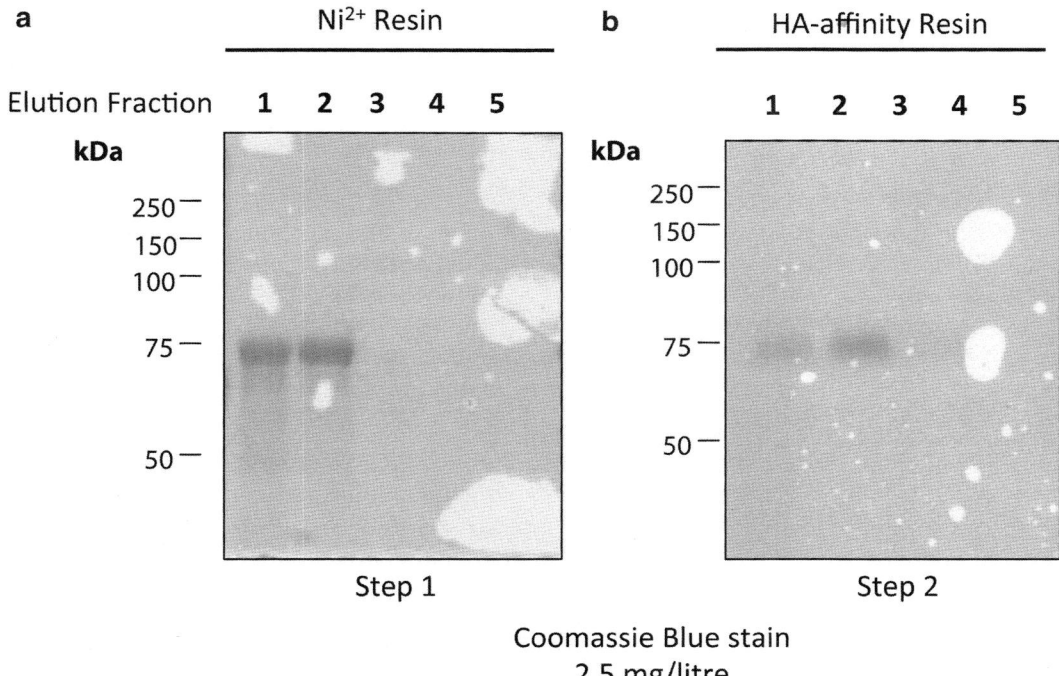

Fig. 3 Coomassie blue stained gel showing versikine elution fractions. Purification from Subheading 3.3.3, **step 8** and Subheading 3.3.4, **step 6** (i.e., Ni2 + -NTA affinity and HA-affinity) are shown in (**a**) and (**b**) respectively. Elution is complete by fraction 3 in both steps

5. Wash resin with 5 column volumes of HA wash buffer 1.

6. Elute bound protein with 5 column volumes HA elution buffer 1, collecting 1 column volume fractions (Fig. 3b).

7. Dialyze each elution fraction into HA wash buffer 1.

8. Take a 30 μl sample from each elution fraction and assess for protein content and purity by analysis using 10 % SDS-PAGE and visualize with Coomassie Brilliant Blue stain.

4   Notes

1. The duration of fixation affects antigen availability and is thus an important variable that affects the results. It is recommended that samples that are to be compared to each other use identical fixation times in equivalent volumes of 4 % paraformaldehyde (PFA) at 4 °C to reduce variability.

2. For sample comparison, all tissue samples should be processed identically (e.g., frozen sections will differ in signal intensity from paraffin-embedded sections).

3. All immunostaining steps are performed in a humidified chamber to prevent evaporation and changes in the concentration of reagents.

4. For secondary antibodies not raised in the goat, an alternative blocking agent is required that is specific to the species providing the secondary antibody. For example, if the secondary antibody was produced in the donkey, then normal donkey serum should be used to block.

5. All antibodies described here were raised in mouse or rabbit. Secondary antibody is dependent on the primary antibody used, e.g., an antibody raised in mouse will require a secondary antibody specific to mouse.

6. Controls are very important for tissue localization by antibodies. The primary antibody should be omitted from the protocol as a negative control. A mid-gestation mouse embryo section (gestational age 9.5–12.5 days) can be used as a positive control. Versican is abundant in the heart, brain, developing limb, and many other locations [2–4, 35–39]. A good positive control for anti-DPEAAE staining is the regressing interdigital web (e.g., forelimb autopod from a 13.5 old embryo) [38].

7. In tissue sections, it does not automatically follow that immunostaining using GAG-domain antibodies solely indicates the presence of intact versican. Following proteolytic cleavage, the GAG domains will be contained in the C-terminal fragment and may linger owing to the multiple interactions of the G3 domains. The fate of versican fragments following proteolysis is not known.

8. Examples of immunostaining with versican antibodies or neo-epitope antibodies can be obtained in the literature [4, 35–38, 42, 43]. Pretreatment of sections using chondroitinase ABC treatment may improve antibody binding.

9. Protein extracts should be processed immediately and kept on ice to prevent degradation.

10. The amount of versican produced by isolated primary cells of different origins is likely to vary and so this step will require optimization for each cell type used. Typically 25 cm^2 monolayer cultures will yield sufficient versican for analysis and quantification.

11. Without prior C′ABC digestion, versican, like other CS-proteoglycan, migrates as a smear at the very top of the gel. Its significantly retarded mobility within the gel results from the high negative charge and large size of the CS chains. These smears are most unlike the sharp bands typically seen on Western blots of unmodified proteins, and may puzzle observers new to proteoglycan analysis; moreover, they are often not reproducible

between successive Western blots, may not transfer uniformly, and a significant proportion may be retained in the stacking gel. Upon C'ABC digestion, the CS chains are removed and the proteoglycan migrates as a better-defined band and the smear is eliminated or greatly attenuated. A mixture of unmodified core protein, and variable modification explain the dispersion of species that constitute the smear. The expected molecular mass for versican V0, V1, and V2 is 372, 265, and 182 kDa respectively, but the observed migration is typically at >350 kDa for V0 and V1, and at ~200 kDa for V2. Versikine has a predicted molecular mass of 49 kDa, but typically migrates at 70 kDa for reasons that are not presently understood [33, 38]. GHAP has a predicted molecular mass of 43 kDa, but migrates as a 64 kDa band [34].

12. StrataClean resin binds all protein and can be used to extract protein from dilute samples for SDS-PAGE analysis (i.e., instead of protein precipitation). If multiple analyses are required, fractions should be aliquoted prior to addition of StrataClean resin. The resin does not need to be washed prior to use and can be added directly from the supplied container. The ethanol will be sufficiently diluted so as to not affect versican/versikine binding. The resin should be loaded directly onto the gel since the bound protein is not removed by SDS-sample buffer (i.e., it needs to be electrophoresed off) [22].

13. The V1 construct in a pSecTagA plasmid [5] contained the entire open reading frame (ORF) but originally had an intervening 3'-untranslated sequence between the stop codon and the epitope tags. Therefore, an Xho I restriction site was inserted to disrupt the stop codon using the QuikChange Mutagenesis kit (Stratagene, Santa Clara, CA), the 3'-untranslated sequence was excised, and the vector religated to render the versican ORF continuous with the myc and (His)6 tags. To generate the V-5GAG construct with C-terminal myc and His6 tags, another Xho I site was placed at the appropriate location within the versican ORF. Mutagenized plasmid was digested with Xho I and the region between the two Xho I sites was separated by agarose electrophoresis followed by re-ligation of the plasmid.

14. The versikine expression plasmid was previously described [38].

15. It is important that all reagents used are sterile and that sterile technique is used throughout. Incubation steps are carried out at 37 °C and in 5 % (v/v) CO_2.

16. Ensure selected stable lines are expressing sufficient protein before expanding colonies. Generally speaking, if expressed protein can be seen clearly on a Coomassie blue stained SDS-PAGE gel of the unconcentrated conditioned medium, expression levels are sufficient to achieve a good final protein yield (~2–3 mg/l). Proteoglycans do not take a Coomassie blue stain and may be stained using silver staining or toluidine blue.

17. This step can often allow sufficient cell recovery to induce the production of more protein; however, this may not always be possible. Collected serum-free medium containing the expressed protein should be processed immediately (being kept on ice between procedures) or frozen until purification to prevent degradation.

18. Carbodiimide reactions cause a gradual increase in pH, which will result in a progressive inhibition of the coupling reaction.

19. It is often desirable to keep all eluted fractions to quantify the amount of HA bound to the column (thus making it possible to estimate the binding capacity of the column). This can be achieved using the meta-hydroxybiphenol reaction [47].

20. Any large-scale protein concentrating system can be used. For example Vivaflow 200 blocks (Membrane 5000PES, Vivascience Ltd).

21. This buffer has threefold actions: It partially unfolds the protein to ensure that all target protein is free of hyaluronan and it reduces nonspecific protein–protein interactions to reduce impurities. It also exposes the His-tag, improving binding to the nickel resin.

22. This step is necessary to remove all chaotropic agent and refold the protein.

23. Samples containing V-5GAG will require C'ABC treatment prior to SDS-PAGE analysis. After purification it is important to characterize the protein to ensure that the protein is functional and has the correct molecular weight. A solid phase hyaluronan-binding assay and mass spectrometry can be used [22].

Acknowledgements

This work was supported by a National Institutes of Health Programs of Excellence in Glycosciences award (NIH PO1 HL107147) and by NIH RO1 HD069747 to S.A. Some of the methods were developed when S.J.F. was the recipient of a Medical Research Council Studentship (UK, G0800127).

References

1. Bode-Lesniewska B et al (1996) Distribution of the large aggregating proteoglycan versican in adult human tissues. J Histochem Cytochem 44(4):303–312

2. Henderson DJ, Copp AJ (1998) Versican expression is associated with chamber specification, septation, and valvulogenesis in the developing mouse heart. Circ Res 83(5):523–532

3. Snow HE et al (2005) Versican expression during skeletal/joint morphogenesis and patterning of muscle and nerve in the embryonic mouse limb. Anat Rec A Discov Mol Cell Evol Biol 282(2):95–105

4. Zimmermann DR et al (1994) Versican is expressed in the proliferating zone in the epidermis and in association with the elastic

network of the dermis. J Cell Biol 124(5): 817–825

5. LeBaron RG, Zimmermann DR, Ruoslahti E (1992) Hyaluronate binding properties of versican. J Biol Chem 267(14):10003–10010

6. Mjaatvedt CH et al (1998) The Cspg2 gene, disrupted in the hdf mutant, is required for right cardiac chamber and endocardial cushion formation. Dev Biol 202(1):56–66

7. Yamamura H et al (1997) A heart segmental defect in the anterior-posterior axis of a transgenic mutant mouse. Dev Biol 186(1):58–72

8. Bratt P et al (1992) Isolation and characterization of bovine gingival proteoglycans versican and decorin. Int J Biochem 24(10):1573–1583

9. Choocheep K et al (2010) Versican facilitates chondrocyte differentiation and regulates joint morphogenesis. J Biol Chem 285(27): 21114–21125

10. Williams DR Jr et al (2005) Limb chondrogenesis is compromised in the versican deficient hdf mouse. Biochem Biophys Res Commun 334(3):960–966

11. Kloeckener-Gruissem B et al (2006) Identification of the genetic defect in the original Wagner syndrome family. Mol Vis 12:350–355

12. Kenagy RD, Plaas AH, Wight TN (2006) Versican degradation and vascular disease. Trends Cardiovasc Med 16(6):209–215

13. Wight TN, Merrilees MJ (2004) Proteoglycans in atherosclerosis and restenosis: key roles for versican. Circ Res 94(9):1158–1167

14. Kischel P et al (2010) Versican overexpression in human breast cancer lesions: known and new isoforms for stromal tumor targeting. Int J Cancer 126(3):640–650

15. Ricciardelli C et al (2009) The biological role and regulation of versican levels in cancer. Cancer Metastasis Rev 28(1–2):233–245

16. Wu Y et al (2005) Versican protects cells from oxidative stress-induced apoptosis. Matrix Biol 24(1):3–13

17. Yang BL et al (2003) Versican G3 domain enhances cellular adhesion and proliferation of bovine intervertebral disc cells cultured in vitro. Life Sci 73(26):3399–3413

18. Yee AJ et al (2007) The effect of versican G3 domain on local breast cancer invasiveness and bony metastasis. Breast Cancer Res 9(4):R47

19. Zimmermann DR, Ruoslahti E (1989) Multiple domains of the large fibroblast proteoglycan, versican. EMBO J 8(10):2975–2981

20. Dours-Zimmermann MT, Zimmermann DR (1994) A novel glycosaminoglycan attachment domain identified in two alternative

splice variants of human versican. J Biol Chem 269(52):32992–32998

21. Matsumoto K et al (2003) Distinct interaction of versican/PG-M with hyaluronan and link protein. J Biol Chem 278(42):41205–41212

22. Seyfried NT et al (2005) Expression and purification of functionally active hyaluronan-binding domains from human cartilage link protein, aggrecan and versican: formation of ternary complexes with defined hyaluronan oligosaccharides. J Biol Chem 280(7):5435–5448

23. Zou K et al (2000) A heparin-binding growth factor, midkine, binds to a chondroitin sulfate proteoglycan, PG-M/versican. Eur J Biochem 267(13):4046–4053

24. Aspberg A et al (1999) Fibulin-1 is a ligand for the C-type lectin domains of aggrecan and versican. J Biol Chem 274(29):20444–20449

25. Aspberg A, Binkert C, Ruoslahti E (1995) The versican C-type lectin domain recognizes the adhesion protein tenascin-R. Proc Natl Acad Sci U S A 92(23):10590–10594

26. Isogai Z et al (2002) Versican interacts with fibrillin-1 and links extracellular microfibrils to other connective tissue networks. J Biol Chem 277(6):4565–4572

27. Evanko SP et al (2007) Hyaluronan-dependent pericellular matrix. Adv Drug Deliv Rev 59(13): 1351–1365

28. Hattori N et al (2011) Pericellular versican regulates the fibroblast-myofibroblast transition: a role for ADAMTS5 protease-mediated proteolysis. J Biol Chem 286(39):34298–34310

29. Kwok JC, Carulli D, Fawcett JW (2010) In vitro modeling of perineuronal nets: hyaluronan synthase and link protein are necessary for their formation and integrity. J Neurochem 114(5):1447–1459

30. Stupka N et al (2013) Versican processing by a disintegrin-like and metalloproteinase domain with thrombospondin-1 repeats proteinases-5 and -15 facilitates myoblast fusion. J Biol Chem 288(3):1907–1917

31. Wu YJ et al (2005) The interaction of versican with its binding partners. Cell Res 15(7): 483–494

32. Apte SS (2009) A disintegrin-like and metalloprotease (reprolysin-type) with thrombospondin type 1 motif (ADAMTS) superfamily: functions and mechanisms. J Biol Chem 284(46):31493–31497

33. Sandy JD et al (2001) Versican V1 proteolysis in human aorta in vivo occurs at the Glu441-Ala442 bond, a site that is cleaved by recombinant ADAMTS-1 and ADAMTS-4. J Biol Chem 276(16):13372–13378

34. Westling J et al (2004) ADAMTS4 (aggrecanase-1) cleaves human brain versican V2 at Glu405-Gln406 to generate glial hyaluronate binding protein. Biochem J 377(Pt 3): 787–795

35. Dupuis LE et al (2011) Altered versican cleavage in ADAMTS5 deficient mice; a novel etiology of myxomatous valve disease. Dev Biol 357(1):152–164

36. Enomoto H, Nelson C, Somerville RPT, Mielke K, Dixon L, Powell K, Apte SS (2010) Cooperation of two ADAMTS metalloproteases in closure of the mouse palate identifies a requirement for versican proteolysis in regulating palatal mesenchyme proliferation. Development 137:4029–4038

37. Kern CB et al (2006) Proteolytic cleavage of versican during cardiac cushion morphogenesis. Dev Dyn 235(8):2238–2247

38. McCulloch DR et al (2009) ADAMTS metalloproteases generate active versican fragments that regulate interdigital web regression. Dev Cell 17(5):687–698

39. Stankunas K et al (2008) Endocardial Brg1 represses ADAMTS1 to maintain the microenvironment for myocardial morphogenesis. Dev Cell 14(2):298–311

40. Brown HM et al (2010) ADAMTS1 cleavage of versican mediates essential structural remodeling of the ovarian follicle and cumulus-oocyte matrix during ovulation in mice. Biol Reprod 83(4):549–557

41. Brown HM et al (2006) Requirement for ADAMTS-1 in extracellular matrix remodeling during ovarian folliculogenesis and lymphangiogenesis. Dev Biol 300(2):699–709

42. Kern CB et al (2007) Versican proteolysis mediates myocardial regression during outflow tract development. Dev Dyn 236(3):671–683

43. Kern CB et al (2010) Reduced versican cleavage due to Adamts9 haploinsufficiency is associated with cardiac and aortic anomalies. Matrix Biol 29(4):304–316

44. Nandadasa S et al (2014) The multiple, complex roles of versican and its proteolytic turnover by ADAMTS proteases during embryogenesis. Matrix Biol 35:34–41

45. Olin AI et al (2001) The proteoglycans aggrecan and Versican form networks with fibulin-2 through their lectin domain binding. J Biol Chem 276(2):1253–1261

46. Olin KL et al (1999) Lipoprotein lipase enhances the binding of native and oxidized low density lipoproteins to versican and biglycan synthesized by cultured arterial smooth muscle cells. J Biol Chem 274(49):34629–34636

47. Filisetti-Cozzi TM, Carpita NC (1991) Measurement of uronic acids without interference from neutral sugars. Anal Biochem 197(1): 157–162

Chapter 47

Analysis of Human Hyaluronan Synthase Gene Transcriptional Regulation and Downstream Hyaluronan Cell Surface Receptor Mobility in Myofibroblast Differentiation

Adam C. Midgley and Timothy Bowen

Abstract

The ubiquitous extracellular glycosaminoglycan hyaluronan (HA) is a polymer composed of repeated disaccharide units of alternating D-glucuronic acid and D-N-acetylglucosamine residues linked via alternating β-1,4 and β-1,3 glycosidic bonds. Emerging data continue to reveal functions attributable to HA in a variety of physiological and pathological contexts. Defining the mechanisms regulating expression of the human hyaluronan synthase (*HAS*) genes that encode the corresponding HA-synthesizing HAS enzymes is therefore important in the context of HA biology in health and disease. We describe here methods to analyze transcriptional regulation of the *HAS* and *HAS2-antisense RNA 1* genes. Elucidation of mechanisms of HA interaction with receptors such as the cell surface molecule CD44 is also key to understanding HA function. To this end, we provide protocols for fluorescent recovery after photobleaching analysis of CD44 membrane dynamics in the process of fibroblast to myofibroblast differentiation, a phenotypic transition that is common to the pathology of fibrosis of large organs such as the liver and kidney.

Key words Hyaluronan, Hyaluronan synthase, Transcriptional regulation, CD44, Fluorescence recovery after photobleaching, Fibroblast

1 Introduction

1.1 Regulation of HAS and HAS2-AS1 Gene Expression

The human hyaluronan synthase (*HAS*) [1–3] and HAS2 antisense RNA 1 (*HAS2-AS1*) [4] genes are located at the following autosomal loci (Table 1).

Dobzhansky reminds us that nothing in biology makes sense except in the light of evolution [5]. From this perspective, it is unclear why the three *HAS* genes, encoding proteins that synthesize the same HA polymer from UDP-*N*-acetylglucosamine and UDP-glucuronic acid [6, 7], have been retained during human evolution. Looking beyond genetic redundancy, identification of functional reasons underlying this retention promises a more complete understanding of HA biology.

Kuberan Balagurunathan et al. (eds.), *Glycosaminoglycans: Chemistry and Biology*, Methods in Molecular Biology, vol. 1229, DOI 10.1007/978-1-4939-1714-3_47, © Springer Science+Business Media New York 2015

Table 1
Human genome sequence data for the *HAS* genes: data shown were extracted from UCSC browser freeze GRCh37/hg19

Gene symbol	NCBI RefSeq[a] (accession number)	Strand	Chromosome	Nucleotides
HAS1	NM_001523.2	–	19	52,216,365–52,227,221
HAS2	NM_005328.2	–	8	122,625,271–122,653,630
HAS2-AS1	NR_002835.2	+	8	122,651,586–122,657,564
HAS3 v1	NM_005329.2	+	16	69,141,443–69,151,570
HAS3 v2	NM_138612.2	+	16	69,140,129–69,152,619
HAS3 v3	NM_001199280.1	+	16	69,139,467–69,151,570

[a]National Center for Biotechnology Information Reference Sequence

Previous work has shown in vitro that the different HAS enzymes synthesize HA of different sizes (e.g., [8]) and have different enzymatic properties (e.g., [8, 9]) with posttranscriptional, translational, and posttranslational mechanisms regulating their expression [3, 4, 10–12]. Basal and cytokine-induced transcriptional regulation of HAS mRNAs and HAS2-AS1 noncoding RNA, a natural antisense to HAS2 and posttranscriptional regulator of its expression, are also potentially highly significant (e.g., [3, 4, 10, 13–17]).

The different promoter sequences found upstream of the *HAS* and *HAS2-AS1* genes have characteristic transcription factor binding profiles conferring the potential to drive cell- and tissue-specific expression (e.g., [3, 4, 10, 14–17]). Further to the identification of three reference HAS3 transcript variant sequences, emerging refinements in sequencing and related technologies may soon reveal evidence of multiple transcripts generated from different promoters at each *HAS* locus.

We provide here a protocol for preparing luciferase reporter vectors containing nested sets of promoter fragments from the above loci with which to analyze transcriptional regulation.

1.2 Cell Surface Mobility of HA Receptor CD44

HA receptor CD44 is a transmembrane receptor and glycoprotein with variants ranging from 78 to 742 amino acids in length. CD44 functions as, but is not limited to, a cell surface receptor for HA. Binding of other ligands, such as collagens and MMPs [18, 19], confers a wide range of cellular CD44 functions. CD44 also acts as a co-receptor, an important function in the myofibroblast differentiation mechanism. In fibroblasts stimulated with cytokine TGF-β1, CD44 association with the epidermal growth factor receptor (EGFR) can phosphorylate EGFR and trigger intracellular signaling cascades resulting in cellular differentiation and proliferation [20–22]. HA is required for CD44 to function as a co-receptor to EGFR via a mechanism whereby pericellular HA bound to CD44 is reorganized by hyaladherin TSG-6 [23] into a cross-linked HA network as a result of TGF-β1-dependent actions, leading to the re-localization of CD44 within the cellular membrane of the activated fibroblast [21].

Changes in CD44 membrane dynamics during myofibroblast differentiation have been determined through the use of fluorescent recovery after photobleaching (FRAP) and laser confocal microscopy [21]. Through analysis of the potential for CD44 to move throughout the cell membrane of fibroblasts and myofibroblasts, in the presence or absence of HA synthesis inhibition, the cellular localization and re-localization of CD44 was determined [21, 24]. This methodology can be used for the investigation of cell surface receptor dynamics and receptor–receptor associations in live-cell studies.

2 Materials

2.1 Luciferase Reporter Vector Analysis of HAS and HAS2-AS1 Gene Promoters

2.1.1 In Silico Analysis and PCR Primer Design

Freely available software and archived data can be used to retrieve target genomic DNA sequences and design oligonucleotide primers for PCR amplification.

1. The promoter sequences upstream of the *HAS* and *HAS2-AS1* genes shown in Table 1 (e.g., HAS2, accession number NM_005328) can be located using resources such as the UCSC (http://genome.ucsc.edu/), Ensembl (http://www.ensembl.org/index.html), or NCBI (http://www.ncbi.nih.gov) genome browsers.

2. Primers can be designed using various primer design software tools. We routinely use Primer3 (http://frodo.wi.mit.edu/cgi-bin/primer3/) and Primer-BLAST (http://www.ncbi.nlm.nhi.gov/tools/primer-blast/) with default parameters [25]. Ideally, GC content is between 50 and 60 % and the primer annealing temperature (T_M) is approximately 60 °C.

2.1.2 PCR and Extraction Kits

Various commercially available PCR thermo-cyclers can be used for generation of amplified promoter fragments. It is recommended to use a thermo-cycler with programmable stages accounting for time and temperature.

1. Endonuclease digestion and vector ligation steps can be completed using a heating block. Heating blocks with constant agitation may yield improved results by limiting reaction solution condensation on microfuge tube lids.

2. PCR purification, gel extraction, mini-prep, and midi-prep kits are commercially available and include optimized step-by-step protocols.

3. Agarose powder, 1× TAE buffer, ethidium bromide (*see* **Note 1**), and flat-bed electrophoresis equipment will be required for the generation of DNA gels. A UV transillumina-tor will be required for visualization of DNA bands.

2.1.3 Cells for Transfection and Transformation

Although multiple cell lines can be used with this technique (*see* **Note 2**), in this methodology fibroblast cells are the preferred choice.

1. Human lung fibroblasts (AG02262) were purchased from Coriell Cell Repositories (Coriell Institute for Medical Research, NJ, USA). The cells are cultured in DMEM/F-12 cell culture medium (*see* **Notes 3** and **4**) and incubated at 37 °C and in 5 % CO_2.

2. Incubating the cells with DMEM/F-12 growth arrest medium for approximately 48 h will ensure cell cycle synchronicity at G_0 arrest, preventing proliferation and over-confluence (*see* **Note 5**).

3. The preferred bacterial cell for transformation is One Shot® competent *E. coli* (Life Technologies). *E. coli* cells are kept frozen at −80 °C until use and freeze-thawing should be avoided when possible.

2.2 Fluorescent Recovery After Photobleaching (FRAP): Analysis of the Influence of HA on CD44 Membrane Dynamics

2.2.1 Cell Treatments

Multiple cell lines can be used with this technique (*see* **Note 6**). In the context of fibrosis however, myofibroblasts, fibroblast cells that differentiate readily in response to key fibrotic mediator TGF-β1, are the preferred choice.

1. Human lung fibroblasts (AG02262) (*see* **Notes 7** and **8**)

2. Cellular growth arrest medium (*see* **Note 9**).

3. Differentiation of fibroblasts to myofibroblasts requires 72 h of TGF-β1 treatment. Spent culture medium is carefully aspirated from the cells and replaced with fresh medium containing 10 ng/ml TGF-β1 (*see* **Note 10**).

4. Treatment with 0.5 μM of HA synthesis inhibitor 4-methylumbelliferone (4MU) (*see* **Note 11**) in combination with TGF-β1 treatment for 72 h can be used to investigate CD44 membrane motility and function in the presence and absence of HA synthesis. Since 4MU is reconstituted in DMSO before further dilution in sterile PBS, use of control cultures with an equal final DMSO concentration to the test cultures is advised (*see* **Note 12**).

5. Antibody treatments are made at least 10 min before laser con-focal analysis (*see* **Note 13**). An anti-human CD44 with FITC conjugate (we routinely use rat mAb to CD44-FITC: IM7, ab19622, Abcam) was used at a final dilution factor of 1:5,000 (*see* **Note 14**). Due to this dilution factor, preparation of a

serum-free medium stock antibody solution is recommended (e.g., 2 μl of antibody in 10 ml of medium followed by equal distribution between culture dishes).

2.2.2 Laser Confocal Microscopy

Various microscopes are suitable for this technique; however it is recommended to use a laser confocal microscope with manual laser intensity, Z-axis focus, and zoom controls, such as the Leica SP2 and Leica RS5 confocal microscope models (*see* **Note 15**).

1. The microscope should have appropriate classes of laser wavelengths that correlate with the excitation and emission wavelengths of the antibody conjugate used (*see* **Note 16**), control panels with adjustable laser intensity dials for these wavelengths (*see* **Note 17**), high-magnification objective lenses (40–63× oil are appropriate), and a heated stage with a 5 % CO_2 incubated chamber (*see* **Note 18**).

2. Ensure the computer connected has appropriate software installed for image capture, time-lapse capture, and data analysis (e.g., Leica LAS AF Confocal Imaging Software).

3. Consult the microscope manual for appropriate software settings and microscope setup, which will vary depending on the objective lens used.

4. Additional useful materials for microscopy include pipettes, an additional stock of serum-free medium, objective lens oil, and water-tight lubricant.

2.3 Solutions

1. 50× TAE buffer: 242 g Tris-base, 57.1 ml 100 % acetic acid, 100 ml 0.5 M sodium EDTA, 1 l dH_2O.

2. 1× TAE buffer: 20 ml 50× TAE buffer, 980 ml dH_2O.

3. DMEM/F-12 cell culture medium: 1:1 DMEM:Ham's F-12 mix containing 2 mM L-glutamine, 100 U/ml penicillin, 100 μg/ml streptomycin, 10 % (v/v) fetal calf serum (FCS).

4. DMEM/F-12 growth arrest medium: 1:1 DMEM:Ham's F-12 mix containing 2 mM L-glutamine, 100 U/ml penicillin, 100 μg/ml streptomycin.

5. LB liquid broth: 1 % (w/v) peptone, 0.5 % (w/v) yeast extract, 0.5 % (w/v) NaCl, 1 l dH_2O.

3 Methods

3.1 Luciferase Reporter Vector Analysis of HAS and HAS2-AS1 Gene Promoters

3.1.1 Primer Design

1. Identify the target promoter sequence directly upstream of the transcription start site and design primers using appropriate computational software.

2. Set the parameters for reverse primer design to include approximately 50 bp of the target mRNA sequence, thus spanning downstream promoter elements. Forward primers are then designed at appropriate increments into the upstream promoter

sequence such that the PCR fragments amplified will span promoter regions of increasing size (e.g., ten primer pairs with product length of approximately 100 bp, 200 bp, 300 bp, 400 bp, 500 bp, 1 kb, 1.5 kb, 2 kb, 2.5 kb, and 3 kb).

3. Addition of appropriate restriction endonuclease sequences to the 5′ end of each primer is required. Using suitable software such as the free online tool at New England Biolabs Ltd. (http://tools.neb.com/NEBcutter2/), obtain a restriction endonuclease map of the sequence. Identify sites present within the multiple cloning site of the luciferase reporter vector but not in the target promoter sequence (*see* **Note 19**).

3.1.2 PCR Amplification and Fragment Purification

1. Lyophilized primers are reconstituted to 200 µM in nuclease-free H_2O, vortexed briefly, then left to stand for 30 min at 4 °C after which vortexing is repeated followed by brief centrifugation. Working dilutions of 10 µM are then made using nuclease-free H_2O (*see* **Note 20**).

2. Promoter fragments are generated from primer pairs using an appropriate PCR thermo-cycler with the following touchdown PCR program. A denaturation step of 94 °C for 5 min, followed by an initial cycle comprising 30 s at 94 °C, 30 s at a T_M of 60 °C (*see* **Note 21**), and 30 s at 72 °C. T_M is then reduced by 0.5 °C per cycle for the next ten cycles, remaining at 55 °C for the concluding 25 cycles.

3. Reactions are performed in a volume of 50 µl comprising 1× PCR buffer containing 1.5 mM $MgCl_2$, 40 ng of genomic DNA (*see* **Note 22**), 20 pmol of each primer (one pair of forward and reverse primers per reaction), 100 µM dNTPs, and 0.5 U of *Taq* polymerase (*see* **Note 23**).

4. If required, sizing of PCR fragments can be tested at this stage by submerged flat-bed electrophoresis.

5. Purification of the PCR product can be performed with commercially available kits; purified products are typically eluted in 30 µl nuclease-free H_2O.

3.1.3 Restriction Endonuclease Digestion and Gel Purification

1. We routinely use the Dual-Luciferase Reporter Assay System (Promega) and pGL-3 firefly luciferase reporter vector (Promega). Use the appropriate restriction endonucleases to digest the sites within the primer sequences flanking the amplified PCR products and to linearize the luciferase reporter vector (e.g., *Kpn*I/*Hind*III for HAS2 with the above system [14]). Incubate 30 µl of purified promoter fragment or 1.5 µg of luciferase reporter vector in a solution containing 4 µl appropriate digestion buffer (*see* **Note 24**), 4 µl 10× BSA, 1 µl *Kpn*I, and 1 µl *Hind*III (restriction endonucleases from New England Biolabs Inc.). Incubate at 37 °C overnight in a heating block.

2. Digestion can be performed using both enzymes simultaneously (double digestion) or sequentially (two-step digestion; *see* **Note 25**).

3. Following endonuclease restriction digestion, gel purification is used to remove PCR product fragments or uncut luciferase reporter vector. Mix 30 μl of digested promoter fragment or vector with loading buffer and subject to flat-bed electrophoresis on a 0.01 % (v/v) ethidium bromide, 1.5 % (w/v) agarose gel (in 1× TAE buffer) following which bands of interest are visualized on a UV transilluminator and excised. Gel purification is performed according to the kit manufacturer's protocol.

4. The purified, linearized reporter vector is routinely treated with 2 μl Shrimp Alkaline Phosphatase (SAP) (1 U/μl; Promega) for 20 min at 37 °C in 1× SAP buffer to dephosphorylate and prevent vector re-ligation. The reaction is then incubated at 65 °C for 20 min to inactivate the SAP.

3.1.4 Ligation and Transformation

1. Use an appropriate DNA ligase (e.g., T4 DNA ligase, commercially available) to ligate the PCR-amplified promoter fragments into the luciferase reporter vector.

2. Ligation is performed using a molar ratio of 1:3 of linearized vector to digested promoter fragment per reaction (*see* **Note 26**). Typically, ligation is performed overnight at 16 °C.

3. Following ligation, the promoter-reporter vectors are transformed into competent One Shot® *E. coli* cells by heat shock.

4. Incubate the *E. coli* and ligation reaction together in a microfuge tube on ice for 30 min, and then incubate in a 42 °C water bath for 45 s.

5. Return the microfuge tube to ice for a further 5 min then transfer the ligation reaction into 400 μl of room temperature super optimal broth (SOC).

6. Incubate the mixture under agitation for 1 h (37 °C, 220 rpm).

7. Spread 200 μl of 1× and 1/10× dilutions of the recovery culture onto pre-warmed and pre-dried ampicillin-containing agar plates (*see* **Note 27**) and incubate at 37 °C overnight.

3.1.5 Colony Selection and Mini-Preps

1. Ensure that clearly defined and independent bacterial colonies have formed on the agar plate.

2. Pick individual colonies, transfer into 5 ml of ampicillin-containing LB liquid broth (*see* **Note 27**), and incubate under agitation for 24 h (37 °C, 220 rpm).

3. Isolate vectors using a suitable mini-prep kit according to the manufacturer's optimized protocol.

4. Quantify vector concentration and estimate purity by spectrophotometry.

3.1.6 Vector Test Digestion and Sequencing

1. To ensure ligated vectors contain appropriately sized inserts, a double-digestion reaction is performed with the same restriction endonucleases used for pre-ligation digestion.

2. Following digestion, run the sample on an agarose gel using flat-bed electrophoresis alongside an appropriately sized double-stranded DNA ladder, and visualize using an UV transilluminator.

3. Following confirmation of inserts of correct size, inserts are sequenced to confirm sequence identity and ensure amplification fidelity.

3.1.7 Midi-Preps

1. Midi-preps are then prepared to provide vector yields suitable for transfection. Colonies positive for reporter vector are grown in 30–50 ml LB broth overnight (37 °C, 220 rpm).

2. The use of a suitable midi-prep kit for vector isolation is recommended following the manufacturer's optimized protocol.

3.1.8 Transfection and Luciferase Analysis

1. Culture the desired cell type to 50–60 % confluence: for fibroblasts, 12-well tissue culture plates are ideal (*see* **Note 28**).

2. Transfect promoter constructs into target cells using an appropriate transfection reagent system according to the manufacturer's optimized protocol; we routinely use Lipofectamine LTX (Life Technologies).

3. Using the Dual-Luciferase Reporter System (Promega), cells are co-transfected with a *Renilla* luciferase reporter vector (Promega; recommended starting concentration ratio 1:4, *Renilla* vector:promoter reporter construct) to control for variation in transfection efficiency.

4. Following transfection and cell treatments, cells are lysed and luciferase luminescence is measured using a suitable plate reader.

3.2 Fluorescent Recovery After Photobleaching (FRAP): Analysis of the Influence of HA on CD44 Membrane Dynamics

3.2.1 Preparation of Cells

1. If using glass coverslips (*see* **Note 29**) go to **step 2**, if using glass-bottom culture plates (*see* **Note 30**) go to **step 3**.

2. Cells are seeded onto circular glass coverslips (22 mm diameter in 35 mm diameter culture plates) that must be sterilized before use by washing in 70 % ethanol and thorough drying before seeding cells. Approximately 7.5×10^5 cells are seeded on top of the coverslips in 2 ml of medium.

3. Approximately 7.5×10^5 cells are seeded onto 35 mm diameter glass-bottom culture plates (In Vitro Scientific, D35-14-1.5-N) in 2 ml of medium (*see* **Note 31**).

4. Once cells have reached desired confluence (approximately 70 % works well for the majority of microscopy and antibody methods) they are growth arrested in serum-free medium for 48 h before further treatment.

5. Serum-free medium solutions containing desired cellular treatments (plain serum-free medium, 10 ng/ml TGF-β1, 0.6 % DMSO controls, or 0.5 μM 4 MU) are then applied to the cells for 24-72 h (*see* **Note 32**).

6. Antibody treatments are made (CD44-FITC in serum-free medium, 1:5,000 dilution) and incubated with the cells for at least 10–15 min at 37 °C, followed by two washes with serum-free medium prior to analysis (*see* **Note 33**).

3.2.2 Microscopy

1. Microscopy is performed on a heated 37 °C stage, and the glass coverslip sealed with lubricant to avoid medium loss. Ideally, an incubator-style stage (37 °C, 5 % CO_2) is used (go to **step 3**); however, if not available, a heated stage will suffice (go to **step 2**).

2. For older model microscopes with heated stages, after applying oil to the objective lens, the edge of the glass coverslips must be made water-tight to avoid medium loss. Carefully remove the glass coverslip from the culture plate and apply a lubricant/sealant to the underside edge before placing it on the stage and above the objective lens. Serum-free medium, warmed to 37 °C, is then carefully pipetted on top of the coverslip. The hydrophobicity of the sealant and surface tension will keep the medium on top of the coverslip, covering the cells sufficiently to prevent desiccation (*see* **Note 34**).

3. For newer model microscopes, following oil application to the objective lens, the glass-bottom culture dish can be placed directly onto the stage and secured.

4. The microscope is initialized, the laser is activated, and the coverslip/culture plate is brought into focus.

5. After focus, ensure there is good antibody coverage, that cells are not over-crowded, and that a clear membrane boundary is defined (*see* **Note 35**).

6. Capture a suitable image of the target cell (*see* **Note 36**) for FRAP analysis.

3.2.3 Photobleaching

1. A 10 μm area of the cell membrane is chosen for photobleaching: if automatic software is unavailable then the area must be selected manually. Once the target area occupies the viewfinder, the laser intensity is adjusted and increased for 5–10 s or until the antibody fluorophore diminishes, using software or manual controls.

2. Upon zooming out, a gap of approximately 10 μm will be visible in the CD44-stained membrane. This technique is more effective using antibodies that stain the entire surface of the cell, as the photobleached region is clearly visible.

3. A suitable post-bleach image is captured, followed by a time-lapse video in which clips are recorded for >100 frames (*see* **Note 37**). Ensure the laser intensity has been returned to normal levels as high intensity laser light will continue to residually photobleach the fluorophore-conjugated antibody.

4. If background noise is high, then 2–4 frame balancing and 2–4 Z-axis scans can help to eliminate interference: refer to software guide or technical manual if required (*see* **Note 38**).

3.2.4 Data Analysis

1. Using the software viewfinder, open the time-lapse clip and highlight the photobleached region of interest (ROI) using the rectangle draw tool (*see* **Note 39**). The region selected should be within the photobleached region covering approximately 10 μm.

2. Select an equally sized region of control (ROC) outside the photobleached zone on another area of the membrane, and a third background region can be selected to eliminate any remaining background noise if necessary. Note that the time-lapse slider beneath the viewfinder can be moved manually or played automatically, allowing visualization of the entire time-lapse.

3. Use the software data analyzer to visualize the time-lapse stack profiles; the light intensity within each region (ROI and ROC) will be displayed as a stack profile histogram of time frame against fluorescent intensity.

4. Export the data as numerical spreadsheets to calculate the fluorescent intensity (FI) ratio, the receptor diffusion constants (D), and mobile fraction of the receptors (MF).

5. To calculate the FI Ratio at each time frame, divide the FI value of the ROI by the FI value of the ROC (FI ratio $= F_{ROI}/F_{ROC}$).

6. To calculate D, use the following formula: $D = (w^2/2t_{1/2})\gamma_D$ where w is the radius of the photobleached area (*see* **Note 40**), $t_{1/2}$ is the half time of fluorescent recovery, and γ_D is a constant that is dependent on experimental conditions (*see* **Note 41**).

7. The MF represents the fraction of receptors recovered into the photobleached zone over the observed time, where $1 = 100\ \%$ and $0 = 0\ \%$ (*see* **Note 42**).

4 Notes

1. Appropriate care should be taken when handling ethidium bromide.

2. This methodology can be applicable to many different types of cells, but transfection protocols and optimization can vary depending on cell type and cell susceptibility. Careful optimization of transfection is essential.

3. The use of antibiotics in cell culture media should be considered carefully and avoided when possible. Ideally, transfection should be performed with cells grown in antibiotic-free medium that has been conditioned for at least 48 h.

4. Cell culture medium is dependent on cell type; the use of different media has no influence on this methodology.

5. Over-confluent cells can reduce the efficiency of transfection; confluence should be limited to 50–60 % during transfection.

6. We describe here the use of fibroblasts, although many types and sizes of cells can be analyzed using FRAP, including both adhesive and nonadhesive cells.

7. The use of antibiotics in cell culture is not usually recommended when culturing primary cell lines and is best avoided if possible.

8. Cell culture medium is dependent on cell type; the use of different media has no influence on this methodology.

9. The recommended level of confluence for antibody experimentation is 70 %; this ensures evenly distributed coverage of antibody to its target antigen across the cell membrane surface.

10. The application of cytokine treatments is dependent on the experimentation of the end-user; dose-response experiments should be performed to optimize cytokine concentration.

11. Before treatment of cells with inhibitors, chemicals, and reagents, cytotoxicity tests should be performed to ensure cell viability after treatment.

12. DMSO controls test for cytotoxicity.

13. For anti-CD44 antibody, 10 min is sufficient for optimal coverage, and this time may vary for different antibodies.

14. Dilution factors for antibodies must be optimized.

15. Various suitable models of laser confocal microscopes are available; consult the technical manual to ensure FRAP can be used.

16. Several fluorophore conjugates can be used for FRAP; always check the suitability of the antibody-fluorophore conjugate for FRAP application. The photostability of the antibody conjugate can gauge the suitability: too low and residual photobleaching will be problematic, too high and the fluorophore will not be susceptible to photobleaching.

17. Ensure the confocal microscope has lasers with appropriate wavelengths for fluorophore excitation and refer to the antibody data sheet for appropriate wavelengths.

18. The use of a heated stage and 5 % CO_2 chamber is preferred. However, if a 5 % CO_2 chamber is unavailable, a 37 °C heated stage will suffice for the time of experimentation. Return culture plates containing cells to a 37 °C, 5 % CO_2, humidified incubator between experiments.

19. The restriction endonuclease recognition sites added to the 5′ ends of PCR primer sequences must be present in the correct orientation in the multiple cloning site of the luciferase reporter vector and absent from the remainder of the vector sequence. Primer and target promoter sequences must be checked to ensure restriction endonuclease recognition sites are absent.

20. Primer solutions can be divided into aliquots and kept in storage at −20 °C until further use.

21. T_M varies depending on primer sequence; PCR primer pairs should ideally have approximately similar T_M values.

22. Human genomic DNA templates may be generated by the end-user, but ethical approval may be required. High molecular weight genomic extracts can be prepared from lymphocytes via phenol:chloroform extraction as we have described previously [26].

23. The efficiency and quantity of thermostable DNA polymerase required will depend on the manufacturer and enzyme type.

24. Use an online tool such as the New England Biolabs Inc. NEBcutter (http://tools.neb.com/NEBcutter2) to determine optimal conditions for double restriction endonuclease digestion.

25. Two-step digestion may require heat-inactivation of the first endonuclease used.

26. Vector and PCR-amplified promoter fragments must be quantified to ensure optimal ligation ratios.

27. The selective antibiotic included in agar/liquid broth media is determined by the antibiotic resistance conferred by the luciferase reporter vector.

28. Tissue culture plate size will vary depending on cell type.

29. Glass coverslips are suitable substitutes for glass bottom culture dishes, and work well with older laser confocal microscope models. These must first be sterilized before cells are seeded. Ensure the diameter of the coverslip is appropriate for the stage of the microscope.

30. Glass bottom culture plates fit the stage brackets of newer models of confocal microscopes, allowing ease of cell culture and analysis without disturbing the cell monolayer. Nonadhesive cells may also be used in these culture plates.

31. Cell seeding density varies depending on cell type and size, and should be optimized.

32. Time courses of cellular treatments should be performed to determine the appropriate treatment time to use for FRAP analysis.

33. Cells are washed following antibody treatment to remove unbound antibody that may give false positive results after photobleaching.

34. If using glass coverslips, ensure cells are sufficiently covered with warm 37 °C culture medium; this avoids desiccation of the cells and helps maintain the temperature of the coverslip.

35. A clearly defined membrane boundary is necessary for accurate FRAP analysis; if the antibody is bound in a globular/clustered pattern then ensure the photobleaching is performed within a region flanked by bound antibody.

36. A pre-photobleached image can be used to calculate the starting FI Ratio (close to 1); a short time-lapse of the pre-photobleached cell can be used to evaluate residual photobleaching.

37. The number of frames that the time-lapse covers is dependent on the cell type and receptor targeted. It is recommended to first perform a 5 min time-lapse to measure recovery within that period and then adjust the time accordingly.

38. Background noise or interference can occur due to light scatter and lead to false positive emission detection. Frame or axis balancing will help to eliminate false emission detection.

39. All confocal software will have the rectangle select/highlight function; refer to the technical manual or software trouble-shooting guides if necessary.

40. Radius is raised to the power of 2 in the given equation due to recovery being 2-directional: recovery is from the membrane at either side of the ROI.

41. For most circular lenses used, the value of γ_D is typically 0.88 [27].

42. The MF can be calculated from the values obtained from determining the FI ratio.

Acknowledgements

The authors are supported by funding from the Medical Research Council, UK, and Kidney Wales Foundation. We thank our colleagues Dr. Robert H. Jenkins and Dr. John Martin for their expert comments on the manuscript.

References

1. Spicer AP, Seldin MF, Olsen AS et al (1997) Chromosomal localization of the human and mouse hyaluronan synthase genes. Genomics 41:493–497

2. Sayo T, Sugiyama Y, Takahashi Y et al (2002) Hyaluronan synthase 3 regulates hyaluronan synthesis in cultured human keratinocytes. J Invest Dermatol 118:43–48

3. Monslow J, Williams JD, Norton N et al (2003) The human hyaluronan synthase genes: genomic structures, proximal promoters and polymorphic microsatellite markers. Int J Biochem Cell Biol 35:1272–1283

4. Chao H, Spicer AP (2005) Natural antisense mRNAs to hyaluronan synthase 2 inhibit hyaluronan biosynthesis and cell proliferation. J Biol Chem 280:27513–27522

5. Dobzhansky T (1964) Biology, molecular and organismic. Am Zool 4:443–452

6. Weigel PH, DeAngelis PL (2007) Hyaluronan synthases: a decade-plus of novel glycosyltransferases. J Biol Chem 282:36777–36781

7. Weigel PH, Hascall VC, Tammi M (1997) Hyaluronan synthases. J Biol Chem 272:13997–14000

8. Itano N, Sawai T, Yoshida M et al (1999) Three isoforms of mammalian hyaluronan synthases have distinct enzymatic properties. J Biol Chem 274:25085–25092

9. Rilla K, Oikari S, Jokela TA et al (2013) Hyaluronan synthase 1 (HAS1) requires higher cellular UDP-GlcNAc concentration than HAS2 and HAS3. J Biol Chem 288:5973–5983

10. Michael DR, Phillips AO, Krupa A et al (2011) The human hyaluronan synthase 2 (HAS2) gene and its natural antisense RNA exhibit coordinated expression in the renal proximal tubular epithelial cell. J Biol Chem 286:19523–19532

11. Tammi RH, Passi AG, Rilla K (2011) Transcriptional and post-translational regulation of hyaluronan synthesis. FEBS J 278:1419–1428

12. Vigetti D, Viola M, Karousou E et al (2014) Metabolic control of hyaluronan synthases. Matrix Biol 35:8–13

13. Yamada Y, Itano N, Hata K-I et al (2004) Differential regulation by IL-1beta and EGF of expression of three different hyaluronan synthases in oral mucosal epithelial cells and fibroblasts and dermal fibroblasts: quantitative analysis using real-time RT-PCR. J Invest Dermatol 122:631–639

14. Monslow J, Williams JD, Guy CA et al (2004) Identification and analysis of the promoter region of the human hyaluronan synthase 2 gene. J Biol Chem 279:20576–20581

15. Monslow J, Williams JD, Fraser DJ et al (2006) Sp1 and Sp3 mediate constitutive transcription of the human hyaluronan synthase 2 gene. J Biol Chem 281:18043–18050

16. Chen L, Neville RD, Michael DR et al (2012) Identification and analysis of the human hyaluronan synthase 1 gene promoter reveals Smad3- and Sp3-mediated transcriptional induction. Matrix Biol 31:373–379

17. Saavalainen K, Tammi MI, Bowen T et al (2007) Integration of the activation of the human hyaluronan synthase 2 gene promoter by common co-factors of the transcription factors RAR and NF-κB. J Biol Chem 282:11530–11539

18. Lauer-Fields JL, Malkar NB, Richet G et al (2003) Melanoma cell CD44 interaction with the alpha 1(IV) 1263–1277 region from basement membrane collagen is modulated by ligand glycosylation. J Biol Chem 278:14321–14330

19. Marrero-Diaz R, Bravo-Cordero JJ, Megías D et al (2009) Polarized MT1-MMP-CD44 interaction and CD44 cleavage during cell retraction reveal an essential role for MT1-MMP in CD44-mediated invasion. Cell Motil Cytoskeleton 66:48–61

20. Meran S, Luo DD, Simpson RM et al (2011) Hyaluronan facilitates transforming growth factor-beta1-dependent proliferation via CD44 and epidermal growth factor receptor interaction. J Biol Chem 286:17618–17630

21. Midgley AC, Rogers M, Hallett MB et al (2013) Transforming growth factor-beta1 (TGF-beta1)-stimulated fibroblast to myofibroblast differentiation is mediated by hyaluronan (HA)-facilitated epidermal growth factor receptor (EGFR) and CD44 co-localization in lipid rafts. J Biol Chem 288:14824–14838

22. Simpson RM, Wells A, Thomas DW et al (2010) Aging fibroblasts resist phenotypic maturation because of impaired hyaluronan-dependent CD44/epidermal growth factor receptor signaling. Am J Pathol 176:1215–1228

23. Webber J, Meran S, Steadman R et al (2009) Hyaluronan orchestrates transforming growth factor-beta1-dependent maintenance of myofibroblast phenotype. J Biol Chem 284:9083–9092

24. Midgley AC, Bowen T, Phillips AO et al (2014) MicroRNA-7 inhibition rescues age-associated loss of EGF receptor and hyaluronan (HA)-dependent differentiation in fibroblasts. Aging Cell 13:235–244

25. Rozen S, Skaletsky HJ (2000) Primer3 on the WWW for general users and for biologist programmers. Methods Mol Biol 132:365–386

26. Speight G, Guy CA, Bowen T et al (1997) Exclusion of CAG/CTG trinucleotide repeat loci which map to chromosome 4 in bipolar disorder and schizophrenia. Am J Med Genet 74:204–206

27. Axelrod D, Koppel DE, Schlessinger J et al (1976) Mobility measurement by analysis of fluorescence photobleaching recovery kinetics. Biophys J 16:1055–1069

INDEX

Kuberan Balagurunathan et al. (eds.), *Glycosaminoglycans: Chemistry and Biology*, Methods in Molecular Biology,
vol. 1229, DOI 10.1007/978-1-4939-1714-3, © Springer Science+Business Media, New York 2015

Printed by Printforce, the Netherlands